STATISTICS FOR
BUSINESS AND
ECONOMICS

SECOND EDITION

Table I Cumulative Probabilities for the Standard Normal Distribution

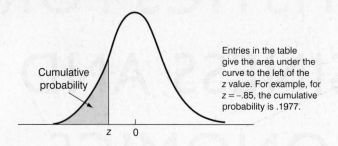

Cumulative probability

Entries in the table give the area under the curve to the left of the z value. For example, for z = −.85, the cumulative probability is .1977.

z	.00	.01	.02	.03	.04	.05	.06	.07	.08	.09
−3.0	.0013	.0013	.0013	.0012	.0012	.0011	.0011	.0011	.0010	.0010
−2.9	.0019	.0018	.0018	.0017	.0016	.0016	.0015	.0015	.0014	.0014
−2.8	.0026	.0025	.0024	.0023	.0023	.0022	.0021	.0021	.0020	.0019
−2.7	.0035	.0034	.0033	.0032	.0031	.0030	.0029	.0028	.0027	.0026
−2.6	.0047	.0045	.0044	.0043	.0041	.0040	.0039	.0038	.0037	.0036
−2.5	.0062	.0060	.0059	.0057	.0055	.0054	.0052	.0051	.0049	.0048
−2.4	.0082	.0080	.0078	.0075	.0073	.0071	.0069	.0068	.0066	.0064
−2.3	.0107	.0104	.0102	.0099	.0096	.0094	.0091	.0089	.0087	.0084
−2.2	.0139	.0136	.0132	.0129	.0125	.0122	.0119	.0116	.0113	.0110
−2.1	.0179	.0174	.0170	.0166	.0162	.0158	.0154	.0150	.0146	.0143
−2.0	.0228	.0222	.0217	.0212	.0207	.0202	.0197	.0192	.0188	.0183
−1.9	.0287	.0281	.0274	.0268	.0262	.0256	.0250	.0244	.0239	.0233
−1.8	.0359	.0351	.0344	.0336	.0329	.0322	.0314	.0307	.0301	.0294
−1.7	.0446	.0436	.0427	.0418	.0409	.0401	.0392	.0384	.0375	.0367
−1.6	.0548	.0537	.0526	.0516	.0505	.0495	.0485	.0475	.0465	.0455
−1.5	.0668	.0655	.0643	.0630	.0618	.0606	.0594	.0582	.0571	.0559
−1.4	.0808	.0793	.0778	.0764	.0749	.0735	.0721	.0708	.0694	.0681
−1.3	.0968	.0951	.0934	.0918	.0901	.0885	.0869	.0853	.0838	.0823
−1.2	.1151	.1131	.1112	.1093	.1075	.1056	.1038	.1020	.1003	.0985
−1.1	.1357	.1335	.1314	.1292	.1271	.1251	.1230	.1210	.1190	.1170
−1.0	.1587	.1562	.1539	.1515	.1492	.1469	.1446	.1423	.1401	.1379
−.9	.1841	.1814	.1788	.1762	.1736	.1711	.1685	.1660	.1635	.1611
−.8	.2119	.2090	.2061	.2033	.2005	.1977	.1949	.1922	.1894	.1867
−.7	.2420	.2389	.2358	.2327	.2296	.2266	.2236	.2206	.2177	.2148
−.6	.2743	.2709	.2676	.2643	.2611	.2578	.2546	.2514	.2483	.2451
−.5	.3085	.3050	.3015	.2981	.2946	.2912	.2877	.2843	.2810	.2776
−.4	.3446	.3409	.3372	.3336	.3300	.3264	.3228	.3192	.3156	.3121
−.3	.3821	.3783	.3745	.3707	.3669	.3632	.3594	.3557	.3520	.3483
−.2	.4207	.4168	.4129	.4090	.4052	.4013	.3974	.3936	.3897	.3859
−.1	.4602	.4562	.4522	.4483	.4443	.4404	.4364	.4325	.4286	.4247
−.0	.5000	.4960	.4920	.4880	.4840	.4801	.4761	.4721	.4681	.4641

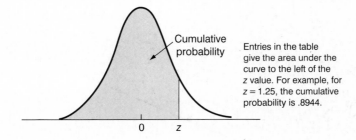

Cumulative probability

Entries in the table give the area under the curve to the left of the z value. For example, for z = 1.25, the cumulative probability is .8944.

z	.00	.01	.02	.03	.04	.05	.06	.07	.08	.09
0.0	.5000	.5040	.5080	.5120	.5160	.5199	.5239	.5279	.5319	.5359
0.1	.5398	.5438	.5478	.5517	.5557	.5596	.5636	.5675	.5714	.5753
0.2	.5793	.5832	.5871	.5910	.5948	.5987	.6026	.6064	.6103	.6141
0.3	.6179	.6217	.6255	.6293	.6331	.6368	.6406	.6443	.6480	.6517
0.4	.6554	.6591	.6628	.6664	.6700	.6736	.6772	.6808	.6844	.6879
0.5	.6915	.6950	.6985	.7019	.7054	.7088	.7123	.7157	.7190	.7224
0.6	.7257	.7291	.7324	.7357	.7389	.7422	.7454	.7486	.7517	.7549
0.7	.7580	.7611	.7642	.7673	.7704	.7734	.7764	.7794	.7823	.7852
0.8	.7881	.7910	.7939	.7967	.7995	.8023	.8051	.8078	.8106	.8133
0.9	.8159	.8186	.8212	.8238	.8264	.8289	.8315	.8340	.8365	.8389
1.0	.8413	.8438	.8461	.8485	.8508	.8531	.8554	.8577	.8599	.8621
1.1	.8643	.8665	.8686	.8708	.8729	.8749	.8770	.8790	.8810	.8830
1.2	.8849	.8869	.8888	.8907	.8925	.8944	.8962	.8980	.8997	.9015
1.3	.9032	.9049	.9066	.9082	.9099	.9115	.9131	.9147	.9162	.9177
1.4	.9192	.9207	.9222	.9236	.9251	.9265	.9279	.9292	.9306	.9319
1.5	.9332	.9345	.9357	.9370	.9382	.9394	.9406	.9418	.9429	.9441
1.6	.9452	.9463	.9474	.9484	.9495	.9505	.9515	.9525	.9535	.9545
1.7	.9554	.9564	.9573	.9582	.9591	.9599	.9608	.9616	.9625	.9633
1.8	.9641	.9649	.9656	.9664	.9671	.9678	.9686	.9693	.9699	.9706
1.9	.9713	.9719	.9726	.9732	.9738	.9744	.9750	.9756	.9761	.9767
2.0	.9772	.9778	.9783	.9788	.9793	.9798	.9803	.9808	.9812	.9817
2.1	.9821	.9826	.9830	.9834	.9838	.9842	.9846	.9850	.9854	.9857
2.2	.9861	.9864	.9868	.9871	.9875	.9878	.9881	.9884	.9887	.9890
2.3	.9893	.9896	.9898	.9901	.9904	.9906	.9909	.9911	.9913	.9913
2.4	.9918	.9920	.9922	.9925	.9927	.9929	.9931	.9932	.9934	.9936
2.5	.9938	.9940	.9941	.9943	.9945	.9946	.9948	.9949	.9951	.9952
2.6	.9953	.9955	.9956	.9957	.9959	.9960	.9961	.9962	.9963	.9964
2.7	.9965	.9966	.9967	.9968	.9969	.9970	.9971	.9972	.9973	.9974
2.8	.9974	.9975	.9976	.9977	.9977	.9978	.9979	.9979	.9980	.9981
2.9	.9981	.9982	.9982	.9983	.9984	.9984	.9985	.9985	.9986	.9986
3.0	.9986	.9987	.9987	.9988	.9988	.9989	.9989	.9989	.9990	.9990

STATISTICS FOR BUSINESS AND ECONOMICS

SECOND EDITION

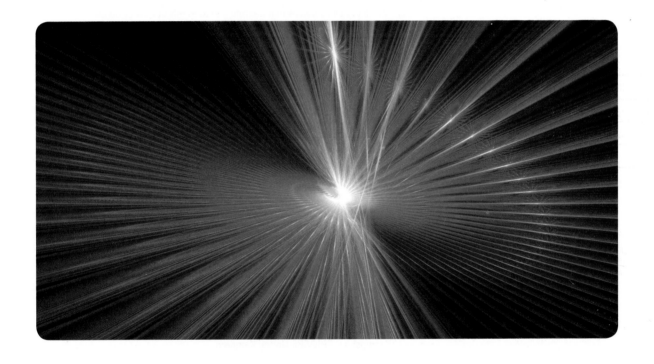

ANDERSON   SWEENEY   WILLIAMS
FREEMAN   SHOESMITH

SOUTH-WESTERN
CENGAGE Learning

Australia • Brazil • Japan • Korea • Mexico • Singapore • Spain • United Kingdom • United States

SOUTH-WESTERN
CENGAGE Learning

Statistics for Business and Economics, 2/e

David R. Anderson, Dennis J. Sweeney, Thomas A. Williams, Jim Freeman and Eddie Shoesmith

Publishing Director: Linden Harris

Publisher: Thomas Rennie

Development Editor: Jennifer Seth

Content Project Editor: Adam Paddon

Senior Production Controller: Paul Herbert

Marketing Manager: Vicky Fielding

Typesetter: Integra Software Services

Cover design: Design Deluxe

Text design: Design Deluxe

For product information and technology assistance, contact **emea.info@cengage.com.**

For permission to use material from this text or product, and for permission queries, email **clsuk.permissions@cengage.com.**

British Library Cataloguing-in-Publication Data
A catalogue record for this book is available from the British Library.

ISBN: 978-1-4080-1810-1

Cengage Learning EMEA
Cheriton House, North Way, Andover, Hampshire, SP10 5BE United Kingdom

Cengage Learning products are represented in Canada by Nelson Education Ltd.

For your lifelong learning solutions, visit **www.cengage.co.uk**

Purchase your next print book, e-book or e-chapter at **www.CengageBrain.com**

Printed by C&C Offset, China
1 2 3 4 5 6 7 8 9 10—12 11 10

'To the memory of my grandparents, Lizzie and Halsey'
JIM FREEMAN

'To my daughter Hannah, and to all my family, past, present and future'
EDDIE SHOESMITH

Brief contents

Contents

Preface and Acknowledgements

The purpose of *Statistics for Business and Economics* is to give students, primarily those in the fields of business, management and economics, a conceptual introduction to the field of statistics and its many applications. The text is applications-oriented and written with the needs of the non-mathematician in mind. The mathematical pre-requisite is knowledge of algebra.

Applications of data analysis and statistical methodology are an integral part of the organization and presentation of the material in the text. The discussion and development of each technique are presented in an application setting, with the statistical results providing insights to problem solution and decision-making.

Although the book is applications oriented, care has been taken to provide sound methodological development and to use notation that is generally accepted for the topic being covered. Hence, students will find that this text provides good preparation for the study of more advanced statistical material. A revised and updated bibliography to guide further study is included as an appendix.

The text introduces the student to the software packages Minitab 15, PASW 17 and Microsoft® Office EXCEL 2007 and emphasizes the role of computer software in the application of statistical analysis. Minitab and PASW are illustrated as they are two of the leading statistical software packages for both education and statistical practice. EXCEL is not a statistical software package, but the wide availability and use of EXCEL makes it important for students to understand the statistical capabilities of this package. Minitab, PASW and EXCEL procedures are provided in end-of-chapter sections so that instructors have the flexibility of using as much computer emphasis as desired for the course.

The international edition

This is the 2nd international edition of *Statistics for Business and Economics*. It is based on the 1st international edition and the 10th United States (US) edition. The US editions have a distinguished history and deservedly high reputation for clarity and soundness of approach, and we maintained the presentation style and readability of those editions in preparing the international edition. We have replaced many of the US-based examples, case studies and exercises with equally interesting and appropriate ones sourced from a wider geographical base, particularly the UK, Ireland, continental Europe, South Africa and the Middle East. We have also made the book slightly more compact by moving some of the exercises and some of the supplementary exercises to the associated website. Other notable changes in this second international edition are summarized here.

Changes in the 2nd International Edition

- **Software sections** In the 2nd international edition, we have updated the software sections to provide step-by-step instructions for the latest versions of the software packages: Minitab 15, PASW 17 and Microsoft® Office EXCEL 2007. The software sections have been relocated at the end of chapters, as in the US editions.

- **Cumulative Standard Normal Distribution Table** In common with the 10th US edition, we use the Cumulative Standard Normal Distribution Table. There is a growing trend for more and more students and practitioners alike to use statistics in an environment that emphasizes modern computer software. Historically, a table was used by everyone because a table was the only source of information about the normal distribution. However, many of today's students are ready and willing to learn about the use of computer software in statistics. Students will find that virtually every computer software package uses the cumulative standard normal distribution. Thus, it is becoming more and more important for introductory statistical texts to use a normal probability table that is consistent with what the student will see when working with statistical software. It is no longer desirable to use one form of the Standard Normal Distribution Table in the text and then use a different type of standard normal distribution calculation when using a software package. Those who are using the Cumulative Normal Distribution Table for the first time will find that, in general, it eases the normal probability calculations. In particular, a cumulative normal probability table makes it easier to compute p-values for hypothesis testing.

- **Upper case notation for random variables** A key change in the 2nd international edition is the use of upper case notation for random variables – lower case notation being reserved, correspondingly, for realizations or values of random variables. This is a widely adopted convention around the world and therefore aligns the text more closely with standard statistical practice.

- **Self-test exercises** Certain exercises are identified as self-test exercises. Completely worked-out solutions for those exercises are provided in Appendix D at the back of the book. Students can attempt the self-test exercises and immediately check the solution to evaluate their understanding of the concepts presented in the chapter.

- **Other Content Revisions** The following additional content revisions appear in the new edition.

  - New examples of times series data are provided in Chapter 1.

  - The EXCEL appendix to Chapter 2 now provides more complete instructions on how to develop a frequency distribution and a histogram for quantitative data.

  - Chapter 17 has been updated with current index numbers.

  - The Solutions Manual now shows the exercise solution steps using the cumulative normal distribution and more details in the explanations about how to compute p-values for hypothesis testing.

  - A number of new case problems have been added. These are in the chapters on Interval Estimation, Inferences about Population Variances, Tests of Goodness of Fit and Independence, Analysis of Variance and Experimental Design, and Decision Analysis. These case problems provide students with the opportunity to analyze somewhat larger data sets and prepare managerial reports based on the results of the analysis.

  - Each chapter begins with a Statistics in Practice article that describes an application of the statistical methodology to be covered in the chapter. New to this edition are Statistics in Practice articles for Chapters 2, 7, 17 and 19, with several other articles substantially updated and revised for this new edition.

  - New examples and exercises have been added throughout the book, based on real data and recent reference sources of statistical information. We believe that the use of real data helps generate more student interest in the material and enables the student to learn about both the statistical methodology and its application.

- To accompany the new exercises and examples, data files are available on the CD-ROM that is packaged with the text. The data sets are available in Minitab, PASW and EXCEL formats. Data set logos are used in the text to identify the data sets that are available on the CD. Data sets for all case problems as well as data sets for larger exercises are included.

Acknowledgements

The authors and publisher acknowledge the contribution of the following reviewers to both editions of this textbook:

- John R. Calvert – Loughborough University
- Naomi Feldman – Ben-Gurion University of the Negev
- Luc Hens – Vesalius College
- Martyn Jarvis – University of Glamorgan
- Alan Matthews – Trinity College Dublin
- Suzanne McCallum – Glasgow University
- Chris Muller – University of Stellenbosch
- Surette Oosthuizen – University of Stellenbosch
- Mark Stevenson – Lancaster University
- Dave Worthington – Lancaster University

We would also like to thank:

- Colin James and the STARS team for kindly providing permission to use some of the datasets and accompanying material which feature on their website (www.stars.ac.uk)
- Ibrahim Wazir at Webster University in Vienna for providing the case problem for Chapter Four

The publisher also thanks the various copyright holders for granting permission to reproduce material throughout the text. Every effort has been made to trace all copyright holders, but if anything has been inadvertently overlooked the publisher will be pleased to make the necessary arrangements at the first opportunity. Please contact the publisher directly.

About the Authors

Jim Freeman is Senior Lecturer in Statistics and Operational Research at Manchester Business School, United Kingdom and Editor of the Operational Research Society's OR Insight journal. He was born in Tewkesbury, Gloucestershire. After taking a first degree in pure mathematics at UCW Aberystwyth, he went on to receive MSc and PhD degrees in Applied Statistics from Bath and Salford universities respectively. In 1992/3 he was Visiting Professor at the University of Alberta. Before joining MBS, he was Statistician at the Distributive Industries Training Board – and prior to that – the Universities Central Council on Admissions. He has taught undergraduate and postgraduate courses in business statistics and operational research courses to students from a wide range of management and engineering backgrounds. For many years he was also responsible for providing introductory statistics courses to staff and research students at the University of Manchester's Staff Teaching Workshop. Through his gaming and simulation interests he has been involved in a significant number of external consultancy projects. In July 2008 he was appointed Editor of the Operational Research Society's *OR Insight* journal.

Eddie Shoesmith is Senior Lecturer in Statistics and Programme Director for undergraduate business and management programmes in the School of Business, University of Buckingham, UK. He was born in Barnsley, Yorkshire. He was awarded an MA (Natural Sciences) at the University of Cambridge, and a BPhil (Economics and Statistics) at the University of York. Prior to taking an academic post at Buckingham, he worked for the UK Government Statistical Service, in the Cabinet Office, for the London Borough of Hammersmith and for the London Borough of Haringey. At Buckingham, before joining the School of Business, he held posts as Dean of Sciences and Head of Psychology. He has taught introductory and intermediate-level applied statistics courses to undergraduate and postgraduate student groups in a wide range of disciplines: business and management, economics, accounting, psychology, biology and social sciences. He has also taught statistics to social and political sciences undergraduates at the University of Cambridge.

David R. Anderson is Professor of Quantitative Analysis in the College of Business Administration at the University of Cincinnati. Born in Grand Forks, North Dakota, he earned his BS, MS and PhD degrees from Purdue University. Professor Anderson has served as Head of the Department of Quantitative Analysis and Operations Management and as Associate Dean of the College of Business Administration. In addition, he was the coordinator of the college's first executive programme. In addition to teaching introductory statistics for business students, Dr Anderson has taught graduate-level courses in regression analysis, multivariate analysis, and management science. He also has taught statistical courses at the Department of Labor in Washington, DC Professor Anderson has been honoured with nominations and awards for excellence in teaching and excellence in service to student organizations. He has co-authored ten textbooks related to decision sciences and actively consults with businesses in the areas of sampling and statistical methods.

Dennis J. Sweeney is Professor of Quantitative Analysis and founder of the Center for Productivity Improvement at the University of Cincinnati. Born in Des Moines, Iowa, he earned BS and BA degrees from Drake University, graduating summa cum laude. He received his MBA and DBA degrees from Indiana University, where he was an NDEA

Fellow. Dr Sweeney has worked in the management science group at Procter & Gamble and has been a visiting professor at Duke University. Professor Sweeney served five years as Head of the Department of Quantitative Analysis and four years as Associate Dean of the College of Business Administration at the University of Cincinnati.

He has published more than 30 articles in the area of management science and statistics. The National Science Foundation, IBM, Procter & Gamble, Federated Department Stores, Kroger and Cincinnati Gas & Electric have funded his research, which has been published in *Management Science, Operations Research, Mathematical Programming, Decision Sciences,* and other journals. Professor Sweeney has coauthored ten textbooks in the areas of statistics, management science, linear programming, and production and operations management.

Thomas A. Williams is Professor of Management Science in the College of Business at Rochester Institute of Technology (RIT). Born in Elmira, New York, he earned his BS degree at Clarkson University. He completed his graduate work at Rensselaer Polytechnic Institute, where he received his MS and PhD degrees.

Before joining the College of Business at RIT, Professor Williams served for seven years as a faculty member in the College of Business Administration at the University of Cincinnati, where he developed the first undergraduate programme in Information Systems. At RIT he was the first chair of the Decision Sciences Department.

Professor Williams is the co-author of 11 textbooks in the areas of management science, statistics, production and operations management, and mathematics. He has been a consultant for numerous *Fortune* 500 companies in areas ranging from the use of elementary data analysis to the development of large-scale regression models.

Walk-through Tour

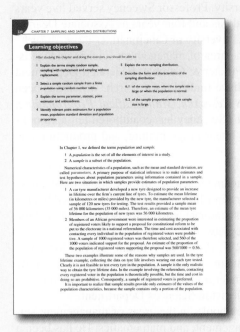

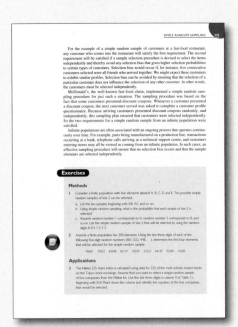

Learning objectives

We have set out clear learning objectives at the start of each chapter in the text, as is now common in texts in the UK and elsewhere. These objectives summarise the core content of each chapter in a list of key points.

Statistics in practice

Each chapter begins with a Statistics in Practice article that describes an application of the statistical methodology to be covered in the chapter.

Exercises

The exercises are split into two parts, Methods and Applications. The Methods exercises require students to use the formulae and make the necessary computations. The Applications exercises require students to use the chapter material in real-world situations. Thus, students first focus on the computational 'nuts and bolts', then move on to the subtleties of statistical application and interpretation. Answers to even-numbered exercises are provided at the back of the textbook, while a full set of answers are provided in the lecturers' Solutions Manual. Supplementary exercises are provided on the textbook's companion website. New Self-test exercises are highlighted throughout by the paper and pencil icon and contain fully-worked solutions in Appendix D.

Notes

Recent US editions have included marginal and end-of-chapter notes. We have not adopted this layout, but have included the important material in the text itself.

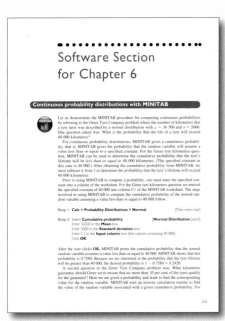

Software sections

Statistical analyses throughout the text provide step-by-step instructions that make it easy for students to use the latest versions of software packages Minitab 15, PASW 17 and Microsoft® Office EXCEL 2007, to conduct the statistical analysis presented in the chapter. Each package is highlighted in the chapter by marginal logos. The software sections have been relocated at the end of chapters, as in the US Editions.

Data sets accompany text

Over 200 data sets are available on the CD-ROM that is packaged with the text. The data sets are available in MINITAB, PASW and EXCEL formats. Data set logos are used in the text to identify the data sets that are available on the CD. Data sets for all case problems as well as data sets for larger exercises are also included on the CD.

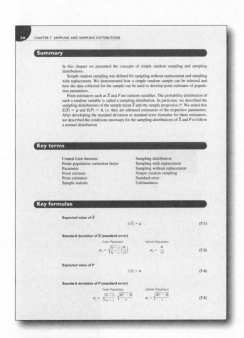

Summaries Each chapter includes a summary to remind students of what they have learnt so far and offering a useful way to review for exams.

Key terms Key terms are highlighted in the text, listed at the end of each chapter and given a full definition in the Glossary at the end of the textbook.

Key formulae Key formulae are listed at the end of each chapter for easy reference.

Case Problems The end-of-chapter case problems provide students with the opportunity to analyse some- what larger data sets and prepare managerial reports based on the results of the analysis.

Accompanying Website

Visit the Statistics for Business and Economics companion website at **www.cengage.co.uk/aswsbe2** to find valuable teaching and learning material including:

For students

- Multiple Choice Questions
- Crossword Puzzles
- Glossary
- Learning Objectives
- Weblinks

For lecturers

- Solutions Manual
- Extra Exercises
- Extra Case Problems
- Solutions to all Case Problems
- PowerPoint Presentation Slides
- Weblinks

All of the web material is available in a variety of VLE cartridges including Blackboard/WebCT and Moodle.

Supplements

Cengage Learning offers various supplements for lecturers and students who use this book. All of these supplements can be found on the companion website: www.cengage.co.uk/aswsbe2

For the lecturer:

- **Solutions manual** – The solutions manual, fully adapted by the authors and downloadable as a PDF document, includes solutions to all exercises that accompany the textbook.
- **PowerPoint presentation slides** – A full set of PowerPoint slides for every chapter, fully adapted by the authors, can save lecturers valuable time in classroom preparation.
- **Extra Exercises** – Hundreds of extra exercises with solutions are included in the solutions manual.
- **Extra case problems** – In addition to those case problems found in the textbook, extra case problems with full solutions are also provided.
- **Solutions to all case problems** – A downloadable PDF document contains full solutions to case problems in the textbook and all extra case problems.
- **Web links** – Carefully selected by the authors, dozens of web links provide valuable guidance on useful websites.
- **ExamView®** – Using thousands of questions created specifically for this textbook, this test bank and test generator allows lecturers to create online, paper and local area network (LAN) tests.

For the student:

- **Multiple choice questions** – Dozens of practice questions for every chapter provide students with a valuable revision tool.
- **Crossword puzzles** – Dozens of crossword puzzles based on chapter content provide a unique learning tool.
- **Glossary** – An electronic glossary provides useful definitions of all key terms.
- **Learning objectives** – Listed for each chapter, they help the student to monitor their understanding and progress through the chapter.
- **Web links** – Carefully selected by the authors, dozens of web links provide valuable guidance on useful websites.

Other supplementary resources:

- Links to an array of further supplementary products such as WebTutor, EasyQuant: Digital Tutor for EXCEL, EasyStat: Digital Tutor for MINITAB, Business Statistics NOW, South Western STAT+ 2.0, Tree Plan and Talk To Us are included on the companion website.
- **Virtual learning environment** – All of the web material is available in a variety of VLE cartridges including Blackboard/WebCT and Moodle.
- **Text choice** – TextChoice (www.textchoice.com) is the home of Cengage Learning's online digital content. It provides the fastest, easiest way for you to create your own learning materials. You may select content from hundreds of our best-selling titles to make a custom text. Please contact your Cengage Learning sales representative to discuss further.

Chapter 1

··

Data and Statistics

Statistics in practice: The *Economist*

Learning objectives

After reading this chapter and doing the exercises, you should be able to:

1 Appreciate the breadth of statistical applications in business and economics.

2 Understand the meaning of the terms elements, variables, and observations as they are used in statistics.

3 Understand the difference between qualitative, quantitative, cross-sectional and time series data.

4 Find out about data sources available for statistical analysis both internal and external to the firm.

5 Appreciate how errors can arise in data.

6 Understand the meaning of descriptive statistics and statistical inference.

7 Distinguish between a population and a sample.

8 Understand the role a sample plays in making statistical inferences about the population.

Frequently, we see the following kinds of statements in newspaper and magazine articles:

- The Ifo World Economic Climate Index fell again substantially in January 2009. The climate indicator stands at 50.1 (1995 = 100); its historically lowest level since introduction in the early 1980s (CESifo, April 2009).

- The IMF projected the global economy would shrink 1.3 per cent in 2009 (*Fin24*, 23 April 2009).

- The Footsie finished the week on a winning streak despite shock figures that showed the economy has contracted by almost 2 per cent already in 2009 (*This is Money*, 25 April 2009).

- China's growth rate fell to 6.1 per cent in the year to the first quarter (*The Economist*, 16 April 2009).

- GM receives further $2 bn in loans (*BBC News*, 24 April 2009).

- Handset shipments to drop by 20 per cent (*In-Stat*, 2009).

The numerical facts in the preceding statements (50.1, 1.3 per cent, 2 per cent, 6.1 per cent, $2 bn, 20 per cent) are called statistics. Thus, in everyday usage, the term *statistics* refers to numerical facts. However, the field, or subject, of statistics involves much more than numerical facts. In a broad sense, **statistics** is the art and science of collecting, analyzing, presenting and interpreting data. Particularly in business and economics, the information provided by collecting, analyzing, presenting and interpreting data gives managers and decision-makers a better understanding of the business and economic environment and thus enables them to make more informed and better decisions. In this text, we emphasize the use of statistics for business and economic decision-making.

Chapter 1 begins with some illustrations of the applications of statistics in business and economics. In Section 1.2 we define the term *data* and introduce the concept of a data set. This section also introduces key terms such as *variables* and *observations*, discusses the difference between quantitative and qualitative data, and illustrates the uses of cross-sectional and time series data. Section 1.3 discusses how data can be obtained from existing sources or through survey and experimental studies designed to obtain new data. The important role that the Internet now plays in obtaining data is also highlighted. The use of data in developing descriptive statistics and in making statistical inferences is described in Sections 1.4 and 1.5.

Statistics in Practice

The *Economist*

Founded in 1843, *The Economist* is an international weekly news and business magazine written for top-level business executives and political decision-makers. The publication aims to provide readers with in-depth analyses of international politics, business news and trends, global economics and culture.

Economist Intelligence Unit website. Reproduced with permission.

The Economist is published by the Economist Group – an international company employing nearly 1000 staff worldwide – with offices in London, Frankfurt, Paris and Vienna; in New York, Boston and Washington DC; and in Hong Kong, mainland China, Singapore and Tokyo.

Between 1998 and 2008 the magazine's worldwide circulation grew by 100 per cent – recently exceeding 180 000 in the UK, 230 000 in continental Europe, 780 000 plus copies in North America and nearly 130 000 in the Asia-Pacific region. It is read in more than 200 countries and with a readership of 4 million, is one of the world's most influential business publications. Along with the *Financial Times*, it is arguably one of the two most successful print publications to be introduced in the US market during the past decade.

Complementing *The Economist* brand within the Economist Brand family, the Economist Intelligence Unit provides access to a comprehensive database of worldwide indicators and forecasts covering more than 200 countries, 45 regions and eight key industries. The Economist Intelligence Unit aims to help executives make informed business decisions through dependable intelligence delivered online, in print, in customized research as well as through conferences and peer interchange.

Alongside the Economist Brand family, the Group manages and runs the CFO and Government brand families for the benefit of senior finance executives and government decision-makers (in Brussels and Washington) respectively.

1.1 Applications in business and economics

In today's global business and economic environment, anyone can access vast amounts of statistical information. The most successful managers and decision-makers understand the information and know how to use it effectively. In this section, we provide examples that illustrate some of the uses of statistics in business and economics.

Accounting

Public accounting firms use statistical sampling procedures when conducting audits for their clients. For instance, suppose an accounting firm wants to determine whether the amount of accounts receivable shown on a client's balance sheet fairly represents the actual amount of accounts receivable. Usually the large number of individual accounts receivable makes reviewing and validating every account too time-consuming and expensive. As common practice in such situations, the audit staff selects a subset of the accounts called a sample. After reviewing the accuracy of the sampled accounts, the auditors draw a conclusion as to whether the accounts receivable amount shown on the client's balance sheet is acceptable.

Finance

Financial analysts use a variety of statistical information to guide their investment recommendations. In the case of stocks, the analysts review a variety of financial data including price/earnings ratios and dividend yields. By comparing the information for an individual stock with information about the stock market averages, a financial analyst can begin to draw a conclusion as to whether an individual stock is over- or under-priced. Similarly, historical trends in stock prices can provide a helpful indication on when investors might consider entering (or re-entering) the market. For example, *Money Week* (3 April 2009) reported a Goldman Sachs analysis that indicated because stocks were unusually cheap at the time, real average returns of up to 6 per cent in the US and 7 per cent in Britain might be possible over the next decade – based on long-term cyclically adjusted price/earnings ratios.

Marketing

Electronic scanners at retail checkout counters collect data for a variety of marketing research applications. For example, data suppliers such as ACNielsen purchase point-of-sale scanner data from grocery stores, process the data and then sell statistical summaries of the data to manufacturers. Manufacturers spend vast amounts per product category to obtain this type of scanner data. Manufacturers also purchase data and statistical summaries on promotional activities such as special pricing and the use of in-store displays. Brand managers can review the scanner statistics and the promotional activity statistics to gain a better understanding of the relationship between promotional activities and sales. Such analyses often prove helpful in establishing future marketing strategies for the various products.

Production

Today's emphasis on quality makes quality control an important application of statistics in production. A variety of statistical quality control charts are used to monitor the output of a production process. In particular, an x-bar chart can be used to monitor the average output. Suppose, for example, that a machine fills containers with 330 g of a soft drink. Periodically, a production worker selects a sample of containers and computes the average number of grams in the sample. This average, or x-bar value, is plotted on an x-bar chart. A plotted value above the chart's upper control limit indicates overfilling, and a plotted value below the chart's lower control limit indicates underfilling. The process is termed 'in control' and allowed to continue as long as the plotted x-bar values fall between the chart's upper and lower control limits. Properly interpreted, an x-bar chart can help determine when adjustments are necessary to correct a production process.

Economics

Economists frequently provide forecasts about the future of the economy or some aspect of it. They use a variety of statistical information in making such forecasts. For instance, in forecasting inflation rates, economists use statistical information on such indicators as the Producer Price Index, the unemployment rate, and manufacturing capacity utilization. Often these statistical indicators are entered into computerized forecasting models that predict inflation rates.

Applications of statistics such as those described in this section are an integral part of this text. Such examples provide an overview of the breadth of statistical applications. To supplement these examples, chapter-opening Statistics in Practice articles obtained from a variety of topical sources are used to introduce the material covered in each chapter. These articles show the importance of statistics in a wide variety of business and economic situations.

1.2 Data

Data are the facts and figures collected, analyzed and summarized for presentation and interpretation. All the data collected in a particular study are referred to as the data set for the study. Table 1.1 shows a data set summarizing information for equity (share) trading at the 22 European Stock Exchanges in March 2009.

Elements, variables and observations

Elements are the entities on which data are collected. For the data set in Table 1.1, each individual European exchange is an element; the element names appear in the first column. With 22 exchanges, the data set contains 22 elements.

A variable is a characteristic of interest for the elements. The data set in Table 1.1 includes the following three variables:

- *Exchanges*: at which the equities were traded.
- *Trades*: number of trades during the month.
- *Turnover*: value of trades (€m) during the month.

Measurements collected on each variable for every element in a study provide the data. The set of measurements obtained for a particular element is called an observation. Referring to Table 1.1, we see that the set of measurements for the first observation (Athens Exchange) is 599 192 and 2009.8. The set of measurements for the second observation (Borsa Italiana) is 5 921 099 and 44 385.9; and so on. A data set with 22 elements contains 22 observations.

Scales of measurement

Data collection requires one of the following scales of measurement: nominal, ordinal, interval or ratio. The scale of measurement determines the amount of information contained in the data and indicates the most appropriate data summarization and statistical analyses.

When the data for a variable consist of labels or names used to identify an attribute of the element, the scale of measurement is considered a nominal scale. For example, referring to the data in Table 1.1, we see that the scale of measurement for the exchange variable is nominal because Athens Exchange, Borsa Italiana . . . Wiener Börse are labels used to identify where the equities are traded. In cases where the scale of measurement is nominal, a numeric code as well as non-numeric labels may be used. For example, to facilitate data collection and to prepare the data for entry into a computer

Table 1.1 European stock exchange monthly statistics domestic equity trading (electronic order book transactions) March 2009

Exchange	Total Trades	Total Turnover
Athens	599 192	2 009.8
Borsa Italiana	5 921 099	44 385.9
Bratislava	111	0.1
Bucharest	79 921	45.3
Budapest	298 871	1 089.6
Bulgarian	14 040	64.4
Cyprus	31 167	76.1
Deutsh Borse	7 642 241	86 994.5
Euronext	15 282 996	116 488
Irish	79 973	549.8
Ljubljana	11 172	35.6
London	16 539 588	114 283.6
Luxembourg	1 152	12.5
Malta	638	1.9
NASDAQ OMX Nordic	4 550 073	40 927.4
Oslo Bars	981 362	9 755.1
Prague	65 153	1 034.8
SIX Swiss	440 578	2 667.1
Spanish (BME)	2 799 329	60 387.6
SWX Europe	n/a	n/a
Warsaw	1 155 379	2 468.6
Wiener Borse	433 545	2 744
TOTAL	**56 927 580**	**486 021.7**

Source: European Stock Exchange monthly statistics (http://www.fese.be/en/?inc=art&id=3)

database, we might use a numeric code by letting 1 denote the Athens Exchange, 2, the Borsa Italiana . . . and 22, Wiener Börse. In this case the numeric values 1, 2, . . . 22 provide the labels used to identify where the stock is traded. The scale of measurement is nominal even though the data appear as numeric values.

The scale of measurement for a variable is called an **ordinal scale** if the data exhibit the properties of nominal data and the order or rank of the data is meaningful. For example, Eastside Automotive sends customers a questionnaire designed to obtain data on the quality of its automotive repair service. Each customer provides a repair service rating of excellent, good or poor. Because the data obtained are the labels – excellent, good or poor – the data have the properties of nominal data. In addition, the data can be ranked, or ordered, with respect to the service quality. Data recorded as excellent indicate the best service, followed by good and then poor. Thus, the scale of measurement is ordinal. Note that the ordinal data can also be recorded using a numeric code. For example, we could use 1 for excellent, 2 for good and 3 for poor to maintain the properties of ordinal data. Thus, data for an ordinal scale may be either non-numeric or numeric.

The scale of measurement for a variable becomes an **interval scale** if the data show the properties of ordinal data and the interval between values is expressed in terms of a fixed unit of measure. Interval data are always numeric. Graduate Management Admission Test (GMAT) scores are an example of interval-scaled data. For example, three students with GMAT scores of 620 550 and 470 can be ranked or ordered in terms of best performance to poorest performance. In addition, the differences between the scores are meaningful. For instance, student one scored $620 - 550 = 70$ points more than student two, while student two scored $550 - 470 = 80$ points more than student three.

The scale of measurement for a variable is a **ratio scale** if the data have all the properties of interval data and the ratio of two values is meaningful. Variables such as distance, height, weight and time use the ratio scale of measurement. This scale requires that a zero value be included to indicate that nothing exists for the variable at the zero point. For example, consider the cost of a car. A zero value for the cost would indicate that the car has no cost and is free. In addition, if we compare the cost of €30 000 for one car to the cost of €15 000 for a second car, the ratio property shows that the first car is €30 000/€15 000 = 2 times, or twice, the cost of the second car.

Qualitative and quantitative data

Data can be further classified as either qualitative or quantitative. **Qualitative data** include labels or names used to identify an attribute of each element. Qualitative data use either the nominal or ordinal scale of measurement and may be non-numeric or numeric. **Quantitative data** require numeric values that indicate how much or how many. Quantitative data are obtained using either the interval or ratio scale of measurement.

A **qualitative variable** is a variable with qualitative data, and a **quantitative variable** is a variable with quantitative data. The statistical analysis appropriate for a particular variable depends upon whether the variable is qualitative or quantitative. If the variable is qualitative, the statistical analysis is rather limited. We can summarize qualitative data by counting the number of observations in each qualitative category or by computing the proportion of the observations in each qualitative category. However, even when the qualitative data use a numeric code, arithmetic operations such as addition, subtraction, multiplication and division do not provide meaningful results. Section 2.1 discusses ways for summarizing qualitative data.

On the other hand, arithmetic operations often provide meaningful results for a quantitative variable. For example, for a quantitative variable, the data may be added and then divided by the number of observations to compute the average value. This average is usually meaningful and easily interpreted. In general, more alternatives for statistical analysis are possible when the data are quantitative. Section 2.2 and Chapter 3 provide ways of summarizing quantitative data.

Cross-sectional and time series data

For purposes of statistical analysis, distinguishing between cross-sectional data and time series data is important. **Cross-sectional data** are data collected at the same or approximately the same point in time. The data in Table 1.1 are cross-sectional because they describe the two variables for the 22 exchanges at the same point in time. **Time series data** are data collected over several time periods. For example, Figure 1.1 provides a graph of the wholesale price (US$) of crude oil per gallon for the period April 2005 – March 2006. It shows a sharp increase in the average price during July and September 2005, followed by a rather stable average price per gallon from November 2005 – March 2006. Most of the statistical methods presented in this text apply to cross-sectional rather than time series data.

Quantitative data that measure how many are discrete. Quantitative data that measure how much are continuous because no separation occurs between the possible data values.

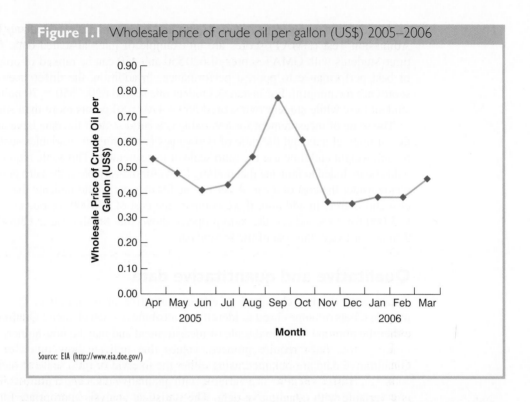

Figure 1.1 Wholesale price of crude oil per gallon (US$) 2005–2006

Source: EIA (http://www.eia.doe.gov/)

1.3 Data sources

Data can be obtained from existing sources or from surveys and experimental studies designed to collect new data.

Existing sources

In some cases, data needed for a particular application already exist. Companies maintain a variety of databases about their employees, customers and business operations. Data on employee salaries, ages and years of experience can usually be obtained from internal personnel records. Other internal records contain data on sales, advertising expenditures, distribution costs, inventory levels and production quantities. Most companies also maintain detailed data about their customers. Table 1.2 shows some of the data commonly available from internal company records.

Organizations that specialize in collecting and maintaining data make available substantial amounts of business and economic data. Companies access these external data sources through leasing arrangements or by purchase. Dun & Bradstreet, Bloomberg and the Economist Intelligence Unit are three sources that provide extensive business database services to clients. ACNielsen built successful businesses collecting and processing data that they sell to advertisers and product manufacturers.

Data are also available from a variety of industry associations and special interest organizations. The European Tour Operators Association and European Travel Commission provide information on tourist trends and travel expenditures by visitors to and from countries in Europe. Such data would be of interest to firms and individuals in the travel industry. The Graduate Management Admission Council maintains data on test scores, student characteristics and graduate management education

Table 1.2 Examples of data available from internal company records

Source	Some of the data typically available
Employee records	Name, address, social security number, salary, number of vacation days, number of sick days and bonus
Production records	Part or product number, quantity produced, direct labour cost and materials cost
Inventory records	Part or product number, number of units on hand, reorder level, economic order quantity and discount schedule
Sales records	Product number, sales volume, sales volume by region and sales volume by customer type
Credit records	Customer name, address, phone number, credit limit and accounts receivable balance
Customer profile	Age, gender, income level, household size, address and preferences

programmes. Most of the data from these types of sources are available to qualified users at a modest cost.

The Internet continues to grow as an important source of data and statistical information. Almost all companies maintain websites that provide general information about the company as well as data on sales, number of employees, number of products, product prices and product specifications. In addition, a number of companies now specialize in making information available over the Internet. As a result, one can obtain access to stock quotes, meal prices at restaurants, salary data and an almost infinite variety of information. Government agencies are another important source of existing data. For instance, Eurostat maintains considerable data on employment rates, wage rates, size of the labour force and union membership. Table 1.3 lists selected governmental agencies and some of the data they provide. Most government agencies that collect and process data also make the results available through a website. For instance, the Eurostat has a wealth of data at its website, epp.eurostat.ec.europa.eu. Figure 1.2 shows the homepage for the Eurostat.

Statistical studies

Sometimes the data needed for a particular application are not available through existing sources. In such cases, the data can often be obtained by conducting a statistical study. Statistical studies can be classified as either *experimental* or *observational*.

In an experimental study, a variable of interest is first identified. Then one or more other variables are identified and controlled so that data can be obtained about how they influence the variable of interest. For example, a pharmaceutical firm might be interested in conducting an experiment to learn about how a new drug affects blood pressure. Blood pressure is the variable of interest in the study. The dosage level of the new drug is another variable that is hoped to have a causal effect on blood pressure. To obtain data about the effect of the new drug, researchers select a sample of individuals. The dosage level of the new drug is controlled, as different groups of individuals are given different dosage levels. Before and after data on blood pressure are collected for each group. Statistical analysis of the experimental data can help determine how the new drug affects blood pressure.

Non-experimental, or observational, statistical studies make no attempt to control the variables of interest. A survey is perhaps the most common type of observational study. For instance, in a personal interview survey, research questions are first identified. Then

Figure 1.2 Eurostat homepage

Table 1.3 Examples of data available from selected European sources

Source	Some of the data available
Europa rates (http://europa.eu)	Travel, VAT (value added tax), euro exchange employment, population and social conditions
Eurostat (http://epp.eurostat.ec.europa.eu/)	Education and training, labour market, living conditions and welfare
European Central Bank (http://www.ecb.int/)	Monetary, financial markets, interest rate and balance of payments statistics, unit labour costs, compensation per employee, labour productivity, consumer prices, construction prices

a questionnaire is designed and administered to a sample of individuals. Some restaurants use observational studies to obtain data about their customers' opinions of the quality of food, service, atmosphere and so on. A questionnaire used by the Lobster Pot Restaurant in Limerick City, Ireland, is shown in Figure 1.3. Note that the customers completing the questionnaire are asked to provide ratings for five variables: food quality, friendliness of service, promptness of service, cleanliness and management. The response categories of excellent, good, satisfactory and unsatisfactory provide ordinal data that enable Lobster Pot's managers to assess the quality of the restaurant's operation.

Managers wanting to use data and statistical analyses as an aid to decision-making must be aware of the time and cost required to obtain the data. The use of existing data sources is desirable when data must be obtained in a relatively short period of time. If important data are not readily available from an existing source, the additional time and cost involved in obtaining the data must be taken into account. In all cases, the decision-maker should consider the contribution of the statistical analysis to the decision-making process. The cost of data acquisition and the subsequent statistical analysis should not exceed the savings generated by using the information to make a better decision.

Data acquisition errors

Managers should always be aware of the possibility of data errors in statistical studies. Using erroneous data can be worse than not using any data at all. An error in data acquisition occurs whenever the data value obtained is not equal to the true or actual value

Figure 1.3 Customer opinion questionnaire used by the Lobster Pot Restaurant, Limerick City, Ireland

The
LOBSTER
Pot
RESTAURANT

W̲e are happy you stopped by the Lobster Pot Restaurant and want to make sure you will come back. So, if you have a little time, we will really appreciate it if you will fill out this card. Your comments and suggestions are extremely important to us. Thank you!

Server's Name _____

	Excellent	Good	Satisfactory	Unsatisfactory
Food Quality	❏	❏	❏	❏
Friendly Service	❏	❏	❏	❏
Prompt Service	❏	❏	❏	❏
Cleanliness	❏	❏	❏	❏
Management	❏	❏	❏	❏

Comments _____

_____

What prompted your vist to us? _____

_____

Please drop in suggestion box at entrance. Thank you.

that would be obtained with a correct procedure. Such errors can occur in a number of ways. For example, an interviewer might make a recording error, such as a transposition in writing the age of a 24-year-old person as 42, or the person answering an interview question might misinterpret the question and provide an incorrect response.

Experienced data analysts take great care in collecting and recording data to ensure that errors are not made. Special procedures can be used to check for internal consistency of the data. For instance, such procedures would indicate that the analyst should review the accuracy of data for a respondent shown to be 22 years of age but reporting 20 years of work experience. Data analysts also review data with unusually large and small values, called outliers, which are candidates for possible data errors. In Chapter 3 we present some of the methods statisticians use to identify outliers.

Errors often occur during data acquisition. Blindly using any data that happen to be available or using data that were acquired with little care can result in misleading information and bad decisions. Thus, taking steps to acquire accurate data can help ensure reliable and valuable decision-making information.

1.4 Descriptive statistics

Most of the statistical information in newspapers, magazines company reports and other publications consists of data that are summarized and presented in a form that is easy for the reader to understand. Such summaries of data, which may be tabular, graphical or numerical, are referred to as **descriptive statistics**.

Refer again to the data set in Table 1.1 showing data on 22 European stock exchanges. Methods of descriptive statistics can be used to provide summaries of the information in this data set. For example, a tabular summary of the data for the six busiest exchanges by trade for the qualitative variable exchange is shown in Table 1.4. A graphical summary of the same data, called a bar graph, is shown in Figure 1.4. These types of tabular and graphical summaries generally make the data easier to interpret. Referring to Table 1.4 and Figure 1.4, we can see easily that the majority of trades are for the London exchange (covering trading in Paris, Brussels, Amsterdam and Lisbon). On a percentage basis, 29.1 per cent of all trades for the 22 European stock exchanges occur through London. Similarly 26.8 per cent occur for Euronext and 18.8 per cent for Deutsche Börse. Note from Table 1.4 that 93 per cent of all trades take place in just six of the 22 European exchanges.

Table 1.4 Per cent frequencies for six busiest exchanges by trades

Exchange	% of Trades
London	29.1
Euronext	26.8
Deutsh Borse	13.4
Borsa Italiana	10.4
NASDAQ OMX Nordic	8.0
Spanish (BME)	4.9
TOTAL	92.6

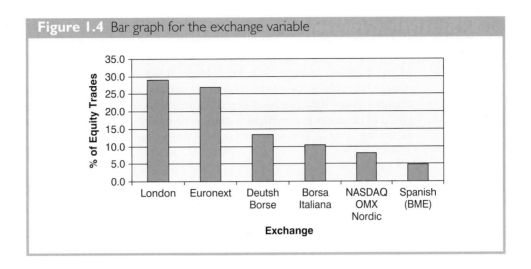

Figure 1.4 Bar graph for the exchange variable

A graphical summary of the data for the quantitative variable turnover for the exchanges, called a histogram, is provided in Figure 1.5. The histogram makes it easy to see that the turnover ranges from €0.0 to €120 000 m, with the highest concentrations between €0 and €30 000 m.

In addition to tabular and graphical displays, numerical descriptive statistics are used to summarize data. The most common numerical descriptive statistic is the average, or mean. Using the data on the variable turnover for the exchanges in Table 1.1, we can compute the average turnover by adding the turnover for the 21 exchanges where turnover has been declared and dividing the sum by 21. Doing so provides an average turnover of €23 144 million. This average demonstrates a measure of the central tendency, or central location, of the data for that variable.

In a number of fields, interest continues to grow in statistical methods that can be used for developing and presenting descriptive statistics. Chapters 2 and 3 devote attention to the tabular, graphical and numerical methods of descriptive statistics.

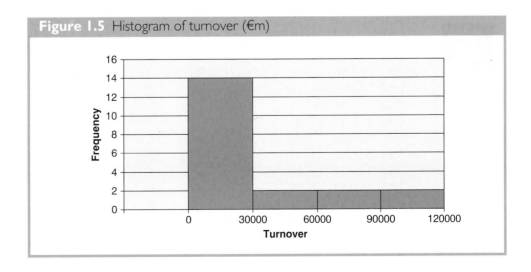

Figure 1.5 Histogram of turnover (€m)

1.5 Statistical inference

Many situations require data for a large group of elements (individuals, companies, voters, households, products, customers and so on). Because of time, cost and other considerations, data can be collected from only a small portion of the group. The larger group of elements in a particular study is called the **population**, and the smaller group is called the **sample**. Formally, we use the following definitions.

> **Population**
>
> A *population* is the set of all elements of interest in a particular study.

> **Sample**
>
> A *sample* is a subset of the population.

The process of conducting a survey to collect data for the entire population is called a **census**. The process of conducting a survey to collect data for a sample is called a **sample survey**. As one of its major contributions, statistics uses data from a sample to make estimates and test hypotheses about the characteristics of a population through a process referred to as **statistical inference**.

NIEVES

Table 1.5 Hours until failure for a sample of 200 light bulbs for the Electronica Nieves example

107	73	68	97	76	79	94	59	98	57
54	65	71	70	84	88	62	61	79	98
66	62	79	86	68	74	61	82	65	98
62	116	65	88	64	79	78	79	77	86
74	85	73	80	68	78	89	72	58	69
92	78	88	77	103	88	63	68	88	81
75	90	62	89	71	71	74	70	74	70
65	81	75	62	94	71	85	84	83	63
81	62	79	83	93	61	65	62	92	65
83	70	70	81	77	72	84	67	59	58
78	66	66	94	77	63	66	75	68	76
90	78	71	101	78	43	59	67	61	71
96	75	64	76	72	77	74	65	82	86
66	86	96	89	81	71	85	99	59	92
68	72	77	60	87	84	75	77	51	45
85	67	87	80	84	93	69	76	89	75
83	68	72	67	92	89	82	96	77	102
74	91	76	83	66	68	61	73	72	76
73	77	79	94	63	59	62	71	81	65
73	63	63	89	82	64	85	92	64	73

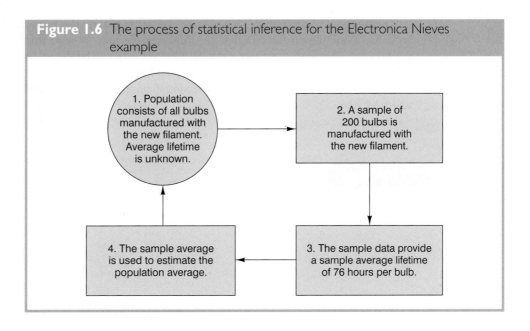

Figure 1.6 The process of statistical inference for the Electronica Nieves example

As an example of statistical inference, let us consider the study conducted by Electronica Nieves. Nieves manufactures a high-intensity light bulb used in a variety of electrical products. In an attempt to increase the useful life of the light bulb, the product design group developed a new light bulb filament. In this case, the population is defined as all light bulbs that could be produced with the new filament. To evaluate the advantages of the new filament, 200 bulbs with the new filament were manufactured and tested. Data collected from this sample showed the number of hours each light bulb operated before the filament burned out or the bulb failed. See Table 1.5.

Suppose Nieves wants to use the sample data to make an inference about the average hours of useful life for the population of all light bulbs that could be produced with the new filament. Adding the 200 values in Table 1.5 and dividing the total by 200 provides the sample average lifetime for the light bulbs: 76 hours. We can use this sample result to estimate that the average lifetime for the light bulbs in the population is 76 hours. Figure 1.6 provides a graphical summary of the statistical inference process for Electronica Nieves.

Whenever statisticians use a sample to estimate a population characteristic of interest, they usually provide a statement of the quality, or precision, associated with the estimate. For the Nieves example, the statistician might state that the point estimate of the average lifetime for the population of new light bulbs is 76 hours with a margin of error of ±4 hours. Thus, an interval estimate of the average lifetime for all light bulbs produced with the new filament is 72 hours to 80 hours. The statistician can also state how confident he or she is that the interval from 72 hours to 80 hours contains the population average.

1.6 Computers and statistical analysis

Because statistical analysis typically involves large amounts of data, analysts frequently use computer software for this work. For instance, computing the average lifetime for the 200 light bulbs in the Electronica Nieves example (see Table 1.5) would be quite tedious without a computer. To facilitate computer usage, the larger data sets in this book are available on the CD that accompanies the text. A logo in the left margin of the text (e.g. Nieves) identifies each of these data sets. The data files are available in MINITAB, PASW and EXCEL formats. In addition, we provide instructions at the end of chapters for carrying out many of the statistical procedures using MINITAB, PASW and EXCEL.

Exercises

1   Discuss the differences between statistics as numerical facts and statistics as a discipline or field of study.

2   Every year *Condé Nast Traveler* conducts an annual survey of subscribers to determine the best new places to stay throughout the world. Table 1.6 shows the ten hotels that were most highly ranked in their 2006 'hot list' survey. Note that (daily) rates quoted are for double rooms and are variously expressed in US dollars, British pounds or euros.

   a.   How many elements are in this data set?
   b.   How many variables are in this data set?
   c.   Which variables are qualitative and which variables are quantitative?
   d.   What type of measurement scale is used for each of the variables?

3   Refer to Table 1.6.

   a.   What is the average number of rooms for the ten hotels?
   b.   If €1 = US$1.3149 = £0.8986 compute the average room rate in euros.

Table 1.6 The ten best new hotels to stay in, in the world

Hot list ranking	Name of property	Country	Room rate	Number of rooms
1	Amangalla, Galle	Sri Lanka	US$574	30
2	Amanwella, Tangalle	Sri Lanka	US$275	30
3	Bairro Alto Hotel, Lisbon	Portugal	€180	55
4	Basico, Playa Del Carmen	Mexico	US$166	15
5	Beit Al Mamlouka	Syria	£75	8
6	Browns Hotel, London	England	£347	117
7	Byblos Art Hotel Villa Amista, Verona	Italy	€270	60
8	Cavas Wine Lodge, Mendoza	Argentina	US$375	14
9	Convento Do Espinheiro Heritage Hotel & Spa, Evora	Portugal	€213	59
10	Cosmopolitan, Toronto	Canada	£150	97

Source: *Condé Nast Traveler*, May 2006 (http://www.cntraveller.co.uk/Special_Features/The_Hot_List_2006/)

   c.   What is the percentage of hotels located in Portugal?
   d.   What is the percentage of hotels with 20 rooms or fewer?

HOTELS

4   Audio systems are typically made up of an MP3 player, a mini disk player, a cassette player, a CD player and separate speakers. The data in Table 1.7 shows the product rating and retail price range for a popular selection of systems. Note that the code Y is used to confirm when a player is included in the system, N when it is not. Output power (watts) details are also provided (Kelkoo Electronics 2006).

   a. How many elements does this data set contain?
   b. What is the population?
   c. Compute the average output power for the sample.

5   Consider the data set for the sample of eight audio systems in Table 1.7.

   a. How many variables are in the data set?
   b. Which of the variables are quantitative and which are qualitative?
   c. What percentage of the audio systems has a four star rating or higher?
   d. What percentage of the audio systems includes an MP3 player?

Table 1.7 A sample of eight audio systems

Brand and model	Product rating (# of stars)	Price (£)	MP3 player	Mini disk player	Cassette player	CD (watts) player	Output
Technics SCEH790	1	320–400	Y	N	Y	Y	360
Yamaha M170	3	162–290	N	N	N	Y	50
Panasonic SCPM29	5	188	Y	N	Y	Y	70
Pure Digital DMX50	3	180–230	N	N	N	Y	80
Sony CMTNEZ3	5	60–100	Y	N	Y	Y	30
Philips FWM589	4	143–200	Y	N	N	Y	400
PHILIPS MCM9	5	93–110	Y	N	Y	Y	100
Samsung MM-C6	5	100–130	Y	N	N	Y	40

Source: Kelkoo (http://audiovisual.kelkoo.co.uk)

6   Columbia House provides CDs to its mail-order club members. A Columbia House Music Survey asked new club members to complete an 11-question survey. Some of the questions asked were:

   a. How many CDs have you bought in the last 12 months?
   b. Are you currently a member of a national mail-order book club? (Yes or No)
   c. What is your age?
   d. Including yourself, how many people (adults and children) are in your household?
   e. What kinds of music are you interested in buying? (15 categories were listed, including hard rock, soft rock, adult contemporary, heavy metal, rap and country.)

Comment on whether each question provides qualitative or quantitative data.

7 The Health & Wellbeing Survey ran over a three-week period (ending 19 October 2007) and 389 respondents took part. The survey asked the respondents to respond to the statement, 'How would you describe your own physical health at this time?' (http://inform. glam.ac.uk/news/2007/10/24/health-wellbeing-staff-survey-results/). Response categories were strongly agree, agree, neither agree or disagree, disagree, and strongly disagree.

a. What was the sample size for this survey?

b. Are the data qualitative or quantitative?

c. Would it make more sense to use averages or percentages as a summary of the data for this question?

d. Of the respondents, 57 per cent agreed with the statement. How many individuals provided this response?

8 State whether each of the following variables is qualitative or quantitative and indicate its measurement scale.

a. Age.

b. Gender.

c. Class rank.

d. Make of car.

e. Number of people favouring closer European integration.

9 Figure 1.7 provides a bar chart summarizing the actual earnings for Volkswagen for the years 2000 to 2008 (Source: *Volkswagen AG Annual Reports 2001–2008*).

a. Are the data qualitative or quantitative?

b. Are the data times series or cross-sectional?

c. What is the variable of interest?

d. Comment on the trend in Volkswagen's earnings over time. Would you expect to see an increase or decrease in 2009?

10 Refer again to the data in Table 1.7 for the audio systems. Are the data cross-sectional or time series? Why?

11 The marketing group at your company developed a new diet soft drink that it claims will capture a large share of the young adult market.

a. What data would you want to see before deciding to invest substantial funds in introducing the new product into the marketplace?

b. How would you expect the data mentioned in part (a) to be obtained?

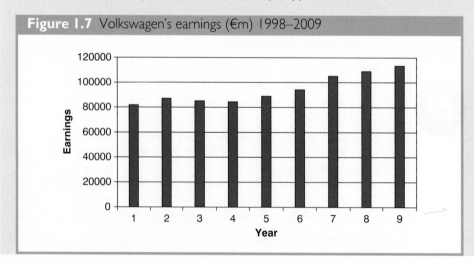

Figure 1.7 Volkswagen's earnings (€m) 1998–2009

12 In a recent study of causes of death in men 60 years of age and older, a sample of 120 men indicated that 48 died as a result of some form of heart disease.

   a. Develop a descriptive statistic that can be used as an estimate of the percentage of men 60 years of age or older who die from some form of heart disease.

   b. Are the data on cause of death qualitative or quantitative?

   c. Discuss the role of statistical inference in this type of medical research.

13 In 2007, 75.4 per cent of *Economist* readers had stayed in a hotel on business in the previous 12 months with 32.4 per cent of readers using first / business class for travel.

   a. What is the population of interest in this study?

   b. Is class of travel a qualitative or quantitative variable?

   c. If a reader had stayed in a hotel on business in the previous 12 months would this be classed as a qualitative or quantitative variable?

   d. Does this study involve cross-sectional or time series data?

   e. Describe any statistical inferences *The Economist* might make on the basis of the survey.

Summary

Statistics is the art and science of collecting, analyzing, presenting and interpreting data. Nearly every college student majoring in business or economics is required to take a course in statistics. We began the chapter by describing typical statistical applications for business and economics.

Data consist of the facts and figures that are collected and analyzed. A set of measurements obtained for a particular element is an observation, Four scales of measurement used to obtain data on a particular variable include nominal, ordinal, interval and ratio. The scale of measurement for a variable is nominal when the data use labels or names to identify an attribute of an element. The scale is ordinal if the data demonstrate the properties of nominal data and the order or rank of the data is meaningful. The scale is interval if the data demonstrate the properties of ordinal data and the interval between values is expressed in terms of a fixed unit of measure. Finally, the scale of measurement is ratio if the data show all the properties of interval data and the ratio of two values is meaningful.

For purposes of statistical analysis, data can be classified as qualitative or quantitative.

Qualitative data use labels or names to identify an attribute of each element. Qualitative data use either the nominal or ordinal scale of measurement and may be non-numeric or numeric.

Quantitative data are numeric values that indicate how much or how many. Quantitative data use either the interval or ratio scale of measurement. Ordinary arithmetic operations are meaningful only if the data are quantitative. Therefore, statistical computations used for quantitative data are not always appropriate for qualitative data. Sources of data – both within organizations and externally – were reviewed and data acquisition errors also discussed.

In Sections 1.4 and 1.5 we introduced the topics of descriptive statistics and statistical inference. Definitions of the population and sample were provided and different types of descriptive statistics – tabular, graphical, and numerical – used to summarize data. The process of statistical inference uses data obtained from a sample to make estimates or test hypotheses about the characteristics of a population.

In the last section of the chapter we noted that computers facilitate statistical analysis. The larger data sets contained in MINITAB, EXCEL and PASW files can be found on the CD that accompanies the text.

Key terms

Census
Cross-sectional data
Data
Data set
Descriptive statistics
Element
Interval scale
Nominal scale
Observation
Ordinal scale
Population

Qualitative data
Qualitative variable
Quantitative data
Quantitative variable
Ratio scale
Sample
Sample survey
Statistical inference
Statistics
Time series data
Variable

Chapter 2

Descriptive Statistics: Tabular and Graphical Presentations

Statistics in practice: YouGov and BrandIndex

Learning objectives

After studying this chapter and doing the exercises, you should be able to construct and interpret a number of different types of tabular and graphical summaries of data.

1 For single qualitative variables: frequency, relative frequency and percentage frequency distributions; bar charts and pie charts.

2 For single quantitative variables: frequency, relative frequency and percentage frequency distributions; cumulative frequency, relative cumulative frequency and percentage cumulative frequency distributions; dot plots, stem-and-leaf plots, histograms and ogives.

3 For pairs of qualitative and quantitative data: cross-tabulations, with row and column percentages.

4 For pairs of quantitative variables: scatter diagrams.

5 You should be able to give an example of Simpson's paradox and explain the relevance of this paradox to the cross-tabulation of variables.

As indicated in Chapter 1, data can be classified as either qualitative or quantitative. **Qualitative data** use labels or names to identify categories of like items. **Quantitative data** are numerical values that indicate how much or how many.

This chapter introduces tabular and graphical methods commonly used to summarize both qualitative and quantitative data. Tabular and graphical summaries of data can be found in annual reports, newspaper articles and research studies. Everyone is exposed to these types of presentations, so it is important to understand how they are prepared and how they should be interpreted. We begin with tabular and graphical methods for summarizing data concerning a single variable. Section 2.3 introduces methods for summarizing data when the relationship between two variables is of interest.

Modern spreadsheet and statistical software packages provide extensive capabilities for summarizing data and preparing graphical presentations. MINITAB, PASW and EXCEL are three packages that are widely available. At the end of this chapter, we show some of their capabilities.

2.1 Summarizing qualitative data

Frequency distribution

We begin with the definition of a **frequency distribution**.

Frequency distribution

A frequency distribution is a tabular summary of data showing the number (frequency) of items in each of several non-overlapping classes.

Statistics in Practice

YouGov and BrandIndex

YouGov is an international research agency that has pioneered Internet polling and research, and is an acknowledged specialist in that area. The agency operates worldwide from five hubs in Europe and North America. One of its leading subscription services is BrandIndex.

To create BrandIndex, YouGov tracks consumers' perceptions, on a daily basis, of over 1000 consumer brands across more than 30 retail sectors. BrandIndex is currently available in the UK, Germany, Turkey and the US. The construction of BrandIndex involves a sophisticated Internet survey design based on large panels of consumers, who are asked to report on seven aspects of their brand perceptions for quite large numbers of consumer brands. Part of the sophistication of the survey design is in organizing and tailoring the items presented to each respondent to prevent question 'overload'.

The seven aspects of brand perception are 'buzz', general impression, quality, value, satisfaction,

YouGov.com website. Reproduced with permission.

recommendation and corporate reputation. Ratings on the latter six of these are averaged to give the BrandIndex rating. 'Buzz' is reported separately, as well as 'mindshare', the proportion of those polled who have offered a view (either positive or negative) on the brand in question.

BrandIndex Turkey was launched in 2007, and is based on consumer responses for about 400 brands. In August 2008, a YouGov report commented on the trends in the BrandIndex Turkey ratings for two coffee shop chains, Starbucks and Kahve Dünyasi (Coffee World). Starbucks opened its first coffee shop in Turkey in 2003, and in 2008 celebrated the opening of its 100th shop. During most of that period, its BrandIndex ratings had shown a positive trend. However, in 2008 it was facing stiff and effective competition from Kahve Dünyasi, an indigenous Turkish coffee shop, whose BrandIndex ratings had outpaced those of Starbucks.

The YouGov website features many headline reports (with a 24-hour delay) such as the one outlined above, and others regularly accompany articles and analysis in the marketing press. Subscribers to the full BrandIndex service can obtain much more detailed figures and reports, and are able to segment the results according to consumer characteristics such as age group, sex and income category. A feature of almost all published BrandIndex reports is a set of charts summarizing, highlighting and comparing trends in the ratings.

In this chapter, you will learn about tabular and graphical methods of descriptive statistics such as frequency distributions, bar charts, histograms, stem-and-leaf displays, cross-tabulations, and others. The goal of these methods is to summarize data so that they can be easily understood and interpreted.

The following example demonstrates the construction and interpretation of a frequency distribution for qualitative data. Coke Classic, Diet Coke, Dr Pepper, Pepsi-Cola and Sprite are five popular soft drinks. The data in Table 2.1 show the soft drinks selected in a sample of 50 soft drink purchases.

To construct a frequency distribution, we count the number of times each soft drink appears in Table 2.1. Coke Classic appears 19 times, Diet Coke appears 8 times and so on. These counts are summarized in the frequency distribution in Table 2.2. The summary offers

Table 2.1 Data from a sample of 50 soft drink purchases

Coke Classic	Sprite	Pepsi-Cola
Diet Coke	Coke Classic	Coke Classic
Pepsi-Cola	Diet Coke	Coke Classic
Diet Coke	Coke Classic	Coke Classic
Coke Classic	Diet Coke	Pepsi-Cola
Coke Classic	Coke Classic	Dr Pepper
Dr Pepper	Sprite	Coke Classic
Diet Coke	Pepsi-Cola	Diet Coke
Pepsi-Cola	Coke Classic	Pepsi-Cola
Pepsi-Cola	Coke Classic	Pepsi-Cola
Coke Classic	Coke Classic	Pepsi-Cola
Dr Pepper	Pepsi-Cola	Pepsi-Cola
Sprite	Coke Classic	Coke Classic
Coke Classic	Sprite	Dr Pepper
Diet Coke	Dr Pepper	Pepsi-Cola
Coke Classic	Pepsi-Cola	Sprite
Coke Classic	Diet Coke	

SOFTDRINK

more insight than the original data shown in Table 2.1. We see that Coke Classic is the leader, Pepsi-Cola is second, Diet Coke is third and Sprite and Dr Pepper are tied for fourth.

Relative frequency and percentage frequency distributions

A frequency distribution shows the number (frequency) of items in each of several non-overlapping classes. We are often interested in the proportion, or percentage, of items in each class. The *relative frequency* of a class equals the fraction or proportion of items belonging to a class. For a data set with n observations, the relative frequency of each class is:

Relative frequency

$$\text{Relative frequency of a class} = \frac{\text{Frequency of the class}}{n} \tag{2.1}$$

The *percentage frequency* of a class is the relative frequency multiplied by 100.

Table 2.2 Frequency distribution of soft drink purchases

Soft drink	Frequency
Coke Classic	19
Diet Coke	8
Dr Pepper	5
Pepsi-Cola	13
Sprite	5
Total	50

Table 2.3 Relative and percentage frequency distributions of soft drink purchases

Soft drink	Relative frequency	Percentage frequency
Coke Classic	0.38	38
Diet Coke	0.16	16
Dr Pepper	0.10	10
Pepsi-Cola	0.26	26
Sprite	0.10	10
Total	1.00	100

A **relative frequency distribution** is a tabular summary showing the relative frequency for each class. A **percentage frequency distribution** summarizes the percentage frequency for each class. Table 2.3 shows these distributions for the soft drink data. The relative frequency for Coke Classic is 19/50 = 0.38, the relative frequency for Diet Coke is 8/50 = 0.16 and so on. From the percentage frequency distribution, we see that 38 per cent of the purchases were Coke Classic, 16 per cent of the purchases were Diet Coke and so on. We can also note that 38 per cent + 26 per cent + 16 per cent = 80 per cent of the purchases were of the top three soft drinks.

Bar charts and pie charts

A **bar chart**, or **bar graph**, is a graphical device for depicting qualitative data summarized in a frequency, relative frequency, or percentage frequency distribution. On one axis of the chart (usually the horizontal axis), we specify the labels for the classes (categories) of data. A frequency, relative frequency or percentage frequency scale can be used for the other axis of the chart (usually the vertical axis). Then, using a bar of fixed width drawn above each class label, we make the length of the bar equal the frequency, relative frequency, or percentage frequency of the class. For qualitative data, the bars should be separated to emphasize the fact that each class is separate. Figure 2.1 shows a bar chart

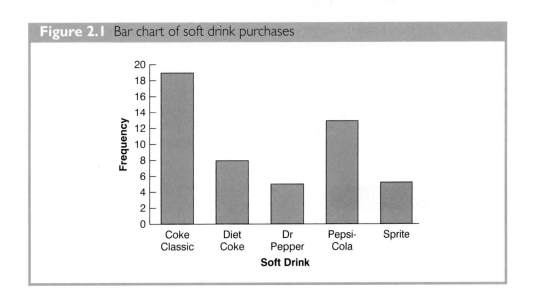

Figure 2.1 Bar chart of soft drink purchases

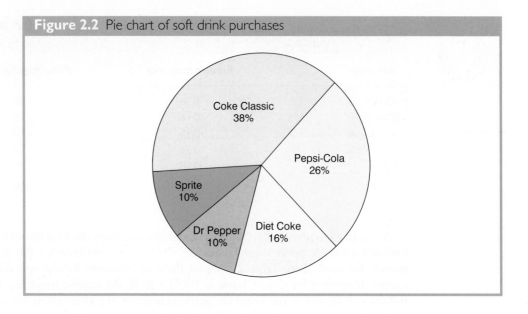

Figure 2.2 Pie chart of soft drink purchases

of the frequency distribution for the 50 soft drink purchases. The graphical presentation shows Coke Classic, Pepsi-Cola and Diet Coke to be the most preferred brands.

A **pie chart** is another way of presenting relative frequency and percentage frequency distributions for qualitative data. We first draw a circle to represent all of the data. Then we use the relative frequencies to subdivide the circle into sectors, or parts, that correspond to the relative frequency for each class. For example, because a circle contains 360 degrees and Coke Classic shows a relative frequency of 0.38, the sector of the pie chart labelled Coke Classic consists of 0.38(360) = 136.8 degrees. The sector of the pie chart labelled Diet Coke consists of 0.16(360) = 57.6 degrees. Similar calculations for the other classes give the pie chart in Figure 2.2. The numerical values shown for each sector can be frequencies, relative frequencies or percentage frequencies.

Often the number of classes in a frequency distribution is the same as the number of categories found in the data, as is the case for the soft drink purchase data in this section. Data that included all soft drinks would require many categories, most of which would have a small number of purchases. Classes with smaller frequencies can be grouped into an aggregate class labelled 'other'. Classes with frequencies of 5 per cent or less would most often be treated in this fashion.

In quality control applications, bar charts are used to identify the most important causes of problems. When the bars are arranged in descending order of height from left to right with the most frequently occurring cause appearing first, the bar chart is called a *Pareto diagram*, named after its founder, Vilfredo Pareto, an Italian economist.

Exercises

Methods

1 The response to a question has three alternatives: A, B and C. A sample of 120 responses provides 60 A, 24 B and 36 C. Construct the frequency and relative frequency distributions.

2   A partial relative frequency distribution is given below.

Class	Relative frequency
A	0.22
B	0.18
C	0.40
D	

   a. What is the relative frequency of class D?
   b. The total sample size is 200. What is the frequency of class D?
   c. Construct the frequency distribution.
   d. Construct the percentage frequency distribution.

3   A questionnaire provides 58 Yes, 42 No and 20 No-opinion answers.

   a. In the construction of a pie chart, how many degrees would be in the sector of the pie showing the Yes answers?
   b. How many degrees would be in the sector of the pie showing the No answers?
   c. Construct a pie chart.
   d. Construct a bar chart.

Applications

4   Figures available on the Broadcasters' Audience Research Board website in October 2008 showed that four of the most popular shows broadcast on terrestrial television in the UK were *The X Factor, Coronation Street, A Touch of Frost* and *Strictly Come Dancing*. Data indicating the favourite show of a sample of 50 viewers follows.

Strictly	Strictly	X Factor	Coronation	X Factor	X Factor	Coronation	X Factor	X Factor	Strictly
Strictly	Frost	Coronation	X Factor	Coronation	Strictly	X Factor	X Factor	X Factor	Coronation
Coronation	X Factor	Frost	X Factor	Coronation	Frost	Strictly	Coronation	Strictly	X Factor
Strictly	Frost	Frost	X Factor	Strictly	Strictly	X Factor	X Factor	Coronation	X Factor
X Factor	Coronation	Coronation	Coronation	X Factor	Strictly	X Factor	Frost	Frost	Strictly

   a. Are these data qualitative or quantitative?
   b. Construct frequency and percentage frequency distributions.
   c. Construct a bar chart and a pie chart.
   d. On the basis of the sample, which television show was the most popular? Which one was second?

5   A Wikipedia article (November 2008) listed the five most common last names in Israel as (in alphabetical order): Biton, Cohen, Levi, Mizrachi and Peretz. A sample of 50 individuals with one of these last names provided the following data.

Cohen	Cohen	Peretz	Cohen	Cohen	Cohen	Levi	Levi	Cohen	Mizrachi
Biton	Levi	Cohen	Peretz	Levi	Levi	Cohen	Cohen	Levi	Levi
Cohen	Cohen	Cohen	Levi	Cohen	Cohen	Mizrachi	Biton	Biton	Cohen
Mizrachi	Levi	Cohen	Cohen	Peretz	Peretz	Cohen	Cohen	Peretz	Mizrachi
Levi	Peretz	Cohen	Cohen	Mizrachi	Cohen	Cohen	Mizrachi	Mizrachi	Cohen

Summarize the data by constructing the following:

   a. Relative and percentage frequency distributions.
   b. A bar chart.
   c. A pie chart.
   d. Based on these data, what are the three most common last names?

6   The flexitime system at Electronics Associates allows employees to begin their working day at 7:00, 7:30, 8:00, 8:30, or 9:00 a.m. The following data represent a sample of the starting times selected by the employees.

7:00	8:30	9:00	8:00	7:30	7:30	8:30	8:30	7:30	7:00
8:30	8:30	8:00	8:00	7:30	8:30	7:00	9:00	8:30	8:00

Summarize the data by constructing the following:

a. A frequency distribution.
b. A percentage frequency distribution.
c. A bar chart.
d. A pie chart.
e. What do the summaries tell you about employee preferences in the flexitime system?

7   A Merrill Lynch Client Satisfaction Survey asked clients to indicate how satisfied they were with their financial consultant. Client responses were coded 1 to 7, with 1 indicating 'not at all satisfied' and 7 indicating 'extremely satisfied'. The following data are from a sample of 60 responses for a particular financial consultant.

CLIENT

5	7	6	6	7	5	5	7	3	6
7	7	6	6	6	5	5	6	7	7
6	6	4	4	7	6	7	6	7	6
5	7	5	7	6	4	7	5	7	6
6	5	3	7	7	6	6	6	6	5
5	6	6	7	7	5	6	4	6	6

a. Comment on why these data are qualitative.
b. Construct a frequency distribution and a relative frequency distribution for the data.
c. Construct a bar chart.
d. On the basis of your summaries, comment on the clients' overall evaluation of the financial consultant.

2.2 Summarizing quantitative data

Frequency distribution

As defined in Section 2.1, a frequency distribution is a tabular summary of data showing the number (frequency) of items in each of several non-overlapping classes. This definition holds for quantitative as well as qualitative data. However, with quantitative data there is usually more work involved in defining the non-overlapping classes to be used in the frequency distribution.

Consider the quantitative data in Table 2.4. These data show the time in days required to complete year-end audits for a sample of 20 clients of Sanderson and Clifford, a small accounting firm. The data are rounded to the nearest day. The three steps necessary to define the classes for a frequency distribution with quantitative data are:

1 Determine the number of non-overlapping classes.
2 Determine the width of each class.
3 Determine the class limits.

AUDIT

Table 2.4 Year-end audit times (in days)									
12	14	19	18	15	15	18	17	20	27
22	23	22	21	33	28	14	18	16	13

We demonstrate these steps by constructing a frequency distribution for the audit time data in Table 2.4.

Number of classes

Classes are formed by specifying ranges that will be used to group the data. As a general guideline, we recommend using between 5 and 20 classes. For a small number of data items, as few as five or six classes may be used to summarize the data. For a larger number of data items, a larger number of classes is usually required. The goal is to use enough classes to show the variation in the data, but not so many classes that some contain only a few data items. Because the number of data items in Table 2.4 is relatively small ($n = 20$), we chose to construct a frequency distribution with five classes.

Width of the classes

The second step is to choose a width for the classes. As a general guideline, we recommend that the width be the same for each class, which reduces the chance of inappropriate interpretations by the user. The choices for the number of classes and the width of classes are not independent decisions. A larger number of classes means a smaller class width and vice versa. To determine an approximate class width, we identify the largest and smallest data values. Then we can use the following expression to determine the approximate class width.

Approximate class width

$$\frac{\text{Largest data value} - \text{Smallest data value}}{\text{Number of classes}} \qquad \text{(2.2)}$$

The approximate class width given by equation (2.2) can be rounded to a more convenient value. For example, an approximate class width of 9.28 might be rounded to 10.

For the year-end audit times, the largest value is 33 and the smallest value is 12. We decided to summarize the data with five classes, so equation (2.2) provides an approximate class width of $(33 - 12)/5 = 4.2$. We decided to round up and use a class width of five days in the frequency distribution.

In practice, the number of classes and the appropriate class width are determined by trial and error. Once a possible number of classes is chosen, equation (2.2) is used to find the approximate class width. The process can be repeated for a different number of classes. Ultimately, the analyst uses judgment to determine the combination of the number of classes and class width that provides a good frequency distribution for summarizing the data. Different people may construct different, but equally acceptable, frequency distributions. The goal is to reveal the natural grouping and variation in the data.

For the audit time data, after deciding to use five classes, each with a width of five days, the next task is to specify the class limits for each of the classes.

Class limits

Class limits must be chosen so that each data item belongs to one and only one class. The *lower class limit* identifies the smallest possible data value assigned to the class. The *upper class limit* identifies the largest possible data value assigned to the class. In constructing frequency distributions for qualitative data, we did not need to specify class limits because each data item naturally fell into a separate class (category). But with quantitative data, such as the audit times in Table 2.4, class limits are necessary to determine where each data value belongs.

Using the audit time data, we selected 10 days as the lower class limit and 14 days as the upper class limit for the first class. This class is denoted 10–14 in Table 2.5. The smallest data value, 12, is included in the 10–14 class. We then selected 15 days as the lower class limit and 19 days as the upper class limit of the next class. We continued defining the lower and upper class limits to obtain a total of five classes: 10–14, 15–19, 20–24, 25–29 and 30–34. The largest data value, 33, is included in the 30–34 class. The difference between the lower class limits of adjacent classes is the class width. Using the first two lower class limits of 10 and 15, we see that the class width is 15 − 10 = 5.

A frequency distribution can now be obtained by counting the number of data values belonging to each class. For example, the data in Table 2.5 show that four values − 12, 14, 14 and 13 − belong to the 10–14 class. The frequency for the 10–14 class is 4. Continuing this counting process for the 15–19, 20–24, 25–29 and 30–34 classes provides the frequency distribution in Table 2.5. Using this frequency distribution, we can observe that:

1 The most frequently occurring audit times are in the class of 15–19 days. Eight of the 20 audit times belong to this class.

2 Only one audit required 30 or more days.

Other comments are possible, depending on the interests of the person viewing the frequency distribution. The value of a frequency distribution is that it provides insights about the data that are not easily obtained by viewing the data in their original unorganized form.

The appropriate values for the class limits with quantitative data depend on the level of accuracy of the data. For instance, with the audit time data, the limits used were integer values because the data were rounded to the nearest day. If the data were rounded to the nearest tenth of a day (e.g. 12.3, 14.4 and so on), the limits would be stated in tenths of days. For example, the first class would be 10.0–14.9. If the data were rounded to the nearest hundredth of a day (e.g. 12.34, 14.45 and so on), the limits would be stated in hundredths of days, e.g. the first class would be 10.00–14.99.

Table 2.5 Frequency distribution for the audit time data

Audit time (days)	Frequency
10–14	4
15–19	8
20–24	5
25–29	2
30–34	1
Total	20

An *open-ended* class requires only a lower class limit or an upper class limit. For example, in the audit time data, suppose two of the audits had taken 58 and 65 days. Rather than continue with the classes of width 5 with classes 35–39, 40–44, 45–49 and so on, we could simplify the frequency distribution to show an open-ended class of '35 or more'. This class would have a frequency count of 2. Most often the open-ended class appears at the upper end of the distribution. Sometimes an open-ended class appears at the lower end of the distribution and occasionally such classes appear at both ends.

Class midpoint

In some applications, we want to know the midpoints of the classes in a frequency distribution for quantitative data. The **class midpoint** is the value halfway between the lower and upper class limits. For the audit time data, the five class midpoints are 12, 17, 22, 27 and 32.

Relative frequency and percentage frequency distributions

We define the relative frequency and percentage frequency distributions for quantitative data in the same manner as for qualitative data. The relative frequency is simply the proportion of the observations belonging to a class. With n observations,

$$\text{Relative frequency of a class} = \frac{\text{Frequency of the class}}{n}$$

The percentage frequency of a class is the relative frequency multiplied by 100.

Based on the class frequencies in Table 2.5 and with $n = 20$, Table 2.6 shows the relative frequency and percentage frequency distributions for the audit time data. Note that 0.40 of the audits, or 40 per cent, required from 15 to 19 days. Only 0.05 of the audits, or 5 per cent, required 30 or more days. Again, additional interpretations and insights can be obtained by using Table 2.6.

Dot plot

One of the simplest graphical summaries of data is a **dot plot**. A horizontal axis shows the range of values for the observations. Each data value is represented by a dot placed above the axis. Figure 2.3 is the dot plot for the audit time data in Table 2.4. The three

Table 2.6 Relative and percentage frequency distributions for the audit time data

Audit time (days)	Relative frequency	Percentage frequency
10–14	0.20	20
15–19	0.40	40
20–24	0.25	25
25–29	0.10	10
30–34	0.05	5
Total	1.00	100

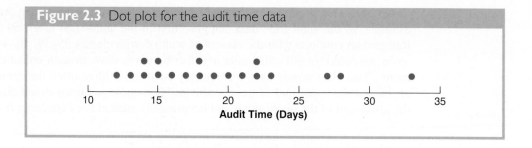

Figure 2.3 Dot plot for the audit time data

dots located above 18 on the horizontal axis indicate that three audit times of 18 days occurred. Dot plots show the details of the data and are useful for comparing the distributions of the data for two or more datasets.

Histogram

A common graphical presentation of quantitative data is a **histogram**. which can be prepared for data previously summarized in a frequency, relative frequency or percentage frequency distribution. The variable of interest is placed on the horizontal axis and the frequency, relative frequency or percentage frequency on the vertical axis. The frequency, relative frequency or percentage frequency of each class is shown by drawing a rectangle whose base is determined by the class limits on the horizontal axis and whose height is the corresponding frequency, relative frequency or percentage frequency.

Figure 2.4 is a histogram for the audit time data. Note that the class with the greatest frequency is shown by the rectangle appearing above the class of 15–19 days. The height of the rectangle shows that the frequency of this class is 8. A histogram for the relative or percentage frequency distribution of this data would look the same as the histogram in Figure 2.4 except that the vertical axis would be labelled with relative or percentage frequency values.

As Figure 2.4 shows, the adjacent rectangles of a histogram touch one another. This is the usual convention for a histogram, unlike a bar chart. Because the classes for the audit time data are stated as 10–14, 15–19, 20–24 and so on, one-unit spaces of 14 to 15, 19 to 20, etc. would seem to be needed between the classes. Eliminating the spaces in the histogram for the audit-time data helps show that, even though the data are rounded

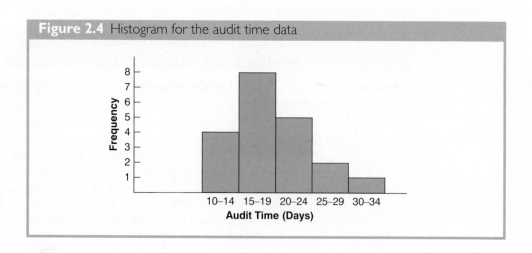

Figure 2.4 Histogram for the audit time data

Figure 2.5 Histograms showing differing levels of skewness

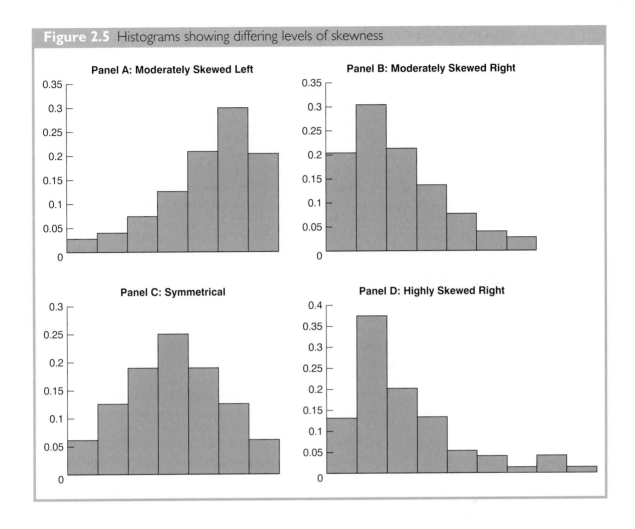

to the nearest full day, all values between the lower limit of the first class and the upper limit of the last class are possible.

One of the most important uses of a histogram is to provide information about the shape, or form, of a distribution. Figure 2.5 contains four histograms constructed from relative frequency distributions. Panel A shows the histogram for a set of data moderately skewed to the left. A histogram is skewed to the left, or negatively skewed, if its tail extends further to the left. A histogram like this might be seen for exam scores. There are no scores above 100 per cent, most of the scores are often above 70 per cent and only a few really low scores occur. Panel B shows the histogram for a set of data moderately skewed to the right. A histogram is skewed to the right, or positively skewed, if its tail extends further to the right. An example of this type of histogram would be for data such as house values. A relatively small number of expensive homes create the skewness in the right tail.

Panel C shows a symmetrical histogram. In a symmetrical histogram, the left tail mirrors the shape of the right tail. Histograms for data found in applications are never perfectly symmetrical, but the histogram for many applications may be roughly symmetrical. Data for IQ scores, heights and weights of people and so on, lead to histograms that are roughly symmetrical. Panel D shows a histogram highly skewed to the right (positively skewed). This histogram was constructed from data on the amount of customer purchases over one day at a women's clothing store. Data from applications in business and economics often lead to histograms that are skewed to the right: for instance, data on housing values, salaries, purchase amounts and so on.

Cumulative distributions

A variation of the frequency distribution that provides another tabular summary of quantitative data is the **cumulative frequency distribution**. The cumulative frequency distribution uses the number of classes, class widths and class limits developed for the frequency distribution. However, rather than showing the frequency of each class, the cumulative frequency distribution shows the number of data items with values *less than or equal to the upper class limit* of each class. The first two columns of Table 2.7 show the cumulative frequency distribution for the audit time data.

Consider the class with the description 'less than or equal to 24'. The cumulative frequency for this class is simply the sum of the frequencies for all classes with data values less than or equal to 24. For the frequency distribution in Table 2.5, the sum of the frequencies for classes 10–14, 15–19 and 20–24 indicates that $4 + 8 + 5 = 17$ data values are less than or equal to 24. The cumulative frequency distribution in Table 2.7 also shows that four audits were completed in 14 days or less and 19 audits were completed in 29 days or less.

A **cumulative relative frequency distribution** shows the proportion of data items and a **cumulative percentage frequency distribution** shows the percentage of data items with values less than or equal to the upper limit of each class. The cumulative relative frequency distribution can be computed either by summing the relative frequencies in the relative frequency distribution, or by dividing the cumulative frequencies by the total number of items. Using the latter approach, we found the cumulative relative frequencies in column 3 of Table 2.7 by dividing the cumulative frequencies in column 2 by the total number of items ($n = 20$). The cumulative percentage frequencies were computed by multiplying the relative frequencies by 100. The cumulative relative and percentage frequency distributions show that 0.85 of the audits, or 85 per cent, were completed in 24 days or less; 0.95 of the audits, or 95 per cent, were completed in 29 days or less and so on.

The last entry in a cumulative frequency distribution always equals the total number of observations. The last entry in a cumulative relative frequency distribution always equals 1.00 and the last entry in a cumulative percentage frequency distribution always equals 100.

Ogive

A graph of a cumulative distribution, called an **ogive**, shows data values on the horizontal axis and either the cumulative frequencies, the cumulative relative frequencies, or the cumulative percentage frequencies on the vertical axis. Figure 2.6 illustrates an ogive for the cumulative frequencies of the audit time data.

Table 2.7 Cumulative frequency, cumulative relative frequency and cumulative percentage frequency distributions for the audit time data

Audit time (days)	Cumulative frequency	Cumulative relative frequency	Cumulative percentage frequency
Less than or equal to 14	4	0.20	20
Less than or equal to 19	12	0.60	60
Less than or equal to 24	17	0.85	85
Less than or equal to 29	19	0.95	95
Less than or equal to 34	20	1.00	100

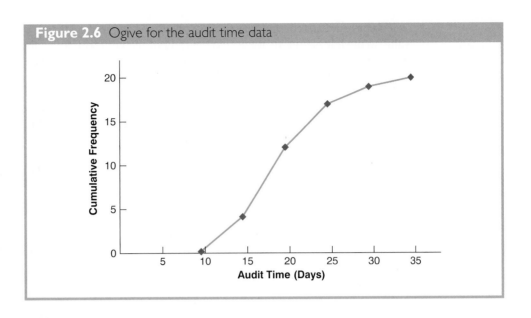

Figure 2.6 Ogive for the audit time data

The ogive is constructed by plotting a point corresponding to the cumulative frequency of each class. Because the classes for the audit time data are 10–14, 15–19, 20–24 and so on, one-unit gaps appear from 14 to 15, 19 to 20 and so on. These gaps are eliminated by plotting points halfway between the class limits. So, 14.5 is used for the 10–14 class, 19.5 is used for the 15–19 class and so on. The 'less than or equal to 14' class with a cumulative frequency of four is shown on the ogive in Figure 2.6 by the point located at 14.5 on the horizontal axis and 4 on the vertical axis. The 'less than or equal to 19' class with a cumulative frequency of 12 is shown by the point located at 19.5 on the horizontal axis and 12 on the vertical axis. Note that one additional point is plotted at the left end of the ogive. This point starts the ogive by showing that no data values fall below the 10–14 class. It is plotted at 9.5 on the horizontal axis and 0 on the vertical axis. The plotted points are connected by straight lines to complete the ogive.

Exploratory data analysis: stem-and-leaf display

Exploratory data analysis techniques consist of simple arithmetic and easy-to-draw graphs that can be used to summarize data quickly. One technique – referred to as a **stem-and-leaf display** – can be used to show both the rank order and shape of a data set simultaneously. To illustrate the stem-and-leaf display, consider the data in Table 2.8. These came from a 150-question aptitude test given to 50 individuals recently interviewed for a position at Hawkins Manufacturing. The data indicate the number of questions answered correctly.

Table 2.8 Number of questions answered correctly on an aptitude test

112	72	69	97	107	73	92	76	86	73
126	128	118	127	124	82	104	132	134	83
92	108	96	100	92	115	76	91	102	81
95	141	81	80	106	84	119	113	98	75
68	98	115	106	95	100	85	94	106	119

To construct a stem-and-leaf display, we first arrange the leading digits of each data value to the left of a vertical line. To the right of the vertical line, on the line corresponding to the appropriate first digit, we record the last digit for each data value as we pass through the observations in the order they were recorded.

```
 6 | 9  8
 7 | 2  3  6  3  6  5
 8 | 6  2  3  |  |  0  4  5
 9 | 7  2  2  6  2  |  5  8  8  5  4
10 | 7  4  8  0  2  6  6  0  6
11 | 2  8  5  9  3  5  9
12 | 6  8  7  4
13 | 2  4
14 | |
```

Sorting the digits on each line into rank order is now relatively simple. This leads to the stem-and-leaf display shown here.

```
 6 | 8  9
 7 | 2  3  3  5  6  6
 8 | 0  |  |  2  3  4  5  6
 9 | |  2  2  2  4  5  5  6  7  8  8
10 | 0  0  2  4  6  6  6  7  8
11 | 2  3  5  5  8  9  9
12 | 4  6  7  8
13 | 2  4
14 | |
```

The numbers to the left of the vertical line (6, 7, 8, 9, 10, 11, 12, 13 and 14) form the *stem*, and each digit to the right of the vertical line is a *leaf*. For example, the first row has a stem value of 6 and leaves of 8 and 9. It indicates that two data values have a first digit of six. The leaves show that the data values are 68 and 69. Similarly, the second row indicates that six data values have a first digit of seven. The leaves show that the data values are 72, 73, 73, 75, 76 and 76. Rotating the page counter-clockwise onto its side provides a picture of the data that is similar to a histogram with classes of 60–69, 70–79, 80–89 and so on.

Although the stem-and-leaf display may appear to offer the same information as a histogram, it has two primary advantages.

1 The stem-and-leaf display is easier to construct by hand for small data sets.

2 Within a class interval, the stem-and-leaf display provides more information than the histogram because the stem-and-leaf shows the actual data.

Just as a frequency distribution or histogram has no absolute number of classes, neither does a stem-and-leaf display have an absolute number of rows or stems. If we believe that our original stem-and-leaf display condensed the data too much, we can stretch the display by using two or more stems for each leading digit. For example, to use two stems for each leading digit, we would place all data values ending in 0, 1, 2, 3 and 4 in one row and all values ending in 5, 6, 7, 8 and 9 in a second row. The following display illustrates this approach. This stretched stem-and-leaf display is similar to a frequency distribution with intervals of 65–69, 70–74, 75–79 and so on.

```
 6 | 8  9
 7 | 2  3  3
 7 | 5  6  6
 8 | 0  1  1  2  3  4
 8 | 5  6
 9 | 1  2  2  2  4
 9 | 5  5  6  7  8  8
10 | 0  0  2  4
10 | 6  6  6  7  8
11 | 2  3
11 | 5  5  8  9  9
12 | 4
12 | 6  7  8
13 | 2  4
14 | 1
```

The preceding example showed a stem-and-leaf display for data with three digits. Stem-and-leaf displays for data with more than three digits are possible. For example, consider the following data on the number of burgers sold by a fast-food restaurant for each of 15 weeks.

| 1565 | 1852 | 1644 | 1766 | 1888 | 1912 | 2044 | 1812 |
| 1790 | 1679 | 2008 | 1852 | 1967 | 1954 | 1733 | |

A stem-and-leaf display of these data follows.

```
       Leaf unit = 10
15 | 6
16 | 4  7
17 | 3  6  9
18 | 1  5  5  8
19 | 1  5  6
20 | 0  4
```

A single digit is used to define each leaf, and only the first three digits of each observation have been used to construct the display. At the top of the display we have specified leaf unit = 10. Consider the first stem (15) and its associated leaf (6). Combining these numbers gives 156. To reconstruct an approximation of the original data value, we must multiply this number by 10, the value of the *leaf unit*: 156 × 10 = 1560. Although it is not possible to reconstruct the exact data value from the display, using a single digit for each leaf enables stem-and-leaf displays to be constructed for data having a large number of digits. Leaf units may be 100, 10, 1, 0.1 and so on. Where the leaf unit is not shown on the display, it is assumed to equal 1.

Exercises

Methods

8   Consider the following data.

14	21	23	21	16	19	22	25	16	16
24	24	25	19	16	19	18	19	21	12
16	17	18	23	25	20	23	16	20	19
24	26	15	22	24	20	22	24	22	20

a. Construct a frequency distribution using classes of 12–14, 15–17, 18–20, 21–23 and 24–26.

b. Construct a relative frequency distribution and a percentage frequency distribution using the classes in (a).

9   Consider the following frequency distribution. Construct a cumulative frequency distribution and a cumulative relative frequency distribution.

Class	Frequency
10–19	10
20–29	14
30–39	17
40–49	7
50–59	2

10   Construct a histogram and an ogive for the data in Exercise 9.

11   Consider the following data.

8.9	10.2	11.5	7.8	10.0	12.2	13.5	14.1	10.0	12.2
6.8	9.5	11.5	11.2	14.9	7.5	10.0	6.0	15.8	11.5

a. Construct a dot plot.

b. Construct a frequency distribution.

c. Construct a percentage frequency distribution.

12   Construct a stem-and-leaf display for the following data.

70   72   75   64   58   83   80   82   76   75   68   65   57   78   85   72

13   Construct a stem-and-leaf display for the following data.

11.3   9.6   10.4   7.5   8.3   10.5   10.0   9.3   8.1   7.7   7.5   8.4   6.3   8.8

Applications

14   A doctor's office staff studied the waiting times for patients who arrive at the office with a request for emergency service. The following data with waiting times in minutes were collected over a one-month period.

2   5   10   12   4   4   5   17   11   8   9   8   12   21   6   8   7   13   18   3

Use classes of 0–4, 5–9 and so on in the following:

a. Show the frequency distribution.

b. Show the relative frequency distribution.

c. Show the cumulative frequency distribution.

d. Show the cumulative relative frequency distribution.

e. What proportion of patients needing emergency service wait nine minutes or less?

15 Data for the numbers of units produced by a production employee during the most recent 20 days are shown here.

160	170	181	156	176	148	198	179	162	150
162	156	179	178	151	157	154	179	148	156

Summarize the data by constructing the following:

a. A frequency distribution.

b. A relative frequency distribution.

c. A cumulative frequency distribution.

d. A cumulative relative frequency distribution.

e. An ogive.

16 The closing prices of 40 company shares (in euros) follow.

SHARES

29.63	34.00	43.25	8.75	37.88	8.63	7.63	30.38
35.25	19.38	9.25	16.50	38.00	53.38	16.63	1.25
48.38	18.00	9.38	9.25	10.00	25.02	18.00	8.00
28.50	24.25	21.63	18.50	33.63	31.13	32.25	29.63
79.38	11.38	38.88	11.50	52.00	14.00	9.00	33.50

a. Construct frequency and relative frequency distributions.

b. Construct cumulative frequency and cumulative relative frequency distributions.

c. Construct a histogram.

d. Using your summaries, make comments and observations about the price of shares.

17 The table below shows the estimated 2009 mid-year population of Zambia, by age group, rounded to the nearest thousand (from the US Census Bureau International Data Base).

Age group	Population (000s)
0 – 4	2005
5 – 9	1749
10 – 14	1591
15 – 19	1440
20 – 24	1253
25 – 29	1022
30 – 34	770
35 – 39	536
40 – 44	369
45 – 49	288
50 – 54	227
55 – 59	186
60 – 64	146
65 – 69	113
70 – 74	83
75 – 79	50
80+	33

a. Construct a percentage frequency distribution.

b. Construct a cumulative percentage frequency distribution.

c. Construct an ogive.

d. Using the ogive, estimate the median age of the population.

COMPUTER

18 The *Nielsen Home Technology Report* provided information about home technology and its usage by individuals aged 12 and older. The following data are the hours of personal computer usage during one week for a sample of 50 individuals.

4.1	1.5	5.9	3.4	5.7	1.6	6.1	3.0	3.7	3.1	4.8	2.0	3.3
11.1	3.5	4.1	4.1	8.8	5.6	4.3	7.1	10.3	6.2	7.6	10.8	0.7
4.0	9.2	4.4	5.7	7.2	6.1	5.7	5.9	4.7	3.9	3.7	3.1	12.1
14.8	5.4	4.2	3.9	4.1	2.8	9.5	12.9	6.1	3.1	10.4		

Summarize the data by constructing the following:

a. A frequency distribution (use a class width of three hours).
b. A relative frequency distribution.
c. A histogram.
d. An ogive.
e. Comment on what the data indicate about personal computer usage at home.

HIGH-LOW

19 The daily high and low temperatures (in degrees Celsius) for 20 cities on one particular day follow.

City	High	Low	City	High	Low
Athens	24	12	Melbourne	19	10
Bangkok	33	23	Montreal	18	11
Cairo	29	14	Paris	25	13
Copenhagen	18	4	Rio de Janeiro	27	16
Dublin	18	8	Rome	27	12
Havana	30	20	Seoul	18	10
Hong Kong	27	22	Singapore	32	24
Johannesburg	16	10	Sydney	20	13
London	23	9	Tokyo	26	15
Manila	34	24	Vancouver	14	6

a. Prepare a stem-and-leaf display for the high temperatures.
b. Prepare a stem-and-leaf display for the low temperatures.
c. Compare the stem-and-leaf displays from parts (a) and (b), and comment on the differences between daily high and low temperatures.
d. Use the stem-and-leaf display from part (a) to determine the number of cities having a high temperature of 25 degrees or above.
e. Provide frequency distributions for both high and low temperature data.

2.3 Cross-tabulations and scatter diagrams

So far in this chapter, we have focused on tabular and graphical methods used to summarize the data for *one variable at a time*. Often a manager or decision-maker requires tabular and graphical methods that will assist in the understanding of the *relationship between two variables*. Cross-tabulation and scatter diagrams are two such methods.

Cross-tabulation

A **cross-tabulation** is a tabular summary of data for two variables. Consider the following data from a consumer restaurant review, based on a sample of 300 restaurants located in a large European city. Table 2.9 shows the data for the first five restaurants. Data on

Table 2.9 Quality rating and meal price for 300 restaurants

RESTAURANT

Restaurant	Quality rating	Meal price (€)
1	Good	18
2	Very Good	22
3	Good	28
4	Excellent	38
5	Very Good	33
⋮	⋮	⋮

a restaurant's quality rating and typical meal price are reported. Quality rating is a qualitative variable with rating categories of good, very good and excellent. Meal price is a quantitative variable that ranges from €10 to €49.

A cross-tabulation of the data is shown in Table 2.10. The left and top margin labels define the classes for the two variables. In the left margin, the row labels (good, very good and excellent) correspond to the three classes of the quality rating variable. In the top margin, the column labels (€10–19, €20–29, €30–39 and €40–49) correspond to the four classes of the meal price variable. Each restaurant in the sample provides a quality rating and a meal price, and so is associated with a cell appearing in one of the rows and one of the columns of the cross-tabulation. For example, restaurant 5 is identified as having a very good quality rating and a meal price of €33. This restaurant belongs to the cell in row 2 and column 3 of Table 2.10. In constructing a cross-tabulation, we simply count the number of restaurants that belong to each of the cells in the cross-tabulation.

We see that the greatest number of restaurants in the sample (64) have a very good rating and a meal price in the €20–29 range. Only two restaurants have an excellent rating and a meal price in the €10–19 range. In addition, note that the right and bottom margins of the cross-tabulation provide the frequency distributions for quality rating and meal price separately. From the frequency distribution in the right margin, we see that data on quality ratings show 84 good restaurants, 150 very good restaurants and 66 excellent restaurants.

Dividing the totals in the right margin of the cross-tabulation by the total for that column provides relative and percentage frequency distributions for the quality rating variable.

Quality rating	Relative frequency	Percentage frequency
Good	0.28	28
Very good	0.50	50
Excellent	0.22	22
Total	1.00	100

Table 2.10 Cross-tabulation of quality rating and meal price for 300 restaurants

Quality rating	Meal price				
	€10–19	€20–29	€30–39	€40–49	Total
Good	42	40	2	0	84
Very good	34	64	46	6	150
Excellent	2	14	28	22	66
Total	78	118	76	28	300

From the percentage frequency distribution we see that 28 per cent of the restaurants were rated good, 50 per cent were rated very good and 22 per cent were rated excellent.

Dividing the totals in the bottom row of the cross-tabulation by the total for that row provides relative and percentage frequency distributions for the meal price variable. Note that in this case the values do not add exactly to 100, because the values being summed are rounded. From the percentage frequency distribution we quickly see that 26 per cent of the meal prices are in the lowest price class (€10–19), 39 per cent are in the next higher class and so on.

Meal price	Relative frequency	Percentage frequency
€10–19	0.26	26
€20–29	0.39	39
€30–39	0.25	25
€40–49	0.09	9
Total	1.00	100

The frequency and relative frequency distributions constructed from the margins of a cross-tabulation provide information about each of the variables individually, but they do not shed any light on the relationship between the variables. The primary value of a cross-tabulation lies in the insight it offers about the relationship between the variables. The cross-tabulation in Table 2.10 reveals that higher meal prices are associated with the higher quality restaurants, and the lower meal prices are associated with the lower quality restaurants.

Converting the entries in a cross-tabulation into row percentages or column percentages can provide more insight into the relationship between the two variables. For row percentages, the results of dividing each frequency in Table 2.10 by its corresponding row total are shown in Table 2.11. Each row of Table 2.11 is a percentage frequency distribution of meal price for one of the quality rating categories. Of the restaurants with the lowest quality rating (good), we see that the greatest percentages are for the less expensive restaurants (50 per cent have €10–19 meal prices and 47.6 per cent have €20–29 meal prices). Of the restaurants with the highest quality rating (excellent), we see that the greatest percentages are for the more expensive restaurants (42.4 per cent have €30–39 meal prices and 33.4 per cent have €40–49 meal prices). Hence, we see that the more expensive meals are associated with the higher quality restaurants.

Cross-tabulation is widely used for examining the relationship between two variables. The final reports for many statistical studies include a large number of cross-tabulations. In the restaurant survey, the cross-tabulation is based on one qualitative variable (quality rating) and one quantitative variable (meal price). Cross-tabulations can also be constructed when both variables are qualitative and when both variables are quantitative.

Table 2.11 Row percentages for each quality rating category

| Quality rating | Meal price | | | | |
	€10–19	€20–29	€30–39	€40–49	Total
Good	50.0	47.6	2.4	0.0	100
Very good	22.7	42.7	30.6	4.0	100
Excellent	3.0	21.2	42.4	33.4	100

When quantitative variables are used, we must first create classes for the values of the variable. For instance, in the restaurant example we grouped the meal prices into four classes (€10–19, €20–29, €30–39 and €40–49).

Simpson's paradox

The data in two or more cross-tabulations are sometimes combined or aggregated to produce a summary cross-tabulation showing how two variables are related. In such cases, we must be careful in drawing conclusions about the relationship between the two variables in the aggregated cross-tabulation. In some cases the conclusions based upon the aggregated cross-tabulation can be completely reversed if we look at the non-aggregated data, something known as **Simpson's paradox**. To provide an illustration, we consider an example involving the analysis of sales success for two sales executives in a mobile telephone company.

The two sales executives are Aaron and Theo. They handle enquiries for renewal of two types of mobile telephone agreement: pre-pay contracts and pay-as-you-go (PAYG) agreements. The cross-tabulation below shows the outcomes for 200 enquiries each for Aaron and Theo, aggregated across the two types of agreement. The cross-tabulation involves two variables: outcome (sale or no sale) and sales executive (Aaron or Theo). It shows the number of sales and the number of no-sales for each executive, along with the column percentages in parentheses next to each value.

	Sales executive		
	Aaron	Theo	Total
Sales	82 (41%)	102 (51%)	184
No-sales	118 (59%)	98 (49%)	216
Total	200 (100%)	200 (100%	400

The column percentages indicate that Aaron's overall sales success rate was 41 per cent, compared with Theo's 51 per cent success rate, suggesting that Theo has the better sales performance. A problem arises with this conclusion, however. The following cross-tabulations show the enquiries handled by Aaron and Theo for the two types of agreement separately.

	Pre-pay					PAYG		
	Aaron	Theo	Total			Aaron	Theo	Total
Sales	56 (35%)	18 (30%)	74		Sales	26 (65%)	84 (60%)	110
No-sales	104 (65%)	42 (70%)	146		No-sales	14 (35%)	56 (40%)	70
Total	160 (100%)	60 (100%)	220		Total	40 (100%)	140 (100%)	180

We see that Aaron achieved a 35 per cent success rate for pre-pay contracts and 65 per cent for PAYG agreements. Theo had a 30 per cent success rate for pre-pay and 60 per cent for PAYG. This comparison suggests that Aaron has a better success rate than Theo for both types of agreement, a result that contradicts the conclusion reached when the data were aggregated across the two types of agreement. This example illustrates Simpson's paradox.

Note that for both sales executives the sales success rate was much higher for PAYG than for pre-pay contracts. Because Theo handled a much higher proportion of PAYG enquiries than Aaron, the aggregated data favoured Theo. When we look at the cross-tabulations for the two types of agreement separately, however, Aaron shows the better record. Hence, for the original cross-tabulation, we see that the *type of agreement* is a hidden variable that should not be ignored when evaluating the records of the sales executives.

Because of Simpson's paradox, we need to be especially careful when drawing conclusions using aggregated data. Before drawing any conclusions about the relationship between two variables shown for a cross-tabulation – or, indeed, any type of display involving two variables (like the scatter diagram illustrated in the next section) – you should consider whether any hidden variable or variables could affect the results.

Scatter diagram and trend line

A **scatter diagram** is a graphical presentation of the relationship between two quantitative variables, and a **trend line** is a line that provides an approximation of the relationship. Consider the advertising/sales relationship for a hi-fi equipment store. On ten occasions during the past three months, the store used weekend television commercials to promote sales at its stores. The managers want to investigate whether a relationship exists between the number of commercials shown and sales at the store during the following week. Sample data for the ten weeks with sales in thousands of euros (€000s) are shown in Table 2.12.

Figure 2.7 shows the scatter diagram and the trend line* for the data in Table 2.12. The number of commercials (x) is shown on the horizontal axis and the sales (y) are shown on the vertical axis. For week 1, $x = 2$ and $y = 50$. A point with those coordinates is plotted on the scatter diagram. Similar points are plotted for the other nine weeks. Note that during two of the weeks one commercial was shown, during two of the weeks two commercials were shown, and so on.

The completed scatter diagram in Figure 2.7 indicates a positive relationship between the number of commercials and sales. Higher sales are associated with a higher number of commercials. The relationship is not perfect in that all points are not on a straight line. However, the general pattern of the points and the trend line suggest that the overall relationship is positive.

Some general scatter diagram patterns and the types of relationships they suggest are shown in Figure 2.8. The top left panel depicts a positive relationship similar to the one

Table 2.12 Sample data for the hi-fi equipment store		
Week	Number of commercials	Sales in €000s
1	2	50
2	5	57
3	1	41
4	3	54
5	4	54
6	1	38
7	5	63
8	3	48
9	4	59
10	2	46

*The equation of the trend line is $y = 4.95x + 36.15$. The slope of the trend line is 4.95 and the y-intercept (the point where the line intersects the y axis) is 36.15. We will discuss in detail the interpretation of the slope and y-intercept for a linear trend line in Chapter 14 when we study simple linear regression.

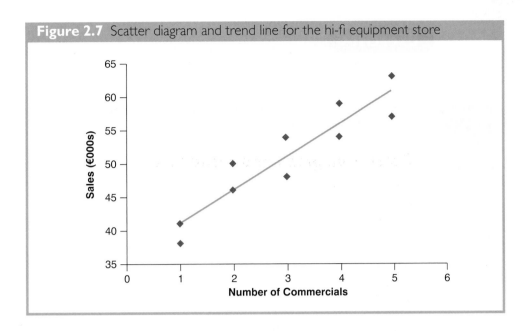

Figure 2.7 Scatter diagram and trend line for the hi-fi equipment store

for the number of commercials and sales example. In the top right panel, the scatter diagram shows no apparent relationship between the variables. The bottom panel depicts a negative relationship where *y* tends to decrease as *x* increases.

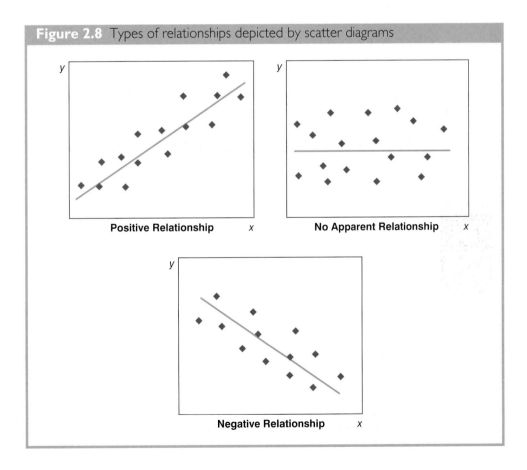

Figure 2.8 Types of relationships depicted by scatter diagrams

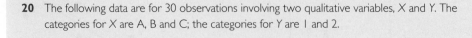

Exercises

Methods

CROSSTAB

20 The following data are for 30 observations involving two qualitative variables, X and Y. The categories for X are A, B and C; the categories for Y are 1 and 2.

Observation	X	Y	Observation	X	Y
1	A	1	16	B	2
2	B	1	17	C	1
3	B	1	18	B	1
4	C	2	19	C	1
5	B	1	20	B	1
6	C	2	21	C	2
7	B	1	22	B	1
8	C	2	23	C	2
9	A	1	24	A	1
10	B	1	25	B	1
11	A	1	26	C	2
12	B	1	27	C	2
13	C	2	28	A	1
14	C	2	29	B	1
15	C	2	30	B	2

a. Construct a cross-tabulation for the data, with X as the row variable and Y as the column variable.
b. Calculate the row percentages.
c. Calculate the column percentages.
d. What is the relationship, if any, between X and Y?

21 The following 20 observations are for two quantitative variables.

SCATTER

Observation	X	Y	Observation	X	Y
1	−22	22	11	−37	48
2	−33	49	12	34	−29
3	2	8	13	9	−18
4	29	−16	14	−33	31
5	−13	10	15	20	−16
6	21	−28	16	−3	14
7	−13	27	17	−15	18
8	−23	35	18	12	17
9	14	−5	19	−20	−11
10	3	−3	20	−7	−22

a. Construct a scatter diagram for the relationship between X and Y.
b. What is the relationship, if any, between X and Y?

Applications

22 Recently, management at Oak Tree Golf Course received a few complaints about the condition of the greens. Several players complained that the greens are too fast. Rather than react to the comments of just a few, the Golf Association conducted a survey of 100 male and 100 female golfers. The survey results are summarized here.

	Male golfers			Female golfers	
	Greens condition			Greens condition	
Handicap	Too fast	Fine	Handicap	Too fast	Fine
Under 15	10	40	Under 15	1	9
15 or more	25	25	15 or more	39	51

a. Combine these two cross-tabulations into one with male, female as the row labels and the column labels too fast and fine. Which group shows the highest percentage saying that the greens are too fast?

b. Refer to the initial cross-tabulations. For those players with low handicaps (better players), which group (male or female) shows the highest percentage saying the greens are too fast?

c. Refer to the initial cross-tabulations. For those players with higher handicaps, which group (male or female) shows the highest percentage saying the greens are too fast?

d. What conclusions can you draw about the preferences of men and women concerning the speed of the greens? Are the conclusions you draw from part (a) as compared with parts (b) and (c) consistent? Explain any apparent inconsistencies.

23 The file 'House Sales' on the accompanying CD contains data for a sample of 50 houses advertised for sale in a regional UK newspaper in autumn 2008. The first five rows of data are shown for illustration below.

Price (£)	Location	House type	Bedrooms	Reception rooms	Bedrooms + Receptions	Garage capacity
234 995	Town	Detached	4	2	6	1
319 000	Town	Detached	4	2	6	1
154 995	Town	Semi-detached	2	1	3	0
349 950	Village	Detached	4	2	6	2
244 995	Town	Detached	3	2	5	1

a. Prepare a cross-tabulation using sale price (rows) and house type (columns). Use classes of 100 000–199 999, 200 000–299 999, etc. for sale price.

b. Compute row percentages and comment on any relationship between the variables.

24 Refer to the data in Exercise 23.

a. Prepare a cross-tabulation using number of bedrooms and house type.

b. Prepare a frequency distribution for number of bedrooms.

c. Prepare a frequency distribution for house type.

d. How has the cross-tabulation helped in preparing the frequency distributions in parts (b) and (c)?

25 The file 'Income Inequality' on the accompanying CD contains data for 29 countries prepared by the Organization for Economic Cooperation & Development (OECD) and published in an article in the *Guardian* newspaper in October 2008. The two variables in the file are the Gini coefficient for each country and the percentage of children in the country estimated to be living in poverty. The Gini coefficient is a widely used measure of income inequality. It varies between 0 and 1, with higher coefficients indicating more inequality. The first five rows of data are shown for illustration below.

INCOME
INEQUALITY

	Child poverty (%)	Income inequality
Turkey	24.6	0.430
Mexico	22.2	0.474
Poland	21.5	0.372
US	20.6	0.381
Spain	17.3	0.319

a. Prepare a scatter diagram using the data on child poverty and income inequality.
b. Comment on the relationship, if any, between the variables.

For additional online summary questions and answers go to the companion website at www.cengage.co.uk/aswsbe2

Summary

A set of data, even if modest in size, is often difficult to interpret directly in the form in which it is gathered. Tabular and graphical methods provide procedures for organizing and summarizing data so that patterns are revealed and the data are more easily interpreted.

Figure 2.9 shows the tabular and graphical methods presented in this chapter.

Frequency distributions, relative frequency distributions, percentage frequency distributions, bar charts and pie charts were presented as tabular and graphical procedures for summarizing qualitative data. Frequency distributions, relative frequency distributions, percentage frequency distributions, histograms, cumulative frequency distributions, cumulative relative frequency distributions, cumulative percentage frequency distributions and ogives were presented as ways of summarizing quantitative data. A stem-and-leaf display provides an exploratory data analysis technique that can be used to summarize quantitative data.

Cross-tabulation was presented as a tabular method for summarizing data for two variables. An example of Simpson's paradox was set out, to illustrate the care that must be taken when interpreting relationships between two variables using aggregated data. The scatter diagram was introduced as a graphical method for showing the relationship between two quantitative variables.

With large data sets, computer software packages are essential in constructing tabular and graphical summaries of data. We show in the Software Section how MINITAB, PASW and EXCEL can be used for this purpose.

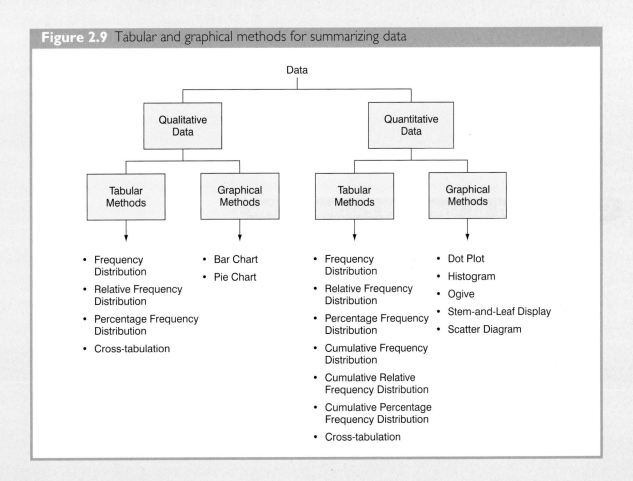

Figure 2.9 Tabular and graphical methods for summarizing data

Key terms

Bar chart
Bar graph
Class midpoint
Cross-tabulation
Cumulative frequency distribution
Cumulative percentage frequency distribution
Cumulative relative frequency distribution
Dot plot
Exploratory data analysis
Frequency distribution

Histogram
Ogive
Percentage frequency distribution
Pie chart
Qualitative data
Quantitative data
Relative frequency distribution
Scatter diagram
Simpson's paradox
Stem-and-leaf display
Trend line

Key formulae

Relative frequency

$$\frac{\text{Frequency of the class}}{n} \tag{2.1}$$

Approximate class width

$$\frac{\text{Largest data value} - \text{Smallest data value}}{\text{Number of classes}} \tag{2.2}$$

Case problem In The Mode Fashion Stores

In The Mode is a chain of women's fashion stores. The chain recently ran a promotion in which discount coupons were sent to customers of related stores. Data collected for a sample of 100 in-store credit card transactions during a single day following the promotion are contained in the file named 'Mode' on the CD accompanying the text. A portion of the data set is shown below. A non-zero amount for the discount variable indicates that the customer brought in the promotional coupons and used them. For a very few customers, the discount amount is actually greater than the sales amount (see, for example, customer 4). In The Mode's management would like to use this sample data to learn about its customer base and to evaluate the promotion involving discount coupons.

Customer	Method of Payment	Items	Discount	Sales	Gender	Marital Status	Age
1	Switch	1	0.00	39.50	Male	Married	32
2	Store Card	1	25.60	102.40	Female	Married	36
3	Store Card	1	0.00	22.50	Female	Married	32
4	Store Card	5	121.10	100.40	Female	Married	28
5	Mastercard	2	0.00	54.00	Female	Married	34
6	Mastercard	1	0.00	44.50	Female	Married	44
7	Store Card	2	19.50	78.00	Female	Married	30
8	Visa	1	0.00	22.50	Female	Married	40
9	Store Card	2	22.48	56.52	Female	Married	46
10	Store Card	1	0.00	44.50	Female	Married	36
11	Store Card	1	0.00	29.50	Female	Married	48
12	Store Card	1	7.90	31.60	Female	Married	40
13	Visa	9	103.60	160.40	Female	Married	40
14	Visa	2	24.50	64.50	Female	Married	46
15	Visa	1	0.00	49.50	Male	Single	24
16	Store Card	2	12.60	71.40	Male	Single	36
17	Store Card	3	53.00	94.00	Female	Single	22
18	Switch	3	0.00	54.50	Female	Married	40
19	Mastercard	2	19.00	38.50	Female	Married	32

Managerial report

Use tabular and graphical descriptive statistics to help management develop a customer profile and to evaluate the promotional campaign. At a minimum, your report should include the following.

1 Percentage frequency distributions for key variables.

2 A bar chart or pie chart showing the percentage of customer purchases possibly attributable to the promotional campaign.

3 A cross-tabulation of type of customer (regular or promotional) versus sales. Comment on any similarities or differences present.

4 A scatter diagram of sales versus discount for only those customers responding to the promotion. Comment on any relationship apparent between sales and discount.

5 A scatter diagram to explore the relationship between sales and customer age.

Clothes on a rail at a women's fashion store. © martin mcelligott.

Software Section
for Chapter 2

MINITAB offers extensive capabilities for constructing tabular and graphical summaries of data. In this section we show how MINITAB can be used to construct several graphical summaries and a cross-tabulation. The graphical methods presented are the dot plot, the histogram and the scatter diagram.

Dot plot

Assume the audit times data of Table 2.4 are in column C1 of a MINITAB worksheet. The following steps will generate a dot plot.

Step 1   **Graph** > **Dotplot**   [Main menu bar]

Step 2   Select **One Y**, **Simple**   [**Dotplots** panel]
                Click **OK**

Step 3   Enter **C1** in the **Graph Variables** box   [**Dotplot – One Y, Simple** panel]
                Click **OK**

Histogram

Again, assume the audit times data are in column C1 of a MINITAB worksheet. The following steps will generate a histogram.

Step 1   **Graph** > **Histogram**   [Main menu bar]

Step 2   Select **Simple**   [**Histogram** panel]
                Click **OK**

Step 3   Enter **C1** in the **Graph Variables** box   [**Histogram – Simple** panel]
                Click **OK**

When the Histogram appears:

Step 4 Position the mouse pointer over any one of the bars, and **Double Click**
Select the **Binning** tab   [**Edit Bars** panel]
Select **Midpoint** for **Interval Type**
Select **Midpoint/Cutpoint positions** for **Interval Definition**
Enter **12:32/5** in the **Midpoint/Cutpoint positions** box*
Click **OK**

Scatter diagram

We use the hi-fi equipment store data in Table 2.12 to demonstrate the construction of a scatter diagram. The weeks are numbered from 1 to 10 in column C1, the data for number of commercials are in column C2, and the data for sales are in column C3 of a MINITAB worksheet. The following steps will generate the scatter diagram shown in Figure 2.7.

Step 1 **Graph** > **Scatterplot**   [Main menu bar]

Step 2 Select **Simple**   [**Scatterplot** panel]
Click **OK**

Step 3 Enter **C3** under **Y Variables**   [**Scatterplot – Simple** panel]
Enter **C2** under **X Variables**
Click **OK**

Cross-tabulation

We use the data from the restaurant review of section 2.4, part of which is shown in Table 2.9, to demonstrate. The restaurants are numbered from 1 to 300 in column C1 of the MINITAB worksheet. The quality ratings are in column C2, and the meal prices are in column C3. MINITAB can create a cross-tabulation only for qualitative variables, so we need to first code the meal price data by specifying a category (class) to which each meal price belongs. The following steps will code the meal price data to create four categories of meal price in column C4: €10–19, €20–29, €30–39 and €40–49.

Step 1 **Data** > **Code** > **Numeric to Text**   [Main menu bar]

Step 2 Enter **C3** in the **Code data from columns** box   [**Code – Numeric to Text** panel]
Enter **C4** in the **Store coded data in columns** box
Enter **10:19** in the first **Original values** box
Enter **€10–19** in the first **New** box
   Repeat the last two operations using 20:29, 30:39 and 40:49 in the second, third and fourth **Original values** boxes, and using €20–29, €30–39 and €40–49 in the second, third and fourth **New** boxes.
Click **OK**

---

*The entry 12:35/5 indicates that 12 is the midpoint of the first class, 32 is the midpoint of the last class, and 5 is the class width.

For each meal price in column C3 the associated meal price category will now appear in column C4. We can now construct a cross-tabulation for quality rating and the meal price categories by using the data in columns C2 and C4. The following steps will create a cross-tabulation containing the same information as shown in Table 2.10.

Step 3 **Stat** > **Tables** > **Cross Tabulation and Chi-Square**   [Main menu bar]

Step 4 Enter **C2** in the **For rows** box   [**Cross Tabulation and Chi-Square** panel]
Enter **C4** in the **For columns** box
Select **Counts** under **Display**
Click **OK**

Tabular and graphical presentations using EXCEL

SOFTDRINK

EXCEL offers extensive capabilities for constructing tabular and graphical summaries of data. In this appendix, we show how EXCEL can be used to construct a frequency distribution, bar chart, pie chart, histogram, scatter diagram and cross-tabulation. We will demonstrate two of EXCEL's most powerful tools for data analysis: creating charts and creating PivotTable Reports.

Frequency distribution and bar chart for qualitative data

In this section we show how EXCEL can be used to construct a frequency distribution and a bar chart for qualitative data. We illustrate each using the data on soft drink purchases in Table 2.1.

Frequency distribution

We begin by showing how the COUNTIF function can be used to construct a frequency distribution. Refer to Figure 2.10 as we describe the steps involved. The formula worksheet (showing the functions and formulae used) is set in the background, and the value worksheet (showing the results obtained using the functions and formulae) appears in the foreground.

The label 'Brand Purchased' and the data for the 50 soft drink purchases are in cells A1:A51. We also entered the labels 'Soft Drink' and 'Frequency' in cells C1:D1. The five soft drink names are entered into cells C2:C6. EXCEL's COUNTIF function can now be used to count the number of times each soft drink appears in cells A2:A51. The following steps are used.

Step 1 Select cell D2

Step 2 Enter =**COUNTIF($A$2:$A$51,C2)**

Step 3 Copy cell D2 to cells D3:D6

The formula worksheet in Figure 2.10 shows the cell formulae inserted by applying these steps. The value worksheet shows the values computed by the cell formulae. This worksheet shows the same frequency distribution that we constructed in Table 2.2.

Figure 2.10 Frequency distribution for soft drink purchases constructed using EXCEL's Countif function

	A	B	C	D
1	Brand Purchased		Soft Drink	Frequency
2	Coke Classic		Coke Classic	=COUNTIF($A$2:$A$51,C2)
3	Diet Coke		Diet Coke	=COUNTIF($A$2:$A$51,C3)
4	Pepsi-Cola		Dr. Pepper	=COUNTIF($A$2:$A$51,C4)
5	Diet Coke		Pepsi-Cola	=COUNTIF($A$2:$A$51,C5)
6	Coke Classic		Sprite	=COUNTIF($A$2:$A$51,C6)
7	Coke Classic			
8	Dr. Pepper			
9	Diet Coke			
10	Pepsi-Cola			
45	Pepsi-Cola			
46	Pepsi-Cola			
47	Pepsi-Cola			
48	Coke Classic			
49	Dr. Pepper			
50	Pepsi-Cola			
51	Sprite			
52				

	A	B	C	D
1	Brand Purchased		Soft Drink	Frequency
2	Coke Classic		Coke Classic	19
3	Diet Coke		Diet Coke	8
4	Pepsi-Cola		Dr. Pepper	5
5	Diet Coke		Pepsi-Cola	13
6	Coke Classic		Sprite	5
7	Coke Classic			
8	Dr. Pepper			
9	Diet Coke			
10	Pepsi-Cola			
45	Pepsi-Cola			
46	Pepsi-Cola			
47	Pepsi-Cola			
48	Coke Classic			
49	Dr. Pepper			
50	Pepsi-Cola			
51	Sprite			
52				

Bar chart

SOFTDRINK

Here we show how EXCEL's chart tools can be used to construct a bar chart for the soft drink data. Refer to the frequency distribution shown in the value worksheet of Figure 2.10. The bar chart that we are going to develop is an extension of this worksheet. The worksheet and the bar chart developed are shown in Figure 2.11. The steps are as follows:

Step 1   Select cells C2:D6

Step 2   Click the **Insert** tab on the Ribbon

Step 3   In the **Charts** group, click **Column**

Figure 2.11 Bar chart of soft drink purchases constructed using EXCEL

	A	B	C	D	E	F	G
1	**Brand Purchased**		**Soft Drink**	**Frequency**			
2	Coke Classic		Coke Classic	19			
3	Diet Coke		Diet Coke	8			
4	Pepsi-Cola		Dr. Pepper	5			
5	Diet Coke		Pepsi-Cola	13			
6	Coke Classic		Sprite	5			
7	Coke Classic						
8	Dr. Pepper						
9	Diet Coke						
10	Pepsi-Cola						
11	Pepsi-Cola						
12	Coke Classic						
13	Dr. Pepper						
14	Sprite						
15	Coke Classic						
16	Diet Coke						
17	Coke Classic						
18	Coke Classic						
19	Sprite						
20	Coke Classic						
21	Diet Coke						
22	Coke Classic						

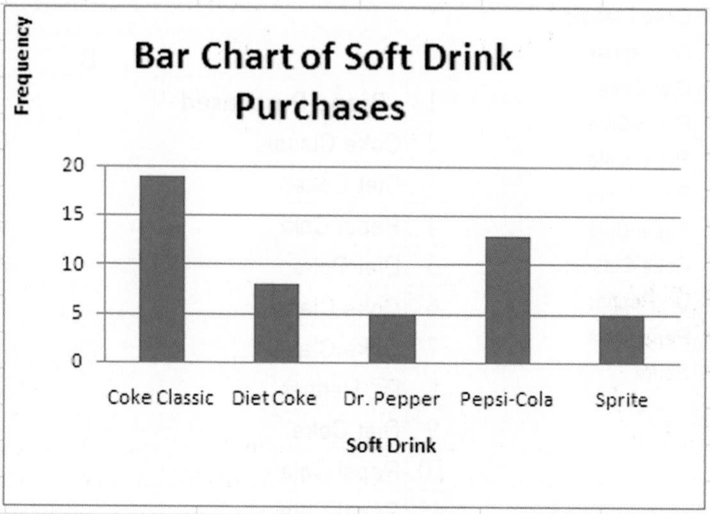

Bar Chart of Soft Drink Purchases

Step 4 When the list of column chart subtypes appears:
Go to the **2-D Column** section
Click **Clustered Column** (the leftmost chart)

Step 5 In the **Chart Layouts** group, click the **More** button (the downward-pointing arrow with a line over it) to display all the options

Step 6 Choose **Layout 9**

Step 7 Select the **Chart Title** and replace it with **Bar Chart of Soft Drink Purchases**

Step 8 Select the **Horizontal (Category) Axis Title** and replace it with **Soft Drink**

Step 9 Select the **Vertical (Value) Axis Title** and replace it with **Frequency**

Step 10 Right-click the **Series 1 Legend Entry**
Click **Delete**

Step 11 Right-click the vertical axis
Click **Format Axis**

Step 12   When the Format Axis panel appears:
Go to the **Axis Options** section
Select **Fixed** for **Major Unit** and enter **5.0** in the corresponding box
Click **Close**

The resulting bar chart is shown in Figure 2.11.*

EXCEL can produce a pie chart for the soft drink data in a similar fashion. The major difference is that in step 3 we would click **Pie** in the **Charts** group. Several styles of pie charts are available.

Frequency distribution and histogram for quantitative data

In this section we show how EXCEL can be used to construct a frequency distribution and a histogram for quantitative data. We illustrate each using the audit time data shown in Table 2.4.

Frequency distribution

Excel's FREQUENCY function can be used to construct a frequency distribution for quantitative data. Refer to Figure 2.12 as we describe the steps involved. The formula worksheet is in the background, and the value worksheet is in the foreground. The label 'Audit Time' is in cell A1 and the data for the 20 audits are in cells A2:A21. Using the procedures described in the text, we make the five classes 10–14, 15–19, 20–24, 25–29, and 30–34. The label 'Audit Time' and the five classes are entered in cells C1:C6. The label 'Upper Limit' and the five class upper limits are entered in cells D1:D6. We also entered the label 'Frequency' in cell E1. EXCEL's FREQUENCY function will be used to show the class frequencies in cells E2:E6. The following steps describe how to develop a frequency distribution for the audit time data.

Step 1   Select cells E2:E6

Step 2   Type, but do not enter, the following formula:
=FREQUENCY(A2:A21,D2:D6)

Step 3   Press CTRL + SHIFT + ENTER and the array formula will be entered into each of the cells E2:E6

The results are shown in Figure 2.12. The values displayed in the cells E2:E6 indicate frequencies for the corresponding classes. Referring to the FREQUENCY function, we see that the range of cells for the upper class limits (D2:D6) provides input to the function. These upper class limits, which EXCEL refers to as *bins,* tell EXCEL which frequency to put into the cells of the output range (E2:E6). For example, the frequency for the class with an upper limit, or bin, of 14 is placed in the first cell (E2), the frequency for the class with an upper limit, or bin, of 19 is placed in the second cell (E3), and so on.

---

* The bar chart in Figure 2.11 can be resized. Resizing an EXCEL chart is not difficult. First, select the chart. Sizing handles will appear on the chart border. Click on the sizing handles and drag them to resize the figure to your preference.

Figure 2.12 Frequency distribution for audit time data constructed using EXCEL's frequency function, plus histogram

	A	B	C	D	E	F	G	H
1	**Audit Time**		**Audit Time**	**Upper Limit**	**Frequency**			
2	12		10-14	14	4			
3	15		15-19	19	8			
4	20		20-24	24	5			
5	22		25-29	29	2			
6	14		30-34	34	1			
7	14							
8	15							
9	27							
10	21							
11	18							
12	19							
13	18							
14	22							
15	33							
16	16							
17	18							
18	17							
19	23							
20	28							
21	13							
22								

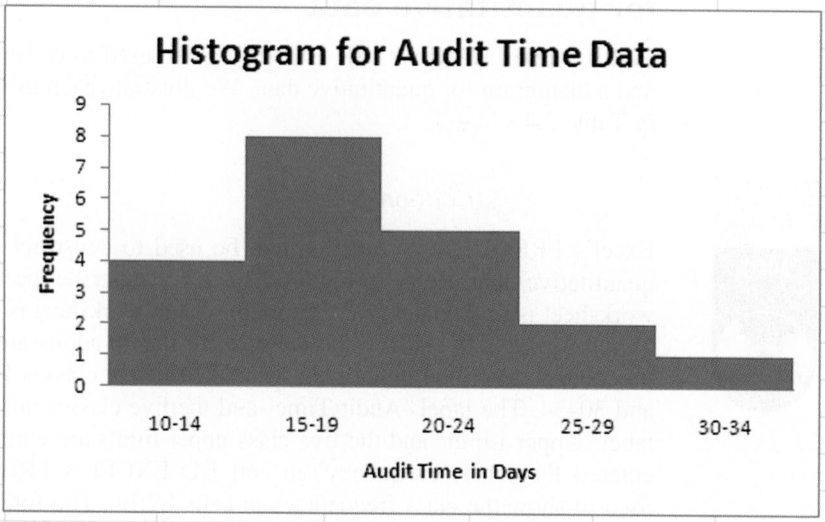

Histogram for Audit Time Data

Histogram

To use EXCEL to construct a histogram for the audit time data, we begin with the frequency distribution as shown in Figure 2.12. The histogram output is also shown in Figure 2.12. The following steps describe how to construct a histogram from a frequency distribution.

Step 1   Select cells C2:C6

Step 2   Press the Ctrl key and also select cells E2:E6

Step 3   Click the **Insert** tab on the Ribbon

Step 4   In the **Charts** group, click **Column**

Step 5   When the list of column chart subtypes appears:
Go to the **2-D Column** section
Click **Clustered Column** (the leftmost chart)

Step 6   In the **Chart Layouts** group, click the **More** button (the downward-pointing arrow with a line over it)

Step 7 Choose **Layout 8**

Step 8 Select the **Chart Title** and replace it with **Histogram for Audit Time Data**

Step 9 Select the **Horizontal (Category) Axis Title** and replace it with **Audit Time in Days**

Step 10 Select the **Vertical (Value) Axis Title** and replace it with **Frequency**

Finally, an interesting aspect of the worksheet in Figure 2.12 is that EXCEL has linked the data in cells A2:A21 to the frequencies in cells E2:E6 and to the histogram. If an edit or revision of the data in cells A2:A21 occurs, the frequencies in cells E2:E6 and the histogram will be updated automatically to display a revised frequency distribution and histogram. Try one or two data edits to see how this automatic updating works.

Scatter diagram

We use the hi-fi equipment store data in Table 2.12 to demonstrate the use of EXCEL to construct a scatter diagram. Refer to Figure 2.13 as we describe the tasks involved. The value worksheet is set in the background, and the scatter diagram produced by EXCEL appears in the foreground. The following steps will produce the scatter diagram.

Step 1 Select cells B2:C11

Step 2 Click the **Insert** tab on the Ribbon

Figure 2.13 Scatter diagram for hi-fi equipment store using EXCEL

	A	B	C	D	E	F	G	H	I
1	Week	No. of Commercials	Sales Volume						
2	1	2	50						
3	2	5	57						
4	3	1	41						
5	4	3	54						
6	5	4							
7	6	1							
8	7	5							
9	8	3							
10	9	4							
11	10	2							
12									
13									
14									
15									
16									
17									
18									
19									
20									

Step 3   In the **Charts** group, click **Scatter**

Step 4   When the list of scatter diagram subtypes appears:
Click **Scatter with only Markers** (the chart in the upper-left corner)

Step 5   In the **Chart Layouts** group, click **Layout 1**

Step 6   Select the **Chart Title** and replace it with **Scatter Diagram for the H-Fi Equipment Store**

Step 7   Select the **Horizontal (Value) Axis Title** and replace it with **Number of Commercials**

Step 8   Select the **Vertical (Value) Axis Title** and replace it with **Sales Volume**

Step 9   Right-click the **Series 1 Legend Entry**
Click **Delete**

Step 10 Right-click the vertical axis
Click **Format Axis**

Step 11 When the Format Axis panel appears:
Go to the **Axis Options** section
Select **Fixed** for **Minimum** and enter **35** in the corresponding box
Select **Fixed** for **Maximum** and enter **65** in the corresponding box
Select **Fixed** for **Major Unit** and enter **5** in the corresponding box
Click **Close**

A trendline can be added to the scatter diagram as follows.

Step 12 Position the mouse pointer over any data point in the scatter diagram and right-click to display a list of options

Step 13 Choose **Add Trendline**

Step 14 When the **Add Format Trendline** dialog box appears:
Go to the **Trendline Options** section
Choose **Linear** in the **Trend/Regression Type** section
Click **Close**

The worksheet in Figure 2.13 shows the scatter diagram with the trendline added.

PivotTable report

EXCEL's PivotTable Report provides a valuable tool for managing data sets involving more than one variable. We will illustrate its use by showing how to develop a cross-tabulation using the restaurant data in Figure 2.14. Labels are entered in row 1, and the data for each of the 300 restaurants are entered into cells A2:C301.

Creating the initial worksheet

The following steps are needed to create a worksheet containing the initial PivotTable Report and PivotTable Field List.

Figure 2.14 EXCEL worksheet containing restaurant data

	A	B	C	D
1	Restaurant	Quality Rating	Meal Price (€)	
2	1	Good	18	
3	2	Very Good	22	
4	3	Good	28	
5	4	Excellent	38	
6	5	Very Good	33	
7	6	Good	28	
8	7	Very Good	19	
9	8	Very Good	11	
10	9	Very Good	23	
11	10	Good	13	
292	291	Very Good	23	
293	292	Very Good	24	
294	293	Excellent	45	
295	294	Good	14	
296	295	Good	18	
297	296	Good	17	
298	297	Good	16	
299	298	Good	15	
300	299	Very Good	38	
301	300	Very Good	31	
302				

Step 1 Click the **Insert** tab on the Ribbon

Step 2 In the **Tables** group, click the icon above PivotTable

Step 3 When the Create PivotTable panel appears:
Choose **Select a table or range**
Enter **A1:C301** in the **Table/Range** box
Select **New Worksheet**
Click **OK**

The resulting PivotTable Field List is shown in Figure 2.15.

Figure 2.15 PivotTable field list

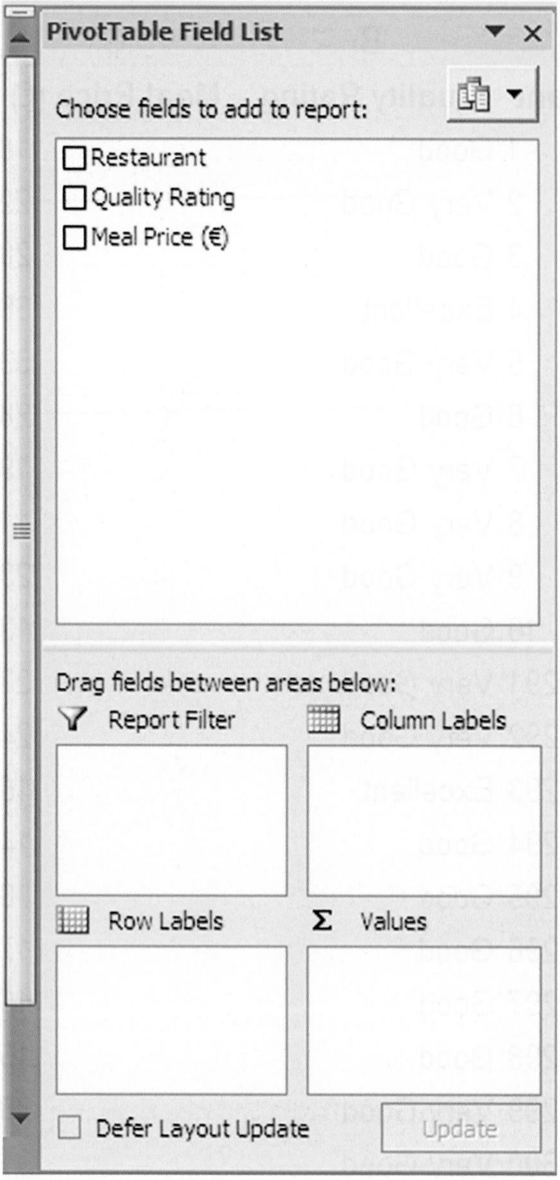

Using the PivotTable Field List

Each column in Figure 2.14 (Restaurant, Quality Rating, and Meal Price) is considered a field by EXCEL. The following steps show how to use EXCEL's PivotTable Field List to move the Quality Rating field to the row section, the Meal Price (€) field to the column section, and the Restaurant field to the values section of the PivotTable report.

Step 1 In the **PivotTable Field List,** go to **Choose Fields to add to report:**
Drag the **Quality Rating** field to the **Row Labels** area
Drag the **Meal Price (€)** field to the **Column Labels** area
Drag the **Restaurant** field to the **Values** area

Figure 2.16 Completed PivotTable field list and a portion of PivotTable Report

	A	B	C	D	AL	AM	AN	AO	AP	AQ	AR
1											
2											
3	Count of Restaurant	Meal Price (€) ▾									
4	Quality Rating ▾	10	11	12	47	48	Grand Total				
5	Excellent				2	2	66				
6	Good	6	4	3			84				
7	Very Good	1	4	3		1	150				
8	Grand Total	7	8	6	2	3	300				
9											

PivotTable Field List ▾ ✕

Choose fields to add to report:

☑ Restaurant
☑ Quality Rating
☑ Meal Price (€)

Drag fields between areas below:

▼ Report Filter   ▦ Column Labels

Meal Price (€) ▾

▦ Row Labels   Σ Values

Quality Rating ▾   Count of Res... ▾

☐ Defer Layout Update   Update

Step 2 Click **Sum of Restaurant** in the **Values** area
Click **Value Field Settings**

Step 3 When the Value Field Settings panel appears:
Under **Summarize value field by,** choose **Count**
Click **OK**

Figure 2.16 shows the completed PivotTable Field List and a portion of the PivotTable Report.

Finalizing the PivotTable Report

To complete the PivotTable Report, the following steps are used to group the columns representing meal prices and place the row labels for quality rating in the proper order.

Step 1 Right-click in cell B4 or in any other cell containing meal prices
Select **Group**

Step 2 When the Grouping panel appears:
Enter **10** in the **Starting at** box
Enter **49** in the **Ending at** box
Enter **10** in the **By** box
Click **OK**

Figure 2.17 Final PivotTable Report

	A	B	C	D	E	F	G
1							
2							
3	Count of Restaurant	Meal Price (€) ▾					
4	Quality Rating ▾	10-19	20-29	30-39	40-49	Grand Total	
5	Good	42	40	2		84	
6	Very Good	34	64	46	6	150	
7	Excellent	2	14	28	22	66	
8	Grand Total	78	118	76	28	300	
9							
10							

Step 3 Right-click on **Excellent** in cell A5
Choose **Move**
Select **Move "Excellent" to END**

Step 4 Close the PivotTable Field List dialog box

The final PivotTable Report is shown in Figure 2.17. Note that it provides the same information as the cross-tabulation shown in Table 2.10.

Tabular and graphical presentations using PASW

PASW offers extensive capabilities for constructing tabular and graphical summaries of data. In this section we show how PASW can be used to construct a histogram, a scatter diagram, and a cross-tabulation.

Histogram

Assume the audit times data of Table 2.4 are in the first column of the PASW Data Editor. The following steps will generate a histogram.

Step 1 **Graph** > **Chart Builder**   [Main menu bar]

Step 2 Under **Gallery**, choose **Histogram**   [**Chart Builder** panel]
Drag and drop the **Simple Histogram** icon into the **Chart Preview** area
Drag and drop the audit times variable to the **X-axis** area in **Chart Preview**
Click **OK**

Scatter diagram

We use the hi-fi equipment store data in Table 2.12 to demonstrate the construction of a scatter diagram. The weeks are numbered from 1 to 10 in the first column of the Data Editor, the data for number of commercials are in column 2 and the data for sales are in column 3. The following steps will generate the scatter diagram shown in Figure 2.7.

Step 1 Graph > Chart Builder   [Main menu bar]

Step 2 Under **Gallery**, choose **Scatter/Dot**   [**Chart Builder** panel]
Drag and drop the **Simple Scatter** icon into the **Chart Preview** area
Drag and drop the sales volume variable to the **Y-axis** area in **Chart Preview**
Drag and drop the number of commercials variable to the **X-axis** area in **Chart Preview**
Click **OK**

Cross-tabulation

We use the data from the restaurant review of section 2.4, part of which is shown in Table 2.9, to demonstrate. The restaurants are numbered from 1 to 300 in the first column of the PASW Data Editor. The quality ratings are in column 2 and the meal prices are in column 3. PASW can create a cross-tabulation only for categorized variables, so we need to first code the meal price data by specifying a category (class) to which each meal price belongs. The following steps will code the meal price data to create four categories of meal price in column 4: €10–19, €20–29, €30–39 and €40–49.

Step 1 Transform > Recode Into Different variables   [Main menu bar]

Step 2 Transfer the meal price variable to the **Input Variable–>Output Variable** box
[**Recode Into Different variables** panel]
Under **Output Variable**, give the new variable a name and label
Click **Change**
Click **Old and New Values**

Step 3 Under **Old Values**, check **Range,** and enter **10** and **19** in the two boxes
[**Recode Into Different variables: Old and New Values** panel]
Under **New Value**, check **Value** and enter **1** in the box
Click **Add**
Step 3 allocates code 1 to the €10–19 meal price range. Repeat this step for the 20–29, 30–39 and 40–49 ranges, allocating them codes 2, 3 and 4 respectively.
Click **Continue**

Step 4 Click **OK**   [**Recode Into Different variables** panel]

The new categorized variable will be added to the Data Editor, in column 4. Appropriate labels can be defined for the codes of this new variable in the **Variables** view of the Data Editor.

We can now construct a cross-tabulation for quality rating and the meal price categories by using the data in columns 2 and 4 of the Data Editor. The following steps will create a cross-tabulation containing the same information as shown in Table 2.10.

Step 5 Analyze > Descriptive Statistics > Crosstabs   [Main menu bar]

Step 6 Transfer the quality rating variable to the **Rows** box   [**Crosstabs** panel]
Transfer the new meal price variable to the **Columns** box
Click **OK**

Chapter 3

··

Descriptive Statistics: Numerical Measures

Learning objectives

After studying this chapter and doing the exercises, you should be able to calculate and interpret the following statistical measures that help to describe the central location, variability and shape of data sets.

1 The mean, median and mode.

2 Percentiles (including quartiles), the range, the interquartile range, the variance, the standard deviation and the coefficient of variation.

3 You should understand the concept of skewness of distribution. You should be able to calculate z-scores and understand their role in identifying data outliers.

4 You should understand the role of Chebyshev's theorem and the empirical rule in estimating the spread of data sets.

5 Five-number summaries and box plots.

6 Scatter diagrams, covariance and Pearson's correlation coefficient.

7 Weighted means.

8 Estimates of mean and standard deviation for grouped data.

In Chapter 2 we discussed tabular and graphical summaries of data. In this chapter, we present several numerical measures for summarizing data.

We start with numerical summary measures for data sets containing a single variable. When a data set contains more than one variable, the same numerical measures can be computed separately for each variable. However, in the two-variable case, we shall also examine measures of the relationship between the variables.

We introduce numerical measures of location, dispersion, shape and association. If the measures are computed for data from a sample, they are called **sample statistics**. If the measures are computed for data from a population, they are called **population parameters**. In statistical inference, a sample statistic is referred to as the **point estimator** of the corresponding population parameter. In Chapter 7 we shall discuss in more detail the process of point estimation. At the end of the present chapter we show how MINITAB, PASW and EXCEL can be used to compute many of the numerical measures described in the chapter.

3.1 Measures of location

Mean

Perhaps the most important measure of location is the **mean**, or average value, for a variable. The mean provides a measure of central location for the data. If the data are from a sample, the mean is denoted by $\bar{x}$. If the data are from a population, the Greek letter μ is used to denote the mean.

Statistics in Practice

TV audience measurement

Television audience levels and audience share are important issues for advertisers and sponsors and, in some countries like the UK where viewers pay a licence fee, also for the government. In recent years in many countries, the number of TV channels available has increased substantially because of the use of digital, satellite and cable services. The Broadcasters' Audience Research Board (BARB) in the UK, for example, lists over 250 channels in its 'multi-channel viewing summary'. Technology also now allows viewers to 'time-shift' their viewing. Accurate audience measurement thereby becomes a more difficult task.

The *Handbook on Radio and Television Audience Research** has a section on data analysis, in which the author

makes the point 'most audience research is quantitative'. He then goes on to describe the various measures that are commonly used in this field, including: 'ratings', 'gross rating points', 'viewing share', 'viewing hours', and 'reach'. Many of the measures involve the use of averages: for example, 'average weekly viewing per person'.

BARB publishes viewing figures regularly on its website, www.barb.co.uk. Figures for the week ending 24 August 2008, for example (the closing week of the Olympic Games in Beijing), showed that 'average daily reach' for all channels was over 36 million viewers, which represented over 70 per cent of the potential viewing audience. Terrestrial TV channels accounted for about two-thirds of this average daily reach. Average weekly viewing was estimated at nearly 26 hours per person.

In this chapter, you will learn how to compute and interpret some of the statistical measures used in reports such as those published by BARB. You will learn about the mean, median and mode, and about other descriptive statistics such as the range, variance, standard deviation, percentiles and correlation. These numerical measures will assist in the understanding and interpretation of data.

© Terraxplorer

Handbook on Radio and Television Audience Research, by Graham Mytton, published by UNICEF/UNESCO/BBC World Service Training Trust, web edition (2007) available at www.cba.org.uk/audience_research/documents/ar_handbook_2007_complete.pdf.

In statistical formulae, it is customary to denote the value of variable X for the first sample observation by x_1, for the second sample observation by x_2, and so on. In general, the value of variable X for the ith observation is denoted by x_i. (As we shall see in Chapters 5 and 6, a common convention in statistics is to *name* variables using capital letters, e.g. X, but to refer to specific values of those variables using small letters, e.g. x.) For a sample with n observations, the formula for the sample mean is as follows.

Sample mean

$$\bar{x} = \frac{\Sigma x_i}{n} \tag{3.1}$$

In (3.1), the numerator is the sum of the values of the n observations. That is,

$$\Sigma x_i = x_1 + x_2 + \cdots + x_n$$

The Greek letter Σ is the summation sign.

To illustrate the computation of a sample mean, consider the following class size data for a sample of five university classes.

$$46 \quad 54 \quad 42 \quad 46 \quad 32$$

We use the notation x_1, x_2, x_3, x_4, x_5 to represent the number of students in each of the five classes.

$$x_1 = 46 \quad x_2 = 54 \quad x_3 = 42 \quad x_4 = 46 \quad x_5 = 32$$

To compute the sample mean, we can write

$$\bar{x} = \frac{\Sigma x_i}{n} = \frac{x_1 + x_2 + x_3 + x_4 + x_5}{n} = \frac{46 + 54 + 42 + 46 + 32}{5} = 44$$

The sample mean class size is 44 students.

Here is a second illustration. Suppose a university careers office has sent a questionnaire to a sample of business school graduates requesting information on monthly starting salaries. Table 3.1 shows the data collected. The mean monthly starting salary for the sample of 12 business school graduates is computed as

$$\bar{x} = \frac{\Sigma x_i}{n} = \frac{x_1 + x_2 + \cdots + x_{12}}{12} = \frac{2020 + 2075 + \cdots + 2040}{12} = \frac{24\,840}{12} = 2070$$

SALARY

Equation (3.1) shows how the mean is computed for a sample with n observations. The formula for computing the mean of a population remains the same, but we use different notation to indicate that we are working with the entire population. We denote the number of observations in a population by N, and the population mean as μ.

Table 3.1 Monthly starting salaries for a sample of 12 business school graduates

Graduate	Monthly starting salary (€)	Graduate	Monthly starting salary (€)
1	2020	7	2050
2	2075	8	2165
3	2125	9	2070
4	2040	10	2260
5	1980	11	2060
6	1955	12	2040

Population mean

$$\mu = \frac{\Sigma x_i}{N} \qquad\qquad (3.2)$$

Median

The **median** is another measure of central location for a variable. The median is the value in the middle when the data are arranged in ascending order (smallest value to largest value).

Median

Arrange the data in ascending order.

a For an odd number of observations, the median is the middle value.

b For even number of observations, the median is the average of the two middle values.

Let us apply this definition to compute the median class size for the sample of five university classes. Arranging the data in ascending order produces the following list.

$$32 \quad 42 \quad 46 \quad 46 \quad 54$$

Because $n = 5$ is odd, the median is the middle value. The median class size is 46 students. This data set contains two observations with values of 46, but each observation is treated separately when we arrange the data in ascending order.

Suppose we also compute the median starting salary for the 12 business school graduates in Table 3.1. We first arrange the data in ascending order.

1955 1980 2020 2040 <u>2050 2060</u> 2070 2075 2125 2165 2260

Middle two values

Because $n = 12$ is even, we identify the middle two values: 2050 and 2060. The median is the average of these values.

$$\text{Median} = \frac{2050 + 2060}{2} = 2055$$

Although the mean is the more commonly used measure of central location, in some situations the median is preferred. The mean is influenced by extremely small and large data values. For example, suppose one of the graduates (see Table 3.1) had a starting salary of €5000 per month (perhaps the individual's family owns the company). If we change the highest monthly starting salary in Table 3.1 from €2260 to €5000 and re-calculate the mean, the sample mean changes from €2070 to €2298. The median of €2055, however, is unchanged, because €2050 and €2060 are still the middle two values.

With the extremely high starting salary included, the median provides a better measure of central location than the mean. When a data set contains extreme values, the median is often the preferred measure of central location.

Another measure sometimes used when extreme values are present is the *trimmed mean*. A percentage of the smallest and largest values are removed from a data set, and the mean of the remaining values is computed. For example, to get the 5 per cent trimmed mean, the smallest 5 per cent and the largest 5 per cent of the data values are removed, and the mean of the remaining values is computed. Using the sample with $n = 12$ starting salaries, $0.05(12) = 0.6$. Rounding this value to 1 indicates that the 5 per cent trimmed mean would remove the smallest data value and the largest data value. The 5 per cent trimmed mean using the 10 remaining observations is 2062.5.

Mode

A third measure of location is the **mode**. The mode is defined as follows.

Mode

The mode is the value that occurs with the greatest frequency.

To illustrate the identification of the mode, consider the sample of five class sizes. The only value that occurs more than once is 46. This value occurs twice, and consequently is the mode. In the sample of starting salaries for the business school graduates, the only monthly starting salary that occurs more than once is €2040, and therefore this value is the mode for that data set.

Situations can arise for which the greatest frequency occurs at two or more different values. In these instances more than one mode exists. If the data contain exactly two modes, we say that the data are *bimodal*. If data contain more than two modes, we say that the data are *multimodal*. In multimodal cases the modes are almost never reported, because listing three or more modes would not be particularly helpful in describing a location for the data.

The mode is an important measure of location for qualitative data. For example, the qualitative data set in Table 2.2 resulted in the following frequency distribution for soft drink purchases.

Soft drink	Frequency
Coke Classic	19
Diet Coke	8
Dr Pepper	5
Pepsi-Cola	13
Sprite	5
Total	50

The mode, or most frequently purchased soft drink, is Coke Classic. For this type of data it obviously makes no sense to speak of the mean or median. The mode provides the information of interest, the most frequently purchased soft drink.

Percentiles

A **percentile** provides information about how the data are spread over the interval from the smallest value to the largest value. For data that do not contain numerous repeated values, the pth percentile divides the data into two parts: approximately p per cent of the observations have values less than the pth percentile; approximately $(100 - p)$ per cent of the observations have values greater than the pth percentile. The pth percentile is formally defined as follows.

> **Percentile**
>
> The pth percentile is a value such that *at least p* per cent of the observations are less than or equal to this value and *at least* $(100 - p)$ per cent of the observations are greater than or equal to this value.

Colleges and universities sometimes report admission test scores in terms of percentiles. For instance, suppose an applicant obtains a raw score of 54 on the verbal portion of an admission test. How this student performed in relation to other students taking the same test may not be readily apparent. However, if the raw score of 54 corresponds to the 70th percentile, we know that approximately 70 per cent of the students scored lower than this individual and approximately 30 per cent of the students scored higher than this individual.

The following procedure can be used to compute the pth percentile.

> **Calculating the pth percentile**
>
> 1 Arrange the data in ascending order (smallest value to largest value).
> 2 Compute an index i
>
> $$i = \left(\frac{p}{100}\right)n$$
>
> where p is the percentile of interest and n is the number of observations.
> 3 a. If i *is not an integer, round up.* The next integer *greater* than i denotes the position of the pth percentile.
>
>   b. If i *is an integer*, the pth percentile is the average of the values in positions i and $i + 1$.

As an illustration, consider the 85th percentile for the starting salary data in Table 3.1.

1 Arrange the data in ascending order.

1955 1980 2020 2040 2040 2050 2060 2070 2075 2125 2165 2260

2 $i = \left(\frac{p}{100}\right)n = \left(\frac{85}{100}\right)12 = 10.2$

3 Because i is not an integer, *round up.* The position of the 85th percentile is the next integer greater than 10.2, the 11th position.

Returning to the data, we see that the 85th percentile is the data value in the 11th position, or 2165.

As another illustration of this procedure, consider the calculation of the 50th percentile for the starting salary data. Applying step 2, we obtain

$$i = \left(\frac{p}{100}\right)n = \left(\frac{50}{100}\right)12 = 6$$

Because i is an integer, step 3(b) states that the 50th percentile is the average of the sixth and seventh data values; that is $(2050 + 2060)/2 = 2055$. Note that the *50th percentile is also the median*.

Quartiles

It is often desirable to divide data into four parts, with each part containing approximately one-quarter, or 25 per cent of the observations. Figure 3.1 shows a data distribution divided into four parts. The division points are referred to as the **quartiles** and are defined as

Q_1 = first quartile, or 25th percentile
Q_2 = second quartile, or 50th percentile (also the median)
Q_3 = third quartile, or 75th percentile

The starting salary data are again arranged in ascending order. We have already identified Q_2, the second quartile (median), as 2055.

1955 1980 2020 2040 2040 2050 2060 2070 2075 2125 2165 2260

The computations of quartiles Q_1 and Q_3 use the rule for finding the 25th and 75th percentiles.

For Q_1,

$$i = \left(\frac{p}{100}\right)n = \left(\frac{25}{100}\right)12 = 3$$

Because i is an integer, step 3(b) indicates that the first quartile, or 25th percentile, is the average of the third and fourth data values; hence,

$$Q_1 = (2020 + 2040)/2 = 2030.$$

For Q_3,

$$i = \left(\frac{p}{100}\right)n = \left(\frac{75}{100}\right)12 = 9$$

Figure 3.1 Location of the quartiles

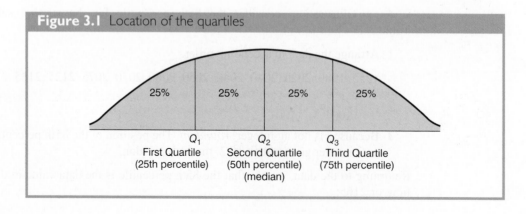

25%	25%	25%	25%

Q_1
First Quartile
(25th percentile)

Q_2
Second Quartile
(50th percentile)
(median)

Q_3
Third Quartile
(75th percentile)

Again, because i is an integer, step 3(b) indicates that the third quartile, or 75th percentile, is the average of the ninth and tenth data values; hence,

$$Q_3 = (2075 + 2125)/2 = 2100.$$

The quartiles divide the starting salary data into four parts, with each part containing 25 per cent of the observations.

1955   1980   2020 | 2040   2040   2050 | 2060   2070   2075 | 2125   2165   2260

$Q_1 = 2030$ $Q_2 = 2055$ $Q_3 = 2100$

(Median)

We defined the quartiles as the 25th, 50th and 75th percentiles. Hence, we computed the quartiles in the same way as percentiles. However, other conventions are sometimes used to compute quartiles and the actual values reported for quartiles may vary slightly depending on the convention used (see the Software Section at the end of the chapter). Nevertheless, the objective of all procedures for computing quartiles is to divide the data into four equal parts.

Exercises

Methods

1   Consider a sample with data values of 10, 20, 12, 17 and 16. Compute the mean and median.

2   Consider a sample with data values of 10, 20, 21, 17, 16 and 12. Compute the mean and median.

3   Consider a sample with data values of 27, 25, 20, 15, 30, 34, 28 and 25. Compute the 20th, 25th, 65th and 75th percentiles.

4   Consider a sample with data values of 53, 55, 70, 58, 64, 57, 53, 69, 57, 68 and 53. Compute the mean, median and mode.

Applications

ENGSAL

5   A sample of 30 Irish engineering graduates had the following starting salaries. Data are in thousands of euros.

36.8	34.9	35.2	37.2	36.2	35.8	36.8	36.1	36.7	36.6
37.3	38.2	36.3	36.4	39.0	38.3	36.0	35.0	36.7	37.9
38.3	36.4	36.5	38.4	39.4	38.8	35.4	36.4	37.0	36.4

a. What is the mean starting salary?
b. What is the median starting salary?
c. What is the mode?
d. What is the first quartile?
e. What is the third quartile?

6   The following data were obtained for the number of minutes spent listening to recorded music for a sample of 30 individuals on one particular day.

88.3	4.3	4.6	7.0	9.2	0.0	99.2	34.9	81.7	0.0
85.4	0.0	17.5	45.0	53.3	29.1	28.8	0.0	98.9	64.5
4.4	67.9	94.2	7.6	56.6	52.9	145.6	70.4	65.1	63.6

a. Compute the mean.
b. Compute the median.
c. Compute the first and third quartiles.
d. Compute and interpret the 40th percentile.

7 miniRank (www.minirank.com) rates the popularity of websites in most countries of the world, using a points system. The 25 most popular sites in Cyprus as listed in November 2008 were as follows (the points scores have been rounded to one decimal place):

Website	Points	Website	Points
www.dart.com.cy	59.2	www.chris-michael.com.cy	8.8
www.dvds.com.cy	21.0	www.music.net.cy	8.7
www.fitness.com.cy	20.5	drivenet.com.cy	8.6
www.airlinetickets.com.cy	20.0	www.prismastore.com.cy	8.6
www.weightloss.com.cy	19.8	www.force.com.cy	8.5
www.cyprus.gov.cy	17.3	www.prisma.com.cy	8.5
www.netcars.com.cy	14.3	www.prismanet.cy	8.5
www.visitcyprus.org.cy	14.3	www.ebos.com.cy	7.3
www.flowershop.com.cy	13.1	www.cytanet.com.cy	6.7
www.netinfo.com.cy	12.5	www.hrdauth.org.cy	6.2
www.interprom.cy	9.5	www.ucy.ac.cy	5.8
www.cyta.com.cy	9.4	www.eplaza.com.cy	5.7
www.drivenet.com.cy	9.1		

a. Compute the mean and median.
b. Do you think it would be better to use the mean or the median as the measure of central location for these data? Explain.
c. Compute the first and third quartiles.
d. Compute and interpret the 85th percentile.

8 Following is a sample of age data for individuals working from home by 'telecommuting'.

18	54	20	46	25	48	53	27	26	37
40	36	42	25	27	33	28	40	45	25

a. Compute the mean and the mode.
b. Suppose the median age of the population of all adults is 35.5 years. Use the median age of the preceding data to comment on whether the at-home workers tend to be younger or older than the population of all adults.
c. Compute the first and third quartiles.
d. Compute and interpret the 32nd percentile.

3.2 Measures of variability

In addition to measures of location, it is often desirable to consider measures of variability, or dispersion. For example, suppose you are a purchasing agent for a large manufacturing firm and that you regularly place orders with two different suppliers. After several months of operation, you find that the mean number of days required to fill orders is ten days for both of the suppliers. The histograms summarizing the number of working days required to fill orders from the suppliers are shown in Figure 3.2. Although the mean number of days is ten for both suppliers, do the two suppliers demonstrate the same degree of reliability in terms of making deliveries on schedule?

Note the dispersion, or variability, in delivery times indicated by the histograms. Which supplier would you prefer?

For most firms, receiving materials and supplies on schedule is important. The seven- or eight-day deliveries shown for J.C. Clark Distributors might be viewed favourably. However, a few of the slow 13- to 15-day deliveries could be disastrous in terms of keeping a workforce busy and production on schedule. This example illustrates a situation in which the variability in the delivery times may be an overriding consideration in selecting a supplier. For most purchasing agents, the lower variability shown for Dawson Supply would make Dawson the preferred supplier.

We turn now to a discussion of some commonly used measures of variability.

Range

The simplest measure of variability is the **range**.

> **Range**
>
> Range = Largest value − Smallest value

Refer to the data on starting salaries for business school graduates in Table 3.1. The largest starting salary is 2260 and the smallest is 1955. The range is $2260 - 1955 = 305$.

Although the range is the easiest of the measures of variability to compute, it is seldom used as the only measure. The reason is that the range is based on only two of the observations and so is highly influenced by extreme values. Suppose one of the graduates received a starting salary of €5000 per month. In this case, the range would be $5000 - 1955 = 3045$ rather than 305. This large value for the range would not be especially descriptive of the variability in the data because 11 of the 12 starting salaries are relatively closely grouped between 1955 and 2165.

Interquartile range

A measure of variability that overcomes the dependency on extreme values is the **interquartile range (IQR)**. This measure of variability is simply the difference between the third quartile, Q_3, and the first quartile, Q_1. In other words, the interquartile range is the range for the middle 50 per cent of the data.

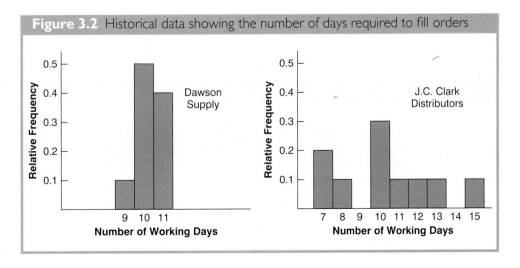

Figure 3.2 Historical data showing the number of days required to fill orders

Interquartile range

$$IQR = Q_3 - Q_1 \tag{3.3}$$

For the data on monthly starting salaries, the quartiles are $Q_3 = 2100$ and $Q_1 = 2030$. The interquartile range is $2100 - 2030 = 70$.

Variance

The **variance** is a measure of variability that uses all the data. The variance is based on the difference between the value of each data value and the mean. This difference is called a *deviation about the mean*. For a sample, a deviation about the mean is written $(x_i - \bar{x})$. For a population, it is written $(x_i - \mu)$. In the computation of the variance, the deviations about the mean are *squared*.

If the data are for a population, the average of the squared deviations is called the *population variance*. The population variance is denoted by the Greek symbol σ^2. For a population of N observations and with μ denoting the population mean, the definition of the population variance is as follows.

Population variance

$$\sigma^2 = \frac{\Sigma(x_i - \mu)^2}{N} \tag{3.4}$$

In most statistical applications, the data being analyzed are for a sample. When we compute a sample variance, we are often interested in using it to estimate the population variance σ^2. Although a detailed explanation is beyond the scope of this text, it can be shown that if the sum of the squared deviations about the sample mean is divided by $n - 1$, and not n, the resulting sample variance provides an unbiased estimate of the population variance.

For this reason, the *sample variance*, denoted by s^2, is defined as follows.

Sample variance

$$s^2 = \frac{\Sigma(x_i - \bar{x})^2}{n - 1} \tag{3.5}$$

Consider the data on class size for the sample of five university classes (Section 3.1). A summary of the data, including the computation of the deviations about the mean and the squared deviations about the mean, is shown in Table 3.2. The sum of squared deviations about the mean is $\Sigma(x_i - \bar{x})^2 = 256$. Hence, with $n - 1 = 4$, the sample variance is

$$s^2 = \frac{\Sigma(x_i - \bar{x})^2}{n - 1} = \frac{256}{4} = 64$$

Table 3.2 Computation of deviations and squared deviations about the mean for the class size data

Number of students in class (x_i)	Mean class size ($\bar{x}$)	Deviation about the Mean ($x_i - \bar{x}$)	Squared deviation about the mean ($x_i - \bar{x}$)^2
46	44	2	4
54	44	10	100
42	44	−2	4
46	44	2	4
32	44	−12	144
Totals		0	256
		$\Sigma(x_i - \bar{x})$	$\Sigma(x_i - \bar{x})^2$

The units associated with the sample variance often cause confusion. Because the values being summed in the variance calculation, $(x_i - \bar{x})^2$, are squared, the units associated with the sample variance are also *squared*. For instance, the sample variance for the class size data is $s^2 = 64$ (students)^2. The squared units make it difficult to obtain an intuitive understanding and interpretation of the numerical value of the variance. We recommend that you think of the variance as a measure useful in comparing the amount of variability for two or more variables. In a comparison of the variables, the one with the larger variance will show the most variability.

As another illustration, consider the starting salaries in Table 3.1 for the 12 business school graduates. In Section 3.1, we showed that the sample mean starting salary was 2070. The computation of the sample variance ($s^2 = 6754.5$) is shown in Table 3.3.

In Tables 3.2 and 3.3 we show both the sum of the deviations about the mean and the sum of the squared deviations about the mean. Note that in both tables, $\Sigma(x_i - \bar{x}) = 0$.

Table 3.3 Computation of the sample variance for the starting salary data

Monthly salary (x_i)	Sample mean ($\bar{x}$)	Deviation about the mean ($x_i - \bar{x}$)	Squared deviation about the mean ($x_i - \bar{x}$)^2
2020	207	−50	2 500
2075	207	5	25
2125	207	55	3 025
2040	207	−30	900
1980	207	−90	8 100
1955	207	−115	13 225
2050	207	−20	400
2165	207	95	9 025
2070	207	0	0
2260	207	190	36 100
2060	207	−10	100
2040	207	−30	900
Totals		0	74 300

Using equation (3.5)

$$s^2 = \frac{\Sigma(x_i - \bar{x})^2}{n - 1} = \frac{74\ 300}{11} = 6754.5$$

The positive deviations and negative deviations cancel each other, causing the sum of the deviations about the mean to equal zero. For any data set, the sum of the deviations about the mean will *always equal zero*.

An alternative formula for the computation of the sample variance is

$$s^2 = \frac{\sum x_i^2 - n\bar{x}^2}{n-1}$$

where

$$\sum x_i^2 = x_1^2 + x_2^2 + \cdots + x_n^2$$

Standard deviation

The **standard deviation** is defined to be the positive square root of the variance. Following the notation we adopted for a sample variance and a population variance, we use s to denote the sample standard deviation and σ to denote the population standard deviation. The standard deviation is derived from the variance in the following way.

Standard deviation	
Population standard deviation $= \sigma = \sqrt{\sigma^2}$	**(3.6)**
Sample standard deviation $= s = \sqrt{s^2}$	**(3.7)**

Recall that the sample variance for the sample of class sizes in five university classes is $s^2 = 64$. Hence the sample standard deviation is

$$s = \sqrt{64} = 8$$

For the data on starting salaries, the sample standard deviation is

$$s = \sqrt{6754.5} = 82.2$$

What is gained by converting the variance to its corresponding standard deviation? Recall that the units associated with the variance are squared. For example, the sample variance for the starting salary data of business school graduates is $s^2 = 6754.5$ (€)^2. Because the standard deviation is the square root of the variance, the units are converted to euros in the standard deviation. Hence, the standard deviation of the starting salary data is €82.2. In other words, the standard deviation is measured in the same units as the original data. For this reason the standard deviation is more easily compared to the mean and other statistics that are measured in the same units as the original data.

Coefficient of variation

In some situations we may be interested in a descriptive statistic that indicates how large the standard deviation is relative to the mean. This measure is called the **coefficient of variation** and is usually expressed as a percentage.

Coefficient of variation

$$\left(\frac{\text{Standard deviation}}{\text{Mean}} \times 100\right)\%$$ **(3.8)**

For the class size data, we found a sample mean of 44 and a sample standard deviation of 8. The coefficient of variation is $[(8/44) \times 100]\% = 18.2\%$. The coefficient of variation tells us that the sample standard deviation is 18.2 per cent of the value of the sample mean. For the starting salary data with a sample mean of 2070 and a sample standard deviation of 82.2, the coefficient of variation, $[(82.2/2070) \times 100]\% = 4.0\%$, tells us the sample standard deviation is only 4.0 per cent of the value of the sample mean. In general, the coefficient of variation is a useful statistic for comparing the variability of variables that have different standard deviations and different means.

Exercises

Methods

9 Consider a sample with data values of 10, 20, 12, 17 and 16. Calculate the range and interquartile range.

10 Consider a sample with data values of 10, 20, 12, 17 and 16. Calculate the variance and standard deviation.

11 Consider a sample with data values of 27, 25, 20, 15, 30, 34, 28 and 25. Calculate the range, interquartile range, variance and standard deviation.

Applications

12 A batsman's cricket scores for six innings were 41, 34, 42, 45, 35 and 37. Using these data as a sample, compute the following descriptive statistics.

   a. Range.
   b. Variance.
   c. Standard deviation.
   d. Coefficient of variation.

13 Dinner bill amounts for set menus at a Dubai restaurant, Al Khayam, show the following frequency distribution. The amounts are in AED (Emirati dirham). Compute the mean, variance and standard deviation.

Dinner bill (AED)	Frequency
30	2
40	6
50	4
60	4
70	2
80	2
Total	20

14   The following data were used to construct the histograms of the number of days required to fill orders for Dawson Supply and for J.C. Clark Distributors (see Figure 3.2).

Dawson Supply days for delivery: 11   10   9   10   11   11   10   11   10   10
Clark Distributors days for delivery: 8   10   13   7   10   11   10   7   15   12

Use the range and standard deviation to support the previous observation that Dawson Supply provides the more consistent and reliable delivery times.

15   Police records show the following numbers of daily crime reports for a sample of days during the winter months and a sample of days during the summer months.

Winter: 18   20   15   16   21   20   12   16   19   20
Summer: 28   18   24   32   18   29   23   38   28   18

a.   Compute the range and interquartile range for each period.
b.   Compute the variance and standard deviation for each period.
c.   Compute the coefficient of variation for each period.
d.   Compare the variability of the two periods.

16   A production department uses a sampling procedure to test the quality of newly produced items. The department employs the following decision rule at an inspection station: if a sample of 14 items has a variance of more than 0.005, the production line must be shut down for repairs. Suppose the following data have just been collected:

3.43   3.45   3.43   3.48   3.52   3.50   3.39
3.48   3.41   3.38   3.49   3.45   3.51   3.50

Should the production line be shut down? Why or why not?

3.3 Measures of distributional shape, relative location and detecting outliers

We described several measures of location and variability for data distributions. It is also often important to have a measure of the shape of a distribution. In Chapter 2 we noted that a histogram offers an excellent graphical display showing the shape of a distribution. An important numerical measure of the shape of a distribution is **skewness**.

Distributional shape

Four histograms constructed from relative frequency distributions are shown in Figure 3.3. The histograms in Panels A and B are moderately skewed. The one in Panel A is skewed to the left: its skewness is −0.85 (negative skewness). The histogram in Panel B is skewed to the right: its skewness is +0.85 (positive skewness). The histogram in Panel C is symmetrical: its skewness is zero. The histogram in Panel D is highly skewed to the right: its skewness is 1.62. The formula used to compute skewness is somewhat complex.* However, the skewness can be easily computed using statistical software (see Software Section at the end of this chapter).

---

*The formula for the skewness of sample data:

$$\text{Skewness} = \frac{n}{(n-1)(n-2)} \Sigma \left(\frac{x_i - \bar{x}}{s^3} \right)$$

Figure 3.3 Histograms showing the skewness for four distributions

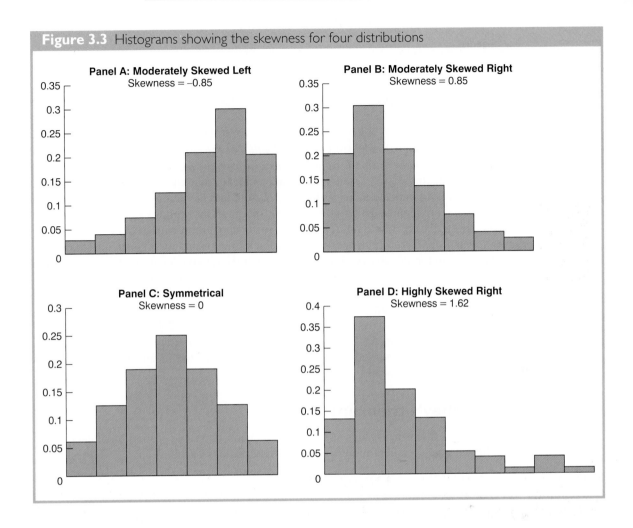

For a symmetrical distribution, the mean and the median are equal. When the data are positively skewed, the mean will usually be greater than the median. When the data are negatively skewed, the mean will usually be less than the median. The data used to construct the histogram in Panel D are customer purchases at a women's fashion store. The mean purchase amount is €77.60 and the median purchase amount is €59.70. The few large purchase amounts pulled up the mean, but the median remains unaffected. The median provides a better measure of typical values when the data are highly skewed.

z-Scores

In addition to measures of location, variability and shape for a data set, we are often also interested in the relative location of data items within a data set. Such measures can help us determine whether a particular item is close to the centre of a data set or far out in one of the tails.

By using both the mean and standard deviation, we can determine the relative location of any observation. Suppose we have a sample of n observations, with the values denoted by $x_1, x_2, \ldots, x_n$. In addition, assume that the sample mean, $\bar{x}$, and the sample standard deviation, s, are already computed. Associated with each value, x_i is another value called its z-**score**. Equation (3.9) shows how the z-score is computed for each x_i.

z-score

$$z_i = \frac{x_i - \bar{x}}{s} \qquad\qquad (3.9)$$

where z_i = the z-score for x, $\bar{x}$ = the sample mean, s = the sample standard deviation.

The z-score is often called the *standardized value* or the *standard score*. The z-score, z_i, represents the *number of standard deviations x_i is from the mean $\bar{x}$*. For example, $z_1 = 1.2$ would indicate that x_1 is 1.2 standard deviations higher than the sample mean. Similarly, $z_2 = -0.5$ would indicate that x_2 is 0.5, or 1/2, standard deviation lower than the sample mean. A z-score greater than zero occurs for observations with a value greater than the mean, and a z-score less than zero occurs for observations with a value less than the mean. A z-score of zero indicates that the value of the observation is equal to the mean.

The z-score for any observation is a measure of the relative location of the observation in a data set. Hence, observations in two different data sets with the same z-score can be said to have the same relative location in terms of being the same number of standard deviations from the mean.

The z-scores for the class size data are computed in Table 3.4. Recall the previously computed sample mean, $\bar{x} = 44$, and sample standard deviation, $s = 8$. The z-score of -1.50 for the fifth observation shows it is farthest from the mean: it is 1.50 standard deviations below the mean.

Chebyshev's theorem

Chebyshev's theorem enables us to make statements about the proportion of data values that lie within a specified number of standard deviations of the mean.

Chebyshev's theorem

At least $(1 - 1/z^2) \times 100\%$ of the data values must be within z standard deviations of the mean, where z is any value greater than 1.

Some of the implications of this theorem, with $z = 2$, 3 and 4 standard deviations, follow.

- At least 75 per cent of the data values must be within $z = 2$ standard deviations of the mean.
- At least 89 per cent of the data values must be within $z = 3$ standard deviations of the mean.

Table 3.4 z-scores for the class size data

Number of students in class (x_i)	Deviation about the mean ($x_i - \bar{x}$)	z-score $\dfrac{(x_i - \bar{x})}{s}$
46	2	2/8 = 0.25
54	10	10/8 = 1.25
42	−2	−2/8 = −0.25
46	2	2/8 = 0.25
32	−12	−12/8 = −1.50

- At least 94 per cent of the data values must be within $z = 4$ standard deviations of the mean.

Suppose that the mid-term test scores for 100 students in a university business statistics course had a mean of 70 and a standard deviation of 5. How many students had test scores between 60 and 80? How many students had test scores between 58 and 82?

For the test scores between 60 and 80, we note that 60 is two standard deviations below the mean and 80 is two standard deviations above the mean. Using Chebyshev's theorem, we see that at least 75 per cent of the observations must have values within two standard deviations of the mean. Hence, at least 75 per cent of the students must have scored between 60 and 80.

For the test scores between 58 and 82, we see that $(58 - 70)/5 = -2.4$, i.e. 58 is 2.4 standard deviations below the mean. Similarly, $(82 - 70)/5 = +2.4$, so 82 is 2.4 standard deviations above the mean. Applying Chebyshev's theorem with $z = 2.4$, we have

$$\left(1 - \frac{1}{z^2}\right) = \left(1 - \frac{1}{(2.4)^2}\right) = 0.826$$

At least 82.6 per cent of the students must have test scores between 58 and 82.

Empirical rule

Chebyshev's theorem applies to any data set, regardless of the shape of the distribution. It could be used, for example, with any of the skewed distributions in Figure 3.3. In many practical applications, however, data sets exhibit a symmetrical mound-shaped or bell-shaped distribution like the one shown in Figure 3.4. When the data are believed to approximate this distribution, the **empirical rule** can be used to determine the percentage of data values that must be within a specified number of standard deviations of the mean. The empirical rule is based on the normal probability distribution, which will be discussed in Chapter 6. The normal distribution is used extensively throughout the text.

Empirical rule

For data with a bell-shaped distribution:

- Approximately 68 per cent of the data values will be within one standard deviation of the mean.
- Approximately 95 per cent of the data values will be within two standard deviations of the mean.
- Almost all of the data values will be within three standard deviations of the mean.

For example, the empirical rule allows us to say that *approximately* 95 per cent of the data values will be within two standard deviations of the mean (Chebyshev's theorem allows us to conclude only that at least 75 per cent of the data values will be in that interval).

For example, liquid detergent cartons are filled automatically on a production line. Filling weights frequently have a bell-shaped distribution. If the mean filling weight is 500 grams and the standard deviation is 7 grams, we can use the empirical rule to draw the following conclusions.

- Approximately 68 per cent of the filled cartons will have weights between 493 and 507 grams (that is, within one standard deviation of the mean).
- Approximately 95 per cent of the filled cartons will have weights between 486 and 514 grams (that is, within two standard deviations of the mean).
- Almost all filled cartons will have weights between 479 and 521 grams (that is, within three standard deviations of the mean).

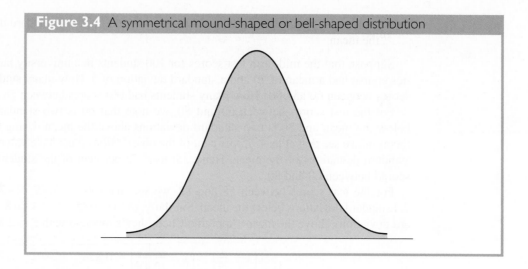

Figure 3.4 A symmetrical mound-shaped or bell-shaped distribution

Detecting outliers

Sometimes a data set will have one or more observations with unusually large or unusually small values. These extreme values are called **outliers**. Experienced statisticians take steps to identify outliers and then review each one carefully. An outlier may be a data value that has been incorrectly recorded. If so, it can be corrected before further analysis. An outlier may also be from an observation that was incorrectly included in the data set. If so, it can be removed. Finally, an outlier may be an unusual data value that has been recorded correctly and belongs in the data set. In such cases it should remain.

Standardized values (z-scores) can be used to identify outliers. The empirical rule allows us to conclude that for data with a bell-shaped distribution, almost all the data values will be within three standard deviations of the mean. Hence, we recommend treating any data value with a z-score less than −3 or greater than +3 as an outlier, if the sample is small or moderately sized. Such data values can then be reviewed for accuracy and to determine whether they belong in the data set.

Refer to the z-scores for the class size data in Table 3.4. The z-score of −1.50 shows the fifth class size is furthest from the mean. However, this standardized value is well within the −3 to +3 guideline for outliers. Hence, the z-scores do not indicate that outliers are present in the class size data.

Exercises

Methods

17 Consider a sample with data values of 10, 20, 12, 17 and 16. Calculate the z-score for each of the five observations.

18 Consider a sample with a mean of 500 and a standard deviation of 100. What are the z-scores for the following data values: 520, 650, 500, 450 and 280?

19 Consider a sample with a mean of 30 and a standard deviation of 5. Use Chebyshev's theorem to determine the percentage of the data within each of the following ranges.

    a. 20 to 40   b. 15 to 45   c. 22 to 38   d. 18 to 42   e. 12 to 48

20 Suppose the data have a bell-shaped distribution with a mean of 30 and a standard deviation of 5. Use the empirical rule to determine the percentage of data within each of the following ranges.

    a. 20 to 40   b. 15 to 45   c. 25 to 35

Applications

21 The results of a survey of 1154 adults showed that on average, adults sleep 6.9 hours per day during the working week. Suppose that the standard deviation is 1.2 hours.

    a. Use Chebyshev's theorem to calculate the percentage of individuals who sleep between 4.5 and 9.3 hours per day.

    b. Use Chebyshev's theorem to calculate the percentage of individuals who sleep between 3.9 and 9.9 hours per day.

    c. Assume that the number of hours of sleep follows a bell-shaped distribution. Use the empirical rule to calculate the percentage of individuals who sleep between 4.5 and 9.3 hours per day. How does this result compare to the value that you obtained using Chebyshev's theorem in part (a)?

22 Suppose that IQ scores have a bell-shaped distribution with a mean of 100 and a standard deviation of 15.

    a. What percentage of people have an IQ score between 85 and 115?

    b. What percentage of people have an IQ score between 70 and 130?

    c. What percentage of people have an IQ score of more than 130?

    d. A person with an IQ score greater than 145 is considered a genius. Does the empirical rule support this statement? Explain.

23 Suppose the average hourly labour cost for car servicing in Johannesburg is ZAR (South African rand) 75.00, and the standard deviation is ZAR 20.00.

    a. What is the z-score for a car service with an hourly labour cost of ZAR 56.00?

    b. What is the z-score for a car service with an hourly labour cost of ZAR 153.00?

    c. Interpret the z-scores in parts (a) and (b). Comment on whether either should be considered an outlier.

24 *Consumer Review* posts reviews and ratings of a variety of products on the Internet. The following is a sample of 20 speaker systems and their ratings, on a scale of 1 to 5, with 5 being best.

    a. Compute the mean and the median.

    b. Compute the first and third quartiles.

    c. Compute the standard deviation.

    d. The skewness of this data is 1.67. Comment on the shape of the distribution.

    e. What are the z-scores associated with Allison One and Omni Audio?

    f. Do the data contain any outliers? Explain.

SPEAKERS

Speaker	Rating	Speaker	Rating
Infinity Kappa 6.1	4.00	ACI Sapphire III	4.67
Allison One	4.12	Bose 501 Series	2.14
Cambridge Ensemble II	3.82	DCM KX-212	4.09
Dynaudio Contour 1.3	4.00	Eosone RSF1000	4.17
Hsu Rsch. HRSW12V	4.56	Joseph Audio RM7si	4.88
Legacy Audio Focus	4.32	Martin Logan Aerius	4.26
26 Mission 73li	4.33	Omni Audio SA 12.3	2.32
PSB 400i	4.50	Polk Audio RT12	4.50
Snell Acoustics D IV	4.64	Sunfire True Subwoofer	4.17
Thiel CS1.5	4.20	Yamaha NS-A636	2.17

3.4 Exploratory data analysis

In Chapter 2 we introduced the stem-and-leaf display as an exploratory data analysis technique. In this section we continue exploratory data analysis by considering five-number summaries and box plots.

Five-number summary

In a **five-number summary** the following five numbers are used to summarize the data.

1 Smallest value
2 First quartile (Q_1)
3 Median (Q_2)
4 Third quartile (Q_3)
5 Largest value

The easiest way to construct a five-number summary is to first place the data in ascending order. Then it is easy to identify the smallest value, the three quartiles and the largest value. The monthly starting salaries shown in Table 3.1 for a sample of 12 business school graduates are repeated here in ascending order.

$$1955 \quad 1980 \quad 2020 \mid 2040 \quad 2040 \quad 2050 \mid 2060 \quad 2070 \quad 2075 \mid 2125 \quad 2165 \quad 2260$$

$$Q_1 = 2030 \qquad\qquad Q_2 = 2055 \qquad\qquad Q_3 = 2100$$

$$\text{(Median)}$$

The median of 2055 and the quartiles $Q_1 = 2030$ and $Q_3 = 2100$ were computed in Section 3.1. The smallest value is 1955 and the largest value is 2260. Hence the five-number summary for the salary data is 1955, 2030, 2055, 2100, 2260. Approximately one-quarter, or 25 per cent, of the observations are between adjacent numbers in a five-number summary.

Box plot

A **box plot** is a graphical summary that is based on a five-number summary. A key to the construction of a box plot is the computation of the median and the quartiles, Q_1 and Q_3. The interquartile range, IQR = $Q_3 - Q_1$, is also used. Figure 3.5 is the box plot for the monthly starting salary data. The steps used to construct the box plot follow.

1 A box is drawn with the ends of the box located at the first and third quartiles. For the salary data, $Q_1 = 2030$ and $Q_3 = 2100$. This box contains the middle 50 per cent of the data.

2 A vertical line is drawn in the box at the location of the median (2055 for the salary data).

3 By using the interquartile range, IQR = $Q_3 - Q_1$, *limits* are located. The limits for the box plot are 1.5(IQR) below Q_1 and 1.5(IQR) above Q_3. For the salary data, IQR = $Q_3 - Q_1$ = 2100 - 2030 = 70. Hence, the limits are 2030 - 1.5(70) = 1925 and 2100 + 1.5(70) = 2205. Data outside these limits are considered *outliers*.

4 The dashed lines in Figure 3.5 are called *whiskers*. The whiskers are drawn from the ends of the box to the smallest and largest values *inside the limits* computed in step 3. Hence the whiskers end at salary values of 1955 and 2165.

5 Finally, the location of each outlier is shown with the symbol.* In Figure 3.5 we see one outlier, 2260.

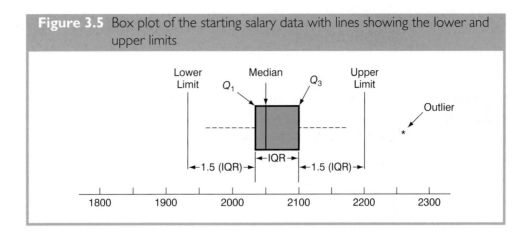

Figure 3.5 Box plot of the starting salary data with lines showing the lower and upper limits

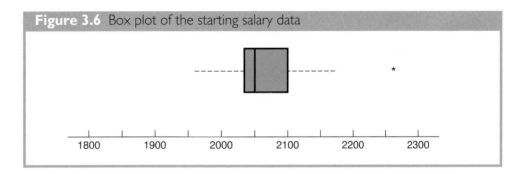

Figure 3.6 Box plot of the starting salary data

In Figure 3.5 we included lines showing the location of the upper and lower limits. These lines were drawn to show how the limits are computed and where they are located for the salary data. Although the limits are always computed, generally they are not drawn on the box plots. Figure 3.6 shows the usual appearance of a box plot for the salary data. Box plots provide another way to identify outliers. But they do not necessarily identify the same values as those with a z-score less than -3 or greater than $+3$. Either, or both, procedures may be used.

Exercises

Methods

25 Consider a sample with data values of 27, 25, 20, 15, 30, 34, 28 and 25. Provide the five-number summary for the data.

26 Construct a box plot for the data in Exercise 25.

27 Prepare the five-number summary and the box plot for the following data: 5, 15, 18, 10, 8, 12, 16, 10, 6.

28 A data set has a first quartile of 42 and a third quartile of 50. Compute the lower and upper limits for the corresponding box plot. Should a data value of 65 be considered an outlier?

Applications

29 Annual sales, in millions of dollars, for 21 pharmaceutical companies follow.

8408	1374	1872	8879	2459	11413	608
14138	6452	1850	2818	1356	10498	7478
4019	4341	739	2127	3653	5794	8305

a. Provide a five-number summary.
b. Compute the lower and upper limits (for the box plot).
c. Do the data contain any outliers?
d. Johnson & Johnson's sales are the largest on the list at $14 138 million. Suppose a data entry error (a transposition) had been made and the sales had been entered as $41 138 million. Would the method of detecting outliers in part (c) identify this problem and allow for correction of the data entry error?
d. Construct a box plot.

30 A goal of management is to help their company earn as much as possible relative to the capital invested. One measure of success is return on equity – the ratio of net income to stockholders' equity. Return on equity percentages are shown here for 25 companies.

9.0	19.6	22.9	41.6	11.4	15.8	52.7	17.3	12.3	5.1
17.3	31.1	9.6	8.6	11.2	12.8	12.2	14.5	9.2	16.6
5.0	30.3	14.7	19.2	6.2					

a. Provide a five-number summary.
b. Compute the lower and upper limits (for the box plot).
c. Do the data contain any outliers? How would this information be helpful to a financial analyst?
d. Construct a box plot.

31 In 2008, stock markets around the world lost value. The website www.owneverystock.com listed the following percentage falls in stock market indices between the start of the year and the beginning of October.

Country	% Fall	Country	% Fall
New Zealand	27.05	Brazil	39.59
Canada	27.30	Japan	39.88
Switzerland	28.42	Sweden	40.35
Mexico	29.99	Egypt	41.57
Australia	31.95	Singapore	41.60
Korea	32.18	Italy	42.88
United Kingdom	32.37	Belgium	43.70
Spain	32.69	India	44.16
Malaysia	32.86	Hong Kong	44.52
Argentina	36.83	Netherlands	44.61
France	37.71	Norway	46.98
Israel	37.84	Indonesia	47.13
Germany	37.85	Austria	50.06
Taiwan	38.79	China	60.24

a. What are the mean and median percentage changes for these countries?
b. What are the first and third quartiles?
c. Do the data contain any outliers? Construct a box plot.
d. What percentile would you report for Belgium?

3.5 Measures of association between two variables

So far we have examined numerical methods used to summarize the data for *one variable at a time*. Often a manager or decision-maker is interested in the *relationship between two variables*. In this section we present covariance and correlation as descriptive measures of the relationship between two variables.

We begin by reconsidering the hi-fi equipment store discussed in Section 2.4. The store's manager wants to determine the relationship between the number of weekend television commercials shown and the sales at the store during the following week. Sample data with sales expressed in €000s were given in Table 2.12, and are repeated here in the first three columns of Table 3.5. It shows ten observations ($n = 10$), one for each week.

The scatter diagram in Figure 3.7 shows a positive relationship, with higher sales (vertical axis) associated with a greater number of commercials (horizontal axis). In fact, the scatter diagram suggests that a straight line could be used as an approximation of the relationship. In the following discussion, we introduce **covariance** as a descriptive measure of the linear association between two variables.

Covariance

For a sample of size n with the observations (x_1, y_1), (x_2, y_2) and so on, the sample covariance is defined as follows:

Sample covariance

$$s_{XY} = \frac{\Sigma(x_i - \bar{x})(y_i - \bar{y})}{n - 1}$$

(3.10)

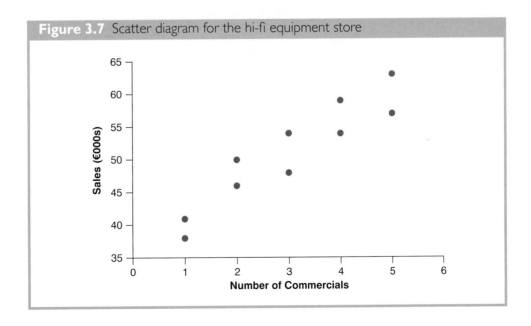

Figure 3.7 Scatter diagram for the hi-fi equipment store

This formula pairs each x_i with a corresponding y_i. We then sum the products obtained by multiplying the deviation of each x_i from its sample mean $\bar{x}$ by the deviation of the corresponding y_i from its sample mean $\bar{y}$. This sum is then divided by $n - 1$.

To measure the strength of the linear relationship between the number of commercials X and the sales volume Y in the hi-fi equipment store problem, we use equation (3.10) to compute the sample covariance. The calculations in Table 3.5 show the computation of

$$\Sigma(x_i - \bar{x})(y_i - \bar{y})$$

Note that $\bar{x} = 30/10 = 3$ and $\bar{y} = 510/10 = 51$. Using equation (3.10), we obtain a sample covariance of

$$s_{XY} = \frac{\Sigma(x_i - \bar{x})(y_i - \bar{y})}{n - 1} = \frac{99}{10 - 1} = 11$$

The formula for computing the covariance of a population of size N is similar to equation (3.10), but we use different notation to indicate that we are working with the entire population.

Population covariance

$$\sigma_{XY} = \frac{\Sigma(x_i - \mu_x)(y_i - \mu_y)}{N}$$

(3.11)

In equation (3.11) we use the notation μ_X for the population mean of X and μ_Y for the population mean of Y. The population covariance σ_{XY} is defined for a population of size N.

Interpretation of the covariance

To aid in the interpretation of the sample covariance, consider Figure 3.8. It is the same as the scatter diagram of Figure 3.7 with a vertical dashed line at $\bar{x} = 3$ and a horizontal dashed line at $\bar{y} = 51$. The lines divide the graph into four quadrants. Points in quadrant I

Table 3.5 Calculations for the sample covariance

Week	Number of commercials x_i	Sales volume (€000s) y_i	$x_i - \bar{x}$	$y_i - \bar{y}$	$(x_i - \bar{x})(y_i - \bar{y})$
1	2	50	−1	−1	1
2	5	57	2	6	12
3	1	41	−2	−10	20
4	3	54	0	3	0
5	4	54	1	3	3
6	1	38	−2	−13	26
7	5	63	2	12	24
8	3	48	0	−3	0
9	4	59	1	8	8
10	2	46	−1	−5	5
Totals	30	510	0	0	99

correspond to $\bar{x}_i$ greater than $\bar{x}$ and y_i greater than $\bar{y}$, points in quadrant II correspond to x_i less than $\bar{x}$ and y_i greater than $\bar{y}$ and so on. Hence, the value of $(x_i - \bar{x})(y_i - \bar{y})$ is positive for points in quadrant I, negative for points in quadrant II, positive for points in quadrant III and negative for points in quadrant IV.

If the value of s_{XY} is positive, the points with the greatest influence on s_{XY} are in quadrants I and III. Hence, a positive value for s_{XY} indicates a positive linear association between X and Y; that is, as the value of X increases, the value of Y increases. If the value of s_{XY} is negative, however, the points with the greatest influence are in quadrants II and IV. Hence, a negative value for s_{XY} indicates a negative linear association between X and Y; that is, as the value of X increases, the value of Y decreases. Finally, if the points are evenly distributed across all four quadrants, the value s_{XY} will be close to zero, indicating no linear association between X and Y. Figure 3.9 shows the values of s_{XY} that can be expected with three different types of scatter diagrams.

Referring again to Figure 3.8, we see that the scatter diagram for the hi-fi equipment store follows the pattern in the top panel of Figure 3.9. As we expect, the value of the sample covariance indicates a positive linear relationship with $s_{XY} = 11$.

From the preceding discussion, it might appear that a large positive value for the covariance indicates a strong positive linear relationship and that a large negative value indicates a strong negative linear relationship. However, one problem with using covariance as a measure of the strength of the linear relationship is that the value of the covariance depends on the units of measurement for X and Y. For example, suppose we are interested in the relationship between height X and weight Y for individuals. Clearly the strength of the relationship should be the same whether we measure height in metres or centimetres (or feet). Measuring the height in centimetres, however, gives us much larger numerical values for $(x_i - \bar{x})$ than when we measure height in metres. Hence, with height measured in centimetres, we would obtain a larger value for the numerator

$$\Sigma(x_i - \bar{x})(y_i - \bar{y})$$

in equation (3.10) – and hence a larger covariance – when in fact the relationship does not change. The **correlation coefficient** is measure of the relationship between two variables that is not affected by the units of measurement for X and Y.

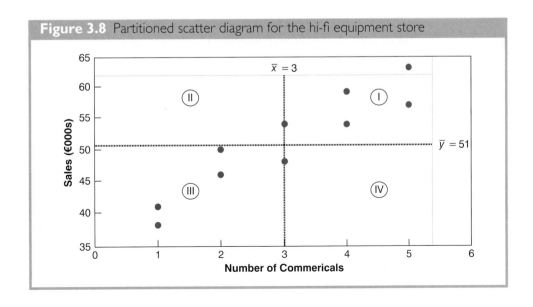

Figure 3.8 Partitioned scatter diagram for the hi-fi equipment store

Correlation coefficient

For sample data, the Pearson product moment correlation coefficient is defined as follows.

Pearson product moment correlation coefficient: sample data

$$r_{XY} = \frac{s_{XY}}{s_X s_Y}$$

(3.12)

where

r_{XY} = sample correlation coefficient
s_{XY} = sample covariance
s_X = sample standard deviation of X
s_Y = sample standard deviation of Y

Equation (3.12) shows that the Pearson product moment correlation coefficient for sample data (commonly referred to more simply as the *sample correlation coefficient*) is computed by dividing the sample covariance by the product of the sample standard deviation of X and the sample standard deviation of Y.

Let us now compute the sample correlation coefficient for the hi-fi equipment store. Using the data in Table 3.5, we can compute the sample standard deviations for the two variables.

$$s_X = \sqrt{\frac{\Sigma(x_i - \bar{x})^2}{n - 1}} = \sqrt{\frac{20}{9}} = 1.49$$

$$s_Y = \sqrt{\frac{\Sigma(y_i - \bar{y})^2}{n - 1}} = \sqrt{\frac{566}{9}} = 7.93$$

Now, because $s_{XY} = 11$, the sample correlation coefficient equals

$$r_{XY} = \frac{s_{XY}}{s_X s_Y} = \frac{11}{(1.49)(7.93)} = +0.93$$

The formula for computing the correlation coefficient for a population, denoted by the Greek letter ρ_{XY} (ρ is rho, pronounced 'row', to rhyme with 'go'), follows.

Pearson product moment correlation coefficient: population data

$$\rho_{XY} = \frac{\sigma_{XY}}{\sigma_X \sigma_Y}$$

(3.13)

where

ρ_{XY} = population correlation coefficient
σ_{XY} = population covariance
σ_X = population standard deviation for X
σ_Y = population standard deviation for Y

The sample correlation coefficient r_{XY} provides an estimate of the population correlation coefficient ρ_{XY}.

Figure 3.9 Interpretation of sample covariance

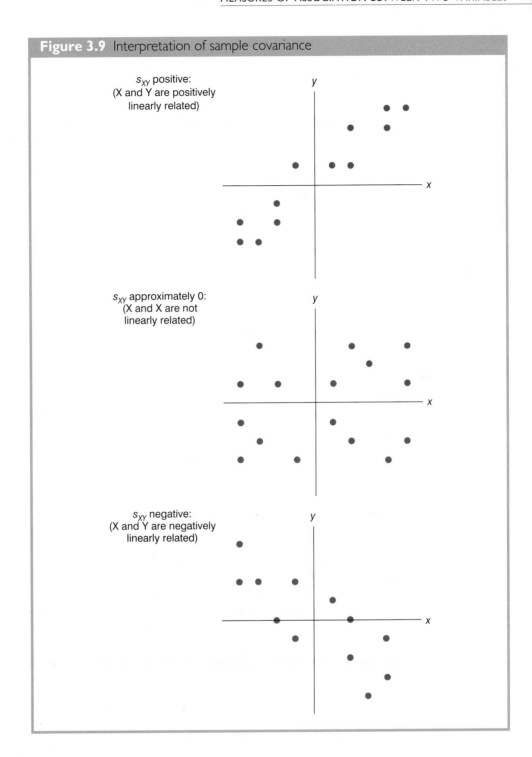

Interpretation of the correlation coefficient

First let us consider a simple example that illustrates the concept of a perfect positive linear relationship. The scatter diagram in Figure 3.10 depicts the relationship between X and Y based on the following sample data.

x_i	y_i
5	10
10	30
15	50

The straight line drawn through the three points shows a perfect linear relationship between X and Y. In order to apply equation (3.12) to compute the sample correlation we must first compute s_{XY}, s_X, and s_Y. Some of the computations are shown in Table 3.6 Using the results in Table 3.6, we find

$$s_{XY} = \frac{\Sigma(x_i - \bar{x})(y_i - \bar{y})}{n - 1} = \frac{200}{2} = 100$$

$$s_X = \sqrt{\frac{\Sigma(x_i - \bar{x})^2}{n - 1}} = \sqrt{\frac{50}{2}} = 5$$

$$s_Y = \sqrt{\frac{\Sigma(y_i - \bar{y})^2}{n - 1}} = \sqrt{\frac{800}{2}} = 20$$

$$r_{XY} = \frac{s_{XY}}{s_X s_Y} = \frac{100}{(5) \times (20)} = +1$$

We see that the value of the sample correlation coefficient is $+1$.

In general, it can be shown that if all the points in a data set fall on a positively sloped straight line, the value of the sample correlation coefficient is $+1$. That is, a sample correlation coefficient of $+1$ corresponds to a perfect positive linear relationship between X and Y. If the points in the data set fall on a straight line with a negative slope, the value of the sample correlation coefficient is -1. That is, a sample correlation coefficient of -1 corresponds to a perfect negative linear relationship between X and Y.

Suppose that a data set indicates a positive linear relationship between X and Y but that the relationship is not perfect. The value of r_{XY} will be less than 1, indicating that the points in the scatter diagram are not all on a straight line. As the points deviate more and more from a perfect positive linear relationship, the value of r_{XY} becomes closer and closer to zero. A value of r_{XY} equal to zero indicates no linear relationship between X and Y, and values of r_{XY} near zero indicate a weak linear relationship.

For the data involving the hi-fi equipment store, recall that $r_{XY} = +0.93$. Therefore, we conclude that a strong positive linear relationship occurs between the number of commercials and sales. More specifically, an increase in the number of commercials is associated with an increase in sales.

In closing, we note that correlation provides a measure of linear association and not necessarily causation. A high correlation between two variables does not mean that one

Table 3.6 Computations used in calculating the sample correlation coefficient

x_i	y_i	$x_i - \bar{x}$	$(x_i - \bar{x})^2$	$y_i - \bar{y}$	$(y_i - \bar{y})^2$	$(x_i - \bar{x})(y_i - \bar{y})$
5	10	−5	25	−20	400	100
10	30	0	0	0	0	0
15	50	5	25	20	400	100
Totals 30	90	0	50	0	800	200
$\bar{x} = 10$	$\bar{y} = 30$					

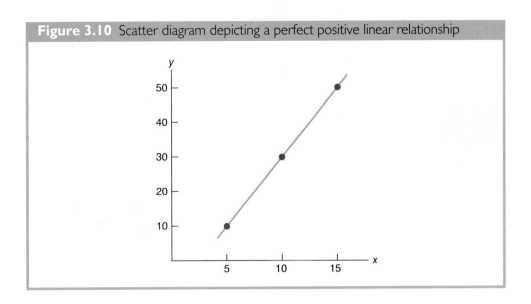

Figure 3.10 Scatter diagram depicting a perfect positive linear relationship

variable causes the other. For instance, we may find that a restaurant's quality rating and its typical meal price are positively correlated. However, increasing the meal price will not cause quality to increase.

Exercises

Methods

32 Five observations taken for two variables follow.

x_i	4	6	11	3	16
y_i	50	50	40	60	30

a. Construct a scatter diagram with the x_i values on the horizontal axis.
b. What does the scatter diagram developed in part (a) indicate about the relationship between the two variables?
c. Compute and interpret the sample covariance.
d. Compute and interpret the sample correlation coefficient.

33 Five observations taken for two variables follow.

x_i	6	11	15	21	27
y_i	6	9	6	17	12

a. Construct a scatter diagram for these data.
b. What does the scatter diagram indicate about a relationship between X and Y?
c. Compute and interpret the sample covariance.
d. Compute and interpret the sample correlation coefficient.

Applications

34 *PCWorld* provided performance scores and ratings for 15 notebook PCs. The performance score is a measure of how fast a PC can run a mix of common business applications as compared to a baseline machine. For example, a PC with a performance score of 200 is twice as fast as the baseline machine. A 100-point scale was used to provide an overall rating for each notebook tested in the study, with higher scores indicating a better rating. The data are shown below.

PCs

Notebook	Performance score	Overall rating
AMS Tech Roadster 15CTA380	115	67
Compaq Armada M700	191	78
Compaq Prosignia Notebook 150	153	79
Dell Inspiron 3700 C466GT	194	80
Dell Inspiron 7500 R500VT	236	84
Dell Latitude Cpi A366XT	184	76
Enpower ENP-313 Pro	184	77
Gateway Solo 9300LS	216	92
HP Pavillion Notebook PC	185	83
IBM ThinkPad I Series 1480	183	78
Micro Express NP7400	189	77
Micron TransPort NX PII-400	202	78
NEC Versa SX	192	78
Sceptre Soundx 5200	141	73
Sony VAIO PCG-F340	187	77

a. Construct a scatter diagram with performance score on the horizontal axis.
b. Is there any relationship between performance score and overall rating? Explain.
c. Compute and interpret the sample covariance.
d. Compute and interpret the sample correlation coefficient.
e. What does the sample correlation coefficient tell you about the relationship between the performance score and the overall rating?

35 The Dow Jones Industrial Average (DJIA) and the Standard & Poor's (S&P) 500 Index are both used as measures of overall movement in the US stock market. The DJIA is based on the price movements of 30 large companies; the S&P 500 is an index composed of 500 stocks. Some say the S&P 500 is a better measure of stock market performance because it is broader based. The index levels of the DJIA and the S&P 500 for 10 weeks beginning with 1 July 2008 are shown below (file 'DowS&P08' on the accompanying CD).

DOWS&P08

Date	DJIA	S&P
1 July	11 382.26	1284.91
8 July	11 384.21	1273.70
15 July	10 962.54	1214.91
22 July	11 602.50	1277.00
29 July	11 397.56	1263.20
5 August	11 615.77	1284.88
12 August	11 642.47	1289.59
19 August	11 348.55	1266.69
26 August	11 412.87	1271.51
2 September	11 516.92	1277.58

a. Compute the sample correlation coefficient for these data.
b. Are they poorly correlated, or do they have a close association?

3.6 The weighted mean and working with grouped data

In Section 3.1, we presented the mean as one of the most important measures of central location. The formula for the mean of a sample with n observations is restated as follows.

$$\bar{x} = \frac{\Sigma x_i}{n} = \frac{x_1 + x_2 + \cdots + x_n}{n}$$ (3.14)

In this formula, each x_i is given equal importance or weight. Although this practice is most common, in some instances the mean is computed by giving each observation a weight that reflects its importance. A mean computed in this manner is referred to as a **weighted mean**.

Weighted mean

The weighted mean is computed as follows:

Weighted mean

$$\bar{x} = \frac{\Sigma w_i x_i}{\Sigma w_i}$$ (3.15)

where

x_i = value of observation i
w_i = weight for observation i

For sample data, equation (3.15) provides the weighted sample mean. For population data, μ replaces $\bar{x}$ and equation (3.15) provides the weighted population mean.

As an example of the need for a weighted mean, consider the following sample of five purchases of a raw material over the past three months. Note that the cost per kilogram has varied from €2.80 to €3.40 and the quantity purchased has varied from 500 to 2750 kilograms.

Purchase	Cost per kilogram (€)	Number of kilograms
1	3.00	1200
2	3.40	500
3	2.80	2750
4	2.90	1000
5	3.25	800

Suppose that a manager asked for information about the mean cost per kilogram of the raw material. Because the quantities ordered vary, we must use the formula for a weighted mean. The five cost-per-kilogram data values are $x_1 = 3.00$, $x_2 = 3.40$, . . . etc. The weighted mean cost per kilogram is found by weighting each cost by its corresponding quantity corresponding quantity. For this example, the weights are

$w_1 = 1200$, $w_2 = 500$, . . . etc. Using equation (3.15), the weighted mean is calculated as follows:

$$\bar{x} = \frac{\Sigma w_i x_i}{\Sigma w_i} = \frac{1200(3.00) + 500(3.40) + 2750(2.80) + 1000(2.90) + 800(3.25)}{1200 + 500 + 2750 + 1000 + 800}$$

$$= \frac{18,500}{6250} = 2.96$$

The weighted mean computation shows that the mean cost per kilogram for the raw material is €2.96. Note that using equation (3.14) rather than the weighted mean formula would have provided misleading results. In this case, the mean of the five cost-per-kilogram values is (3.00 + 3.40 + 2.80 + 2.90 + 3.25)/5 = 15.35/5 = €3.07, which overstates the actual mean cost per kilogram purchased.

When observations vary in importance, the analyst must choose the weight that best reflects the importance of each observation in the determination of the mean. The choice of weights for a particular weighted mean computation depends upon the application.

Grouped data

In most cases, measures of location and variability are computed by using the individual data values. Sometimes, however, data are available only in a grouped or frequency distribution form. We show how the weighted mean formula can be used to obtain approximations of the mean, variance and standard deviation for **grouped data**.

In Section 2.2 we provided a frequency distribution of the time in days required to complete year-end audits for the small accounting firm of Sanderson and Clifford. The frequency distribution of audit times based on a sample of 20 clients is shown again in the first two columns of Table 3.7. Based on this frequency distribution, what is the sample mean audit time?

To compute the mean using only the grouped data, we treat the midpoint of each class as being representative of the items in the class. Let M_i denote the midpoint for class i and let f_i denote the frequency of class i. The weighted mean formula (3.15) is then used with the data values denoted as M_i and the weights given by the frequencies f_i. In this case, the denominator of equation (3.15) is the sum of the frequencies, which is the sample size n. That is,

$$\Sigma f_i = n$$

Hence, the equation for the sample mean for grouped data is as follows.

Sample mean for grouped data

$$\bar{x} = \frac{\Sigma f_i M_i}{n} \tag{3.16}$$

where

M_i = the midpoint for class i

f_i = the frequency for class i

n = the sample size

Table 3.7 Computation of the sample mean audit time for grouped data

Audit time (days)	Frequency (f_i)	Class midpoint (M_i)	($f_i M_i$)
10–14	4	12	48
15–19	8	17	136
20–24	5	22	110
25–29	2	27	54
30–34	1	32	32
Totals	20		380

$$\text{Sample mean } \bar{x} = \frac{\Sigma f_i M_i}{n} = \frac{380}{20} = 19 \text{ days}$$

With the class midpoints, M_i, halfway between the class limits, the first class of 10–14 in Table 3.7 has a midpoint at $(10 + 14)/2 = 12$. The five class midpoints and the weighted mean computation for the audit time data are summarized in Table 3.7. The sample mean audit time is 19 days.

To compute the variance for grouped data, we use a slightly altered version of the formula for the variance provided in equation (3.5). In equation (3.5), the squared deviations of the data about the sample mean $\bar{x}$ were written $(x_i - \bar{x})^2$. However, with grouped data, the values are not known. In this case, we treat the class midpoint, M_i, as being representative of the x_i values in the corresponding class. The squared deviations about the sample mean, $(x_i - \bar{x})^2$, are replaced by $(M_i - \bar{x})^2$. Then, just as we did with the sample mean calculations for grouped data, we weight each value by the frequency of the class, f_i. The sum of the squared deviations about the mean for all the data is approximated by

$$\Sigma f_i (M_i - \bar{x})^2$$

The term $n - 1$ rather than n appears in the denominator in order to make the sample variance the estimate of the population variance. The following formula is used to obtain the sample variance for grouped data.

Sample variance for grouped data

$$s^2 = \frac{\Sigma f_i (M_i - \bar{x})^2}{n - 1} \tag{3.17}$$

The calculation of the sample variance for audit times based on the grouped data from Table 3.7 is shown in Table 3.8. The sample variance is 30. The standard deviation for grouped data is simply the square root of the variance for grouped data. For the audit time data, the sample standard deviation is

$$s = \sqrt{30} = 5.48$$

Note that formulae (3.16) and (3.17) are for a sample. Population summary measures are computed similarly. The grouped data formulae for a population mean and variance follow.

Table 3.8 Computation of the sample variance of audit times for grouped data

Audit time (days)	Class midpoint (M_i)	Frequency (f_i)	Deviation ($M_i - \bar{x}$)	Squared deviation ($M_i - \bar{x}$)^2	$f_i(M_i - \bar{x})^2$
10–14	12	4	−7	49	196
15–19	17	8	−2	4	32
20–24	22	5	3	9	45
25–29	27	2	8	64	128
30–34	32	1	13	169	169
Total		20			570
					$\Sigma f_i(M_i - \bar{x})^2$

$$\text{Sample variance } s^2 = \frac{\Sigma f_i(M_i - \bar{x})^2}{n - 1} = \frac{570}{19} = 30$$

Population mean for grouped data

$$\mu = \frac{\Sigma f_i M_i}{N} \tag{3.18}$$

Population variance for grouped data

$$\sigma^2 = \frac{\Sigma f_i(M_i - \mu)^2}{N} \tag{3.19}$$

Exercises

Methods

36 Consider the following data and corresponding weights.

x_i	Weight
3.2	6
2.0	3
2.5	2
5.0	8

a. Compute the weighted mean.
b. Compute the sample mean of the four data values without weighting. Note the difference in the results provided by the two computations.

37 Consider the sample data in the following frequency distribution.

Class	Midpoint	Frequency
3–7	5	4
8–12	10	7
13–17	15	9
18–22	20	5

a. Compute the sample mean.
b. Compute the sample variance and sample standard deviation.

Applications

38 *Bloomberg Personal Finance* (July/August 2001) included the following companies in its recommended investment portfolio. For a portfolio value of €25 000, the recommended euro amounts allocated to each stock are shown.

Company	Portfolio (€)	Estimated growth rate (%)	Dividend yield (%)
Citigroup	3000	15	1.21
General Electric	5500	14	1.48
Kimberley-Clark	4200	12	1.72
Oracle	3000	25	0.00
Pharmacia	3000	20	0.96
SBC Communications	3800	12	2.48
WorldCom	2500	35	0.00

a. Using the portfolio euro amounts as the weights, what is the weighted average estimated growth rate for the portfolio?
b. What is the weighted average dividend yield for the portfolio?

39 A petrol station recorded the following frequency distribution for the number of litres of petrol sold per car in a sample of 680 cars.

Petrol (litres)	Frequency
1–15	74
16–30	192
31–45	280
46–60	105
61–75	23
76–90	6
Total	680

Compute the mean, variance and standard deviation for these grouped data. If the petrol station expects to serve petrol to about 120 cars on a given day, estimate the total number of litres of petrol that will be sold.

For additional online summary questions and answers go to the companion website at www.cengage.co.uk/aswsbe2

Summary

In this chapter we introduced several descriptive statistics that can be used to summarize the location, variability and shape of a data distribution. Unlike the tabular and graphical procedures in Chapter 2, the measures introduced in this chapter summarize the data in terms of numerical values. When the numerical values obtained are for a sample, they are called sample statistics. When the numerical values obtained are for a population, they are called population parameters. In statistical inference, the sample statistic is referred to as the point estimator of the population parameter. Some of the notation used for sample statistics and population parameters follow.

	Sample statistic	Population parameter
Mean	$\bar{x}$	μ
Variance	s^2	σ^2
Standard deviation	s	σ
Covariance	s_{XY}	σ_{XY}
Correlation	r_{XY}	ρ_{XY}

As measures of central location, we defined the mean, median and mode. Then the concept of percentiles was used to describe other locations in the data set. Next, we presented the range, interquartile range, variance, standard deviation and coefficient of variation as measures of variability or dispersion. Our primary measure of the shape of a data distribution was the skewness. Negative values indicate a data distribution skewed to the left. Positive values indicate a data distribution skewed to the right. We showed how to calculate z-scores, and indicated how they can be used to identify outlying observations. We then described how the mean and standard deviation could be used, applying Chebyshev's theorem and the empirical rule, to provide more information about the distribution of data and to identify outliers.

In Section 3.4 we showed how to develop a five-number summary and a box plot to provide simultaneous information about the location, variability and shape of the distribution.

Section 3.5 introduced covariance and the correlation coefficient as measures of association between two variables.

In Section 3.6, we showed how to compute a weighted mean and how to calculate a mean, variance and standard deviation for grouped data.

Key terms

Box plot	Mean
Chebyshev's theorem	Median
Coefficient of variation	Mode
Correlation coefficient	Outlier
Covariance	Percentile
Empirical rule	Point estimator
Five-number summary	Population parameter
Grouped data	Quartiles
Interquartile range (IQR)	Range

Sample statistic   Variance
Skewness   Weighted mean
Standard deviation   z-score

Key formulae

Sample mean

$$\bar{x} = \frac{\Sigma x_i}{n}$$ (3.1)

Population mean

$$\mu = \frac{\Sigma x_i}{N}$$ (3.2)

Interquartile range

$$IQR = Q_3 - Q_1$$ (3.3)

Population variance

$$\sigma^2 = \frac{\Sigma (x_i - \mu)^2}{N}$$ (3.4)

Sample variance

$$s^2 = \frac{\Sigma (x_i - \bar{x})^2}{n - 1}$$ (3.5)

Standard deviation

Population standard deviation $= \sigma = \sqrt{\sigma^2}$ (3.5)

Sample standard deviation $= s = \sqrt{s^2}$ (3.7)

Coefficient of variation

$$\left(\frac{\text{Standard deviation}}{\text{Mean}} \times 100 \right)\%$$ (3.8)

z-score

$$z_i = \frac{x_i - \bar{x}}{s}$$ (3.9)

Sample covariance

$$s_{XY} = \frac{\Sigma (x_i - \bar{x})(y_i - \bar{y})}{n - 1}$$ (3.10)

Population covariance

$$\sigma_{XY} = \frac{\Sigma(x_i - \mu_X)(y_i - \mu_Y)}{N} \qquad (3.11)$$

Pearson product moment correlation coefficient: sample data

$$r_{XY} = \frac{s_{XY}}{s_X s_Y} \qquad (3.12)$$

Pearson product moment correlation coefficient: population data

$$\rho_{xy} = \frac{\sigma_{XY}}{\sigma_X \sigma_Y} \qquad (3.13)$$

Weighted mean

$$\bar{x} = \frac{\Sigma w_i x_i}{\Sigma w_i} \qquad (3.15)$$

Sample mean for grouped data

$$\bar{x} = \frac{\Sigma f_i M_i}{n} \qquad (3.16)$$

Sample variance for grouped data

$$s^2 = \frac{\Sigma f_i (M_i - \bar{x})^2}{n - 1} \qquad (3.17)$$

Population mean for grouped data

$$\mu = \frac{\Sigma f_i M_i}{N} \qquad (3.18)$$

Population variance for grouped data

$$\sigma^2 = \frac{\Sigma f_i (M_i - \mu)^2}{N} \qquad (3.19)$$

Case problem Company profiles

The file 'Companies' on the CD accompanying the text contains a data set compiled in late March 2006. It comprises figures relating to samples of companies whose shares are traded on the stock exchanges in Germany, France, Ireland, South Africa and Israel. The data contained in the file are:

Name of company

Country of stock exchange where the shares are traded

Code for industrial/commercial sector in which the company operates

Market value of company (expressed in £ million) in early 2006

Price change (per cent) in a recent 12-month period
Dividend yield (per cent), from the most recently available company accounts
Market to book value (per cent), from the most recently available company accounts
Return on investment (per cent), from the most recently available company accounts

Percentage of company share in 'free float' (i.e. available for trading)
'Volatility' (standard deviation of share price over last 12 months, divided by mean over last 12 months, multiplied by 40)

A screenshot of the first few rows of data is shown below.

Company name	Country	Sector	Market value (£ million)	12-month price change (%)	Dividend yield (%)	Market to book value	Return on investment (%)	% of shares in free float	Volatility
Adidas-Salomon	BD	PERSONAL GOODS	5750.63	36.7	0.79	3.02	23.50	100	5
Allianz	BD	NON-LIFE INSURANCE	38271.64	38.8	1.29	1.49	9.04	100	6
Altana	BD	PHARMACEUTICALS	4838.53	2.4	1.91	3.93	27.03	50	3
BASF	BD	CHEMICALS	24746.79	15.7	2.65	1.92	12.16	95	4
Bayer	BD	CHEMICALS	17050.88	32.4	2.82	1.48	4.94	95	5
BMW	BD	AUTOMB	19462.39	26.9	1.37	1.28	13.76	53	3
Commerxbank	BD	BANKS	14454.39	90.6	0.79	0.93	4.32	92	9
Continental	BD	AUTOMOBILES	9037.81	55.1	0.89	2.23	33.98	84	6
Daimler Chrysler	BD	AUTOMOBILES	32925.26	37.5	3.19	1.20	8.49	93	7

Managerial Report

1 Using suitable descriptive statistics, produce summaries for each of the numerical variables in the file. For each variable, identify outliers as well as summarizing the overall characteristics of the data distribution.

Trading floor of stock exchange in stuttgart, Germany © Stefan Kiefer imagebroker.net

2 Produce a cross-tabulation of country by industrial/commercial sector, with percentages appropriate to comparing the industrial/commercial breakdown of companies between countries.

3 Investigate whether there are any differences between countries in average company market value. Similarly, investigate whether there are differences between countries in average market to book value and in return on investment.

4 Investigate whether there is any relationship between price change over the last 12 months and 'volatility'? Would you expect a relationship, given the way volatility has been calculated? Similarly, investigate whether there is any relationship between volatility and the percentage of the company's shares that is in 'free float'.

Software Section for Chapter 3

Descriptive statistics using MINITAB

Table 3.1 listed the starting salaries for 12 business school graduates. Panel A of Figure 3.11 shows the descriptive statistics obtained by using MINITAB to summarize these data. Definitions of the headings in Panel A follow.

SALARY

N	number of data values	Min	minimum data value
N*	number of missing data values	Q1	first quartile
Mean	mean	Median	median
SE Mean	standard error of mean	Q3	third quartile
StDev	standard deviation	Max	maximum data value

The label SE Mean refers to the *standard error of the mean*, which is computed by dividing the standard deviation by the square root of the number of data values. This statistic is discussed in Chapter 7 when we introduce the topics of sampling and sampling distributions. Although the range, interquartile range, variance and coefficient of variation do not appear on the MINITAB output, these values can be easily computed from the results in Figure 3.11 as follows.

$$\text{Range} = \text{Max} - \text{Min}$$
$$\text{IQR} = \text{Q3} - \text{Q1}$$
$$\text{Variance} = (\text{StDev})^2$$
$$\text{Coefficient of Variation} = (\text{StDev/Mean}) \times 100$$

Note that MINITAB's quartiles Q1 = 2025 and Q3 = 2112.5 are slightly different from the quartiles Q_1 = 2030 and Q_3 = 2100 computed in Section 3.1. The different conventions* used to identify the quartiles explain this difference. The values provided by one convention may not be identical to the values by another convention, but the differences tend to be negligible so far as interpretation is concerned.

The statistics in Figure 3.11 are generated as follows. The starting salary data are in column C2 of a MINITAB worksheet.

---

*With the n observations arranged in ascending order (smallest value to largest value), MINITAB uses the positions given by $(n + 1)/4$ and $3(n + 1)/4$ to locate Q_1 and Q_3, respectively. When a position is fractional, MINITAB interpolates between the two adjacent ordered data values to determine the corresponding quartile.

Step 1 **Stat > Basic Statistics > Display Descriptive Statistics**   [Main menu bar]

Step 2 Enter **C2** in the **Variables** box   [**Descriptive Statistics** panel]
Click **OK**

Panel B of Figure 3.11 is a MINITAB box plot. The box drawn from the first to third quartiles contains the middle 50 per cent of the data. The line within the box locates the median. The asterisk indicates an outlier at 2260. The following steps generate the box plot.

Step 1 **Graph > Boxplot**   [Main menu bar]

Step 2 Select **Simple**   [**Boxplot** panel]
Click **OK**

Step 3 Enter **C2** in the **Graph variables** box   [**Boxplot – One Y, Simple** panel]
Click **OK**

The skewness measure also does not appear as part of MINITAB's standard descriptive statistics output. However, we can include it in the descriptive statistics display by following these steps.

Figure 3.11 Descriptive statistics and box plot provided by MINITAB

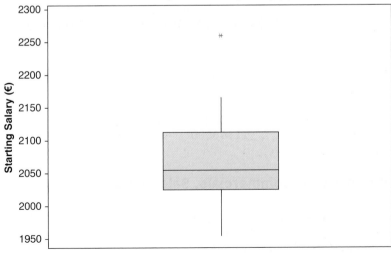

Panel A

Descriptive Statistics: Starting Salary (€)

Variable	N	N*	Mean	SE Mean	StDev	Minimum	Q1	Median
Starting Salary (€)	12	0	2070.0	23.7	82.2	1955.0	2025.0	2055.0

Variable	Q3	Maximum
Starting Salary (€)	2112.5	2260.0

Panel B

Boxplot of Starting Salary (€)

Step 1 **Stat > Basic Statistics > Display Descriptive Statistics**   [Main menu bar]

Step 2 Enter **C2** in the **Variables** box   [**Descriptive Statistics** panel]
Click the **Statistics** button

Step 3 Check **Skewness**   [**Descriptive Statistics – Statistics** panel]
Click **OK**

Step 4 Click **OK**   [**Descriptive Statistics** panel]

The skewness measure of 1.07 will then appear in your Session window.

Figure 3.12 shows the covariance and correlation output that MINITAB provided for the hi-fi equipment store data in Table 3.5. In the covariance portion of the figure, No. of Commercia denotes the number of weekend television commercials and Sales Volume denotes the sales during the following week. The value in column No. of Commercia and row Sales Volume, 11.00, is the sample covariance as computed in Section 3.5. The value in column No. of Commercia and row No. of Commercia, 2.22, is the sample variance for the number of commercials and the value in column Sales Volume and row Sales Volume, 62.89, is the sample variance for sales. The sample correlation coefficient, 0.93, is shown in the correlation portion of the output. The interpretation and use of the *p*-value provided in the output are discussed in Chapter 9.

To obtain the information in Figure 3.12, we entered the data for the number of commercials into column C2 and the data for sales volume into column C3 of a MINITAB worksheet. The steps necessary to generate the covariance output are:

Step 1 **Stat** > **Basic Statistics** > **Covariance**   [Main menu bar]
Enter **C2 C3** in the **Variables** box   [**Covariance** panel]
Click **OK**

To obtain the correlation output in Figure 3.12, only one change is necessary in the steps for obtaining the covariance: on the **Basic Statistics** menu (step 1), choose **Correlation** rather than **Covariance**.

Figure 3.12 Covariance and correlation provided by MINITAB for the number of commercials and sales data

```
Covariances: No. of Commercials, Sales Volume

                    No. of Commercia      Sales Volume
   No. of Commercia          2.22222
   Sales Volume             11.00000          62.88889


Correlations: No. of Commercials, Sales Volume

Pearson correlation of No. of Commercials and Sales Volume = 0.930
P-Value = 0.000
```

Descriptive statistics using EXCEL

We show how EXCEL can be used to generate several measures of location and variability for a single variable and to generate the covariance and correlation coefficient as measures of association between two variables.

Using EXCEL Functions

EXCEL provides functions for computing the mean, median, mode, sample variance, and sample standard deviation. We illustrate the use of these EXCEL functions by computing the mean, median, mode, sample variance and sample standard deviation for the starting salary data in Table 3.1. Refer to Figure 3.13 as we describe the steps involved. The data are entered in column B.

EXCEL's **AVERAGE** function can be used to compute the mean by entering the following formula into cell E1:

$$= \text{AVERAGE(B2:B13)}$$

**START
SALARY**

Similarly, the formulae =**MEDIAN**(B2:B13), =**MODE**(B2:B13), =**VAR**(B2:B13), and =**STDEV**(B2:B13) are entered into cells E2:E5, respectively, to compute the median, mode, variance, and standard deviation. The worksheet in the foreground shows that the values computed using the EXCEL functions are the same as we computed earlier in the chapter.

Figure 3.13 Using EXCEL functions for computing the mean, median, mode, variance and standard deviation

	A	B	C	D	E
1	Graduate	Starting Salary (€)		Mean	=AVERAGE(B2:B13)
2	1	2020		Median	=MEDIAN(B2:B13)
3	2	2075		Mode	=MODE(B2:B13)
4	3	2125		Variance	=VAR(B2:B13)
5	4	2040		Standard Deviation	=STDEV(B2:B13)
6	5	1980			
7	6	1955			
8	7	2050			
9	8	2165			
10	9	2070			
11	10	2260			
12	11	2060			
13	12	2040			
14					

	A	B	C	D	E
1	Graduate	Starting Salary (€)		Mean	2070
2	1	2020		Median	2055
3	2	2075		Mode	2040
4	3	2125		Variance	6754.5
5	4	2040		Standard Deviation	82.2
6	5	1980			
7	6	1955			
8	7	2050			
9	8	2165			
10	9	2070			
11	10	2260			
12	11	2060			
13	12	2040			
14					

EXCEL also provides functions for computing the covariance and correlation coefficient. You must be careful when using these functions because the covariance function treats the data as a population and the correlation function treats the data as a sample. So the result obtained using EXCEL's covariance function must be adjusted to provide the sample covariance. We show here how these functions can be used to compute the sample covariance and the sample correlation coefficient for the stereo and sound equipment store data in Table 3.7. Refer to Figure 3.14 as we present the steps involved.

EXCEL's covariance function, **COVAR**, can be used to compute the population covariance by entering the following formula into cell F1:

$$=COVAR(B2:B11,C2:C11)$$

Similarly, the formula =**CORREL**(B2:B11,C2:C11) is entered into cell F2 to compute the sample correlation coefficient. The worksheet in the foreground shows the values computed using the EXCEL functions. Note that the value of the sample correlation coefficient (0.93) is the same as computed using equation (3.12). However, the result provided by the EXCEL COVAR function, 9.9, was obtained by treating the data as a population. We must adjust the EXCEL result of 9.9 to obtain the sample covariance. The adjustment is rather simple. First, note that the formula for the population covariance, equation (3.11), requires dividing by the total number of observations in the data set. But the formula for the sample covariance, equation (3.10), requires dividing by the total number of observations minus 1. So, to use the EXCEL result of 9.9 to compute the sample covariance, we simply multiply 9.9 by $n/(n - 1)$. Because $n = 10$, we obtain

$$s_{XY} = \left(\frac{10}{9}\right)9.9 = 11$$

The sample covariance for the stereo and sound equipment data is 11.

Figure 3.14 Using EXCEL functions for computing covariance and correlation

	A	B	C	D	E	F
1	Week	No. of Commercials	Sales Volume		Population Covariance	=COVAR(B2:B11,C2:C11)
2	1	2	50		Sample Correlation	=CORREL(B2:B11,C2:C11)
3	2	5	57			
4	3	1	41			
5	4	3	54			
6	5	4	54			
7	6	1	38			
8	7	5	63			
9	8	3	48			
10	9	4	59			
11	10	2	46			
12						

	A	B	C	D	E	F
1	Week	No. of Commercials	Sales Volume	Population Covariance		9.9
2	1	2	50	Sample Correlation		0.93
3	2	5	57			
4	3	1	41			
5	4	3	54			
6	5	4	54			
7	6	1	38			
8	7	5	63			
9	8	3	48			
10	9	4	59			
11	10	2	46			
12						

Using EXCEL's descriptive statistics tool

As we already demonstrated, EXCEL provides statistical functions to compute descriptive statistics for a data set. These functions can be used to compute one statistic at a time (e.g. mean, variance, etc.). EXCEL also provides a set of Data Analysis Tools. One of these tools, called Descriptive Statistics, allows the user to compute a variety of descriptive statistics at once. We show here how it can be used to compute descriptive statistics for the starting salary data in Table 3.1. Refer to Figure 3.15 as we describe the steps involved.

Step 1   Click the **Data** tab on the Ribbon

Step 2   In the **Analysis** group, click **Data Analysis**

Step 3   Choose **Descriptive Statistics**   [**Data Analysis** panel]
Click **OK**

Step 4   Enter **B1:B13** in the **Input Range** box   [**Descriptive Statistics** panel]
Select **Grouped By Columns**
Check **Labels in First Row**
Select **Output Range**
Enter **D1** in the **Output Range** box
Check **Summary statistics**
Click **OK**

Cells D1:E15 of Figure 3.15 show the descriptive statistics provided by EXCEL. The boldface entries are the descriptive statistics we covered in this chapter. The descriptive statistics that are not boldface are either covered subsequently in the text or discussed in more advanced texts.

Figure 3.15 EXCEL's descriptive statistics tool output

	A	B	C	D	E	F
1	**Graduate**	**Starting Salary (€)**		*Starting Salary (€)*		
2	1	2020				
3	2	2075		**Mean**	2070	
4	3	2125		Standard Error	23.72507	
5	4	2040		**Median**	2055	
6	5	1980		**Mode**	2040	
7	6	1955		**Standard Deviation**	82.18604	
8	7	2050		**Sample Variance**	6754.545	
9	8	2165		Kurtosis	1.639058	
10	9	2070		**Skewness**	1.070021	
11	10	2260		**Range**	305	
12	11	2060		**Minimum**	1955	
13	12	2040		**Maximum**	2260	
14				**Sum**	24840	
15				**Count**	12	
16						

START
SALARY

Descriptive statistics using PASW

In PASW, a limited set of descriptive statistics can be produced as follows:

Step 1   Analyze > Descriptive Statistics > Descriptives   [Main menu bar]

Step 2   Transfer the variable(s) to be analyzed to the **Variables** box   [**Descriptives** panel]
Click **OK**

The default PASW output for the graduate starting salaries data is shown in the first part of Figure 3.16. As you can see there, PASW calculates the mean, the standard deviation, the minimum and the maximum. The variance, the range and the skewness can be added to these defaults by using the **Options** button on the **Descriptives** panel.

To produce the median and quartiles, a different PASW routine is required:

Step 1   Analyze > Descriptive Statistics > Frequencies   [Main menu bar]

Step 2   Transfer the variable(s) to be analyzed to the **Variables** box   [**Frequencies** panel]
Click **Statistics**

Step 3   Check the statistics you wish to calculate   [**Frequencies:Statistics** panel]
Click **Continue**

Step 4   Remove the check in the **Display frequency tables** box   [**Frequencies** panel]
Click **OK**

Output for the starting salaries data is shown in the second part of Figure 3.16. User-defined percentiles can also be produced using this routine, by making the appropriate choices on the **Frequencies:Statistics** dialogue panel.

Note that PASW's quartiles (25th percentile = 2025 and 75th percentile = 2112.5) are slightly different from the quartiles $Q_1 = 2030$ and $Q_3 = 2100$ computed in Section 3.1. The different conventions* used to identify the quartiles explain this difference. The values provided by one convention may not be identical to the values by another convention, but any differences tend to be negligible for interpretation purposes.

Figure 3.17 is a box plot produced by PASW for the graduate starting salaries data. The box drawn from the first to third quartiles contains the middle 50 per cent of the data. The line within the box locates the median. The small open circle indicates an outlier at 2260 (identified as the 10th data value). The following steps generate the box plot.

Step 1   Graphs > Legacy Dialogs > Boxplot   [Main menu bar]

Step 2   Select **Simple**   [**Boxplot** panel]
Check **Summaries of separate variables**
Click **Define**

Step 3   Transfer the variable(s)   [**Define Simple Boxplot:Summaries of Separate**
to be analyzed to the   **Variables** panel]
Boxes represent box
Click **OK**

---

*With the n observations arranged in ascending order (smallest value to largest value), PASW uses the positions given by $(n + 1)/4$ and $3(n + 1)/4$ to locate Q_1 and Q_3, respectively. When a position is fractional, PASW interpolates between the two adjacent ordered data values to determine the corresponding quartile.

Figure 3.16 Descriptive statistics provided by PASW

Descriptive Statistics

	N	Minimum	Maximum	Mean	Std. Deviation
Starting Salary (€)	12	1955	2260	2070.00	82.186
Valid N (listwise)	12				

Statistics

Starting Salary (€)

N	Valid	12
	Missing	0
Mean		2070.00
Median		2055.00
Mode		2040
Std. Deviation		82.186
Variance		6754.545
Range		305
Minimum		1955
Maximum		2260
Percentiles	25	2025.00
	50	2055.00
	75	2112.50

Figure 3.17 Box plot provided by PASW

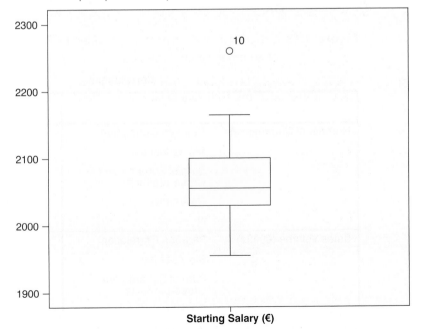

Starting Salary (€)

STEREO

Figure 3.18 shows the covariance and correlation output that PASW provided for the hi-fi equipment store data in Table 3.5. The bottom left and top right panels in the table are identical and each shows the sample correlation coefficient (0.930) and the sample covariance (11.00). Also shown, in the row labelled Sum of Squares and Cross-products, is the numerator in the variance calculation

$$\Sigma(x_i - \bar{x})(y_i - \bar{y}) = 99$$

The interpretation and use of the figure in the row labelled Sig. (2-tailed), and the asterisked note below the table, are discussed in Chapter 9.

The top left panel in the table shows the sample variance for the number of commercials (2.22), and the numerator in the variance calculation

$$\Sigma(x_i - \bar{x})^2 = 20$$

Similarly, the bottom right panel shows the sample variance for the sales volume (62.9), and the numerator in the variance calculation

$$\Sigma(y_i - \bar{y})^2 = 566$$

To obtain the information in Figure 3.18, we entered the data for the number of commercials into the second column of the PASW Data Editor and the data for sales volume into the third column.

Step 1 Analyze > Correlate > Bivariate   [Main menu bar]

Step 2 Transfer the two variables to the **Variables** box   [**Bivariate Correlations** panel]
Under **Correlation Coefficients**, ensure that the **Pearson** box is checked
Click **Options**

Step 3 Check the **Cross-product**   [**Bivariate Correlations:Options** panel]
deviations and covariances box
Click **Continue**
Click **OK**

Figure 3.18 Covariance and correlation provided by PASW for the number of commercials and sales data

Correlations

		Number of Commercials	Sales Volume (€000s)
Number of Commercials	Pearson Correlation	1	.930**
	Sig. (2-tailed)		.000
	Sum of Squares and Cross-products	20.000	99.000
	Covariance	2.222	11.000
	N	10	10
Sales Volume (€000s)	Pearson Correlation	.930**	1
	Sig. (2-tailed)	.000	
	Sum of Squares and Cross-products	99.000	566.000
	Covariance	11.000	62.889
	N	10	10

**. Correlation is significant at the 0.01 level (2-tailed).

Chapter 4

···

Introduction to Probability

Learning objectives

After reading this chapter and doing the exercises, you should be able to:

1 Appreciate the role probability information plays in the decision-making process.

2 Understand probability as a numerical measure of the likelihood of occurrence.

3 Appreciate the three methods commonly used for assigning probabilities and understand when they should be used.

4 Use the laws that are available for computing the probabilities of events.

5 Understand how new information can be used to revise initial (prior) probability estimates using Bayes' theorem.

Managers often base their decisions on an analysis of uncertainties such as the following:

1 What are the chances that sales will decrease if we increase prices?
2 What is the likelihood a new assembly method will increase productivity?
3 How likely is it that the project will be finished on time?
4 What is the chance that a new investment will be profitable?

Probability is a numerical measure of the likelihood that an event will occur. Thus, probabilities can be used as measures of the degree of uncertainty associated with the four events previously listed. If probabilities are available, we can determine the likelihood of each event occurring.

Probability values are always assigned on a scale from 0 to 1. A probability near zero indicates an event is unlikely to occur; a probability near 1 indicates an event is almost certain to occur. Other probabilities between 0 and 1 represent degrees of likelihood that an event will occur. For example, if we consider the event 'rain tomorrow', we understand that when the weather report indicates 'a near-zero probability of rain', it means almost no chance of rain. However, if a 0.90 probability of rain is reported, we know that rain is likely to occur. A 0.50 probability indicates that rain is just as likely to occur as not. Figure 4.1 depicts the view of probability as a numerical measure of the likelihood of an event occurring.

Figure 4.1 Probability as a numerical measure of the likelihood of an event occurring

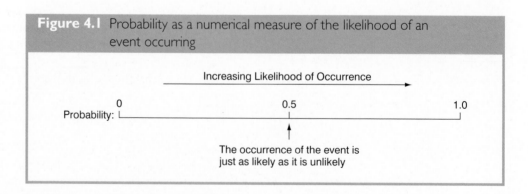

Statistics in Practice

Combating junk e-mail

Junk e-mail remains a major Internet scourge. In November 2005 it was estimated 67.39 per cent of electronic mail worldwide was spam (unsolicited commercial e-mail), more than half of it originating in China, South Korea and the USA[1]. Spam is inextricably linked to the spread of Internet viruses on the web and indeed a significant proportion of spam is now virus-generated[2]. In January 2005, 21 per cent of spam was porn[3]. Spam is time-consuming to deal with and an increasing brake on further e-mail take-up and usage.

Various initiatives have been undertaken to help counter the problem. However determining which messages are 'good' and which 'spam' is difficult to establish even with the most sophisticated spam filters (spam-busters). One of the earliest techniques for dealing with spam was the Naïve Bayes method which exploits the probability relationship:

© Young Hian Lim.

$$P(\text{spam} \mid \text{message}) = \frac{P(\text{message} \mid \text{spam})\,P(\text{spam})}{P(\text{message})}$$

where $P(\text{message}) = P(\text{message} \mid \text{spam})\,P(\text{spam})$
$+ P(\text{message} \mid \text{good})\,P(\text{good})$

Here:

P(spam) is the prior probability a message is spam based on past experience,

P(message | spam) is estimated from a training corpus (a set of messages known to be good or spam) on the (naïve) assumption that every word in the message is independent of every other so that

$P(\text{message} \mid \text{spam}) = P(\text{first word} \mid \text{spam})\,P(\text{second word} \mid \text{spam}) \ldots P(\text{last word} \mid \text{spam})$

Similarly

$P(\text{message} \mid \text{good}) = P(\text{first word} \mid \text{good})\,P(\text{second word} \mid \text{good}) \ldots P(\text{last word} \mid \text{good})$

Advantages of Naïve Bayes are its simplicity and ease of implementation. Indeed it is often found to be very effective – even compared to methods based on more complex modelling procedures.

[1] www.commtouch.com//Site/News_Events/
[2] www.messagelabs.com/
[3] www.aladdin.com/home/csrt/statistics/statistics_2005.asp

4.1　Experiments, counting rules and assigning probabilities

We define an **experiment** as a process that generates well-defined outcomes. On any single repetition of an experiment, one and only one of the possible experimental outcomes will occur. Several examples of experiments and their associated outcomes follow.

Experiment	Experimental outcomes
Toss a coin	Head, tail
Select a part for inspection	Defective, non-defective
Conduct a sales call	Purchase, no purchase
Role a die	1, 2, 3, 4, 5, 6
Play a football game	Win, lose, draw

By specifying all possible experimental outcomes, we identify the **sample space** for an experiment.

> **Sample space**
>
> The sample space for an experiment is the set of all experimental outcomes.

An experimental outcome is also called a **sample point** to identify it as an element of the sample space.

Consider the first experiment in the preceding table – tossing a coin. The upward face of the coin – a head or a tail – determines the experimental outcomes (sample points). If we let S denote the sample space, we can use the following notation to describe the sample space.

$$S = \{\text{Head, Tail}\}$$

The sample space for the second experiment in the table – selecting a part for inspection – can be described as follows:

$$S = \{\text{Defective, Non-defective}\}$$

Both of the experiments just described have two experimental outcomes (sample points). However, suppose we consider the fourth experiment listed in the table – rolling a die. The possible experimental outcomes, defined as the number of dots appearing on the upward face of the die, are the six points in the sample space for this experiment.

$$S = \{1, 2, 3, 4, 5, 6\}$$

Counting rules, combinations and permutations

Being able to identify and count the experimental outcomes is a necessary step in assigning probabilities. We now discuss three useful counting rules.

Multiple-step experiments

The first counting rule applies to multiple-step experiments. Consider the experiment of tossing two coins. Let the experimental outcomes be defined in terms of the pattern of heads and tails appearing on the upward faces of the two coins. How many experimental outcomes are possible for this experiment? The experiment of tossing two coins can be thought of as a two-step experiment in which step 1 is the tossing of the first coin and step 2 is the tossing of the second coin. If we use H to denote a head and T to denote a tail, (H, H) indicates the experimental outcome with a head on the first coin and a head on the second coin. Continuing this notation, we can describe the sample space (S) for this coin-tossing experiment as follows:

$$S = \{(H, H), (H, T), (T, H), T, T)\}$$

Thus, we see that four experimental outcomes are possible. In this case, we can easily list all of the experimental outcomes.

The counting rule for multiple-step experiments makes it possible to determine the number of experimental outcomes without listing them.

A counting rule for multiple-step experiments

If an experiment can be described as a sequence of k steps with n_1 possible outcomes on the first step, n_2 possible outcomes on the second step, and so on, then the total number of experimental outcomes is given by

$$n_1 \times n_2 \times \ldots \times n_k.$$

Viewing the experiment of tossing two coins as a sequence of first tossing one coin ($n_1 = 2$) and then tossing the other coin ($n_2 = 2$), we can see from the counting rule that there are $2 \times 2 = 4$ distinct experimental outcomes. They are $S = \{(H, H), (H, T), (T, H), (T, T)\}$. The number of experimental outcomes in an experiment involving tossing six coins is $2 \times 2 \times 2 \times 2 \times 2 \times 2 = 64$.

A **tree diagram** is a graphical representation that helps in visualizing a multiple-step experiment. Figure 4.2 shows a tree diagram for the experiment of tossing two coins. The sequence of steps moves from left to right through the tree. Step 1 corresponds to tossing the first coin, and step 2 corresponds to tossing the second coin. For each step, the two possible outcomes are head or tail. Note that for each possible outcome at step 1 two branches correspond to the two possible outcomes at step 2. Each of the points on the right end of the tree corresponds to an experimental outcome. Each path through the tree from the leftmost node to one of the nodes at the right side of the tree corresponds to a unique sequence of outcomes.

Let us now see how the counting rule for multiple-step experiments can be used in the analysis of a capacity expansion project for Kristof Projects Limited (KPL). KPL is starting a project designed to increase the generating capacity of one of its plants in southern Norway. The project is divided into two sequential stages or steps: stage 1 (design) and stage 2 (construction). Even though each stage will be scheduled and controlled as closely as possible, management cannot predict beforehand the exact time required to complete each stage of the project. An analysis of similar construction projects revealed possible completion times for the design stage of 2, 3 or 4 months and possible completion times for the construction stage of 6, 7 or 8 months. In addition, because of the critical need for additional electrical power, management set a goal of ten months for the completion of the entire project.

Figure 4.2 Tree diagram for the experiment of tossing two coins

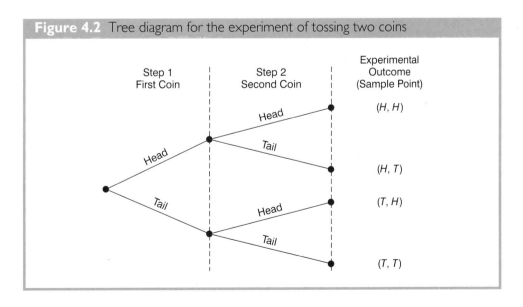

Table 4.1 Experimental outcomes (sample points) for the KPL project

Completion Time (months)

Stage 1 Design	Stage 2 Construction	Notation for experimental outcome	Total project completion time (months)
2	6	(2, 6)	8
2	7	(2, 7)	9
2	8	(2, 8)	10
3	6	(3, 6)	9
3	7	(3, 7)	10
3	8	(3, 8)	11
4	6	(4, 6)	10
4	7	(4, 7)	11
4	8	(4, 8)	12

Because this project has three possible completion times for the design stage (step 1) and three possible completion times for the construction stage (step 2), the counting rule for multiple-step experiments can be applied here to determine a total of $3 \times 3 = 9$ experimental outcomes. To describe the experimental outcomes, we use a two-number notation; for instance, (2, 6) indicates that the design stage is completed in two months and the construction stage is completed in 6 months. This experimental outcome results in a total of $2 + 6 = 8$ months to complete the entire project. Table 4.1 summarizes the nine experimental outcomes for the KPL problem. The tree diagram in Figure 4.3 shows how the nine outcomes (sample points) occur.

The counting rule and tree diagram help the project manager identify the experimental outcomes and determine the possible project completion times. We see that the project will be completed in 8 to 12 months, with six of the nine experimental outcomes providing the desired completion time of ten months or less. Even though identifying the experimental outcomes may be helpful, we need to consider how probability values can be assigned to the experimental outcomes before making an assessment of the probability that the project will be completed within the desired ten months.

Combinations

A second useful counting rule allows one to count the number of experimental outcomes when the experiment involves selecting n objects from a (usually larger) set of N objects. It is called the counting rule for combinations.

Counting rule for combinations

The number of combinations of N objects taken n at a time is

$$^N C_n = \binom{N}{n} = \frac{N!}{n!(N - n)!} \qquad (4.1)$$

where

$$N! = N(N - 1)(N - 2) \ldots (2)(1)$$
$$n! = n(n - 1)(n - 2) \ldots (2)(1)$$

and, by definition

$$0! = 1$$

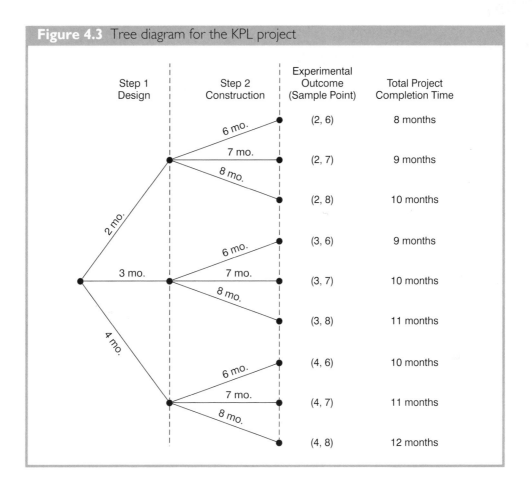

Figure 4.3 Tree diagram for the KPL project

The notation ! means *factorial:* for example, 5 factorial is $5! = 5 \times 4 \times 3 \times 2 \times 1 = 120$.

Consider a quality control procedure in which an inspector randomly selects two of five parts to test for defects. In a group of five parts, how many combinations of two parts can be selected? The counting rule in equation (4.1) shows that with $N = 5$ and $n = 2$, we have

$$^5C_2 = \binom{5}{2} = \frac{5 \times 4 \times 3 \times 2 \times 1}{(2 \times 1) \times (3 \times 2 \times 1)} = \frac{120}{12} = 10$$

Thus, ten outcomes are possible for the experiment of randomly selecting two parts from a group of five. If we label the five parts as A, B, C, D and E, the ten combinations or experimental outcomes can be identified as AB, AC, AD, AE, BC, BD, BE, CD, CE and DE.

As another example, consider that the Spanish Lotto 6-49 system uses the random selection of six integers from a group of 49 to determine the weekly lottery winner. The counting rule for combinations, equation (4.1), can be used to determine the number of ways six different integers can be selected from a group of 49.

$$\binom{49}{6} = \frac{49!}{6!(49-6)!} = \frac{49!}{6!43!} = \frac{49 \times 48 \times 47 \times 46 \times 45 \times 44}{6 \times 5 \times 4 \times 3 \times 2 \times 1} = 13\,983\,816$$

The counting rule for combinations tells us that more than 13 million experimental outcomes are possible in the lottery drawing. An individual who buys a lottery ticket has one chance in 13 983 816 of winning.

Permutations

A third counting rule that is sometimes useful is the counting rule for permutations. It allows one to compute the number of experimental outcomes when n objects are to be selected from a set of N objects where the order of selection is important. The same n objects selected in a different order is considered a different experimental outcome.

Counting rule for permutations

The number of permutations of N objects taken at n is given by

$$^N P_n = n! \binom{N}{n} = \frac{N!}{(N-n)!} \qquad (4.2)$$

The counting rule for permutations closely relates to the one for combinations; however, an experiment results in more permutations than combinations for the same number of objects because every selection of n objects can be ordered in $n!$ different ways.

As an example, consider again the quality control process in which an inspector selects two of five parts to inspect for defects. How many permutations may be selected? The counting rule in equation (4.2) shows that with $N = 5$ and $n = 2$, we have

$$^5 P_2 = \frac{5!}{(5-2)!} = \frac{5!}{3!} = \frac{5 \times 4 \times 3 \times 2 \times 1}{3 \times 2 \times 1} = \frac{120}{6} = 20$$

Thus, 20 outcomes are possible for the experiment of randomly selecting two parts from a group of five when the order of selection must be taken into account. If we label the parts A, B, C, D and E, the 20 permutations are AB, BA, AC, CA, AD, DA, AE, EA, BC, CB, BD, DB, BE, EB, CD, DC, CE, EC, DE and ED.

Assigning probabilities

Now let us see how probabilities can be assigned to experimental outcomes. The three approaches most frequently used are the classical, relative frequency, and subjective methods. Regardless of the method used, two **basic requirements for assigning probabilities** must be met.

Basic requirements for assigning probabilities

1 The probability assigned to each experimental outcome must be between 0 and 1, inclusively. If we let E_i denote the ith experimental outcome and $P(E_i)$ its probability, then this requirement can be written as

$$0 \leq P(E_i) \leq 1) \text{ for all } i \qquad (4.3)$$

2 The sum of the probabilities for all the experimental outcomes must equal 1.0. For n experimental outcomes, this requirement can be written as

$$P(E_1) + P(E_2) + \ldots + P(E_n) = 1 \qquad (4.4)$$

The **classical method** of assigning probabilities is appropriate when all the experimental outcomes are equally likely. If n experimental outcomes are possible, a probability of $1/n$ is assigned to each experimental outcome. When using this approach, the two basic requirements for assigning probabilities are automatically satisfied.

For example, consider the experiment of tossing a fair coin; the two experimental outcomes – head and tail – are equally likely. Because one of the two equally likely outcomes is a head, the probability of observing a head is 1/2 or 0.50. Similarly, the probability of observing a tail is also 1/2 or 0.50.

As another example, consider the experiment of rolling a die. It would seem reasonable to conclude that the six possible outcomes are equally likely, and hence each outcome is assigned a probability of 1/6. If $P(1)$ denotes the probability that one dot appears on the upward face of the die, then $P(1) = 1/6$. Similarly, $P(2) = 1/6$, $P(3) = 1/6$, $P(4) = 1/6$, $P(5) = 1/6$ and $P(6) = 1/6$. Note that these probabilities satisfy the two basic requirements of equations (4.3) and (4.4) because each of the probabilities is greater than or equal to zero and they sum to 1.0.

The **relative frequency method** of assigning probabilities is appropriate when data are available to estimate the proportion of the time the experimental outcome will occur if the experiment is repeated a large number of times. As an example consider a study of waiting times in the X-ray department for a local hospital. A clerk recorded the number of patients waiting for service at 9:00 a.m. on 20 successive days, and obtained the following results.

Number Waiting	Number of days outcome occurred
0	2
1	5
2	6
3	4
4	3
	Total = 20

These data show that on two of the 20 days, zero patients were waiting for service; on five of the days, one patient was waiting for service and so on. Using the relative frequency method, we would assign a probability of 2/20 = 0.10 to the experimental outcome of zero patients waiting for service, 5/20 = 0.25 to the experimental outcome of one patient waiting, 6/20 = 0.30 to two patients waiting, 4/20 = 0.20 to three patients waiting, and 3/20 = 0.15 to four patients waiting. As with the classical method, using the relative frequency method automatically satisfies the two basic requirements of equations (4.3) and (4.4).

The **subjective method** of assigning probabilities is most appropriate when one cannot realistically assume that the experimental outcomes are equally likely and when little relevant data are available. When the subjective method is used to assign probabilities to the experimental outcomes, we may use any information available, such as our experience or intuition. After considering all available information, a probability value that expresses our *degree of belief* (on a scale from 0 to 1) that the experimental outcome will occur is specified. Because subjective probability expresses a person's degree of belief, it is personal. Using the subjective method, different people can be expected to assign different probabilities to the same experimental outcome.

The subjective method requires extra care to ensure that the two basic requirements of equations (4.3) and (4.4) are satisfied. Regardless of a person's degree of belief, the probability value assigned to each experimental outcome must be between 0 and 1, inclusive, and the sum of all the probabilities for the experimental outcomes must equal 1.0.

Consider the case in which Tomas and Margit Elsbernd make an offer to purchase a house. Two outcomes are possible:

$$E_1 = \text{their offer is accepted}$$
$$E_2 = \text{their offer is rejected}$$

Margit believes that the probability their offer will be accepted is 0.8; thus, Margit would set $P(E_1) = 0.8$ and $P(E_2) = 0.2$. Tomas, however, believes that the probability that their offer will be accepted is 0.6; hence, Tomas would set $P(E_1) = 0.6$ and $P(E_2) = 0.4$. Note that Tomas' probability estimate for E_1 reflects a greater pessimism that their offer will be accepted.

Both Margit and Tomas assigned probabilities that satisfy the two basic requirements. The fact that their probability estimates are different emphasizes the personal nature of the subjective method.

Even in business situations where either the classical or the relative frequency approach can be applied, managers may want to provide subjective probability estimates. In such cases, the best probability estimates often are obtained by combining the estimates from the classical or relative frequency approach with subjective probability estimates.

Probabilities for the KPL project

To perform further analysis on the KPL project, we must develop probabilities for each of the nine experimental outcomes listed in Table 4.1. On the basis of experience and judgment, management concluded that the experimental outcomes were not equally likely. Hence, the classical method of assigning probabilities could not be used. Management then decided to conduct a study of the completion times for similar projects undertaken by KPL over the past three years. The results of a study of 40 similar projects are summarized in Table 4.2.

After reviewing the results of the study, management decided to employ the relative frequency method of assigning probabilities. Management could have provided subjective probability estimates, but felt that the current project was quite similar to the 40 previous projects. Thus, the relative frequency method was judged best.

Table 4.2 Completion results for 40 KPL projects

Completion times (months) Stage 1 Design	Stage 2 Construction	Sample point	Number of past projects having these completion times
2	6	(2, 6)	6
2	7	(2, 7)	6
2	8	(2, 8)	2
3	6	(3, 6)	4
3	7	(3, 7)	8
3	8	(3, 8)	2
4	6	(4, 6)	2
4	7	(4, 7)	4
4	8	(4, 8)	6
			Total = 40

Table 4.3 Probability assignments for the KPL project based on the relative frequency method

Sample point	Project completion time	Probability of sample point
(2, 6)	8 months	$P(2, 6) = 6/40 = 0.15$
(2, 7)	9 months	$P(2, 7) = 6/40 = 0.15$
(2, 8)	10 months	$P(2, 8) = 2/40 = 0.05$
(3, 6)	9 months	$P(3, 6) = 4/40 = 0.10$
(3, 7)	10 months	$P(3, 7) = 8/40 = 0.20$
(3, 8)	11 months	$P(3, 8) = 2/40 = 0.05$
(4, 6)	10 months	$P(4, 6) = 2/40 = 0.05$
(4, 7)	11 months	$P(4, 7) = 4/40 = 0.10$
(4, 8)	12 months	$P(4, 8) = 6/40 = 0.15$
	Total	1.00

In using the data in Table 4.2 to compute probabilities, we note that outcome (2, 6) – stage 1 completed in two months and stage 2 completed in six months – occurred six times in the 40 projects. We can use the relative frequency method to assign a probability of 6/40 = 0.15 to this outcome. Similarly, outcome (2, 7) also occurred in six of the 40 projects, providing a 6/40 = 0.15 probability. Continuing in this manner, we obtain the probability assignments for the sample points of the KPL project shown in Table 4.3. Note that $P(2, 6)$ represents the probability of the sample point (2, 6), $P(2, 7)$ represents the probability of the sample point (2, 7) and so on.

Exercises

Methods

1   An experiment has three steps with three outcomes possible for the first step, two outcomes possible for the second step, and four outcomes possible for the third step. How many experimental outcomes exist for the entire experiment?

2   How many ways can three items be selected from a group of six items? Use the letters A, B, C, D, E, and F to identify the items, and list each of the different combinations of three items.

3   How many permutations of three items can be selected from a group of six? Use the letters A, B, C, D, E, and F to identify the items, and list each of the permutations of items B, D, and F.

4   Consider the experiment of tossing a coin three times.

   a. Develop a tree diagram for the experiment.
   b. List the experimental outcomes.
   c. What is the probability for each experimental outcome?

5   Suppose an experiment has five equally likely outcomes: E_1, E_2, E_3, E_4, E_5. Assign probabilities to each outcome and show that the requirements in equations (4.3) and (4.4) are satisfied. What method did you use?

6   An experiment with three outcomes has been repeated 50 times, and it was learned that E_1 occurred 20 times, E_2 occurred 13 times, and E_3 occurred 17 times. Assign probabilities to the outcomes. What method did you use?

7   A decision-maker subjectively assigned the following probabilities to the four outcomes of an experiment: $P(E_1) = 0.10$, $P(E_2) = 0.15$, $P(E_3) = 0.40$, and $P(E_4) = 0.20$. Are these probability assignments valid? Explain.

Applications

8   Applications for zoning changes in a large metropolitan city go through a two-step process: a review by the planning commission and a final decision by the city council. At step 1 the planning commission reviews the zoning change request and makes a positive or negative recommendation concerning the change. At step 2 the city council reviews the planning commission's recommendation and then votes to approve or to disapprove the zoning change. Suppose the developer of an apartment complex submits an application for a zoning change. Consider the application process as an experiment.

   a. How many sample points are there for this experiment? List the sample points.
   b. Construct a tree diagram for the experiment.

9   Simple random sampling uses a sample of size n from a population of size N to obtain data that can be used to make inferences about the characteristics of a population. Suppose that, from a population of 50 bank accounts, we want to take a random sample of four accounts in order to learn about the population. How many different random samples of four accounts are possible?

10   A company that franchises coffee houses conducted taste tests for a new coffee product. Four blends were prepared, then randomly chosen individuals were asked to taste the blends and state which one they liked best. Results of the taste test for 100 individuals are given.

Blend	Number choosing
1	20
2	30
3	35
4	15

   a. Define the experiment being conducted. How many times was it repeated?
   b. Prior to conducting the experiment, it is reasonable to assume preferences for the four blends are equal. What probabilities would you assign to the experimental outcomes prior to conducting the taste test? What method did you use?
   c. After conducting the taste test, what probabilities would you assign to the experimental outcomes? What method did you use?

11   A company that manufactures toothpaste is studying five different package designs. Assuming that one design is just as likely to be selected by a consumer as any other design, what selection probability would you assign to each of the package designs? In an actual experiment, 100 consumers were asked to pick the design they preferred. The following data were obtained. Do the data confirm the belief that one design is just as likely to be selected as another? Explain.

Design times	Number of preferred
1	5
2	15
3	30
4	40
5	10

4.2 Events and their probabilities

In the introduction to this chapter we used the term *event* much as it would be used in everyday language. Then, in Section 4.1 we introduced the concept of an experiment and its associated experimental outcomes or sample points. Sample points and events provide the foundation for the study of probability. We must now introduce the formal definition of an **event** as it relates to sample points. Doing so will provide the basis for determining the probability of an event.

> **Event**
>
> An event is a collection of sample points.

For example, let us return to the KPL project and assume that the project manager is interested in the event that the entire project can be completed in ten months or less. Referring to Table 4.3, we see that six sample points – (2, 6), (2, 7), (2, 8), (3, 6), (3, 7) and (4, 6) – provide a project completion time of ten months or less. Let C denote the event that the project is completed in 10 months or less; we write

$$C = \{(2, 6), (2, 7), (2, 8), (3, 6), (3, 7), (4, 6)\}$$

Event C is said to occur if *any one* of these six sample points appears as the experimental outcome.

Other events that might be of interest to KPL management include the following.

L = The event that the project is completed in *less* than ten months
M = The event that the project is completed in *more* than ten months

Using the information in Table 4.3, we see that these events consist of the following sample points.

$$L = \{(2, 6), (2, 7), (3, 6)\}$$
$$M = \{3, 8), (4, 7), (4, 8)\}$$

A variety of additional events can be defined for the KPL project, but in each case the event must be identified as a collection of sample points for the experiment.

Given the probabilities of the sample points shown in Table 4.3, we can use the following definition to compute the probability of any event that KPL management might want to consider.

> **Probability of an event**
>
> The probability of any event is equal to the sum of the probabilities of the sample points for the event.

Using this definition, we calculate the probability of a particular event by adding the probabilities of the sample points (experimental outcomes) that make up the event. We can now compute the probability that the project will take ten months or less to complete. Because this event is given by $C = \{(2, 6), (2, 7), (2, 8), (3, 6), (3, 7), (4, 6)\}$, the probability of event C, denoted $P(C)$, is given by

$$P(C) = P(2, 6) + P(2, 7) + P(2, 8) + P(3, 6) + P(3, 7) + P(4, 6)$$
$$= 0.15 + 0.15 + 0.05 + 0.10 + 0.20 + 0.05 = 0.70$$

Similarly, because the event that the project is completed in less than ten months is given by $L = \{(2, 6), (2, 7), (3, 6)\}$, the probability of this event is given by

$$P(L) = P(2, 6) + P(2, 7) + (3, 6)$$
$$= 0.15 + 0.15 + 0.10 = 0.40$$

Finally, for the event that the project is completed in more than ten months, we have $M = \{(3, 8), (4, 7), (4, 8)\}$ and thus

$$P(M) = P(3,8) + P(4, 7) + (4, 8)$$
$$= 0.05 + 0.10 + 0.15 = 0.30$$

Using these probability results, we can now tell KPL management that there is a 0.70 probability that the project will be completed in ten months or less, a 0.40 probability that the project will be completed in less than ten months, and a 0.30 probability that the project will be completed in more than ten months. This procedure of computing event probabilities can be repeated for any event of interest to the KPL management.

Any time that we can identify all the sample points of an experiment and assign probabilities to each, we can compute the probability of an event using the definition. However, in many experiments the large number of sample points makes the identification of the sample points, as well as the determination of their associated probabilities, extremely cumbersome, if not impossible. In the remaining sections of this chapter, we present some basic probability relationships that can be used to compute the probability of an event without knowledge of all the sample point probabilities.

Exercises

Methods

12 An experiment has four equally likely outcomes: E_1, E_2, E_3, and E_4.

   a. What is the probability that E_2 occurs?
   b. What is the probability that any two of the outcomes occur (e.g. E_1 or E_3)?
   c. What is the probability that any three of the outcomes occur (e.g. E_1 or E_2 or E_4)?

13 Consider the experiment of selecting a playing card from a deck of 52 playing cards. Each card corresponds to a sample point with a 1/52 probability.

   a. List the sample points in the event an ace is selected.
   b. List the sample points in the event a club is selected.
   c. List the sample points in the event a face card (jack, queen, or king) is selected.
   d. Find the probabilities associated with each of the events in parts (a), (b) and (c).

14 Consider the experiment of rolling a pair of dice. Suppose that we are interested in the sum of the face values showing on the dice.

   a. How many sample points are possible? (*Hint:* Use the counting rule for multiple-step experiments.)
   b. List the sample points.
   c. What is the probability of obtaining a value of 7?
   d. What is the probability of obtaining a value of 9 or greater?
   e. Because each roll has six possible even values (2, 4, 6, 8, 10 and 12) and only five possible odd values (3, 5, 7, 9 and 11), the dice should show even values more often than odd values. Do you agree with this statement? Explain.
   f. What method did you use to assign the probabilities requested?

Applications

15 Refer to the KPL sample points and sample point probabilities in Tables 4.2 and 4.3.

   a. The design stage (stage 1) will run over budget if it takes four months to complete. List the sample points in the event the design stage is over budget.
   b. What is the probability that the design stage is over budget?
   c. The construction stage (stage 2) will run over budget if it takes eight months to complete. List the sample points in the event the construction stage is over budget.
   d. What is the probability that the construction stage is over budget?
   e. What is the probability that both stages are over budget?

16 Suppose that a manager of a large apartment complex provides the following subjective probability estimates about the number of vacancies that will exist next month.

Vacancies	Probability
0	0.10
1	0.15
2	0.30
3	0.20
4	0.15
5	0.10

Provide the probability of each of the following events.
   a. No vacancies.
   b. At least four vacancies.
   c. Two or fewer vacancies.

17 A survey of 50 college students about the number of extracurricular activities resulted in the data shown.

   a. Let A be the event that a student participates in at least one activity. Find $P(A)$.
   b. Let B be the event that a student participates in three or more activities. Find $P(B)$.

c. What is the probability that a student participates in exactly two activities?

Number of activities	Frequency
0	8
1	20
2	12
3	6
4	3
5	1

4.3 Some basic relationships of probability

Complement of an event

Given an event A, the **complement of A** is defined to be the event consisting of all sample points that are *not* in A. The complement of A is denoted by $\overline{A}$. Figure 4.4 is a diagram, known as a **Venn diagram**, which illustrates the concept of a complement. The rectangular area represents the sample space for the experiment and as such contains all possible sample points. The circle represents event A and contains only the sample points that belong to A. The shaded region of the rectangle contains all sample points not in event A, and is by definition the complement of A.

In any probability application, either event A or its complement $\overline{A}$ must occur. Therefore, we have

$$P(A) + P(\overline{A}) = 1$$

Solving for $P(A)$, we obtain the following result.

Computing probability using the complement

$$P(A) = 1 - P(\overline{A}) \tag{4.5}$$

Figure 4.4 Complement of event A is shaded

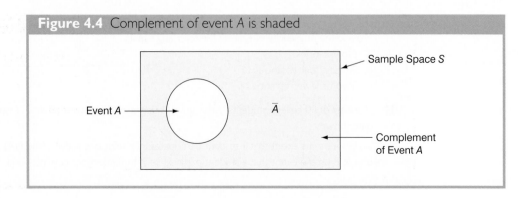

Equation (4.5) shows that the probability of an event A can be computed easily if the probability of its complement, $P(\overline{A})$, is known.

As an example, consider the case of a sales manager who, after reviewing sales reports, states that 80 per cent of new customer contacts result in no sale. By allowing A to denote the event of a sale and $\overline{A}$ to denote the event of no sale, the manager is stating that $P(\overline{A}) = 0.80$. Using equation (4.5), we see that

$$P(A) = 1 - P(\overline{A}) = 1 - 0.80 = 0.20$$

We can conclude that a new customer contact has a 0.20 probability of resulting in a sale.

In another example, a purchasing agent states a 0.90 probability that a supplier will send a shipment that is free of defective parts. Using the complement, we can conclude that there is a $1 - 0.90 = 0.10$ probability that the shipment will contain defective parts.

Addition law

The addition law is helpful when we are interested in knowing the probability that at least one of two events occurs. That is, with events A and B we are interested in knowing the probability that event A or event B or both occur.

Before we present the addition law, we need to discuss two concepts related to the combination of events: the *union* of events and the *intersection* of events. Given two events A and B, the **union of A and B** is defined as follows.

Union of two events

The *union* of A and B is the event containing *all* sample points belonging to A or B or both. The union is denoted by $A \cup B$.

The Venn diagram in Figure 4.5 depicts the union of events A and B. Note that the two circles contain all the sample points in event A as well as all the sample points in event B.

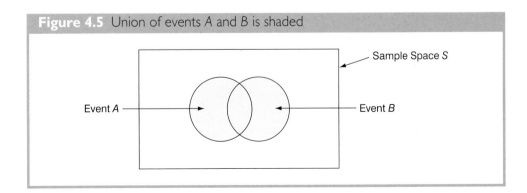

Figure 4.5 Union of events A and B is shaded

The fact that the circles overlap indicates that some sample points are contained in both *A* and *B*.

The definition of the **intersection of *A* and *B*** follows.

> **Intersection of two events**
>
> Given two events *A* and *B*, the *intersection* of *A* and *B* is the event containing the sample points belonging to *both A and B*. The intersection is denoted by *A* ∩ *B*.

The Venn diagram depicting the intersection of events *A* and *B* is shown in Figure 4.6. The area where the two circles overlap is the intersection; it contains the sample points that are in both *A* and *B*.

The **addition law** provides a way to compute the probability that event *A* or event *B* or both occur. In other words, the addition law is used to compute the probability of the union of two events. The addition law is written as follows.

> **Addition law**
>
> $$P(A \cup B) = P(A) + P(B) - P(A \cap B) \qquad (4.6)$$

To understand the addition law intuitively, note that the first two terms in the addition law, $P(A) + P(B)$, account for all the sample points in $A \cup B$. However, because the sample points in the intersection $A \cap B$ are in both *A* and *B*, when we compute $P(A) + P(B)$, we are in effect counting each of the sample points in $A \cup B$ twice. We correct for this over-counting by subtracting $P(A \cap B)$.

As an example of an application of the addition law, consider the case of a small assembly plant with 50 employees. Each worker is expected to complete work assignments on time and in such a way that the assembled product will pass a final inspection. On occasion, some of the workers fail to meet the performance standards by completing work late or assembling a defective product. At the end of a performance evaluation period, the production manager found that 5 of the 50 workers completed work late, 6 of the 50 workers assembled a defective product, and 2 of the 50 workers both completed work late *and* assembled a defective product.

Let

$$L = \text{the event that the work is completed}$$
$$D = \text{the event that the assembled product is defective}$$

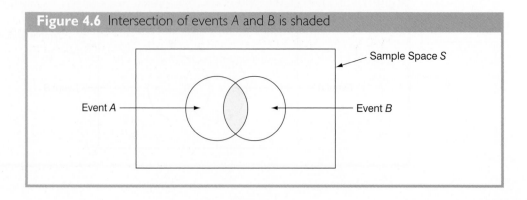

Figure 4.6 Intersection of events *A* and *B* is shaded

The relative frequency information leads to the following probabilities.

$$P(L) = \frac{5}{50} = 0.10$$

$$P(D) = \frac{6}{50} = 0.12$$

$$P(L \cap D) = \frac{2}{50} = 0.04$$

After reviewing the performance data, the production manager decided to assign a poor performance rating to any employee whose work was either late or defective; thus the event of interest is $L \cup D$. What is the probability that the production manager assigned an employee a poor performance rating?

Using equation (4.6), we have

$$P(L \cup D) = P(L) + P(D) - P(L \cap D)$$
$$= 0.10 + 0.12 - 0.04 = 0.18$$

This calculation tells us that there is a 0.18 probability that a randomly selected employee received a poor performance rating.

As another example of the addition law, consider a recent study conducted by the personnel manager of a major computer software company. The study showed that 30 per cent of the employees who left the firm within two years did so primarily because they were dissatisfied with their salary, 20 per cent left because they were dissatisfied with their work assignments, and 12 per cent of the former employees indicated dissatisfaction with *both* their salary and their work assignments. What is the probability that an employee who leaves within two years does so because of dissatisfaction with salary, dissatisfaction with the work assignment, or both?

Let

S = the event that the employee leaves because of salary
W = the event that the employee leaves because of work assignment

We have $P(S) = 0.30$, $P(W) = 0.20$, and $P(S \cap W) = 0.12$. Using equation (4.6), we have

$$P(S) + P(W) - P(S \cap W) = 0.30 + 0.20 - 0.12 = 0.38$$

We find a 0.38 probability that an employee leaves for salary or work assignment reasons.

Before we conclude our discussion of the addition law, let us consider a special case that arises for **mutually exclusive events**.

> **Mutually exclusive events**
>
> Two events are said to be mutually exclusive if the events have no sample points in common.

Events A and B are mutually exclusive if, when one event occurs, the other cannot occur. Thus, a requirement for A and B to be mutually exclusive is that their intersection must contain no sample points. The Venn diagram depicting two mutually exclusive events A and B is shown in Figure 4.7. In this case $P(A \cap B) = 0$ and the addition law can be written as follows.

Figure 4.7 Mutually exclusive events

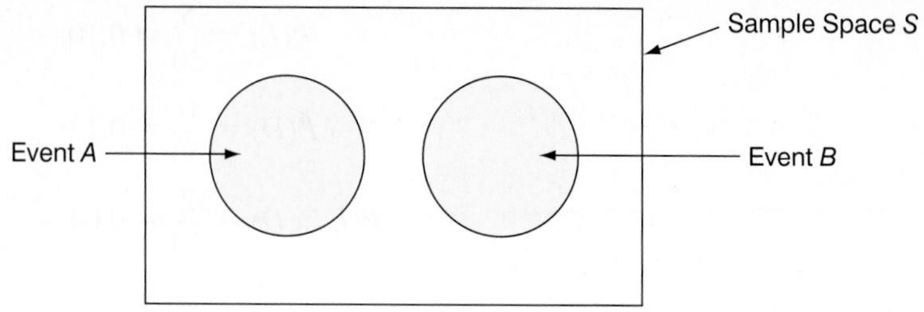

Addition law for mutually exclusive events

$$P(A \cup B) = P(A) + P(B)$$

Exercises

Methods

18 Suppose that we have a sample space with five equally likely experimental outcomes: E_1, E_2, E_3, E_4, E_5. Let

$$A = \{E_1, E_2\}$$
$$B = \{E_3, E_4\}$$
$$C = \{E_2, E_3, E_5\}$$

a. Find $P(A)$, $P(B)$, and $P(C)$.
b. Find $P(A \cup B)$. Are A and B mutually exclusive?
c. Find $\overline{A}$, $\overline{C}$, $P(\overline{A})$, and $P(\overline{C})$.
d. Find $A \cup \overline{B}$ and $P(A \cup \overline{B})$.
e. Find $P(B \cup C)$.

19 Suppose that we have a sample space $S = \{E_1, E_2, E_3, E_4, E_5, E_6, E_7\}$, where $E_1, E_2, \ldots, E_7$ denote the sample points. The following probability assignments apply: $P(E_1) = 0.05$, $P(E_2) = 0.20$, $P(E_3) = 0.20$, $P(E_4) = 0.25$, $P(E_5) = 0.15$, $P(E_6) = 0.10$, and $P(E_7) = 0.05$. Let

$$A = \{E_1, E_2\}$$
$$B = \{E_3, E_4\}$$
$$C = \{E_2, E_3, E_5\}$$

a. Find $P(A)$, $P(B)$, and $P(C)$.
b. Find $A \cup B$ and $P(A \cup B)$.
c. Find $A \cap B$ and $P(A \cap B)$.
d. Are events A and C mutually exclusive?
e. Find $\overline{B}$ and $P(\overline{B})$.

Applications

20 A survey of magazine subscribers showed that 45.8 per cent rented a car during the past 12 months for business reasons, 54 per cent rented a car during the past 12 months for personal reasons, and 30 per cent rented a car during the past 12 months for both business and personal reasons.

    a. What is the probability that a subscriber rented a car during the past 12 months for business or personal reasons?

    b. What is the probability that a subscriber did not rent a car during the past 12 months for either business or personal reasons?

4.4 Conditional probability

Often, the probability of an event is influenced by whether a related event already occurred. Suppose we have an event A with probability $P(A)$. If we obtain new information and learn that a related event, denoted by B, already occurred, we will want to take advantage of this information by calculating a new probability for event A. This new probability of event A is called a **conditional probability** and is written $P(A \mid B)$. We use the notation $\mid$ to indicate that we are considering the probability of event A *given* the condition that event B has occurred. Hence, the notation $P(A \mid B)$ reads 'the probability of A given B'.

Consider the situation of the promotion status of male and female police officers of a regional police force in France. The police force consists of 1200 officers, 960 men and 240 women. Over the past two years, 324 officers on the police force received promotions. The specific breakdown of promotions for male and female officers is shown in Table 4.4.

After reviewing the promotion record, a committee of female officers raised a discrimination case on the basis that 288 male officers had received promotions but only 36 female officers had received promotions. The police administration argued that the relatively low number of promotions for female officers was due not to discrimination, but to the fact that relatively few females are members of the police force. Let us show how conditional probability could be used to analyze the discrimination charge.

Let

$$M = \text{event an officer is a man}$$
$$W = \text{event an officer is a woman}$$
$$A = \text{event an officer is promoted}$$
$$\overline{A} = \text{event an officer is not promoted}$$

Table 4.4 Promotion status of police officers over the past two years

	Men	Women	Total
Promoted	288	36	324
Not Promoted	672	204	876
Totals	960	240	1200

Dividing the data values in Table 4.4 by the total of 1200 officers enables us to summarize the available information with the following probability values.

$P(M \cap A) = 288/1200 = 0.24 =$ probability that a randomly selected officer is a man *and* is promoted

$P(M \cap \overline{A}) = 672/1200 = 0.56 =$ probability that a randomly selected officer is a man *and* not promoted

$P(W \cap A) = 36/1200 = 0.03 =$ probability that a randomly selected officer is a woman *and* is promoted

$P(W \cap \overline{A}) = 204/1200 = 0.17 =$ probability that a randomly selected officer is a woman *and* is not promoted

Because each of these values gives the probability of the intersection of two events, the probabilities are called **joint probabilities**. Table 4.5 is referred to as a *joint probability table*.

The values in the margins of the joint probability table provide the probabilities of each event separately. That is, $P(M) = 0.80$, $P(W) = 0.20$, $P(A) = 0.27$, and $P(\overline{A}) = 0.73$. These probabilities are referred to as **marginal probabilities** because of their location in the margins of the joint probability table. We note that the marginal probabilities are found by summing the joint probabilities in the corresponding row or column of the joint probability table. For instance, the marginal probability of being promoted is $P(A) = P(M \cap A) + P(W \cap A) = 0.24 + 0.03 = 0.27$. From the marginal probabilities, we see that 80 per cent of the force is male, 20 per cent of the force is female, 27 per cent of all officers received promotions, and 73 per cent were not promoted.

Consider the probability that an officer is promoted given that the officer is a man. In conditional probability notation, we are attempting to determine $P(A \mid M)$. By definition, $P(A \mid M)$ tells us that we are concerned only with the promotion status of the 960 male officers. Because 288 of the 960 male officers received promotions, the probability of being promoted given that the officer is a man is $288/960 = 0.30$. In other words, given that an officer is a man, that officer has a 30 per cent chance of receiving a promotion over the past two years.

This procedure was easy to apply because the values in Table 4.4 show the number of officers in each category. We now want to demonstrate how conditional probabilities such as $P(A \mid M)$ can be computed directly from related event probabilities rather than the frequency data of Table 4.4.

Table 4.5 Joint probability table for promotions

Joint probabilities appear in the body of the table	Men (M)	Women (W)	Totals
Promoted (A)	0.24	0.03	0.27
Not Promoted ($\overline{A}$)	0.56	0.17	0.73
Totals	0.80	0.20	1.00

Marginal probabilities appear in the margins of the table.

We have shown that $P(A \mid M) = 288/960 = 0.30$. Let us now divide both the numerator and denominator of this fraction by 1200, the total number of officers in the study.

$$P(A \mid M) = \frac{288}{960} = \frac{288/1200}{960/1200} = \frac{0.24}{0.80} = 0.30$$

We now see that the conditional probability $P(A \mid M)$ can be computed as 0.24/0.80. Refer to the joint probability table (Table 4.5). Note in particular that 0.24 is the joint probability of A and M; that is, $P(A \cap M) = 0.24$. Also note that 0.80 is the marginal probability that a randomly selected officer is a man; that is, $P(M) = 0.80$. Thus, the conditional probability $P(A \mid M)$ can be computed as the ratio of the joint probability $P(A \cap M)$ to the marginal probability $P(M)$.

$$P(A \mid M) = \frac{P(A \cap M)}{P(M)} = \frac{0.24}{0.80} = 0.30$$

The fact that conditional probabilities can be computed as the ratio of a joint probability to a marginal probability provides the following general formula for conditional probability calculations for two events A and B.

Conditional probability

$$P(A \mid B) = \frac{P(A \cap B)}{P(B)} \qquad (4.7)$$

or

$$P(B \mid A) = \frac{P(A \cap B)}{P(A)} \qquad (4.8)$$

The Venn diagram in Figure 4.8 is helpful in obtaining an intuitive understanding of conditional probability. The circle on the right shows that event B has occurred; the portion of the circle that overlaps with event A denotes the event $(A \cap B)$. We know that once event B has occurred, the only way that we can also observe event A is for the event $(A \cap B)$ to occur. Thus, the ratio $P(A \cap B)/P(B)$ provides the conditional probability that we will observe event A given that event B has already occurred.

Let us return to the issue of discrimination against the female officers. The marginal probability in row 1 of Table 4.5 shows that the probability of promotion of an officer is $P(A) = 0.27$ (regardless of whether that officer is male or female). However, the critical issue in the discrimination case involves the two conditional probabilities $P(A \mid M)$ and $P(A \mid W)$. That is, what is the probability of a promotion *given* that the officer is a man, and what is the probability of a promotion *given* that the officer is a woman? If these two probabilities are equal, a discrimination argument has no basis because the chances of a promotion are the same for male and female officers. However, a difference in the two conditional probabilities will support the position that male and female officers are treated differently in promotion decisions.

We already determined that $P(A \mid M) = 0.30$. Let us now use the probability values in Table 4.5 and the basic relationship of conditional probability in equation (4.7) to

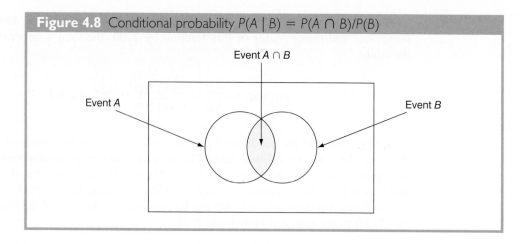

Figure 4.8 Conditional probability $P(A \mid B) = P(A \cap B)/P(B)$

compute the probability that an officer is promoted given that the officer is a woman; that is, $P(A \mid W)$. Using equation (4.7), with W replacing B, we obtain

$$P(A \mid W) = \frac{P(A \cap W)}{P(W)} = \frac{0.03}{0.20} = 0.15$$

What conclusion do you draw? The probability of a promotion given that the officer is a man is 0.30, twice the 0.15 probability of a promotion given that the officer is a woman. Although the use of conditional probability does not in itself prove that discrimination exists in this case, the conditional probability values support the argument presented by the female officers.

Independent events

In the preceding illustration, $P(A) = 0.27$, $P(A \mid M) = 0.30$, and $P(A \mid W) = 0.15$. We see that the probability of a promotion (event A) is affected or influenced by whether the officer is a man or a woman. Particularly, because $P(A \mid M) \neq P(A)$, we would say that events A and M are dependent events. That is, the probability of event A (promotion) is altered or affected by knowing that event M (the officer is a man) exists. Similarly, with $P(A \mid W) \neq P(A)$, we would say that events A and W are *dependent events*. However, if the probability of event A is not changed by the existence of event M – that is, $P(A \mid M) = P(A)$ – we would say that events A and M are **independent events**. This situation leads to the following definition of the independence of two events.

Independent events

Two events A and B are independent if

$$P(A \mid B) = P(A) \tag{4.9}$$

or

$$P(B \mid A) = P(B) \tag{4.10}$$

Otherwise, the events are dependent.

Multiplication law

Whereas the addition law of probability is used to compute the probability of a union of two events, the multiplication law is used to compute the probability of the intersection of two events. The multiplication law is based on the definition of conditional probability. Using equations (4.7) and (4.8) and solving for $P(A \cap B)$, we obtain the **multiplication law**.

Multiplication law

$$P(A \cap B) = P(A)P(B \mid A) \qquad\qquad (4.11)$$

or

$$P(A \cap B) = P(B)P(A \mid B) \qquad\qquad (4.12)$$

To illustrate the use of the multiplication law, consider a newspaper circulation department where it is known that 84 per cent of the households in a particular neighbourhood subscribe to the daily edition of the paper. If we let D denote the event that a household subscribes to the daily edition, $P(D) = 0.84$. In addition, it is known that the probability that a household that already holds a daily subscription also subscribes to the Sunday edition (event S) is 0.75; that is, $P(S \mid D) = 0.75$.

What is the probability that a household subscribes to both the Sunday and daily editions of the newspaper? Using the multiplication law, we compute the desired $P(S \cap D)$ as

$$P(S \cap D) = P(D)P(S \mid D) = 0.84 \times 0.75 = 0.63$$

We now know that 63 per cent of the households subscribe to both the Sunday and daily editions.

Before concluding this section, let us consider the special case of the multiplication law when the events involved are independent. Recall that events A and B are independent whenever $P(A \mid B) = P(A)$ or $P(B \mid A) = P(B)$. Hence, using equations (4.11) and (4.12) for the special case of independent events, we obtain the following multiplication law.

Multiplication law for independent events

$$P(A \cap B) = P(A)P(B) \qquad\qquad (4.13)$$

To compute the probability of the intersection of two independent events, we simply multiply the corresponding probabilities. Note that the multiplication law for independent events provides another way to determine whether A and B are independent. That is, if $P(A \cap B) = P(A)P(B)$, then A and B are independent; if $P(A \cap B) \neq P(A)P(B)$, then A and B are dependent.

As an application of the multiplication law for independent events, consider the situation of a service station manager who knows from past experience that 80 per cent of the customers use a credit card when they purchase petrol. What is the probability that the next two customers purchasing petrol will each use a credit card? If we let

$A =$ the event that the first customer uses a credit card
$B =$ the event that the second customer uses a credit card

then the event of interest is $A \cap B$. Given no other information, we can reasonably assume that A and B are independent events. Thus,

$$P(A \cap B) = P(A)P(B) = 0.80 \times 0.80 = 0.64$$

To summarize this section, we note that our interest in conditional probability is motivated by the fact that events are often related. In such cases, we say the events are dependent and the conditional probability formulae in equations (4.7) and (4.8) must be used to compute the event probabilities. If two events are not related, they are independent; in this case neither event's probability is affected by whether the other event occurred.

Exercises

Methods

21 Suppose that we have two events, A and B, with $P(A) = 0.50$, $P(B) = 0.60$, and $P(A \cap B) = 0.40$.

    a. Find $P(A \mid B)$.

    b. Find $P(B \mid A)$.

    c. Are A and B independent? Why or why not?

22 Assume that we have two events, A and B, that are mutually exclusive. Assume further that we know $P(A) = 0.30$ and $P(B) = 0.40$.

    a. What is $P(A \cap B)$?

    b. What is $P(A \mid B)$?

    c. A student in statistics argues that the concepts of mutually exclusive events and independent events are really the same, and that if events are mutually exclusive they must be independent. Do you agree with this statement? Use the probability information in this problem to justify your answer.

    d. What general conclusion would you make about mutually exclusive and independent events given the results of this problem?

Applications

23 A Paris nightclub obtains the following data on the age and marital status of 140 customers.

	Marital status	
Age	Single	Married
Under 30	77	14
30 or over	28	21

    a. Develop a joint probability table for these data.

    b. Use the marginal probabilities to comment on the age of customers attending the club.

    c. Use the marginal probabilities to comment on the marital status of customers attending the club.

    d. What is the probability of finding a customer who is single and under the age of 30?

    e. If a customer is under 30, what is the probability that he or she is single?

    f. Is marital status independent of age? Explain, using probabilities.

24. In a survey of MBA students, the following data were obtained on 'students' first reason for application to the school in which they matriculated'.

		Reason for application			
		School quality	School cost or convenience	Other	Totals
Enrolment status	Full time	421	393	76	890
	Part time	400	593	46	1039
	Totals	821	986	122	1929

a. Develop a joint probability table for these data.
b. Use the marginal probabilities of school quality, school cost or convenience, and other to comment on the most important reason for choosing a school.
c. If a student goes full time, what is the probability that school quality is the first reason for choosing a school?
d. If a student goes part time, what is the probability that school quality is the first reason for choosing a school?
e. Let A denote the event that a student is full time and let B denote the event that the student lists school quality as the first reason for applying. Are events A and B independent? Justify your answer.

25. A sample of convictions and compensation orders issued at a number of Scottish courts was followed up to see whether the offender had paid the compensation to the victim. Details by gender of offender are as follows:

Offender gender	Payment outcome		
	Paid in full	Part paid	Nothing paid
Male	754	62	61
Female	157	7	6

a. What is the probability that no compensation was paid?
b. What is the probability that the offender was not male given that compensation was part paid?

26 A purchasing agent placed rush orders for a particular raw material with two different suppliers, A and B. If neither order arrives in four days, the production process must be shut down until at least one of the orders arrives. The probability that supplier A can deliver the material in four days is 0.55. The probability that supplier B can deliver the material in four days is 0.35.

a. What is the probability that both suppliers will deliver the material in four days? Because two separate suppliers are involved, we are willing to assume independence.
b. What is the probability that at least one supplier will deliver the material in four days?
c. What is the probability that the production process will be shut down in four days because of a shortage of raw material (that is, both orders are late)?

4.5 Bayes' theorem

In the discussion of conditional probability, we indicated that revising probabilities when new information is obtained is an important phase of probability analysis. Often, we begin the analysis with initial or **prior probability** estimates for specific events of interest. Then,

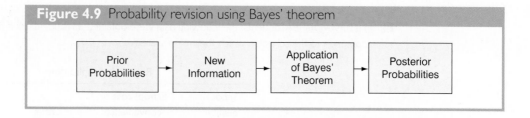

Figure 4.9 Probability revision using Bayes' theorem

from sources such as a sample, a special report, or a product test, we obtain additional information about the events. Given this new information, we update the prior probability values by calculating revised probabilities, referred to as **posterior probabilities**. **Bayes' theorem** provides a means for making these probability calculations. The steps in this probability revision process are shown in Figure 4.9.

As an application of Bayes' theorem, consider a manufacturing firm that receives shipments of parts from two different suppliers. Let A_1 denote the event that a part is from supplier 1 and A_2 denote the event that a part is from supplier 2. Currently, 65 per cent of the parts purchased by the company are from supplier 1 and the remaining 35 per cent are from supplier 2. Hence, if a part is selected at random, we would assign the prior probabilities $P(A_1) = 0.65$ and $P(A_2) = 0.35$.

The quality of the purchased parts varies with the source of supply. Historical data suggest that the quality ratings of the two suppliers are as shown in Table 4.6. If we let G denote the event that a part is good and B denote the event that a part is bad, the information in Table 4.6 provides the following conditional probability values.

$$P(G\,|\,A_1) = 0.98 \quad P(B\,|\,A_1) = 0.02$$
$$P(G\,|\,A_2) = 0.95 \quad P(B\,|\,A_2) = 0.05$$

The tree diagram in Figure 4.10 depicts the process of the firm receiving a part from one of the two suppliers and then discovering that the part is good or bad as a two-step experiment. We see that four experimental outcomes are possible; two correspond to the part being good and two correspond to the part being bad.

Each of the experimental outcomes is the intersection of two events, so we can use the multiplication rule to compute the probabilities. For instance,

$$P(A_1, G) = P(A_1 \cap G) = P(A_1)\,P(G\,|\,A_1) = 0.05$$

The process of computing these joint probabilities can be depicted in what is called a probability tree (see Figure 4.11). From left to right through the tree, the probabilities for each branch at step 1 are prior probabilities and the probabilities for each branch at step 2 are conditional probabilities. To find the probabilities of each experimental outcome, we simply multiply the probabilities on the branches leading to the outcome. Each of these joint probabilities is shown in Figure 4.11 along with the known probabilities for each branch.

Table 4.6 Historical quality levels of two suppliers

	Percentage good parts	Percentage bad parts
Supplier 1	98	2
Supplier 2	95	5

Figure 4.10 Tree diagram for two-supplier example

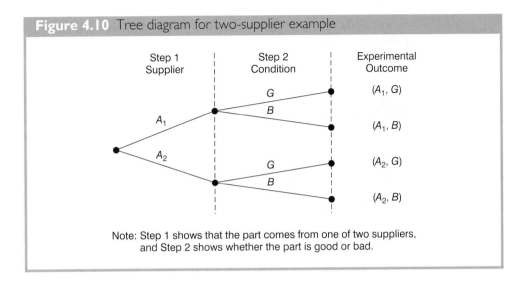

Note: Step 1 shows that the part comes from one of two suppliers,
and Step 2 shows whether the part is good or bad.

Suppose now that the parts from the two suppliers are used in the firm's manufacturing process and that a machine breaks down because it attempts to process a bad part. Given the information that the part is bad, what is the probability that it came from supplier 1 and what is the probability that it came from supplier 2? With the information in the probability tree (Figure 4.11), Bayes' theorem can be used to answer these questions.

Letting B denote the event that the part is bad, we are looking for the posterior probabilities $P(A_1 \mid B)$ and $P(A_2 \mid B)$. From the law of conditional probability, we know that

$$P(A_1 \mid B) = \frac{P(A_1 \cap B)}{P(B)} \qquad (4.14)$$

Referring to the probability tree, we see that

$$P(A_1 \cap B) = P(A_1)P(B \mid A_1) \qquad (4.15)$$

Figure 4.11 Probability tree for two-supplier example

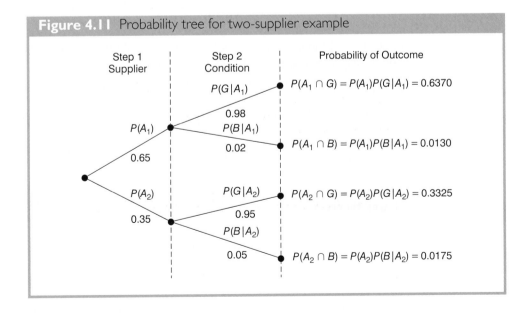

To find $P(B)$, we note that event B can occur in only two ways: $(A_1 \cap B)$ and $(A_2 \cap B)$. Therefore, we have

$$
\begin{aligned}
P(B) &= P(A_1 \cap B) + P(A_2 \cap B) \\
&= P(A_1)P(B \mid A_1) + P(A_2) P(B \mid A_2)
\end{aligned}
\tag{4.16}
$$

Substituting from equations (4.15) and (4.16) into equation (4.14) and writing a similar result for $P(A_2 \mid B)$, we obtain Bayes' theorem for the case of two events.

Bayes' theorem (two-event case)

$$
P(A_1 \mid B) = \frac{P(A_1)P(B \mid A_1)}{P(A_1)P(B \mid A_1) + P(A_2)P(B \mid A_2)}
\tag{4.17}
$$

$$
P(A_2 \mid B) = \frac{P(A_2)P(B \mid A_1)}{P(A_1)P(B \mid A_1) + P(A_2)P(B \mid A_2)}
\tag{4.18}
$$

Using equation (4.17) and the probability values provided in the example, we have

$$
\begin{aligned}
P(A_1 \mid B) &= \frac{P(A_1)P(B \mid A_1)}{P(A_1)P(B \mid A_1) + P(A_2)P(B \mid A_2)} \\
&= \frac{0.65 \times 0.02}{0.65 \times 0.02 + 0.35 \times 0.05} = \frac{0.0130}{0.0130 + 0.0175} \\
&= \frac{0.0130}{0.0305} = 0.4262
\end{aligned}
$$

In addition, using equation (4.18), we find $P(A_2 \mid B)$.

$$
\begin{aligned}
P(A_2 \mid B) &= \frac{0.35 \times 0.05}{0.65 \times 0.02 + 0.35 \times 0.05} \\
&= \frac{.0175}{0.0130 + 0.0175} = \frac{0.0175}{0.0305} = 0.5738
\end{aligned}
$$

Note that in this application we started with a probability of 0.65 that a part selected at random was from supplier 1. However, given information that the part is bad, the probability that the part is from supplier 1 drops to 0.4262. In fact, if the part is bad, it has better than a 50–50 chance that it came from supplier 2; that is, $P(A_2 \mid B) = 0.5738$.

Bayes' theorem is applicable when the events for which we want to compute posterior probabilities are mutually exclusive and their union is the entire sample space.* For the case of n mutually exclusive events $A_1, A_2, \ldots, A_n$, whose union is the entire sample space, Bayes' theorem can be used to compute any posterior probability $P(A_i \mid B)$ as shown here.

Bayes' theorem

$$
P(A_i \mid B) = \frac{P(A_i)P(B \mid A_i)}{P(A_1)P(B \mid A_1) + P(A_2)P(B \mid A_2) + \ldots + P(A_n)P(B \mid A_n)}
\tag{4.19}
$$

*If the union of events is the entire sample space, the events are said to be collectively exhaustive.

With prior probabilities $P(A_1)$, $P(A_2)$, . . . , $P(A_n)$ and the appropriate conditional probabilities $P(B \mid A_1)$, $P(B \mid A_2)$, . . . , $P(B \mid A_n)$, equation (4.19) can be used to compute the posterior probability of the events A_1, A_2, . . . , A_n.

Tabular approach

A tabular approach is helpful in conducting the Bayes' theorem calculations. Such an approach is shown in Table 4.7 for the parts supplier problem. The computations shown there are done in the following steps.

Step 1 Prepare the following three columns:
Column 1 – The mutually exclusive events A_i for which posterior probabilities are desired.
Column 2 – The prior probabilities $P(A_i)$ for the events.
Column 3 – The conditional probabilities $P(B \mid A_i)$ of the new information B given each event.

Step 2 In column 4, compute the joint probabilities $P(A_i \cap B)$ for each event and the new information B by using the multiplication law. These joint probabilities are found by multiplying the prior probabilities in column 2 by the corresponding conditional probabilities in column 3; that is, $P(A_i \cap B) = P(A_i)P(B \mid A_i)$.

Step 3 Sum the joint probabilities in column 4. The sum is the probability of the new information, $P(B)$. Thus we see in Table 4.7 that there is a 0.0130 probability that the part came from supplier 1 and is bad and a 0.0175 probability that the part came from supplier 2 and is bad. Because these are the only two ways in which a bad part can be obtained, the sum $0.0130 + 0.0175$ shows an overall probability of 0.0305 of finding a bad part from the combined shipments of the two suppliers.

Step 4 In column 5, compute the posterior probabilities using the basic relationship of conditional probability.

$$P(A_i \mid B) = \frac{P(A_i \cap B)}{P(B)}$$

Note that the joint probabilities $P(A_i \cap B)$ are in column (4) and the probability $P(B)$ is the sum of column (4).

Table 4.7 Tabular approach to Bayes' theorem calculations for the two-supplier problem

(1) Events A_i	(2) Prior probabilities $P(A_i)$	(3) Conditional probabilities $P(B \mid A_i)$	(4) Joint probabilities $P(A_i \cap B)$	(5) Posterior probabilities $P(A_i \mid B)$
A_1	0.65	0.02	0.0130	0.0130/0.0305 = 0.4262
A_2	0.35	0.05	0.0175	0.0175/0.0305 = 0.5738
			$P(B) = 0.0305$	1.0000

Exercises

Methods

27 The prior probabilities for events A_1 and A_2 are $P(A_1) = 0.40$ and $P(A_2) = 0.60$. It is also known that $P(A_1 \cap A_2) = 0$. Suppose $P(B \mid A_1) = 0.20$ and $P(B \mid A_2) = 0.05$.

   a. Are A_1 and A_2 mutually exclusive? Explain.
   b. Compute $P(A_1 \cap B)$ and $P(A_2 \cap B)$.
   c. Compute $P(B)$.
   d. Apply Bayes' theorem to compute $P(A_1 \mid B)$ and $P(A_2 \mid B)$.

28 The prior probabilities for events A_1, A_2, and A_3 are $P(A_1) = 0.20$, $P(A_2) = 0.50$ and $P(A_3) = 0.30$. The conditional probabilities of event B given A_1, A_2, and A_3 are $P(B \mid A_1) = 0.50$, $P(B \mid A_2) = 0.40$ and $P(B \mid A_3) = 0.30$.

   a. Compute $P(B \cap A_1)$, $P(B \cap A_2)$ and $P(B \cap A_3)$.
   b. Apply Bayes' theorem, equation (4.19), to compute the posterior probability $P(A_2 \mid B)$.
   c. Use the tabular approach to applying Bayes' theorem to compute $P(A_1 \mid B)$, $P(A_2 \mid B)$ and $P(A_3 \mid B)$.

Applications

29 A consulting firm submitted a bid for a large research project. The firm's management initially felt they had a 50–50 chance of getting the project. However, the agency to which the bid was submitted subsequently requested additional information on the bid. Past experience indicates that for 75 per cent of the successful bids and 40 per cent of the unsuccessful bids the agency requested additional information.

   a. What is the prior probability of the bid being successful (that is, prior to the request for additional information)?
   b. What is the conditional probability of a request for additional information given that the bid will ultimately be successful?
   c. Compute the posterior probability that the bid will be successful given a request for additional information.

30 A local bank reviewed its credit card policy with the intention of recalling some of its credit cards. In the past approximately 5 per cent of cardholders defaulted, leaving the bank unable to collect the outstanding balance. Hence, management established a prior probability of 0.05 that any particular cardholder will default. The bank also found that the probability of missing a monthly payment is 0.20 for customers who do not default. Of course, the probability of missing a monthly payment for those who default is 1.

   a. Given that a customer missed one or more monthly payments, compute the posterior probability that the customer will default.
   b. The bank would like to recall its card if the probability that a customer will default is greater than 0.20. Should the bank recall its card if the customer misses a monthly payment? Why or why not?

31 In 2006, there were 3172 fatalities recorded on Britain's roads, 169 of which were for children (Department of Transport, 2007). Correspondingly, serious injuries totalled 28 390 of which 25 625 were for adults.

   a. What is the probability of a serious injury given the victim was a child?
   b. What is the probability that the victim was an adult given a fatality occurred?

...s-tabulation shows industry type and Price/Earnings (P/E) ratio for
... the consumer products and banking industries.

	P/E ratio					
	5–9	10–14	15–19	20–24	25–29	Total
...er	4	10	18	10	8	50
	14	14	12	6	4	50
	18	24	30	16	12	100

... obability that a company had a P/E greater than 9 and belonged to the
... stry?

... obability that a company with a P/E in the range 15–19 belonged to the
... y?

... t advisory service has a number of analysts who prepare detailed studies
... anies. On the basis of these studies the analysts make 'buy' or 'sell'
... on the companies' shares. The company classes an excellent analyst as
... orrect 80 per cent of the time, a good analyst as who will be correct
... time, and a poor analyst who will be correct 40 per cent of the time.
... , the advisory service hired Mr Smith who came with considerable
... he research department of another firm. At the time he was hired it was
thought that the probability was 0.90 that he was an excellent analyst, 0.09 that he was a
good analyst and 0.01 that he was a poor analyst. In the past two years he has made ten
recommendations of which only three have been correct.

Assuming that each recommendation is an independent event what probability would
you assign to Mr Smith being:

a. An excellent analyst?
b. A good analyst?
c. A poor analyst?

34 An electronic component is produced by four production lines in a manufacturing operation.
The components are costly, are quite reliable and are shipped to suppliers in 50-component
lots. Because testing is destructive, most buyers of the components test only a small number
before deciding to accept or reject lots of incoming components. All four production lines
usually only produce 1 per cent defective components which are randomly dispersed in the
output. Unfortunately, production line 1 suffered mechanical difficulty and produced 10 per
cent defectives during the month of April. This situation became known to the manufacturer
after the components had been shipped. A customer received a lot in April and tested five
components. Two failed. What is the probability that this lot came from production line 1?

**For additional online summary questions and answers go
to the companion website at www.cengage.co.uk/aswsbe2**

Summary

In this chapter we introduced basic probability concepts and illustrated how probability analysis can be used to provide helpful information for decision-making. We described how probability can be interpreted as a numerical measure of the likelihood that an event will occur and reviewed classical, relative frequency and subjective methods for deriving probabilities. In addition, we saw that the probability of an event can be computed either by summing the probabilities of the experimental outcomes (sample points) comprising the event or by using the relationships established by the addition, conditional probability, and multiplication laws of probability. For cases in which new information is available, we showed how Bayes' theorem can be used to obtain revised or posterior probabilities.

Key terms

Addition law

Basic requirements for assigning
    probabilities

Bayes' theorem

Classical method

Complement of A

Conditional probability

Event

Experiment

Independent events

Intersection of A and B

Joint probability

Marginal probability

Multiplication law

Mutually exclusive events

Posterior probabilities

Prior probabilities

Probability

Relative frequency method

Sample point

Sample space

Subjective method

Tree diagram

Union of A and B

Venn diagram

Key formulae

Counting rule for combinations

$$^NC_n = \binom{N}{n} = \frac{N!}{n!(N-n)!}$$

(4.1)

Counting rule for permutations

$$^NP_n = n!\binom{N}{n} = \frac{N!}{(N-n)!}$$

(4.2)

Computing probability using the complement

$$P(A) = 1 - P(\bar{A})$$

(4.5)

Addition law

$$P(A \cup B) = P(A) + P(B) - P(A \cap B) \tag{4.6}$$

Conditional probability

$$P(A \mid B) = \frac{P(A \cap B)}{P(B)} \tag{4.7}$$

$$P(B \mid A) = \frac{P(A \cap B)}{P(A)} \tag{4.8}$$

Multiplication law

$$P(A \cap B) = P(B)P(A \mid B) \tag{4.11}$$

$$P(A \cap B) = P(A)P(B \mid A) \tag{4.12}$$

Multiplication law for independent events

$$P(A \cap B) = P(A)P(B) \tag{4.13}$$

Bayes' theorem

$$P(A_i \mid B) = \frac{P(A_i)\, P(B \mid A_i)}{P(A_1)\, P(B \mid A_1) + P(A_2)\, P(B \mid A_2) + \ldots + P(A_n)\, P(B \mid A_n)} \tag{4.19}$$

Case problem BAC and the Alcohol Test

In 2005, 6.7 per cent of accidents with injuries in Austria were caused by drunk drivers. The police in Wachau, Austria, a region which is famous for its wine production, is interested in buying equipment for testing drivers' blood alcohol levels. The law in Austria requires that the driver's licence be withdrawn if the driver is found to have more than 0.05 per cent BAC (Blood Alcohol Concentration).

Due to the large number of factors that come into play regarding the consumption and reduction (burn off) rates of different people, there is no blood alcohol calculator that is 100 per cent accurate. Factors include the sex (male/female) of the drinker, differing metabolism rates, various health issues and the combination of medications being taken, drinking frequency, amount of food in the stomach and small intestine and when it was eaten, elapsed time and many others. The best that can be done is a rough estimate of the BAC level based on known inputs.

There are three types of equipment available with the following conditions:

1 The Saliva Screen is a disposable strip which can be used once – this is the cheapest method.

2 The Alcometer™ is an instrument attached to a container into which the driver breathes, with the Alcometer™ then measuring the BAC concentration through an analysis of the driver's breath. The drawback to the Alcometer is that it can only detect the alcohol level correctly if it is used within two hours of alcohol consumption. It is less effective if used beyond this two-hour period.

Type	False positive	False negative
Saliva Alcohol Screen	0.020	0.03
Alcometer™	0.015	0.02
Intoximeter	0.020	0.01

3 The Intoximeter is the most expensive of the three and it works through a blood sample of the driver. The advantage for this is that it can test the BAC up to 12 hours after alcohol consumption. False positive is the situation where the test indicates a high BAC level in a driver that actually does not have such a level. The false negative is when the test indicates a low level of BAC when the driver is actually highly intoxicated.

Police records show that the percentage of drivers (late night) that drink heavily and drive ranges between 6 per cent on weekdays and 10 per cent on the weekend.

Close up of a police breathalyser being used on a man to test blood alcohol levels.
© Jack Sullivan/Alamy.

Managerial report

Carry out an appropriate probability analysis of this information on behalf of the police and advise them accordingly. (Note that it would be particularly helpful if you could assess the effectiveness of the different equipment types separately for weekdays and weekends.)

———
*Case problem provided by Dr Ibrahim Wazir, Webster University, Vienna

Chapter 5

Discrete Probability Distributions

Learning objectives

After reading this chapter and doing the exercises, you should be able to:

1 Understand the concepts of a random variable and a probability distribution.

2 Distinguish between discrete and continuous random variables.

3 Compute and interpret the expected value, variance and standard deviation for a discrete random variable.

4 Compute and work with probabilities involving a binomial probability distribution.

5 Compute and work with probabilities involving a Poisson probability distribution.

6 Know when and how to use the hypergeometric probability distribution.

In this chapter we continue the study of probability by introducing the concepts of random variables and probability distributions. The focus of this chapter is discrete probability distributions. Three special discrete probability distributions – the binomial, Poisson and hypergeometric – are covered.

5.1 Random variables

In Chapter 4 we defined the concept of an experiment and its associated experimental outcomes. A random variable provides a means for describing experimental outcomes using numerical values. Random variables must assume numerical values.

Random variable

A random variable is a numerical description of the outcome of an experiment.

In effect, a random variable associates a numerical value with each possible experimental outcome. The particular numerical value of the random variable depends on the outcome of the experiment. A random variable can be classified as being either *discrete* or *continuous* depending on the numerical values it assumes.

Discrete random variables

A random variable that may assume either a finite number of values or an infinite sequence of values such as 0, 1, 2, . . . is referred to as a **discrete random variable**. For example, consider the experiment of an accountant taking the chartered accountancy (CA) examination.

Statistics in Practice

Improving the performance reliability of combat aircraft

Modern combat aircraft are expensive to acquire and maintain. In today's post-cold war world the emphasis is therefore on deploying as few aircraft as are required and for these to be made to perform as reliably as possible in conflict and peace-keeping situations. Different strategies have been considered by manufacturers for improving the performance reliability of aircraft. One such is to reduce the incidence of faults

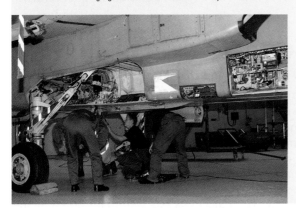

A British Tornado undergoing maintenance. © ALAN OLIVER/Alamy.

per flying hour to improve the aircraft's survival time. For example, the Tornado averages 800 faults per 1000 flying hours but if this rate could be halved, the mean operational time between faults would double. Another strategy is to build 'redundancy' into the design. In practice this would involve the aircraft carrying additional engines which would only come into use if one of the operational engines failed. To determine the number of additional engines required, designers have relied on the Poisson distribution. Calculations based on this distribution show that an aircraft with two engines would need at least four redundant engines to achieve a target maintenance-free operating period (MFOP) of 150 hours. Given each engine weighs over a tonne, occupies a space of at least $2\,m^3$ and costs some €3 m clearly this has enormous implications for those wishing to pursue this solution further.

Source: Kumar U D, Knezivic J and Crocker (1999) Maintenance-free operating period – an alternative measure to MTBF and failure rate for specifying reliability. *Reliability Engineering & System Safety* Vol 64 pp 127–131.

The examination has four parts. We can define a random variable as $X =$ the number of parts of the CA examination passed. It is a discrete random variable because it may assume the finite number of values 0, 1, 2, 3 or 4.

As another example of a discrete random variable, consider the experiment of cars arriving at a tollbooth. The random variable of interest is $X =$ the number of cars arriving during a one-day period. The possible values for X come from the sequence of integers 0, 1, 2 and so on. Hence, X is a discrete random variable assuming one of the values in this infinite sequence. Although the outcomes of many experiments can naturally be described by numerical values, others cannot. For example, a survey question might ask an individual to recall the message in a recent television commercial. This experiment would have two possible outcomes: the individual cannot recall the message and the individual can recall the message.

We can still describe these experimental outcomes numerically by defining the discrete random variable X as follows: let $X = 0$ if the individual cannot recall the message and $X = 1$ if the individual can recall the message. The numerical values for this random variable are arbitrary (we could use 5 and 10), but they are acceptable in terms of the definition of a random variable – namely, X is a random variable because it provides a numerical description of the outcome of the experiment.

Table 5.1 Examples of discrete random variables

Experiment	Random variable (X)	Possible values for the random variable
Contact five customers	Number of customers who place an order	0, 1, 2, 3, 4, 5
Inspect a shipment of 50 radios	Number of defective radios	0, 1, 2, . . . , 49, 50
Operate a restaurant for one day	Number of customers	0, 1, 2, 3, . . .
Sell a car	Gender of the customer	0 if male; 1 if female

Table 5.1 provides some additional examples of discrete random variables. Note that in each example the discrete random variable assumes a finite number of values or an infinite sequence of values such as 0, 1, 2, ... These types of discrete random variables are discussed in detail in this chapter.

Continuous random variables

A random variable that may assume any numerical value in an interval or collection of intervals is called a **continuous random variable**. Experimental outcomes based on measurement scales such as time, weight, distance, and temperature can be described by continuous random variables. For example, consider an experiment of monitoring incoming telephone calls to the claims office of a major insurance company. Suppose the random variable of interest is $X =$ the time between consecutive incoming calls in minutes. This random variable may assume any value in the interval $X \geq 0$. Actually, an infinite number of values are possible for X, including values such as 1.26 minutes, 2.751 minutes, 4.3333 minutes and so on. As another example, consider a 90-kilometre section of the A8 Autobahn in Germany. For an emergency ambulance service located in Stuttgart, we might define the random variable as $X =$ number of kilometres to the location of the next traffic accident along this section of the A8. In this case, X would be a continuous random variable assuming any value in the interval $0 \leq X \leq 90$. Additional examples of continuous random variables are listed in Table 5.2. Note that each example describes a random variable that may assume any value in an interval of values. Continuous random variables and their probability distributions will be the topic of Chapter 6.

Table 5.2 Examples of continuous random variables

Experiment	Random variable (X)	Possible values for the random variable
Operate a bank	Time between customer arrivals	$X \geq 0$ in minutes
Fill a soft drink can (max = 350 g)	Number of grams	$0 \leq X \leq 350$
Construct a new library	Percentage of project complete after six months	$0 \leq X \leq 100$
Test a new chemical process	Temperature when the desired reaction takes place (min 65° C; max 100° C)	$65 \leq X \leq 100$

Exercises

Methods

1   Consider the experiment of tossing a coin twice.

   a. List the experimental outcomes.
   b. Define a random variable that represents the number of heads occurring on the two tosses.
   c. Show what value the random variable would assume for each of the experimental outcomes.
   d. Is this random variable discrete or continuous?

2   Consider the experiment of a worker assembling a product.

   a. Define a random variable that represents the time in minutes required to assemble the product.
   b. What values may the random variable assume?
   c. Is the random variable discrete or continuous?

Applications

3   Three students have interviews scheduled for summer employment. In each case the interview results in either an offer for a position or no offer. Experimental outcomes are defined in terms of the results of the three interviews.

   a. List the experimental outcomes.
   b. Define a random variable that represents the number of offers made. Is the random variable continuous?
   c. Show the value of the random variable for each of the experimental outcomes.

4   Suppose we know home mortgage rates for 12 Danish lending institutions. Assume that the random variable of interest is the number of lending institutions in this group that offers a 30-year fixed rate of 1.5 per cent or less. What values may this random variable assume?

5   To perform a certain type of blood analysis, lab technicians must perform two procedures. The first procedure requires either 1 or 2 separate steps, and the second procedure requires either 1, 2 or 3 steps.

   a. List the experimental outcomes associated with performing the blood analysis.
   b. If the random variable of interest is the total number of steps required to do the complete analysis (both procedures), show what value the random variable will assume for each of the experimental outcomes.

6   Listed is a series of experiments and associated random variables. In each case, identify the values that the random variable can assume and state whether the random variable is discrete or continuous.

Experiment	Random variable (X)
a. Take a 20-question examination	Number of questions answered correctly
b. Observe cars arriving at a tollbooth for one hour	Number of cars arriving at tollbooth
c. Audit 50 tax returns	Number of returns containing errors
d. Observe an employee's work	Number of non-productive hours in an eight-hour workday
e. Weigh a shipment of goods	Number of kilograms

5.2 Discrete probability distributions

The **probability distribution** for a random variable describes how probabilities are distributed over the values of the random variable. For a discrete random variable X, the probability distribution is defined by a **probability function**, denoted by $p(x) = p(X = x)$ for all possible values, x. The probability function provides the probability for each value of the random variable. Consider the sales of cars at DiCarlo Motors in Sienna, Italy. Over the past 300 days of operation, sales data show 54 days with no cars sold, 117 days with one car sold, 72 days with two cars sold, 42 days with three cars sold, 12 days with four cars sold, and three days with five cars sold. Suppose we consider the experiment of selecting a day of operation at DiCarlo Motors and define the random variable of interest as X = the number of cars sold during a day. From historical data, we know X is a discrete random variable that can assume the values 0, 1, 2, 3, 4 or 5. In probability function notation, $p(0)$ provides the probability of 0 cars sold, $p(1)$ provides the probability of one car sold and so on. Because historical data show 54 of 300 days with no cars sold, we assign the value 54/300 = 0.18 to $p(0)$, indicating that the probability of no cars being sold during a day is 0.18. Similarly, because 117 of 300 days had one car sold, we assign the value 117/300 = 0.39 to $p(1)$, indicating that the probability of exactly one car being sold during a day is 0.39. Continuing in this way for the other values of the random variable, we compute the values for $p(2)$, $p(3)$, $p(4)$, and $p(5)$ as shown in Table 5.3, the probability distribution for the number of cars sold during a day at DiCarlo Motors.

A primary advantage of defining a random variable and its probability distribution is that once the probability distribution is known, it is relatively easy to determine the probability of a variety of events that may be of interest to a decision-maker. For example, using the probability distribution for DiCarlo Motors as shown in Table 5.3, we see that the most probable number of cars sold during a day is one with a probability of $p(1) = 0.39$. In addition, there is a $p(3) + p(4) + p(5) = 0.14 + 0.04 + 0.01 = 0.19$ probability of selling three or more cars during a day. These probabilities, plus others the decision-maker may ask about, provide information that can help the decision-maker understand the process of selling cars at DiCarlo Motors.

In the development of a probability function for any discrete random variable, the following two conditions must be satisfied.

Required conditions for a discrete probability function

$$p(x) \geq 0 \tag{5.1}$$

$$\sum p(x) = 1 \tag{5.2}$$

Table 5.3 shows that the probabilities for the random variable X satisfy equation (5.1); $p(x)$ is greater than or equal to 0 for all values of x. In addition, because the probabilities sum to 1, equation (5.2) is satisfied. Thus, the DiCarlo Motors probability function is a valid discrete probability function.

We can also present probability distributions graphically. In Figure 5.1 the values of the random variable X for DiCarlo Motors are shown on the horizontal axis and the probability associated with these values is shown on the vertical axis. In addition to tables and graphs, a formula that gives the probability function, $p(x)$, for every value of $X = x$ is often used to describe probability distributions. The simplest example of a discrete probability distribution given by a formula is the **discrete uniform probability distribution.** Its probability function is defined by equation (5.3).

Table 5.3 Probability distribution for the number of cars sold during a day at DiCarlo Motors

x	p(x)
0	0.18
1	0.39
2	0.24
3	0.14
4	0.04
5	0.01
	Total 1.00

Discrete uniform probability function

$$p(x) = 1/n \qquad (5.3)$$

where

n = the number of values the random variable may assume

For example, suppose that for the experiment of rolling a die we define the random variable X to be the number of dots on the upward face. There are $n = 6$ possible values for the random variable; $X = 1, 2, 3, 4, 5, 6$. Thus, the probability function for this discrete uniform random variable is

$$p(x) = 1/6 \qquad x = 1, 2, 3, 4, 5, 6$$

Figure 5.1 Graphical representation of the probability distribution for the number of cars sold during a day at DiCarlo Motors

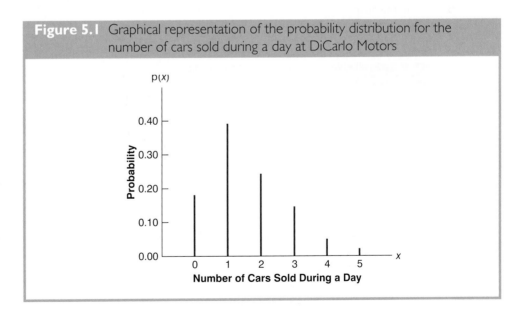

The possible values of the random variable and the associated probabilities are shown.

x	p(x)
1	1/6
2	1/6
3	1/6
4	1/6
5	1/6
6	1/6

As another example, consider the random variable X with the following discrete probability distribution.

x	p(x)
1	1/10
2	2/10
3	3/10
4	4/10

This probability distribution can be defined by the formula

$$p(x) = \frac{x}{10} \qquad \text{for } x = 1, 2, 3 \text{ or } 4$$

Evaluating p(x) for a given value of the random variable will provide the associated probability. For example, using the preceding probability function, we see that $p(2) = 2/10$ provides the probability that the random variable assumes a value of 2. The more widely used discrete probability distributions generally are specified by formulae. Three important cases are the binomial, Poisson and hypergeometric distributions; these are discussed later in the chapter.

Exercises

Methods

7   The probability distribution for the random variable X follows.

x	p(x)
20	0.20
25	0.15
30	0.25
35	0.40

a. Is this probability distribution valid? Explain.
b. What is the probability that $X = 30$?

c. What is the probability that X is less than or equal to 25?

d. What is the probability that X is greater than 30?

Applications

8   The following data were collected by counting the number of operating rooms in use at a general hospital over a 20-day period. On three of the days only one operating room was used, on five of the days two were used, on eight of the days three were used, and on four days all four of the hospital's operating rooms were used.

a. Use the relative frequency approach to construct a probability distribution for the number of operating rooms in use on any given day.

b. Draw a graph of the probability distribution.

c. Show that your probability distribution satisfies the required conditions for a valid discrete probability distribution.

9   Table 5.4 shows the percent frequency distributions of job satisfaction scores for a sample of information systems (IS) senior executives and IS middle managers. The scores range from a low of 1 (very dissatisfied) to a high of 5 (very satisfied).

Table 5.4 Percent frequency distribution of job satisfaction scores for information systems executives and middle managers		
Job satisfaction	IS senior executives (%)	IS middle score managers (%)
1	5	4
2	9	10
3	3	12
4	42	46
5	41	28

a. Develop a probability distribution for the job satisfaction score of a senior executive.

b. Develop a probability distribution for the job satisfaction score of a middle manager.

c. What is the probability a senior executive will report a job satisfaction score of 4 or 5?

d. What is the probability a middle manager is very satisfied?

e. Compare the overall job satisfaction of senior executives and middle managers.

10 A technician services mailing machines at companies in the Berne area. Depending on the type of malfunction, the service call can take 1, 2, 3 or 4 hours. The different types of malfunctions occur at about the same frequency.

a. Develop a probability distribution for the duration of a service call.

b. Draw a graph of the probability distribution.

c. Show that your probability distribution satisfies the conditions required for a discrete probability function.

d. What is the probability a service call will take three hours?

e. A service call has just come in, but the type of malfunction is unknown. It is 3:00 p.m. and service technicians usually get off at 5:00 p.m. What is the probability the service technician will have to work overtime to fix the machine today?

11   A college admissions tutor subjectively assessed a probability distribution for X, the number of entering students, as follows.

x	p(x)
1000	0.15
1100	0.20
1200	0.30
1300	0.25
1400	0.10

   a. Is this probability distribution valid? Explain.
   b. What is the probability of 1200 or fewer entering students?

12   A psychologist determined that the number of sessions required to obtain the trust of a new patient is either 1, 2 or 3. Let X be a random variable indicating the number of sessions required to gain the patient's trust. The following probability function has been proposed.

$$p(x) = \frac{x}{6} \quad \text{for } x = 1, 2, \text{ or } 3$$

   a. Is this probability function valid? Explain.
   b. What is the probability that it takes exactly two sessions to gain the patient's trust?
   c. What is the probability that it takes at least two sessions to gain the patient's trust?

13   The following table is a partial probability distribution for the MRA Company's projected profits (X = profit in €'000s) for the first year of operation (the negative value denotes a loss).

x	p(x)
−100	0.10
0	0.20
50	0.30
100	0.25
150	0.10
200	

   a. What is the proper value for p(200)? What is your interpretation of this value?
   b. What is the probability that MRA will be profitable?
   c. What is the probability that MRA will make at least €100 000?

5.3 Expected value and variance

Expected value

The **expected value**, or mean, of a random variable is a measure of the central location for the random variable. The formula for the expected value of a discrete random variable X follows.

> **Expected value of a discrete random variable**
>
> $$E(X) = \mu = \Sigma x \, p(x) \qquad \qquad \textbf{(5.4)}$$

Both the notations $E(X)$ and μ are used to denote the expected value of a random variable. Equation (5.4) shows that to compute the expected value of a discrete random variable, we must multiply each value of the random variable by the corresponding probability $p(x)$ and then add the resulting products. Using the DiCarlo Motors car sales example from Section 5.2, we show the calculation of the expected value for the number of cars sold during a day in Table 5.5. The sum of the entries in the $xp(x)$ column shows that the expected value is 1.50 cars per day. We therefore know that although sales of 0, 1, 2, 3, 4 or 5 cars are possible on any one day, over time DiCarlo can anticipate selling an average of 1.50 cars per day. Assuming 30 days of operation during a month, we can use the expected value of 1.50 to forecast average monthly sales of $30(1.50) = 45$ cars.

Variance

Even though the expected value provides the mean value for the random variable, we often need a measure of variability, or dispersion. Just as we used the variance in Chapter 3 to summarize the variability in data, we now use **variance** to summarize the variability in the values of a random variable. The formula for the variance of a discrete random variable follows.

> **Variance of a discrete random variable**
>
> $$\mathrm{Var}(X) = \sigma^2 = \Sigma (x - \mu)^2 p(x) \qquad \qquad \textbf{(5.5)}$$

Table 5.5 Calculation of the expected value for the number of cars sold during a day at DiCarlo Motors

x	$p(x)$	$xp(x)$
0	0.18	0 (0.18) = 0.00
1	0.39	1 (0.39) = 0.39
2	0.24	2 (0.24) = 0.48
3	0.14	3 (0.14) = 0.42
4	0.04	4 (0.04) = 0.16
5	0.01	5 (0.01) = 0.05
		1.50

$$E(X) = \mu = \Sigma xp(x)$$

Table 5.6 Calculation of the variance for the number of cars sold during a day at DiCarlo Motors

x	x − μ	(x − μ)²	p(x)	(x − μ)²p(x)
0	0 − 1.50 = −1.50	2.25	0.18	2.25 × 0.18 = 0.4050
1	1 − 1.50 = −0.50	0.25	0.39	0.25 × 0.39 = 0.0975
2	2 − 1.50 = 0.50	0.25	0.24	0.25 × 0.24 = 0.0600
3	3 − 1.50 = 1.50	2.25	0.14	2.25 × 0.14 = 0.3150
4	4 − 1.50 = 2.50	6.25	0.04	6.25 × 0.04 = 0.2500
5	5 − 1.50 = 3.50	12.25	0.01	12.25 × 0.01 = 0.1225
				1.2500

$$\sigma^2 = \Sigma(x - \mu)^2 p(x)$$

As equation (5.5) shows, an essential part of the variance formula is the deviation, $x - \mu$, which measures how far a particular value of the random variable is from the expected value, or mean, μ. In computing the variance of a random variable, the deviations are squared and then weighted by the corresponding value of the probability function. The sum of these weighted squared deviations for all values of the random variable is referred to as the *variance*. The notations Var(X) and σ^2 are both used to denote the variance of a random variable.

The calculation of the variance for the probability distribution of the number of cars sold during a day at DiCarlo Motors is summarized in Table 5.6. We see that the variance is 1.25. The **standard deviation**, σ, is defined as the positive square root of the variance. Thus, the standard deviation for the number of cars sold during a day is

$$\sigma = \sqrt{1.25} = 1.118$$

The standard deviation is measured in the same units as the random variable ($\sigma = 1.118$ cars) and therefore is often preferred in describing the variability of a random variable. The variance σ^2 is measured in squared units and is thus more difficult to interpret.

Exercises

Methods

14   The following table provides a probability distribution for the random variable X.

x	p(x)
3	0.25
6	0.50
9	0.25

a. Compute $E(X)$, the expected value of X.
b. Compute σ^2, the variance of X.
c. Compute σ, the standard deviation of X.

15 The following table provides a probability distribution for the random variable Y.

y	p(y)
2	0.20
4	0.30
7	0.40
8	0.10

a. Compute $E(Y)$.
b. Compute $Var(Y)$ and σ.

Applications

16 A local ambulance service handles 0 to 5 service calls on any given day. The probability distribution for the number of service calls is as follows.

Number of service calls	Probability
0	0.10
1	0.15
2	0.30
3	0.20
4	0.15
5	0.10

a. What is the expected number of service calls?
b. What is the variance in the number of service calls? What is the standard deviation?

17 A certain machinist works an eight-hour shift. An efficiency expert wants to assess the value of this machinist where value is defined as value added minus the machinist's labour cost. The value added for the work the machinist does is €30 per item and the machinist earns €16 per hour. From past records, the machinist's output per shift is known to have the following probability distribution:

Output/shift	Probability
5	0.2
6	0.4
7	0.3
8	0.1

a. What is the expected monetary value of the machinist to the company per shift?
b. What is the corresponding variance value?

18 The probability distribution for damage claims paid by a national car insurance company on collision waiver insurance follows.

Payment (€)	Probability
0	0.90
600	0.04
1500	0.03
3000	0.01
6000	0.01
9000	0.01

a. Use the expected collision waiver payment to determine the collision insurance premium that would enable the company to break even.

b. The insurance company charges an annual rate of €390 for the collision waiver coverage. What is the expected value of the collision policy for a policyholder? (*Hint:* It is the expected payments from the company minus the cost of coverage.) Why does the policyholder purchase a collision waiver policy with this expected value?

19 The following probability distributions of job satisfaction scores for a sample of information systems (IS) senior executives and IS middle managers range from a low of 1 (very dissatisfied) to a high of 5 (very satisfied).

Job satisfaction score	Probability	
	IS senior executives	IS middle managers
1	0.05	0.04
2	0.09	0.10
3	0.03	0.12
4	0.42	0.46
5	0.41	0.28

a. What is the expected value of the job satisfaction score for senior executives?

b. What is the expected value of the job satisfaction score for middle managers?

c. Compute the variance of job satisfaction scores for executives and middle managers.

d. Compute the standard deviation of job satisfaction scores for both probability distributions.

e. Compare the overall job satisfaction of senior executives and middle managers.

20 The demand for a product of Cobh Industries varies greatly from month to month. The probability distribution in the following table, based on the past two years of data, shows the company's monthly demand.

Unit Demand	Probability
300	0.20
400	0.30
500	0.35
600	0.15

a. If the company bases monthly orders on the expected value of the monthly demand, what should Cobh's monthly order quantity be for this product?

b. Assume that each unit demanded generates €70 in revenue and that each unit ordered costs €50. How much will the company gain or lose in a month if it places an order based on your answer to part (a) and the actual demand for the item is 300 units?

5.4 Binomial probability distribution

The binomial probability distribution is a discrete probability distribution that provides many applications. It is associated with a multiple-step experiment that we call the binomial experiment.

A binomial experiment

A **binomial experiment** exhibits the following four properties.

Properties of a binomial experiment

1 The experiment consists of a sequence of n identical trials.
2 Two outcomes are possible on each trial. We refer to one outcome as a *success* and the other outcome as a *failure*.
3 The probability of a success, denoted by π, does not change from trial to trial. Consequently, the probability of a failure, denoted by $1 - \pi$, does not change from trial to trial.
4 The trials are independent.

If properties 2, 3 and 4 are present, we say the trials are generated by a Bernoulli process. If, in addition, property 1 is present, we say we have a binomial experiment. Figure 5.2 depicts one possible sequence of successes and failures for a binomial experiment involving eight trials.

In a binomial experiment, our interest is in the *number of successes occurring in the n trials*. If we let X denote the number of successes occurring in the n trials, we see that X can assume the values of 0, 1, 2, 3, . . . , n. Because the number of values is finite, X is a *discrete* random variable. The probability distribution associated with this random variable is called the **binomial probability distribution**. For example, consider the experiment of tossing a coin five times and on each toss observing whether the coin lands with a head or a tail on its upward face. Suppose we want to count the number of heads appearing over the five tosses. Does this experiment show the properties of a binomial experiment? What is the random variable of interest? Note that:

1 The experiment consists of five identical trials; each trial involves the tossing of one coin.

2 Two outcomes are possible for each trial: a head or a tail. We can designate head a success and tail a failure.

3 The probability of a head and the probability of a tail are the same for each trial, with $\pi = 0.5$ and $1 - \pi = 0.5$.

4 The trials or tosses are independent because the outcome on any one trial is not affected by what happens on other trials or tosses.

Figure 5.2 One possible sequence of successes and failures for an eight-trial binomial experiment

Property 1: The experiment consists of $n = 8$ identical trials.

Property 2: Each trial results in either success (S) or failure (F).

Trials ⟶	1	2	3	4	5	6	7	8
Outcomes ⟶	S	F	F	S	S	F	S	S

Thus, the properties of a binomial experiment are satisfied. The random variable of interest is $X =$ the number of heads appearing in the five trials. In this case, X can assume the values of 0, 1, 2, 3, 4 or 5.

As another example, consider an insurance salesperson who visits ten randomly selected families. The outcome associated with each visit is classified as a success if the family purchases an insurance policy and a failure if the family does not. From past experience, the salesperson knows the probability that a randomly selected family will purchase an insurance policy is 0.10. Checking the properties of a binomial experiment, we observe that:

1 The experiment consists of ten identical trials; each trial involves contacting one family.

2 Two outcomes are possible on each trial: the family purchases a policy (success) or the family does not purchase a policy (failure).

3 The probabilities of a purchase and a non-purchase are assumed to be the same for each sales call, with $\pi = 0.10$ and $1 - \pi = 0.90$.

4 The trials are independent because the families are randomly selected.

Because the four assumptions are satisfied, this example is a binomial experiment. The random variable of interest is the number of sales obtained in contacting the ten families. In this case, X can assume the values of 0, 1, 2, 3, 4, 5, 6, 7, 8, 9 and 10.

Property 3 of the binomial experiment is called the *stationarity assumption* and is sometimes confused with property 4, independence of trials. To see how they differ, consider again the case of the salesperson calling on families to sell insurance policies. If, as the day wore on, the salesperson got tired and lost enthusiasm, the probability of success (selling a policy) might drop to 0.05, for example, by the tenth call. In such a case, property 3 (stationarity) would not be satisfied, and we would not have a binomial experiment. Even if property 4 held – that is, the purchase decisions of each family were made independently – it would not be a binomial experiment if property 3 was not satisfied.

In applications involving binomial experiments, a special mathematical formula, called the *binomial probability function,* can be used to compute the probability of x successes in the n trials. We will show in the context of an illustrative problem how the formula can be developed.

Marrine clothing store problem

Let us consider the purchase decisions of the next three customers who enter the Marrine Clothing Store. On the basis of past experience, the store manager estimates the probability that any one customer will make a purchase is 0.30. What is the probability that two of the next three customers will make a purchase?

Using a tree diagram (Figure 5.3), we see that the experiment of observing the three customers each making a purchase decision has eight possible outcomes. Using S to denote success (a purchase) and F to denote failure (no purchase), we are interested in experimental outcomes involving two successes in the three trials (purchase decisions). Next, let us verify that the experiment involving the sequence of three purchase decisions can be viewed as a binomial experiment. Checking the four requirements for a binomial experiment, we note that:

1 The experiment can be described as a sequence of three identical trials, one trial for each of the three customers who will enter the store.

2 Two outcomes – the customer makes a purchase (success) or the customer does not make a purchase (failure) – are possible for each trial.

Figure 5.3 Tree diagram for the Marrine clothing store problem

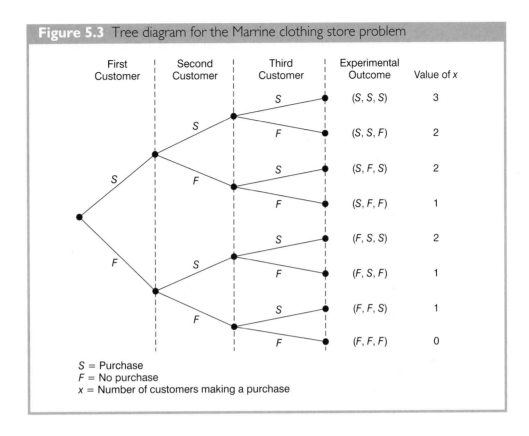

S = Purchase
F = No purchase
x = Number of customers making a purchase

3   The probability that the customer will make a purchase (0.30) or will not make a purchase (0.70) is assumed to be the same for all customers.

4   The purchase decision of each customer is independent of the decisions of the other customers.

Hence, the properties of a binomial experiment are present.

The number of experimental outcomes resulting in exactly x successes in n trials can be computed using the following formula.*

Number of experimental outcomes providing exactly x successes in n trials

$$\binom{n}{x} = \frac{n!}{x!(n-x)!} \tag{5.6}$$

where

$$n! = n(n-1)(n-2)\ldots(2)(1)$$

and, by definition,

$$0! = 1$$

---

*This formula, introduced in Chapter 4, determines the number of combinations of n objects selected x at a time. For the binomial experiment, this combinatorial formula provides the number of experimental outcomes (sequences of n trials) resulting in x successes.

Now let us return to the Marrine Clothing Store experiment involving three customer purchase decisions. Equation (5.6) can be used to determine the number of experimental outcomes involving two purchases; that is, the number of ways of obtaining $X = 2$ successes in the $n = 3$ trials. From equation (5.6) we have

$$\binom{n}{x} = \binom{3}{2} = \frac{3!}{2! \times (3-2)!} = \frac{3 \times 2 \times 1}{(2 \times 1) \times (1)} = \frac{6}{2} = 3$$

Equation (5.6) shows that three of the experimental outcomes yield two successes. From Figure 5.3 we see these three outcomes are denoted by (S, S, F), (S, F, S) and (F, S, S). Using equation (5.6) to determine how many experimental outcomes have three successes (purchases) in the three trials, we obtain

$$\binom{n}{x} = \binom{3}{3} = \frac{3!}{3! \times (3-2)!} = \frac{3 \times 2 \times 1}{(3 \times 2 \times 1) \times (1)} = \frac{6}{6} = 1$$

From Figure 5.3 we see that the one experimental outcome with three successes is identified by (S, S, S).

We know that equation (5.6) can be used to determine the number of experimental outcomes that result in X successes. If we are to determine the probability of x successes in n trials, however, we must also know the probability associated with each of these experimental outcomes. Because the trials of a binomial experiment are independent, we can simply multiply the probabilities associated with each trial outcome to find the probability of a particular sequence of successes and failures.

The probability of purchases by the first two customers and no purchase by the third customer, denoted (S, S, F), is given by

$$\pi\pi(1 - \pi)$$

With a 0.30 probability of a purchase on any one trial, the probability of a purchase on the first two trials and no purchase on the third is given by

$$0.30 \times 0.30 \times 0.70 = 0.30^2 \times 0.70 = 0.063$$

Two other experimental outcomes also result in two successes and one failure. The probabilities for all three experimental outcomes involving two successes follow.

Trial outcomes				
1st customer	2nd customer	3rd customer	Experimental outcome	Probability of experimental outcome
Purchase	Purchase	No purchase	(S, S, F)	$\pi\pi(1 - \pi) = \pi^2(1 - \pi)$ $= (0.30)^2(0.70) = 0.063$
Purchase	No purchase	Purchase	(S, F, S)	$\pi(1 - \pi)\pi = \pi^2(1 - \pi)$ $= (0.30)^2(0.70) = 0.063$
No purchase	Purchase	Purchase	(F, S, S)	$(1 - \pi)\pi\pi = \pi^2(1 - \pi)$ $= (0.30)^2(0.70) = 0.063$

Observe that all three experimental outcomes with two successes have exactly the same probability. This observation holds in general. In any binomial experiment, all sequences of trial outcomes yielding x successes in n trials have the *same probability* of occurrence.

The probability of each sequence of trials yielding X successes in n trials follows.

$$\text{Probability of a particular}$$
$$\text{sequence of trial outcomes} = \pi^x(1 - \pi)^{(n - x)} \qquad \textbf{(5.7)}$$
$$\text{with } X \text{ successes in } n \text{ trials}$$

For the Marrine Clothing Store, this formula shows that any experimental outcome with two successes has a probability of $\pi^2(1 - \pi)^{(3 - 2)} = \pi^2(1 - \pi)^1 = (0.30)^2 (0.70)^1 = 0.063$. Combining equations (5.6) and (5.7) we obtain the following **binomial probability function**.

Binomial probability function

$$p(x) = \binom{n}{x}\pi^x (1 - \pi)^{(n - x)} \qquad \textbf{(5.8)}$$

where $p(x)$ = the probability of x successes in n trials

$\quad n$ = the number of trials

$$\binom{n}{x} = \frac{n!}{x!(n - x)!}$$

$\quad \pi$ = the probability of a success on any one trial

$1 - \pi$ = the probability of a failure on any one trial

In the Marrine Clothing Store example, we can use this function to compute the probability that no customer makes a purchase, exactly one customer makes a purchase, exactly two customers make a purchase, and all three customers make a purchase. The calculations are summarized in Table 5.7, which gives the probability distribution of the number of customers making a purchase. Figure 5.4 is a graph of this probability distribution.

The binomial probability function can be applied to *any* binomial experiment. If we are satisfied that a situation demonstrates the properties of a binomial experiment and if we know the values of n and π, we can use equation (5.8) to compute the probability of x successes in the n trials.

If we consider variations of the Marrine experiment, such as ten customers rather than three entering the store, the binomial probability function given by equation (5.8) is still

---

Table 5.7 Probability distribution for the number of customers making a purchase

x	$p(x)$
0	$\dfrac{3!}{0!3!}(0.30)^0(0.70)^3 = 0.343$
1	$\dfrac{3!}{1!2!} (0.30)^1(0.70)^2 = 0.441$
2	$\dfrac{3!}{2!1!}(0.30)^2(0.70)^1 = 0.189$
3	$\dfrac{3!}{3!0!} (0.30)^3(0.70)^0 = \dfrac{0.027}{1.000}$

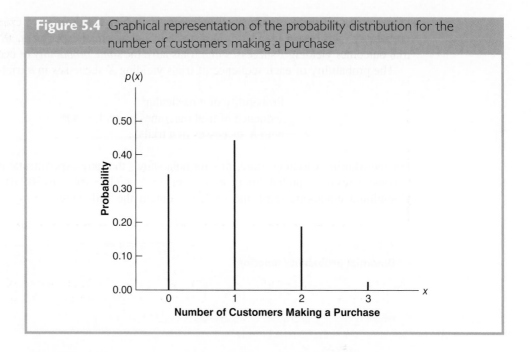

Figure 5.4 Graphical representation of the probability distribution for the number of customers making a purchase

applicable. Suppose we have a binomial experiment with $n = 10$, $x = 4$, and $\pi = 0.30$. The probability of making exactly four sales to 10 customers entering the store is

$$p(4) = \frac{10!}{4!6!} (0.30)^4 (0.70)^6 = 0.2001$$

Using tables of binomial probabilities

Tables have been developed that give the probability of x successes in n trials for a binomial experiment. The tables are generally easy to use and quicker than equation (5.8). Table 5 of Appendix B provides such a table of binomial probabilities. To use this table, we must specify the values of n, π and x for the binomial experiment of interest. For example the probability of $x = 3$ successes in a binomial experiment with $n = 10$ and $\pi = 0.40$ can be seen to be 0.2150. You can use equation (5.8) to verify that you would obtain the same answer if you used the binomial probability function directly.

Now let us use the same table to verify the probability of four successes in ten trials for the Marrine Clothing Store problem. Note that the value of $p(4) = 0.2001$ can be read directly from the table of binomial probabilities, with $n = 10$, $x = 4$ and $\pi = 0.30$.

Even though the tables of binomial probabilities are relatively easy to use, it is impossible to have tables that show all possible values of n and π that might be encountered in a binomial experiment. However, with today's calculators, using equation (5.8) to calculate the desired probability is not difficult, especially if the number of trials is not large. In the exercises, you should practice using equation (5.8) to compute the binomial probabilities unless the problem specifically requests that you use the binomial probability table.

Statistical software packages such as MINITAB, PASW and spreadsheet packages such as EXCEL also provide a capability for computing binomial probabilities. Consider the Marrine Clothing Store example with $n = 10$ and $\pi = 0.30$. Figure 5.5 shows the binomial probabilities generated by MINITAB for all possible values of x. Note that these

Figure 5.5 MINITAB output showing binomial probabilities for the Marrine clothing store problem

```
          x   P ( X = x )
          0     0. 028248
          1     0. 121061
          2     0. 233474
          3     0. 266828
          4     0. 200121
          5     0. 102919
          6     0. 036757
          7     0. 009002
          8     0. 001447
          9     0. 000138
         10     0. 000006
```

values are the same as those found in the $\pi = 0.30$ column of Table 5 of Appendix B. At the end of the chapter, details are given on how to generate the output in Figure 5.5 using first MINITAB, then EXCEL and finally PASW.

Expected value and variance for the binomial distribution

In Section 5.3 we provided formulae for computing the expected value and variance of a discrete random variable. In the special case where the random variable has a binomial distribution with a known number of trials n and a known probability of success π, the general formulae for the expected value and variance can be simplified. The results follow.

Expected value and variance for the binomial distribution

$$E(X) = \mu = n\pi \tag{5.9}$$

$$\text{Var}(X) = \sigma^2 = n\pi(1 - \pi) \tag{5.10}$$

For the Marrine Clothing Store problem with three customers, we can use equation (5.9) to compute the expected number of customers who will make a purchase.

$$E(X) = n\pi = 3 \times 0.30 = 0.9$$

Suppose that for the next month the Marrine Clothing Store forecasts 1000 customers will enter the store. What is the expected number of customers who will make a purchase? The answer is $\mu = n\pi = 1000 \times 0.3 = 300$. Thus, to increase the expected number of purchases, Marrine's must induce more customers to enter the store and/or somehow increase the probability that any individual customer will make a purchase after entering.

For the Marrine Clothing Store problem with three customers, we see that the variance and standard deviation for the number of customers who will make a purchase are

$$\sigma^2 = n\pi(1 - \pi) = 3 \times 0.3 \times 0.7 = 0.63$$
$$\sigma = \sqrt{.63} = 0.79$$

For the next 1000 customers entering the store, the variance and standard deviation for the number of customers who will make a purchase are

$$\sigma^2 = n\pi(1 - \pi) = 1000 \times 0.3 \times 0.7 = 210$$
$$\sigma = \sqrt{210} = 14.49$$

Exercises

Methods

21 Consider a binomial experiment with two trials and $\pi = 0.4$.

   a. Draw a tree diagram for this experiment (see Figure 5.3).
   b. Compute the probability of one success, $p(1)$.
   c. Compute $p(0)$.
   d. Compute $p(2)$.
   e. Compute the probability of at least one success.
   f. Compute the expected value, variance, and standard deviation.

22 Consider a binomial experiment with $n = 10$ and $\pi = 0.10$.

   a. Compute $p(0)$.
   b. Compute $p(2)$.
   c. Compute $P(x \leq 2)$.
   d. Compute $P(x \geq 1)$.
   e. Compute $E(X)$.
   f. Compute $\text{Var}(X)$ and σ.

23 Consider a binomial experiment with $n = 20$ and $\pi = 0.70$.

   a. Compute $p(12)$.
   b. Compute $p(16)$.
   c. Compute $P(X \geq 16)$.
   d. Compute $P(X \leq 15)$.
   e. Compute $E(X)$.
   f. Compute $\text{Var}(X)$ and σ.

Applications

24 When a new machine is functioning properly, only 3 per cent of the items produced are defective. Assume that we will randomly select two parts produced on the machine and that we are interested in the number of defective parts found.

   a. Describe the conditions under which this situation would be a binomial experiment.
   b. Draw a tree diagram similar to Figure 5.3 showing this problem as a two-trial experiment.
   c. How many experimental outcomes result in exactly one defect being found?
   d. Compute the probabilities associated with finding no defects, exactly one defect, and two defects.

25 Military radar and missile detection systems are designed to warn a country of an enemy attack. A reliability question is whether a detection system will be able to identify an attack and issue a warning. Assume that a particular detection system has a 0.90 probability of detecting a missile attack. Use the binomial probability distribution to answer the following questions.

   a. What is the probability that a single detection system will detect an attack?

b. If two detection systems are installed in the same area and operate independently, what is the probability that at least one of the systems will detect the attack?

c. If three systems are installed, what is the probability that at least one of the systems will detect the attack?

d. Would you recommend that multiple detection systems be used? Explain.

26 A firm bills its accounts at a 1 per cent discount for payment within ten days and the full amount is due after ten days. In the past 30 per cent of all invoices have been paid within ten days. If the firm sends out eight invoices during the first week of January, what is the probability that:

a. No one receives the discount?

b. Everyone receives the discount?

c. No more than three receive the discount?

d. At least two receive the discount?

27 For the special case of a binomial random variable, we stated that the variance could be computed using the formula $\sigma^2 = n\pi(1 - \pi)$. For the Marrine Clothing Store problem with $n = 3$ and $\pi = 0.3$ we found $\sigma^2 = n\pi(1 - \pi) = 3 \times 0.3 \times 0.7 = 0.63$. Use the general definition of variance for a discrete random variable, equation (5.5), and the probabilities in Table 5.7 to verify that the variance is in fact 0.63.

5.5 Poisson probability distribution

In this section we consider a discrete random variable that is often useful in estimating the number of occurrences over a specified interval of time or space. For example, the random variable of interest might be the number of arrivals at a car wash in one hour, the number of repairs needed in 10 kilometres of highway, or the number of leaks in 100 kilometres of pipeline.

If the following two properties are satisfied, the number of occurrences is a random variable described by the **Poisson probability distribution**.

Properties of a Poisson experiment

1 The probability of an occurrence is the same for any two intervals of equal length.

2 The occurrence or non-occurrence in any interval is independent of the occurrence or non-occurrence in any other interval.

The **Poisson probability function** is defined by equation (5.11).

Poisson probability function

$$p(x) = \frac{\mu^x e^{-\mu}}{x!}$$ (5.11)

where

$p(x) =$ the probability of x occurrences in an interval

$\mu =$ expected value or mean number of occurrences in an interval

$e = 2.71828$

Before we consider a specific example to see how the Poisson distribution can be applied, note that the number of occurrences, x, has no upper limit. It is a discrete random variable that may assume an infinite sequence of values ($x = 0, 1, 2, \ldots$).

An example involving time intervals

Suppose that we are interested in the number of arrivals at the payment kiosk of a car park during a 15-minute period on weekday mornings. If we can assume that the probability of a car arriving is the same for any two time periods of equal length and that the arrival or non-arrival of a car in any time period is independent of the arrival or non-arrival in any other time period, the Poisson probability function is applicable. Suppose these assumptions are satisfied and an analysis of historical data shows that the average number of cars arriving in a 15-minute period of time is ten; in this case, the following probability function applies.

$$p(x) = \frac{10^x\, e^{-10}}{x!}$$

The random variable here is X = number of cars arriving in any 15-minute period.

If management wanted to know the probability of exactly five arrivals in 15 minutes, we would set $X = 5$ and thus obtain

$$\text{Probability of exactly five arrivals in 15 minutes} = p(5) = \frac{10^5 e^{-10}}{5!} = 0.0378$$

Although this probability was determined by evaluating the probability function with $\mu = 10$ and $x = 5$, it is often easier to refer to a table for the Poisson distribution. The table provides probabilities for specific values of x and μ. We include such a table as Table 7 of Appendix B. Note that to use the table of Poisson probabilities, we need know only the values of x and μ. From this table we see that the probability of five arrivals in a 15-minute period is found by locating the value in the row of the table corresponding to $x = 5$ and the column of the table corresponding to $\mu = 10$. Hence, we obtain p(5) = 0.0378.

In the preceding example, the mean of the Poisson distribution is $\mu = 10$ arrivals per 15-minute period. A property of the Poisson distribution is that the mean of the distribution and the variance of the distribution are *equal*. Thus, the variance for the number of arrivals during 15-minute periods is $\sigma^2 = 10$. The standard deviation is $\sigma = \sqrt{10} = 3.16$.

Our illustration involves a 15-minute period, but other time periods can be used. Suppose we want to compute the probability of one arrival in a three-minute period. Because ten is the expected number of arrivals in a 15-minute period, we see that $10/15 = 2/3$ is the expected number of arrivals in a one-minute period and that $2/3 \times 3$ minutes $= 2$ is the expected number of arrivals in a three-minute period. Thus, the probability of x arrivals in a three-minute time period with $\mu = 2$ is given by the following Poisson probability function.

$$p(x) = \frac{2^x e^{-2}}{x!}$$

The probability of one arrival in a three-minute period is calculated as follows:

$$\text{Probability of exactly one arrival in three minutes} = P(1) = \frac{2^1 e^{-2}}{1!} = 0.2707$$

Earlier we computed the probability of five arrivals in a 15-minute period; it was 0.0378. Note that the probability of one arrival in a three-minute period (0.2707) is not the same. When computing a Poisson probability for a different time interval, we must first convert the mean arrival rate to the time period of interest and then compute the probability.

An example involving length or distance intervals

Consider an application not involving time intervals in which the Poisson distribution is useful. Suppose we are concerned with the occurrence of major defects in a highway, one month after resurfacing. We will assume that the probability of a defect is the same for any two highway intervals of equal length and that the occurrence or non-occurrence of a defect in any one interval is independent of the occurrence or non-occurrence of a defect in any other interval. Hence, the Poisson distribution can be applied.

Suppose that major defects one month after resurfacing occur at the average rate of two per kilometre. Let us find the probability of no major defects in a particular three-kilometre section of the highway. Because we are interested in an interval with a length of three kilometres, $\mu = 2$ defects/kilometre $\times$ 3 kilometres $= 6$ represents the expected number of major defects over the three-kilometre section of highway. Using equation (5.11), the probability of no major defects is $p(0) = 6^0 e^{-6}/0! = 0.0025$. Thus, it is unlikely that no major defects will occur in the three-kilometre section. Equivalently there is a $1 - 0.0025 = 0.9975$ probability of at least one major defect in the three-kilometre highway section.

Exercises

Methods

28 Consider a Poisson distribution with $\mu = 3$.

    a. Write the appropriate Poisson probability function.
    b. Compute $p(2)$.
    c. Compute $p(1)$.
    d. Compute $P(X \geq 2)$.

29 Consider a Poisson distribution with a mean of two occurrences per time period.

    a. Write the appropriate Poisson probability function.
    b. What is the expected number of occurrences in three time periods?
    c. Write the appropriate Poisson probability function to determine the probability of x occurrences in three time periods.
    d. Compute the probability of two occurrences in one time period.
    e. Compute the probability of six occurrences in three time periods.
    f. Compute the probability of five occurrences in two time periods.

Applications

30 Phone calls arrive at the rate of 48 per hour at the reservation desk for Regional Airways.

    a. Compute the probability of receiving three calls in a five-minute interval of time.
    b. Compute the probability of receiving exactly ten calls in 15 minutes.

c. Suppose no calls are currently on hold. If the agent takes five minutes to complete the current call, how many callers do you expect to be waiting by that time? What is the probability that none will be waiting?

d. If no calls are currently being processed, what is the probability that the agent can take three minutes for personal time without being interrupted by a call?

31 During the period of time that a local university takes phone-in registrations, calls come in at the rate of one every two minutes.

a. What is the expected number of calls in one hour?

b. What is the probability of three calls in five minutes?

c. What is the probability of no calls in a five-minute period?

32 Airline passengers arrive randomly and independently at the passenger-screening facility at a major international airport. The mean arrival rate is ten passengers per minute.

a. Compute the probability of no arrivals in a one-minute period.

b. Compute the probability that three or fewer passengers arrive in a one-minute period.

c. Compute the probability of no arrivals in a 15-second period.

d. Compute the probability of at least one arrival in a 15-second period.

5.6 Hypergeometric probability distribution

The **hypergeometric probability distribution** is closely related to the binomial distribution. The two probability distributions differ in two key ways. With the hypergeometric distribution, the trials are not independent; and the probability of success changes from trial to trial.

In the usual notation for the hypergeometric distribution, r denotes the number of elements in the population of size N labelled success, and $N - r$ denotes the number of elements in the population labelled failure. The **hypergeometric probability function** is used to compute the probability that in a random selection of n elements, selected without replacement, we obtain x elements labelled success and $n - x$ elements labelled failure. For this outcome to occur, we must obtain x successes from the r successes in the population and $n - x$ failures from the $N - r$ failures. The following hypergeometric probability function provides $p(x)$, the probability of obtaining x successes in a sample of size n.

Hypergeometric probability function

$$p(x) = \frac{\binom{r}{x}\binom{N-r}{n-x}}{\binom{N}{n}} \tag{5.12}$$

where

$p(x)$ = probability of x successes in n trials

n = number of trials

N = number of elements in the population

r = number of elements in the population labelled success

Note that $\binom{N}{n}$ represents the number of ways a sample of size n can be selected from a population of size N; $\binom{r}{x}$ represents the number of ways that x successes can be selected from a total of r successes in the population; and $\binom{N - r}{n - x}$ represents the number of ways that $n - x$ failures can be selected from a total of $N - r$ failures in the population.

To illustrate the computations involved in using equation (5.12), consider the following quality control application. Electric fuses produced by Warsaw Electric are packaged in boxes of 12 units each. Suppose an inspector randomly selects three of the 12 fuses in a box for testing. If the box contains exactly five defective fuses, what is the probability that the inspector will find exactly one of the three fuses defective? In this application, $n = 3$ and $N = 12$. With $r = 5$ defective fuses in the box the probability of finding $x = 1$ defective fuse is

$$p(1) = \frac{\binom{5}{1}\binom{7}{2}}{\binom{12}{3}} = \frac{\dfrac{5!}{1!4!}\dfrac{7!}{2!3!}}{\dfrac{12!}{3!9!}} = \frac{5 \times 21}{220} = 0.4733$$

Now suppose that we wanted to know the probability of finding *at least* 1 defective fuse. The easiest way to answer this question is to first compute the probability that the inspector does not find any defective fuses. The probability of $x = 0$ is

$$p(0) = \frac{\binom{5}{0}\binom{7}{3}}{\binom{12}{3}} = \frac{\dfrac{5!}{0!5!}\dfrac{7!}{3!4!}}{\dfrac{12!}{3!9!}} = \frac{1 \times 35}{220} = 0.1591$$

With a probability of zero defective fuses $p(0) = 0.1591$, we conclude that the probability of finding at least one defective fuse must be $1 - 0.1591 = 0.8409$. Thus, there is a reasonably high probability that the inspector will find at least one defective fuse.

The mean and variance of a hypergeometric distribution are as follows.

Expected value for the hypergeometric distribution

$$E(X) = \mu = n\left(\frac{r}{N}\right) \tag{5.13}$$

Variance for the hypergeometric distribution

$$\text{Var}(X) = \sigma^2 = n\left(\frac{r}{N}\right)\left(1 - \frac{r}{N}\right)\left(\frac{N - n}{N - 1}\right) \tag{5.14}$$

In the preceding example $n = 3$, $r = 5$, and $N = 12$. Thus, the mean and variance for the number of defective fuses is

$$\mu = n\left(\frac{r}{N}\right) = 3\left(\frac{5}{12}\right) = 1.25$$

$$\sigma^2 = n\left(\frac{r}{N}\right)\left(1 - \frac{r}{N}\right)\left(\frac{N - n}{N - 1}\right) = 3\left(\frac{5}{12}\right)\left(1 - \frac{5}{12}\right)\left(\frac{12 - 3}{12 - 1}\right) = 0.60$$

The standard deviation is $\sigma = \sqrt{0.60} = 0.77$

Exercises

Methods

33 Suppose $N = 10$ and $r = 3$. Compute the hypergeometric probabilities for the following values of n and x.

a. $n = 4, x = 1$.
b. $n = 2, x = 2$.
c. $n = 2, x = 0$.
d. $n = 4, x = 2$.

34 Suppose $N = 15$ and $r = 4$. What is the probability of $x = 3$ for $n = 10$?

Applications

35 Blackjack, or Twenty-one as it is frequently called, is a popular gambling game played in Monte Carlo casinos. A player is dealt two cards. Face cards (jacks, queens and kings) and tens have a point value of ten. Aces have a point value of one or 11. A 52-card deck contains 16 cards with a point value of ten (jacks, queens, kings and tens) and four aces.

a. What is the probability that both cards dealt are aces or ten-point cards?
b. What is the probability that both of the cards are aces?
c. What is the probability that both of the cards have a point value of ten?
d. A blackjack is a ten-point card and an ace for a value of 21. Use your answers to parts (a), (b) and (c) to determine the probability that a player is dealt blackjack. (*Hint:* Part (d) is not a hypergeometric problem. Develop your own logical relationship as to how the hypergeometric probabilities from parts (a), (b) and (c) can be combined to answer this question.)

36 An electronic device has two components and works only when both components function. The probability a component functions is π independently of the other. A firm buys four devices.

a. What is the probability that three or more work?
b. What value must π take to ensure a probability 0.75 that three or more devices work?

37 A shipment of ten items has two defective and eight non-defective items. In the inspection of the shipment, a sample of items will be selected and tested. If a defective item is found, the shipment of ten items will be rejected.

a. If a sample of three items is selected, what is the probability that the shipment will be rejected?
b. If a sample of four items is selected, what is the probability that the shipment will be rejected?
c. If a sample of five items is selected, what is the probability that the shipment will be rejected?
d. If management would like a 0.90 probability of rejecting a shipment with two defective and eight non-defective items, how large a sample would you recommend?

**For additional online summary questions and answers go
to the companion website at www.cengage.co.uk/aswsbe2**

Summary

A random variable provides a numerical description of the outcome of an experiment. The probability distribution for a random variable describes how the probabilities are distributed over the values the random variable can assume. A variety of examples are used to distinguish between discrete and continuous random variables. For any discrete random variable X, the probability distribution is defined by a probability function, denoted by $p(x)=p(X=x)$, which provides the probability associated with each value of the random variable. From the probability function, the expected value, variance, and standard deviation for the random variable can be computed and relevant interpretations of these terms are provided.

Particular attention was devoted to the binomial distribution which can be used to determine the probability of x successes in n trials whenever the experiment has the following properties:

1 The experiment consists of a sequence of n identical trials.

2 Two outcomes are possible on each trial, one called success and the other failure.

3 The probability of a success π does not change from trial to trial. Consequently, the probability of failure, $1 - \pi$, does not change from trial to trial.

4 The trials are independent.

Formulae were also presented for the probability function, mean and variance of the binomial distribution.

The Poisson distribution can be used to determine the probability of obtaining x occurrences over an interval of time or space. The necessary assumptions for the Poisson distribution to apply in a given situation are that:

1 The probability of an occurrence of the event is the same for any two intervals of equal length.

2 The occurrence or non-occurrence of the event in any interval is independent of the occurrence or non-occurrence of the event in any other interval.

A third discrete probability distribution, the hypergeometric, was introduced in Section 5.6. Like the binomial, it is used to compute the probability of x successes in n trials. But, in contrast to the binomial, the probability of success changes from trial to trial.

Key terms

Binomial experiment
Binomial probability distribution
Binomial probability function
Continuous random variable
Discrete random variable
Discrete uniform probability
  distribution
Expected value
Hypergeometric probability
  distribution

Hypergeometric probability function
Poisson probability distribution
Poisson probability function
Probability distribution
Probability function
Random variable
Standard deviation
Variance

Key formulae

Discrete uniform probability function

$$p(x) = 1/n \qquad (5.3)$$

where

$$n = \text{the number of values the random variable may assume}$$

Expected value of a discrete random variable

$$E(X) = \mu = \Sigma x \, p(x) \qquad (5.4)$$

Variance of a discrete random variable

$$\text{Var}(X) = \sigma^2 = \Sigma (x - \mu)^2 p(x) \qquad (5.5)$$

Number of experimental outcomes providing exactly x successes in n trials

$$\binom{n}{x} = \frac{n!}{x!(n-x)!} \qquad (5.6)$$

Binomial probability function

$$p(x) = \binom{n}{x} \pi^x (1 - \pi)^{(n-x)} \qquad (5.8)$$

Expected value for the binomial distribution

$$E(X) = \mu = n\pi \qquad (5.9)$$

Variance for the binomial distribution

$$\text{Var}(X) = \sigma^2 = n\pi(1 - \pi) \qquad (5.10)$$

Poisson probability punction

$$p(x) = \frac{\mu^x e^{-\mu}}{x!} \qquad (5.11)$$

Hypergeometric probability function

$$p(x) = \frac{\binom{r}{x}\binom{N-r}{n-x}}{\binom{N}{n}} \qquad (5.12)$$

Expected value for the hypergeometric distribution

$$E(X) = \mu = n\left(\frac{r}{N}\right) \qquad (5.13)$$

Variance for the hypergeometric distribution

$$\text{Var}(X) = \sigma^2 = n\left(\frac{r}{N}\right)\left(1 - \frac{r}{N}\right)\left(\frac{N-n}{N-1}\right) \qquad (5.14)$$

Case problem Adapting a Bingo Game

Gaming Machines International (GMI) is investigating the adaptation of one of its bingo machine formats to allow for a bonus game facility. With the existing setup, the player has to select seven numbers from the series 1 to 80. Fifteen numbers are then drawn randomly from the 80 available and prizes awarded, according to how many of the 15 coincide with the player's selection, as follows:

Number of 'hits',	Payoff
0	0
1	0
2	0
3	1
4	10
5	100
6	1 000
7	100 000

Gaming machines at a casino. © Jean Miele/CORBIS.

With the new 'two ball bonus draw' feature players effectively have the opportunity to improve their prize by buying an extra two balls. Note however that the bonus draw is only expected to be available to players who have scored 2, 3, 4 or 5 hits in the main game.

Managerial report

1 Determine the probability characteristics of GMI's original bingo game and calculate the player's expected payoff.

2 Derive corresponding probability details for the proposed bonus game. What is the probability of the player scoring

   a 0 hits
   b 1 hit
   c 2 hits

   with the extra two balls?

3 Use the results obtained from two to revise the probability distribution found for one. Hence calculate the player's expected payoff in the enhanced game. Comment on how much the player might be charged for the extra gamble.

Software Section for Chapter 5

Statistical packages such as MINITAB offer a relatively easy and efficient procedure for computing binomial probabilities. In this section, we show the step-by-step procedure for determining the binomial probabilities for the Marrine Clothing Store problem in Section 5.4. Recall that the desired binomial probabilities are based on $n = 10$ and $\pi = 0.30$. Before beginning the MINITAB routine, the user must enter the desired values of the random variable X into a column of the worksheet. We entered the values 0, 1, 2, . . . , 10 in column 1 (see Figure 5.5) to generate the entire binomial probability distribution. The MINITAB steps to obtain the required binomial probabilities are as follows:

Step 1   [Main menu bar]
        Calc > Probability Distributions > Binomial

Step 2   [**Binomial Distribution** panel]
        Select **Probability**
        Enter 10 in the **Number of trials** box
        Enter 0.3 in the **Probability of success** box
        Enter C1 in the **Input column** box
        Click **OK**

The MINITAB output with the binomial probabilities will appear as shown in Figure 5.5. MINITAB provides Poisson and hypergeometric probabilities in a similar manner. For instance to compute Poisson probabilities, the only differences are in step 2, where the **Poisson** option would be selected and the **Mean** entered afterwards rather than the number of trials and the probability of success.

EXCEL provides functions for computing probabilities for the binomial, Poisson and hypergeometric distributions introduced in this chapter. The EXCEL function for computing binomial probabilities is BINOMDIST. It has four arguments: x (the number of successes), n (the number of trials), π (the probability of success), and cumulative. FALSE is used for the fourth argument (cumulative) if we want the probability of x successes, and TRUE is used for the fourth argument if we want the cumulative probability of x or fewer successes. Here we show how to compute the probabilities of 0 through 10 successes in the Marrine Clothing Store problem of Section 5.4 (see Figure 5.5).

As we describe the worksheet development refer to Figure 5.6; the formula worksheet is set in the background, and the value worksheet appears in the foreground. We entered the number of trials (10) into cell B1, the probability of success into cell B2, and the values for the random variable into cells B5:B15. The following steps will generate the required desired probabilities.

Step 1 Use the BINOMDIST function to compute the probability for $x = 0$ by entering the following formula into cell C5:

$$=BINOMDIST(B5, \$B\$1, \$B\$2, FALSE)$$

Step 2 Copy the formula in cell C5 into cells C6:C15.

Figure 5.6 EXCEL worksheet for computing binomial probabilities

Check Box	B	C	D	E	F
1 Number of Trials (n)	10				
2 Probability of Success (π)	0.3				
3					
4	x	p(π)			
5	0	=BINOMDIST(B5,$B$1,$B$2,FALSE)			
6	1	=BINOMDIST(B6,$B$1,$B$2,FALSE)			
7	2	=BINOMDIST(B7,$B$1,$B$2,FALSE)			
8	3	=BINOMDIST(B8,$B$1,$B$2,FALSE)			
9	4	=BINOMDIST(B9,$B$1,$B$2,FALSE)			
10	5	=BINOMDIST(B10,$B$1,$B$2,FALSE)			
11	6	=BINOMDIST(B11,$B$1,$B$2,FALSE)			
12	7	=BINOMDIST(B12,$B$1,$B$2,FALSE)			
13	8	=BINOMDIST(B13,$B$1,$B$2,FALSE)			
14	9	=BINOMDIST(B14,$B$1,$B$2,FALSE)			
15	10	=BINOMDIST(B15,$B$1,$B$2,FALSE)			
16					

Check Box	B	C	D
1 Number of Trials (n)	10		
2 Probability of Success (π)	0.3		
3			
4	x	p(π)	
5	0	0.0282	
6	1	0.1211	
7	2	0.2335	
8	3	0.2668	
9	4	0.2001	
10	5	0.1029	
11	6	0.0368	
12	7	0.0090	
13	8	0.0014	
14	9	0.0041	
15	10	0.0000	
16			

The value worksheet in the foreground of Figure 5.6 shows that the probabilities obtained are the same as in Figure 5.5. Poisson and hypergeometric probabilities can be computed in a similar fashion. The POISSON and HYPERGEOMDIST functions are used.

EXCEL's Insert Function dialogue box can help the user in entering the proper arguments for these functions (see Appendix 2.2).

Discrete probability distributions with PASW

In this section, we show the step-by-step procedure in PASW for determining the binomial probabilities for the Marrine Clothing Store problem in Section 5.4. Recall that the desired binomial probabilities are based on $n = 10$ and $\pi = 0.30$. Before beginning the PASW routine, first, the data must be entered in a PASW worksheet. We entered the values 0, 1, 2, . . . , 10 in the leftmost column to generate the entire binomial probability distribution. This is automatically labelled by the system V1. The latter variable name can then be changed to x in 'Variable View' mode. The following steps will generate the required probabilities:

Step 1 [Main menu bar]
Transform > Compute Variable

Step 2 [**Compute Variable** panel]
Enter p(x) in **Target variable** box
Click on **Function group:**
Select **PDF & Noncentral PDF**
Click on **Pdf.Binom** (which has three arguments)
Click on x (for argument 1)
Enter 10 (for argument 2)
Enter .3 (for argument 3)
Click **OK**

PASW provides Poisson probabilities in a similar manner. The only differences are that **Pdf.Poisson** (which has two arguments) would be selected instead of **Pdf.Binomial** at step 2 above and the **Mean** would be entered afterwards as the second argument. Analogously for the hypogeomentric distribution Pdf.Hyper would be selected and then the Total entered followed by Sample and Hits corresponding to the N, n and r values in equation (5.12) respectively.

Chapter 6

....

Continuous Probability Distributions

Learning objectives

After reading this chapter and doing the exercises, you should be able to:

1 Understand the difference between how probabilities are computed for discrete and continuous random variables.

2 Compute probability values for a continuous uniform probability distribution and be able to compute the expected value and variance for such a distribution.

3 Compute probabilities using a normal probability distribution. Understand the role of the standard normal distribution in this process.

4 Use the normal distribution to approximate binomial probabilities.

5 Compute probabilities using an exponential probability distribution.

6 Understand the relationship between the Poisson and exponential probability distributions.

In this chapter we turn to the study of continuous random variables. Specifically, we discuss three continuous probability distributions: the uniform, the normal and the exponential. A fundamental difference separates discrete and continuous random variables in terms of how probabilities are computed. For a discrete random variable, the probability function $p(x)$ provides the probability that the random variable assumes a particular value. With continuous random variables the counterpart of the probability function is the **probability density function**, denoted by $f(x)$. The difference is that the probability density function does not directly provide probabilities. However, the area under the graph of $f(x)$ corresponding to a given interval does provide the probability that the continuous random variable X assumes a value in that interval. So when we compute probabilities for continuous random variables we are computing the probability that the random variable assumes any value in an interval.

One of the implications of the definition of probability for continuous random variables is that the probability of any particular value of the random variable is zero, because the area under the graph of $f(x)$ at any particular point is zero. In Section 6.1 we demonstrate these concepts for a continuous random variable that has a uniform distribution.

Much of the chapter is devoted to describing and showing applications of the normal distribution. The main importance of normal distribution is its extensive use in statistical inference. The chapter closes with a discussion of the exponential distribution.

6.1 Uniform probability distribution

Consider the random variable X representing the flight time of an aeroplane travelling from Graz to Stansted. Suppose the flight time can be any value in the interval from 120 minutes to 140 minutes. Because the random variable X can assume any value in that interval, X is a continuous rather than a discrete random variable. Let us assume that sufficient actual flight data are available to conclude that the probability of a flight time within any 1-minute interval is the same as the probability of a flight time within any other 1-minute interval contained in the larger interval from 120 to 140 minutes. With every 1-minute interval being equally likely, the random variable X is said to have a **uniform probability distribution**.

Statistics in Practice

Assessing the effectiveness of new medical procedures

Clinical trials are a vital and commercially very important, application of statistics, typically involving the random assignment of patients to two experimental groups. One group receives the treatment of interest, the second a placebo (a dummy treatment that has no effect). To assess the evidence that the probability of success with the treatment will be better than that with the placebo, frequencies a, b, c and d can be collected for a predetermined number of trials according to the following two-way table:

	Treatment	Placebo
Success	a	b
Failure	c	d

This woman is a test participant for a clinical trial to test the effectiveness of a new drug.
© Karen Kasmauski/Scinece/Faction/Corbis.

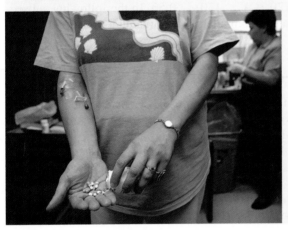

and the quantity ('log odds ratio') $X = \log (a/c/b/d)$ calculated. Clearly the larger the value of X obtained the greater the evidence that the treatment is better than the placebo.

In the particular case that the treatment has no effect the distribution of X can be shown to align very closely to a normal distribution with a mean of zero:

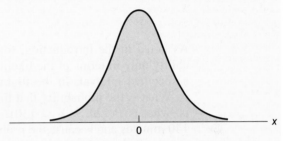

Thus, as values of X fall increasingly to the right of the zero mean this should signify stronger and stronger support for the belief in the treatment's relative effectiveness.

Intriguingly, this formulation was adapted by Copas (2005) recently to cast doubt on the findings of a recent study linking passive smoking to an increased risk of lung cancer.

Source: Copas John (2005) The downside of publication. *Significance* Vol 2 Issue 4 pp 154–157.

If x is any number lying in the range that the random variable X can take then the probability density function, which defines the uniform distribution for the flight-time random variable, is

$$f(x) = \begin{cases} 1/20 & \text{for } 120 \leq x \leq 140 \\ 0 & \text{elsewhere} \end{cases}$$

Figure 6.1 is a graph of this probability density function. In general, the uniform probability density function for a random variable X is defined by the following formula.

Uniform probability density function

$$f(x) = \begin{cases} \dfrac{1}{b-a} & \text{for } a \leq x \leq b \\ 0 & \text{elsewhere} \end{cases} \qquad (6.1)$$

For the flight-time random variable, $a = 120$ and $b = 140$.

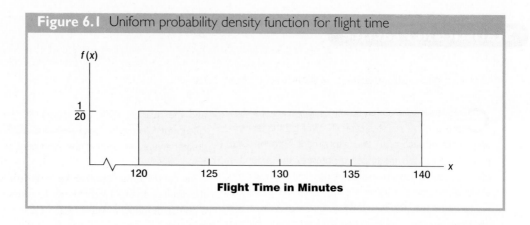

Figure 6.1 Uniform probability density function for flight time

As noted in the introduction, for a continuous random variable, we consider probability only in terms of the likelihood that a random variable assumes a value within a specified interval. In the flight time example, an acceptable probability question is: What is the probability that the flight time is between 120 and 130 minutes? That is, what is $P(120 \leq X \leq 130)$? Because the flight time must be between 120 and 140 minutes and because the probability is described as being uniform over this interval, we feel comfortable saying $P(120 \leq X \leq 130) = 0.50$. In the following subsection we show that this probability can be computed as the area under the graph of $f(x)$ from 120 to 130 (see Figure 6.2).

Area as a measure of probability

Let us make an observation about the graph in Figure 6.2. Consider the area under the graph of $f(x)$ in the interval from 120 to 130. The area is rectangular, and the area of a rectangle is simply the width multiplied by the height. With the width of the interval equal to $130 - 120 = 10$ and the height equal to the value of the probability density function $f(x) = 1/20$, we have area = width × height = $10 \times 1/20 = 10/20 = 0.50$.

What observation can you make about the area under the graph of $f(x)$ and probability? They are identical! Indeed, this observation is valid for all continuous random variables.

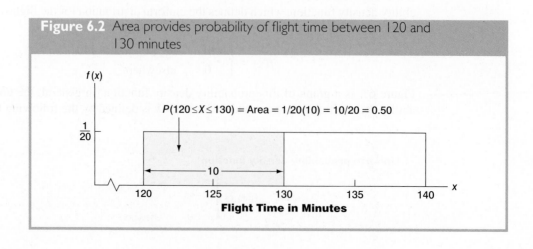

Figure 6.2 Area provides probability of flight time between 120 and 130 minutes

Once a probability density function $f(x)$ is identified, the probability that X takes a value x between some lower value x_1 and some higher value x_2 can be found by computing the area under the graph of $f(x)$ over the interval from x_1 to x_2.

Given the uniform distribution for flight time and using the interpretation of area as probability, we can answer any number of probability questions about flight times. For example, what is the probability of a flight time between 128 and 136 minutes? The width of the interval is $136 - 128 = 8$. With the uniform height of $f(x) = 1/20$, we see that $P(128 \leq X \leq 136) = 8 \times 1/20 = 0.40$. Note that $P(120 \leq X \leq 140) = 20 \times 1/20 = 1$; that is, the total area under the graph of $f(x)$ is equal to 1. This property holds for all continuous probability distributions and is the analogue of the condition that the sum of the probabilities must equal 1 for a discrete probability function. For a continuous probability density function, we must also require that $f(x) \geq 0$ for all values of x. This requirement is the analogue of the requirement that $p(x) \geq 0$ for discrete probability functions.

Two major differences stand out between the treatment of continuous random variables and the treatment of their discrete counterparts.

I We no longer talk about the probability of the random variable assuming a particular value. Instead, we talk about the probability of the random variable assuming a value within some given interval.

2 The probability of the random variable assuming a value within some given interval from x_1 to x_2 is defined to be the area under the graph of the probability density function between x_1 and x_2. It implies that the probability of a continuous random variable assuming any particular value exactly is zero, because the area under the graph of $f(x)$ at a single point is zero.

The calculation of the expected value and variance for a continuous random variable is analogous to that for a discrete random variable. However, because the computational procedure involves integral calculus, we leave the derivation of the appropriate formulae to more advanced texts.

For the uniform continuous probability distribution introduced in this section, the formulae for the expected value and variance are

$$E(X) = \frac{a + b}{2}$$

$$Var(X) = \frac{(b - a)^2}{12}$$

In these formulae, a is the smallest value and b is the largest value that the random variable may assume.

Applying these formulae to the uniform distribution for flight times from Graz to Stansted, we obtain

$$E(X) = \frac{(120 + 140)}{2} = 130$$

$$Var(X) = \frac{(140 - 120)^2}{12} = 33.33$$

The standard deviation of flight times can be found by taking the square root of the variance. Thus, $\sigma = 5.77$ minutes.

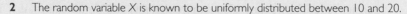

Exercises

Methods

1   The random variable X is known to be uniformly distributed between 1.0 and 1.5.

    a. Show the graph of the probability density function.
    b. Compute $P(X = 1.25)$.
    c. Compute $P(1.0 \leq X \leq 1.25)$.
    d. Compute $P(1.20 < X < 1.5)$.

2   The random variable X is known to be uniformly distributed between 10 and 20.

    a. Show the graph of the probability density function.
    b. Compute $P(X < 15)$.
    c. Compute $P(12 \leq X \leq 18)$.
    d. Compute $E(X)$.
    e. Compute $Var(X)$.

Applications

3   Alpha Airlines quotes a flight time of 2 hours, 30 minutes for its flights from Seville to Milan.
    Suppose we believe that actual flight times are uniformly distributed between 2 hours,
    20 minutes and 2 hours, 50 minutes.

    a. Show the graph of the probability density function for flight time.
    b. What is the probability that the flight will be no more than 5 minutes late?
    c. What is the probability that the flight will be more than 10 minutes late?
    d. What is the expected flight time?

4   Most computer languages include a function that can be used to generate random numbers.
    In EXCEL, the RAND function can be used to generate random numbers between 0 and 1.
    If we let X denote a random number generated using RAND, then X is a continuous random
    variable with the following probability density function.

$$f(x) = \begin{cases} 1 & \text{for } 0 \leq x \leq 1 \\ 0 & \text{elsewhere} \end{cases}$$

    a. Graph the probability density function.
    b. What is the probability of generating a random number between 0.25 and 0.75?
    c. What is the probability of generating a random number with a value less than or equal
        to 0.30?
    d. What is the probability of generating a random number with a value greater than 0.60?

5   The driving distance for the top 100 golfers on the PGA tour is between 284.7 and
    310.6 metres (*Golfweek*, 29 March 2003). Assume that the driving distance for these golfers is
    uniformly distributed over this interval.

    a. Give a mathematical expression for the probability density function of driving distance.
    b. What is the probability the driving distance for one of these golfers is less than
        290 metres?
    c. What is the probability the driving distance for one of these golfers is at least
        300 metres?
    d. What is the probability the driving distance for one of these golfers is between 290 and
        305 metres?
    e. How many of these golfers drive the ball at least 290 metres?

6   The label on a bottle of liquid detergent shows contents to be 12 grams per bottle. The production operation fills the bottle uniformly according to the following probability density function.

$$f(x) = \begin{cases} 8 & \text{for } 11.975 \leq x \leq 12.100 \\ 0 & \text{elsewhere} \end{cases}$$

a. What is the probability that a bottle will be filled with between 12 and 12.05 grams?
b. What is the probability that a bottle will be filled with 12.02 or more grams?
c. Quality control accepts a bottle that is filled to within 0.02 grams of the number of grams shown on the container label. What is the probability that a bottle of this liquid detergent will fail to meet the quality control standard?

7   Suppose we are interested in bidding on a piece of land and we know there is one other bidder. The seller announced that the highest bid in excess of €10 000 will be accepted. Assume that the competitor's bid X is a random variable that is uniformly distributed between €10 000 and €15 000.

a. Suppose you bid €12 000. What is the probability that your bid will be accepted?
b. Suppose you bid €14 000. What is the probability that your bid will be accepted?
c. What amount should you bid to maximize the probability that you get the property?
d. Suppose you know someone who is willing to pay you €16 000 for the property. Would you consider bidding less than the amount in part (c)? Why or why not?

6.2 Normal probability distribution

The most important probability distribution for describing a continuous random variable is the **normal probability distribution**. The normal distribution has been used in a wide variety of practical applications in which the random variables are heights and weights of people, test scores, scientific measurements, amounts of rainfall, and so on. It is also widely used in statistical inference, which is the major topic of the remainder of this book. In such applications, the normal distribution provides a description of the likely results obtained through sampling.

Normal curve

The form, or shape, of the normal distribution is illustrated by the bell-shaped normal curve in Figure 6.3. The probability density function that defines the bell-shaped curve of the normal distribution follows.

Figure 6.3 Bell-shaped curve for the normal distribution

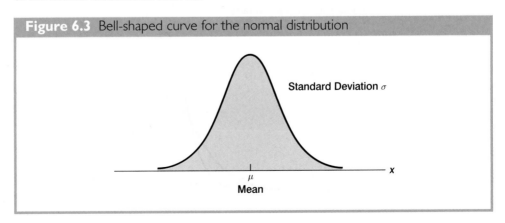

Standard Deviation σ

μ
Mean

x

Normal probability density function

$$f(x) = \frac{1}{\sigma\sqrt{2\pi}}\, e^{-(x-\mu)^2/2\sigma^2}$$

(6.2)

where

μ = mean
σ = standard deviation
π = 3.14159
e = 2.71828

We make several observations about the characteristics of the normal distribution.

1 The entire family of normal distributions is differentiated by its mean μ and its standard deviation σ.

2 The highest point on the normal curve is at the mean, which is also the median and mode of the distribution.

3 The mean of the distribution can be any numerical value: negative, zero, or positive. Three normal distributions with the same standard deviation but three different means (-10, 0, and 20) are shown here.

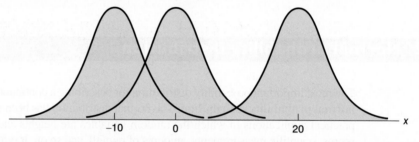

4 The normal distribution is symmetric, with the shape of the curve to the left of the mean a mirror image of the shape of the curve to the right of the mean. The tails of the curve extend to infinity in both directions and theoretically never touch the horizontal axis. Because it is symmetric, the normal distribution is not skewed; its skewness measure is zero.

5 The standard deviation determines how flat and wide the curve is. Larger values of the standard deviation result in wider, flatter curves, showing more variability in the data. Two normal distributions with the same mean but with different standard deviations are shown here.

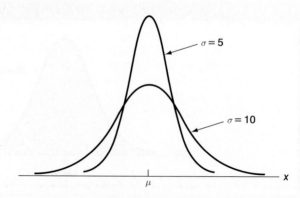

6   Probabilities for the normal random variable are given by areas under the curve. The total area under the curve for the normal distribution is 1. Because the distribution is symmetric, the area under the curve to the left of the mean is 0.50 and the area under the curve to the right of the mean is 0.50.

7   The percentage of values in some commonly used intervals are:

   a.   68.3 per cent of the values of a normal random variable are within plus or minus one standard deviation of its mean.
   b.   95.4 per cent of the values of a normal random variable are within plus or minus two standard deviations of its mean.
   c.   99.7 per cent of the values of a normal random variable are within plus or minus three standard deviations of its mean.

Figure 6.4 shows properties (a), (b) and (c) graphically.

Standard normal probability distribution

A random variable that has a normal distribution with a mean of zero and a standard deviation of one is said to have a **standard normal probability distribution**. The letter Z is commonly used to designate this particular normal random variable. Figure 6.5 is the graph of the standard normal distribution. It has the same general appearance as other normal distributions, but with the special properties of $\mu = 0$ and $\sigma = 1$.

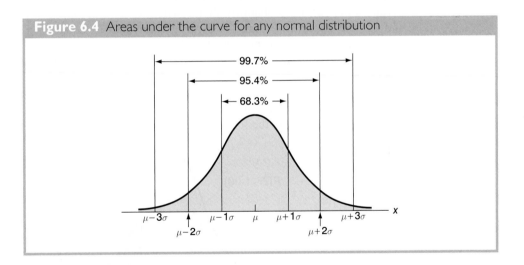

Figure 6.4 Areas under the curve for any normal distribution

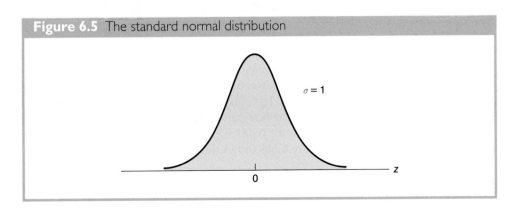

Figure 6.5 The standard normal distribution

Because $\mu = 0$ and $\sigma = 1$, the formula for the standard normal probability density function is a simpler version of equation (6.2).

Standard normal density function

$$f(z) = \frac{1}{\sqrt{2\pi}}\, e^{-z^2/2}$$

As with other continuous random variables, probability calculations with any normal distribution are made by computing areas under the graph of the probability density function. Thus, to find the probability that a normal random variable is within any specific interval, we must compute the area under the normal curve over that interval.

For the standard normal distribution, areas under the normal curve have been computed and are available in tables that can be used to compute probabilities. Such a table appears on the two pages inside the front cover of the text. The table on the left-hand page contains areas, or cumulative probabilities, for z values less than or equal to the mean of zero. The table on the right-hand page contains areas, or cumulative probabilities, for z values greater than or equal to the mean of zero.

The three types of probabilities we need to compute include (1) the probability that the standard normal random variable Z will be less than or equal to a given value; (2) the probability that Z will take a value between two given values; and (3) the probability that Z will be greater than or equal to a given value. To see how the cumulative probability table for the standard normal distribution can be used to compute these three types of probabilities, let us consider some examples.

We start by showing how to compute the probability that Z is less than or equal to 1.00; that is, $P(Z \le 1.00)$. This cumulative probability is the area under the normal curve to the left of $z = 1.00$ in the following graph.

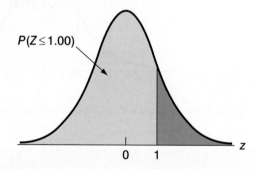

Refer to the right-hand page of the standard normal probability table inside the front cover of the text. The cumulative probability corresponding to $z = 1.00$ is the table value located at the intersection of the row labeled 1.0 and the column labeled .00. First we find 1.0 in the left column of the table and then find .00 in the top row of the table. By looking in the body of the table, we find that the 1.0 row and the .00 column intersect at the value of 0.8413; thus, $P(Z \le 1.00) = 0.8413$. The following excerpt from the probability table shows these steps.

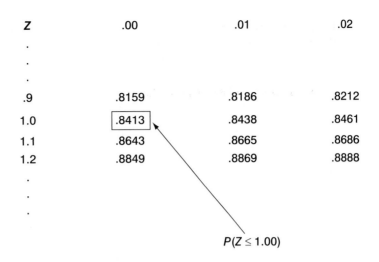

Z	.00	.01	.02
.			
.			
.			
.9	.8159	.8186	.8212
1.0	.8413	.8438	.8461
1.1	.8643	.8665	.8686
1.2	.8849	.8869	.8888
.			
.			
.			

$P(Z \leq 1.00)$

To illustrate the second type of probability calculation we show how to compute the probability that Z is in the interval between -0.50 and 1.25; that is, $P(-0.50 \leq Z \leq 1.25)$. The following graph shows this area, or probability.

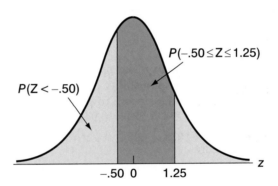

Three steps are required to compute this probability. First, we find the area under the normal curve to the left of $z = 1.25$. Second, we find the area under the normal curve to the left of $z = -0.50$. Finally, we subtract the area to the left of $z = -0.50$ from the area to the left of $z = 1$ to find $P(-0.5 \leq Z \leq 1.25)$.

To find the area under the normal curve to the left of $z = 1.25$, we first locate the 1.2 row in the standard normal probability table and then move across to the .05 column. Because the table value in the 1.2 row and the .05 column is 0.8944, $P(Z \leq 1.25) = 0.8944$. Similarly, to find the area under the curve to the left of $z = -0.50$ we use the left-hand page of the table to locate the table value in the -0.5 row and the .00 column; with a table value of 0.3085, $P(Z \leq -0.50) = 0.3085$. Thus, $P(-0.50 \leq Z \leq 1.25) = P(Z \leq 1.25) - P(Z \leq -0.50) = 0.8944 - 0.3085 = 0.5859$.

Let us consider another example of computing the probability that Z is in the interval between two given values. Often it is of interest to compute the probability that a normal random variable assumes a value within a certain number of standard deviations of the mean. Suppose we want to compute the probability that the standard normal random variable is within one standard deviation of the mean; that is, $P(-1.00 \leq Z \leq 1.0)$

To compute this probability we must find the area under the curve between -1.0 and 1.00. Earlier we found that $P(Z \leq 1.00) = 0.8413$. Referring again to the table inside the front cover of the book, we find that the area under the curve to the left of $z = -1.00$ is 0.1587, so $P(Z \leq -1.00) = 0.1587$. Therefore $P(-1.00 \leq Z \leq 1.00) = P(Z \leq 1.00) - P(Z \leq -1.00) = 0.8413 - 0.1587 = 0.6826$. This probability is shown graphically in the following figure.

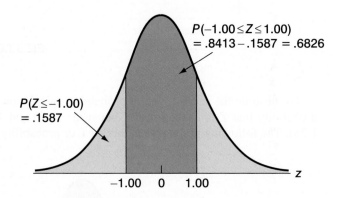

To illustrate how to make the third type of probability computation, suppose we want to compute the probability of obtaining a z value of at least 1.58; that is, $P(Z \geq 1.58)$. The value in the $z = 1.5$ row and the $.08$ column of the cumulative normal table is 0.9429; thus, $P(Z < 1.58) = 0.9429$. However, because the total area under the normal curve is 1, $P(Z \geq 1.58) = 1 - 0.9429 = 0.0571$. This probability is shown in the following figure.

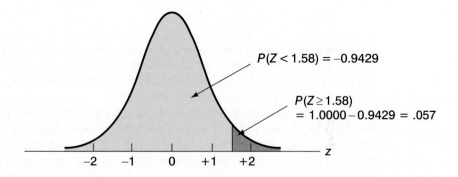

In the preceding illustrations, we showed how to compute probabilities given specified z values. In some situations, we are given a probability and are interested in working backward to find the corresponding z value. Suppose we want to find a z value such that the probability of obtaining a larger z value is 0.10. The following figure shows this situation graphically.

This problem is the inverse of those in the preceding examples. Previously, we specified the z value of interest and then found the corresponding probability, or area.

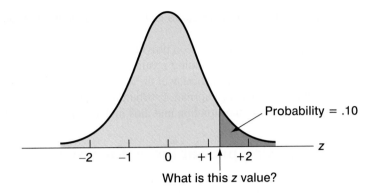

In this example, we are given the probability, or area, and asked to find the corresponding z value. To do so, we use the standard normal probability table somewhat differently.

z	.06	.07	.08	.09
.				
.				
.				
1.0	.8554	.8577	.8599	.8621
1.1	.8770	.8790	.8810	.8830
1.2	.8962	.8980	.8997	.9015
1.3	.9131	.9147	.9162	.9177
1.4	.9279	.9292	.9306	.9319
.				
.				
.				

Cumulative probability value
closest to 0.9000

Recall that the standard normal probability table gives the area under the curve to the left of a particular z value. We have been given the information that the area in the upper tail of the curve is 0.10. Hence, the area under the curve to the left of the unknown z value must equal 0.9000. Scanning the body of the table, we find 0.8997 is the cumulative probability value closest to 0.9000. The section of the table providing this result follows. Reading the z value from the left-most column and the top row of the table, we find that the corresponding z value is 1.28. Thus, an area of approximately 0.9000 (actually 0.8997) will be to the left of $z = 1.28$.* In terms of the question originally asked, the probability is approximately 0.10 that the z value will be larger than 1.28.

The examples illustrate that the table of areas for the standard normal distribution can be used to find probabilities associated with values of the standard normal random variable Z. Two types of questions can be asked. The first type of question specifies a

---

*We could use interpolation in the body of the table to get a better approximation of the z value that corresponds to an area of 0.9000. Doing so provides one more decimal place of accuracy and yields a z value of 1.282. However, in most practical situations, sufficient accuracy is obtained by simply using the table value closest to the desired probability.

value, or values, for z and asks us to use the table to determine the corresponding areas, or probabilities.

The second type of question provides an area, or probability, and asks us to use the table to determine the corresponding z value. Thus, we need to be flexible in using the standard normal probability table to answer the desired probability question. In most cases, sketching a graph of the standard normal distribution and shading the appropriate area or probability helps to visualize the situation and aids in determining the correct answer.

Computing probabilities for any normal distribution

The reason for discussing the standard normal distribution so extensively is that probabilities for all normal distributions are computed by using the standard normal distribution. That is, when we have a normal distribution with any mean μ and any standard deviation σ, we answer probability questions about the distribution by first converting to the standard normal distribution. Then we can use the standard normal probability table and the appropriate z values to find the desired probabilities. The formula used to convert any normal random variable X with mean μ and standard deviation σ to the standard normal distribution follows.

Converting to the standard normal distribution

$$Z = \frac{X - \mu}{\sigma}$$

(6.3)

A value of X equal to the mean μ results in $z = (\mu - \mu)/\sigma = 0$. Thus, we see that a value of X equal to the mean μ of X corresponds to a value of Z at the mean 0 of Z. Now suppose that x is one standard deviation greater than the mean; that is, $x = \mu + \sigma$. Applying equation (6.3), we see that the corresponding z value $= [(\mu + \sigma) - \mu]/\sigma = \sigma/\sigma = 1$. Thus, a value of X that is one standard deviation above the mean μ of X corresponds to a z value $= 1$. In other words, we can interpret Z as the number of standard deviations that the normal random variable X is from its mean μ.

To see how this conversion enables us to compute probabilities for any normal distribution, suppose we have a normal distribution with $\mu = 10$ and $\sigma = 2$. What is the probability that the random variable X is between 10 and 14? Using equation (6.3) we see that at $x = 10$, $z = (x - \mu)/\sigma = (10 - 10)/2 = 0$ and that at $x = 14$, $z = (14 - 10)/2 = 4/2 = 2$. Thus, the answer to our question about the probability of X being between 10 and 14 is given by the equivalent probability that Z is between 0 and 2 for the standard normal distribution.

In other words, the probability that we are seeking is the probability that the random variable X is between its mean and two standard deviations greater than the mean. Using $z = 2.00$ and standard normal probability table, we see that $P(Z \leq 2) = 0.9772$. Because $P(Z \leq 0) = 0.5000$ we can compute $P(0.00 \leq Z \leq 2.00) = P(Z \leq 2) - P(Z \leq 0) = 0.9772 - 0.5000 = 0.4772$. Hence the probability that X is between 10 and 14 is 0.4772.

Greer Tyre Company problem

We turn now to an application of the normal distribution. Suppose the Greer Tyre Company just developed a new steel-belted radial tyre that will be sold through a national chain of discount stores. Because the tyre is a new product, Greer's managers believe that the kilometres guarantee offered with the tyre will be an important factor in the acceptance of the product. Before finalizing the kilometres guarantee policy, Greer's managers want probability information about the number of kilometres the tyres will last.

From actual road tests with the tyres, Greer's engineering group estimates the mean number of kilometres the tyre will last is $\mu = 36\ 500$ kilometres and that the standard deviation is $\sigma = 5000$. In addition, the data collected indicate a normal distribution is a reasonable assumption. What percentage of the tyres can be expected to last more than 40 000 kilometres? In other words, what is the probability that the number of kilometres the tyre lasts will exceed 40 000? This question can be answered by finding the area of the darkly shaded region in Figure 6.6. At $x = 40\ 000$, we have

$$z = \frac{x - \mu}{\sigma} = \frac{40\ 000 - 36\ 500}{5000} = \frac{3500}{5000} = 0.70$$

Refer now to the bottom of Figure 6.6. We see that a value of $x = 40\ 000$ on the Greer Tyre normal distribution corresponds to a value of $z = 0.70$ on the standard normal distribution. Using the standard normal probability table, we see that the area to the left of $z = 0.70$ is 0.7580. Referring again to Figure 6.6, we see that the area to the left of $x = 40\ 000$ on the Greer Tyre normal distribution is the same. Thus, $1.000 - 0.7580 = 0.2420$ is the probability that X will exceed 40 000. We can conclude that about 24.2 per cent of the tyres will last longer than 40 000 kilometres.

Let us now assume that Greer is considering a guarantee that will provide a discount on replacement tyres if the original tyres do not exceed the number of kilometres stated in the guarantee. What should the guaranteed number of kilometres be if Greer wants no more than 10 per cent of the tyres to be eligible for the discount guarantee? This question is interpreted graphically in Figure 6.7.

According to Figure 6.7, the area under the curve to the left of the unknown guaranteed number of kilometers must be 0.10. So we must find the z value that cuts off an area of 0.10 in the left tail of a standard normal distribution. Using the standard normal probability table, we see that $z = -1.28$ cuts off an area of 0.10 in the lower tail. Hence

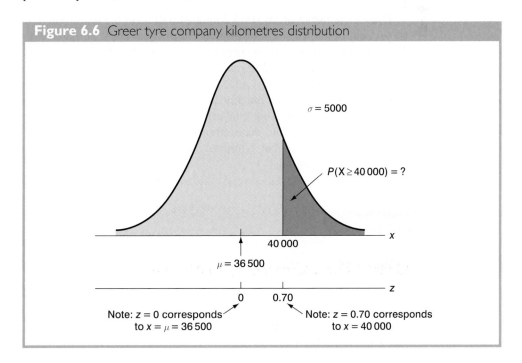

Figure 6.6 Greer tyre company kilometres distribution

$\sigma = 5000$

$P(X \geq 40\ 000) = ?$

40 000

$\mu = 36\ 500$

0   0.70   z

Note: z = 0 corresponds to x = μ = 36 500

Note: z = 0.70 corresponds to x = 40 000

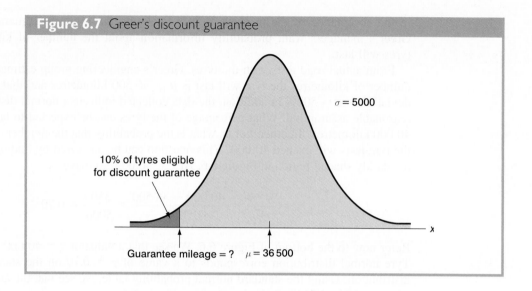

Figure 6.7 Greer's discount guarantee

$z = -1.28$ is the value of the standard normal variable corresponding to the desired number of kilometres guarantee on the Greer Tyre normal distribution. To find the value of X corresponding to $z = -1.28$, we have

$$z = \frac{x - \mu}{\sigma} = -1.28$$

$$x - \mu = -1.28\sigma$$

$$x = \mu - 1.28\sigma$$

With $\mu = 36\,500$ and $\sigma = 5000$,

$$x = 36\,500 - 1.28 \times 5000 = 30\,100$$

Thus, a guarantee of 30 100 kilometres will meet the requirement that approximately 10 per cent of the tyres will be eligible for the guarantee. Perhaps, with this information, the firm will set its tyre kilometres guarantee at 30 000 kilometres.

Again, we see the important role that probability distributions play in providing decision-making information. Namely, once a probability distribution is established for a particular application, it can be used quickly and easily to obtain probability information about the problem. Probability does not establish a decision recommendation directly, but it provides information that helps the decision-maker better understand the risks and uncertainties associated with the problem. Ultimately, this information may assist the decision-maker in reaching a good decision.

Exercises

Methods

8   Using Figure 6.4 as a guide, sketch a normal curve for a random variable X that has a mean of $\mu = 100$ and a standard deviation of $\sigma = 10$. Label the horizontal axis with values of 70, 80, 90, 100, 110, 120 and 130.

9   A random variable is normally distributed with a mean of $\mu = 50$ and a standard deviation of $\sigma = 5$.

   a. Sketch a normal curve for the probability density function. Label the horizontal axis with values of 35, 40, 45, 50, 55, 60 and 65. Figure 6.4 shows that the normal curve almost touches the horizontal axis at three standard deviations below and at three standard deviations above the mean (in this case at 35 and 65).

   b. What is the probability the random variable will assume a value between 45 and 55?

   c. What is the probability the random variable will assume a value between 40 and 60?

10 Draw a graph for the standard normal distribution. Label the horizontal axis at values of -3, -2, -1, 0, 1, 2 and 3. Then use the table of probabilities for the standard normal distribution to compute the following probabilities.

   a. $P(0 \le Z \le 1)$
   b. $P(0 \le Z \le 1.5)$
   c. $P(0 < Z < 2)$
   d. $P(0 < Z < 2.5)$

11 Given that Z is a standard normal random variable, compute the following probabilities.

   a. $P(-1 \le Z \le 0)$
   b. $P(-1.5 \le Z \le 0)$
   c. $P(-2 < Z < 0)$
   d. $P(-2.5 \le Z \le 0)$
   e. $P(-3 \le Z \le 0)$

12 Given that Z is a standard normal random variable, compute the following probabilities.

   a. $P(0 \le Z \le 0.83)$
   b. $P(-1.57 \le Z \le 0)$
   c. $P(Z > 0.44)$
   d. $P(Z \ge -0.23)$
   e. $P(Z < 1.20)$
   f. $P(Z \le -0.71)$

13 Given that Z is a standard normal random variable, compute the following probabilities.

   a. $P(-1.98 \le Z \le 0.49)$
   b. $P(0.52 \le Z \le 1.22)$
   c. $P(-1.75 \le Z \le -1.04)$

14 Given that Z is a standard normal random variable, find z for each situation.

   a. The area between 0 and z is 0.4750.
   b. The area between 0 and z is 0.2291.
   c. The area to the right of z is 0.1314.
   d. The area to the left of z is 0.6700.

15 Given that Z is a standard normal random variable, find z for each situation.

   a. The area to the left of z is 0.2119.
   b. The area between $-z$ and z is 0.9030.
   c. The area between $-z$ and z is 0.2052.
   d. The area to the left of z is 0.9948.
   e. The area to the right of z is 0.6915.

16 Given that Z is a standard normal random variable, find z for each situation.

   a. The area to the right of z is 0.01.

   b. The area to the right of z is 0.025.

   c. The area to the right of z is 0.05.

   d. The area to the right of z is 0.10.

Applications

17 The time a salesperson takes to travel from customer A to customer B varies but can be described by a normal probability function with mean 45 minutes and standard deviation 6 minutes.

   a. What proportion of the journeys takes less than 35 minutes?

   b. What proportion of the journeys takes over 60 minutes?

   c. How long should the salesperson allow for a journey if they want to be 70 per cent sure of not being late?

18 The holdings of clients of a successful on-line stockbroker are normally distributed with a mean of £20 000 and standard deviation of £1 500. To increase its business, the stockbroker is looking to email special promotions to the top 20 per cent of its clientele based on the value of their holdings. What is the minimum holding of this group?

19 A company has been involved in developing a new pesticide. Tests show that the average proportion, p, of insects killed by administration of x units of the insecticide is given by $p = P(X \leq x)$ where the probability $P(X \leq x)$ relates to a normal distribution with unknown mean and standard deviation.

   a. Given that $x = 10$ when $p = 0.4$ and that $x = 15$ when $p = 0.9$, determine the dose that will be lethal to 50 per cent of the insect population on average.

   b. If a dose of 17.5 units is administered to each of 100 insects, how many will be expected to die?

6.3 Normal approximation of binomial probabilities

In Section 5.4 we presented the discrete binomial distribution. Recall that a binomial experiment consists of a sequence of n identical independent trials with each trial having two possible outcomes, a success or a failure. The probability of a success on a trial is the same for all trials and is denoted by π. The binomial random variable is the number of successes in the n trials, and probability questions pertain to the probability of x successes in the n trials. When the number of trials becomes large, evaluating the binomial probability function by hand or with a calculator is difficult. In addition, the binomial tables in Appendix B do not include values of n greater than 20. Hence, when we encounter a binomial distribution problem with a large number of trials, we may want to approximate the binomial distribution. In cases where the number of trials is greater than 20, $n\pi \geq 5$, and $n(1 - \pi) \geq 5$, the normal distribution provides an easy-to-use approximation of binomial probabilities.

When using the normal approximation to the binomial, we set $\mu = n\pi$ and $\sigma = \sqrt{n\pi(1 - \pi)}$ in the definition of the normal curve. Let us illustrate the normal approximation to the binomial by supposing that a particular company has a history of making errors in 10 per cent of its invoices. A sample of 100 invoices has been taken,

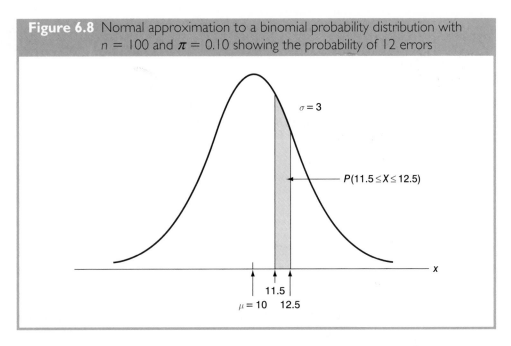

Figure 6.8 Normal approximation to a binomial probability distribution with $n = 100$ and $\pi = 0.10$ showing the probability of 12 errors

and we want to compute the probability that 12 invoices contain errors. That is, we want to find the binomial probability of 12 successes in 100 trials.

In applying the normal approximation to the binomial, we set $\mu = n\pi = 100 \times 0.1 = 10$ and $\sigma = \sqrt{n\pi(1 - \pi)} = \sqrt{100 \times 0.1 \times 0.9} = 3$. A normal distribution with $\mu = 10$ and $\sigma = 3$ is shown in Figure 6.8.

Recall that, with a continuous probability distribution, probabilities are computed as areas under the probability density function. As a result, the probability of any single value for the random variable is zero. Thus to approximate the binomial probability of 12 successes, we must compute the area under the corresponding normal curve between 11.5 and 12.5. The 0.5 that we add and subtract from 12 is called a **continuity correction factor**. It is introduced because a continuous distribution is being used to approximate a discrete distribution. Thus, $P(X = 12)$ for the *discrete* binomial distribution is approximated by $P(11.5 \leq X \leq 12.5)$ for the *continuous* normal distribution.

Converting to the standard normal distribution to compute $P(11.5 \leq X \leq 12.5)$, we have

$$z = \frac{x - \mu}{\sigma} = \frac{12.5 - 10.0}{3} = 0.83 \text{ at } X = 12.5$$

and

$$z = \frac{x - \mu}{\sigma} = \frac{11.5 - 10.0}{3} = 0.50 \text{ at } X = 11.5$$

Using the standard normal probability table, we find that the area under the curve (in Figure 6.8) to the left of 12.5 is 0.7967. Similarly, the area under the curve to the left of 11.5 is 0.6915. Therefore, the area between 11.5 and 12.5 is $0.7967 - 0.6915 = 0.1052$. The normal approximation to the probability of 12 successes in 100 trials is 0.1052.

For another illustration, suppose we want to compute the probability of 13 or fewer errors in the sample of 100 invoices. Figure 6.9 shows the area under the normal curve that approximates this probability. Note that the use of the continuity correction factor

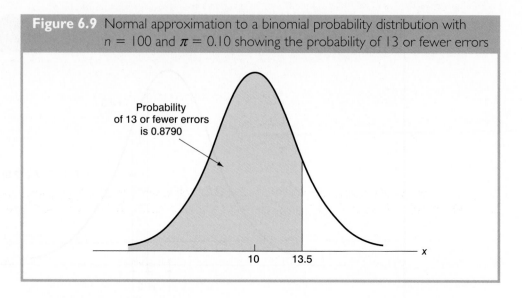

Figure 6.9 Normal approximation to a binomial probability distribution with $n = 100$ and $\pi = 0.10$ showing the probability of 13 or fewer errors

results in the value of 13.5 being used to compute the desired probability. The z value corresponding to $x = 13.5$ is

$$z = \frac{13.5 - 10.0}{3} = 1.17$$

The standard normal probability table shows that the area under the standard normal curve to the left of 1.17 is 0.8790. The area under the normal curve approximating the probability of 13 or fewer errors is given by the heavily shaded portion of the graph in Figure 6.9.

Exercises

Methods

20 A binomial probability distribution has $\pi = 0.20$ and $n = 100$.

    a. What is the mean and standard deviation?

    b. Is this a situation in which binomial probabilities can be approximated by the normal probability distribution? Explain.

    c. What is the probability of exactly 24 successes?

    d. What is the probability of 18 to 22 successes?

    e. What is the probability of 15 or fewer successes?

21 Assume a binomial probability distribution has $\pi = 0.60$ and $n = 200$.

    a. What is the mean and standard deviation?

    b. Is this a situation in which binomial probabilities can be approximated by the normal probability distribution? Explain.

    c. What is the probability of 100 to 110 successes?

    d. What is the probability of 130 or more successes?

    e. What is the advantage of using the normal probability distribution to approximate the binomial probabilities? Use part (d) to explain the advantage.

Applications

22 A hotel in Nice has 120 rooms. In the spring months, hotel room occupancy is approximately 75 per cent.

  a. What is the probability that at least half of the rooms are occupied on a given day?
  b. What is the probability that 100 or more rooms are occupied on a given day?
  c. What is the probability that 80 or fewer rooms are occupied on a given day?

6.4 Exponential probability distribution

The **exponential probability distribution** may be used for random variables such as the time between arrivals at a car wash, the time required to load a truck, the distance between major defects in a highway and so on. The exponential probability density function follows.

Exponential probability density function

$$f(x) = \frac{1}{\mu}\, e^{-x/\mu} \qquad \text{for } x \geq 0, \mu > 0 \tag{6.4}$$

As an example of the exponential distribution, suppose that $X =$ the time it takes to load a truck at the Schips loading dock follows such a distribution. If the mean, or average, time to load a truck is 15 minutes ($\mu = 15$), the appropriate probability density function is

$$f(x) = \frac{1}{15}\, e^{-x/15}$$

Figure 6.10 is the graph of this probability density function.

Computing probabilities for the exponential distribution

As with any continuous probability distribution, the area under the curve corresponding to an interval provides the probability that the random variable assumes a value in that interval. In the Schips loading dock example, the probability that loading a truck will take six minutes or less ($X \leq 6$) is defined to be the area under the curve in Figure 6.10 from $x = 0$ to $x = 6$. Similarly, the probability that loading a truck will take 18 minutes or less ($X \leq 18$) is the area under the curve from $x = 0$ to $x = 18$. Note also that the probability that loading a truck will take between 6 minutes and 18 minutes ($6 \leq X \leq 18$) is given by the area under the curve from $x = 6$ to $x = 18$.

To compute exponential probabilities such as those just described, we use the following formula. It provides the cumulative probability of obtaining a value for the exponential random variable of less than or equal to some specific value denoted by x_0.

Exponential distribution: cumulative probabilities

$$P(X \leq x_0) = 1 - e^{-x_0/\mu} \tag{6.5}$$

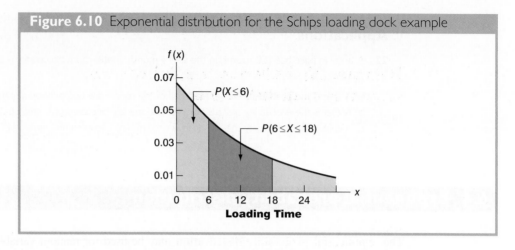

Figure 6.10 Exponential distribution for the Schips loading dock example

For the Schips loading dock example, X = loading time and $\mu = 15$, which gives us

$$P(X \leq x_0) = 1 - e^{-x_0/15}$$

Hence, the probability that loading a truck will take six minutes or less is

$$P(X \leq 6) = 1 - e^{-6/15} = 0.3297$$

Figure 6.11 shows the area or probability for a loading time of six minutes or less. Using equation (6.5), we calculate the probability of loading a truck in 18 minutes or less.

$$P(X \leq 18) = 1 - e^{-18/15} = 0.6988$$

Thus, the probability that loading a truck will take between six minutes and 18 minutes is equal to $0.6988 - 0.3297 = 0.3691$. Probabilities for any other interval can be computed similarly.

In the preceding example, the mean time it takes to load a truck is $\mu = 15$ minutes. A property of the exponential distribution is that the mean of the distribution and the

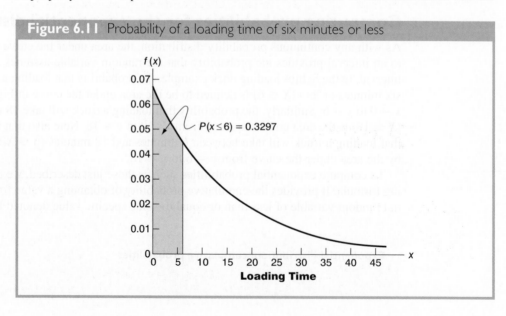

Figure 6.11 Probability of a loading time of six minutes or less

standard deviation of the distribution are *equal*. Thus, the standard deviation for the time it takes to load a truck is $\sigma = 15$ minutes. The variance is $\sigma^2 = (15)^2 = 225$.

Relationship between the Poisson and exponential distributions

In Section 5.5 we introduced the Poisson distribution as a discrete probability distribution that is often useful in examining the number of occurrences of an event over a specified interval of time or space. Recall that the Poisson probability function is

$$p(x) = \frac{\mu^x e^{-\mu}}{x!}$$

where

μ = expected value or mean number of occurrences over a specified interval

The continuous exponential probability distribution is related to the discrete Poisson distribution. If the Poisson distribution provides an appropriate description of the number of occurrences per interval, the exponential distribution provides a description of the length of the interval between occurrences.

To illustrate this relationship, suppose the number of cars that arrive at a car wash during one hour is described by a Poisson probability distribution with a mean of ten cars per hour. The Poisson probability function that gives the probability of X arrivals per hour is

$$p(x) = \frac{10^x e^{-10}}{x!}$$

Because the average number of arrivals is ten cars per hour, the average time between cars arriving is

$$\frac{1 \text{ hour}}{10 \text{ cars}} = 0.1 \text{ hour/car}$$

Thus, the corresponding exponential distribution that describes the time between the arrivals has a mean of $\mu = 0.1$ hour per car; as a result, the appropriate exponential probability density function is

$$f(x) = \frac{1}{0.1} e^{-x/0.1} = 10 e^{-10x}$$

Exercises

Methods

23 Consider the following exponential probability density function.

$$f(x) = \frac{1}{8} e^{-x/8} \qquad \text{for } x \geq 0$$

a. Find $P(X \leq 6)$.
b. Find $P(X \leq 4)$.

c. Find $P(X \geq 6)$.

d. Find $P(4 \leq X \leq 6)$.

24 Consider the following exponential probability density function.

$$f(x) = \frac{1}{3} e^{-x/3} \qquad \text{for } x \geq 0$$

a. Write the formula for $P(X \leq x_0)$.

b. Find $P(X \leq 2)$.

c. Find $P(X \geq 3)$.

d. Find $P(X \leq 5)$.

e. Find $P(2 \leq X \leq 5)$.

Applications

25 The lifetime (hours) of an electronic device is a random variable with the following exponential probability density function.

$$f(x) = \frac{1}{50} e^{-x/50} \qquad \text{for } x \geq 0$$

a. What is the mean lifetime of the device?

b. What is the probability that the device will fail in the first 25 hours of operation?

c. What is the probability that the device will operate 100 or more hours before failure?

26 The time between arrivals of vehicles at a particular intersection follows an exponential probability distribution with a mean of 12 seconds.

a. Sketch this exponential probability distribution.

b. What is the probability that the arrival time between vehicles is 12 seconds or less?

c. What is the probability that the arrival time between vehicles is 6 seconds or less?

d. What is the probability of 30 or more seconds between vehicle arrivals?

27 According to Barron's 1998 Primary Reader Survey, the average annual number of investment transactions for a subscriber is 30 (*www.barronsmag.com*, 28 July 2000). Suppose the number of transactions in a year follows the Poisson probability distribution.

a. Show the probability distribution for the time between investment transactions.

b. What is the probability of no transactions during the month of January for a particular subscriber?

c. What is the probability that the next transaction will occur within the next half month for a particular subscriber?

For additional online summary questions and answers go to the companion website at www.cengage.co.uk/aswsbe2

Summary

This chapter extended the discussion of probability distributions to the case of continuous random variables. The major conceptual difference between discrete and continuous probability distributions involves the method of computing probabilities. With discrete distributions, the probability function p(x) provides the probability that the random variable X assumes various values. With continuous distributions, the probability density function f(x) does not provide probability values directly. Instead, probabilities are given by areas under the curve or graph of f(x). Three continuous probability distributions – the uniform, normal and exponential distributions were the particular focus – with detailed examples showing how probabilities could be straightforwardly computed. In addition, relationships between the binomial and normal distributions and Poisson and exponential distribution were established and related probability results, exploited.

Key terms

Continuity correction factor

Exponential probability distribution

Normal probability distribution

Probability density function

Standard normal probability
   distribution

Uniform probability distribution

Key formulae

Uniform Probability Density Function

$$f(x) = \begin{cases} \dfrac{1}{b-a} & \text{for } a \leq x \leq b \\[2mm] 0 & \text{elsewhere} \end{cases} \tag{6.1}$$

Normal Probability Density Function

$$f(x) = \frac{1}{\sigma \sqrt{2\pi}} e^{-(x-\mu)^2/2\sigma^2} \tag{6.2}$$

Converting to the Standard Normal Distribution

$$Z = \frac{X - \mu}{\sigma} \tag{6.3}$$

Exponential Probability Density Function

$$f(x) = \frac{1}{\mu} e^{-x/\mu} \quad \text{for } x \geq 0, \mu > 0 \tag{6.4}$$

Exponential Distribution: Cumulative Probabilities

$$P(X \leq x_0) = 1 - e^{-x_0/\mu} \qquad\qquad (6.5)$$

Case problem 1 Prix-Fischer Toys

Prix-Fischer Toys sells a variety of new and innovative children's toys. Management learned that the pre-holiday season is the best time to introduce a new toy, because many families use this time to look for new ideas for December holiday gifts. When Prix-Fischer discovers a new toy with good market potential, it chooses an October market entry date.

In order to get toys in its stores by October, Prix-Fischer places one-time orders with its manufacturers in June or July of each year. Demand for children's toys can be highly volatile. If a new toy catches on, a sense of shortage in the market place often increases the demand to high levels and large profits can be realised. However, new toys can also flop, leaving Prix-Fischer stuck with high levels of inventory that must be sold at reduced prices. The most important question the company faces is deciding how many units of a new toy should be purchased to meet anticipated sales demand. If too few are purchased, sales will be lost; if too many are purchased, profits will be reduced because of low prices realised in clearance sales.

For the coming season, Prix-Fischer plans to introduce a new talking bear product called Chattiest Teddy. As usual, Prix-Fischer faces the decision of how many Chattiest Teddy units to order for the coming holiday season. Members of the management team suggested order quantities of 15 000, 18 000, 24 000 or 28 000 units. The wide range of order quantities suggested, indicate considerable disagreement concerning the market potential. The product management team asks you for an analysis of the stock-out probabilities for various order quantities, an estimate of the profit potential, and to help make an order quantity recommendation.

Prix-Fischer expects to sell Chattiest Teddy for €24 based on a cost of €16 per unit. If inventory remains after the holiday season, Prix-Fischer will sell all surplus inventory for €5 per unit. After reviewing the sales history of similar products, Prix-Fischer's senior sales forecaster predicted an expected demand of 20 000 units with a 0.90 probability that demand would be between 10 000 units and 30 000 units.

Managerial Report

Prepare a managerial report that addresses the following issues and recommends an order quantity for the Chattiest Teddy product.

1 Use the sales forecaster's prediction to describe a normal probability distribution that can be used to approximate the demand distribution. Sketch the distribution and show its mean and standard deviation.

2 Compute the probability of a stock-out for the order quantities suggested by members of the management team.

3 Compute the projected profit for the order quantities suggested by the management team under three scenarios: worst case in which sales = 10 000 units, most likely case in which sales = 20 000 units, and best case in which sales = 30 000 units.

4 One of Prix-Fischer's managers felt that the profit potential was so great that the order quantity should have a 70 per cent chance of meeting demand and only a 30 per cent chance of any stock-outs. What quantity would be ordered under this policy, and what is the projected profit under the three sales scenarios?

5 Provide your own recommendation for an order quantity and note the associated profit projections. Provide a rationale for your recommendation.

The children's toy, Talking Teady Bears. © H Tuller.

Case problem 2 Queuing patterns in a retail furniture store

The assistant manager of one of the larger stores in a retail chain selling furniture and household appliances has recently become interested in using quantitative techniques in the store operation. To help resolve a longstanding queuing problem, data have been collected on the time between customer arrivals and the time that a given number of customers were in a particular store department. Relevant details are summarised in Tables 6.1 and 6.2 respectively. Corresponding data on service times per customer are tabulated in Table 6.3.

Table 6.1 Time between arrivals (during a four hour period)

Time between arrivals (in minutes)	frequency
0.0 < 0.2	31
0.2 < 0.4	32
0.4 < 0.6	23
0.6 < 0.8	21
0.8 < 1.0	19
1.0 < 1.2	11
1.2 < 1.4	14
1.4 < 1.6	8
1.6 < 1.8	6
1.8 < 2.0	9
2.0 < 2.2	6
2.2<2.4	4
2.4<2.6	5
2.6 < 2.8	4
2.8 < 3.0	4
3.0 < 3.2	3
More than 3.2	10

Table 6.2 Time that n customers were in the department (during a four-hour period)

Number of customers, n	Time (in minutes)
0	16.8
1	35.5
2	52
3	49
4	29.3
5	18.8
6	13.6
7	9.6
8	5.8
More than 8	9.6

Table 6.3 Service time (from a sample of 31)

Service time (in minutes)	Frequency
Less than 1	5
1 < 2	7
2 < 3	6
3 < 4	4
4 < 5	2
5 < 6	3
6 < 7	1
7 < 8	1
8 < 9	1
9 < 10	0
More than 10	1

Queuing up to enter an Ikea home furnishing store in Beijing, China. © Lou Linwei/Alamy.

In order to arrive at an appropriate solution strategy for the department's queuing difficulties, the manager has come to you for advice on possible statistical patterns that might apply to this information.

1 By plotting the arrival and service patterns shown in Table 6.1 and 6.3 show that they can each be reasonably represented by an exponential distribution.

2. If λ = mean arrival rate for the queuing system here and μ, the mean service rate for each

channel, estimate the mean arrival time $(1/\lambda)$ and mean service time $(1/\mu)$ respectively.

3. If k = the number of service channels and the mean service time for the system (store) is greater than the mean arrival rate (i.e. $k\mu > \lambda$) then the following formulae can be shown to apply to the system 'in the steady state' - subject to certain additional mathematical assumptions[1]:

a. The probability there are no customers in the system

$$P(0) = \cfrac{1}{\sum_{n=0}^{K-1} \cfrac{(\lambda / \mu)^n}{n!} + \cfrac{(\lambda / \mu)^k}{k!} \cdot \cfrac{k\mu}{(k\mu - \lambda)}}$$

b. The average number of customers in the queue

$$L_q = \frac{(\lambda / \mu)^k \lambda\mu}{(k - 1)!(k\mu - \lambda)^2} P(0)$$

c. The average number of customers in the store

$$L = L_q + \frac{\lambda}{\mu}$$

[1]The queue has two or more channels; the mean service rate μ is the same for each channel; arrivals wait in a single queue and then move to the first open channel for service; the queue discipline is first-come, first-served (FCFS).

d. The average time a customer spends in the queue

$$W_q = \frac{L_q}{\lambda}$$

e. The average time a customer spends in the store

$$W = W_q + \frac{1}{\mu}$$

f. The probability of n customers in the system

$$P(n) = \frac{(\lambda + \mu)^n}{n!} P(0) \qquad \text{for } n \le k$$

$$P(n) = \frac{(\lambda + \mu)^n}{k!k^{(n-k)}} P(0) \qquad \text{for } n > k$$

According to this model, what is the smallest value that k can take? If this is the number of channels that the retailer currently operates, estimate the above operating characteristics for the store. How would these values change if the k channels were increased by 1 or 2 extra channels? Discuss what factors might influence the retailer in arriving at an appropriate value of k.

Software Section for Chapter 6

Continuous probability distributions with MINITAB

Let us demonstrate the MINITAB procedure for computing continuous probabilities by referring to the Greer Tyre Company problem where the number of kilometres that a tyre lasts was described by a normal distribution with $\mu = 36\,500$ and $\sigma = 5000$. One question asked was: What is the probability that the life of a tyre will exceed 40 000 kilometres?

For continuous probability distributions, MINITAB gives a cumulative probability; that is, MINITAB gives the probability that the random variable will assume a value less than or equal to a specified constant. For the Greer tyre kilometres question, MINITAB can be used to determine the cumulative probability that the tyre's lifetime will be *less than or equal* to 40 000 kilometres. (The specified constant in this case is 40 000.) After obtaining the cumulative probability from MINITAB, we must subtract it from 1 to determine the probability that the tyre's lifetime will exceed 40 000 kilometres.

Prior to using MINITAB to compute a probability, one must enter the specified constant into a column of the worksheet. For the Greer tyre kilometres question we entered the specified constant of 40 000 into column C1 of the MINITAB worksheet. The steps involved in using MINITAB to compute the cumulative probability of the normal random variable assuming a value less than or equal to 40 000 follow.

Step 1   **Calc > Probability Distributions > Normal**   [Main menu bar]

Step 2   Select **Cumulative probability**   [**Normal Distribution** panel]
           Enter 36500 in the **Mean** box
           Enter 5000 in the **Standard deviation** box
           Enter C1 in the **Input column** box (the column containing 40 000)
           Click **OK**

After the user clicks **OK**, MINITAB prints the cumulative probability that the normal random variable assumes a value less than or equal to 40 000. MINITAB shows that this probability is 0.7580. Because we are interested in the probability that the tyre lifetime will be greater than 40 000, the desired probability is $1 - 0.7580 = 0.2420$.

A second question in the Greer Tyre Company problem was: What kilometres guarantee should Greer set to ensure that no more than 10 per cent of the tyres qualify for the guarantee? Here we are given a probability and want to find the corresponding value for the random variable. MINITAB uses an inverse calculation routine to find the value of the random variable associated with a given cumulative probability. For step 1, we must enter the cumulative probability into a column of the MINITAB worksheet (say C1). In this case, the desired cumulative probability is 0.10. Then, step 2 of the MINITAB procedure is as already listed. In step 2, we select **Inverse cumulative probability** instead of **Cumulative probability** and complete the remaining parts of the step.

MINITAB then displays the kilometres guarantee of 30 092 kilometres. MINITAB is capable of computing probabilities for other continuous probability distributions, including the exponential probability distribution. To compute exponential probabilities, follow the procedure shown previously for the normal probability distribution but choose **Exponential** instead of **Normal**. Again the same entries follow on as before, with the exception that entering the standard deviation is not required. Output for inverse cumulative probabilities is identical to that described for the normal probability distribution.

Continuous probability distributions with EXCEL

EXCEL provides the capability for computing probabilities for several continuous probability distributions, including the normal and exponential probability distributions. In this appendix, we describe how EXCEL can be used to compute probabilities for any normal distribution. The procedures for the exponential and other continuous distributions are similar to the one we describe for the normal distribution.

Let us return to the Greer Tyre Company problem where number of kilometres the tyre lasted was described by a normal distribution with $\mu = 36\ 500$ and $\sigma = 5000$. Assume we are interested in the probability that tyre's lifetime will exceed 40 000 kilometres. EXCEL's NORMDIST function provides cumulative probabilities for a normal distribution. The general form of the function is NORMDIST $(x, \mu, \sigma, \text{cumulative})$. For the fourth argument, TRUE is specified if a cumulative probability is desired. Thus, to compute the cumulative probability that the tyre's lifetime will be less than or equal to 40 000 kilometres we would enter the following formula into any cell of an EXCEL worksheet:

$$=\text{NORMDIST}(40000,36500,5000,\text{TRUE})$$

At this point, 0.7580 will appear in the cell where the formula was entered, indicating that the probability of tyre's lifetime being less than or equal to 40 000 kilometres is 0.7580. Therefore, the probability that tyre's lifetime will exceed 40 000 kilometres is $1 - 0.7580 = 0.2420$. EXCEL's NORMINV function uses an inverse computation to find the x value corresponding to a given cumulative probability. For instance, suppose we want to find the guaranteed number of kilometres Greer should offer so that no more than 10 per cent of the tyres will be eligible for the guarantee. We would enter the following formula into any cell of an EXCEL worksheet:

$$=\text{NORMINV}(.1,36500,5000)$$

At this point, 30092 will appear in the cell where the formula was entered indicating that the probability of a tyre not lasting 30 092 kilometres is 0.10. The EXCEL function for computing exponential probabilities is EXPONDIST. Using it is straightforward. But if one needs help specifying the proper values for the arguments, EXCEL's Insert Function tool can be used (see Appendix 2.2).

Continuous probability distributions with PASW

In this appendix, we demonstrate the use of PASW for computing probabilities for the Greer Tyre Company problem where the number of kilometres that a tyre lasts was described by a normal distribution with μ = 36 500 and σ = 5000. One question asked was: What is the probability that the life of a tyre will exceed 40 000 kilometres?

For continuous probability distributions, PASW gives a cumulative probability; that is, PASW gives the probability that the random variable will assume a value less than or equal to a specified constant. For the Greer tyre kilometres question, PASW can be used to determine the cumulative probability that the tyre's lifetime will be *less than or equal* to 40 000 kilometres. (The specified constant in this case is 40 000.) After obtaining the cumulative probability from PASW, we must subtract it from 1 to determine the probability that the tyre's lifetime will exceed 40 000 kilometres.

Before beginning the PASW routine, first, the data must be entered in a PASW worksheet. For the Greer tyre kilometres question we entered the specified constant of 40 000 into column 1. This is automatically labelled by the system VAR00001. The latter variable name can then be changed to x in 'Variable View' mode. The steps involved in using PASW to compute the cumulative probability of the normal random variable assuming a value less than or equal to 40 000 follow.

Step 1 **Transform > Compute Variable**   [Main menu bar]

Step 2 Enter p(x) in **Target variable** box   [**Compute Variable** panel]
Choose **Function group**
Select **CDF & Noncentral CDF**
Click on **Cdf.Normal** (which has three arguments)
Select the **Numeric Expression** box
Click on x (for argument 1)
Enter 36500 (for argument 2)
Enter 5000 (for argument 3)
Click **OK**

After the user clicks **OK**, PASW provides the cumulative probability that the normal random variable assumes a value less than or equal to 40 000. PASW shows that this probability is 0.7580. Because we are interested in the probability that the tyre lifetime will be greater than 40 000, the desired probability is $1 - 0.7580 = 0.2420$.

A second question in the Greer Tyre Company problem was: What kilometres guarantee should Greer set to ensure that no more than 10 per cent of the tyres qualify for the guarantee? Here we are given a probability and want to find the corresponding value for the random variable. PASW uses an inverse calculation routine to find the value of the random variable associated with a given cumulative probability. First, we must enter the cumulative probability into the leftmost column of the PASW worksheet. In this case, the desired cumulative probability is 0.10. This is automatically labelled by the system var00001. The latter variable name can then be changed to p in 'Variable View' mode.

Then, the first three steps of the PASW procedure are as already listed. In step 2, we enter x in Target Variable box, Select Inverse DF instead of CDF & Noncentral CDF, click on 'p' for the first argument and repeat the entries for arguments 2 and 3.

PASW then provides the kilometres guarantee of 30 100 kilometres.

PASW is capable of computing probabilities for other continuous probability distributions, including the exponential probability distribution. To compute exponential probabilities, follow the procedure shown previously for the normal probability distribution but choose **Cdf.Exp** instead of **Cdf.Normal**. Again the same entries follow on as before, with the exception that entering the standard deviation is not required. Output for inverse cumulative probabilities is identical to that described for the normal probability distribution but using **Idf.Exp** instead of **Idf.Normal**.

Chapter 7

···

Sampling and Sampling Distributions

Learning objectives

After studying this chapter and doing the exercises, you should be able to:

1 Explain the terms simple random sample, sampling with replacement and sampling without replacement.

2 Select a simple random sample from a finite population using random number tables.

3 Explain the terms parameter, statistic, point estimator and unbiasedness.

4 Identify relevant point estimators for a population mean, population standard deviation and population proportion.

5 Explain the term sampling distribution.

6 Describe the form and characteristics of the sampling distribution:

6.1 of the sample mean, when the sample size is large or when the population is normal.

6.2 of the sample proportion when the sample size is large.

In Chapter 1, we defined the terms *population* and *sample*.

1 A *population* is the set of all the elements of interest in a study.

2 A *sample* is a subset of the population.

Numerical characteristics of a population, such as the mean and standard deviation, are called **parameters**. A primary purpose of statistical inference is to make estimates and test hypotheses about population parameters using information contained in a sample. Here are two situations in which samples provide estimates of population parameters.

1 A car tyre manufacturer developed a new tyre designed to provide an increase in lifetime over the firm's current line of tyres. To estimate the mean lifetime (in kilometres or miles) provided by the new tyre, the manufacturer selected a sample of 120 new tyres for testing. The test results provided a sample mean of 56 000 kilometres (35 000 miles). Therefore, an estimate of the mean tyre lifetime for the population of new tyres was 56 000 kilometres.

2 Members of an African government were interested in estimating the proportion of registered voters likely to support a proposal for constitutional reform to be put to the electorate in a national referendum. The time and cost associated with contacting every individual in the population of registered voters were prohibitive. A sample of 1000 registered voters was therefore selected, and 560 of the 1000 voters indicated support for the proposal. An estimate of the proportion of the population of registered voters supporting the proposal was 560/1000 = 0.56.

These two examples illustrate some of the reasons why samples are used. In the tyre lifetime example, collecting the data on tyre life involves wearing out each tyre tested. Clearly it is not feasible to test every tyre in the population. A sample is the only realistic way to obtain the tyre lifetime data. In the example involving the referendum, contacting every registered voter in the population is theoretically possible, but the time and cost in doing so are prohibitive. Consequently, a sample of registered voters is preferred.

It is important to realize that sample results provide only *estimates* of the values of the population characteristics, because the sample contains only a portion of the population.

Statistics in Practice

Copyright and Public Lending Right

How would you feel if the size of your income was determined each year by a sampling procedure? This is the situation that often exists, for at least part of annual income, for musicians and other artists who receive copyright payments for the performance, broadcasting or other use of their work. Even in this 21st century world of large databases and sophisticated communication, it is not always possible, or it is too costly, to maintain 100 per cent checks on what is being broadcast over TV, radio and Internet, so an alternative is to sample.

In a similar vein, many book authors receive payments through a Public Lending Right (PLR) scheme. This is particularly so for authors of fiction, or of popular non-fiction, whose books are available for loan in public libraries. A PLR scheme is intended to compensate authors for potential loss of income because their books are available in public libraries, and are therefore borrowed rather than bought by readers. Most of the current working PLR schemes are in Northern Europe (Denmark was the first country to establish a scheme, in 1946), though Australia, Canada, New Zealand and Israel also have schemes.

The UK PLR scheme was set up in 1979. From the outset, it was decided that it would be too costly to try and collect data from all libraries in the UK. Data on lending are therefore collected from a sample of libraries. In the early days of the UK scheme, the sample included only 16 branch libraries, but now the sample contains at least 30 library authorities. This is around 15 per cent of all library authorities in the UK (there are over 200). The 30 or so sampled library authorities typically cover more than 1000 branch libraries out of the 4000 to 5000 branch libraries in total in the UK. At least seven of the authorities are replaced each year in the sample, and no authority is allowed to remain in the sample for longer than four years.

The National Library of Scotland, Edinburgh. © David Robertson/Alamy.

The examples from copyright and PLR are cases where the sampling schemes involved can influence the income of individuals – the copyright holders or authors. The website for the UK PLR scheme acknowledges, for example, that authors of books with a 'local interest' – local history, say – are likely to qualify for PLR payments only if the library sample for the year contains library authorities in the relevant geographical area.

A great deal of the information on which companies and governments make important decisions are based on sample data. This chapter examines the basis and practicalities of scientific sampling.

A sample mean provides an estimate of a population mean, and a sample proportion provides an estimate of a population proportion. Some estimation error can be expected. This chapter provides the basis for determining how large the estimation error might be. With proper sampling methods, the sample results will provide 'good' estimates of the population parameters.

In this chapter we show how simple random sampling can be used to select a sample from a population. We then show how data obtained from a simple random sample can be used to compute estimates of a population mean, a population standard deviation, and a population proportion. In addition, we introduce the important concept of a sampling distribution. Knowledge of the appropriate sampling distribution enables us to make statements about how close the sample estimates are to the corresponding population parameters.

7.1   The EAI sampling problem

EAI

The head of personnel services for E-Applications & Informatics plc (EAI) has been given the task of developing a profile of the company's 2500 managers. The characteristics to be identified include the mean annual salary for the managers and the proportion of managers who have completed the company's management training programme. The 2500 managers are the population for this study. We can find the annual salary and training programme status for each individual by referring to the firm's personnel records. The data file containing this information for all 2500 managers in the population is on the CD that accompanies the text, in the file EAI.

Using the EAI data set and the formulae presented in Chapter 3, we calculate the population mean and the population standard deviation for the annual salary data.

$$\text{Population mean: } \mu = \text{€}51\ 800$$
$$\text{Population standard deviation: } \sigma = \text{€}4000$$

The data for training programme status show that 1500 of the 2500 managers completed the training programme. Let π denote the proportion of the population that completed the training programme: $\pi = 1500/2500 = 0.60$. The population mean annual salary ($\mu = \text{€}51\ 800$), the population standard deviation of annual salary ($\sigma = \text{€}4000$), and the population proportion that completed the training programme ($\pi = 0.60$) are parameters of the population of EAI managers.

Now, suppose the necessary information on all the EAI managers was *not* readily available in the company's database. How can the firm's head of personnel services obtain estimates of the population parameters by using a sample of managers, rather than all 2500 managers in the population? Suppose a sample of 30 managers will be used. Clearly, the time and the cost of developing a profile would be substantially less for 30 managers than for the entire population. If the head of personnel could be assured that a sample of 30 managers would provide adequate information about the population of 2500 managers, working with a sample would be preferable to working with the entire population. Often the cost of collecting information from a sample is substantially less than from a population, especially when personal interviews must be conducted to collect the information.

First we consider how we can identify a sample of 30 managers.

7.2 Simple random sampling

Several methods can be used to select a sample from a population. One of the most common is **simple random sampling**. The definition of a simple random sample and the process of selecting such a sample depend on whether the population is *finite* or *infinite*. We first consider sampling from a finite population, because the EAI sampling problem involves a finite population of 2500 managers.

Sampling from a finite population

A simple random sample of size n from a finite population of size N is defined as follows.

Simple random sample (finite population)

A simple random sample of size n from a finite population of size N is a sample selected such that each possible sample of size n has the same probability of being selected.

One procedure for selecting a simple random sample from a finite population is to choose the elements for the sample one at a time in such a way that, at each step, each of the elements remaining in the population has the same probability of being selected.

To select a simple random sample from the population of EAI managers, we first assign each manager a number. We can assign the managers the numbers 1 to 2500 in the order their names appear in the EAI personnel file. Next, we refer to the table of random numbers shown in Table 7.1. Using the first row of the table, each digit, 6, 3, 2, . . . , is a random digit with an equal chance of occurring. The random numbers in the table are shown in groups of five for readability. Because the largest number in the population list, 2500, has four digits, we shall select random numbers from the table in groups of four digits. We may start the selection of random numbers anywhere in the table and move systematically in a direction of our choice. We shall use the first row of Table 7.1 and move from left to right. The first seven four-digit random numbers are

6327   1599   8671   7445   1102   1514   1807

Table 7.1 Random numbers

63271	59986	71744	51102	15141	80714	58683	93108	13554	79945
88547	09896	95436	79115	08303	01041	20030	63754	08459	28364
55957	57243	83865	09911	19761	66535	40102	26646	60147	15702
46276	87453	44790	67122	45573	84358	21625	16999	13385	22782
55363	07449	34835	15290	76616	67191	12777	21861	68689	03263
69393	92785	49902	58447	42048	30378	87618	26933	40640	16281
13186	29431	88190	04588	38733	81290	89541	70290	40113	08243
17726	28652	56836	78351	47327	18518	92222	55201	27340	10493
36520	64465	05550	30157	82242	29520	69753	72602	23756	54935
81628	36100	39254	56835	37636	02421	98063	89641	64953	99337
84649	48968	75215	75498	49539	74240	03466	49292	36401	45525
63291	11618	12613	75055	43915	26488	41116	64531	56827	30825
70502	53225	03655	05915	37140	57051	48393	91322	25653	06543
06426	24771	59935	49801	11082	66762	94477	02494	88215	27191
20711	55609	29430	70165	45406	78484	31639	52009	18873	96927
41990	70538	77191	25860	55204	73417	83920	69468	74972	38712
72452	36618	76298	26678	89334	33938	95567	29380	75906	91807
37042	40318	57099	10528	09925	89773	41335	96244	29002	46453
53766	52875	15987	46962	67342	77592	57651	95508	80033	69828
90585	58955	53122	16025	84299	53310	67380	84249	25348	04332
32001	96293	37203	64516	51530	37069	40261	61374	05815	06714
62606	64324	46354	72157	67248	20135	49804	09226	64419	29457
10078	28073	85389	50324	14500	15562	64165	06125	71353	77669
91561	46145	24177	15294	10061	98124	75732	00815	83452	97355
13091	98112	53959	79607	52244	63303	10413	63839	74762	50289

These four-digit numbers are equally likely, because the numbers in the table are random. We use them to give each manager in the population an equal chance of being included in the random sample.

The first number, 6327, is greater than 2500. We discard it because it does not correspond to one of the numbered managers in the population. The second number, 1599, is between 1 and 2500. So the first manager selected for the random sample is number 1599 on the list of EAI managers. Continuing this process, we ignore the numbers 8671 and 7445 (greater than 2500) before identifying managers numbered 1102, 1514 and 1807 to be included in the random sample. This process continues until the simple random sample of 30 EAI managers has been obtained.

It is possible that a random number already used may appear again in the table before the sample of 30 EAI managers has been fully selected. Because we do not want to select a manager more than once, any previously used random numbers are ignored. Selecting a sample in this manner is referred to as **sampling without replacement**. If we selected a sample such that previously used random numbers are acceptable, and specific managers could be included in the sample two or more times, we would be **sampling with replacement**. Sampling with replacement is a valid way of identifying a simple random sample, but sampling without replacement is used more often. When we refer to simple random sampling, we shall assume that the sampling is without replacement.

Computer-generated random numbers can also be used to implement the random sample selection process. EXCEL, MINITAB and PASW all provide functions for generating random numbers.

The number of different simple random samples of size n that can be selected from a finite population of size N is

$$\frac{N!}{n!(N-n)!}$$

In this formula, $N!$ and $n!$ are the factorial computations discussed in Chapter 4. For the EAI problem with $N = 2500$ and $n = 30$, this expression can be used to show that approximately 2.75×10^{69} different simple random samples of 30 EAI managers can be obtained.

Sampling from an infinite population

In some situations, the population is either infinite, or so large that for practical purposes it must be treated as infinite. For example, suppose that a fast-food restaurant would like to obtain a profile of its customers by selecting a simple random sample of customers and asking each customer to complete a short questionnaire. The ongoing process of customer visits to the restaurant can be viewed as coming from an infinite population. In practice, a population being studied is usually considered infinite if it involves an ongoing process that makes listing or counting every element in the population impossible. The definition of a simple random sample from an infinite population follows.

Simple random sample (infinite population)

A simple random sample from an infinite population is a sample selected such that the following conditions are satisfied.

1　Each element selected comes from the population.
2　Each element is selected independently.

For the example of a simple random sample of customers at a fast-food restaurant, any customer who comes into the restaurant will satisfy the first requirement. The second requirement will be satisfied if a sample selection procedure is devised to select the items independently and thereby avoid any selection bias that gives higher selection probabilities to certain types of customers. Selection bias would occur if, for instance, five consecutive customers selected were all friends who arrived together. We might expect these customers to exhibit similar profiles. Selection bias can be avoided by ensuring that the selection of a particular customer does not influence the selection of any other customer. In other words, the customers must be selected independently.

McDonald's, the well-known fast-food chain, implemented a simple random sampling procedure for just such a situation. The sampling procedure was based on the fact that some customers presented discount coupons. Whenever a customer presented a discount coupon, the next customer served was asked to complete a customer profile questionnaire. Because arriving customers presented discount coupons randomly, and independently, this sampling plan ensured that customers were selected independently. So the two requirements for a simple random sample from an infinite population were satisfied.

Infinite populations are often associated with an ongoing process that operates continuously over time. For example, parts being manufactured on a production line, transactions occurring at a bank, telephone calls arriving at a technical support centre, and customers entering stores may all be viewed as coming from an infinite population. In such cases, an effective sampling procedure will ensure that no selection bias occurs and that the sample elements are selected independently.

Exercises

Methods

1   Consider a finite population with five elements labeled A, B, C, D and E. Ten possible simple random samples of size 2 can be selected.

   a.   List the ten samples beginning with AB, AC and so on.
   b.   Using simple random sampling, what is the probability that each sample of size 2 is selected?
   c.   Assume random number 1 corresponds to A, random number 2 corresponds to B, and so on. List the simple random sample of size 2 that will be selected by using the random digits 8 0 5 7 5 3 2.

2   Assume a finite population has 350 elements. Using the last three digits of each of the following five-digit random numbers (601, 022, 448, . . .), determine the first four elements that will be selected for the simple random sample.

   98601   73022   83448   02147   34229   27553   84147   93289   14209

Applications

3   The Nikkei 225 share index is calculated using data for 225 of the most actively traded stocks on the Tokyo stock exchange. Assume that you want to select a simple random sample of five companies from the Nikkei list. Use the last three digits in column 9 of Table 7.1, beginning with 554. Read down the column and identify the numbers of the five companies that would be selected.

4   A student union is interested in estimating the proportion of students who favour a mandatory 'pass–fail' grading policy for optional courses. A list of names and addresses of the 645 students enrolled during the current semester is available from the registrar's office. Using three-digit random numbers in row 10 of Table 7.1 and moving across the row from left to right, identify the first ten students who would be selected using simple random sampling. The three-digit random numbers begin with 816, 283 and 610.

5   Assume that we want to identify a simple random sample of 12 of the 372 doctors practising in a particular city. The doctors' names are available from the local health authority. Use the eighth column of five-digit random numbers in Table 7.1 to identify the 12 doctors for the sample. Ignore the first two random digits in each five-digit grouping of the random numbers. This process begins with random number 108 and proceeds down the column of random numbers.

6   Indicate whether the following populations should be considered finite or infinite.

   a. All registered voters in Ireland.
   b. All television sets that could be produced by the Johannesburg factory of the TV-M Company.
   c. All orders that could be processed by a mail-order firm.
   d. All emergency telephone calls that could come into a local police station.
   e. All components that Fibercon plc produced on the second shift on 17 May 2009.

7.3 Point estimation

Let us return to the EAI problem. A simple random sample of 30 managers and the corresponding data on annual salary and management training programme participation are shown in Table 7.2. The notation x_1, x_2 and so on is used to denote the annual salary of the first manager in the sample, the annual salary of the second manager in the sample and so on. Participation in the management training programme is indicated by Yes in the management training programme column.

To estimate the value of a population parameter, we compute a corresponding characteristic of the sample, referred to as a **sample statistic**. For example, to estimate the population mean μ and the population standard deviation σ for the annual salary of EAI managers, we use the data in Table 7.2 to calculate the corresponding sample statistics: the sample mean and the sample standard deviation. Using the formulae presented in Chapter 3, the sample mean is

$$\bar{x} = \frac{\Sigma x_i}{n} = \frac{1\ 554\ 420}{30} = 51\ 814\ (\text{\euro})$$

and the sample standard deviation is

$$s = \sqrt{\frac{\Sigma (x_i - \bar{x})^2}{n} - 1} = \sqrt{\frac{325\ 009\ 260}{29}} = 3348\ (\text{\euro})$$

Annual salary (€)	Management training programme	Annual salary (€)	Management training programme
$x_1 = 49\,094.30$	Yes	$x_{16} = 51\,766.00$	Yes
$x_2 = 53\,263.90$	Yes	$x_{17} = 52\,541.30$	No
$x_3 = 49\,643.50$	Yes	$x_{18} = 44\,980.00$	Yes
$x_4 = 49\,894.90$	Yes	$x_{19} = 51\,932.60$	Yes
$x_5 = 47\,621.60$	No	$x_{20} = 52\,973.00$	Yes
$x_6 = 55\,924.00$	Yes	$x_{21} = 45\,120.90$	Yes
$x_7 = 49\,092.30$	Yes	$x_{22} = 51\,753.00$	Yes
$x_8 = 51\,404.40$	Yes	$x_{23} = 54\,391.80$	No
$x_9 = 50\,957.70$	Yes	$x_{24} = 50\,164.20$	No
$x_{10} = 55\,109.70$	Yes	$x_{25} = 52\,973.60$	No
$x_{11} = 45\,922.60$	Yes	$x_{26} = 50\,241.30$	No
$x_{12} = 57\,268.40$	No	$x_{27} = 52\,793.90$	No
$x_{13} = 55\,688.80$	Yes	$x_{28} = 50\,979.40$	Yes
$x_{14} = 51\,564.70$	No	$x_{29} = 55\,860.90$	Yes
$x_{15} = 56\,188.20$	No	$x_{30} = 57\,309.10$	No

Table 7.2 Annual salary and training programme status for a simple random sample of 30 EAI managers

To estimate π, the proportion of managers in the population who completed the management training programme, we use the corresponding sample proportion. Let m denote the number of managers in the sample who completed the management training programme. The data in Table 7.2 show that $m = 19$. So, with a sample size of $n = 30$, the sample proportion is

$$p = \frac{m}{n} = \frac{19}{30} = 0.63$$

These computations are an example of the statistical procedure called *point estimation*. We refer to the sample mean as the **point estimator** of the population mean μ, the sample standard deviation as the point estimator of the population standard deviation σ, and the sample proportion as the point estimator of the population proportion π. The numerical value obtained for the sample mean, sample standard deviation or sample proportion is called a **point estimate**. For the simple random sample of 30 EAI managers shown in Table 7.2, €51 814 is the point estimate of μ, €3348 is the point estimate of σ and 0.63 is the point estimate of π. Table 7.3 summarizes the sample results and compares the point estimates to the actual values of the population parameters.

The point estimates in Table 7.3 differ somewhat from the corresponding population parameters. This difference is to be expected because a sample, rather than a census of the entire population, is being used to obtain the point estimates. In the next chapter, we shall show how to construct an interval estimate in order to provide information about how close the point estimate is to the population parameter.

Table 7.3 Summary of point estimates obtained from a simple random sample of 30 EAI managers

Population parameter	Parameter value	Point estimator	Point estimate
Population mean annual salary	$\mu = €51\ 800$	Sample mean annual salary	$\bar{x} = €51\ 814$
Population standard deviation for annual salary	$\sigma = €4\ 000$	Sample standard deviation for annual salary	$s = €3\ 348$
Population proportion who have completed the management training programme	$\pi = 0.60$	Sample proportion who have completed the management training programme	$p = 0.63$

Exercises

Methods

7 The following data are from a simple random sample.

$$5 \qquad 8 \qquad 10 \qquad 7 \qquad 10 \qquad 14$$

a. Calculate a point estimate of the population mean.
b. Calculate a point estimate of the population standard deviation.

8 A survey question for a sample of 150 individuals yielded 75 Yes responses, 55 No responses, and 20 No Opinion responses.

a. Calculate a point estimate of the proportion in the population who respond Yes.
b. Calculate a point estimate of the proportion in the population who respond No.

Applications

9 A simple random sample of five months of sales data provided the following information:

Month:	1	2	3	4	5
Units sold:	94	100	85	94	92

a. Calculate a point estimate of the population mean number of units sold per month.
b. Calculate a point estimate of the population standard deviation.

10 The data set Mutual Fund contains data on a sample of 40 mutual funds. These were randomly selected from 283 funds featured in *Business Week*. Use the data set to answer the following questions.

a. Compute a point estimate of the proportion of the *Business Week* mutual funds that are load funds.
b. Compute a point estimate of the proportion of the funds that are classified as high risk.
c. Compute a point estimate of the proportion of the funds that have a below-average risk rating.

11 In an ICM poll for the *Guardian* newspaper in October 2008, during the turbulence in the world's financial markets, respondents were asked to what extent they felt they and their families would be affected financially. The opinions of the 1007 adult respondents were:

98	Suffer a great deal
320	Suffer quite a lot

426	Suffer a little
132	Not suffer at all
31	Don't know

Calculate point estimates of the following population parameters.

a. The proportion of all adults who feel they would suffer a little.
b. The proportion of all adults who feel they would not suffer at all.
c. The proportion of all adults who feel they would suffer quite a lot or a great deal.

12 Many drugs used to treat cancer are expensive. *BusinessWeek* reported on the cost per treatment of Herceptin, a drug used to treat breast cancer. Typical treatment costs (in dollars) for Herceptin are provided by a simple random sample of 10 patients.

4376	5578	2717	4920	4495
4798	6446	4119	4237	3814

a. Calculate a point estimate of the mean cost per treatment with Herceptin.
b. Calculate a point estimate of the standard deviation of the cost per treatment with Herceptin.

7.4 Introduction to sampling distributions

For the simple random sample of 30 EAI managers shown in Table 7.2, the point estimate of μ is $\bar{x} = $€51 814 and the point estimate of π is $p = 0.63$. Suppose we select another simple random sample of 30 EAI managers and obtain the following point estimates:

$$\text{Sample mean: } \bar{x} = \text{€}52\ 670$$
$$\text{Sample proportion: } p = 0.70$$

Note that different values of the sample mean and sample proportion were obtained. A second simple random sample of 30 EAI managers cannot be expected to provide exactly the same point estimates as the first sample.

Now, suppose we repeat the process of selecting a simple random sample of 30 EAI managers over and over again, each time computing the values of the sample mean and sample proportion. Table 7.4 contains a portion of the results obtained for 500 simple random samples, and Table 7.5 shows the frequency and relative frequency distributions for the 500 values. Figure 7.1 shows the relative frequency histogram for the values.

In Chapter 5 we defined a random variable as a numerical description of the outcome of an experiment. If we consider selecting a simple random sample as an experiment, the sample mean is a numerical description of the outcome of the experiment. So, the sample mean is a random variable. In accordance with the naming conventions for random variables described in Chapters 5 and 6 (i.e. use of capital letters for names of random variables), we denote this random variable $\bar{X}$. Just like other random variables, $\bar{X}$ has a mean or expected value, a standard deviation, and a probability distribution. Because the various possible values of $\bar{X}$ are the result of different simple random samples, the probability distribution of $\bar{X}$ is called the **sampling distribution** of $\bar{X}$. Knowledge of this sampling distribution will enable us to make probability statements about how close the sample mean is to the population mean μ.

Let us return to Figure 7.1. We would need to enumerate every possible sample of 30 managers and compute each sample mean to completely determine the sampling

Table 7.4 Values $\bar{x}$ and p from 500 simple random samples of 30 EAI managers

Sample number	Sample mean ($\bar{x}$)	Sample proportion (p)
1	51 814	0.63
2	52 670	0.70
3	51 780	0.67
4	51 588	0.53
.	.	.
.	.	.
.	.	.
500	51 752	0.50

distribution of $\overline{X}$. However, the histogram of 500 $\bar{x}$ values gives an approximation of this sampling distribution. From the approximation we observe the bell-shaped appearance of the distribution. We note that the largest concentration of the $\bar{x}$ values and the mean of the 500 $\bar{x}$ values is near the population mean $\mu = €51\ 800$. We shall describe the properties of the sampling distribution of $\overline{X}$ more fully in the next section.

The 500 values of the sample proportion are summarized by the relative frequency histogram in Figure 7.2. As in the case of the sample mean, the sample proportion is a random variable, which we denote P. If every possible sample of size 30 were selected from the population and if a value p were computed for each sample, the resulting probability distribution would be the sampling distribution of P. The relative frequency histogram of the 500 sample values in Figure 7.2 provides a general idea of the appearance of the sampling distribution of P.

In practice, we select only one simple random sample from the population for estimating population characteristics. We repeated the sampling process 500 times in this section simply to illustrate that many different samples are possible and that the different samples generate a variety of values $\bar{x}$ and p for the sample statistics $\overline{X}$ and P. The probability distribution of any particular sample statistic is called the sampling distribution of the statistic. In Section 7.5 we show the characteristics of the sampling distribution of $\overline{X}$. In Section 7.6 we show the characteristics of the sampling distribution of P. The ability to understand the material in subsequent chapters depends heavily on the ability to understand and use the sampling distributions presented in this chapter.

Table 7.5 Frequency distribution of $\overline{X}$ values from 500 simple random samples of 30 EAI managers

Mean annual salary (€)	Frequency	Relative frequency
49 500.00–49 999.99	2	0.004
50 000.00–50 499.99	16	0.032
50 500.00–50 999.99	52	0.104
51 000.00–51 499.99	101	0.202
51 500.00–51 999.99	133	0.266
52 000.00–52 499.99	110	0.220
52 500.00–52 999.99	54	0.108
53 000.00–53 499.99	26	0.052
53 500.00–53 999.99	6	0.012
Totals	500	1.000

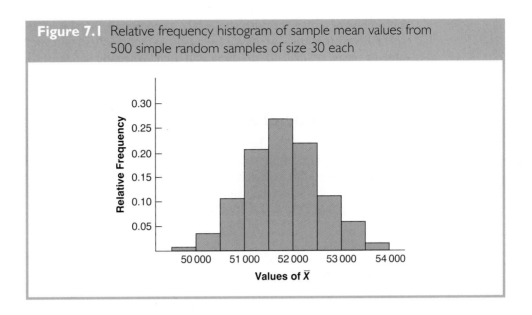

Figure 7.1 Relative frequency histogram of sample mean values from 500 simple random samples of size 30 each

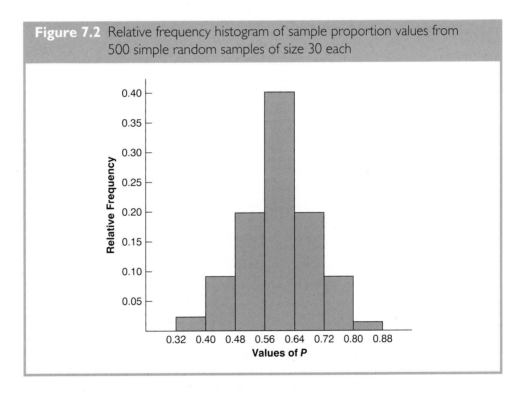

Figure 7.2 Relative frequency histogram of sample proportion values from 500 simple random samples of size 30 each

7.5 Sampling distribution of $\overline{X}$

Sampling distribution of $\overline{X}$

The sampling distribution of $\overline{X}$ is the probability distribution of all possible values of the sample mean.

This section describes the properties of the sampling distribution of $\overline{X}$. Just as with other probability distributions we have studied, the sampling distribution of $\overline{X}$ has an expected value or mean, a standard deviation and a characteristic shape or form. We begin by considering the expected value of $\overline{X}$.

Expected value of $\overline{X}$

We are often interested in the mean of all possible values of $\overline{X}$ that can be generated by the various possible simple random samples. This is known as the expected value of $\overline{X}$. Let $E(\overline{X})$ represent the expected value of $\overline{X}$, and μ represent the mean of the population from which we are selecting a simple random sample. It can be shown that with simple random sampling, $E(\overline{X})$ and μ are equal.

Expected value of $\overline{X}$

$$E(\overline{X}) = \mu \tag{7.1}$$

where

$E(\overline{X})$ = the expected value of $\overline{X}$

μ = the mean of the population from which the sample is selected

In Section 7.1 we saw that the mean annual salary for the population of EAI managers is $\mu = 51\ 800$. So according to equation (7.1), the mean of all possible sample means for the EAI study is also €51 800.

When the expected value of a point estimator equals the population parameter, we say the point estimator is an **unbiased** estimator of the population parameter.

Unbiasedness

The sample statistic Q is an unbiased estimator of the population parameter θ if

$$E(Q) = \theta$$

where $E(Q)$ is the expected value of the sample statistic Q.

Figure 7.3 shows the cases of unbiased and biased point estimators. In the illustration showing the unbiased estimator, the mean of the sampling distribution is equal to the value of the population parameter. The estimation errors balance out in this case, because sometimes the value of the point estimator may be less than θ and other times it may be greater than θ.

In the case of a biased estimator, the mean of the sampling distribution is less than or greater than the value of the population parameter. In the illustration in Panel B of Figure 7.3, $E(Q)$ is greater than θ; the sample statistic has a high probability of overestimating the value of the population parameter. The amount of the bias is shown in the figure.

Equation (7.1) shows that $\overline{X}$ is an unbiased estimator of the population mean μ.

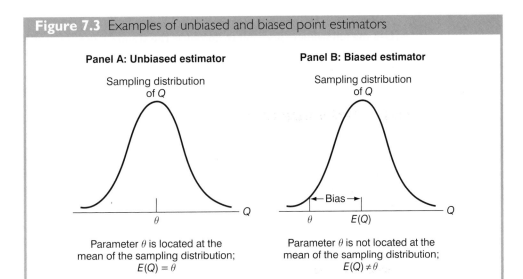

Figure 7.3 Examples of unbiased and biased point estimators

Panel A: Unbiased estimator

Sampling distribution of Q

Parameter θ is located at the mean of the sampling distribution; $E(Q) = \theta$

Panel B: Biased estimator

Sampling distribution of Q

Parameter θ is not located at the mean of the sampling distribution; $E(Q) \neq \theta$

Standard deviation of $\overline{X}$

It can be shown that with simple random sampling, the standard deviation of $\overline{X}$ depends on whether the population is finite or infinite. We use the following notation.

$\sigma_{\overline{x}}$ = the standard deviation of $\overline{X}$
σ = the standard deviation of the population
n = the sample size
N = the population size

Standard deviation of $\overline{X}$

Finite Population	Infinite Population	
$\sigma_{\overline{x}} = \sqrt{\dfrac{N-n}{N-1}}\left(\dfrac{\sigma}{\sqrt{n}}\right)$	$\sigma_{\overline{x}} = \dfrac{\sigma}{\sqrt{n}}$	(7.2)

In comparing the two formulae in (7.2), we see that the factor

$$\sqrt{(N-n)/(N-1)}$$

is required for the finite population case but not for the infinite population case. This factor is commonly referred to as the **finite population correction factor**. In many practical sampling situations, we find that the population involved, although finite, is 'large', whereas the sample size is relatively 'small'. In such cases the finite population correction factor is close to 1. As a result, the difference between the values of the standard deviation of $\overline{X}$ for the finite and infinite population cases becomes negligible. Then,

$$\sigma_{\overline{x}} = \sigma/\sqrt{n}$$

becomes a good approximation to the standard deviation of $\overline{X}$ even though the population is finite. This observation leads to the following general guideline, or rule of thumb, for computing the standard deviation of $\overline{X}$.

Use the following expression to compute the standard deviation of $\overline{X}$

$$\sigma_{\overline{X}} = \frac{\sigma}{\sqrt{n}} \qquad\qquad (7.3)$$

whenever

1 The population is infinite; or
2 The population is finite *and* the sample size is less than or equal to 5 per cent of the population size; that is, $n/N \leq 0.05$.

In cases where $n/N > 0.05$, the finite population version of formula (7.2) should be used in the computation of $\sigma_{\overline{X}}$. Unless otherwise noted, throughout the text we shall assume that the population size is 'large', $n/N \leq 0.05$, and expression (7.3) can be used to compute $\sigma_{\overline{X}}$.

To compute $\sigma_{\overline{X}}$, we need to know σ, the standard deviation of the population. To further emphasize the difference between $\sigma_{\overline{X}}$ and σ, we refer to $\sigma_{\overline{X}}$ as the **standard error** of the mean. The term standard error is used throughout statistical inference to refer to the standard deviation of a point estimator. Later we shall see that the value of the standard error of the mean is helpful in determining how far the sample mean may be from the population mean.

We return to the EAI example and compute the standard error of the mean associated with simple random samples of 30 EAI managers. In Section 7.1 we saw that the standard deviation of annual salary for the population of 2500 EAI managers is $\sigma = 4000$. In this case, the population is finite, with $N = 2500$. However, with a sample size of 30, we have $n/N = 30/2500 = 0.012$. Because the sample size is less than 5 per cent of the population size, we can ignore the finite population correction factor and use equation (7.3) to compute the standard error.

$$\sigma_{\overline{X}} = \frac{\sigma}{\sqrt{n}} = \frac{4000}{\sqrt{30}} = 730.3$$

Form of the sampling distribution of $\overline{X}$

The preceding results concerning the expected value and standard deviation for the sampling distribution of $\overline{X}$ are applicable for any population. The final step in identifying the characteristics of the sampling distribution of $\overline{X}$ is to determine the form or shape of the sampling distribution. We shall consider two cases: (1) the population has a normal distribution; and (2) the population does not have a normal distribution.

Population has a normal distribution

In many situations it is reasonable to assume that the population from which we are selecting a simple random sample has a normal, or nearly normal, distribution. When the population has a normal distribution, the sampling distribution of $\overline{X}$ is normally distributed for any sample size.

Population does not have a normal distribution

When the population from which we are selecting a simple random sample does not have a normal distribution, the **central limit theorem** is helpful in identifying the shape of the sampling distribution of $\overline{X}$.

Figure 7.4 Illustration of the central limit theorem for three populations

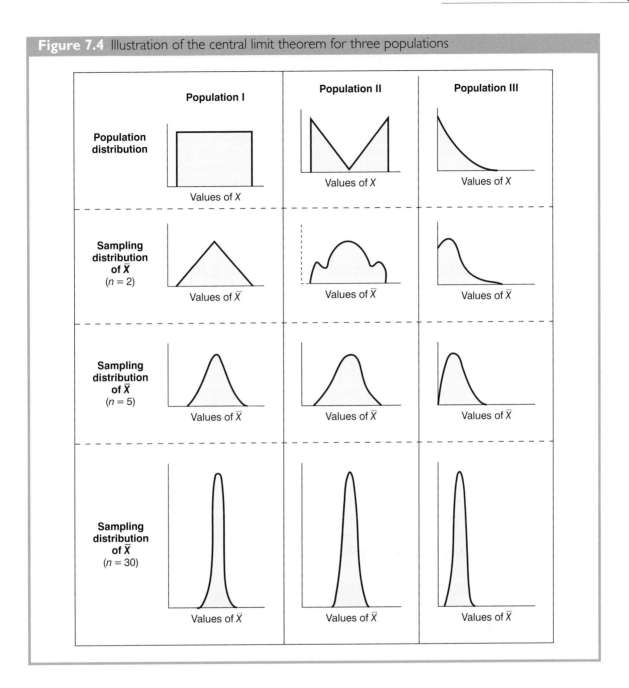

Central limit theorem

In selecting simple random samples of size *n* from a population, the sampling distribution of the sample mean $\overline{X}$ can be approximated by a *normal distribution* as the sample size becomes large.

Figure 7.4 shows how the central limit theorem works for three different populations. Each column refers to one of the populations. The top panel of the figure shows that none of the populations is normally distributed. When the samples are of size 2, we see that the sampling distribution begins to take on an appearance different from that of the population distribution. For samples of size 5, we see all three sampling distributions beginning to take on a bell-shaped appearance. Finally, the samples of size 30 show all three sampling distributions to be approximately normally distributed. For sufficiently large samples, the sampling distribution of $\overline{X}$ can be approximated by a normal distribution. How large must the sample size be before we can assume that the central limit theorem applies? Studies of the sampling distribution of $\overline{X}$ for a variety of populations and a variety of sample sizes have indicated that, for most applications, the sampling distribution of $\overline{X}$ can be approximated by a normal distribution whenever the sample size is 30 or more.

The theoretical proof of the central limit theorem requires independent observations in the sample. This condition is met for infinite populations and for finite populations where sampling is done with replacement. Although the central limit theorem does not directly address sampling without replacement from finite populations, general statistical practice applies the findings of the central limit theorem when the population size is large.

Sampling distribution of $\overline{X}$ for the EAI problem

For the EAI problem, we previously showed that $E(\overline{X}) = $ €51 800 and $\sigma_{\overline{x}} = $ €730.3. At this point, we do not have any information about the population distribution; it may or may not be normally distributed. If the population has a normal distribution, the sampling distribution of $\overline{X}$ is normally distributed. If the population does not have a normal distribution, the simple random sample of 30 managers and the central limit theorem enable us to conclude that the sampling distribution can be approximated by a normal distribution. In either case, we can proceed with the conclusion that the sampling distribution can be described by the normal distribution shown in Figure 7.5.

Practical value of the sampling distribution of $\overline{X}$

We are interested in the sampling distribution of $\overline{X}$ because it can be used to provide probability information about the difference between the sample mean and the population mean. Suppose the head of personnel services believes the sample mean will be an acceptable estimate of the population mean if the sample mean is within €500 of the population mean. It is not possible to guarantee that the sample mean will be within €500 of the population mean. Indeed, Table 7.5 and Figure 7.1 show that some of the 500 sample means differed by more

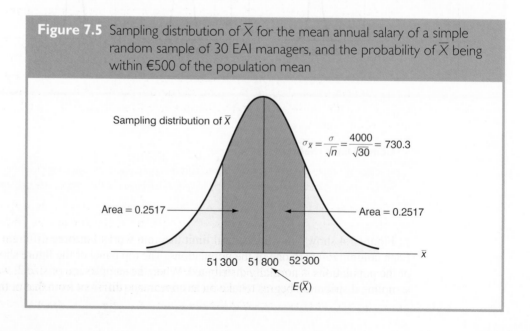

Figure 7.5 Sampling distribution of $\overline{X}$ for the mean annual salary of a simple random sample of 30 EAI managers, and the probability of $\overline{X}$ being within €500 of the population mean

Sampling distribution of $\overline{X}$

$$\sigma_{\overline{x}} = \frac{\sigma}{\sqrt{n}} = \frac{4000}{\sqrt{30}} = 730.3$$

Area = 0.2517

Area = 0.2517

51 300   51 800   52 300

$E(\overline{X})$

than €2000 from the population mean. So we must think of the head of personnel's request in probability terms. What is the probability that the sample mean computed using a simple random sample of 30 EAI managers will be within €500 of the population mean?

We can answer this question using the sampling distribution of $\overline{X}$. Refer to Figure 7.5. With a population mean of €51 800, the personnel manager wants to know the probability that $\overline{X}$ is between €51 300 and €52 300. The darkly shaded area of the sampling distribution shown in Figure 7.5 gives this probability. Because the sampling distribution is normally distributed, with mean 51 800 and standard error of the mean 730.3, we can use the table of areas for the standard normal distribution to find the area or probability. At $\overline{X} = 51\ 300$, we have

$$z = \frac{51\ 300 - 51\ 800}{730.3} = -0.68$$

Referring to the standard normal distribution table, we find the cumulative probability for $z = -0.68$ is 0.2483. Similar calculations for $\overline{X} = 52\ 300$ show a cumulative probability for $z = +0.68$ of 0.7517. So the probability of the sample mean is between 51 300 and 52 300 is $0.7517 - 0.2483 = 0.5034$.

These computations show that a simple random sample of 30 EAI managers has a 0.5034 probability of providing a sample mean that is within €500 of the population mean. Hence, there is a $1 - 0.5034 = 0.4966$ probability that the difference between $\overline{X}$ and μ will be more than €500. In other words, a simple random sample of 30 EAI managers has roughly a 50/50 chance of providing a sample mean within the allowable €500. Perhaps a larger sample size should be considered. We explore this possibility by considering the relationship between the sample size and the sampling distribution of $\overline{X}$.

Relationship between sample size and the sampling distribution of $\overline{X}$

Suppose that in the EAI sampling problem we select a simple random sample of 100 EAI managers instead of the 30 originally considered. Intuitively, it would seem that, with more data provided by the larger sample size, the sample mean based on $n = 100$ should provide a better estimate of the population mean than the sample mean based on $n = 30$. To see how much better, let us consider the relationship between the sample size and the sampling distribution of $\overline{X}$.

First note that $E(\overline{X}) = \mu$, i.e. $\overline{X}$ is an unbiased estimator of μ, regardless of the sample size n. However, the standard error of the mean, $\sigma_{\overline{X}}$, is related to the square root of the sample size. When the sample size increases, the standard error of the mean $\sigma_{\overline{X}}$ decreases. With $n = 30$, the standard error of the mean for the EAI problem is 730.3. With the increase in the sample size to $n = 100$, the standard error of the mean is decreased to

$$\sigma_{\overline{X}} = \frac{\sigma}{\sqrt{n}} = \frac{4000}{\sqrt{100}} = 400$$

The sampling distributions of $\overline{X}$ with $n = 30$ and $n = 100$ are shown in Figure 7.6. Because the sampling distribution with $n = 100$ has a smaller standard error, the values of $\overline{X}$ have less variation and tend to be closer to the population mean than the values of $\overline{X}$ with $n = 30$.

We can use the sampling distribution of $\overline{X}$ for $n = 100$ to compute the probability that a simple random sample of 100 EAI managers will provide a sample mean within €500 of the population mean. Because the sampling distribution is normal, with mean 51 800

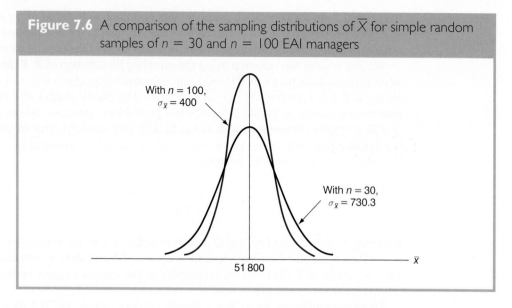

Figure 7.6 A comparison of the sampling distributions of $\overline{X}$ for simple random samples of $n = 30$ and $n = 100$ EAI managers

and standard error of the mean 400, we can use the standard normal distribution table to find the area or probability. At $\overline{X} = 51\,300$ (Figure 7.7), we have

$$z = \frac{51300 - 51800}{400} = -1.25$$

Referring to the standard normal probability distribution table, we find a cumulative probability for $z = -1.25$ of 0.1056. With a similar calculation for $\overline{X} = 52\,300$, we see that the probability of the sample mean being between 51 300 and 52 300 is $0.8944 - 0.1056 = 0.7888$. By increasing the sample size from 30 to 100 EAI managers, we have increased the probability of obtaining a sample mean within €500 of the population mean from 0.5034 to 0.7888.

The important point in this discussion is that as the sample size is increased, the standard error of the mean decreases. As a result, the larger sample size provides a higher probability that the sample mean is within a specified distance of the population mean.

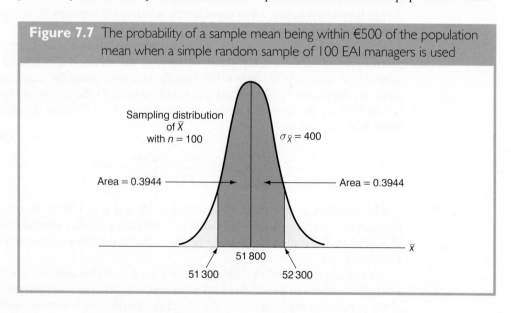

Figure 7.7 The probability of a sample mean being within €500 of the population mean when a simple random sample of 100 EAI managers is used

In presenting the sampling distribution of $\overline{X}$ for the EAI problem, we have taken advantage of the fact that the population mean $\mu = 51\ 800$ and the population standard deviation $\sigma = 4000$ were known. However, usually the values μ and σ that are needed to determine the sampling distribution of $\overline{X}$ will be unknown. In Chapter 8 we shall show how the sample mean $\overline{X}$ and the sample standard deviation S are used when μ and σ are unknown.

Exercises

Methods

13 A population has a mean of 200 and a standard deviation of 50. A simple random sample of size 100 will be taken and the sample mean will be used to estimate the population mean.

    a. What is the expected value of $\overline{X}$?
    b. What is the standard deviation of $\overline{X}$?
    c. Sketch the sampling distribution of $\overline{X}$.
    d. What does the sampling distribution of $\overline{X}$ show?

14 A population has a mean of 200 and a standard deviation of 50. Suppose a simple random sample of size 100 is selected and is used to estimate μ.

    a. What is the probability that the sample mean will be within ± 5 of the population mean?
    b. What is the probability that the sample mean will be within ± 10 of the population mean?

15 Assume the population standard deviation is $\sigma = 25$. Compute the standard error of the mean, $\sigma_{\overline{X}}$, for sample sizes of 50, 100, 150 and 200. What can you say about the size of the standard error of the mean as the sample size is increased?

16 Suppose a simple random sample of size 50 is selected from a population with $\sigma = 25$. Find the value of the standard error of the mean in each of the following cases (use the finite population correction factor if appropriate).

    a. The population size is infinite.
    b. The population size is $N = 50\ 000$.
    c. The population size is $N = 5000$.
    d. The population size is $N = 500$.

Applications

17 Refer to the EAI sampling problem. Suppose a simple random sample of 60 managers is used.

    a. Sketch the sampling distribution of $\overline{X}$ when simple random samples of size 60 are used.
    b. What happens to the sampling distribution of $\overline{X}$ if simple random samples of size 120 are used?
    c. What general statement can you make about what happens to the sampling distribution of $\overline{X}$ as the sample size is increased? Does this generalization seem logical? Explain.

18 In the EAI sampling problem (see Figure 7.5), we showed that for $n = 30$, there was a 0.5034 probability of obtaining a sample mean within $\pm €500$ of the population mean.

    a. What is the probability that $\overline{X}$ is within €500 of the population mean if a sample of size 60 is used?
    b. Answer part (a) for a sample of size 120.

19 The Automobile Association in the UK gave the average price of unleaded petrol as
    106.4 p per litre in November 2008. Assume this price is the population mean, and that the
    population standard deviation is $\sigma = 4.5$ p.

    a. What is the probability that the mean price for a sample of 30 petrol stations is within
        1.0 p of the population mean?
    b. What is the probability that the mean price for a sample of 50 petrol stations is within
        1.0 p of the population mean?
    c. What is the probability that the mean price for a sample 100 petrol stations is within
        1.0 p of the population mean?
    d. Would you recommend a sample size of 30, 50 or 100 to have at least a 0.95 probability
        that the sample mean is within 1.0 p of the population mean?

20 According to *Golf Digest*, the average score for male golfers is 95 and the average score for
    female golfers is 106. Use these values as population means. Assume that the population
    standard deviation is $\sigma = 14$ strokes for both men and women. A simple random sample of
    30 male golfers and another simple random sample of 45 female golfers are taken.

    a. Sketch the sampling distribution of $\overline{X}$ for male golfers.
    b. What is the probability that the sample mean is within 3 strokes of the population mean
        for the sample of male golfers?
    c. What is the probability that the sample mean is within 3 strokes of the population mean
        for the sample of female golfers?
    d. In which case is the probability higher (b or c)? Why?

21 A researcher reports survey results by stating that the standard error of the mean is 20. The
    population standard deviation is 500.

    a. How large was the sample?
    b. What is the probability that the point estimate was within ±25 of the population mean?

22 To estimate the mean age for a population of 4000 employees, a simple random sample of
    40 employees is selected.

    a. Would you use the finite population correction factor in calculating the standard error of
        the mean? Explain.
    b. If the population standard deviation is $\sigma = 8.2$, compute the standard error both with
        and without the finite population correction factor. What is the rationale for ignoring the
        finite population correction factor whenever $n/N \leq 0.05$?
    c. What is the probability that the sample mean age of the employees will be within ±2
        years of the population mean age?

7.6 Sampling distribution of *P*

The sample proportion P is a point estimator of the population proportion π. The formula
for computing the sample proportion is

$$p = \frac{m}{n}$$

where
 $m =$ the number of elements in the sample that possess the characteristic of interest
 $n =$ sample size.

As noted in Section 7.4, the sample proportion P is a random variable and its probability distribution is called the sampling distribution of P.

Sampling distribution of P

The sampling distribution of P is the probability distribution of all possible values of the sample proportion P.

To determine how close the sample proportion is to the population proportion π, we need to understand the properties of the sampling distribution of P: the expected value of P, the standard deviation of P, and the shape of the sampling distribution of P.

Expected value of P

The expected value of P, the mean of all possible values of P, is equal to the population proportion π. P is an unbiased estimator of π.

Expected value of P

$$E(P) = \pi \qquad (7.4)$$

where

$$E(P) = \text{the expected value of } P$$
$$\pi = \text{the population proportion}$$

In Section 7.1 we noted that $\pi = 0.60$ for the EAI population, where π is the proportion of the population of managers who participated in the company's management training programme. The expected value of P for the EAI sampling problem is therefore 0.60.

Standard deviation of P

Just as we found for the standard deviation of $\overline{X}$, the standard deviation of P depends on whether the population is finite or infinite.

Standard deviation of P

Finite Population	Infinite Population
$\sigma_P = \sqrt{\dfrac{N-n}{N-1}}\sqrt{\dfrac{\pi(1-\pi)}{n}}$	$\sigma_P = \sqrt{\dfrac{\pi(1-\pi)}{n}}$ (7.5)

Comparing the two formulae in (7.5), we see that the only difference is the use of the finite population correction factor

$$\sqrt{(N-n)/(N-1)}$$

As was the case with the sample mean, the difference between the expressions for the finite population and the infinite population becomes negligible if the size of the finite population is large in comparison to the sample size. We follow the same rule of thumb that we recommended for the sample mean. That is, if the population is finite with $n/N \leq 0.05$, we shall use

$$\sigma_P = \sqrt{\pi(1 - \pi)/n}.$$

However, if the population is finite with $n/N > 0.05$, the finite population correction factor should be used. Again, unless specifically noted, throughout the text we shall assume that the population size is large in relation to the sample size and so the finite population correction factor is unnecessary.

In Section 7.5 we used the term standard error of the mean to refer to the standard deviation of $\overline{X}$. We stated that in general the term standard error refers to the standard deviation of a point estimator. Accordingly, for proportions we use *standard error of the proportion* to refer to the standard deviation of P.

Let us now return to the EAI example and compute the standard error of the proportion associated with simple random samples of 30 EAI managers. For the EAI study we know that the population proportion of managers who participated in the management training programme is $\pi = 0.60$. With $n/N = 30/2500 = 0.012$, we can ignore the finite population correction factor when we compute the standard error of the proportion. For the simple random sample of 30 managers, σ_P is

$$\sigma_P = \sqrt{\frac{\pi(1 - \pi)}{n}} = \sqrt{\frac{0.60(1 - 0.60)}{30}} = 0.0894$$

Form of the sampling distribution of P

The sample proportion is $p = m/n$. For a simple random sample from a large population, the value of m is a binomial random variable indicating the number of elements in the sample with the characteristic of interest. Because n is a constant, the probability of m/n is the same as the binomial probability of m, which means that the sampling distribution of P is also a discrete probability distribution and that the probability for each value of m/n is the same as the probability of m.

In Chapter 6 we also showed that a binomial distribution can be approximated by a normal distribution whenever the sample size is large enough to satisfy the following two conditions:

$$n\pi \geq 5 \quad \text{and} \quad n(1 - \pi) \geq 5$$

Assuming these two conditions are satisfied, the probability of m in the sample proportion, $p = m/n$, can be approximated by a normal distribution. And because n is a constant, the sampling distribution of P can also be approximated by a normal distribution. This approximation is stated as follows:

> The sampling distribution of P can be approximated by a normal distribution whenever $n\pi \geq 5$ and $n(1 - \pi) \geq 5$.

In practical applications, when an estimate of a population proportion is needed, we find that sample sizes are almost always large enough to permit the use of a normal approximation for the sampling distribution of P.

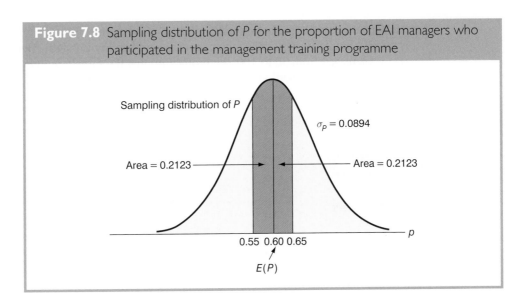

Figure 7.8 Sampling distribution of P for the proportion of EAI managers who participated in the management training programme

Recall that for the EAI sampling problem the population proportion of managers who participated in the training programme is $\pi = 0.60$. With a simple random sample of size 30, we have $n\pi = 30(0.60) = 18$ and $n(1 - \pi) = 30(0.40) = 12$. Consequently, the sampling distribution of P can be approximated by the normal distribution shown in Figure 7.8.

Practical value of the sampling distribution of P

The practical value of the sampling distribution of P is that it can be used to provide probability information about the difference between the sample proportion and the population proportion. For instance, suppose that in the EAI problem the head of personnel services wants to know the probability of obtaining a value of P that is within 0.05 of the population proportion of EAI managers who participated in the training programme. That is, what is the probability of obtaining a sample with a sample proportion P between 0.55 and 0.65? The darkly shaded area in Figure 7.8 shows this probability. Using the fact that the sampling distribution of P can be approximated by a normal distribution with a mean of 0.60 and a standard error of the proportion of $\sigma_p = 0.0894$, we find that the standard normal random variable corresponding to $p = 0.55$ has a value of $z = (0.55 - 0.60)/0.0894 = -0.56$. Referring to the standard normal distribution table, we see that the cumulative probability for $z = -0.56$ is 0.2877. Similarly, for $p = 0.56$ we find a cumulative probability of 0.7123. Hence, the probability of selecting a sample that provides a sample proportion P within 0.05 of the population proportion π is 0.7123 − 0.2877 = 0.4246.

If we consider increasing the sample size to $n = 100$, the standard error of the proportion becomes

$$\sigma_P == \sqrt{\frac{0.60(1 - 0.60)}{100}} = 0.049$$

The probability of the sample proportion being within 0.05 of the population proportion can now be calculated. We can again use the standard normal distribution table to find the area or probability. At $p = 0.55$, we have $z = (0.55 - 0.60)/0.049 = -1.02$. Referring to the standard normal distribution table, we see that the cumulative probability for $z = -1.02$ is 0.1539. Similarly, at $p = 0.65$ the cumulative probability is 0.8461. Hence, if the sample size is increased from 30 to 100, the probability that the sample proportion is within 0.05 of the population proportion π will increase to $0.8461 - 0.1539 = 0.6922$.

Exercises

Methods

23 A simple random sample of size 100 is selected from a population with $\pi = 0.40$.

   a. What is the expected value of P?
   b. What is the standard error of P?
   c. Sketch the sampling distribution of P.

24 Assume that the population proportion is 0.55. Compute the standard error of the sample proportion, σ_p, for sample sizes of 100, 200, 500 and 1000. What can you say about the size of the standard error of the proportion as the sample size is increased?

25 The population proportion is 0.30. What is the probability that a sample proportion will be within ±0.04 of the population proportion for each of the following sample sizes?

   a. $n = 100$
   b. $n = 200$
   c. $n = 500$
   d. $n = 1000$
   e. What is the advantage of a larger sample size?

Applications

26 The Chief Executive Officer of Dunkley Distributors plc believes that 30 per cent of the firm's orders come from first-time customers. A simple random sample of 100 orders will be used to estimate the proportion of first-time customers.

   a. Assume that the CEO is correct and $\pi = 0.30$. Describe the sampling distribution of the sample proportion P for this study?
   b. What is the probability that the sample proportion P will be between 0.20 and 0.40?
   c. What is the probability that the sample proportion P will be between 0.25 and 0.35?

27 The UK Office for National Statistics reported that in 2007, 61 per cent of households in the UK had Internet access. Use a population proportion $\pi = 0.61$ and assume that a sample of 300 households will be selected.

   a. Sketch the sampling distribution of P, the sample proportion of households that have Internet access.
   b. What is the probability that the sample proportion P will be within ±0.03 of the population proportion?
   c. Answer part (b) for sample sizes of 600 and 1000.

28 Advertisers contract with Internet service providers and search engines to place ads on websites. They pay a fee based on the number of potential customers who click on their ads. Unfortunately, click fraud – i.e. someone clicking on an ad solely for the purpose of driving up

advertising revenue – has become a problem. Forty per cent of advertisers claim they have been a victim of click fraud. Suppose a simple random sample of 380 advertisers is taken to learn about how they are affected by this practice.

a. What is the probability the sample proportion will be within ±0.04 of the population proportion experiencing click fraud?

b. What is the probability the sample proportion will be greater than 0.45?

29 In July 2005, the Pew Research Center released the results of a worldwide survey of attitudes towards Islamic extremism. In Pakistan, 52 per cent of respondents considered that Islamic extremism was a threat to their country. Assume the population proportion was $\pi = 0.52$, and that P is the sample proportion expressing approval in a simple random sample of 800 adults from the population.

a. Sketch the sampling distribution of P.

b. What is the probability that the sample proportion will be within ±0.02 of the population proportion?

c. Answer part (b) for a sample of 1600 adults.

30 A market research firm conducts telephone surveys with a 40 per cent historical response rate. What is the probability that in a new sample of 400 telephone numbers, at least 150 individuals will cooperate and respond to the questions? In other words, what is the probability that the sample proportion will be at least 150/400 = 0.375?

31 Laura Jeffrey is a successful sales representative for a major publisher of university textbooks. Historically, Laura secures a book adoption on 25 per cent of her sales calls. Assume that her sales calls for one month are taken as a sample of all possible sales calls, and that a statistical analysis of the data estimates the standard error of the sample proportion to be 0.0625.

a. How large was the sample used in this analysis? That is, how many sales calls did Laura make during the month?

b. Let P indicate the sample proportion of book adoptions obtained during the month. Sketch the sampling distribution P.

c. Using the sampling distribution of P, compute the probability that Laura will obtain book adoptions on 30 per cent or more of her sales calls during a one-month period.

Summary

In this chapter we presented the concepts of simple random sampling and sampling distributions.

Simple random sampling was defined for sampling without replacement and sampling with replacement. We demonstrated how a simple random sample can be selected and how the data collected for the sample can be used to develop point estimates of population parameters.

Point estimators such as $\overline{X}$ and P are random variables. The probability distribution of such a random variable is called a sampling distribution. In particular, we described the sampling distributions of the sample mean $\overline{X}$ and the sample proportion P. We stated that $E(\overline{X}) = \mu$ and $E(P) = \pi$, i.e. they are unbiased estimators of the respective parameters. After developing the standard deviation or standard error formulae for these estimators, we described the conditions necessary for the sampling distributions of $\overline{X}$ and P to follow a normal distribution.

Key terms

Central limit theorem

Finite population correction factor

Parameter

Point estimate

Point estimator

Sample statistic

Sampling distribution

Sampling with replacement

Sampling without replacement

Simple random sampling

Standard error

Unbiasedness

Key formulae

Expected value of $\overline{X}$

$$E(\overline{X}) = \mu \tag{7.1}$$

Standard deviation of $\overline{X}$ (standard error)

Finite Population

Infinite Population

$$\sigma_{\overline{X}} = \sqrt{\frac{N-n}{N-1}}\left(\frac{\sigma}{\sqrt{n}}\right) \qquad\qquad \sigma_{\overline{X}} = \frac{\sigma}{\sqrt{n}} \tag{7.2}$$

Expected value of P

$$E(P) = \pi \tag{7.4}$$

Standard deviation of P (standard error)

Finite Population

Infinite Population

$$\sigma_P = \sqrt{\frac{N-n}{N-1}}\sqrt{\frac{\pi(1-\pi)}{n}} \qquad\qquad \sigma_P = \sqrt{\frac{\pi(1-\pi)}{n}} \tag{7.5}$$

Software Section for Chapter 7

Random sampling using MINITAB

If a list of the elements in a population is available in a MINITAB worksheet, MINITAB can be used to select a simple random sample. For example, a list of the top 100 golfers in the official world rankings, as at July 2008, is given in the MINITAB file 'Golfers. MTW'. Column 1 contains the ranking, column 2 the name and country of the golfer, column 3 the golfer's points average, and column 4 the number of events over which the average has been calculated. The first five rows in the data set are shown in Table 7.6. Suppose that you would like to select a simple random sample of 20 golfers from the top 100. The following steps can be used to select the sample.

Step 1   **Calc > Random Data > Sample From Columns**   [Main menu bar]

Step 2   Enter **20** in the **Number of rows to sample** box
                                                                    [**Sample From Columns** panel]

      Enter **C1–C4** in the **From columns** box
      Enter **C5–C8** in the **Store samples in** box
      Click **OK**

The random sample of 20 golfers appears in columns C5–C8.

Table 7.6 Points averages for the world's top golfers, July 2008

		Average points	Events
1	Tiger Woods, USA	9.52	40
2	Phil Mickelson, USA	8.45	47
3	Sergio Garcia, Esp	6.97	50
4	Geoff Ogilvy, Aus	6.38	47
5	Kenny Perry, USA	5.66	52

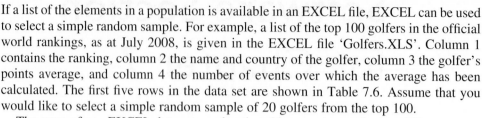

Random sampling using EXCEL

If a list of the elements in a population is available in an EXCEL file, EXCEL can be used to select a simple random sample. For example, a list of the top 100 golfers in the official world rankings, as at July 2008, is given in the EXCEL file 'Golfers.XLS'. Column 1 contains the ranking, column 2 the name and country of the golfer, column 3 the golfer's points average, and column 4 the number of events over which the average has been calculated. The first five rows in the data set are shown in Table 7.6. Assume that you would like to select a simple random sample of 20 golfers from the top 100.

The rows of any EXCEL data set can be placed in a random order by adding an extra column to the data set and filling the column with random numbers using the =**RAND()** function. Then using EXCEL's sorting capability on the random number column, the rows of the data set will be reordered randomly. The random sample of size n appears in the first n rows of the reordered data set. In the Golfers data set, labels are in row 1 and the 100 golfers are in rows 2 to 101. The following steps can be used to select a simple random sample of 20 golfers.

Step 1 Enter =**RAND( )** in cell E2

Step 2 Copy cell E2 to cells E3:E101

Step 3 Select any cell in Column E

Step 4 Click the **Home** tab on the Ribbon

Step 5 In the **Editing** group, click **Sort & Filter**

Step 6 Click **Sort Smallest to Largest**

The random sample of 20 golfers appears in rows 2 to 21 of the reordered data set. The random numbers in column E are no longer necessary and can be deleted.

Random sampling using PASW

If a list of the elements in a population is available in a PASW data file, PASW can be used to select a simple random sample. For example, a list of the top 100 golfers in the official world rankings, as at July 2008, is given in the PASW data file 'Golfers.SAV'. Column 1 contains the ranking, column 2 the name and country of the golfer, column 3 the golfer's points average, and column 4 the number of events over which the average has been calculated. The first five rows in the data set are shown in Table 7.6. Suppose that you would like to select a simple random sample of 20 golfers from the top 100. The following steps can be used to select the sample.

Step 1 **Data > Select Cases**   [Main menu bar]

Step 2 Select **Random sample of cases**   [**Select Cases** panel]
            Click on the **Sample** button

Step 3 Specify **Exactly 20 cases from the first 100 cases**

[**Select Cases:Random Sample** panel]

Click **Continue** to return to the **Select Cases** panel

Step 4 Select **Deleted** if you want to create a file   [**Select Cases** panel]
containing only the 20 sampled golfers
Click **OK**

If you opt to delete the non-selected cases, the 20 randomly selected cases can be saved in a new data file.

Chapter 8

Interval Estimation

Learning objectives

After reading this chapter and doing the exercises, you should be able to:

1 Explain the purpose of an interval estimate of a population parameter.

2 Explain the terms margin of error, confidence interval, confidence level and confidence coefficient.

3 Construct confidence intervals for a population mean:

   3.1 When the population standard deviation is known, using the normal distribution.

   3.2 When the population standard deviation is unknown, using the *t* distribution.

4 Construct large-sample confidence intervals for a population proportion.

5 Calculate the sample size required to construct a confidence interval with a given margin of error for a population mean, when the population standard deviation is known.

6 Calculate the sample size required to construct a confidence interval with a given margin of error for a population proportion.

In Chapter 7, we stated that a point estimator is a sample statistic used to estimate a population parameter. For example, the sample mean is a point estimator of the population mean, and the sample proportion is a point estimator of the population proportion. Because a point estimator cannot be expected to provide the exact value of the population parameter, an **interval estimate** is often computed, by adding and subtracting a **margin of error**. The purpose of an interval estimate is to provide information about how close the point estimate is to the value of the population parameter. The general form of an interval estimate is:

$$\text{Point estimate} \pm \text{Margin of error}$$

In this chapter we show how to compute interval estimates of a population mean μ and a population proportion π. The interval estimates have the same general form:

$$\text{Population mean: } \bar{x} \pm \text{Margin of error}$$
$$\text{Population proportion: } p \pm \text{Margin of error}$$

The sampling distributions of $\overline{X}$ and P play key roles in computing these interval estimates.

8.1 Population mean: σ known

To construct an interval estimate of a population mean, either the population standard deviation σ or the sample standard deviation s must be used to compute the margin of error. Although σ is rarely known exactly, historical data sometimes permit us to obtain a good estimate of the population standard deviation prior to sampling. In such cases, the population standard deviation can, for all practical purposes, be considered known.

Statistics in Practice

How accurate are opinion polls and market research surveys?

Synovate is the market research arm of Aegis Group plc, a company quoted on the London Stock Exchange. Synovate South Africa claims on its website to be the largest marketing research company in Southern Africa, and boasts over 110 blue chip clients. Among the recent research studies featured on the Synovate South Africa website* (late 2008) were three very different characters: a survey of motor vehicle quality in which more than 55 000 vehicle owners were interviewed, a global survey of 18–24 year olds involving around 12 000 young people, and a large survey looking at male beauty in 12 markets across the world.

The survey of 18–24-year-olds was done in 26 countries. The results for South Africa came from an online panel of 420 'techno-savvy' young South Africans. They revealed, for example, that 88 per cent of young South Africans owned a mobile phone, that 59 per cent used the social networking site Facebook, and that stereos still outnumbered iPods (69 per cent to 54 per cent).

Screenshot of Facebook social networking website. © NetPics/Alamy.

The global male beauty survey involved about 10 000 respondents (roughly half men, half women) spread across 213 countries. Among South African men, 78 per cent believed themselves to be sexy, whereas 66 per cent of French men thought themselves not sexy.

But how accurate are estimates like these based on sample evidence?

ICM is one of the largest market research and opinion polling organizations in the UK. The *Guardian* newspaper, Land Rover, and high street names like Laura Ashley, B&Q and Ladbrokes are among ICM's clients. ICM posts results of many of its polls and surveys on its website.** The issue of survey accuracy and margin of error also features on the ICM website. On one website page, there is an interactive 'ready-reckoner' that will calculate the margin of error for any given percentage result, like those above, and for any given sample size. For example, in respect of the 59 per cent of Young South Africans using Facebook, based on a sample size of 420, the ICM ready-reckoner calculates the 'accuracy at 95 per cent confidence level' to be plus or minus 4.7 percentage points. In other words, this implies that we can be 95 per cent confident that the percentage of all South African 18–24 year olds who were using Facebook in late 2008 was between 54.3 per cent and 63.7 per cent.

In this chapter, you will learn the basis for these margins of error, the confidence level of 95 per cent associated with them, and the calculations that underlie the ICM's ready-reckoner.

*www.synovate.com/southafrica
**www.icmresearch.co.uk

We refer to such cases as the *σ* **known** case. In this section we show how a simple random sample can be used to construct an interval estimate of a population mean for the *σ* known case.

Consider the monthly customer service survey conducted by CJW Company Limited, who provide a website for taking customer orders and providing follow-up service. The company prides itself on providing easy online ordering, timely delivery and prompt response to customer enquiries. Good customer service is critical to the company's ongoing success.

CJW's quality assurance team uses a customer service survey to measure satisfaction with its website and online customer service. Each month, the team sends a questionnaire to a random sample of customers who placed an order or requested service during the previous month. The questionnaire asks customers to rate their satisfaction with such things as ease of placing orders, timely delivery, accurate order filling and technical advice. The team summarizes each customer's questionnaire by computing an overall satisfaction score x that ranges from 0 (worst possible score) to 100 (best possible score). A sample mean customer satisfaction score is then computed.

The sample mean satisfaction score provides a point estimate of the mean satisfaction score μ for the population of all CJW customers. With this regular measure of customer service, CJW can promptly take corrective action if a low customer satisfaction score results. The company conducted this satisfaction survey for a number of months, and consistently obtained an estimate near 12 for the standard deviation of satisfaction scores. Based on these historical data, CJW now assumes a known value of $\sigma = 12$ for the population standard deviation. The historical data also indicate that the population of satisfaction scores follows an approximately normal distribution.

During the most recent month, the quality assurance team surveyed 100 customers ($n = 100$) and obtained a sample mean satisfaction score of $\bar{x} = 72$. This provides a point estimate of the population mean satisfaction score μ. We show how to compute the margin of error for this estimate and construct an interval estimate of the population mean.

Margin of error and the interval estimate

In Chapter 7 we showed that the sampling distribution of the sample mean $\bar{X}$ can be used to compute the probability that $\bar{X}$ will be within a given distance of μ. In the CJW example, the historical data show that the population of satisfaction scores is normally distributed with a standard deviation of $\sigma = 12$. So, using what we learned in Chapter 7, we can conclude that the sampling distribution of $\bar{X}$ follows a normal distribution with a standard error of

$$\sigma_{\bar{x}} = \sigma/\sqrt{n} = 12/\sqrt{100} = 1.2$$

This sampling distribution is shown in Figure 8.1.* The sampling distribution of $\bar{X}$ provides information about the possible differences between $\bar{X}$ and μ.

Using the table of cumulative probabilities for the standard normal distribution, we find that 95 per cent of the values of any normally distributed random variable are within ± 1.96 standard deviations of the mean. So, 95 per cent of the $\bar{X}$ values must be within $\pm 1.96\sigma_{\bar{x}}$ of the mean μ. In the CJW example, we know that the sampling distribution of $\bar{X}$ is normal with a standard error of $\sigma_{\bar{x}} = 1.2$. Because $\pm 1.96\sigma_{\bar{x}} = \pm 1.96(1.2) = \pm 2.35$, we conclude that 95 per cent of all $\bar{X}$ values obtained using a sample size of $n = 100$ will be within ± 2.35 units of the population mean μ. See Figure 8.1.

In the introduction to this chapter we said that the general form of an interval estimate of the population mean μ is $\bar{x} \pm$ Margin of error. For the CJW example, suppose we set the margin of error equal to 2.35 and compute the interval estimate of μ using $\bar{x} \pm 2.35$. To provide an interpretation for this interval estimate, let us consider the values of $\bar{x} \pm 2.35$ that could be obtained if we took three different simple random samples, each consisting of 100 CJW customers.

---

*The population of satisfaction scores has a normal distribution, so we can conclude that the sampling distribution of $\bar{X}$ is a normal distribution. If the population did not have a normal distribution, we could rely on the central limit theorem, and the sample size of $n = 100$, to conclude that the sampling distribution of $\bar{X}$ is approximately normal. In either case, the sampling distribution would appear as shown in Figure 8.1.

Figure 8.1 Sampling distribution of the sample mean satisfaction score from simple random samples of 100 customers, also showing the location of sample means that are within ±2.35 units of μ, and intervals calculated from selected sample means at locations $\bar{x}_1$, $\bar{x}_2$ and $\bar{x}_3$.

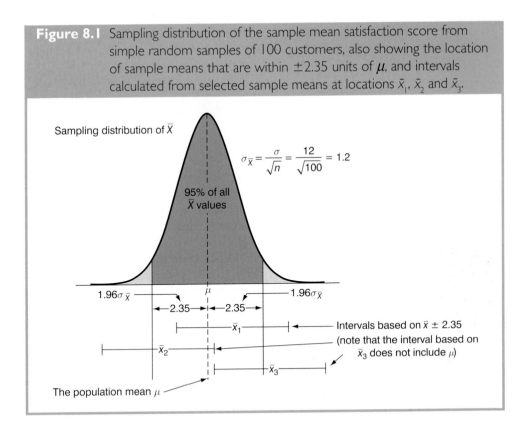

The first sample mean might turn out to have the value shown as $\bar{x}_1$ in Figure 8.1. In this case, the interval formed by subtracting 2.35 from $\bar{x}_1$ and adding 2.35 to $\bar{x}_1$ includes the population mean μ. Now consider what happens if the second sample mean turns out to have the value shown as $\bar{x}_2$ in Figure 8.1. Although $\bar{x}_2$ differs from $\bar{x}_1$, we see that the interval formed by $\bar{x}_2 \pm 2.35$ also includes the population mean μ. However, consider what happens if the third sample mean turns out to have the value shown as $\bar{x}_3$ in Figure 8.1. In this case, because $\bar{x}_3$ falls in the upper tail of the sampling distribution and is further than 2.35 units from μ, the interval $\bar{x}_3 \pm 2.35$ does not include the population mean μ.

Any sample mean that is within the darkly shaded region of Figure 8.1 will provide an interval estimate that contains the population mean μ. Because 95 per cent of all possible sample means are in the darkly shaded region, 95 per cent of all intervals formed by subtracting 2.35 from $\bar{x}$ and adding 2.35 to $\bar{x}$ will include the population mean μ.

The general form of an interval estimate of a population mean for the σ known case is:

Interval estimate of a population mean: σ known

$$\bar{x} \pm z_{\alpha/2}\frac{\sigma}{\sqrt{n}} \tag{8.1}$$

where $(1 - \alpha)$ is the confidence coefficient and $z_{\alpha/2}$ is the z value providing an area $\alpha/2$ in the upper tail of the standard normal probability distribution.

Let us use expression (8.1) to construct a 95 per cent confidence interval for the CJW problem. For a 95 per cent confidence interval, the confidence coefficient is $(1 - \alpha) = 0.95$

and so $\alpha = 0.05$. As we saw above, an area of $\alpha/2 = 0.05/2 = 0.025$ in the upper tail gives $z_{0.025} = 1.96$. With the CJW sample mean $\bar{x} = 72$, $\sigma = 12$, and a sample size $n = 100$, we obtain

$$72 \pm 1.96\frac{12}{\sqrt{100}} = 72 \pm 2.35$$

The specific interval estimate of μ based on the data from the most recent month is $72 - 2.35 = 69.65$, to $72 + 2.35 = 74.35$. Because 95 per cent of all the intervals constructed using $\bar{x} \pm 2.35$ will contain the population mean, we say that we are 95 per cent confident that the interval 69.65 to 74.35 includes the population mean μ. We say that this interval has been established at the 95 per cent **confidence level**. The value 0.95 is referred to as the **confidence coefficient**, and the interval 69.65 to 74.35 is called the 95 per cent **confidence interval**.

Although a 95 per cent confidence level is frequently used, other confidence levels such as 90 per cent and 99 per cent may be considered. Values of $z_{\alpha/2}$ for the most commonly used confidence levels are shown in Table 8.1. Using these values and expression (8.1), the 90 per cent confidence interval for the CJW problem is

$$72 \pm 1.645\frac{12}{\sqrt{100}} = 72 \pm 1.97$$

At 90 per cent confidence, the margin of error is 1.97 and the confidence interval is $72 - 1.97 = 70.03$, to $72 + 1.97 = 73.97$. Similarly, the 99 per cent confidence interval is

$$72 \pm 2.576\frac{12}{\sqrt{100}} = 72 \pm 3.09$$

At 99 per cent confidence, the margin of error is 3.09 and the confidence interval is $72 - 3.09 = 68.93$, to $72 + 3.09 = 75.09$.

Comparing the results for the 90 per cent, 95 per cent and 99 per cent confidence levels, we see that in order to have a higher degree of confidence, the margin of error and consequently the width of the confidence interval must be larger.

Practical advice

If the population follows a normal distribution, the confidence interval provided by expression (8.1) is exact. Therefore, if expression (8.1) were used repeatedly to generate 95 per cent confidence intervals, 95 per cent of the intervals generated (in the long run) would contain the population mean. If the population does not follow a normal distribution, the confidence interval provided by expression (8.1) will be approximate. In this

Table 8.1 Values of $z_{\alpha/2}$ for the most commonly used confidence levels

Confidence level	α	$\alpha/2$	$z_{\alpha/2}$
90%	0.10	0.05	1.645
95%	0.05	0.025	1.960
99%	0.01	0.005	2.576

case, the quality of the approximation depends on both the distribution of the population and the sample size.

In most applications, a sample size of $n \geq 30$ is adequate when using expression (8.1) to construct an interval estimate of a population mean. If the population is not normally distributed but is roughly symmetrical, sample sizes as small as 15 can be expected to provide good approximate confidence intervals. With smaller sample sizes, expression (8.1) should be used only if the analyst believes, or is willing to assume, that the population distribution is at least approximately normal.

Exercises

Methods

1   A simple random sample of 40 items results in a sample mean of 25. The population standard deviation is $\sigma = 5$.

   a. What is the value of the standard error of the mean, $\sigma_{\bar{x}}$?
   b. At 95 per cent confidence, what is the margin of error for estimating the population mean?

2   A simple random sample of 50 items from a population with $\sigma = 6$ results in a sample mean of 32.

   a. Construct a 90 per cent confidence interval for the population mean.
   b. Construct a 95 per cent confidence interval for the population mean.
   c. Construct a 99 per cent confidence interval for the population mean.

3   A simple random sample of 60 items results in a sample mean of 80. The population standard deviation is $\sigma = 15$.

   a. Compute the 95 per cent confidence interval for the population mean.
   c. Assume that the same sample mean was obtained from a sample of 120 items. Construct a 95 per cent confidence interval for the population mean.
   c. What is the effect of a larger sample size on the interval estimate?

4   A 95 per cent confidence interval for a population mean was reported to be 152 to 160. If $\sigma = 15$, what sample size was used in this study?

Applications

5   In an effort to estimate the mean amount spent per customer for dinner at a Johannesburg restaurant, data were collected for a sample of 49 customers. Assume a population standard deviation of 40 South African Rand (ZAR).

   a. At 95 per cent confidence, what is the margin of error?
   b. If the sample mean is ZAR186, what is the 95 per cent confidence interval for the population mean?

6   A survey of small businesses with websites found that the average amount spent on a site was €11 500 per year. Given a sample of 60 businesses and a population standard deviation of $\sigma = €4\,000$, what is the margin of error in estimating the population mean spend per year? Use 95 per cent confidence.

7   The UNITE/MORI 2004 survey of student experiences estimated that university students in the UK paid on average £54 per week in accommodation costs. Assume that this average is based on a sample of 755 students and that the population standard deviation for weekly accommodation costs is £11.

    a. Construct a 90 per cent confidence interval estimate of the population mean.
    b. Construct a 95 per cent confidence interval estimate of the population mean.
    c. Construct a 99 per cent confidence interval estimate of the population mean.
    d. Discuss what happens to the width of the confidence interval as the confidence level is increased. Does this result seem reasonable? Explain.

8.2 Population mean: σ unknown

If a good estimate of the population standard deviation σ cannot be obtained prior to sampling, we must use the sample standard deviation s to estimate σ. This is the σ **unknown** case. When s is used to estimate σ, the margin of error and the interval estimate for the population mean are based on a probability distribution known as the **t distribution**. Although the mathematical development of the t distribution is based on the assumption of a normal distribution for the population from which we are sampling, research shows that the t distribution can be successfully applied in many situations where the population deviates from normal. Later in this section we provide guidelines for using the t distribution if the population is not normally distributed.

The t distribution is a family of similar probability distributions, with a specific t distribution depending on a parameter known as the **degrees of freedom**. The t distribution with one degree of freedom is unique, as is the t distribution with two degrees of freedom, with three degrees of freedom, and so on. As the number of degrees of freedom increases, the difference between the t distribution and the standard normal distribution becomes smaller and smaller. Figure 8.2 shows t distributions with 10 and 20 degrees of freedom and their relationship to the standard normal probability distribution. Note

Figure 8.2 Comparison of the standard normal distribution with t distributions having 10 and 20 degrees of freedom

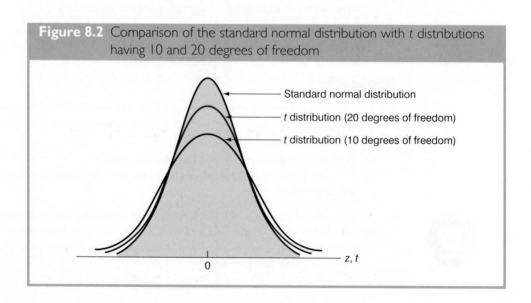

Standard normal distribution

t distribution (20 degrees of freedom)

t distribution (10 degrees of freedom)

z, t

0

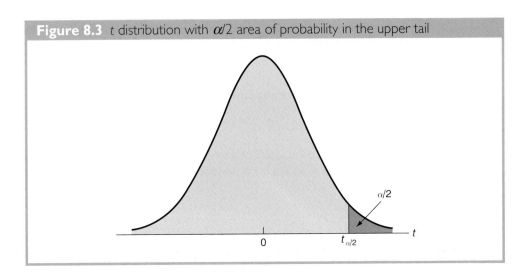

Figure 8.3 *t* distribution with α/2 area of probability in the upper tail

that, the higher the degrees of freedom, the lower is the variability, and the greater the resemblance to the standard normal distribution. Note also that the mean of the *t* distribution is zero.

We place a subscript on *t* to indicate the area in the upper tail of the *t* distribution. For example, just as we used $z_{0.025}$ to indicate the *z* value providing a 0.025 area in the upper tail of a standard normal distribution, we will use $t_{0.025}$ to indicate a 0.025 area in the upper tail of a *t* distribution. In general, we will use the notation $t_{\alpha/2}$ to represent a *t* value with an area of α/2 in the upper tail of the *t* distribution. See Figure 8.3.

Table 2 of Appendix B is a table for the *t* distribution. Each row in the table corresponds to a separate *t* distribution with the degrees of freedom shown. For example, for a *t* distribution with 10 degrees of freedom, $t_{0.025} = 2.228$. Similarly, for a *t* distribution with 20 degrees of freedom, $t_{0.025} = 2.086$. As the degrees of freedom continue to increase, $t_{0.025}$ approaches $z_{0.025} = 1.96$. In fact, the standard normal distribution *z* values can be found in the infinite degrees of freedom row (labelled ∞) of the *t* distribution table. If the degrees of freedom exceed 100, the infinite degrees of freedom row can be used to approximate the actual *t* value. In other words, for more than 100 degrees of freedom, the standard normal *z* value provides a good approximation to the *t* value.

William Sealy Gosset, writing under the name 'Student', was the originator of the *t* distribution. Gosset, an Oxford graduate in mathematics, worked for the Guinness Brewery in Dublin, Ireland. The distribution is sometimes referred to as 'Student's *t* distribution'.

Margin of error and the interval estimate

In Section 8.1 we showed that an interval estimate of a population mean for the σ known case is

$$\bar{x} \pm z_{\alpha/2} \frac{\sigma}{\sqrt{n}}$$

To compute an interval estimate of μ for the σ unknown case, the sample standard deviation *s* is used to estimate σ and $z_{\alpha/2}$ is replaced by the *t* distribution value $t_{\alpha/2}$. The margin of error is then $\pm t_{\alpha/2}s/\sqrt{n}$, and the general expression for an interval estimate of a population mean when σ is unknown is:

Interval estimate of a population mean: σ unknown

$$\bar{x} \pm t_{\alpha/2}\frac{s}{\sqrt{n}} \qquad (8.2)$$

where s is the sample standard deviation, $(1 - \alpha)$ is the confidence coefficient, and $t_{\alpha/2}$ is the t value providing an area of $\alpha/2$ in the upper tail of the t distribution with $n - 1$ degrees of freedom*.

Consider a study designed to estimate the mean credit card debt for a defined population of households. A sample of $n = 85$ households provided the credit card balances in the file 'Balance' on the accompanying CD. The first few rows of this data set are shown in the EXCEL screenshot in Figure 8.4 below. For this situation, no previous estimate of the population standard deviation σ is available. As a consequence, the sample data must be used to estimate both the population mean and the population standard deviation. Using the data in the 'Balance' file, we compute the sample mean $\bar{x} = 5900$ (€) and the sample standard deviation $s = 3058$ (€).

Figure 8.4 First few data rows and summary statistics for credit card balances

BALANCE

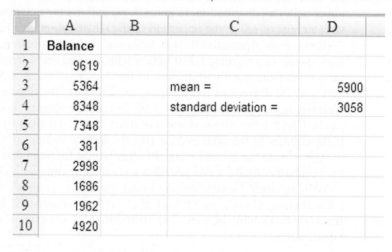

	A	B	C	D
1	Balance			
2	9619			
3	5364		mean =	5900
4	8348		standard deviation =	3058
5	7348			
6	381			
7	2998			
8	1686			
9	1962			
10	4920			

---

*The reason the number of degrees of freedom associated with the t value in expression (8.2) is $n - 1$ concerns the use of s as an estimate of the population standard deviation. The expression for the sample standard deviation is $s = \sqrt{\Sigma(x_i - \bar{x})^2/(n - 1)}$. Degrees of freedom refers to the number of independent pieces of information that go into the computation of $\Sigma(x_i - \bar{x})^2$. The n pieces of information involved in computing $\Sigma(x_i - \bar{x})^2$ are as follows: $x_1 - \bar{x}, x_2 - \bar{x}, \ldots, x_n - \bar{x}$. In Section 3.2 we indicated that $\Sigma(x_i - \bar{x}) = 0$. Hence, only $n - 1$ of the $x_i - \bar{x}$ values are independent; that is, if we know $n - 1$ of the values, the remaining value can be determined exactly by using the condition that $\Sigma(x_i - \bar{x}) = 0$. So $n - 1$ is the number of degrees of freedom associated with $\Sigma(x_i - \bar{x})^2$ and hence the number of degrees of freedom for the t distribution in expression (8.2).

With 95 per cent confidence and $n - 1 = 84$ degrees of freedom, Table 2 in Appendix B gives $t_{0.025} = 1.989$. We can now use expression (8.2) to compute an interval estimate of the population mean.

$$5900 \pm 1.989\frac{3058}{\sqrt{85}} = 5900 \pm 660$$

The point estimate of the population mean is €5900, the margin of error is €660, and the 95 per cent confidence interval is $5900 - 660 = €5240$ to $5900 + 660 = €6560$. We are 95 per cent confident that the population mean credit card balance for all households in the defined population is between €5240 and €6560.

The procedures used by MINITAB, EXCEL and PASW to construct confidence intervals for a population mean are described at the end of the chapter.

Practical advice

If the population follows a normal distribution, the confidence interval provided by expression (8.2) is exact and can be used for any sample size. If the population does not follow a normal distribution, the confidence interval provided by expression (8.2) will be approximate. In this case, the quality of the approximation depends on both the distribution of the population and the sample size.

In most applications, a sample size of $n \geq 30$ is adequate when using expression (8.2) to construct an interval estimate of a population mean. However, if the population distribution is highly skewed or contains outliers, the sample size should be 50 or more. If the population is not normally distributed but is roughly symmetrical, sample sizes as small as 15 can be expected to provide good approximate confidence intervals. With smaller sample sizes, expression (8.2) should only be used if the analyst is confident that the population distribution is at least approximately normal.

Using a small sample

In the following example we construct an interval estimate for a population mean when the sample size is small. As we have already noted, an understanding of the distribution of the population becomes a factor in deciding whether the interval estimation procedure provides acceptable results.

Scheer Industries is considering a new computer-assisted programme to train maintenance employees to do machine repairs. In order to fully evaluate the programme, the director of manufacturing requested an estimate of the population mean time required for maintenance employees to complete the computer-assisted training.

A sample of 20 employees is selected, with each employee in the sample completing the training programme. Data on the training time in days for the 20 employees are shown in Table 8.2. A histogram of the sample data appears in Figure 8.5. What can we say about the distribution of the population based on this histogram? First, the sample data do not support with certainty the conclusion that the distribution of the population is normal, but we do not see any evidence of skewness or outliers. Therefore, using the guidelines in the previous subsection, we conclude that an interval estimate based on the t distribution appears acceptable for the sample of 20 employees.

Table 8.2 Training time in days for a sample of 20 Scheer Industries employees

52	59	54	42
44	50	42	48
55	54	60	55
44	62	62	57
45	46	43	56

We compute the sample mean and sample standard deviation as follows.

$$\bar{x} = \frac{\Sigma x_i}{n} = \frac{1030}{20} = 51.5 \text{ days}$$

$$s = \sqrt{\frac{\Sigma (x_i - \bar{x})^2}{n - 1}} = \sqrt{\frac{889}{20 - 1}} = 6.84 \text{ days}$$

For a 95 per cent confidence interval, we use Table 2 from Appendix B and $n - 1 = 19$ degrees of freedom to obtain $t_{0.025} = 2.093$. Expression (8.2) provides the interval estimate of the population mean .

$$51.5 \pm 2.093 \left(\frac{6.84}{\sqrt{20}} \right) = 51.5 \pm 3.2$$

The point estimate of the population mean is 51.5 days. The margin of error is 3.2 days and the 95 per cent confidence interval is $51.5 - 3.2 = 48.3$ days to $51.5 + 3.2 = 54.7$ days.

Using a histogram of the sample data to learn about the distribution of a population is not always conclusive, but in many cases it provides the only information available. The histogram, along with judgment on the part of the analyst, can often be used to decide if expression (8.2) can be used to construct the interval estimate.

Figure 8.5 Histogram of training times for the Scheer Industries sample

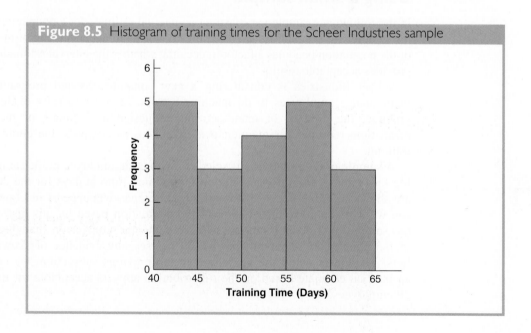

Summary of interval estimation procedures

We provided two approaches to constructing an interval estimate of a population mean. For the σ known case, σ and the standard normal distribution are used in expression (8.1) to compute the margin of error and to develop the interval estimate. For the σ unknown case, the sample standard deviation s and the t distribution are used in expression (8.2) to compute the margin of error and to develop the interval estimate.

A summary of the interval estimation procedures for the two cases is shown in Figure 8.6. In most applications, a sample size of $n \geq 30$ is adequate. If the population has a normal or approximately normal distribution, however, smaller sample sizes may be used. For the σ unknown case a sample size of $n \geq 50$ is recommended if the population distribution is believed to be highly skewed or has outliers.

Figure 8.6 Summary of interval estimation procedures for a population mean

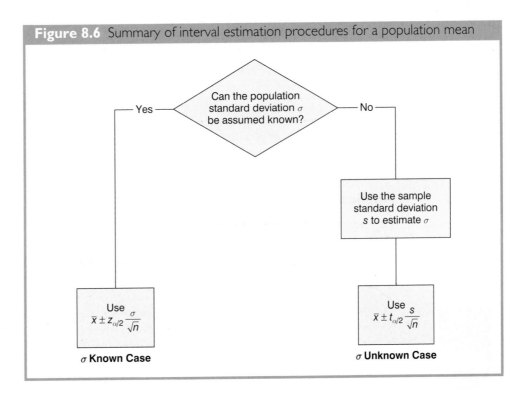

Exercises

Methods

8   For a t distribution with 16 degrees of freedom, find the area, or probability, in each region.

   a. To the right of 2.120
   b. To the left of 1.337
   c. To the left of -1.746
   d. To the right of 2.583
   e. Between -2.120 and 2.120
   f. Between -1.746 and 1.746

9 Find the *t* value(s) for each of the following cases.

    a. Upper tail area of 0.025 with 12 degrees of freedom

    b. Lower tail area of 0.05 with 50 degrees of freedom

    c. Upper tail area of 0.01 with 30 degrees of freedom

    d. Where 90 per cent of the area falls between these two *t* values with 25 degrees of freedom

    e. Where 95 per cent of the area falls between these two *t* values with 45 degrees of freedom

10 The following sample data are from a normal population: 10, 8, 12, 15, 13, 11, 6, 5.

    a. What is the point estimate of the population mean?

    b. What is the point estimate of the population standard deviation?

    c. With 95 per cent confidence, what is the margin of error for the estimation of the population mean?

    d. What is the 95 per cent confidence interval for the population mean?

11 A simple random sample with $n = 54$ provided a sample mean of 22.5 and a sample standard deviation of 4.4.

    a. Construct a 90 per cent confidence interval for the population mean.

    b. Construct a 95 per cent confidence interval for the population mean.

    c. Construct a 99 per cent confidence interval for the population mean.

    d. What happens to the margin of error and the confidence interval as the confidence level is increased?

Applications

12 Sales personnel for Skillings Distributors submit weekly reports listing the customer contacts made during the week. A sample of 65 weekly reports showed a sample mean of 19.5 customer contacts per week. The sample standard deviation was 5.2. Provide 90 per cent and 95 per cent confidence intervals for the population mean number of weekly customer contacts for the sales personnel.

13 Consumption of alcoholic beverages by young women of drinking age has been increasing in the UK, Europe and the US (*The Wall Street Journal*, 15 February, 2006). Data (annual consumption in litres) consistent with the findings reported in *The Wall Street Journal* article are shown for a sample of 20 European young women.

266	82	199	174	97
170	222	115	130	169
164	102	113	171	0
93	0	93	110	130

Assuming the population is roughly symmetrically distributed, construct a 95 per cent confidence interval for the mean annual consumption of alcoholic beverages by young European women.

14 The International Air Transport Association surveys business travellers to develop quality ratings for international airports. The maximum possible rating is ten. Suppose a simple random sample of business travellers is selected and each traveller is asked to provide a rating for Singapore Changi International Airport. The ratings obtained from the sample of 50 business travellers follow. Construct a 95 per cent confidence interval estimate of the population mean rating for Changi.

6	4	6	8	7	7	6	3	3	8	10	4	8	7	8	7	5
9	5	8	4	3	8	5	5	4	4	4	8	4	5	6	2	5
9	9	8	4	8	9	9	5	9	7	8	3	10	8	9	6	

15 Suppose a survey of 40 first-time home buyers finds that the mean of annual household income is €40 000 and the sample standard deviation is €15 300.

    a. At 95 per cent confidence, what is the margin of error for estimating the population mean household income?

    b. What is the 95 per cent confidence interval for the population mean annual household income for first-time home buyers?

16 Thirty fast-food restaurants including McDonald's and Burger King were visited during the summer of 2009. During each visit, the customer went to the drive-through and ordered a basic meal such as a burger, fries and drink. The time between pulling up to the order kiosk and receiving the filled order was recorded. The times in minutes for the 30 visits are as follows:

| 0.9 | 1.0 | 1.2 | 2.2 | 1.9 | 3.6 | 2.8 | 5.2 | 1.8 | 2.1 | 6.8 | 1.3 | 3.0 | 4.5 | 2.8 |
| 2.3 | 2.7 | 5.7 | 4.8 | 3.5 | 2.6 | 3.3 | 5.0 | 4.0 | 7.2 | 9.1 | 2.8 | 3.6 | 7.3 | 9.0 |

    a. Provide a point estimate of the population mean drive-through time at fast-food restaurants.

    b. At 95 per cent confidence, what is the margin of error?

    c. What is the 95 per cent confidence interval estimate of the population mean?

    d. Discuss skewness that may be present in this population. What suggestion would you make for a repeat of this study?

17 A survey by Accountemps asked a sample of 200 executives to provide data on the number of minutes per day office workers waste trying to locate mislabelled, misfiled or misplaced items. Data consistent with this survey are contained in the data set 'ActTemps'.

    a. Use 'ActTemps' to develop a point estimate of the number of minutes per day office workers waste trying to locate mislabelled, misfiled or misplaced items.

    b. What is the sample standard deviation?

    c. What is the 95 per cent confidence interval for the mean number of minutes wasted per day?

8.3 Determining the sample size

In providing practical advice in the two preceding sections, we commented on the role of the sample size in providing good approximate confidence intervals when the population is not normally distributed. In this section, we focus on another aspect of the sample size issue. We describe how to choose a sample size large enough to provide a desired margin of error. To understand how this process is done, we return to the σ known case presented in Section 8.1. Using expression (8.1), the interval estimate is $\bar{x} \pm z_{\alpha/2}\sigma/\sqrt{n}$. We see that $z_{\alpha/2}$, the population standard deviation σ, and the sample size n combine to determine the margin of error. Once we select a confidence coefficient $1 - \alpha$, $z_{\alpha/2}$ can be determined. Then, if we have a value for σ, we can determine the sample size n needed to provide any desired margin of error. Let $E =$ the desired margin of error.

$$E = z_{\alpha/2}\frac{\sigma}{\sqrt{n}}$$

Solving for $\sqrt{n}$, we have

$$\sqrt{n} = \frac{z_{\alpha/2}\sigma}{E}$$

Squaring both sides of this equation, we obtain the following expression for the sample size.

Sample size for an interval estimate of a population mean

$$n = \frac{(z_{\alpha/2})^2\sigma^2}{E^2} \tag{8.3}$$

This sample size provides the desired margin of error at the chosen confidence level.

In equation (8.3), E is the margin of error that the user is willing to accept, and the value of $z_{\alpha/2}$ follows directly from the confidence level to be used in constructing the interval estimate. Although user preference must be considered, 95 per cent confidence is the most frequently chosen value ($z_{0.025} = 1.96$). Equation (8.3) can be used to provide a good sample size recommendation. However, judgment on the part of the analyst should be used to determine whether the final sample size should be adjusted upward.

Finally, use of equation (8.3) requires a value for the population standard deviation σ. However, even if σ is unknown, we can use equation (8.3) provided we have a preliminary or *planning value* for σ. In practice, one of the following procedures can be chosen.

1 Use an estimate of the population standard deviation computed from data of previous studies as the planning value for σ.

2 Use a pilot study to select a preliminary sample. The sample standard deviation from the preliminary sample can be used as the planning value for σ.

3 Use judgment or a 'best guess' for the value of σ. For example, we might begin by estimating the largest and smallest data values in the population. The difference between the largest and smallest values provides an estimate of the range for the data. The range divided by four is often suggested as a rough approximation of the standard deviation and hence an acceptable planning value for σ.

Consider the following example. A previous study that investigated the cost of renting cars in Ireland found a mean cost of approximately €80 per day for renting a family saloon. Suppose that the organization that conducted this study would like to conduct a new study in order to estimate the population mean daily rental cost for a family saloon in Ireland. In designing the new study, the project director specifies that the population mean daily rental cost be estimated with a margin of error of €2 and a 95 per cent level of confidence.

The project director specified a desired margin of error of $E = 2$, and the 95 per cent level of confidence indicates $z_{0.025} = 1.96$. We only need a planning value for the population standard deviation σ to compute the required sample size. At this point, an analyst reviewed the sample data from the previous study and found that the sample standard deviation for the daily rental cost was €9.65. Using 9.65 as the planning value for σ, we obtain

$$n = \frac{(z_{\alpha/2})^2\sigma^2}{E^2} = \frac{(1.96)^2(9.65)^2}{2^2} = 89.43$$

The sample size for the new study needs to be at least 89.43 family saloon rentals in order to satisfy the project director's €2 margin-of-error requirement. In cases where the computed n is not an integer, we round up to the next integer value; hence, the recommended sample size is 90 family saloon rentals.

Exercises

Methods

18 How large a sample should be selected to provide a 95 per cent confidence interval with a margin of error of 10? Assume that the population standard deviation is 40.

19 The range for a set of data is estimated to be 36.

   a. What is the planning value for the population standard deviation?

   b. At 95 per cent confidence, how large a sample would provide a margin of error of 3?

   c. At 95 per cent confidence, how large a sample would provide a margin of error of 2?

Applications

20 Refer to the Scheer Industries example in Section 8.2. Use 6.82 days as a planning value for the population standard deviation.

   a. Assuming 95 per cent confidence, what sample size would be required to obtain a margin of error of 1.5 days?

   b. If the precision statement was made with 90 per cent confidence, what sample size would be required to obtain a margin of error of 2 days?

21 Suppose you are interested in estimating the average cost of staying for one night in a double room in a three-star hotel in France (outside Paris). Using €30.00 as the planning value for the population standard deviation, what sample size is recommended for each of the following cases? Use €3 as the desired margin of error.

   a. A 90 per cent confidence interval estimate of the population mean cost.

   b. A 95 per cent confidence interval estimate of the population mean cost.

   c. A 99 per cent confidence interval estimate of the population mean cost.

   d. When the desired margin of error is fixed, what happens to the sample size as the confidence level is increased? Would you recommend a 99 per cent confidence level be used? Discuss.

22 During the third quarter of 2005, the price/earnings (P/E) ratio for stocks listed on the Hong Kong Stock Exchange generally ranged from 1 to 100. Assume that we want to estimate the population mean P/E ratio for all stocks listed on the exchange. How many stocks should be included in the sample if we want a margin of error of 3? Use 95 per cent confidence.

23 Fuel consumption tests are conducted for a particular model of car. If a 98 per cent confidence interval with a margin of error of 0.2 litres per 100 km is desired, how many cars should be used in the test? Assume that preliminary tests indicate the standard deviation is 0.5 litres per 100 km.

24 In developing patient appointment schedules, a medical centre wants to estimate the mean time that a staff member spends with each patient. How large a sample should be taken if the desired margin of error is 2 minutes at a 95 per cent level of confidence? How large a sample should be taken for a 99 per cent level of confidence? Use a planning value for the population standard deviation of 8 minutes.

8.4 Population proportion

In the introduction to this chapter we said that the general form of an interval estimate of a population proportion π is: $p \pm$ Margin of error. The sampling distribution of the sample proportion of plays a key role in computing the margin of error for this interval estimate.

In Chapter 7 we said that the sampling distribution of the sample proportion P can be approximated by a normal distribution whenever $n\pi \geq 5$ and $n(1 - \pi) \geq 5$. Figure 8.7 shows the normal approximation of the sampling distribution of P. The mean of the sampling distribution of P is the population proportion π, and the standard error of P is

$$\sigma_P = \sqrt{\frac{\pi(1 - \pi)}{n}} \tag{8.4}$$

Because the sampling distribution of P is normally distributed, if we choose $z_{\alpha/2}\sigma_P$ as the margin of error in an interval estimate of a population proportion, we know that $100(1 - \alpha)$ per cent of the intervals generated will contain the true population proportion. But σ_p cannot be used directly in the computation of the margin of error because π will not be known; π is what we are trying to estimate. So, p is substituted for π and the margin of error for an interval estimate of a population proportion is given by

$$\text{Margin of error} = z_{\alpha/2}\sqrt{\frac{p(1 - p)}{n}} \tag{8.5}$$

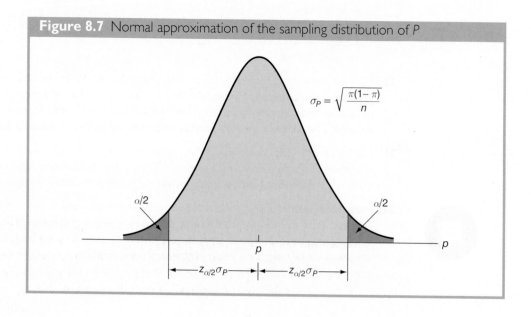

Figure 8.7 Normal approximation of the sampling distribution of P

The general expression for an interval estimate of a population proportion is:

Interval estimate of a population proportion

$$p \pm z_{\alpha/2} \sqrt{\frac{p(1-p)}{n}}$$

(8.6)

where $1 - \alpha$ is the confidence coefficient and $z_{\alpha/2}$ is the z value providing an area of $\alpha/2$ in the upper tail of the standard normal distribution.

TEETIMES

Consider the following example. A national survey of 900 women golfers was conducted to learn how women golfers view their treatment at golf courses. (The data are available in the file 'TeeTimes' on the CD.) The survey found that 396 of the women golfers were satisfied with the availability of tee times. So, the point estimate of the proportion of the population of women golfers who are satisfied with the availability of tee times is 396/900 = 0.44. Using expression (8.6) and a 95 per cent confidence level,

$$p \pm z_{\alpha/2} \sqrt{\frac{p(1-p)}{n}} = 0.44 \pm 1.96 \sqrt{\frac{0.44(1-0.44)}{900}} = 0.44 \pm 0.0324$$

The margin of error is 0.0324 and the 95 per cent confidence interval estimate of the population proportion is 0.408 to 0.472. Using percentages, the survey results enable us to state that with 95 per cent confidence between 40.8 per cent and 47.2 per cent of all women golfers are satisfied with the availability of tee times.

Determining the sample size

The rationale for the sample size determination in developing interval estimates of π is similar to the rationale used in Section 8.3 to determine the sample size for estimating a population mean.

Previously in this section we said that the margin of error associated with an interval estimate of a population proportion is $z_{\alpha/2}\sqrt{p(1-p)/n}$. The margin of error is based on the values of $z_{\alpha/2}$, the sample proportion p, and the sample size n. Larger sample sizes provide a smaller margin of error and better precision. Let E denote the desired margin of error.

$$E = z_{\alpha/2} \sqrt{\frac{p(1-p)}{n}}$$

Solving this equation for n provides a formula for the sample size that will provide a margin of error of size E.

$$n = \frac{(z_{\alpha/2})^2 \, p(1-p)}{E^2}$$

Note, however, that we cannot use this formula to compute the sample size that will provide the desired margin of error because p will not be known until after we select the sample. What we need, then, is a planning value for p that can be used to make the computation. Using p^* to denote the planning value for p, the following formula can be used to compute the sample size that will provide a margin of error of size E.

Sample size for an interval estimate of a population proportion

$$n = \frac{(z_{\alpha/2})^2 \, p^*(1 - p^*)}{E^2} \qquad\qquad (8.7)$$

In practice, the planning value can be chosen by one of the following procedures.

1 Use the sample proportion from a previous sample of the same or similar units.

2 Use a pilot study to select a preliminary sample. The sample proportion from this sample can be used as the planning value, $p.*$

3 Use judgment or a 'best guess' for the value of $p.*$

4 If none of the preceding alternatives apply, use a planning value of $p* = 0.50$.

Let us return to the survey of women golfers and assume that the company is interested in conducting a new survey to estimate the current proportion of the population of women golfers who are satisfied with the availability of tee times. How large should the sample be if the survey director wants to estimate the population proportion with a margin of error of 0.025 at 95 per cent confidence? With $E = 0.025$ and $z_{\alpha/2} = 1.96$, we need a planning value p^* to answer the sample size question. Using the previous survey result of $p = 0.44$ as the planning value p^*, equation (8.7) shows that

$$n = \frac{(z_{\alpha/2})^2 \, p^*(1 - p^*)}{E^2} = \frac{(1.96)^2(0.44)(1 - 0.44)}{(0.025)^2} = 1514.5$$

The sample size must be at least 1514.5 women golfers to satisfy the margin of error requirement. Rounding up to the next integer value indicates that a sample of 1515 women golfers is recommended to satisfy the margin of error requirement.

The fourth alternative suggested for selecting a planning value p^* is to use $p^* = 0.50$. This value of p^* is frequently used when no other information is available. To understand why, note that the numerator of equation (8.7) shows that the sample size is proportional to the quantity $p^*(1 - p^*)$. A larger value for this quantity will result in a larger sample size. Table 8.3 gives some possible values of $p^*(1 - p^*)$. Note that the largest value occurs when $p^* = 0.50$. So, in case of any uncertainty about an appropriate planning value, we know that $p^* = 0.50$ will provide the largest sample size recommendation. In effect, we play it safe by recommending the largest possible sample size. If the sample proportion turns out to be different from the 0.50 planning value, the margin of error will

Table 8.3 Some possible values for $p^*(1 - p^*)$

p^*	$p^*(1 - p^*)$	
0.10	(0.10)(0.90) = 0.09	
0.30	(0.30)(0.70) = 0.21	
0.40	(0.40)(0.60) = 0.24	
0.50	(0.50)(0.50) = 0.25	← ——————Largest value for $p^*(1 - p^*)$
0.60	(0.60)(0.40) = 0.24	
0.70	(0.70)(0.30) = 0.21	
0.90	(0.90)(0.10) = 0.09	

be smaller than anticipated. Hence, in using $p^* = 0.50$, we guarantee that the sample size will be sufficient to obtain the desired margin of error.

In the survey of women golfers example, a planning value of $p^* = 0.50$ would have provided the sample size

$$n = \frac{(z_{\alpha/2})^2 p^*(1 - p^*)}{E^2} = \frac{(1.96)^2(0.5)(1 - 0.5)}{(0.025)^2} = 1536.6$$

A slightly larger sample size of 1537 women golfers would be recommended.

Exercises

Methods

25 A simple random sample of 400 individuals provides 100 Yes responses.

   a. What is the point estimate of the proportion of the population that would provide Yes responses?
   b. What is your estimate of the standard error of the sample proportion?
   c. Compute a 95 per cent confidence interval for the population proportion.

26 A simple random sample of 800 elements generates a sample proportion $p = 0.70$.

   a. Provide a 90 per cent confidence interval for the population proportion.
   b. Provide a 95 per cent confidence interval for the population proportion.

27 In a survey, the planning value for the population proportion is $p^* = 0.35$. How large a sample should be taken to provide a 95 per cent confidence interval with a margin of error of 0.05?

28 At 95 per cent confidence, how large a sample should be taken to obtain a margin of error of 0.03 for the estimation of a population proportion? Assume that past data are not available for developing a planning value for p.

Applications

29 A survey of 611 office workers investigated telephone answering practices, including how often each office worker was able to answer incoming telephone calls and how often incoming telephone calls went directly to voice mail. A total of 281 office workers indicated that they never need voice mail and are able to take every telephone call.

   a. What is the point estimate of the proportion of the population of office workers who are able to take every telephone call?
   b. At 90 per cent confidence, what is the margin of error?
   c. What is the 90 per cent confidence interval for the proportion of the population of office workers who are able to take every telephone call?

30 The French market research and polling company CSA carried out surveys in March 2005 to investigate job satisfaction among professionally qualified employees of private companies. A total of 629 professionals were involved in the surveys, of whom 195 said that they were dissatisfied with their employer's recognition of their professional experience.

   a. What is the point estimate of the proportion of the population of employees who were dissatisfied with their employer's recognition of their professional experience?

b. At 95 per cent confidence, what is the margin of error?

c. What is the 95 per cent confidence interval for the proportion of the population of employees who were dissatisfied with their employer's recognition of their professional experience?

31 According to a *BusinessWeek* report in early 2006, data from Thomson Financial showed that the majority of companies reporting profits had beaten estimates. A sample of 162 companies showed 104 beat estimates, 29 matched estimates, and 29 fell short.

a. What is the point estimate of the proportion that fell short of estimates?

b. Determine the margin of error and provide a 95 per cent confidence interval for the proportion that beat estimates.

c. How large a sample is needed if the desired margin of error is 0.05?

32 In early December 2008, the Palestinian Center for Policy and Survey Research carried out an opinion poll among adults in the West Bank and Gaza Strip. Respondents were asked their opinion about the chance of an independent Palestinian state being established alongside Israel in the next five years. Among the 1270 respondents, 34.6 per cent felt there was no chance of this happening.

a. Provide a 95 per cent confidence interval for the population proportion of adults who thought there was no chance of an independent Palestinian state being established alongside Israel in the next five years.

b. Provide a 99 per cent confidence interval for the population proportion of adults who thought there was no chance of an independent Palestinian state being established alongside Israel in the next five years.

c. What happens to the margin of error as the confidence is increased from 95 per cent to 99 per cent?

33 In a survey conducted by ICM Research in the UK towards the end of December 2008, 710 out of 1000 adults interviewed said that, if there were to be a referendum, they would vote for the UK not to join the European currency (the euro). What is the margin of error and what is the interval estimate of the population proportion of British adults who would vote for the UK not to join the European currency? Use 95 per cent confidence.

34 A well-known bank credit card firm wishes to estimate the proportion of credit card holders who carry a non-zero balance at the end of the month and incur an interest charge. Assume that the desired margin of error is 0.03 at 98 per cent confidence.

a. How large a sample should be selected if it is anticipated that roughly 70 per cent of the firm's cardholders carry a non-zero balance at the end of the month?

b. How large a sample should be selected if no planning value for the proportion could be specified?

For additional online summary questions and answers go to the companion website at www.cengage.co.uk/aswsbe2

Summary

In this chapter we introduced the idea of an interval estimate of a population parameter. A point estimator may or may not provide a good estimate of a population parameter. The use of an interval estimate provides a measure of the precision of an estimate. A common form of interval estimate is a confidence interval.

We presented methods for computing confidence intervals of a population mean and a population proportion. Both are of the form: point estimate $\pm$ margin of error. The confidence interval has a confidence coefficient associated with it.

We presented interval estimates for a population mean for two cases. In the σ known case, historical data or other information is used to make an estimate of σ prior to taking a sample. Analysis of new sample data then proceeds based on the assumption that σ is known. In the σ unknown case, the sample data are used to estimate both the population mean and the population standard deviation. In the σ known case, the interval estimation procedure is based on the assumed value of σ and the use of the standard normal distribution. In the σ unknown case, the interval estimation procedure uses the sample standard deviation s and the t distribution.

In both cases the quality of the interval estimates obtained depends on the distribution of the population and the sample size. Practical advice about the sample size necessary to obtain good approximations was included in Sections 8.1 and 8.2.

The general form of the interval estimate for a population proportion is $p \pm$ margin of error. In practice the sample sizes used for interval estimates of a population proportion are generally large. Consequently, the interval estimation procedure is based on the standard normal distribution.

We explained how the expression for margin of error can be used to calculate the sample size required to achieve a desired margin of error at a given level of confidence. We did this for two cases: estimating a population mean when the population standard deviation is known, and estimating a population proportion.

Key terms

Confidence coefficient
Confidence interval
Confidence level
Degrees of freedom
Interval estimate

Margin of error
σ known
σ unknown
t distribution

Key formulae

Interval estimate of a population mean: σ known

$$\bar{x} \pm z_{\alpha/2} \frac{\sigma}{\sqrt{n}} \tag{8.1}$$

Interval estimate of a population mean: σ unknown

$$\bar{x} \pm t_{\alpha/2} \frac{s}{\sqrt{n}} \tag{8.2}$$

Sample size for an interval estimate of a population mean

$$n = \frac{(z_{\alpha/2})^2 \sigma^2}{E^2} \qquad (8.3)$$

Interval estimate of a population proportion

$$p \pm z_{\alpha/2}\sqrt{\frac{p(1-p)}{n}} \qquad (8.6)$$

Sample size for an interval estimate of a population proportion

$$n = \frac{(z_{\alpha/2})^2 p^*(1-p^*)}{E^2} \qquad (8.7)$$

Case problem 1 International bank

The manager of a city-centre branch of a well-known international bank commissioned a customer satisfaction survey. The survey investigated three areas of customer satisfaction: their experience waiting for service at a till, their experience being served at the till, and their experience of self-service facilities at the branch. Within each of these categories, respondents to the survey were asked to give ratings on a number of aspects of the bank's service. These ratings were then summed to give an overall satisfaction rating in each of the three areas of service. The summed ratings are scaled such that they lie between 0 and 100, with 0 representing extreme dissatisfaction and 100 representing extreme satisfaction. The data file for this case study ('IntnlBank' on the accompanying CD) contains the 0–100 ratings for the three areas of service, together with particulars of respondents' gender and whether they would recommend the bank to other

people (a simple Yes/No response was required to this question). A table containing the first few rows of the data file is shown below.

Waiting	Service	Self-service	Gender	Recommend
55	65	50	male	no
50	80	88	male	no
30	40	44	male	no
65	60	69	male	yes
55	65	63	male	no
40	60	56	male	no
15	65	38	male	yes
45	60	56	male	no
55	65	75	male	no
50	50	69	male	yes

People using automated self service machines at a main bank branch. © david pearson/Alamy.

Managerial report

1 Use descriptive statistics to summarize each of the five variables in the data file (the three service ratings, customer gender and customer recommendation).

2 Calculate a 95 per cent confidence interval estimate of the mean service rating for the population of customers of the branch, for each of the three service areas. Provide a managerial interpretation of each interval estimate.

3 Calculate a 95 per cent confidence interval estimate of the proportion of the branch's customers who would recommend the bank, and a 95 per cent confidence interval estimate of the proportion of

the branch's customers who are female. Provide a managerial interpretation of each interval estimate.

4 Suppose the branch manager required an estimate of the percentage of branch customers who would recommend the branch within a margin of error of 3 percentage points. Using 95 per cent confidence, how large should the sample size be?

5 Suppose the branch manager required an estimate of the percentage of branch customers who are female within a margin of error of 5 percentage points. Using 95 per cent confidence, how large should the sample size be?

Case Problem 2 *Young Professional* Magazine

Young *Professional* magazine was developed for a target audience of recent university graduates who are in their first 10 years in a business/professional career. In its two years of publication, the magazine has been fairly successful. Now the publisher is interested in expanding the magazine's advertising base. Potential advertisers continually ask about the demographics and interests of subscribers to *Young Professional*. To collect this information, the magazine commissioned a survey to develop a profile of its subscribers. The survey results will be used to help the magazine choose articles of interest and provide advertisers with a profile of subscribers. As a new employee of the magazine, you have been asked to help analyze the survey results.

Some of the survey questions follow (these are not necessarily in the order they were asked in the survey):

1 What is your age?

2 Are you: Male _____ Female _____

Young business man and woman reading Young Professional Magazine. © Marcin Balcerzak.

3 Do you plan to make any real estate purchases in the next two years? Yes _____ No _____

4 What is the approximate total value of financial investments, exclusive of your home, owned by you or members of your household?

5 How many stock/bond/mutual fund transactions have you made in the past year?

6 Do you have broadband access to the Internet at home? Yes _____ No _____

7 Please indicate your total household income last year.

8 Do you have children? Yes _____ No _____

The file entitled Professional contains the responses to these questions. The file is on the CD accompanying the text.

Managerial Report

PROFES-SIONAL

Prepare a managerial report summarizing the results of the survey. In addition to statistical summaries, discuss how the magazine might use these results to attract advertisers. You might also comment on how the survey results could be used by the magazine's editors to identify topics that would be of interest to readers. Your report should address the following issues, but do not limit your analysis to just these areas.

1 Develop appropriate descriptive statistics to summarize the data.

2 Develop 95 per cent confidence intervals for the mean age and household income of subscribers.

3 Develop 95 per cent confidence intervals for the proportion of subscribers who have broadband access at home and the proportion of subscribers who have children.

4 Would *Young Professional* be a good advertising outlet for online brokers? Justify your conclusion with statistical data.

5 Would this magazine be a good place to advertize for companies selling educational software and computer games for young children?

6 Comment on the types of articles you believe would be of interest to readers of *Young Professional*.

Software Section
for Chapter 8

Interval estimation using MINITAB

We describe the use of MINITAB in constructing confidence intervals for a population mean and a population proportion.

Population mean: σ known

We illustrate using the CJW example in Section 8.1 (file 'CJW.MTW' on the accompanying CD). The satisfaction scores for the sample of 100 customers are in column C1 of a MINITAB worksheet. The population standard deviation $\sigma = 20$ is assumed known. The following steps can be used to compute a 95 per cent confidence interval estimate of the population mean.

Step 1 **Stat > Basic Statistics > 1-Sample Z** [Main menu bar]

Step 2 Enter **C1** in the **Samples in columns** box
[**1-Sample Z (Test and Confidence Interval)** panel]
Enter **20** in the **Standard deviation** box
Click **OK**
  The MINITAB default is a 95 per cent confidence level. To specify a different confidence level such as 90 per cent:

Step 2 Enter **C1** in the **Samples in columns** box
[**1-Sample Z (Test and Confidence Interval)** panel]
Enter **20** in the **Standard deviation** box
Select **Options**

Step 3 Enter **90** in the **Confidence level** box [**1-Sample Z - Options** panel]
Click **OK**

Step 4 Click **OK** [**1-Sample Z (Test and Confidence Interval)** panel]

BALANCE

Population mean: σ unknown

We illustrate using the credit card balance data for a sample of 85 households that was an example in section 8.2 (file 'Balance.MTW' on the accompanying CD). The data are in column C1 of a MINITAB worksheet. In this case the population standard deviation σ will be estimated by the sample standard deviation s. The following steps can be used to compute a 90 per cent confidence interval estimate of the population mean. The dialogue panels involved are quite similar to those above (but in this case do not involve inputting the value for the standard deviation).

Step 1   **Stat > Basic Statistics > 1-Sample t**   [Main menu bar]

Step 2   Enter **C1** in the **Samples in columns**
                                    [**1-Sample t (Test and Confidence Interval)** panel]
          Click **OK**
            The MINITAB default is a 95 per cent confidence level. To specify a different
            confidence level such as 90 per cent:

Step 2   Enter **C1** in the **Samples in columns** box
                                    [**1-Sample t (Test and Confidence Interval)** panel]
          Select **Options**

Step 3   Enter **90** in the **Confidence level** box   [**1-Sample t - Options** panel]
          Click **OK**

Step 4   Click **OK**   [**1-Sample t (Test and Confidence Interval)** panel]

The results of the MINITAB interval estimation procedure are shown in Figure 8.8. The sample of 85 households provides a sample mean credit card balance of €5900, a sample standard deviation of €3058, an estimate (after rounding) of the standard error of the mean of €332, and a 90 per cent confidence interval of €5348 to €6452.

Population proportion

TEETIMES

We illustrate using the survey data for women golfers presented in Section 8.4 (file 'TeeTimes.MTW' on the accompanying CD). The data are in column C1 of a MINITAB worksheet. Individual responses are recorded as Yes if the golfer is satisfied with the availability of tee times and No otherwise. The following steps can be used to compute

---

Figure 8.8 MINITAB confidence interval for the credit card balance survey

Results for: Balance.MTW

One-Sample T: Balance

Variable	N	Mean	StDev	SE Mean	90% CI
Balance	85	5900	3058	332	(5348, 6452)

a 95 per cent confidence interval estimate of the proportion of women golfers who are satisfied with the availability of tee times. The main dialogue panel is quite similar to those for the population mean procedures described above.

Step 1 **Stat > Basic Statistics > 1 Proportion** [Main menu bar]

Step 2 Enter **C1** in the **Samples in columns**
                                   [**1 Proportion (Test and Confidence Interval)** panel]
    Select **Options**

Step 3 Check **Use test and interval based on normal distribution**
                                         [**1 Proportion - Options** panel]
    (The MINITAB default is a 95 per cent confidence level. To specify a different confidence level, enter the appropriate figure in the **Confidence Level** box)
    Click **OK**

Step 4 Click **OK** [**1 Proportion (Test and Confidence Interval)** panel]

MINITAB's **1 Proportion** routine uses an alphabetical ordering of the responses and selects the *second response* for the population proportion of interest. In the women golfers example, MINITAB uses the alphabetical ordering No–Yes and then provides the confidence interval for the proportion of Yes responses. Because Yes was the response of interest, the MINITAB output was fine. However, if MINITAB's alphabetical ordering does not provide the response of interest, select any cell in the column and use the sequence: **Editor > Column > Value Order**. It will provide you with the option of entering a user-specified order. You must list the response of interest second in the **define-an-order** box.

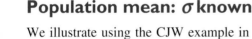

Interval estimation using EXCEL

We describe the use of EXCEL in constructing confidence intervals for a population mean (there is no inbuilt routine for a population proportion).

Population mean: σ known

We illustrate using the CJW example in Section 8.1 (file 'CJW.XLS' on the accompanying CD). The population standard deviation $\sigma = 20$ is assumed known. The satisfaction scores for the sample of 100 customers are in column A of an EXCEL worksheet. The following steps can be used to compute the margin of error for an estimate of the population mean. We begin by using EXCEL's Descriptive Statistics Tool described in Chapter 3.

Step 1 Click the **Data** tab on the Ribbon

Step 2 In the **Analysis** group, click **Data Analysis**

Step 3 Choose **Descriptive Statistics** from the list of Analysis Tools

Step 4 Enter **A1:A101** in the **Input Range** box   **[Descriptive Statistics** panel]
    Select **Grouped by Columns**
    Select **Labels in First Row**
    Select **Output Range**
    Enter **C1** in the **Output Range** box
    Select **Summary Statistics**
    Click **OK**

The summary statistics will appear in columns C and D. Continue by computing the margin of error using EXCEL's Confidence function as follows:

Step 5 Select cell C16 and enter the label **Margin of Error**

Step 6 Select cell D16 and enter the EXCEL formula = **CONFIDENCE(.05,20,100)**
    The three parameters of the Confidence function are
    Alpha $= 1 -$ confidence coefficient $= 1 - 0.95 = 0.05$
    The population standard deviation $= 20$
    The sample size $= 100$ (*Note:* This parameter appears as Count in cell D15.)

The point estimate of the population mean is in cell D3 and the margin of error is in cell D16. The point estimate (82) and the margin of error (3.92) allow the confidence interval for the population mean to be easily computed.

Population mean: σ unknown

BALANCE

We illustrate using the credit card balance data for a sample of 85 households that was an example in section 8.2 (file 'Balance.XLS' on the accompanying CD). The data are in column A of an EXCEL worksheet. The following steps can be used to compute the point estimate and the margin of error for an interval estimate of a population mean. We will use EXCEL's Descriptive Statistics Tool described in Chapter 3.

Step 1 Click the **Data** tab on the Ribbon

Step 2 In the **Analysis** group, click **Data Analysis**

Step 3 Choose **Descriptive Statistics** from the list of Analysis Tools
    Click **OK**

Step 4 Enter **A1:A86** in the **Input Range** box   **[Descriptive Statistics** panel]
    Select **Grouped by Columns**
    Check **Labels in First Row**
    Select **Output Range**
    Enter **C1** in the Output Range box
    Check **Summary Statistics**
    Check **Confidence Level for Mean**
    Enter **95** in the **Confidence Level for Mean** box
    Click **OK**

The summary statistics will appear in columns C and D. The point estimate of the population mean appears in cell D3. The margin of error, labelled 'Confidence Level (95.0 per cent)', appears in cell D16. The point estimate (€5900) and the margin of error (€660)

Figure 8.9 Interval estimation of the population mean credit card balance using EXCEL

	A	B	C	D
1	Balance		*Balance*	
2	9619			
3	5364		Mean	5900
4	8348		Standard Error	331.68667
5	7348		Median	5759
6	381		Mode	8047
7	2998		Standard Deviation	3058
8	1686		Sample Variance	9351363.8
9	1962		Kurtosis	0.2327214
10	4920		Skewness	0.4076447
11	5047		Range	14061
12	6921		Minimum	381
13	5759		Maximum	14442
14	8047		Sum	501500
15	3924		Count	85
16	3470		Confidence Level(95.0%)	659.5953
17	5994			
81	5938			
82	5266			
83	10658			
84	3910			
85	7503			
86	1582			

allow the confidence interval for the population mean to be easily computed. The output from this EXCEL procedure is shown in Figure 8.9.

Interval estimation using PASW

We describe the use of PASW in constructing confidence intervals for a population mean in the σ unknown condition. There are no inbuilt routines in PASW for the σ known condition, nor for a population proportion.

Population mean: σ unknown

We illustrate using the credit card balance data for a sample of 85 households that was an example in section 8.2 (file 'Balance.SAV' on the accompanying CD). The data are in the first column of the data file. The following steps can be used to compute the point estimate and the margin of error for an interval estimate of a population mean.

Figure 8.10 PASW confidence interval for the credit card balance survey

One-Sample Statistics

	N	Mean	Std. Deviation	Std. Error Mean
Balance	85	5900.00	3058.000	331.687

One-Sample Test

	Test Value = 0					
					95% Confidence Interval of the Difference	
	t	df	Sig. (2-tailed)	Mean Difference	Lower	Upper
Balance	17.788	84	.000	5900.000	5240.40	6559.60

Step 1 Analyze > Compare Means > One-Sample T Test   [Main menu bar]

Step 2 Transfer the Balance variable to the **Test Variable(s)** box

[**One-Sample T Test** panel]

The PASW default is a 95 per cent confidence level. To specify a different confidence level, click **Options**

Step 3 Enter the appropriate figure in the **Confidence Interval** box

[**One-Sample T Test: Options** panel]

Click **Continue**

Step 4 Click **OK**   [**One-Sample T Test** panel]

PASW produces two tables, shown in Figure 8.10. These include the sample mean (€5900), the sample standard deviation (€3058), the estimated standard error of the mean (€331.7) and the confidence interval (this is labelled as a confidence interval for 'the Difference'). The second table also includes the result of a hypothesis test (we deal with the hypothesis test in Chapter 9).

Chapter 9

Hypothesis Tests

Statistics in practice: Monitoring the quality of latex condoms

Learning objectives

After studying this chapter and doing the exercises, you should be able to:

1 Set up appropriate null and alternative hypotheses for testing research hypotheses, testing the validity of a claim and testing in decision-making situations.

2 Give an account of the logical steps involved in a statistical hypothesis test.

3 Explain the meaning of the terms null hypothesis, alternative hypothesis, Type I error, Type II error, level of significance, p-value and critical value in statistical hypothesis testing.

4 Construct and interpret hypothesis tests for a population mean:

   4.1 When the population standard deviation is known.

   4.2 When the population standard deviation is unknown.

5 Construct and interpret hypothesis tests for a population proportion.

6 Explain the relationship between the construction of hypothesis tests and confidence intervals.

7 Calculate the probability of a Type II error for a hypothesis test of a population mean when the population standard deviation is known.

8 Estimate the sample size required for a hypothesis test of a population mean when the population standard deviation is known.

In Chapters 7 and 8 we showed how a sample could be used to develop point and interval estimates of population parameters. In this chapter we continue the discussion of statistical inference by showing how hypothesis testing can be used to determine whether a statement about the value of a population parameter should or should not be rejected.

In hypothesis testing we begin by making a tentative assumption about a population parameter. This tentative assumption is called the **null hypothesis** and is denoted by H_0. We then define another hypothesis, called the **alternative hypothesis**. which is the opposite of what is stated in the null hypothesis. We denote the alternative hypothesis by H_1. The hypothesis testing procedure uses data from a sample to assess the two competing statements indicated by H_0 and H_1.

This chapter shows how hypothesis tests can be conducted about a population mean and a population proportion. We begin by providing examples of approaches to formulating null and alternative hypotheses.

9.1 Developing null and alternative hypotheses

In some applications it may not be obvious how the null and alternative hypotheses should be formulated. Care must be taken to structure the hypotheses appropriately so that the conclusion from the hypothesis test provides the information the researcher or decision-maker wants. Learning to formulate hypotheses correctly will take practice. The examples in this section show a variety of forms for H_0 and H_1 depending upon the application. Guidelines for establishing the null and alternative hypotheses will be given for three types of situations in which hypothesis testing procedures are commonly used.

Statistics in Practice

Monitoring the quality of latex condoms

Many consumer products are required by law to meet specifications set out in documents known as *standards*. This is particularly the case when there are issues of consumer safety, such as with electrical goods, children's toys or furniture (fire resistance). In less safety-critical cases, the standards may be permissive rather than obligatory, but manufacturers will often conform to the standards, and tell consumers so, as an assurance of quality. Many standards are established internationally and are embodied in documents published by the International Standards Organization (ISO). Companies based in the UK and other EU countries usually operate according to ISO standards.

The humble latex condom is the subject of ISO standard *4074:2002*. This lays down a range of specifications, relating to materials, dimensions, packaging and performance criteria including, for obvious reasons, freedom from holes. For an outline of quality testing procedures, read the relevant pages at www.durex.com, for example. *ISO 4074:2002* makes reference to other standards documents, including frequent references to *ISO 4859-1*, which lays down specifications for the

Detail of an integrated circuit with a sticker of 'quality control' showing that the product meets the specifications set out in the standards. © Pedro Antonio.

sampling schemes that must be used to ensure quality, such as in respect of freedom from holes. For the latter characteristic, *ISO 4074:2002* specifies an acceptable quality level (AQL) of no more than 0.25 per cent defective condoms in any manufacturing batch: in other words, a probability of no more than 1 in 400 that any particular condom will be defective.

As an example of the sampling specifications, suppose *ISO 4859-1* requires that, from a batch of 10 000 condoms, a random sample of 200 should be taken and examined individually for freedom from holes. This is likely to be a destructive test. Suppose *ISO 4859-1* then stipulates that the whole batch from which the sample was drawn can be declared satisfactory, in respect of freedom from holes, only if the sample contains no more than one defective condom. If the sample does contain more than one defective condom, the whole batch must be scrapped, or further tests of quality must be done to gather further information about the overall quality of the batch.

A statistician would refer to this decision procedure as a hypothesis test. The working hypothesis is that the batch conforms to the AQL specified in *ISO 4074:2002*. If the sampled batch contains more than one defective condom, this hypothesis is rejected. Otherwise the hypothesis is accepted. In making the decision on the basis of sample evidence, the quality controller is taking two risks. One risk is that a batch meeting the AQL requirement will be incorrectly rejected. The second risk is that a batch not meeting the AQL requirement will be incorrectly accepted. The sampling schemes laid down in *ISO 4859-1* are intended to clarify and restrict the level of risk involved, and to strike a sensible balance between the two types of risk.

In this chapter you will learn about the logic of statistical hypothesis testing.

Testing research hypotheses

Consider a particular model of car that currently attains an average fuel consumption of 7 litres of fuel per 100 kilometres of driving. A product research group develops a new fuel injection system specifically designed to decrease the fuel consumption. To evaluate the new system, several will be manufactured, installed in cars, and subjected to research-controlled driving tests. Here the product research group is looking for evidence to conclude that the new system *decreases* the mean fuel consumption. In this case, the research hypothesis is that the new fuel injection system will provide a mean litres-per-100 km rating below 7; that is, $\mu < 7$. As a general guideline, a research hypothesis should be stated as the *alternative hypothesis*. Hence, the appropriate null and alternative hypotheses for the study are:

$$H_0: \mu \geq 7$$
$$H_1: \mu < 7$$

If the sample results indicate that H_0 cannot be rejected, researchers cannot conclude that the new fuel injection system is better. Perhaps more research and subsequent testing should be conducted. However, if the sample results indicate that H_0 can be rejected, researchers can make the inference that $H_1: \mu < 7$ is true. With this conclusion, the researchers gain the statistical support necessary to state that the new system decreases the mean fuel consumption. Production with the new system should be considered.

In research studies such as these, the null and alternative hypotheses should be formulated so that the rejection of H_0 supports the research conclusion. The research hypothesis therefore should be expressed as the alternative hypothesis.

Testing the validity of a claim

As an illustration of testing the validity of a claim, consider the situation of a manufacturer of soft drinks who states that bottles of its products contain an average of at least 1.5 litres. A sample of bottles will be selected, and the contents will be measured to test the manufacturer's claim. In this type of hypothesis testing situation, we generally assume that the manufacturer's claim is true unless the sample evidence is contradictory. Using this approach for the soft-drink example, we would state the null and alternative hypotheses as follows.

$$H_0: \mu \geq 1.5$$
$$H_1: \mu < 1.5$$

If the sample results indicate H_0 cannot be rejected, the manufacturer's claim will not be challenged. However, if the sample results indicate H_0 can be rejected, the inference will be made that $H_1: \mu < 1.5$ is true. With this conclusion, statistical evidence indicates that the manufacturer's claim is incorrect and that the soft-drink bottles are being filled with a mean less than the claimed 1.5 litres. Appropriate action against the manufacturer may be considered.

In any situation that involves testing the validity of a claim, the null hypothesis is generally based on the assumption that the claim is true. The alternative hypothesis is then formulated so that rejection of H_0 will provide statistical evidence that the stated assumption is incorrect. Action to correct the claim should be considered whenever H_0 is rejected.

Testing in decision-making situations

In testing research hypotheses or testing the validity of a claim, action is taken if H_0 is rejected. In many instances, however, action must be taken both when H_0 cannot be rejected and when H_0 can be rejected. In general, this type of situation occurs when a decision-maker must choose between two courses of action, one associated with the null hypothesis and another associated with the alternative hypothesis. The quality-testing scenario outlined in the Statistics in Practice at the beginning of the chapter is an example of this.

Suppose that, on the basis of a sample of parts from a shipment just received, a quality control inspector must decide whether to accept the shipment or to return the shipment to the supplier because it does not meet specifications. The specifications for a particular part require a mean length of two centimetres per part. If the mean length is greater or less than the two-centimetre standard, the parts will cause quality problems in the assembly operation. In this case, the null and alternative hypotheses would be formulated as follows.

$$H_0: \mu = 2$$
$$H_1: \mu \neq 2$$

If the sample results indicate H_0 cannot be rejected, the quality control inspector will have no reason to doubt that the shipment meets specifications, and the shipment will be accepted. However, if the sample results indicate H_0 should be rejected, the conclusion will be that the parts do not meet specifications. In this case, the quality control inspector will have sufficient evidence to return the shipment to the supplier. We see that for these types of situations, action is taken both when H_0 cannot be rejected and when H_0 can be rejected.

Summary of forms for null and alternative hypotheses

The hypothesis tests in this chapter involve one of two population parameters: the population mean and the population proportion. Depending on the situation, hypothesis tests about a population parameter may take one of three forms: two include inequalities in the null hypothesis, the third uses only an equality in the null hypothesis. For hypothesis tests involving a population mean, we let μ_0 denote the hypothesized value and choose one of the following three forms for the hypothesis test.

$$H_0: \mu \geq \mu_0 \quad H_0: \mu \leq \mu_0 \quad H_0: \mu = \mu_0$$
$$H_1: \mu < \mu_0 \quad H_1: \mu > \mu_0 \quad H_1: \mu \neq \mu_0$$

For reasons that will be clear later, the first two forms are called one-tailed tests. The third form is called a two-tailed test.

In many situations, the choice of H_0 and H_1 is not obvious and judgment is necessary to select the proper form. However, as the preceding forms show, the equality part of the expression (either $\geq$, $\leq$ or $=$) *always* appears in the null hypothesis. In selecting the proper form of H_0 and H_1, keep in mind that the alternative hypothesis is often what the test is attempting to establish. Hence, asking whether the user is looking for evidence to support $\mu < \mu_0$, $\mu > \mu_0$ or $\mu \neq \mu_0$ will help determine H_1. The following exercises are designed to provide practice in choosing the proper form for a hypothesis test involving a population mean.

Exercises

I   The manager of the Costa Resort Hotel stated that the mean weekend guest bill is €600 or less. A member of the hotel's accounting staff noticed that the total charges for guest bills have been increasing in recent months. The accountant will use a sample of weekend guest bills to test the manager's claim.

a. Which form of the hypotheses should be used to test the manager's claim? Explain.

$$H_0: \mu \geq 600 \qquad H_0: \mu \leq 600 \qquad H_0: \mu = 600$$
$$H_1: \mu < 600 \qquad H_1: \mu > 600 \qquad H_1: \mu \neq 600$$

b. What conclusion is appropriate when H_0 cannot be rejected?
c. What conclusion is appropriate when H_0 can be rejected?

2 The manager of a car dealership is considering a new bonus plan designed to increase sales volume. Currently, the mean sales volume is 14 cars per month. The manager wants to conduct a research study to see whether the new bonus plan increases sales volume. To collect data on the plan, a sample of sales personnel will be allowed to sell under the new bonus plan for a one-month period.

a. Formulate the null and alternative hypotheses most appropriate for this research situation.
b. Comment on the conclusion when H_0 cannot be rejected.
c. Comment on the conclusion when H_0 can be rejected.

3 A production line operation is designed to fill cartons with laundry detergent to a mean weight of 0.75 kg. A sample of cartons is periodically selected and weighed to determine whether under-filling or over-filling is occurring. If the sample data lead to a conclusion of under-filling or over-filling, the production line will be shut down and adjusted to obtain proper filling.

a. Formulate the null and alternative hypotheses that will help in deciding whether to shut down and adjust the production line.
b. Comment on the conclusion and the decision when H_0 cannot be rejected.
c. Comment on the conclusion and the decision when H_0 can be rejected.

4 Because of high production-changeover time and costs, a director of manufacturing must convince management that a proposed manufacturing method reduces costs before the new method can be implemented. The current production method operates with a mean cost of €320 per hour. A research study will measure the cost of the new method over a sample production period.

a. Formulate the null and alternative hypotheses most appropriate for this study.
b. Comment on the conclusion when H_0 cannot be rejected.
c. Comment on the conclusion when H_0 can be rejected.

9.2 Type I and Type II errors

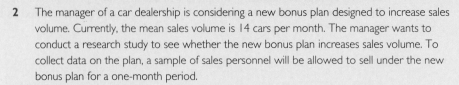

The null and alternative hypotheses are competing statements about the population. Either the null hypothesis H_0 is true or the alternative hypothesis H_1 is true, but not both. Ideally the hypothesis testing procedure should lead to the acceptance of H_0 when H_0 is true and the rejection of H_0 when H_1 is true. Unfortunately, the correct conclusions are not always possible. Because hypothesis tests are based on sample information, we must allow for the possibility of errors. Table 9.1 illustrates the two kinds of errors that can be made in hypothesis testing.

The first row of Table 9.1 shows what can happen if the conclusion is to accept H_0. If H_0 is true, this conclusion is correct. However, if H_1 is true, we make a **Type II error**; that is, we accept H_0 when it is false. The second row of Table 9.1 shows what can happen if the conclusion is to reject H_0. If H_0 is true, we make a **Type I error**; that is, we reject H_0 when it is true. However, if H_1 is true, rejecting H_0 is correct.

Table 9.1 Errors and correct conclusions in hypothesis testing

		Population condition	
		H_0 true	H_1 true
Conclusion	Accept H_0	Correct conclusion	Type II error
	Reject H_0	Type I error	Correct conclusion

Recall the hypothesis testing illustration discussed in Section 9.1 in which a product research group developed a new fuel injection system designed to decrease the fuel consumption of a particular car. With the current model achieving an average of 7 litres of fuel per 100 km, the hypothesis test was formulated as follows.

$$H_0: \mu \geq 7$$
$$H_1: \mu < 7$$

The alternative hypothesis, $H_1: \mu < 7$, indicates that the researchers are looking for sample evidence to support the conclusion that the population mean fuel consumption with the new fuel injection system is less than 7.

In this application, the Type I error of rejecting H_0 when it is true corresponds to the researchers claiming that the new system reduces fuel consumption ($\mu < 7$) when in fact the new system is no better than the current system. In contrast, the Type II error of accepting H_0 when it is false corresponds to the researchers concluding that the new system is no better than the current system ($\mu \geq 7$) when in fact the new system reduces fuel consumption.

For the fuel consumption hypothesis test, the null hypothesis is $H_0: \mu \geq 7$. Suppose the null hypothesis is true as an equality; that is, $\mu = 7$. The probability of making a Type I error when the null hypothesis is true as an equality is called the **level of significance**. For the fuel efficiency hypothesis test, the level of significance is the probability of rejecting $H_0: \mu \geq 7$ when $\mu = 7$. Because of the importance of this concept, we now restate the definition of level of significance.

Level of significance

The level of significance is the probability of making a Type I error when the null hypothesis is true as an equality.

The Greek symbol α (alpha) is used to denote the level of significance. In practice, the person conducting the hypothesis test specifies the level of significance. By selecting α, that person is controlling the probability of making a Type I error. If the cost of making a Type I error is high, small values of α are preferred. If the cost of making a Type I error is not too high, larger values of α are typically used. Common choices for α are 0.05 and 0.01. Applications of hypothesis testing that only control for the Type I error are often called *significance tests*. Most applications of hypothesis testing are of this type.

Although most applications of hypothesis testing control for the probability of making a Type I error, they do not always control for the probability of making a Type II error. Hence, if we decide to accept H_0, we cannot determine how confident we can be with that decision. Because of the uncertainty associated with making a Type II error, statisticians often recommend that we use the statement 'do not reject H_0' instead of 'accept H_0'. Using the statement 'do not reject H_0' carries the recommendation to withhold both judgment and action. In effect, by not directly accepting H_0, the statistician avoids the risk of making a Type II error. Whenever the probability of making a Type II error has not been determined and controlled, we will not make the statement 'accept H_0'. In such cases, the two conclusions possible are: *do not reject H_0* or *reject H_0*.

Although controlling for a Type II error in hypothesis testing is not common, it can be done. In Sections 9.7 and 9.8 we shall illustrate procedures for determining and controlling the probability of making a Type II error. If proper controls have been established for this error, action based on the 'accept H_0' conclusion can be appropriate.

Exercises

5 The label on a container of orange juice claims that the orange juice contains an average of 1 gram of fat or less. Answer the following questions for a hypothesis test that could be used to test the claim on the label.

a. Formulate the appropriate null and alternative hypotheses.

b. What is the Type I error in this situation? What are the consequences of making this error?

c. What is the Type II error in this situation? What are the consequences of making this error?

6 Carpetland salespersons average €5000 per week in sales. The company's chief executive officer proposes a remuneration plan with new selling incentives. The CEO hopes that the results of a trial selling period will enable him to conclude that the remuneration plan increases the average sales per salesperson.

a. Formulate the appropriate null and alternative hypotheses.

b. What is the Type I error in this situation? What are the consequences of making this error?

c. What is the Type II error in this situation? What are the consequences of making this error?

7 Suppose a new production method will be implemented if a hypothesis test supports the conclusion that the new method reduces the mean operating cost per hour.

a. State the appropriate null and alternative hypotheses if the mean cost for the current production method is €320 per hour.

b. What is the Type I error in this situation? What are the consequences of making this error?

c. What is the Type II error in this situation? What are the consequences of making this error?

9.3 Population mean: σ known

In Chapter 8 we said that the σ known case corresponds to applications in which historical data and/or other information are available that enable us to obtain a good estimate of the population standard deviation prior to sampling. In such cases the population standard deviation can, for all practical purposes, be considered known. In this section we show how to conduct a hypothesis test about a population mean for the σ known case. The methods presented in this section are exact if the sample is selected from a population that is normally distributed. In cases where it is not reasonable to assume the population is normally distributed, these methods are still applicable if the sample size is large enough. We provide some practical advice concerning the population distribution and the sample size at the end of this section.

One-tailed test

One-tailed tests about a population mean take one of the following two forms.

Lower tail test	Upper tail test
$H_0: \mu \geq \mu_0$	$H_0: \mu \leq \mu_0$
$H_1: \mu < \mu_0$	$H_1: \mu > \mu_0$

Consider an example. Trading Standards Officers periodically conduct statistical studies to test the claims that manufacturers make about their products. For example, suppose the label on a large bottle of Cola states that the bottle contains three litres of Cola. European legislation acknowledges that the bottling process cannot guarantee exactly three litres of Cola in each bottle, even if the mean filling volume for the population of all bottles filled is three litres per bottle. However, if the population mean filling volume is at least three litres per bottle, the rights of consumers will be protected. The legislation interprets the label information on a large bottle of Cola as a claim that the population mean filling weight is at least three litres per bottle. We shall show how a Trading Standards Officer can check the claim by conducting a lower tail hypothesis test.

The first step is to formulate the null and alternative hypotheses for the test. If the population mean filling volume is at least three litres per bottle, the manufacturer's claim is correct. This establishes the null hypothesis for the test. However, if the population mean weight is less than three litres per bottle, the manufacturer's claim is incorrect. This establishes the alternative hypothesis. With μ denoting the population mean filling volume, the null and alternative hypotheses are as follows:

$$H_0: \mu \geq 3$$
$$H_1: \mu < 3$$

Note that the hypothesized value of the population mean is $\mu_0 \geq 3$. If the sample data indicate that H_0 cannot be rejected, the statistical evidence does not support the conclusion that a labelling violation has occurred. Hence, no action should be taken against the manufacturer. However, if the sample data indicate H_0 can be rejected, we shall conclude that the alternative hypothesis, $H_1: \mu < 3$, is true. In this case a conclusion of under-filling and a charge of a labelling violation against the manufacturer would be justified.

Suppose a sample of 36 bottles is selected and the sample mean is computed as an estimate of the population mean μ. If the value of the sample mean is less than three litres, the sample results will cast doubt on the null hypothesis. What we want to know is how much less than three litres the sample mean must be before we would be willing to declare the difference significant and risk making a Type I error by falsely accusing the manufacturer of a labelling violation. A key factor in addressing this issue is the value the decision-maker selects for the level of significance.

As noted in the preceding section, the level of significance, denoted by α, is the probability of making a Type I error by rejecting H_0 when the null hypothesis is true as an equality. The decision-maker must specify the level of significance. If the cost of making a Type I error is high, a small value should be chosen for the level of significance. If the cost is not high, a larger value is more appropriate. Suppose that in the Cola bottling study, the Trading Standards Officer made the following statement: 'If the manufacturer is meeting its weight specifications at $\mu = 3$, I would like a 99 per cent chance of not taking any action against the manufacturer. Although I do not want to accuse the manufacturer wrongly of under-filling its product, I am willing to risk a 1 per cent chance of making such an error'. From the Standards Officer's statement, we set the level of significance for the hypothesis test at $\alpha = 0.01$. Hence, we must design the hypothesis test so that the probability of making a Type I error when $\mu = 3$ is 0.01.

For the Cola bottling study, by developing the null and alternative hypotheses and specifying the level of significance for the test, we carry out the first two steps required in conducting every hypothesis test. We are now ready to perform the third step of hypothesis testing: collect the sample data and compute the value of what is called a test statistic.

Test statistic

For the Cola bottling study, previous Trading Standards tests show that the population standard deviation can be assumed known with a value of $\sigma = 0.18$. In addition, these tests also show that the population of filling weights can be assumed to have a normal distribution. From the study of sampling distributions in Chapter 7 we know that if the population from which we are sampling is normally distributed, the sampling distribution of the sample mean $\overline{X}$ will also be normal in shape. Hence, for the Cola bottling study, the sampling distribution of $\overline{X}$ is normal. With a known value of $\sigma = 0.18$ and a sample size of $n = 36$, Figure 9.1 shows the sampling distribution of $\overline{X}$ when the null hypothesis is true as an equality; that is, when $\mu = \mu_0 = 3$. In constructing sampling distributions for hypothesis tests, it is assumed that H_0 is satisfied as an equality. Note that the standard error of is given by

$$\sigma_{\overline{X}} = \sigma/\sqrt{n} = 0.18/\sqrt{36} = 0.03$$

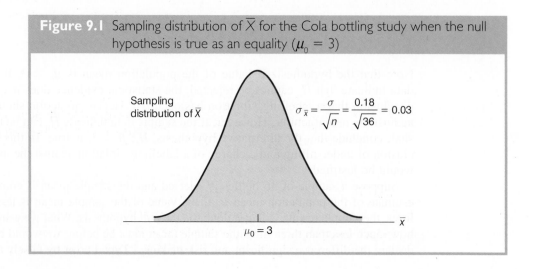

Figure 9.1 Sampling distribution of $\overline{X}$ for the Cola bottling study when the null hypothesis is true as an equality ($\mu_0 = 3$)

Because $\overline{X}$ has a normal sampling distribution, the sampling distribution of

$$Z = \frac{\overline{X} - \mu_0}{\sigma_{\overline{X}}} = \frac{\overline{X} - 3}{0.03}$$

is a standard normal distribution. A value $z = -1$ means that $\bar{x}$ is one standard error below the mean, a value $z = -2$ means that $\bar{x}$ is two standard errors below the mean, and so on. We can use the standard normal distribution table to find the lower tail probability corresponding to any z value. For instance, the standard normal table shows that the cumulative probability for $z = -3.00$ is 0.0014. This is the probability of obtaining a value that is three or more standard errors below the mean. As a result, the probability of obtaining a value $\bar{x}$ that is 3 or more standard errors below the hypothesized population mean $\mu_0 = 3$ is also 0.0014. Such a result is unlikely if the null hypothesis is true.

For hypothesis tests about a population mean for the σ known case, we use the standard normal random variable Z as a **test statistic** to determine whether $\bar{x}$ deviates from the hypothesized value μ_0 enough to justify rejecting the null hypothesis. The test statistic used in the σ known case is as follows (note that $\sigma_{\overline{X}} = \sigma/\sqrt{n}$).

Test statistic for hypothesis tests about a population mean: σ known

$$z = \frac{\overline{x} - \mu_0}{\sigma/\sqrt{n}} \qquad (9.1)$$

The key question for a lower tail test is: How small must the test statistic z be before we choose to reject the null hypothesis? Two approaches can be used to answer this question.

The first approach uses the value z from expression (9.1) to compute a probability called a *p*-**value**. The *p*-value measures the support (or lack of support) provided by the sample for the null hypothesis, and is the basis for determining whether the null hypothesis should be rejected given the level of significance. The second approach requires that we first determine a value for the test statistic called the **critical value**. For a lower tail test, the critical value serves as a benchmark for determining whether the value of the test statistic is small enough to reject the null hypothesis. We begin with the *p*-value approach.

p-value approach

The *p*-value approach has become the preferred method of determining whether the null hypothesis can be rejected, especially when using computer software packages such as MINITAB, PASW and EXCEL. We begin with a formal definition for a *p*-value.

---

p-value

The p-value is a probability, computed using the test statistic, that measures the support (or lack of support) provided by the sample for the null hypothesis.

Because a *p*-value is a probability, it ranges from 0 to 1. A small *p*-value indicates a sample result that is unusual given the assumption that H_0 is true. Small *p*-values lead to rejection of H_0, whereas large *p*-values indicate the null hypothesis should not be rejected.

Two steps are required to use the *p*-value approach. First, we must use the value of the test statistic to compute the *p*-value. The method used to compute a *p*-value depends on whether the test is lower tail, upper tail, or a two-tailed test. For a lower tail test, the *p*-value is the probability of obtaining a value for the test statistic at least as small as that provided by the sample. To compute the *p*-value for the lower tail test in the σ known case, we must find the area under the standard normal curve to the left of the test statistic. After computing the *p*-value, we must then decide whether it is small enough to reject the null hypothesis. As we will show, this involves comparing it to the level of significance.

We now illustrate the *p*-value approach by computing the *p*-value for the Cola bottling lower tail test. Suppose the sample of 36 cola bottles provides a sample mean of $\bar{x} = 2.92$ litres. Is $\bar{x} = 2.92$ small enough to cause us to reject H_0? Because this test is a lower tail test, the *p*-value is the area under the standard normal curve to the left of the test statistic. Using $\bar{x} = 2.92$, $\sigma = 0.18$, and $n = 36$, we compute the value z of the test statistic:

$$z = \frac{\bar{x} - \mu_0}{\sigma/\sqrt{n}} = \frac{2.92 - 3}{0.18/\sqrt{36}} = -2.67$$

The *p*-value is the probability that the test statistic Z is less than or equal to -2.67 (the area under the standard normal curve to the left of $z = -2.67$).

Using the standard normal distribution table, we find that the cumulative probability for $z = -2.67$, which in this case is the *p*-value, is 0.00382. Figure 9.2 shows that $\bar{x} = 2.92$

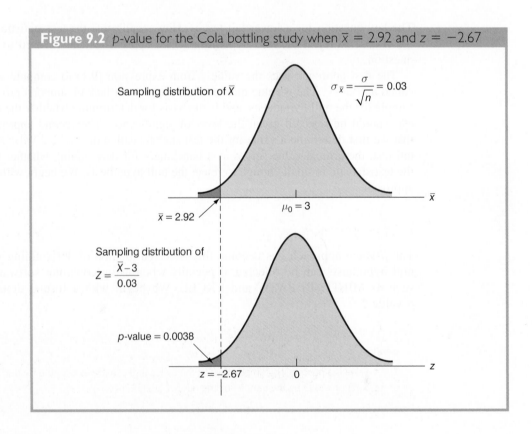

Figure 9.2 *p*-value for the Cola bottling study when $\bar{x} = 2.92$ and $z = -2.67$

corresponds to $z = -2.67$ and a p-value $= 0.0038$. This p-value indicates a small probability of obtaining a sample mean of $\bar{x} = 2.92$ or smaller when sampling from a population with $\mu = 3$. This p-value does not provide much support for the null hypothesis, but is it small enough to cause us to reject H_0? The answer depends upon the level of significance for the test.

As noted previously, the Trading Standards Officer selected a value of 0.01 for the level of significance. The selection of $\alpha = 0.01$ means that the director is willing to accept a probability of 0.01 of rejecting the null hypothesis when it is true as an equality ($\mu_0 = 3$). The sample of 36 bottles in the Cola bottling study resulted in a p-value $= 0.0038$, which means that the probability of obtaining a value of $\bar{x} = 2.92$ or less when the null hypothesis is true as an equality is 0.0038. Because 0.0038 is less than $\alpha = 0.01$ we reject H_0. Therefore, we find sufficient statistical evidence to reject the null hypothesis at the 0.01 level of significance.

We can now state the general rule for determining whether the null hypothesis can be rejected when using the p-value approach. For a level of significance α, the rejection rule using the p-value approach is as follows:

Rejection rule using p-value

Reject H_0 if p-value $\leq \alpha$

In the Cola bottling test, the p-value of 0.0038 resulted in the rejection of the null hypothesis. The basis for rejecting H_0 is a comparison of the p-value to the level of significance ($\alpha = 0.01$) specified by the Trading Standards Officer. However, the observed p-value of 0.0038 means that we would reject H_0 for any value $\alpha \geq 0.0038$. For this reason, the p-value is also called the *observed level of significance* or the *attained level of significance*.

Different decision-makers may express different opinions concerning the cost of making a Type I error and may choose a different level of significance. By providing the p-value as part of the hypothesis testing results, another decision-maker can compare the reported p-value to his or her own level of significance and possibly make a different decision with respect to rejecting H_0. The smaller the p-value, the greater the evidence against H_0, and the more the evidence in favour of H_1. Here are some guidelines statisticians suggest for interpreting small p-values.

- Less than 0.01 – Very strong evidence to conclude H_1 is true.
- Between 0.01 and 0.05 – Moderately strong evidence to conclude H_1 is true.
- Between 0.05 and 0.10 – Weak evidence to conclude H_1 is true.
- Greater than 0.10 – Insufficient evidence to conclude H_1 is true.

Critical value approach

For a lower tail test, the critical value is the value of the test statistic that corresponds to an area of α (the level of significance) in the lower tail of the sampling distribution of the test statistic. In other words, the critical value is the largest value of the test statistic that will result in the rejection of the null hypothesis. Let us return to the Cola bottling example and see how this approach works.

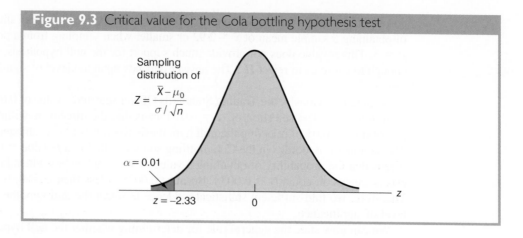

Figure 9.3 Critical value for the Cola bottling hypothesis test

In the σ known case, the sampling distribution for the test statistic Z is a standard normal distribution. Therefore, the critical value is the value of the test statistic that corresponds to an area of $\alpha = 0.01$ in the lower tail of a standard normal distribution. Using the standard normal distribution table, we find that $z = -2.33$ gives an area of 0.01 in the lower tail (see Figure 9.3). So if the sample results in a value of the test statistic that is less than or equal to -2.33, the corresponding p-value will be less than or equal to 0.01; in this case, we should reject the null hypothesis. Hence, for the Cola bottling study the critical value rejection rule for a level of significance of 0.01 is

$$\text{Reject } H_0 \text{ if } z \leq -2.33$$

In the Cola bottling example, $\bar{x} = 2.92$ and the test statistic is $z = -2.67$. Because $z = -2.67 < -2.33$, we can reject H_0 and conclude that the Cola manufacturer is under-filling bottles.

We can generalize the rejection rule for the critical value approach to handle any level of significance. The rejection rule for a lower tail test follows.

Rejection rule for a lower tail test: critical value approach

$$\text{Reject } H_0 \text{ if } z \leq -z_\alpha$$

where $-z_\alpha$ is the critical value; that is, the z value that provides an area of α in the lower tail of the standard normal distribution.

The p-value approach and the critical value approach will always lead to the same rejection decision. That is, whenever the p-value is less than or equal to α, the value of the test statistic will be less than or equal to the critical value. The advantage of the p-value approach is that the p-value tells us *how* significant the results are (the observed level of significance). If we use the critical value approach, we only know that the results are significant at the stated level of significance α.

Computer procedures for hypothesis testing provide the p-value, so it is rapidly becoming the preferred method of doing hypothesis tests. If you do not have access to a computer, you may prefer to use the critical value approach. For some probability distributions it is easier to use statistical tables to find a critical value than to use the tables to compute the p-value. This topic is discussed further in the next section.

At the beginning of this section, we said that one-tailed tests about a population mean take one of the following two forms:

Lower tail test	Upper tail test
$H_0: \mu \geq \mu_0$	$H_0: \mu \leq \mu_0$
$H_1: \mu < \mu_0$	$H_1: \mu > \mu_0$

We used the Cola bottling study to illustrate how to conduct a lower tail test. We can use the same general approach to conduct an upper tail test. The test statistic is still computed using equation (9.1). But, for an upper tail test, the p-value is the probability of obtaining a value for the test statistic at least as large as that provided by the sample. To compute the p-value for the upper tail test in the σ known case, we must find the area under the standard normal curve to the right of the test statistic. Using the critical value approach causes us to reject the null hypothesis if the value of the test statistic is greater than or equal to the critical value z_α. In other words, we reject H_0 if $z \geq z_\alpha$.

Two-tailed test

In hypothesis testing, the general form for a **two-tailed test** about a population mean is as follows:

$$H_0: \mu = \mu_0$$
$$H_1: \mu \neq \mu_0$$

In this subsection we show how to conduct a two-tailed test about a population mean for the σ known case. As an illustration, we consider the hypothesis testing situation facing MaxFlight, a manufacturer of golf equipment who use a high technology manufacturing process to produce golf balls with an average driving distance of 295 metres. Sometimes, however, the process gets out of adjustment and produces golf balls with average distances different from 295 metres. When the average distance falls below 295 metres, the company worries about losing sales because the golf balls do not provide as much distance as advertised. However, some of the national golfing associations impose equipment standards for professional competition and when the average driving distance exceeds 295 metres, MaxFlight's golf balls may be rejected for exceeding the overall distance standard concerning carry and roll.

MaxFlight's quality control programme involves taking periodic samples of 50 golf balls to monitor the manufacturing process. For each sample, a hypothesis test is done to determine whether the process has fallen out of adjustment. Let us develop the null and alternative hypotheses. We begin by assuming that the process is functioning correctly; that is, the golf balls being produced have a mean driving distance of 295 metres. This assumption establishes the null hypothesis. The alternative hypothesis is that the mean driving distance is not equal to 295 yards. With a hypothesized value of $\mu_0 = 295$, the null and alternative hypotheses for the MaxFlight hypothesis test are as follows:

$$H_0: \mu = 295$$
$$H_1: \mu \neq 295$$

If the sample mean is significantly less than 295 metres or significantly greater than 295 metres, we will reject H_0. In this case, corrective action will be taken to adjust the manufacturing process. On the other hand, if $\bar{x}$ does not deviate from the hypothesized mean $\mu_0 = 295$ by a significant amount, H_0 will not be rejected and no action will be taken to adjust the manufacturing process.

The quality control team selected $\alpha = 0.05$ as the level of significance for the test. Data from previous tests conducted when the process was known to be in adjustment show that the population standard deviation can be assumed known with a value of $\sigma = 12$. With a sample size of $n = 50$, the standard error of the sample mean is

$$\sigma_{\bar{x}} = \frac{\sigma}{\sqrt{n}} = \frac{12}{\sqrt{50}} = 1.7$$

GOLFTEST

Because the sample size is large, the central limit theorem (see Chapter 7) allows us to conclude that the sampling distribution of $\bar{X}$ can be approximated by a normal distribution. Figure 9.4 shows the sampling distribution of $\bar{X}$ for the Maxflight hypothesis test with a hypothesized population mean of $\mu_0 = 295$.

Suppose that a sample of 50 golf balls is selected and that the sample mean is 297.6 metres. This sample mean suggests that the population mean may be larger than 295 metres. Is this value $\bar{x} = 297.6$ sufficiently larger than 295 to cause us to reject H_0 at the 0.05 level of significance? In the previous section we described two approaches that can be used to answer this question: the p-value approach and the critical value approach.

p-value approach

Recall that the p-value is a probability, computed using the test statistic, that measures the support (or lack of support) provided by the sample for the null hypothesis. For a two-tailed test, values of the test statistic in *either* tail show a lack of support for the null hypothesis. For a two-tailed test, the p-value is the probability of obtaining a value for the test statistic *at least as unlikely* as that provided by the sample. Let us see how the p-value is computed for the MaxFlight hypothesis test.

First we compute the value of the test statistic. For the σ known case, the test statistic Z is a standard normal random variable. Using equation (9.1) with $\bar{x} = 297.6$, the value of the test statistic is

$$z = \frac{\bar{x} - \mu_0}{\sigma/\sqrt{n}} = \frac{297.6 - 295}{12/\sqrt{50}} = 1.53$$

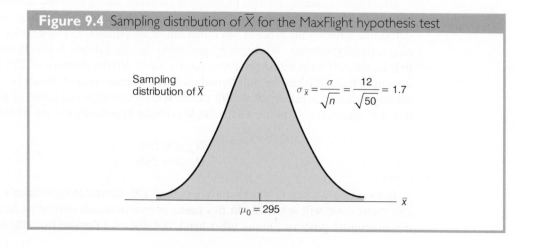

Figure 9.4 Sampling distribution of $\bar{X}$ for the MaxFlight hypothesis test

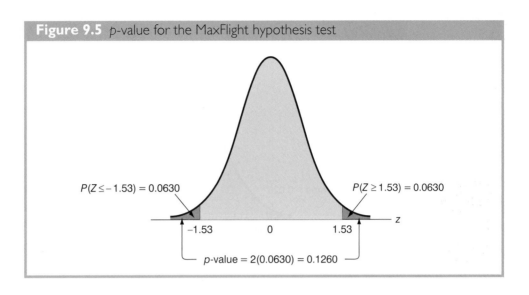

Figure 9.5 p-value for the MaxFlight hypothesis test

$P(Z \leq -1.53) = 0.0630$ $P(Z \geq 1.53) = 0.0630$

-1.53 0 1.53 z

p-value = 2(0.0630) = 0.1260

Now to compute the p-value we must find the probability of obtaining a value for the test statistic *at least as unlikely* as $z = 1.53$. Clearly values ≥ 1.53 are *at least as unlikely*. But, because this is a two-tailed test, values ≤ -1.53 are also *at least as unlikely* as the value of the test statistic provided by the sample. Referring to Figure 9.5, we see that the two-tailed p-value in this case is given by $P(Z \leq -1.53) + P(Z \geq 1.53)$. Because the normal curve is symmetrical, we can compute this probability by finding the area under the standard normal curve to the left of $z = -1.53$ and doubling it. The table of cumulative probabilities for the standard normal distribution shows that the area to the left of $z = -1.53$ is 0.0630. Doubling this, we find the p-value for the MaxFlight two-tailed hypothesis test is $2(0.0630) = 0.1260$.

Next we compare the p-value to the level of significance to see if the null hypothesis should be rejected. With a level of significance of $\alpha = 0.05$, we do not reject H_0 because the p-value $= 0.1260 > 0.05$. Because the null hypothesis is not rejected, no action will be taken to adjust the MaxFlight manufacturing process.

The computation of the p-value for a two-tailed test may seem a bit confusing as compared to the computation of the p-value for a one-tailed test. But it can be simplified by following these three steps.

1 Compute the value of the test statistic z.

2 If the value of the test statistic is in the upper tail ($z > 0$), find the area under the standard normal curve to the right of z. If the value of the test statistic is in the lower tail, find the area under the standard normal curve to the left of z.

3 Double the tail area, or probability, obtained in step 2 to obtain the p-value.

In practice, the computation of the p-value is done automatically when using computer software such as MINITAB, PASW and EXCEL.

Critical value approach

Now let us see how the test statistic can be compared to a critical value to make the hypothesis testing decision for a two-tailed test. Figure 9.6 shows that the critical values for the test will occur in both the lower and upper tails of the standard normal distribution. With a level of significance of $\alpha = 0.05$, the area in each tail beyond the critical values is $\alpha/2 = 0.05/2 = 0.025$. Using the table of probabilities for the standard

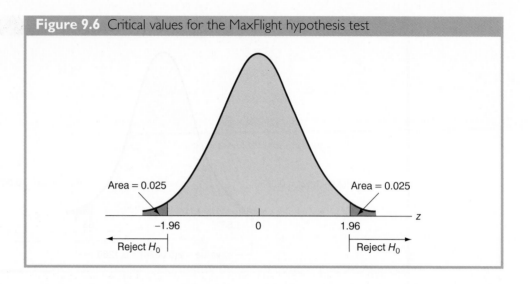

Figure 9.6 Critical values for the MaxFlight hypothesis test

normal distribution, we find the critical values for the test statistic are $-z_{0.025} = -1.96$ and $z_{0.025} = 1.96$. Using the critical value approach, the two-tailed rejection rule is

$$\text{Reject } H_0 \text{ if } z \leq -1.96 \text{ or if } z \geq 1.96$$

Because the value of the test statistic for the MaxFlight study is $z = 1.53$, the statistical evidence will not permit us to reject the null hypothesis at the 0.05 level of significance.

Summary and practical advice

We presented examples of a lower tail test and a two-tailed test about a population mean. Based upon these examples, we can now summarize the hypothesis testing procedures about a population mean for the σ known case as shown in Table 9.2. Note that μ_0 is the hypothesized value of the population mean. The hypothesis testing steps followed in the two examples presented in this section are common to every hypothesis test.

Steps of hypothesis testing

Step 1 Formulate the null and alternative hypotheses.

Step 2 Specify the level of significance α.

Step 3 Collect the sample data and compute the value of the test statistic.

p-value approach

Step 4 Use the value of the test statistic to compute the p-value.

Step 5 Reject H_0 if the p-value $\leq \alpha$.

Critical value approach

Step 4 Use the level of significance α to determine the critical value and the rejection rule.

Step 5 Use the value of the test statistic and the rejection rule to determine whether to reject H_0.

Table 9.2 Summary of hypothesis tests about a population mean: σ known case

	Lower tail test	Upper tail test	Two-tailed test
Hypotheses	$H_0: \mu \geq \mu_0$ $H_1: \mu < \mu_0$	$H_0: \mu \leq \mu_0$ $H_1: \mu > \mu_0$	$H_0: \mu = \mu_0$ $H_1: \mu \neq \mu_0$
Test statistic	$z = \dfrac{\bar{x} - \mu_0}{\sigma/\sqrt{n}}$	$z = \dfrac{\bar{x} - \mu_0}{\sigma/\sqrt{n}}$	$z = \dfrac{\bar{x} - \mu_0}{\sigma/\sqrt{n}}$
Rejection rule: p-value approach	Reject H_0 if p-value $\leq \alpha$	Reject H_0 if p-value $\leq \alpha$	Reject H_0 if p-value $\leq \alpha$
Rejection rule: critical value approach	Reject H_0 if $z \leq -z_\alpha$	Reject H_0 if $z \geq z_\alpha$	Reject H_0 if $z \leq -z_{\alpha/2}$ or if $z \geq z_{\alpha/2}$

Practical advice about the sample size for hypothesis tests is similar to the advice we provided about the sample size for interval estimation in Chapter 8. In most applications, a sample size of $n \geq 30$ is adequate when using the hypothesis testing procedure described in this section. In cases where the sample size is less than 30, the distribution of the population from which we are sampling becomes an important consideration. If the population is normally distributed, the hypothesis testing procedure that we described is exact and can be used for any sample size. If the population is not normally distributed but is at least roughly symmetrical, sample sizes as small as 15 can be expected to provide acceptable results. With smaller sample sizes, the hypothesis testing procedure presented in this section should only be used if the analyst believes, or is willing to assume, that the population is at least approximately normally distributed.

Relationship between interval estimation and hypothesis testing

We close this section by discussing the relationship between interval estimation and hypothesis testing. In Chapter 8 we showed how to construct a confidence interval estimate of a population mean. For the σ known case, the confidence interval estimate of a population mean corresponding to a $1 - \alpha$ confidence coefficient is given by

$$\bar{x} \pm z_{\alpha/2} \frac{\sigma}{\sqrt{n}} \tag{9.2}$$

Doing a hypothesis test requires us first to formulate the hypotheses about the value of a population parameter. In the case of the population mean, the two-tailed test takes the form

$$H_0: \mu = \mu_0$$
$$H_1: \mu \neq \mu_0$$

where μ_0 is the hypothesized value for the population mean. Using the two-tailed critical value approach, we do not reject H_0 for values of the sample mean that are within $-z_{\alpha/2}$ and $+z_{\alpha/2}$ standard errors of μ_0. Hence, the do-not-reject region for the sample mean in a two-tailed hypothesis test with a level of significance of α is given by

$$\mu_0 \pm z_{\alpha/2} \frac{\sigma}{\sqrt{n}} \tag{9.3}$$

A close look at expression (9.2) and expression (9.3) provides insight about the relationship between the estimation and hypothesis testing approaches to statistical inference. Note in particular that both procedures require the computation of the values $z_{\alpha/2}$ and $\sigma/\sqrt{n}$. Focusing on α, we see that a confidence coefficient of $(1 - \alpha)$ for interval estimation corresponds to a level of significance of α in hypothesis testing. For example, a 95 per cent confidence interval corresponds to a 0.05 level of significance for hypothesis testing. Furthermore, expressions (9.2) and (9.3) show that, because $z_{\alpha/2}(\sigma/\sqrt{n})$ is the plus or minus value for both expressions, if $\bar{x}$ is in the do-not-reject region defined by (9.3), the hypothesized value μ_0 will be in the confidence interval defined by (9.2). Conversely, if the hypothesized value μ_0 is in the confidence interval defined by (9.2), the sample mean will be in the do-not-reject region for the hypothesis $H_0: \mu = \mu_0$ as defined by (9.3). These observations lead to the following procedure for using a confidence interval to conduct a two-tailed hypothesis test.

A confidence interval approach to testing a hypothesis of the form

$$H_0: \mu = \mu_0$$
$$H_1: \mu \neq \mu_0$$

1 Select a simple random sample from the population and use the value of the sample mean to construct the confidence interval for the population mean μ.

$$\bar{x} \pm z_{\alpha/2} \frac{\sigma}{\sqrt{n}}$$

2 If the confidence interval contains the hypothesized value μ_0, do not reject H_0. Otherwise, reject H_0.

We return to the MaxFlight hypothesis test, which resulted in the following two-tailed test.

$$H_0: \mu = 295$$
$$H_1: \mu \neq 295$$

To test this hypothesis with a level of significance of $\alpha = 0.05$, we sampled 50 golf balls and found a sample mean distance of $\bar{x} = 297.6$ yards. Recall that the population standard deviation is $\sigma = 12$. Using these results with $z_{0.025} = 1.96$, we find that the 95 per cent confidence interval estimate of the population mean is

$$\bar{x} \pm z_{\alpha/2} \frac{\sigma}{\sqrt{n}} = 297.6 \pm 1.96 \frac{12}{\sqrt{50}} = 297.6 \pm 3.3$$

This finding enables the quality control manager to conclude with 95 per cent confidence that the mean distance for the population of golf balls is between 294.3 and 300.9 metres. Because the hypothesized value for the population mean, $\mu_0 = 295$, is in this interval, the conclusion from the hypothesis test is that the null hypothesis, $H_0: \mu = 295$, cannot be rejected.

Note that this discussion and example pertain to two-tailed hypothesis tests about a population mean. However, the same confidence interval and two-tailed hypothesis testing relationship exists for other population parameters. The relationship can also be extended to one-tailed tests about population parameters. Doing so, however, requires the development of one-sided confidence intervals.

Exercises

Note to student: Some of the exercises that follow ask you to use the p-value approach and others ask you to use the critical value approach. Both methods will provide the same hypothesis testing conclusion. We provide exercises with both methods to give you practice using both. In later sections and in following chapters, we will generally emphasize the p-value approach as the preferred method, but you may select either based on personal preference.

Methods

8   Consider the following hypothesis test:

$$H_0: \mu \geq 20$$
$$H_1: \mu < 20$$

A sample of 50 gave a sample mean of 19.4. The population standard deviation is 2.

a. Compute the value of the test statistic.
b. What is the p-value?
c. Using $\alpha = 0.05$, what is your conclusion?
d. What is the rejection rule using the critical value? What is your conclusion?

9   Consider the following hypothesis test:

$$H_0: \mu = 15$$
$$H_1: \mu \neq 15$$

A sample of 50 provided a sample mean of 14.15. The population standard deviation is 3.

a. Compute the value of the test statistic.
b. What is the p-value?
c. At $\alpha = 0.05$, what is your conclusion?
d. What is the rejection rule using the critical value? What is your conclusion?

10 Consider the following hypothesis test:

$$H_0: \mu \leq 50$$
$$H_1: \mu > 50$$

A sample of 60 is used and the population standard deviation is 8. Use the critical value approach to state your conclusion for each of the following sample results. Use $\alpha = 0.05$.

a. $\bar{x} = 52.5$
b. $\bar{x} = 51.0$
c. $\bar{x} = 51.8$

Applications

11 Suppose that the mean length of the working week for a population of workers has been previously reported as 39.2 hours. We would like to take a current sample of workers to see whether the mean length of a working week has changed from the previously reported 39.2 hours.

a. State the hypotheses that will help us determine whether a change occurred in the mean length of a working week.
b. Suppose a current sample of 112 workers provided a sample mean of 38.5 hours. Use a population standard deviation $\sigma = 4.8$ hours. What is the p-value?
c. At $\alpha = 0.05$, can the null hypothesis be rejected? What is your conclusion?
d. Repeat the preceding hypothesis test using the critical value approach.

12 Suppose the national mean sales price for new two-bedroom houses is £181 900. A sample of 40 new two-bedroom house sales in the north-east of England showed a sample mean of £166 400. Use a population standard deviation of £33 500.

a. Formulate the null and alternative hypotheses that can be used to determine whether the sample data support the conclusion that the population mean sales price for new two-bedroom houses in the north-east is less than the national mean of £181 900.

b. What is the value of the test statistic?

c. What is the p-value?

d. At $\alpha = 0.01$, what is your conclusion?

13 Fowler Marketing Research bases charges to a client on the assumption that telephone surveys can be completed in a mean time of 15 minutes or less per interview. If a longer mean interview time is necessary, a premium rate is charged. Suppose a sample of 35 interviews shows a sample mean of 17 minutes. Use $\sigma = 4$ minutes. Is the premium rate justified?

a. Formulate the null and alternative hypotheses for this application.

b. Compute the value of the test statistic.

c. What is the p-value?

d. At $\alpha = 0.01$, what is your conclusion?

14 CCN and ActMedia provided a television channel targeted to individuals waiting in supermarket checkout lines. The channel showed news, short features, and advertisements. The length of the programme was based on the assumption that the population mean time a shopper stands in a supermarket checkout line is eight minutes. A sample of actual waiting times will be used to test this assumption and determine whether actual mean waiting time differs from this standard.

a. Formulate the hypotheses for this application.

b. A sample of 120 shoppers showed a sample mean waiting time of eight and a half minutes. Assume a population standard deviation $\sigma = 3.2$ minutes. What is the p-value?

c. At $\alpha = 0.05$, what is your conclusion?

d. Compute a 95 per cent confidence interval for the population mean. Does it support your conclusion?

15 In November and December 2008, during global economic upheavals, research companies affiliated to the Worldwide Independent Network of Market Research carried out polls in 17 countries to assess people's views on the economic outlook. One of the questions asked respondents to rate their trust in their government's management of the financial situation, on a 0 to 10 scale (10 being maximum trust). Suppose the worldwide population mean on this trust question was 5.2, and we are interested in the question of whether the population mean in Germany was different from this worldwide mean.

a. State the hypotheses that could be used to address this question.

b. In the Germany survey, respondents gave the government a mean trust score of 4.0. Suppose the sample size in Germany was 1050, and the population standard deviation score was $\sigma = 2.9$. What is the 95 per cent confidence interval estimate of the population mean trust score for Germany?

c. Use the confidence interval to conduct a hypothesis test. Using $\alpha = 0.05$, what is your conclusion?

16 A production line operates with a mean filling weight of 500 grams per container. Over-filling or under-filling presents a serious problem and when detected requires the operator to shut down the production line to readjust the filling mechanism. From past data, a population standard deviation $\sigma = 25$ grams is assumed. A quality control inspector selects a sample of

30 items every hour and at that time makes the decision of whether to shut down the line for readjustment. The level of significance is $\alpha = 0.05$.

a. State the hypotheses in the hypothesis test for this quality control application.
b. If a sample mean of 510 grams were found, what is the p-value? What action would you recommend?
c. If a sample mean of 495 grams were found, what is the p-value? What action would you recommend?
d. Use the critical value approach. What is the rejection rule for the preceding hypothesis testing procedure? Repeat parts (b) and (c). Do you reach the same conclusion?

9.4 Population mean: σ unknown

In this section we describe how to do hypothesis tests about a population mean for the σ unknown case. In the σ unknown case, the sample must be used to compute estimates of both μ and σ. The sample mean is used as an estimate of μ and the sample standard deviation is used as an estimate of σ. The steps of the hypothesis testing procedure for the σ unknown case are the same as those for the σ known case described in Section 9.3. But, with σ unknown, the computation of the test statistic and p-value are a little different. For the σ known case, the sampling distribution of the test statistic has a standard normal distribution. For the σ unknown case, however, the sampling distribution of the test statistic has slightly more variability because the sample is used to compute estimates of both μ and σ.

In Section 8.2 we showed that an interval estimate of a population mean for the σ unknown case is based on a probability distribution known as the t distribution. Hypothesis tests about a population mean for the σ unknown case are also based on the t distribution. The test statistic has a t distribution with $n - 1$ degrees of freedom.

Test statistic for hypothesis tests about a population mean: σ unknown

$$t = \frac{\bar{x} - \mu_0}{s/\sqrt{n}} \qquad (9.4)$$

In Chapter 8 we said that the t distribution is based on an assumption that the population from which we are sampling has a normal distribution. However, research shows that this assumption can be relaxed considerably when the sample size is large enough. We provide some practical advice concerning the population distribution and sample size at the end of the section.

One-tailed test

Consider an example of a one-tailed test about a population mean for the σ unknown case. A business travel magazine wants to classify international airports according to the mean rating for the population of business travellers. A rating scale with a low score of 0 and a high score of 10 will be used, and airports with a population mean rating greater than seven will be designated as superior service airports. The magazine staff surveyed a sample of 60 business travellers at each airport to obtain the ratings data. Suppose the sample for Munich Airport

AIRRATING

provided a sample mean rating of $\bar{x} = 7.25$ and a sample standard deviation of $s = 1.052$. Do the data indicate that Munich should be designated as a superior service airport?

We want to develop a hypothesis test for which the decision to reject H_0 will lead to the conclusion that the population mean rating for Munich Airport is *greater* than seven. Accordingly, an upper tail test with $H_1: \mu > 7$ is required. The null and alternative hypotheses for this upper tail test are as follows:

$$H_0: \mu \leq 7$$
$$H_1: \mu > 7$$

We will use $\alpha = 0.05$ as the level of significance for the test.

Using expression (9.4) with $\bar{x} = 7.25$, $s = 1.052$, and $n = 60$, the value of the test statistic is

$$t = \frac{\bar{x} - \mu_0}{s/\sqrt{n}} = \frac{7.25 - 7}{1.052/\sqrt{60}} = 1.84$$

The sampling distribution of t has $n - 1 = 60 - 1 = 59$ degrees of freedom. Because the test is an upper tail test, the p-value is the area under the curve of the t distribution to the right of $t = 1.84$.

The t distribution table provided in most textbooks will not contain sufficient detail to determine the exact p-value, such as the p-value corresponding to $t = 1.84$. For instance, using Table 2 in Appendix B, the t distribution with 59 degrees of freedom provides the following information.

Area in upper tail	0.20	0.10	0.05	0.025	0.01	0.005
t value (59 df)	0.848	1.296	1.671	2.001	2.391	2.662

$t = 1.84$

We see that $t = 1.84$ is between 1.671 and 2.001. Although the table does not provide the exact p-value, the values in the 'Area in upper tail' row show that the p-value must be less than 0.05 and greater than 0.025. With a level of significance of $\alpha = 0.05$, this placement is all we need to know to make the decision to reject the null hypothesis and conclude that Munich should be classified as a superior service airport. Computer packages such as MINITAB, PASW and EXCEL can easily determine the exact p-value associated with the test statistic $t = 1.84$. Each of these packages will show that the p-value is 0.035 for this example. A p-value $= 0.035 < 0.05$ leads to the rejection of the null hypothesis and to the conclusion Munich should be classified as a superior service airport.

The critical value approach can also be used to make the rejection decision. With $\alpha = 0.05$ and the t distribution with 59 degrees of freedom, $t_{0.05} = 1.671$ is the critical value for the test. The rejection rule is therefore

$$\text{Reject } H_0 \text{ if } t \geq 1.671$$

With the test statistic $t = 1.84 > 1.671$, H_0 is rejected and we can conclude that Munich can be classified as a superior service airport.

Two-tailed test

To illustrate how to do a two-tailed test about a population mean for the σ unknown case, let us consider the hypothesis testing situation facing Mega Toys. The company manufactures and distributes its products through more than 1000 retail outlets. In planning production levels for the coming winter season, Mega Toys must decide how many units of each product to produce prior to knowing the actual demand at the retail level. For this year's most important new toy, Mega Toys' marketing director is expecting demand to average 40 units per retail outlet. Prior to making the final production decision based upon this estimate, Mega Toys decided to survey a sample of 25 retailers in order to develop more information about the demand for the new product. Each retailer was provided with information about the features of the new toy along with the cost and the suggested selling price. Then each retailer was asked to specify an anticipated order quantity.

With μ denoting the population mean order quantity per retail outlet, the sample data will be used to conduct the following two-tailed hypothesis test:

$$H_0: \mu = 40$$
$$H_1: \mu \neq 40$$

If H_0 cannot be rejected, Mega Toys will continue its production planning based on the marketing director's estimate that the population mean order quantity per retail outlet will be $\mu = 40$ units. However, if H_0 is rejected, Mega Toys will immediately re-evaluate its production plan for the product. A two-tailed hypothesis test is used because Mega Toys wants to re-evaluate the production plan if the population mean quantity per retail outlet is less than anticipated or greater than anticipated. Because no historical data are available (it is a new product), the population mean and the population standard deviation must both be estimated using $\bar{x}$ and s from the sample data.

The sample of 25 retailers provided a mean of $\bar{x} = 37.4$ and a standard deviation of $s = 11.79$ units. Before going ahead with the use of the t distribution, the analyst constructed a histogram of the sample data in order to check on the form of the population distribution. The histogram of the sample data showed no evidence of skewness or any extreme outliers, so the analyst concluded that the use of the t distribution with $n - 1 = 24$ degrees of freedom was appropriate. Using equation (9.4) with $\bar{x} = 37.4$, $\mu_0 = 40$, $s = 11.79$, and $n = 25$, the value of the test statistic is

ORDERS

$$t = \frac{\bar{x} - \mu_0}{s / \sqrt{n}} = \frac{37.4 - 40}{11.79 / \sqrt{25}} = -1.10$$

Because we have a two-tailed test, the p-value is two times the area under the curve for the t distribution to the left of $t = -1.10$. Using Table 2 in Appendix B, the t distribution table for 24 degrees of freedom provides the following information.

Area in upper tail	0.20	0.10	0.05	0.025	0.01	0.005
t value (24 df)	0.858	1.318	1.711	2.064	2.492	2.797

$t = 1.10$

The t distribution table only contains positive t values. Because the t distribution is symmetrical, however, we can find the area under the curve to the right of $t = 1.10$ and double it to find the p-value. We see that $t = 1.10$ is between 0.858 and 1.318. From the 'Area in upper tail' row, we see that the area in the tail to the right of $t = 1.10$ is between 0.20 and 0.10. Doubling these amounts, we see that the p-value must be between 0.40 and 0.20. With a level of significance of $\alpha = 0.05$, we now know that the

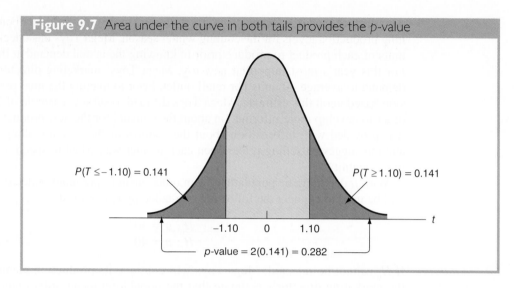

Figure 9.7 Area under the curve in both tails provides the *p*-value

$P(T \leq -1.10) = 0.141$

$P(T \geq 1.10) = 0.141$

−1.10   0   1.10

t

p-value = 2(0.141) = 0.282

p-value is greater than α. Therefore, H_0 cannot be rejected. Sufficient evidence is not available to conclude that Mega Toys should change its production plan for the coming season. Using MINITAB, PASW or EXCEL, we find that the exact *p*-value is 0.282. Figure 9.7 shows the two areas under the curve of the *t* distribution corresponding to the exact *p*-value.

The test statistic can also be compared to the critical value to make the two-tailed hypothesis testing decision. With $\alpha = 0.05$ and the *t* distribution with 24 degrees of freedom, $-t_{0.025} = -2.064$ and $t_{0.025} = 2.064$ are the critical values for the two-tailed test. The rejection rule using the test statistic is

$$\text{Reject } H_0 \text{ if } t \leq -2.064 \text{ or if } t \geq 2.064$$

Based on the test statistic $t = -1.10$, H_0 cannot be rejected. This result indicates that Mega Toys should continue its production planning for the coming season based on the expectation that $\mu = 40$ or do further investigation amongst its retailers.

Summary and practical advice

Table 9.3 provides a summary of the hypothesis testing procedures about a population mean for the σ unknown case. The key difference between these procedures and the ones for the σ known case are that *s* is used, instead of σ, in the computation of the test statistic. For this reason, the test statistic follows the *t* distribution.

The applicability of the hypothesis testing procedures of this section is dependent on the distribution of the population being sampled and the sample size. When the population is normally distributed, the hypothesis tests described in this section provide exact results for any sample size. When the population is not normally distributed, the procedures are approximations. Nonetheless, we find that sample sizes greater than 50 will provide good results in almost all cases. If the population is approximately normal, small sample sizes (e.g. $n < 15$) can provide acceptable results. In situations where the population cannot be approximated by a normal distribution, sample sizes of $n \geq 15$ will provide acceptable results as long as the population is not significantly skewed and does not contain outliers. If the population is significantly skewed or contains outliers, samples sizes approaching 50 are a good idea.

Table 9.3 Summary of hypothesis tests about a population mean: σ unknown case

	Lower tail test	Upper tail test	Two-tailed test
Hypotheses	$H_0: \mu \geq \mu_0$ $H_1: \mu < \mu_0$	$H_0: \mu \leq \mu_0$ $H_1: \mu > \mu_0$	$H_0: \mu = \mu_0$ $H_1: \mu \neq \mu_0$
Test statistic	$t = \dfrac{\bar{x} - \mu_0}{s/\sqrt{n}}$	$t = \dfrac{\bar{x} - \mu_0}{s/\sqrt{n}}$	$t = \dfrac{\bar{x} - \mu_0}{s/\sqrt{n}}$
Rejection rule: p-value approach	Reject H_0 if p-value $\leq \alpha$	Reject H_0 if p-value $\leq \alpha$	Reject H_0 if p-value $\leq \alpha$
Rejection rule: critical value approach	Reject H_0 if $t \leq -t_\alpha$	Reject H_0 if $t \geq t_\alpha$	Reject H_0 if $t \leq -t_{\alpha/2}$ or if $t \geq t_{\alpha/2}$

Exercises

Methods

17 Consider the following hypothesis test:

$$H_0: \mu \leq 12$$
$$H_1: \mu > 12$$

A sample of 25 provided a sample mean $\bar{x} = 14$ and a sample standard deviation $s = 4.32$.
a. Compute the value of the test statistic.
b. What does the t distribution table (Table 2 in Appendix B) tell you about the p-value?
c. At $\alpha = 0.05$, what is your conclusion?
d. What is the rejection rule using the critical value? What is your conclusion?

18 Consider the following hypothesis test:

$$H_0: \mu = 18$$
$$H_1: \mu \neq 18$$

A sample of 48 provided a sample mean $\bar{x} = 17$ and a sample standard deviation $s = 4.5$.
a. Compute the value of the test statistic.
b. What does the t distribution table (Table 2 in Appendix B) tell you about the p-value?
c. At $\alpha = 0.05$, what is your conclusion?
d. What is the rejection rule using the critical value? What is your conclusion?

19 Consider the following hypothesis test:

$$H_0: \mu \geq 45$$
$$H_1: \mu < 45$$

A sample of size 36 is used. Using $\alpha = 0.01$, identify the p-value and state your conclusion for each of the following sample results.

a. $\bar{x} = 44$ and $s = 5.2$
b. $\bar{x} = 43$ and $s = 4.6$
c. $\bar{x} = 46$ and $s = 5.0$

Applications

20 Grolsch lager, like some of its competitors, can be bought in handy 300 ml bottles. If a bottle such as Grolsch is marked as containing 300 ml, legislation requires that the production batch from which the bottle came must have a mean fill volume of at least 300 ml.

a. Formulate hypotheses that could be used to determine whether the mean fill volume for a production batch satisfies the legal requirement of being at least 300 ml.
b. Suppose you take a random sample of 30 bottles from a lager-bottling production line and find that the mean fill for the sample of 30 bottles is 299.5 ml, with a sample standard deviation of 1.9 ml. What is the p-value?
c. At $\alpha = 0.01$, what is your conclusion?

21 AOL Time Warner Inc.'s CNN has been the long-time ratings leader of cable television news in the US. Nielsen Media Research indicated that the mean CNN viewing audience was 600 000 viewers per day during 2002 (*The Wall Street Journal*, 10 March 2003). Assume that for a sample of 40 days during the first half of 2003, the daily audience was 612 000 viewers with a sample standard deviation of 65 000 viewers.

a. What are the hypotheses if CNN management would like information on any change in the CNN viewing audience?
b. What is the p-value?
c. Select your own level of significance. What is your conclusion?
d. What recommendation would you make to CNN management in this application?

22 The *Guardian* newspaper in the UK runs a fantasy football competition called FantasyChairman, in which competitors choose a squad of players and a manager, with the objective of increasing the valuation of the squad over the season. At the beginning of the 2005/06 competition, the mean valuation of strikers in the FantasyChairman list was £4 671 264.

a. Formulate the null and alternative hypotheses that could be used by a football pundit to determine whether mid-fielders have a higher mean valuation than strikers.
b. A random sample of 30 mid-fielders from the FantasyChairman list had a mean valuation at the start of the 2005/06 competition of £5 803 333, with a sample standard deviation of £2 460 810. On average, by how much did the valuation of mid-fielders exceed that of strikers?
c. At $\alpha = 0.05$, what is your conclusion?

23 Most new models of car sold in the European Union have to undergo an official test for fuel consumption. The test is in two parts: an urban cycle and an extra-urban cycle. The urban cycle is carried out under laboratory conditions, over at total distance of 4 km at an average speed of 19 km per hour. Consider a new car model for which the official fuel consumption figure for the urban cycle is published as 11.8 litres of fuel per 100 km. A consumer affairs organization is interested in examining whether this published figure is truly indicative of urban driving.

a. State the hypotheses that would enable the consumer affairs organization to conclude that the model's fuel consumption is more than the published 11.8 litres per 100 km.
b. A sample of 50 mileage tests with the new model of car showed a sample mean of 12.10 litres per 100 km and a sample standard deviation of 0.92 litre per 100 km. What is the p-value?
c. What conclusion should be drawn from the sample results? Use $\alpha = 0.01$.
d. Repeat the preceding hypothesis test using the critical value approach.

24 Joan's Nursery specializes in custom-designed landscaping for residential areas. The estimated labour cost associated with a particular landscaping proposal is based on the number of plantings of trees, shrubs, and so on to be used for the project. For cost-estimating purposes, managers use two hours of labour time for the planting of a medium-sized tree. Actual times from a sample of ten plantings during the past month follow (times in hours).

1.7   1.5   2.6   2.2   2.4   2.3   2.6   3.0   1.4   2.3

With a 0.05 level of significance, test to see whether the mean tree-planting time differs from two hours.

a. State the null and alternative hypotheses.
b. Compute the sample mean.
c. Compute the sample standard deviation.
d. What is the p-value?
e. What is your conclusion?

9.5 Population proportion

In this section we show how to conduct a hypothesis test about a population proportion π. Using π_0 to denote the hypothesized value for the population proportion, the three forms for a hypothesis test about a population proportion are as follows.

$$H_0: \pi \geq \pi_0 \qquad H_0: \pi \leq \pi_0 \qquad H_0: \pi = \pi_0$$
$$H_1: \pi < \pi_0 \qquad H_1: \pi > \pi_0 \qquad H_1: \pi \neq \pi_0$$

The first form is called a lower tail test, the second form is called an upper tail test, and the third form is called a two-tailed test.

Hypothesis tests about a population proportion are based on the difference between the sample proportion p and the hypothesized population proportion π_0. The methods used to do the hypothesis test are similar to those used for hypothesis tests about a population mean. The only difference is that we use the sample proportion and its standard error to compute the test statistic. The p-value approach or the critical value approach is then used to determine whether the null hypothesis should be rejected.

WOMENGYM

Let us consider an example involving a situation faced by Aspire gymnasium. Over the past year, 20 per cent of the users of Aspire were women. In an effort to increase the proportion of women users, Aspire implemented a special promotion designed to attract women. One month after the promotion was implemented, the gym manager requested a statistical study to determine whether the proportion of women users at Aspire had increased. Because the objective of the study is to determine whether the proportion of women users increased, an upper tail test with $H_1: \pi > 0.20$ is appropriate. The null and alternative hypotheses for the Aspire hypothesis test are as follows:

$$H_0: \pi \leq 0.20$$
$$H_1: \pi > 0.20$$

If H_0 can be rejected, the test results will give statistical support for the conclusion that the proportion of women users increased and the promotion was beneficial. The gym manager specified that a level of significance of $\alpha = 0.05$ be used in carrying out this hypothesis test.

The next step of the hypothesis testing procedure is to select a sample and compute the value of an appropriate test statistic. To show how this step is done for the Aspire upper tail test, we begin with a general discussion of how to compute the value of the test statistic for any form of a hypothesis test about a population proportion. The sampling distribution of P, the point estimator of the population parameter π, is the basis for developing the test statistic.

When the null hypothesis is true as an equality, the expected value of P equals the hypothesized value π_0; that is, $E(P) = \pi_0$. The standard error of P is given by

$$\sigma_P = \sqrt{\frac{\pi_0(1 - \pi_0)}{n}}$$

In Chapter 7 we said that if $n\pi \geq 5$ and $n(1 - \pi) \geq 5$, the sampling distribution of P can be approximated by a normal distribution.* Under these conditions, which usually apply in practice, the quantity

$$Z = \frac{P - \pi_0}{\sigma_P} \tag{9.5}$$

has a standard normal probability distribution, with $\sigma_P = \sqrt{\pi_0(1 - \pi_0)/n}$. Expression (9.5) gives the test statistic used to conduct hypothesis tests about a population proportion.

> **Test statistic for hypothesis tests about a population proportion**
>
> $$z = \frac{p - \pi_0}{\sqrt{\dfrac{\pi_0(1 - \pi_0)}{n}}} \tag{9.6}$$

We can now compute the test statistic for the Aspire hypothesis test. Suppose a random sample of 400 gym users was selected and that 100 of the users were women. The proportion of women users in the sample is

$$p = 100/400 = 0.25$$

Using expression (9.6), the value of the test statistic is

$$z = \frac{p - \pi_0}{\sqrt{\dfrac{\pi_0(1 - \pi_0)}{n}}} = \frac{0.25 - 0.20}{\sqrt{\dfrac{0.20(1 - 0.20)}{400}}} = \frac{0.05}{0.02} = 2.50$$

Because the Aspire hypothesis test is an upper tail test, the p-value is the probability that Z is greater than or equal to $z = 2.50$. That is, it is the area under the standard normal curve to the right of $z = 2.50$. Using the table of cumulative probabilities for the standard normal distribution, we find that the p-value for the Aspire test is therefore $(1 - 0.9938) = 0.0062$. Figure 9.8 shows this p-value calculation.

Recall that the gym manager specified a level of significance of $\alpha = 0.05$. A p-value $= 0.0062 < 0.05$ gives sufficient statistical evidence to reject H_0 at the 0.05 level of significance.

---

*In most applications involving hypothesis tests of a population proportion, sample sizes are large enough to use the normal approximation. The exact sampling distribution of P is discrete with the probability for each value of P given by the binomial distribution. So hypothesis testing is more complicated for small samples when the normal approximation cannot be used.

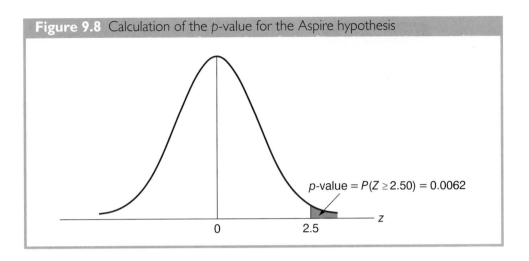

Figure 9.8 Calculation of the p-value for the Aspire hypothesis

p-value $= P(Z \geq 2.50) = 0.0062$

z

0   2.5

The test provides statistical support for the conclusion that the special promotion increased the proportion of women users at the Aspire gymnasium.

The decision whether to reject the null hypothesis can also be made using the critical value approach. The critical value corresponding to an area of 0.05 in the upper tail of a standard normal distribution is $z_{0.05} = 1.645$. Hence, the rejection rule using the critical value approach is to reject H_0 if $z \geq 1.645$. Because $z = 2.50 > 1.645$, H_0 is rejected.

Again, we see that the p-value approach and the critical value approach lead to the same hypothesis testing conclusion, but the p-value approach provides more information. With a p-value $= 0.0062$, the null hypothesis would be rejected for any level of significance greater than or equal to 0.0062.

Summary of hypothesis tests about a population proportion

The procedure used to conduct a hypothesis test about a population proportion is similar to the procedure used to conduct a hypothesis test about a population mean. Although we only illustrated how to conduct a hypothesis test about a population proportion for an upper tail test, similar procedures can be used for lower tail and two-tailed tests. Table 9.4 provides a summary of the hypothesis tests about a population proportion.

Table 9.4 Summary of hypothesis tests about a population proportion

	Lower tail test	Upper tail test	Two-tailed test
Hypotheses	$H_0: \pi \geq \pi_0$ $H_1: \pi < \pi_0$	$H_0: \pi \leq \pi_0$ $H_1: \pi > \pi_0$	$H_0: \pi = \pi_0$ $H_1: \pi \neq \pi_0$
Test statistic	$z = \dfrac{p - \pi_0}{\sqrt{\dfrac{\pi_0(1 - \pi_0)}{n}}}$	$z = \dfrac{p - \pi_0}{\sqrt{\dfrac{\pi_0(1 - \pi_0)}{n}}}$	$z = \dfrac{p - \pi_0}{\sqrt{\dfrac{\pi_0(1 - \pi_0)}{n}}}$
Rejection rule: p-value approach	Reject H_0 if p-value $\leq \alpha$	Reject H_0 if p-value $\leq \alpha$	Reject H_0 if p-value $\leq \alpha$
Rejection rule: critical value approach	Reject H_0 if $z \leq$ $-z_\alpha$	Reject H_0 if $z \geq z_\alpha$	Reject H_0 if $z \leq -z_{\alpha/2}$ or if $z \geq z_{\alpha/2}$

Exercises

Methods

25 Consider the following hypothesis test:

$$H_0: \pi = 0.20$$
$$H_1: \pi \neq 0.20$$

A sample of 400 provided a sample proportion $p = 0.175$.

a. Compute the value of the test statistic.
b. What is the p-value?
c. At $\alpha = 0.05$, what is your conclusion?
d. What is the rejection rule using the critical value? What is your conclusion?

26 Consider the following hypothesis test:

$$H_0: \pi \geq 0.75$$
$$H_1: \pi < 0.75$$

A sample of 300 items was selected. At $\alpha = 0.05$, compute the p-value and state your conclusion for each of the following sample results.

a. $p = 0.68$
b. $p = 0.72$
c. $p = 0.70$
d. $p = 0.77$

Applications

27 An airline promotion to business travellers is based on the assumption that two-thirds of business travellers use a laptop computer on overnight business trips.

a. State the hypotheses that can be used to test the assumption.
b. What is the sample proportion from an American Express sponsored survey that found 355 of 546 business travellers use a laptop computer on overnight business trips?
c. What is the p-value?
d. Use $\alpha = 0.05$. What is your conclusion?

EAGLE

28 Eagle Outfitters is a chain of stores specializing in outdoor clothing and camping gear. They are considering a promotion that involves sending discount coupons to all their credit card customers by direct mail. This promotion will be considered a success if more than 10 per cent of those receiving the coupons use them. Before going nationwide with the promotion, coupons were sent to a sample of 100 credit card customers.

a. Formulate hypotheses that can be used to test whether the population proportion of those who will use the coupons is sufficient to go national.
b. The file 'Eagle' contains the sample data. Compute a point estimate of the population proportion.
c. Use $\alpha = 0.05$ to conduct your hypothesis test. Should Eagle go national with the promotion?

29 Before the Iraqi election in January 2005, an Abu Dhabi TV/Zogby International poll asked a sample of Iraqi adults whether they would prefer an Islamic or a secular government.

a. Formulate the hypotheses that can be used to help determine whether more than 50 per cent of the adult population would prefer a secular government.

b. Suppose that, of 805 respondents to the poll, 475 expressed a preference for a secular government. What is the sample proportion? What is the p-value?

c. At $\alpha = 0.01$, what is your conclusion?

30 A study by *Consumer Reports* showed that 64 per cent of supermarket shoppers believe supermarket brands to be as good as national name brands. To investigate whether this result applies to its own product, the manufacturer of a national name-brand ketchup asked a sample of shoppers whether they believed that supermarket ketchup was as good as the national brand ketchup.

a. Formulate the hypotheses that could be used to determine whether the percentage of supermarket shoppers who believe that the supermarket ketchup was as good as the national brand ketchup differed from 64 per cent.

b. If a sample of 100 shoppers showed 52 stating that the supermarket brand was as good as the national brand, what is the p-value?

c. At $\alpha = 0.05$, what is your conclusion?

d. Should the national brand ketchup manufacturer be pleased with this conclusion? Explain.

31 Microsoft Outlook is the most widely used email manager. A Microsoft executive claims that Microsoft Outlook is used by at least 75 per cent of Internet users. A sample of Internet users will be used to test this claim.

a. Formulate the hypotheses that can be used to test the claim.

b. A Merrill Lynch study reported that Microsoft Outlook is used by 72 per cent of Internet users. Assume that the report was based on a sample size of 300 Internet users. What is the p-value?

c. At $\alpha = 0.05$, should the executive's claim of at least 75 per cent be rejected?

32 In the Kenyan presidential election in December 2002, Mwai Kibaki, representing Narc, was elected with 63 per cent of the vote ($\pi = 0.63$). A month before the election, an opinion poll had estimated the proportion of support for each candidate. Did Kibaki's support change during the last month of the election campaign?

a. Formulate the null and alternative hypotheses.

b. Suppose the November opinion poll had a random sample of 3000 potential voters, and that 68.2 per cent expressed support for Kibaki. What is the p-value? Use $\alpha = 0.05$.

c. What is your conclusion?

9.6 Hypothesis testing and decision-making

In Section 9.1 we discussed three types of situations in which hypothesis testing is used:

1 Testing research hypotheses.

2 Testing the validity of a claim.

3 Testing in decision-making situations.

In the first two situations, action is taken only when the null hypothesis H_0 is rejected and the alternative hypothesis H_1 is concluded to be true. In the third situation – decision-making – it is necessary to take action when the null hypothesis is not rejected as well as when it is rejected.

The hypothesis testing procedures presented so far have limited applicability in a decision-making situation because it is not considered appropriate to accept H_0 and take action based on the conclusion that H_0 is true. The reason for not taking action when the test results indicate *do not reject* H_0 is that the decision to accept H_0 exposes the decision-maker to the risk of making a Type II error; that is, accepting H_0 when it is false. With the hypothesis testing procedures described in the preceding sections, the probability of a Type I error is controlled by establishing a level of significance for the test. However, the probability of making the Type II error is not controlled.

Clearly, in certain decision-making situations the decision-maker may want – and in some cases may be forced – to take action with both the conclusion *do not reject* H_0 and the conclusion *reject* H_0. A good illustration of this situation is lot-acceptance sampling, a topic we will discuss in more depth in Chapter 20. For example, a quality control manager must decide to accept a shipment of batteries from a supplier or to return the shipment because of poor quality. Assume that design specifications require batteries from the supplier to have a mean useful life of at least 120 hours. To evaluate the quality of an incoming shipment, a sample of 36 batteries will be selected and tested. On the basis of the sample, a decision must be made to accept the shipment of batteries or to return it to the supplier because of poor quality. Let μ denote the mean number of hours of useful life for batteries in the shipment. The null and alternative hypotheses about the population mean follow.

$$H_0: \mu \geq 120$$
$$H_1: \mu < 120$$

If H_0 is rejected, the alternative hypothesis is concluded to be true. This conclusion indicates that the appropriate action is to return the shipment to the supplier. However, if H_0 is not rejected, the decision-maker must still determine what action should be taken. Therefore, without directly concluding that H_0 is true, but merely by not rejecting it, the decision-maker will have made the decision to accept the shipment as being of satisfactory quality.

In such decision-making situations, it is recommended that the hypothesis testing procedure be extended to control the probability of making a Type II error. Because a decision will be made and action taken when we do not reject H_0, knowledge of the probability of making a Type II error will be helpful. In Sections 9.7 and 9.8 we explain how to compute the probability of making a Type II error and how the sample size can be adjusted to help control the probability of making a Type II error.

9.7 Calculating the probability of Type II errors

In this section we show how to calculate the probability of making a Type II error for a hypothesis test about a population mean. We illustrate the procedure by using the lot-acceptance example described in Section 9.6. The null and alternative hypotheses about the mean number of hours of useful life for a shipment of batteries are $H_0: \mu \geq 120$ and $H_1: \mu < 120$. If H_0 is rejected, the decision will be to return the shipment to the supplier because the mean hours of useful life are less than the specified 120 hours. If H_0 is not rejected, the decision will be to accept the shipment.

Suppose a level of significance of $\alpha = 0.05$ is used to conduct the hypothesis test. The test statistic in the σ known case is

$$z = \frac{x - \mu_0}{\sigma/\sqrt{n}} = \frac{x - 120}{\sigma/\sqrt{n}}$$

Based on the critical value approach and $z_{0.05}$ = 1.645, the rejection rule for the lower tail test is to reject H_0 if $z \leq -1.645$. Suppose a sample of 36 batteries will be selected and based upon previous testing the population standard deviation can be assumed known with a value of $\sigma = 12$ hours. The rejection rule indicates that we will reject H_0 if

$$z = \frac{\bar{x} - 120}{12 \, / \, \sqrt{36}} \leq -1.645$$

Solving for $\bar{x}$ in the preceding expression indicates that we will reject H_0 if

$$\bar{x} \leq 120 - 1.645 \left(\frac{12}{\sqrt{36}} \right) = 116.71$$

Rejecting H_0 when $\bar{x} \leq 116.71$ means we will accept the shipment whenever $\bar{x} > 116.71$. We are now ready to compute probabilities associated with making a Type II error. We make a Type II error whenever the true shipment mean is less than 120 hours and we decide to accept H_0: $\mu \geq 120$. To compute the probability of making a Type II error, we must therefore select a value of μ less than 120 hours. For example, suppose the shipment is considered to be of poor quality if the batteries have a mean life of $\mu = 112$ hours. If $\mu = 112$, what is the probability of accepting H_0: $\mu \geq 120$ and hence committing a Type II error? This probability is the probability that the sample mean $\bar{x}$ is greater than 116.71 when $\mu = 112$.

Figure 9.9 shows the sampling distribution of the sample mean when the mean is $\mu = 112$. The shaded area in the upper tail gives the probability of obtaining $\bar{x} > 116.71$. Using the standard normal distribution, we see that at $\bar{x} = 116.71$

$$z = \frac{x - \mu_0}{\sigma/\sqrt{n}} = \frac{116.71 - 112}{12/\sqrt{36}} = 2.36$$

The standard normal distribution table shows that with $z = 2.36$, the area in the upper tail is $1 - 0.0909 = 0.0091$. Denoting the probability of making a Type II error as β, we see if $\mu = 112$, $\beta = 0.0091$. If the mean of the population is 112 hours, the probability of making a Type II error is only 0.0091.

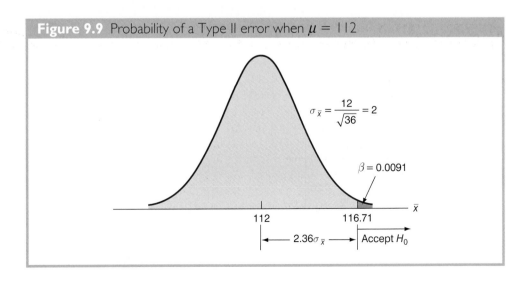

Figure 9.9 Probability of a Type II error when $\mu = 112$

Table 9.5 Probability of making a Type II error for the lot-acceptance hypothesis test

Value of μ	$z = \dfrac{116.71 - \mu}{12/\sqrt{36}}$	Probability of a Type II Error (β)	Power $(1 - \beta)$
112	2.36	0.0091	0.9909
114	1.36	0.0869	0.9131
115	0.86	0.1949	0.8051
116.71	0.00	0.5000	0.5000
117	−0.15	0.5596	0.4404
118	−0.65	0.7422	0.2578
119.999	−1.645	0.9500	0.050

We can repeat these calculations for other values of μ less than 120. Doing so will show a different probability of making a Type II error for each value of μ. For example, suppose the shipment of batteries has a mean useful life of $\mu = 115$ hours. Because we will accept H_0 whenever $\bar{x} > 116.71$, the z value for $\mu = 115$ is given by

$$z = \frac{\bar{x} - \mu_0}{\sigma/\sqrt{n}} = \frac{116.71 - 115}{12/\sqrt{36}} = 0.86$$

From the standard normal distribution table, we find that the area in the upper tail of the standard normal distribution for $z = 0.86$ is $1 - 0.8051 = 0.1949$. The probability of making a Type II error is $\beta = 0.1949$ when the true mean is $\mu = 115$.

In Table 9.5 we show the probability of making a Type II error for a variety of values of μ less than 120. Note that as μ increases towards 120, the probability of making a Type II error increases towards an upper bound of 0.95. However, as μ decreases to values further below 120, the probability of making a Type II error diminishes. This pattern is what we should expect. When the true population mean μ is close to the null hypothesis value of $\mu = 120$, the probability is high that we will make a Type II error. However, when the true population mean μ is far below the null hypothesis value of 120, the probability is low that we will make a Type II error.

Figure 9.10 Power curve for the lot-acceptance hypothesis test

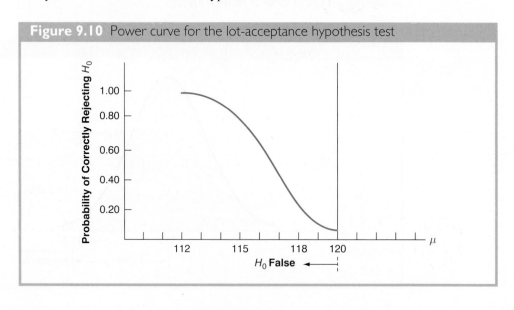

The probability of correctly rejecting H_0 when it is false is called the **power** of the test. For any particular value of μ, the power is $1 - \beta$; that is, the probability of correctly rejecting the null hypothesis is 1 minus the probability of making a Type II error. Values of power are also listed in Table 9.5. On the basis of these values, the power associated with each value of μ is shown graphically in Figure 9.10. Such a graph is called a **power curve**. Note that the power curve extends over the values of μ for which the null hypothesis is false. The height of the power curve at any value of μ indicates the probability of correctly rejecting H_0 when H_0 is false. Another graph, called the *operating characteristic curve,* is sometimes used to provide information about the probability of making a Type II error. The operating characteristic curve shows the probability of accepting H_0 and thus provides β for the values of μ where the null hypothesis is false. The probability of making a Type II error can be read directly from this graph.

In summary, the following step-by-step procedure can be used to compute the probability of making a Type II error in hypothesis tests about a population mean.

1 Formulate the null and alternative hypotheses.

2 Use the level of significance α and the critical value approach to determine the critical value and the rejection rule for the test.

3 Use the rejection rule to solve for the value of the sample mean corresponding to the critical value of the test statistic.

4 Use the results from step 3 to state the values of the sample mean that lead to the acceptance of H_0. These values define the acceptance region for the test.

5 Use the sampling distribution of $\overline{X}$ for a value of μ satisfying the alternative hypothesis, and the acceptance region from step 4, to compute the probability that the sample mean will be in the acceptance region. This probability is the probability of making a Type II error at the chosen value of μ.

Exercises

Methods

33 Consider the following hypothesis test.

$$H_0: \mu \geq 10$$
$$H_1: \mu < 10$$

The sample size is 120 and the population standard deviation is assumed known with $\sigma = 5$. Use $\alpha = 0.05$.

a. If the population mean is 9, what is the probability that the sample mean leads to the conclusion *do not reject H_0?*

b. What type of error would be made if the actual population mean is 9 and we conclude that $H_0: \mu \geq 10$ is true?

c. What is the probability of making a Type II error if the actual population mean is 8?

34 Consider the following hypothesis test.

$$H_0: \mu = 20$$
$$H_1: \mu \neq 20$$

A sample of 200 items will be taken and the population standard deviation is $\sigma = 10$. Use $\alpha = 0.05$. Compute the probability of making a Type II error if the population mean is:

a. $\mu = 18.0$
b. $\mu = 22.5$
c. $\mu = 21.0$

Applications

35 Fowler Marketing Research bases charges to a client on the assumption that telephone survey interviews can be completed within 15 minutes or less. If more time is required, a premium rate is charged. With a sample of 35 interviews, a population standard deviation of four minutes, and a level of significance of 0.01, the sample mean will be used to test the null hypothesis $H_0: \mu \leq 15$.

a. What is your interpretation of the Type II error for this problem? What is its impact on the firm?
b. What is the probability of making a Type II error when the actual mean time is $\mu = 17$ minutes?
c. What is the probability of making a Type II error when the actual mean time is $\mu = 18$ minutes?
d. Sketch the general shape of the power curve for this test.

36 Refer to Exercise 35. Assume the firm selects a sample of 50 interviews and repeat parts (b) and (c). What observation can you make about how increasing the sample size affects the probability of making a Type II error?

37 *Young Adult* magazine states the following hypotheses about the mean age of its subscribers.

$$H_0: \mu = 28$$
$$H_1: \mu \neq 28$$

a. What would it mean to make a Type II error in this situation?
b. The population standard deviation is assumed known at $\sigma = 6$ years and the sample size is 100. With $\alpha = 0.05$, what is the probability of accepting H_0 for μ equal to 26, 27, 29 and 30?
c. What is the power at $\mu = 26$? What does this result tell you?

38 Sparr Investments specializes in tax-deferred investment opportunities for its clients. Recently Sparr offered a payroll deduction investment scheme for the employees of a particular company. Sparr estimates that the employees are currently averaging €100 or less per month in tax-deferred investments. A sample of 40 employees will be used to test Sparr's hypothesis about the current level of investment activity among the population of employees. Assume the employee monthly tax-deferred investment amounts have a standard deviation of €75 and that a 0.05 level of significance will be used in the hypothesis test.

a. What would it mean to make a Type II error in this situation?
b. What is the probability of the Type II error if the actual mean employee monthly investment is €120?
c. What is the probability of the Type II error if the actual mean employee monthly investment is €130?
d. Assume a sample size of 80 employees is used and repeat parts (b) and (c).

9.8 Determining the sample size for hypothesis tests about a population mean

Assume that a hypothesis test is to be conducted about the value of a population mean. The level of significance specified by the user determines the probability of making a Type I error for the test. By controlling the sample size, the user can also control the probability of making a Type II error. Let us show how a sample size can be determined for the following lower tail test about a population mean.

$$H_0: \mu \geq \mu_0$$
$$H_1: \mu < \mu_0$$

The upper panel of Figure 9.11 is the sampling distribution of $\overline{X}$ when H_0 is true with $\mu = \mu_0$. For a lower tail test, the critical value of the test statistic is denoted $-z_\alpha$. In the upper panel of the figure the vertical line, labelled c, is the corresponding value of $\overline{x}$. Note that, if we reject H_0 when $\overline{x} = c$, the probability of a Type I error will be α. With z_α representing the z value corresponding to an area of α in the upper tail of the standard normal distribution, we compute c using the following formula:

$$c = \mu_0 - z_\alpha \frac{\sigma}{\sqrt{n}} \qquad (9.7)$$

The lower panel of Figure 9.11 is the sampling distribution of $\overline{X}$ when the alternative hypothesis is true with $\mu = \mu_1 < \mu_0$. The shaded region shows β, the probability of a Type II error that the decision-maker will be exposed to if the null hypothesis is accepted when $\overline{x} > c$. With z_β representing the z value corresponding to an area of β in the upper tail of the standard normal distribution, we compute c using the following formula:

$$c = \mu_1 + z_\beta \frac{\sigma}{\sqrt{n}} \qquad (9.8)$$

Now what we want to do is to select a value for c so that when we reject or do not reject H_0, the probability of a Type I error is equal to the chosen value of α and the probability of a Type II error is equal to the chosen value of β. Therefore, both equations (9.7) and (9.8) must provide the same value for c. Hence, the following equation must be true.

$$\mu_0 - z_\alpha \frac{\sigma}{\sqrt{n}} = \mu_1 + z_\beta \frac{\sigma}{\sqrt{n}}$$

To determine the required sample size, we first solve for $\sqrt{n}$ as follows.

$$\mu_0 - \mu_1 = z_\alpha \frac{\sigma}{\sqrt{n}} + z_\beta \frac{\sigma}{\sqrt{n}} = \frac{(z_\alpha + z_\beta)\sigma}{\sqrt{n}}$$

and

$$\sqrt{n} = \frac{(z_\alpha + z_\beta)\sigma}{(\mu_0 - \mu_1)}$$

Squaring both sides of the expression provides the following sample size formula for a one-tailed hypothesis test about a population mean.

Figure 9.11   Determining the sample size for specified levels of the Type I (α) and Type II (β) errors

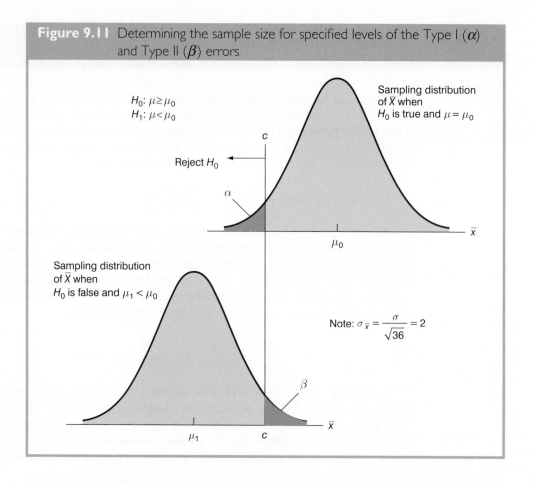

Sample size for a one-tailed hypothesis test about a population mean

$$n = \frac{(z_\alpha + z_\beta)^2 \, \sigma^2}{(\mu_0 - \mu_1)^2} \qquad\qquad \textbf{(9.9)}$$

$z_\alpha = z$ value providing an area of α in the upper tail of a standard normal distribution

$z_\beta = z$ value providing an area of β in the upper tail of a standard normal distribution

$\sigma =$ the population standard deviation

$\mu_0 =$ the value of the population mean in the null hypothesis

$\mu_1 =$ the value of the population mean used for the Type II error

Although the logic of equation (9.9) was developed for the hypothesis test shown in Figure 9.11, it holds for any one-tailed test about a population mean. Note that in a two-tailed hypothesis test about a population mean, $z_{\alpha/2}$ is used instead of z_α in equation (9.9).

Let us return to the lot-acceptance example from Sections 9.6 and 9.7. The design specification for the shipment of batteries indicated a mean useful life of at least 120 hours for the batteries. Shipments were rejected if H_0: $\mu \geq 120$ was rejected. Let us assume that the quality control manager makes the following statements about the allowable probabilities for the Type I and Type II errors.

Type I error statement: If the mean life of the batteries in the shipment is $\mu = 120$, I am willing to risk an $\alpha = 0.05$ probability of rejecting the shipment.

Type II error statement: If the mean life of the batteries in the shipment is five hours under the specification (i.e. $\mu = 115$), I am willing to risk a $\beta = 0.10$ probability of accepting the shipment.

These statements are based on the judgment of the manager. Someone else might specify different restrictions on the probabilities. However, statements about the allowable probabilities of both errors must be made before the sample size can be determined.

In the example, $\alpha = 0.05$ and $\beta = 0.10$. Using the standard normal probability distribution, we have $z_{0.05} = 1.645$ and $z_{0.10} = 1.28$. From the statements about the error probabilities, we note that $\mu_0 = 120$ and $\mu_1 = 115$. Finally, the population standard deviation was assumed known at $\sigma = 12$. By using equation (9.9), we find that the recommended sample size for the lot-acceptance example is

$$n = \frac{(1.645 + 1.28)^2(12)^2}{(120 - 115)^2} = 49.3$$

Rounding up, we recommend a sample size of 50.

Because both the Type I and Type II error probabilities have been controlled at allowable levels with $n = 50$, the quality control manager is now justified in using the *accept* H_0 and *reject* H_0 statements for the hypothesis test. The accompanying inferences are made with allowable probabilities of making Type I and Type II errors.

We can make three observations about the relationship among α, β and the sample size n.

1 Once two of the three values are known, the other can be computed.

2 For a given level of significance α, increasing the sample size will reduce β.

3 For a given sample size, decreasing α will increase β, whereas increasing α will decrease β.

The third observation should be kept in mind when the probability of a Type II error is not being controlled. It suggests that one should not choose unnecessarily small values for the level of significance α. For a given sample size, choosing a smaller level of significance means more exposure to a Type II error. Inexperienced users of hypothesis testing often think that smaller values of α are always better. They are better if we are concerned only about making a Type I error. However, smaller values of α have the disadvantage of increasing the probability of making a Type II error.

Exercises

Methods

39 Consider the following hypothesis test.

$$H_0: \mu \geq 10$$
$$H_1: \mu < 10$$

The sample size is 120 and the population standard deviation is 5. Use $\alpha = 0.05$. If the actual population mean is 9, the probability of a Type II error is 0.2912. Suppose the researcher wants to reduce the probability of a Type II error to 0.10 when the actual population mean is 9. What sample size is recommended?

40 Consider the following hypothesis test.

$$H_0: \mu = 20$$
$$H_1: \mu \neq 20$$

The population standard deviation is 10. Use $\alpha = 0.05$. How large a sample should be taken if the researcher is willing to accept a 0.05 probability of making a Type II error when the actual population mean is 22?

Applications

41 A special industrial battery must have a life of at least 400 hours. A hypothesis test is to be conducted with a 0.02 level of significance. If the batteries from a particular production run have an actual mean use life of 385 hours, the production manager wants a sampling procedure that only 10 per cent of the time would show erroneously that the batch is acceptable. What sample size is recommended for the hypothesis test? Use 30 hours as an estimate of the population standard deviation.

42 *Young Adult* magazine states the following hypotheses about the mean age of its subscribers.

$$H_0: \mu = 28$$
$$H_1: \mu \neq 28$$

If the manager conducting the test will permit a 0.15 probability of making a Type II error when the true mean age is 29, what sample size should be selected? Assume $\sigma = 6$ and a 0.05 level of significance.

43 $H_0: \mu = 120$ and $H_1: \mu \neq 120$ are used to test whether a bath soap production process is meeting the standard output of 120 bars per batch. Use a 0.05 level of significance for the test and a planning value of 5 for the standard deviation.

a. If the mean output drops to 117 bars per batch, the firm wants to have a 98 per cent chance of concluding that the standard production output is not being met. How large a sample should be selected?

b. With your sample size from part (a), what is the probability of concluding that the process is operating satisfactorily for each of the following actual mean outputs: 117, 118, 119, 121, 122 and 123 bars per batch? That is, what is the probability of a Type II error in each case?

For additional online summary questions and answers go to the companion website at www.cengage.co.uk/aswsbe2

Summary

Hypothesis testing uses sample data to determine whether a statement about the value of a population parameter should or should not be rejected. The hypotheses are two competing statements about a population parameter. One is called the null hypothesis (H_0), and the other is called the alternative hypothesis (H_1). In Section 9.1 we provided guidelines for formulating hypotheses for three situations frequently encountered in practice.

In all hypothesis tests, a relevant test statistic is calculated using sample data. The test statistic can be used to compute a p-value for the test. A p-value is a probability that measures the support (or lack of support) provided by the sample for the null hypothesis. If the p-value is less than or equal to the level of significance α, the null hypothesis can be rejected.

Conclusions can also be drawn by comparing the value of the test statistic to a critical value. For lower tail tests, the null hypothesis is rejected if the value of the test statistic is less than or equal to the critical value. For upper tail tests, the null hypothesis is rejected if the value of the test statistic is greater than or equal to the critical value. Two-tailed tests consist of two critical values: one in the lower tail of the sampling distribution and one in the upper tail. In this case, the null hypothesis is rejected if the value of the test statistic is less than or equal to the critical value in the lower tail or greater than or equal to the critical value in the upper tail.

We illustrated the relationship between hypothesis testing and interval construction in Section 9.3.

When historical data or other information provides a basis for assuming that the population standard deviation is known, the hypothesis testing procedure is based on the standard normal distribution. When σ is unknown, the sample standard deviation s is used to estimate σ and the hypothesis testing procedure is based on the t distribution.

In the case of hypothesis tests about a population proportion, the hypothesis testing procedure uses a test statistic based on the standard normal distribution.

Extensions of hypothesis testing procedures to include an analysis of the Type II error were also presented. In Section 9.7 we showed how to compute the probability of making a Type II error. In Section 9.8 we showed how to determine a sample size that will control for both the probability of making a Type I error and a Type II error.

Key terms

Alternative hypothesis
Critical value
Level of significance
Null hypothesis
One-tailed test
p-value

Power
Power curve
Test statistic
Two-tailed test
Type I error
Type II error

Key formulae

Test statistic for hypothesis tests about a population mean: σ known

$$z = \frac{\bar{x} - \mu_0}{\sigma / \sqrt{n}}$$

(9.1)

Test statistic for hypothesis tests about a population mean: σ unknown

$$t = \frac{\bar{x} - \mu_0}{s/\sqrt{n}}$$ **(9.4)**

Test statistic for hypothesis tests about a population proportion

$$z = \frac{p - \pi_0}{\sqrt{\dfrac{\pi_0(1 - \pi_0)}{n}}}$$ **(9.6)**

Sample size for a one-tailed hypothesis test about a population mean

$$n = \frac{(z_\alpha + z_\beta)^2 \, \sigma^2}{(\mu_0 - \mu_1)^2}$$ **(9.9)**

In a two-tailed test, replace z_α with $z_{\alpha/2}$.

Case problem Quality Associates

Quality Associates, a consulting firm, advises its clients about sampling and statistical procedures that can be used to control their manufacturing processes. In one particular application, a client gave Quality Associates a sample of 800 observations taken during a time in which that client's process was operating satisfactorily. The sample standard deviation for these data was 0.21; hence, with so much data, the population standard deviation was assumed to be 0.21. Quality Associates then suggested that random samples of size 30 be taken periodically to monitor the process on an ongoing basis. By analyzing the new samples, the client could quickly learn whether the process was operating satisfactorily. When the process

was not operating satisfactorily, corrective action could be taken to eliminate the problem. The design specification indicated the mean for the process should be 12. The hypothesis test suggested by Quality Associates follows.

$$H_0: \mu = 12$$
$$H_1: \mu \neq 12$$

Corrective action will be taken any time H_0 is rejected.

The data set 'Quality' on the accompanying CD contains data from four samples, each of size 30, collected at hourly intervals during the first day of operation of the new statistical control procedure.

Managerial report

1 Conduct a hypothesis test for each sample at the 0.01 level of significance and determine what action, if any, should be taken. Provide the test statistic and p-value for each test.

2 Compute the standard deviation for each of the four samples. Does the assumption of 0.21 for the population standard deviation appear reasonable?

3 Compute limits for the sample mean $\bar{X}$ around $\mu = 12$ such that, as long as a new sample mean is within those limits, the process will be considered to

Quality control inspector checking that an electrical transformer meets standard requirements. © Edward Todd.

be operating satisfactorily. If $\overline{X}$ exceeds the upper limit or if is below the lower limit, corrective action will be taken. These limits are referred to as upper and lower control limits for quality control purposes.

4 Discuss the implications of changing the level of significance to a larger value. What mistake or error could increase if the level of significance is increased?

Sample 1	Sample 2	Sample 3	Sample 4
11.55	11.62	11.91	12.02
11.62	11.69	11.36	12.02
11.52	11.59	11.75	12.05
11.75	11.82	11.95	12.18
11.90	11.97	12.14	12.11
11.64	11.71	11.72	12.07
11.80	11.87	11.61	12.05
12.03	12.10	11.85	11.64
11.94	12.01	12.16	12.39
11.92	11.99	11.91	11.65
12.13	12.20	12.12	12.11
12.09	12.16	11.61	11.90
11.93	12.00	12.21	12.22
12.21	12.28	11.56	11.88
12.32	12.39	11.95	12.03
11.93	12.00	12.01	12.35
11.85	11.92	12.06	12.09
11.76	11.83	11.76	11.77
12.16	12.23	11.82	12.20
11.77	11.84	12.12	11.79
12.00	12.07	11.60	12.30
12.04	12.11	11.95	12.27
11.98	12.05	11.96	12.29
12.30	12.37	12.22	12.47
12.18	12.25	11.75	12.03
11.97	12.04	11.96	12.17
12.17	12.24	11.95	11.94
11.85	11.92	11.89	11.97
12.30	12.37	11.88	12.23
12.15	12.22	11.93	12.25

Software Section
for Chapter 9

Hypothesis testing using MINITAB

We describe the use of MINITAB to conduct hypothesis tests about a population mean and a population proportion. MINITAB provides both hypothesis testing and interval estimation results simultaneously, so the routines illustrated here were also used in Chapter 8.

Population mean: σ known

We illustrate using the MaxFlight golf ball distance example in Section 9.3. The data are in column C1 of a MINITAB worksheet (file 'GolfTest.MTW' on the accompanying CD). The population standard deviation $\sigma = 12$ is assumed known and the level of significance is $\alpha = 0.05$. The following steps can be used to test the hypothesis H_0: $\mu = 295$ versus H_1: $\mu \neq 295$.

Step 1 **Stat > Basic Statistics > 1-Sample Z**　　　　　　　　[Main menu bar]

Step 2 Enter **C1** in the **Samples in columns** box
　　　　　　　　　　　　　　　[**1-Sample Z (Test and Confidence Interval)** panel]
　　　　Enter **20** in the **Standard deviation** box
　　　　Check the **Perform Hypothesis Test** box
　　　　Enter **295** in the **Hypothesized mean** box
　　　　Click **Options**

Step 3 Enter **95** in the **Confidence level** box　　　　[**1-Sample Z - Options** panel]
　　　　Select **not equal** on the **Alternative** menu
　　　　Click **OK**

Step 4 Click **OK**　　　　[**1-Sample Z (Test and Confidence Interval)** panel]

In addition to the hypothesis testing results, MINITAB provides a 95 per cent confidence interval for the population mean. The MINITAB output is shown below as Figure 9.12. The procedure can be easily modified for a one-tailed hypothesis test by selecting the **less than** or **greater than** option on the **Alternative** drop-down menu (**Step 3**).

Population mean: σ unknown

The ratings that 60 business travellers gave for Munich Airport are entered in column C1 of a MINITAB worksheet (file 'AirRating.MTW' on the accompanying CD). The level of significance for the test is $\alpha = 0.05$, and the population standard deviation σ will be

Figure 9.12 MINITAB output for the MaxFlight hypothesis test

```
Results for: GolfTest.MTW

One-Sample Z: Metres

Test of mu = 295 vs not = 295
The assumed standard deviation = 20


Variable   N    Mean   StDev  SE Mean       95% CI          Z      P
Metres    50  297.60   11.30     2.83  (292.06, 303.14)  0.92  0.358
```

estimated by the sample standard deviation s. The following steps can be used to test the hypothesis H_0: $\mu \leq 7$ against H_1: $\mu > 7$.

Step 1 **Stat > Basic Statistics > 1-Sample t**   [Main menu bar]

Step 2 Enter **C1** in the **Samples in columns** box
                                [**1-Sample t (Test and Confidence Interval)** panel]
             Check the **Perform Hypothesis Test** box
             Enter **7** in the **Hypothesized mean** box
             Click **Options**

Step 3 Enter **95** in the **Confidence level** box   [**1-Sample t - Options** panel]
             Select **greater than** on the **Alternative** menu
             Click **OK**

Step 4 Click **OK**   [**1-Sample t (Test and Confidence Interval)** panel]

The MINITAB results are shown below in Figure 9.13. The Munich Airport rating study involved a 'greater than' alternative hypothesis. The preceding steps can be easily modified for other hypothesis tests by selecting the **less than** or **not equal** options on the **Alternative** drop-down menu (**Step 3**).

Figure 9.13 MINITAB output for the Munich rating hypothesis test

```
Results for: AirRating.MTW

One-Sample T: Rating

Test of mu = 7 vs > 7


                                        95% Lower
Variable   N    Mean   StDev  SE Mean     Bound      T      P
Rating    60   7.250   1.052    0.136     7.023   1.84  0.035
```

Population proportion

We illustrate using the Aspire gymnasium example in Section 9.5. The data with responses Female and Male are in column C1 of a MINITAB worksheet (file 'WomenGym' on the accompanying CD). MINITAB uses an alphabetical ordering of the responses and selects the *second response* for the population proportion of interest. In this example, MINITAB by default uses the ordering Female-Male and gives results for the population proportion of Male responses. Because Female is the response of interest, we change MINITAB's ordering as follows. Select any cell in the column and use the sequence:

Step 1 **Editor > Column > Value Order**   [Main menu bar]

Step 2 Choose **User-specified order**   [**Value Order for C1 (Gym User)** panel]
Enter the responses **Male Female** in the **Define-an-order (one value per line)** box
Click **OK**

Then proceed as follows to test the hypothesis $H_0: \pi \le 0.2$ against $H_1: \pi > 0.2$. The MINITAB results are shown in Figure 9.14.

Step 3 **Stat > Basic Statistics > 1 Proportion**   [Main menu bar]

Step 4 Enter **C1** in the **Samples in columns** box
                          [**1 Proportion (Test and Confidence Interval)** panel]
Check the **Perform Hypothesis Test** box
Enter **0.20** in the **Test proportion** box
Select **Options**

Step 3 Check **Use test and interval based on normal distribution**
                          [**1 Proportion - Options** panel]
Enter **95** in the **Confidence Level** box
Select **greater than** on the **Alternative** menu

Figure 9.14 MINITAB output for the Aspire gymnasium hypothesis test

Results for: WomenGym.MTW

Test and CI for One Proportion: Gym User

Test of p = 0.2 vs p > 0.2

Event = Female

				95% Lower		
Variable	X	N	Sample p	Bound	Z-Value	P-Value
Gym User	100	400	0.250000	0.214388	2.50	0.006

Using the normal approximation.

Click **OK**

Step 4 Click **OK**   **[1 Proportion (Test and Confidence Interval)** panel]

Hypothesis testing using **EXCEL**

EXCEL does not provide inbuilt routines for the hypothesis tests presented in this chapter. To handle these situations, we present EXCEL worksheets that we designed to test hypotheses about a population mean and a population proportion. The worksheets are easy to use and can be modified to handle any sample data. The worksheets are available on the CD that accompanies this book.

Population mean: σ known

We illustrate using the MaxFlight golf ball distance example in Section 9.3. The data are in column A of an EXCEL worksheet. The population standard deviation $\sigma = 12$ is assumed known and the level of significance is $\alpha = 0.05$. The following steps can be used to test the hypothesis $H_0: \mu = 295$ versus $H_1: \mu \neq 295$. Refer to Figure 9.15 as we describe the procedure. The data are entered into cells A2:A51. The following steps are necessary to use the template for this data set.

Step 1 Enter the data range **A2:A51** into the $=$ COUNT cell formula in cell D4

Step 2 Enter the data range **A2:A51** into the $=$ AVERAGE cell formula in cell D5

Step 3 Enter the population standard deviation $\sigma = $ **12** into cell D6

Step 4 Enter the hypothesized value for the population mean **295** into cell D8

The remaining cell formulae automatically provide the standard error, the value of the test statistic z, and three p-values. Because the alternative hypothesis ($\mu_0 \neq 295$) indicates a two-tailed test, the p-value (Two Tail) in cell D15 is used to make the rejection decision. With p-value $= 0.1255 > \alpha = 0.05$, the null hypothesis cannot be rejected. The p-values in cells D13 or D14 would be used if the hypotheses involved a one-tailed test.

This template can be used to do hypothesis test computations for other applications. For example, to conduct a hypothesis test for a new data set, enter the new sample data into column A of the worksheet. Modify the formulas in cells D4 and D5 to correspond to the new data range. Enter the population standard deviation into cell D6 and the hypothesized value for the population mean into cell D8 to obtain the results. If the new sample data have already been summarized, the new sample data do not have to be entered into the worksheet. In this case, enter the sample size into cell D4, the sample mean into cell D5, the population standard deviation into cell D6, and the hypothesized value for the population mean into cell D8 to obtain the results. The worksheet in Figure 9.15 is available in the file Hyp Sigma Known on the CD that accompanies this book.

Figure 9.15 EXCEL worksheet for hypothesis tests about a population mean with σ known

	A	B	C	D	E
1	Metres		Hypothesis Test About a Population Mean		
2	303		With σ Known		
3	282				
4	289		Sample Size	=COUNT(A2:A51)	
5	298		Sample Mean	=AVERAGE(A2:A51)	
6	283		Population Std. Deviation	12	
7	317				
8	297		Hypothesized Value	295	
9	308				
10	317		Standard Error	=D6/SQRT(D4)	
11	293		Test Statistic z	=(D5-D8)/D10	
12	284				
13	290		p-value (Lower Tail)	=NORMSDIST(D11)	
14	304		p-value (Upper Tail)	=1-D13	
15	290		p-value (Two Tail)	=2*(MIN(D13,D14))	
16	311				
17	305				
49	303				
50	301				
51	292				
52					

	A	B	C	D	E
1	Metres		Hypothesis Test About a Population Mean		
2	303		With σ Known		
3	282				
4	289		Sample Size	50	
5	298		Sample Mean	297.6	
6	283		Population Std. Deviation	12	
7	317				
8	297		Hypothesized Value	295	
9	308				
10	317		Standard Error	1.70	
11	293		Test Statistic z	1.53	
12	284				
13	290		p-value (Lower Tail)	0.9372	
14	304		p-value (Upper Tail)	0.0628	
15	290		p-value (Two Tail)	0.1255	
16	311				
17	305				
48	292				
49	303				
50	301				
51	292				
52					

HYP SIGMA UNKNOWN

Population mean: σ unknown

We illustrate using the Munich Airport rating example in Section 9.4. The data are entered into cells A2:A61 of an EXCEL worksheet. The population standard deviation σ is unknown and will be estimated by the sample standard deviation s. The level of significance is $\alpha = 0.05$. The following steps are necessary to use the template for this data set, to test the hypothesis $H_0: \mu \leq 7$ versus $H_1: \mu > 7$.

Step 1 Enter the data range **A2:A61** into the $=$ COUNT cell formula in cell D4

Step 2 Enter the data range **A2:A61** into the $=$ AVERAGE cell formula in cell D5

Step 3 Enter the data range **A2:A61** into the $=$ STDEV cell formula in cell D6

Step 4 Enter the hypothesized value for the population mean **7** into cell D8

The remaining cell formulae automatically provide the standard error, the value of the test statistic t, the number of degrees of freedom, and three p-values. Because the alternative hypothesis ($\mu > 7$) indicates an upper tail test, the p-value (Upper Tail) in cell D15 is used to make the decision. With p-value $= 0.0353 < \alpha = 0.05$, the null hypothesis is rejected. The p-values in cells D14 or D16 would be used if the hypotheses involved a lower tail test or a two-tailed test.

This template can be used to do hypothesis test computations for other applications. For instance, to conduct a hypothesis test for a new data set, enter the new sample data into column A of the worksheet and modify the formulae in cells D4, D5, and D6 to correspond to the new data range. Enter the hypothesized value for the population mean into cell D8 to obtain the results. If the new sample data have already been summarized, the new sample data do not have to be entered into the worksheet. In this case, enter the sample size into cell D4, the sample mean into cell D5, the sample standard deviation into cell D6, and the hypothesized value for the population mean into cell D8 to obtain the results. The worksheet is available in the file Hyp Sigma Unknown on the CD that accompanies this book.

Population proportion

We illustrate using the Aspire gymnasium survey data presented in Section 9.5. The level of significance is $\alpha = 0.05$. The data of Male or Female user are in column A of an EXCEL worksheet. The data are entered into cells A2:A401. The following steps can be used to test the hypothesis H_0: $\pi \leq 0.20$ versus H_1: $\pi > 0.20$.

Step 1 Enter the data range **A2:A401** into the = COUNTA cell formula in cell D3

Step 2 Enter **Female** as the response of interest in cell D4

Step 3 Enter the data range **A2:A401** into the = COUNTIF cell formula in cell D5

Step 4 Enter the hypothesized value for the population proportion **0.20** into cell D8

The remaining cell formulae automatically provide the standard error, the value of the test statistic z, and three p-values. Because the alternative hypothesis ($\pi > 0.20$) indicates an upper tail test, the p-value (Upper Tail) in cell D14 is used to make the decision. With p-value $= 0.0062 < \alpha = 0.05$, the null hypothesis is rejected. The p-values in cells D13 or D15 would be used if the hypothesis involved a lower tail test or a two-tailed test.

This template can be used to do hypothesis test computations for other applications. For instance, to conduct a hypothesis test for a new data set, enter the new sample data into column A of the worksheet. Modify the formulae in cells D3 and D5 to correspond to the new data range. Enter the response of interest into cell D4 and the hypothesized value for the population proportion into cell D8 to obtain the results. If the new sample data have already been summarized, the new sample data do not have to be entered into the worksheet. In this case, enter the sample size into cell D3, the sample proportion into cell D6, and the hypothesized value for the population proportion into cell D8 to obtain the results. There is a worksheet available in the file 'Hypothesis p' on the CD that accompanies this book.

Hypothesis testing using PASW

We describe the use of PASW to construct a hypothesis test for a population mean in the σ unknown condition. There are no inbuilt routines in PASW for the σ known condition, nor for a population proportion.

Population mean: σ unknown

The One-Sample T Test routine in PASW constructs both a confidence interval and a hypothesis test.

Step 1 **Analyze > Compare Means > One-Sample T Test**   [Main menu bar]

Step 2   Transfer the Rating variable to the **Test Variable(s)** box

[**One-Sample T Test** panel]

Enter **7** in the **Test Value** box

Click **OK**

The routine was illustrated in Chapter 8 using the credit card balance data for a sample of 85 households. The PASW results were displayed in Figure 8.10. Similar results are shown here in Figure 9.16 for the Munich Airport ratings, which are in the first column of the PASW data file ('AirRating.SAV' on the accompanying CD). The PASW routine constructs a two-tailed test. The p-value for a one-tailed test can be computed as half the two-tailed p-value shown in the output: $0.071/2 = 0.035$.

Figure 9.16 PASW output for the Munich Airport rating hypothesis test

One-Sample Statistics

	N	Mean	Std. Deviation	Std. Error Mean
Rating	60	7.25	1.052	.136

One-Sample Test

	Test Value = 7					
					95% Confidence Interval of the Difference	
	t	df	Sig. (2-tailed)	Mean Difference	Lower	Upper
Rating	1.841	59	.071	.250	-.02	.52

Chapter 10

··

Statistical Inference About Means and Proportions with Two Populations

Learning objectives

After studying this chapter and doing the exercises, you should be able to:

1 Construct and interpret confidence intervals and hypothesis tests for the difference between two population means, given independent samples from the two populations:

   1.1 When the standard deviations of the two populations are known.

   1.2 When the standard deviations of the two populations are unknown.

2 Construct and interpret confidence intervals and hypothesis tests for the difference between two population means, given matched samples from the two populations.

3 Construct and interpret confidence intervals and hypothesis tests for the difference between two population proportions, given independent samples from the two populations.

In Chapters 8 and 9 we showed how to construct interval estimates and do hypothesis tests for situations involving a single population mean and a single population proportion. In this chapter we continue our discussion of statistical inference by showing how interval estimates and hypothesis tests can be developed for situations involving two populations, when the difference between the two population means or the two population proportions is of prime importance. For example, we may want to construct an interval estimate of the difference between the mean starting salary for a population of men and the mean starting salary for a population of women. Or we may want to conduct a hypothesis test to determine whether any difference is present between the proportion of defective parts in a population of parts produced by supplier A and the proportion of defective parts in a population of parts produced by supplier B. We begin our discussion of statistical inference about two populations by showing how to construct interval estimates and do hypothesis tests about the difference between the means of two populations when the standard deviations of the two populations are assumed known.

10.1 Inferences about the difference between two population means: σ_1 and σ_2 known

Let μ_1 denote the mean of population 1 and μ_2 denote the mean of population 2. We shall focus on inferences about the difference between the means: $\mu_1 - \mu_2$. To make an inference about this difference, we select a simple random sample of n_1 units from population 1 and a second simple random sample of n_2 units from population 2. The two samples, taken separately and independently, are referred to as independent simple random samples. In this section, we assume that information is available such that the two population standard deviations, σ_1 and σ_2, can be assumed known prior to collecting the samples. We refer to this situation as the σ_1 and σ_2 known case. In the following example we show how to compute a margin of error and develop an interval estimate of the difference between the two population means when σ_1 and σ_2 are known.

Statistics in Practice

Fisons Corporation

Fisons plc is a major company that manufactures pharmaceuticals, scientific equipment and horticultural products. The company's pharmaceutical division uses extensive statistical procedures to test and develop new drugs. The testing process usually consists of three stages: (1) pre-clinical testing, (2) testing for long-term usage and safety, and (3) clinical efficacy testing. At each successive stage, the chance that a drug will pass the rigorous tests decreases; however, the cost of further testing increases dramatically. Industry surveys indicate that on average the research and development for one new drug costs over €200 million and takes 12 years. Hence, it is important to eliminate unsuccessful new drugs in the early stages of the testing process, as well as identify promising ones for

Scientist doing pre-clinical testing on a new pharmaceutical drug. © Firefly Productions/ First Collection.

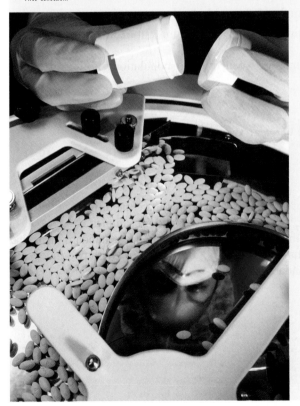

further testing. Statistics plays a major role in pharmaceutical research, where government regulations are stringent.

In pre-clinical testing, a two- or three-population statistical study is typically used to determine whether a new drug shows promise. The populations may consist of the new drug, a control, and a standard drug. The pre-clinical testing process begins when a new drug is sent to the pharmacology group for evaluation of efficacy – the capacity of the drug to produce the desired effects. As part of the process, a statistician is asked to design an experiment that can be used to test the new drug. The design must specify the sample size and the statistical methods of analysis. In a two-population study, one sample is used to obtain data on the efficacy of the new drug (population 1) and a second sample is used to obtain data on the efficacy of a standard drug (population 2). Depending on the intended use, the new and standard drugs are tested in such disciplines as neurology, cardiology and immunology. In most studies, the statistical method involves hypothesis testing for the difference between the means of the new drug population and the standard drug population. If a new drug lacks efficacy or produces undesirable effects in comparison with the standard drug, the new drug is rejected. Only new drugs that show promising comparisons with the standard drugs are forwarded to the long-term usage and safety testing programme.

Further data collection and multi-population studies are conducted in the long-term usage and safety testing programme and in the clinical testing programme. In the UK, the Medicines and Healthcare Products Regulatory Agency (MHRA) requires that statistical methods be defined prior to such testing to avoid data-related biases. In addition, to avoid human biases, some of the clinical trials are double or triple blind. That is, neither the participant nor the investigator knows what drug is administered to whom.

In this chapter you will learn how to construct interval estimates and do hypothesis tests about means and proportions with two populations. Techniques will be presented for analyzing independent random samples as well as matched samples.

Interval estimation of $\mu_1 - \mu_2$

Suppose a retailer such as Currys (selling TVs, DVD players, computers, photographic equipment and so on) operates two stores in Dublin: one is in the inner city and the other is in an out-of-town shopping centre. The regional manager noticed that products that sell well in one store do not always sell well in the other. The manager believes this may be attributable to differences in customer demographics at the two locations. Customers may differ in age, education, income and so on. Suppose the manager asks us to investigate the difference between the mean ages of the customers who shop at the two stores.

Let us define population 1 as all customers who shop at the inner-city store and population 2 as all customers who shop at the suburban store.

μ_1 = mean of population 1 (i.e. the mean age of all customers who shop at the inner-city store)

μ_2 = mean of population 2 (i.e. the mean age of all customers who shop at the out-of-town store)

The difference between the two population means is $\mu_1 - \mu_2$. To estimate $\mu_1 - \mu_2$, we shall select a simple random sample of n_1 customers from population 1 and a simple random sample of n_2 customers from population 2. We then compute the two sample means.

$\bar{x}_1$ = sample mean age for the simple random sample of n_1 inner-city customers

$\bar{x}_2$ = sample mean age for the simple random sample of n_2 out-of-town customers

The point estimator of the difference between the two populations is the difference between the sample means.

Point estimator of the difference between two population means

$$\bar{X}_1 - \bar{X}_2$$

(10.1)

Figure 10.1 provides an overview of the process used to estimate the difference between two population means based on two independent simple random samples.

As with other point estimators, the point estimator $\bar{X}_1 - \bar{X}_2$ has a standard error that describes the variation in the sampling distribution of the estimator. With two independent simple random samples, the standard error of $\bar{X}_1 - \bar{X}_2$ is as follows:

Standard error of $\bar{X}_1 - \bar{X}_2$

$$\sigma_{\bar{X}_1 - \bar{X}_2} = \sqrt{\frac{\sigma_1^2}{n_1} + \frac{\sigma_2^2}{n_2}}$$

(10.2)

If both populations have a normal distribution, or if the sample sizes are large enough that the central limit theorem enables us to conclude that the sampling distributions of $\bar{X}_1$ and $\bar{X}_2$ can be approximated by a normal distribution, the sampling distribution of $\bar{X}_1 - \bar{X}_2$ will have a normal distribution with mean given by $\mu_1 - \mu_2$.

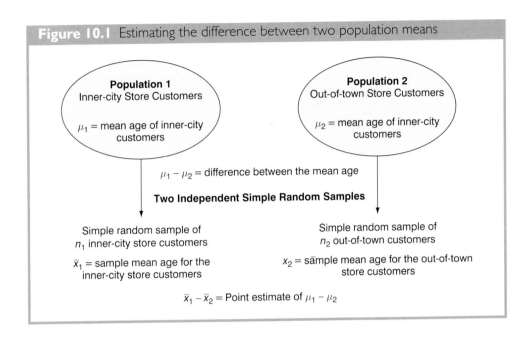

Figure 10.1 Estimating the difference between two population means

Population 1
Inner-city Store Customers

μ_1 = mean age of inner-city customers

Population 2
Out-of-town Store Customers

μ_2 = mean age of inner-city customers

$\mu_1 - \mu_2$ = difference between the mean age

Two Independent Simple Random Samples

Simple random sample of n_1 inner-city store customers

$\bar{x}_1$ = sample mean age for the inner-city store customers

Simple random sample of n_2 out-of-town customers

x_2 = sample mean age for the out-of-town store customers

$\bar{x}_1 - \bar{x}_2$ = Point estimate of $\mu_1 - \mu_2$

As we showed in Chapter 8, an interval estimate is given by a point estimate $\pm$ a margin of error. In the case of estimation of the difference between two population means, an interval estimate will take the form $(\bar{x}_1 - \bar{x}_2) \pm$ margin of error. When the sampling distribution of $\bar{X}_1 - \bar{X}_2$ is a normal distribution, we can write the margin of error as follows:

$$\text{Margin of error} = z_{\alpha/2}\sigma_{\bar{X}_1 - \bar{X}_2} = z_{\alpha/2}\sqrt{\frac{\sigma_1^2}{n_1} + \frac{\sigma_2^2}{n_2}} \qquad (10.3)$$

Therefore the interval estimate of the difference between two population means is as follows:

Interval estimate of the difference between two population means: σ_1 and σ_2 known

$$(\bar{x}_1 - \bar{x}_2) \pm z_{\alpha/2}\sqrt{\frac{\sigma_1^2}{n_1} + \frac{\sigma_2^2}{n_2}} \qquad (10.4)$$

where $1 - \alpha$ is the confidence coefficient.

Let us return to the example of the Dublin retailer. Based on data from previous customer demographic studies, the two population standard deviations are known with $\sigma_1 = 9$ years and $\sigma_2 = 10$ years. The data collected from the two independent simple random samples of the retailer's customers provided the following results.

	Inner city store	Out-of-town store
Sample size	$n_1 = 36$	$n_2 = 49$
Sample mean	$\bar{x}_1 = 40$ years	$\bar{x}_2 = 35$ years

Using expression (10.1), we find that the point estimate of the difference between the mean ages of the two populations is $\bar{x}_1 - \bar{x}_2 = 40 - 35 = 5$ years. We estimate that the customers at the inner-city store have a mean age five years greater than the mean age of the out-of-town customers. We can now use expression (10.4) to compute the margin of error and provide the interval estimate of $\mu_1 - \mu_2$. Using 95 per cent confidence and $z_{\alpha/2} = z_{0.025} = 1.96$, we have

$$(\bar{x}_1 - \bar{x}_2) \pm z_{\alpha/2}\sqrt{\frac{\sigma_1^2}{n_1} + \frac{\sigma_2^2}{n_2}} = (40 - 35) \pm 1.96\sqrt{\frac{9^2}{36} + \frac{10^2}{49}} = 5 \pm 4.1$$

The margin of error is 4.1 years and the 95 per cent confidence interval estimate of the difference between the two population means is $5 - 4.1 = 0.9$ years to $5 + 4.1 = 9.1$ years.

Hypothesis tests about $\mu_1 - \mu_2$

Let us consider hypothesis tests about the difference between two population means. Using D_0 to denote the hypothesized difference between μ_1 and μ_2, the three forms for a hypothesis test are as follows:

$$H_0: \mu_1 - \mu_2 \geq D_0 \qquad H_0: \mu_1 - \mu_2 \leq D_0 \qquad H_0: \mu_1 - \mu_2 = D_0$$
$$H_1: \mu_1 - \mu_2 < D_0 \qquad H_1: \mu_1 - \mu_2 > D_0 \qquad H_1: \mu_1 - \mu_2 \neq D_0$$

In many applications, $D_0 = 0$. Using the two-tailed test as an example, when $D_0 = 0$ the null hypothesis is $H_0: \mu_1 - \mu_2 = 0$, i.e. the null hypothesis is that μ_1 and μ_2 are equal. Rejection of H_0 leads to the conclusion that $H_1: \mu_1 - \mu_2 \neq 0$ is true, i.e. μ_1 and μ_2 are not equal.

The steps for doing hypothesis tests presented in Chapter 9 are applicable here. We must choose a level of significance, compute the value of the test statistic and find the p-value to determine whether the null hypothesis should be rejected. With two independent simple random samples, we showed that the point estimator $\bar{X}_1 - \bar{X}_2$ has a standard error $\sigma_{\bar{X}_1 - \bar{X}_2}$ given by expression (10.2), and the distribution of $\bar{X}_1 - \bar{X}_2$ can be described by a normal distribution. In this case, the test statistic for the difference between two population means when σ_1 and σ_2 are known is as follows.

Test statistic for hypothesis tests about $\mu_1 - \mu_2$: σ_1 and σ_2 known

$$z = \frac{(\bar{x}_1 - \bar{x}_2) - D_0}{\sqrt{\dfrac{\sigma_1^2}{n_1} + \dfrac{\sigma_2^2}{n_2}}} \qquad\qquad (10.5)$$

Here is an example. As part of a study to evaluate differences in education quality between two training centres, a standardized examination is given to individuals who are trained at the centres. The difference between the mean examination scores is used to assess quality differences between the centres. The population means for the two centres are as follows.

$\mu_1 = $ the mean examination score for the population of individuals trained at centre A
$\mu_2 = $ the mean examination score for the population of individuals trained at centre B

EXAMSCORES

We begin with the tentative assumption that no difference exists between the average training quality provided at the two centres. Hence, in terms of the mean examination scores, the null hypothesis is that $\mu_1 - \mu_2 = 0$. If sample evidence leads to the rejection of this hypothesis, we shall conclude that the mean examination scores differ for the two populations. This conclusion indicates a quality differential between the two centres and suggests that a follow-up study investigating the reason for the differential may be warranted. The null and alternative hypotheses for this two-tailed test are written as follows.

$$H_0: \mu_1 - \mu_2 = 0$$
$$H_1: \mu_1 - \mu_2 \neq 0$$

The standardized examination given previously in a variety of settings always resulted in an examination score standard deviation near 10 points. We shall use this information to assume that the population standard deviations are known with $\sigma_1 = 10$ and $\sigma_2 = 10$. An $\alpha = 0.05$ level of significance is specified for the study.

Independent simple random samples of $n_1 = 30$ individuals from training centre A and $n_2 = 40$ individuals from training centre B are taken. The respective sample means are $\bar{x}_1 = 82$ and $\bar{x}_2 = 78$. Do these data suggest a difference between the population means at the two training centres? To help answer this question, we compute the test statistic using equation (10.5).

$$z = \frac{(\bar{x}_1 - \bar{x}_2) - D_0}{\sqrt{\dfrac{\sigma_1^2}{n_1} + \dfrac{\sigma_2^2}{n_2}}} = \frac{(82 - 78) - 0}{\sqrt{\dfrac{(10)^2}{30} + \dfrac{(10)^2}{40}}} = 1.66$$

Next let us compute the p-value for this two-tailed test. Because the test statistic z is in the upper tail, we first compute the area under the curve to the right of $z = 1.66$. Using the standard normal distribution table, the cumulative probability for $z = 1.66$ is 0.9515, so the area in the upper tail of the distribution is $1 - 0.9515 = 0.0485$. Because this test is a two-tailed test, we must double the tail area: p-value $= 2(0.0485) = 0.0970$. Following the usual rule to reject H_0 if p-value $\leq \alpha$, we see that the p-value of 0.0970 does not allow us to reject H_0 at the 0.05 level of significance. The sample results do not provide sufficient evidence to conclude that the training centres differ in quality.

In this chapter we shall use the p-value approach to hypothesis testing as described in Chapter 9. However, if you prefer, the test statistic and the critical value rejection rule may be used. With $\alpha = 0.05$ and $z_{\alpha/2} = z_{0.025} = 1.96$, the rejection rule using the critical value approach would be to reject H_0 if $z \leq -1.96$ or if $z \geq 1.96$. With $z = 1.66$, we reach the same 'do not reject H_0' conclusion.

In the preceding example, we demonstrated a two-tailed hypothesis test about the difference between two population means. Lower tail and upper tail tests can also be considered. These tests use the same test statistic as given in equation (10.5). The procedure for computing the p-value and the rejection rules for these one-tailed tests are the same as those presented in Chapter 9.

Practical advice

In most applications of the interval estimation and hypothesis testing procedures presented in this section, random samples with $n_1 \geq 30$ and $n_2 \geq 30$ are adequate. In cases where either or both sample sizes are less than 30, the distributions of the populations become important considerations. In general, with smaller sample sizes, it is more important for the analyst to be satisfied that it is reasonable to assume the distributions of the two populations are at least approximately normal.

Exercises

Methods

1   Consider the following results for two independent random samples taken from two populations.

Sample 1	Sample 2
$n_1 = 50$	$n_2 = 35$
$\bar{x}_1 = 13.6$	$\bar{x}_2 = 11.6$
$\sigma_1 = 2.2$	$\sigma_2 = 3.0$

   a. What is the point estimate of the difference between the two population means?
   b. Provide a 90 per cent confidence interval for the difference between the two population means.
   c. Provide a 95 per cent confidence interval for the difference between the two population means.

2   Consider the following hypothesis test.

$$H_0: \mu_1 - \mu_2 \leq 0$$
$$H_1: \mu_1 - \mu_2 > 0$$

The following results are for two independent samples taken from the two populations.

Sample 1	Sample 2
$n_1 = 40$	$n_2 = 50$
$\bar{x}_1 = 25.2$	$\bar{x}_2 = 22.8$
$\sigma_1 = 5.2$	$\sigma_2 = 6.0$

   a. What is the value of the test statistic?
   b. What is the p-value?
   c. With $\alpha = 0.05$, what is your hypothesis testing conclusion?

3   Consider the following hypothesis test.

$$H_0: \mu_1 - \mu_2 = 0$$
$$H_1: \mu_1 - \mu_2 \neq 0$$

The following results are for two independent samples taken from the two populations.

Sample 1	Sample 2
$n_1 = 80$	$n_2 = 70$
$\bar{x}_1 = 104$	$\bar{x}_2 = 106$
$\sigma_1 = 8.4$	$\sigma_2 = 7.6$

   a. What is the value of the test statistic?
   b. What is the p-value?
   c. With $\alpha = 0.05$, what is your hypothesis testing conclusion?

Applications

4   A study of wage differentials between men and women reported that one of the reasons wages for men are higher than wages for women is that men tend to have more years of work experience than women. Assume that the following sample summaries show the years of experience for each group.

Men	Women
$n_1 = 100$	$n_2 = 85$
$\bar{x}_1 = 14.9$ years	$\bar{x}_2 = 10.3$ years
$\sigma_1 = 5.2$ years	$\sigma_2 = 3.8$ years

a. What is the point estimate of the difference between the two population means?
b. At 95 per cent confidence, what is the margin of error?
c. What is the 95 per cent confidence interval estimate of the difference between the two population means?

5   The Dublin retailer age study (used as an example above) provided the following data on the ages of customers from independent random samples taken at the two store locations.

Inner-city store	Out-of-town store
$n_1 = 36$	$n_2 = 49$
$\bar{x}_1 = 40$ years	$\bar{x}_2 = 35$ years
$\sigma_1 = 9$ years	$\sigma_2 = 10$ years

a. State the hypotheses that could be used to detect a difference between the population mean ages at the two stores.
b. What is the value of the test statistic?
c. What is the p-value
d. At $\alpha = 0.05$, what is your conclusion?

6   According to a report in *USA Today* on 13 February 2006, the average expenditure on Valentine's Day (14 February) was expected to be about $101. Do male and female consumers differ in the amounts they spend? The average expenditure in a sample survey of 40 male consumers was $135.67, and the average expenditure in a sample survey of 30 female consumers was $68.64. Based on past surveys, the standard deviation for male consumers is assumed to be $35, and the standard deviation for female consumers is assumed to be $20.

a. What is the point estimate of the difference between the population mean expenditure for males and the population mean expenditure for females?
b. At 99 per cent confidence, what is the margin of error?
c. Construct a 99 per cent confidence interval for the difference between the two population means.

10.2 Inferences about the difference between two population means: σ_1 and σ_2 unknown

In this section we extend the discussion of inferences about the difference between two population means to the case when the two population standard deviations, σ_1 and σ_2, are unknown. In this case, we shall use the sample standard deviations, s_1 and s_2, to estimate the unknown population standard deviations. When we use the sample standard deviations, the interval estimation and hypothesis testing procedures will be based on the t distribution rather than the standard normal distribution.

Interval estimation of $\mu_1 - \mu_2$

In the following example we show how to compute a margin of error and construct an interval estimate of the difference between two population means when σ_1 and σ_2 are unknown. The Union Bank is conducting a study designed to identify differences between cheque account practices by customers at two of its branches. A simple random sample of 28 cheque accounts is selected from the Northern Branch and an independent simple random sample of 22 cheque accounts is selected from the Eastern Branch. The current cheque account balance is recorded for each of the accounts. A summary of the account balances follows:

CHEQACCT

	Northern	Eastern
Sample size	$n_1 = 28$	$n_2 = 22$
Sample mean	$\bar{x}_1 = €1025$	$\bar{x}_2 = €910$
Sample standard deviation	$s_1 = €150$	$s_2 = €125$

The Union Bank would like to estimate the difference between the mean cheque account balance maintained by the population of Northern customers and the population of Eastern customers. Let us develop the margin of error and an interval estimate of the difference between these two population means.

In Section 10.1, we provided the following interval estimate for the case when the population standard deviations, σ_1 and σ_2, are known.

$$(\bar{x}_1 - \bar{x}_2) \pm z_{\alpha/2}\sqrt{\frac{\sigma_1^2}{n_1} + \frac{\sigma_2^2}{n_2}}$$

With σ_1 and σ_2 unknown, we shall use the sample standard deviations s_1 and s_2 to estimate σ_1 and σ_2 and replace $z_{\alpha/2}$ with $t_{\alpha/2}$. As a result, the interval estimate of the difference between two population means is given by the following expression:

Interval estimate of the difference between two population means: σ_1 and σ_2 unknown

$$(\bar{x}_1 - \bar{x}_2) \pm t_{\alpha/2}\sqrt{\frac{s_1^2}{n_1} + \frac{s_2^2}{n_2}} \qquad (10.6)$$

where $1 - \alpha$ is the confidence coefficient.

In this expression, the use of the t distribution is an approximation, but it provides excellent results and is relatively easy to use. The only difficulty that we encounter in using expression (10.6) is determining the appropriate degrees of freedom for $t_{\alpha/2}$. Statistical software packages compute the appropriate degrees of freedom automatically. The formula used is as follows:

Degrees of freedom for the t distribution using two independent random samples

$$df = \frac{\left(\dfrac{s_1^2}{n_1} + \dfrac{s_2^2}{n_2}\right)^2}{\left(\dfrac{1}{n_1 - 1}\right)\left(\dfrac{s_1^2}{n_1}\right)^2 + \left(\dfrac{1}{n_2 - 1}\right)\left(\dfrac{s_2^2}{n_2}\right)^2} \qquad (10.7)$$

Let us return to the Union Bank example and show how to use expression (10.6) to provide a 95 per cent confidence interval estimate of the difference between the population mean cheque account balances at the two branches. The sample data show $n_1 = 28$, $\bar{x}_1 = €1025$, and $s_1 = €150$ for the Northern Branch, and $n_2 = 22$, $\bar{x}_2 = €910$, and $s_2 = €125$ for the Eastern Branch. The calculation for degrees of freedom for $t_{\alpha/2}$ is as follows:

$$df = \frac{\left(\dfrac{s_1^2}{n_1} + \dfrac{s_2^2}{n_2}\right)^2}{\left(\dfrac{1}{n_1 - 1}\right)\left(\dfrac{s_1^2}{n_1}\right)^2 + \left(\dfrac{1}{n_2 - 1}\right)\left(\dfrac{s_2^2}{n_2}\right)^2} = \frac{\left(\dfrac{150^2}{28} + \dfrac{125^2}{22}\right)^2}{\left(\dfrac{1}{28 - 1}\right)\left(\dfrac{150^2}{28}\right)^2 + \left(\dfrac{1}{22 - 1}\right)\left(\dfrac{125^2}{22}\right)^2} = 47.8$$

We round the non-integer degrees of freedom *down* to 47 to provide a larger t-value and a more conservative interval estimate. Using the t distribution table with 47 degrees of freedom, we find $t_{0.025} = 2.012$. Using expression (10.6), we construct the 95 per cent confidence interval estimate of the difference between the two population means as follows.

$$(\bar{x}_1 - \bar{x}_2) \pm t_{\alpha/2}\sqrt{\frac{s_1^2}{n_1} + \frac{s_2^2}{n_2}} = (1025 - 910) \pm 2.012\sqrt{\frac{150^2}{28} + \frac{125^2}{22}} = 115 \pm 78$$

The point estimate of the difference between the population mean cheque account balances at the two branches is €115. The margin of error is €78, and the 95 per cent confidence interval estimate of the difference between the two population means is $115 - 78 = €37$ to $115 + 78 = €193$.

The computation of the degrees of freedom (equation (10.7)) is cumbersome if you are doing the calculation by hand, but it is easily implemented with a computer software package. However, note that the expressions s_1^2/n_1 and s_2^2/n_2 appear in both expression (10.6) and equation (10.7). These values only need to be computed once in order to evaluate both (10.6) and (10.7).

Hypothesis tests about $\mu_1 - \mu_2$

Let us now consider hypothesis tests about the difference between the means of two populations when the population standard deviations σ_1 and σ_2 are unknown. Letting D_0 denote the hypothesized difference between μ_1 and μ_2, Section 10.1 showed that the test statistic used for the case where σ_1 and σ_2 are known is as follows. The test statistic, z, follows the standard normal distribution.

$$z = \frac{(\bar{x}_1 - \bar{x}_2) - D_0}{\sqrt{\dfrac{\sigma_1^2}{n_1} + \dfrac{\sigma_2^2}{n_2}}}$$

When σ_1 and σ_2 are unknown, we use s_1 as an estimator of σ_1 and s_2 as an estimator of σ_2. Substituting these sample standard deviations for σ_1 and σ_2 gives the following test statistic when σ_1 and σ_2 are unknown.

Test statistic for hypothesis tests about $\mu_1 - \mu_2$: σ_1 and σ_2 unknown

$$t = \frac{(\bar{x}_1 - \bar{x}_2) - D_0}{\sqrt{\dfrac{s_1^2}{n_1} + \dfrac{s_2^2}{n_2}}} \tag{10.8}$$

The degrees of freedom for t are given by equation (10.7).

Consider an example, involving a new computer software package developed to help systems analysts reduce the time required to design, develop, and implement an information system. To evaluate the benefits of the new software package, a random sample of 24 systems analysts is selected. Each analyst is given specifications for a hypothetical information system. Then 12 of the analysts are instructed to produce the information system by using current technology. The other 12 analysts are trained in the use of the new software package and then instructed to use it to produce the information system.

This study involves two populations: a population of systems analysts using the current technology and a population of systems analysts using the new software package. In terms of the time required to complete the information system design project, the population means are as follow.

μ_1 = the mean project completion time for systems analysts using the current technology

μ_2 = the mean project completion time for systems analysts using the new software package

The researcher in charge of the new software evaluation project hopes to show that the new software package will provide a shorter mean project completion time, i.e. the researcher is looking for evidence to conclude that μ_2 is less than μ_1. In this case, the difference between the two population means, $\mu_1 - \mu_2$, will be greater than zero. The research hypothesis $\mu_1 - \mu_2 > 0$ is stated as the alternative hypothesis. The hypothesis test becomes

SOFTWARE TEST

$$H_0: \mu_1 - \mu_2 \leq 0$$
$$H_1: \mu_1 - \mu_2 > 0$$

We shall use $\alpha = 0.05$ as the level of significance. Suppose that the 24 analysts complete the study with the results shown in Table 10.1.

Table 10.1 Completion time data and summary statistics for the software testing study

	Current technology	New software
	300	274
	280	220
	344	308
	385	336
	372	198
	360	300
	288	315
	321	258
	376	318
	290	310
	301	332
	283	263
Summary statistics		
Sample size	$n_1 = 12$	$n_2 = 12$
Sample mean	$\bar{x}_1 = 325$ hours	$\bar{x}_2 = 286$ hours
Sample standard deviation	$s_1 = 40$	$s_2 = 44$

... test statistic in equation (10.8), we have

$$\frac{6) - 0}{\dfrac{44^2}{12}} = 2.27$$

10.7), we have

$$\frac{\left(\dfrac{40^2}{12} + \dfrac{44^2}{12}\right)^2}{\left(\dfrac{40^2}{12}\right)^2 + \left(\dfrac{1}{12 - 1}\right)\left(\dfrac{44^2}{12}\right)^2} = 21.8$$

h 21 degrees of freedom. This row of

)5	0.025	0.01	0.005
721	2.080	2.518	2.831

$t = 2.27$

With an upper tail test, the p-value is the area in the upper tail to the right of $t = 2.27$. From the above results, we see that the p-value is between 0.025 and 0.01. Hence, the p-value is less than $\alpha = 0.05$ and H_0 is rejected. The sample results enable the researcher to conclude that $\mu_1 - \mu_2 > 0$, or $\mu_1 > \mu_2$. The research study supports the conclusion that the new software package provides a smaller population mean completion time.

Practical advice

The interval estimation and hypothesis testing procedures presented in this section are robust and can be used with relatively small sample sizes. In most applications, equal or nearly equal sample sizes such that the total sample size $n_1 + n_2$ is at least 20 can be expected to provide very good results even if the populations are not normal. Larger sample sizes are recommended if the distributions of the populations are highly skewed or contain outliers. Smaller sample sizes should only be used if the analyst is satisfied that the distributions of the populations are at least approximately normal.

Another approach sometimes used to make inferences about the difference between two population means when σ_1 and σ_2 are unknown is based on the assumption that the two population standard deviations are equal. You will find this approach as an option in MINITAB, PASW and EXCEL. Under the assumption of equal population variances, the two sample standard deviations are combined to provide the following 'pooled' sample variance s^2:

$$s^2 = \frac{(n_1 - 1)s_1^2 + (n_2 - 1)s_2^2}{n_1 + n_2 - 2}$$

The t test statistic becomes

$$t = \frac{(\bar{x}_1 - \bar{x}_2) - D_0}{s\sqrt{\dfrac{1}{n_1} + \dfrac{1}{n_2}}}$$

and has $n_1 + n_2 - 2$ degrees of freedom. At this point, the computation of the p-value and the interpretation of the sample results are identical to the procedures discussed earlier in this section. A difficulty with this procedure is that the assumption that the two population standard deviations are equal is usually difficult to verify. Unequal population standard deviations are frequently encountered. Using the pooled procedure may not provide satisfactory results especially if the sample sizes n_1 and n_2 are quite different. The t procedure that we presented in this section does not require the assumption of equal population standard deviations and can be applied whether the population standard deviations are equal or not. It is a more general procedure and is recommended for most applications.

Exercises

Methods

7 Consider the following results for independent random samples taken from two populations.

Sample 1	Sample 2
$n_1 = 20$	$n_2 = 30$
$\bar{x}_1 = 22.5$	$\bar{x}_2 = 20.1$
$s_1 = 2.5$	$s_2 = 4.8$

a. What is the point estimate of the difference between the two population means?
b. What are the degrees of freedom for the t distribution?
c. At 95 per cent confidence, what is the margin of error?
d. What is the 95 per cent confidence interval for the difference between the two population means?

8 Consider the following hypothesis test.

$$H_0: \mu_1 - \mu_2 = 0$$
$$H_1: \mu_1 - \mu_2 \neq 0$$

The following results are from independent samples taken from two populations.

Sample 1	Sample 2
$n_1 = 35$	$n_2 = 40$
$\bar{x}_1 = 13.6$	$\bar{x}_2 = 10.1$
$s_1 = 5.2$	$s_2 = 8.5$

a. What is the value of the test statistic?
b. What are the degrees of freedom for the t distribution?
c. What is the p-value?
d. At $\alpha = 0.05$, what is your conclusion?

9 Consider the following data for two independent random samples taken from two normal populations.

Sample 1	10	7	13	7	9	8
Sample 2	8	7	8	4	6	9

a. Compute the two sample means.
b. Compute the two sample standard deviations.
c. What is the point estimate of the difference between the two population means?
d. What is the 90 per cent confidence interval estimate of the difference between the two population means?

Applications

10 The International Air Transport Association surveyed business travellers to determine ratings of various international airports. The maximum possible score was 10. Suppose 50 business travellers were asked to rate airport L and 50 other business travellers were asked to rate airport M. The rating scores follow.

Airport L

10 9 6 7 8 7 9 8 10 7 6 5 7 3 5 6 8 7 10 8 4 7 8 6 9
 9 5 3 1 8 9 6 8 5 4 6 10 9 8 3 2 7 9 5 3 10 3 5 10 8

Airport M

6 4 6 8 7 7 6 3 3 8 10 4 8 7 8 7 5 9 5 8 4 3 8 5 5
4 4 4 8 4 5 6 2 5 9 9 8 4 8 9 9 5 9 7 8 3 10 8 9 6

Construct a 95 per cent confidence interval estimate of the difference between the mean ratings of the airports L and M.

11 Suppose independent random samples of 15 unionized women and 20 non-unionized women in a skilled manufacturing job provide the following hourly wage rates (€).

Union workers

| 22.40 | 18.90 | 16.70 | 14.05 | 16.20 | 20.00 | 16.10 | 16.30 | 19.10 | 16.50 |
| 18.50 | 19.80 | 17.00 | 14.30 | 17.20 | | | | | |

Non-union workers

| 17.60 | 14.40 | 16.60 | 15.00 | 17.65 | 15.00 | 17.55 | 13.30 | 11.20 | 15.90 |
| 19.20 | 11.85 | 16.65 | 15.20 | 15.30 | 17.00 | 15.10 | 14.30 | 13.90 | 14.50 |

a. What is the point estimate of the difference between mean hourly wages for the two populations?
b. Develop a 95 per cent confidence interval estimate of the difference between the two population means.
c. Does there appear to be any difference in the mean wage rate for these two groups? Explain.

12 The College Board provided comparisons of Scholastic Aptitude Test (SAT) scores based on the highest level of education attained by the test taker's parents. A research hypothesis was that students whose parents had attained a higher level of education would on average score higher on the SAT. During 2003, the overall mean SAT verbal score was 507 (*The World Almanac 2004*). SAT verbal scores for independent samples of students follow. The first sample shows the SAT verbal test scores for students whose parents are college graduates with a bachelor's degree. The second sample shows the SAT verbal test scores for students whose parents are high school graduates but do not have a college degree.

Student's Parents			
College Grads		High School Grads	
485	487	442	492
534	533	580	478
650	526	479	425
554	410	486	485
550	515	528	390
572	578	524	535
497	448		
592	469		

a. Formulate the hypotheses that can be used to determine whether the sample data support the hypothesis that students show a higher population mean verbal score on the SAT if their parents attained a higher level of education.
b. What is the point estimate of the difference between the means for the two populations?
c. Compute the p-value for the hypothesis test.
d. At $\alpha = 0.05$, what is your conclusion?

13 Periodically, Merrill Lynch customers are asked to evaluate Merrill Lynch financial consultants and services (2000 Merrill Lynch Client Satisfaction Survey). Higher ratings on the client satisfaction survey indicate better service, with 7 the maximum service rating. Independent samples of service ratings for two financial consultants are summarized here. Consultant A has ten years of experience while consultant B has one year of experience. Use $\alpha = 0.05$

and test to see whether the consultant with more experience has the higher population mean service rating.

	Consultant A	Consultant B
	$n_1 = 16$	$n_2 = 10$
	$\bar{x}_1 = 6.82$	$\bar{x}_2 = 6.25$
	$s_1 = 0.64$	$s_2 = 0.75$

a. State the null and alternative hypotheses.
b. Compute the value of the test statistic.
c. What is the p-value?
d. What is your conclusion?

14 Safegate Foods is redesigning the checkouts in its supermarkets throughout the country and is considering two designs. Tests on customer checkout times conducted at two stores where the two new systems have been installed result in the following summary of the data.

	System A	System B
	$n_1 = 120$	$n_2 = 100$
	$\bar{x}_1 = 4.1$ minutes	$\bar{x}_2 = 3.4$ minutes
	$s_1 = 2.2$ minutes	$s_2 = 1.5$ minutes

Test at the 0.05 level of significance to determine whether the population mean checkout times of the two systems differ. Which system is preferred?

15 Samples of final examination scores for two statistics classes with different instructors provided the following results.

	Instructor A	Instructor B
	$n_1 = 12$	$n_2 = 15$
	$\bar{x}_1 = 72$	$\bar{x}_2 = 78$
	$s_1 = 8$	$s_2 = 10$

With $\alpha = 0.05$, test whether these data are sufficient to conclude that the population mean grades for the two classes differ.

16 Educational testing companies provide tutoring, classroom learning, and practice tests in an effort to help students perform better on tests such as the Scholastic Aptitude Test (SAT). The test preparation companies claim that their courses will improve SAT score performances by an average of 120 points (*The Wall Street Journal*, 23 January 2003). A researcher is uncertain of this claim and believes that 120 points may be an overstatement in an effort to encourage students to take the test preparation course. In an evaluation study of one test preparation service, the researcher collects SAT score data for 35 students who took the test preparation course and 48 students who did not take the course.

	Course	No course
Sample mean	1058	983
Sample standard deviation	90	105

a. Formulate the hypotheses that can be used to test the researcher's belief that the improvement in SAT scores may be less than the stated average of 120 points.
b. Use $\alpha = 0.05$ and the data above. What is your conclusion?
c. What is the point estimate of the improvement in the average SAT scores provided by the test preparation course? Provide a 95 per cent confidence interval estimate of the improvement.
d. What advice would you have for the researcher after seeing the confidence interval?

10.3 Inferences about the difference between two population means: matched samples

Suppose employees at a manufacturing company can use two different methods to perform a production task. To maximize production output, the company wants to identify the method with the smaller population mean completion time. Let μ_1 denote the population mean completion time for production method 1 and μ_2 denote the population mean completion time for production method 2. With no preliminary indication of the preferred production method, we begin by tentatively assuming that the two production methods have the same population mean completion time. The null hypothesis is $H_0: \mu_1 - \mu_2 = 0$. If this hypothesis is rejected, we can conclude that the population mean completion times differ. In this case, the method providing the smaller mean completion time would be recommended. The null and alternative hypotheses are written as follows.

$$H_0: \mu_1 - \mu_2 = 0$$
$$H_1: \mu_1 - \mu_2 \neq 0$$

In choosing the sampling procedure that will be used to collect production time data and test the hypotheses, we consider two alternative designs. One is based on **independent samples** and the other is based on **matched samples**.

1 *Independent sample design*: A simple random sample of workers is selected and each worker in the sample uses method 1. A second independent simple random sample of workers is selected and each worker in this sample uses method 2. The test of the difference between population means is based on the procedures in Section 10.2.

2 *Matched sample design*: One simple random sample of workers is selected. Each worker first uses one method and then uses the other method. The order of the two methods is assigned randomly to the workers, with some workers performing method 1 first and others performing method 2 first. Each worker provides a pair of data values, one value for method 1 and another value for method 2.

In the matched sample design the two production methods are tested under similar conditions (i.e. with the same workers). Hence this design often leads to a smaller sampling error than the independent sample design. The primary reason is that in a matched sample design, variation between workers is eliminated because the same workers are used for both production methods.

Let us demonstrate the analysis of a matched sample design by assuming it is the method used to test the difference between population means for the two production

| | Completion time for | Completion time for | Difference in |
Worker	Method 1 (minutes)	Method 2 (minutes)	completion times (d_i)
1	6.0	5.4	0.6
2	5.0	5.2	−0.2
3	7.0	6.5	0.5
4	6.2	5.9	0.3
5	6.0	6.0	0.0
6	6.4	5.8	0.6

Table 10.2 Task completion times for a matched sample design

MATCHED

methods. A random sample of six workers is used. The data on completion times for the six workers are given in Table 10.2. Note that each worker provides a pair of data values, one for each production method. Also note that the last column contains the difference in completion times d_i for each worker in the sample.

The key to the analysis of the matched sample design is to realize that we consider only the column of differences. Therefore, we have six data values (0.6, −0.2, 0.5, 0.3, 0.0, 0.6) that will be used to analyze the difference between population means of the two production methods.

Let μ_d = the mean of the *difference* values for the population of workers. With this notation, the null and alternative hypotheses are rewritten as follows.

$$H_0: \mu_d = 0$$
$$H_1: \mu_d \neq 0$$

If H_0 is rejected, we can conclude that the population mean completion times differ. The d notation is a reminder that the matched sample provides *difference* data. The sample mean and sample standard deviation for the six difference values in Table 10.2 follow. Other than the use of the d notation, the formulae for the sample mean and sample standard deviation are the same ones used previously in the text.

$$\bar{d} = \frac{\Sigma d_i}{n} = \frac{1.8}{6} = 0.30$$

$$s_d = \sqrt{\frac{\Sigma (d_i - \bar{d})^2}{n-1}} = \sqrt{\frac{0.56}{5}} = 0.335$$

With the small sample of $n = 6$ workers, we need to make the assumption that the population of differences has a normal distribution. This assumption is necessary so that we may use the t distribution for hypothesis testing and interval estimation procedures. Sample size guidelines for using the t distribution were presented in Chapters 8 and 9. Based on this assumption, the following test statistic has a t distribution with $n - 1$ degrees of freedom.

Test statistic for hypothesis test involving matched samples

$$t = \frac{\bar{d} - \mu_d}{s_d/\sqrt{n}} \tag{10.9}$$

Let us use equation (10.9) to test the hypotheses $H_0: \mu_d = 0$ and $H_1: \mu_d \neq 0$, using $\alpha = 0.05$. Substituting the sample results $d = 0.30$, $s_d = 0.335$, and $n = 6$ into equation (10.9), we compute the value of the test statistic.

$$t = \frac{\bar{d} - \mu_d}{s_d/\sqrt{n}} = \frac{0.30 - 0}{0.335/\sqrt{6}} = 2.20$$

Now let us compute the p-value for this two-tailed test. Because $t = 2.20 > 0$, the test statistic is in the upper tail of the t distribution. With $t = 2.20$, the area in the upper tail to the right of the test statistic can be found by using the t distribution table with degrees of freedom $= n - 1 = 6 - 1 = 5$. Information from the 5 degrees of freedom row of the t distribution table is as follows:

Area in upper tail	0.20	0.10	0.05	0.025	0.01	0.005
t value (5 df)	0.920	1.476	2.015	2.571	3.365	4.032

$t = 2.20$

We see that the area in the upper tail is between 0.05 and 0.025. Because this test is a two-tailed test, we double these values to conclude that the p-value is between 0.10 and 0.05. This p-value is greater than $\alpha = 0.05$, so the null hypothesis $H_0: \mu_d = 0$ is not rejected. MINITAB, EXCEL and PASW show the p-value as 0.080.

In addition we can obtain an interval estimate of the difference between the two population means by using the single population methodology of Chapter 8. At 95 per cent confidence, the calculation follows.

$$\bar{d} \pm t_{0.025} \frac{s_d}{\sqrt{n}} = 0.30 \pm 2.527 \left(\frac{0.335}{\sqrt{6}} \right) = 0.3 \pm 0.35$$

The margin of error is 0.35 and the 95 per cent confidence interval for the difference between the population means of the two production methods is -0.05 minutes to 0.65 minutes.

In the example presented in this section, workers performed the production task with first one method and then the other method. This example illustrates a matched sample design in which each sampled element (worker) provides a pair of data values. It is also possible to use different but 'similar' elements to provide the pair of data values. For example, a worker at one location could be matched with a similar worker at another location (similarity based on age, education, gender, experience, etc.). The pairs of workers would provide the difference data that could be used in the matched sample analysis. A matched sample procedure for inferences about two population means generally provides better precision than the independent samples approach, therefore it is the recommended design. However, in some applications matching is not feasible, or perhaps the time and cost associated with matching are excessive. In such cases, the independent samples design should be used.

Exercises

Methods

17 Consider the following hypothesis test.

$$H_0: \mu_d \leq 0$$
$$H_1: \mu_d > 0$$

The following data are from matched samples taken from two populations.

| | Population | |
Element	1	2
1	21	20
2	28	26
3	18	18
4	20	20
5	26	24

a. Compute the difference value for each element.
b. Compute $\bar{d}$.
c. Compute the standard deviation s_d
d. Conduct a hypothesis test using $\alpha = 0.05$. What is your conclusion?

18 The following data are from matched samples taken from two populations.

| | Population | |
Element	1	2
1	11	8
2	7	8
3	9	6
4	12	7
5	13	10
6	15	15
7	15	14

a. Compute the difference value for each element.
b. Compute $\bar{d}$.
c. Compute the standard deviation s_d
d. What is the point estimate of the difference between the two population means?
e. Provide a 95 per cent confidence interval for the difference between the two population means.

Applications

19 In recent years, a growing array of entertainment options competes for consumer time. By 2004, cable television and radio surpassed broadcast television, recorded music, and the daily newspaper to become the two entertainment media with the greatest usage (*The Wall Street Journal*, 26 January 2004). Researchers used a sample of 15 individuals and collected data on the hours per week spent watching cable television and hours per week spent listening to the radio.

Individual	Television	Radio	Individual	Television	Radio
1	22	25	9	21	21
2	8	10	10	23	23
3	25	29	11	14	15
4	22	19	12	14	18
5	12	13	13	14	17
6	26	28	14	16	15
7	22	23	15	24	23
8	19	21			

TVRADIO

a. What is the sample mean number of hours per week spent watching cable television? What is the sample mean number of hours per week spent listening to radio? Which medium has the greater usage?

b. Use a 0.05 level of significance and test for a difference between the population mean usage for cable television and radio. What is the p-value?

20 A market research firm used a sample of individuals to rate the purchase potential of a particular product before and after the individuals saw a new television commercial about the product. The purchase potential ratings were based on a 0 to 10 scale, with higher values indicating a higher purchase potential. The null hypothesis stated that the mean rating 'after' would be less than or equal to the mean rating 'before'. Rejection of this hypothesis would show that the commercial improved the mean purchase potential rating. Use $\alpha = 0.05$ and the following data to test the hypothesis and comment on the value of the commercial.

| | Purchase rating | | | Purchase rating | |
Individual	After	Before	Individual	After	Before
1	6	5	5	3	5
2	6	4	6	9	8
3	7	7	7	7	5
4	4	3	8	6	6

21 StreetInsider.com reported 2002 earnings per share data for a sample of major companies (12 February 2003). Prior to 2002, financial analysts predicted the 2002 earnings per share for these same companies (*Barron's*, 10 September 2001). Use the following data to comment on differences between actual and estimated earnings per share.

Company	Actual	Predicted
AT & T	1.29	0.38
American Express	2.01	2.31
Citigroup	2.59	3.43
Coca Cola	1.60	1.78
DuPont	1.84	2.18
Exxon-Mobil	2.72	2.19
General Electric	1.51	1.71
Johnson & Johnson	2.28	2.18
McDonald's	0.77	1.55
Wal-Mart	1.81	1.74

EARNINGS

a. Use $\alpha = 0.05$ and test for any difference between the population mean actual and population mean estimated earnings per share. What is the p-value? What is your conclusion?

b. What is the point estimate of the difference between the two means? Did the analysts tend to underestimate or overestimate the earnings?

c. At 95 per cent confidence, what is the margin of error for the estimate in part (b)? What would you recommend based on this information?

22 A survey was made of Book-of-the-Month-Club members to ascertain whether members spend more time watching television than they do reading. Assume a sample of 15 respondents provided the following data on weekly hours of television watching and

weekly hours of reading. Using a 0.05 level of significance, can you conclude that Book-of-the-Month-Club members spend more hours per week watching television than reading?

Respondent	Television	Reading	Respondent	Television	Reading
1	10	6	9	4	7
2	14	16	10	8	8
3	16	8	11	16	5
4	18	10	12	5	10
5	15	10	13	8	3
6	14	8	14	19	10
7	10	14	15	11	6
8	12	14			

10.4 Inferences about the difference between two population proportions

Let π_1 denote the proportion for population 1 and π_2 denote the proportion for population 2. We next consider inferences about the difference between the two population proportions: $\pi_1 - \pi_2$. To make an inference about this difference, we shall select two independent random samples consisting of n_1 units from population 1 and n_2 units from population 2.

Interval estimation of $\pi_1 - \pi_2$

In the following example, we show how to compute a margin of error and construct an interval estimate of the difference between two population proportions.

An accountancy firm specializing in the preparation of income tax returns is interested in comparing the quality of work at two of its regional offices. The firm will be able to estimate the proportion of erroneous returns by randomly selecting samples of tax returns prepared at each office and verifying the sample returns' accuracy. The difference between these proportions is of particular interest.

π_1 = proportion of erroneous returns for population 1 (office 1)
π_2 = proportion of erroneous returns for population 2 (office 2)
P_1 = sample proportion for a simple random sample from population 1
P_2 = sample proportion for a simple random sample from population 2

The difference between the two population proportions is given by $\pi_1 - \pi_2$. The point estimator of $\pi_1 - \pi_2$ is as follows.

Point estimator of the difference between two population proportions

$$P_1 - P_2 \tag{10.10}$$

The point estimator of the difference between two population proportions is the difference between the sample proportions of two independent simple random samples.

As with other point estimators, the point estimator $P_1 - P_2$ has a sampling distribution that reflects the possible values of $P_1 - P_2$ if we repeatedly took two independent random samples. The mean of this sampling distribution is $\pi_1 - \pi_2$ and the standard error of $P_1 - P_2$ is as follows:

Standard error of $P_1 - P_2$

$$\sigma_{P_1 - P_2} = \sqrt{\frac{\pi_1(1 - \pi_1)}{n_1} + \frac{\pi_2(1 - \pi_2)}{n_2}}$$

(10.11)

If the sample sizes are large enough that $n_1\pi_1$, $n_1(1 - \pi_1)$, $n_2\pi_2$ and $n_2(1 - \pi_2)$ are all greater than or equal to five, the sampling distribution of $P_1 - P_2$ can be approximated by a normal distribution.

As we showed previously, an interval estimate is given by a point estimate ± a margin of error. In the estimation of the difference between two population proportions, an interval estimate will take the form $p_1 - p_2$ ± margin of error. With the sampling distribution of $P_1 - P_2$ approximated by a normal distribution, we would like to use $z_{\alpha/2}\,\sigma_{P_1 - P_2}$ as the margin of error. However, $\sigma_{P_1 - P_2}$ given by equation (10.11) cannot be used directly because the two population proportions, π_1 and π_2, are unknown. Using the sample proportion p_1 to estimate π_1 and the sample proportion p_2 to estimate π_2, the margin of error is as follows.

$$\text{Margin of error} = z_{\alpha/2}\sqrt{\frac{p_1(1 - p_1)}{n_1} + \frac{p_2(1 - p_2)}{n_2}}$$

(10.12)

The general form of an interval estimate of the difference between two population proportions is as follows.

Interval estimate of the difference between two population proportions

$$(p_1 - p_2) \pm z_{\alpha/2}\sqrt{\frac{p_1(1 - p_1)}{n_1} + \frac{p_2(1 - p_2)}{n_2}}$$

(10.13)

where $1 - \alpha$ is the confidence coefficient.

Returning to the tax returns example, we find that independent simple random samples from the two offices provide the following information.

Office 1	Office 2
$n_1 = 250$	$n_1 = 300$
Number of returns with errors = 35	Number of returns with errors = 27

TAX PREP

The sample proportions for the two offices are:

$$p_1 = \frac{35}{250} = 0.14 \qquad p_2 = \frac{27}{300} = 0.09$$

The point estimate of the difference between the proportions of erroneous tax returns for the two populations is $p_1 - p_2 = 0.14 - 0.09 = 0.05$. We estimate that Office 1 has a 0.05, or 5 percentage points, greater error rate than Office 2.

Expression (10.13) can now be used to provide a margin of error and interval estimate of the difference between the two population proportions. Using a 90 per cent confidence interval with $z_{\alpha/2} = z_{0.05} = 1.645$, we have

$$(p_1 - p_2) \pm z_{\alpha/2} \sqrt{\frac{p_1(1 - p_1)}{n_1} + \frac{p_2(1 - p_2)}{n_2}}$$

$$= (0.14 - 0.09) \pm 1.645 \sqrt{\frac{0.14(1 - 0.14)}{250} + \frac{0.09(1 - 0.09)}{300}} = 0.05 \pm 0.045$$

The margin of error is 0.045, and the 90 per cent confidence interval is 0.005 to 0.095.

Hypothesis tests about $\pi_1 - \pi_2$

Let us now consider hypothesis tests about the difference between the proportions of two populations. The three forms for a hypothesis test are as follows:

$$\begin{array}{ccc} H_0: \pi_1 - \pi_2 \geq 0 & H_0: \pi_1 - \pi_2 \leq 0 & H_0: \pi_1 - \pi_2 = 0 \\ H_1: \pi_1 - \pi_2 < 0 & H_1: \pi_1 - \pi_2 > 0 & H_1: \pi_1 - \pi_2 \neq 0 \end{array}$$

When we assume H_0 is true as an equality, we have $\pi_1 - \pi_2 = 0$, which is the same as saying that the population proportions are equal, $\pi_1 = \pi_2$. We shall base the test statistic on the sampling distribution of the point estimator $P_1 - P_2$.

In expression (10.11), we showed that the standard error of $P_1 - P_2$ is given by

$$\sigma_{P_1 - P_2} = \sqrt{\frac{\pi_1(1 - \pi_1)}{n_1} + \frac{\pi_2(1 - \pi_2)}{n_2}}$$

Under the assumption that H_0 is true as an equality, the population proportions are equal and $\pi_1 = \pi_2 = \pi$. In this case, $\sigma_{P_1 - P_2}$ becomes

---

Standard error of $P_1 - P_2$ when $\pi_1 = \pi_2 = \pi$

$$\sigma_{P_1 - P_2} = \sqrt{\frac{\pi(1 - \pi)}{n_1} + \frac{\pi(1 - \pi)}{n_2}} = \sqrt{\pi(1 - \pi)\left(\frac{1}{n_1} + \frac{1}{n_2}\right)} \qquad (10.14)$$

With π unknown, we pool, or combine, the point estimates from the two samples (p_1 and p_2) to obtain a single point estimate of π as follows:

Pooled estimate of π when $\pi_1 = \pi_2 = \pi$

$$p = \frac{n_1 p_1 + n_2 p_2}{n_1 + n_2} \qquad (10.15)$$

This **pooled estimate of π** is a weighted average of p_1 and p_2.

Substituting p for π in equation (10.14), we obtain an estimate of $\sigma_{p_1 - p_2}$, which is used in the test statistic. The general form of the test statistic for hypothesis tests about the difference between two population proportions is the point estimator divided by the estimate of $\sigma_{p_1 - p_2}$:

Test statistic for hypothesis tests about $\pi_1 - \pi_2$

$$z = \frac{(p_1 - p_2)}{\sqrt{p(1 - p)\left(\frac{1}{n_1} + \frac{1}{n_2}\right)}}$$

(10.16)

This test statistic applies to large sample situations where $n_1\pi_1$, $n_1(1 - \pi_1)$, $n_2\pi_2$ and $n_2(1 - \pi_2)$ are all greater than or equal to five.

Let us return to the tax returns example and assume that the firm wants to use a hypothesis test to determine whether the error proportions differ between the two offices. A two-tailed test is required. The null and alternative hypotheses are as follows:

$$H_0: \pi_1 - \pi_2 = 0$$
$$H_1: \pi_1 - \pi_2 \neq 0$$

If H_0 is rejected, the firm can conclude that the error rates at the two offices differ. We shall use $\alpha = 0.10$ as the level of significance.

The sample data previously collected showed $p_1 = 0.14$ for the $n_1 = 250$ returns sampled at Office 1 and $p_2 = 0.09$ for the $n_2 = 300$ returns sampled at Office 2. The pooled estimate of π is

$$p = \frac{n_1 p_1 + n_2 p_2}{n_1 + n_2} = \frac{250(0.14) + 300(0.09)}{250 + 300} = 0.1127$$

Using this pooled estimate and the difference between the sample proportions, the value of the test statistic is as follows.

$$z = \frac{(p_1 - p_2)}{\sqrt{p(1 - p)\left(\frac{1}{n_1} + \frac{1}{n_2}\right)}} = \frac{(0.14 - 0.09)}{\sqrt{0.1127(1 - 0.1127)\left(\frac{1}{250} + \frac{1}{300}\right)}} = 1.85$$

To compute the p-value for this two-tailed test, we first note that $z = 1.85$ is in the upper tail of the standard normal distribution. Using the standard normal distribution table, we find the area in the upper tail for $z = 1.85$ is $1 - 0.9678 = 0.0322$. Doubling this area for a two-tailed test, we find the p-value $= 2(0.0322) = 0.0644$. With the p-value less than $\alpha = 0.10$, H_0 is rejected at the 0.10 level of significance. The firm can conclude that the error rates differ between the two offices. This hypothesis test conclusion is consistent with the earlier interval estimation results that showed the interval estimate of the difference between the population error rates at the two offices to be 0.005 to 0.095, with Office 1 having the higher error rate.

Exercises

Methods

23 Consider the following results for independent samples taken from two populations.

Sample 1	Sample 2
$n_1 = 400$	$n_2 = 300$
$p_1 = 0.48$	$p_2 = 0.36$

a. What is the point estimate of the difference between the two population proportions?
b. Construct a 90 per cent confidence interval for the difference between the two population proportions.
c. Construct a 95 per cent confidence interval for the difference between the two population proportions.

24 Consider the hypothesis test

$$H_0: \pi_1 - \pi_2 \leq 0$$
$$H_1: \pi_1 - \pi_2 > 0$$

The following results are for independent samples taken from the two populations.

Sample 1	Sample 2
$n_1 = 200$	$n_2 = 300$
$p_1 = 0.22$	$p_2 = 0.1$

a. What is the p-value?
b. With $\alpha = 0.05$, what is your hypothesis testing conclusion?

Applications

25 In November and December 2008, research companies affiliated to the Worldwide Independent Network of Market Research carried out polls in 17 countries to assess people's views on the economic outlook. In the Canadian survey, conducted by Léger Marketing, 61 per cent of the sample of 1511 people thought the economic situation would worsen over the next three months. In the UK survey, conducted by ICM Research, 78 per cent of the sample of 1050 felt that economic conditions would worsen over that period. Provide a 95 per cent confidence interval estimate for the difference between the population proportions in the two countries. What is your interpretation of the interval estimate?

26 The Anwar Sadat Chair for Peace and Development carried out an opinion poll among adults in six African and Arab states in May 2004. The results show that 69 per cent of 400 respondents in Jordan felt that the war in Iraq had brought less democracy to the country, compared with 57 per cent of 700 respondents in Lebanon who had that view. Construct a 95 per cent confidence interval for the difference between the proportion of Jordanian adults who held this view and the proportion of Lebanese adults who held this view.

27 In a test of the quality of two television commercials, each commercial was shown in a separate test area six times over a one-week period. The following week a telephone survey was conducted to identify individuals who had seen the commercials. Those individuals were asked to state the primary message in the commercials. The following results were recorded.

	Commercial A	Commercial B
Number who saw commercial	150	200
Number who recalled message	63	60

a. Use $\alpha = 0.05$ and test the hypothesis that there is no difference in the recall proportions for the two commercials.

b. Compute a 95 per cent confidence interval for the difference between the recall proportions for the two populations.

28 UNITE/MORI published annual 'Student Experience Reports' from 2001 to 2005, based on face-to-face interviews carried out at a sample of UK universities. In 2001, it was reported that 74 per cent of 1103 respondents strongly agreed with the statement that 'going to university is a worthwhile experience'. The 2005 report says that 66 per cent of 1065 respondents strongly agreed with this statement. Test the hypothesis $\pi_1 - \pi_2 = 0$ with $\alpha = 0.05$. What is the p-value. What is your conclusion?

29 A large car insurance company selected samples of single and married male policyholders and recorded the number who made an insurance claim over the preceding three-year period.

Single policyholders	Married policyholders
$n_1 = 400$	$n_2 = 900$
Number making claims = 76	Number making claims = 90

a. Use $\alpha = 0.05$ and test to determine whether the claim rates differ between single and married male policyholders.

b. Provide a 95 per cent confidence interval for the difference between the proportions for the two populations.

For additional online summary questions and answers go to the companion website at www.cengage.co.uk/aswsbe2

Summary

In this chapter we discussed procedures for constructing interval estimates and doing hypothesis tests involving two populations. First, we showed how to make inferences about the difference between two population means when independent simple random samples are selected. We considered the case where the population standard deviations, σ_1 and σ_2, could be assumed known. The standard normal distribution z was used to develop the interval estimate and served as the test statistic for hypothesis tests. We then considered the case where the population standard deviations were unknown and estimated by the sample standard deviations s_1 and s_2. In this case, the t distribution was used to develop the interval estimate and served as the test statistic for hypothesis tests.

Inferences about the difference between two population means were then discussed for the matched sample design. In the matched sample design each element provides a pair of data values, one from each population. The difference between the paired data values is then used in the statistical analysis. The matched sample design is generally preferred to the independent sample design, when it is feasible, because the matched-samples procedure often improves the precision of the estimate.

Finally, interval estimation and hypothesis testing about the difference between two population proportions were discussed. Statistical procedures for analyzing the difference between two population proportions are similar to the procedures for analyzing the difference between two population means.

Key terms

Independent samples

Matched samples

Pooled estimator of π

Key formulae

Point estimator of the difference between two population means

$$\overline{X}_1 - \overline{X}_2 \tag{10.1}$$

Standard error of $\overline{X}_1 - \overline{X}_2$

$$\sigma_{\overline{X}_1 - \overline{X}_2} = \sqrt{\frac{\sigma_1^2}{n_1} + \frac{\sigma_2^2}{n_2}} \tag{10.2}$$

Interval estimate of the difference between two population means: σ_1 and σ_2 known

$$(\overline{X}_1 - \overline{X}_2) \pm z_{\alpha/2}\sqrt{\frac{\sigma_1^2}{n_1} + \frac{\sigma_2^2}{n_2}} \tag{10.4}$$

Test statistic for hypothesis tests about $\mu_1 - \mu_2$: σ_1 and σ_2 known

$$z = \frac{(\bar{x}_1 - \bar{x}_2) - D_0}{\sqrt{\dfrac{\sigma_1^2}{n_1} + \dfrac{\sigma_2^2}{n_2}}}$$

(10.5)

Interval estimate of the difference between two population means: σ_1 and σ_2 unknown

$$(\bar{x}_1 - \bar{x}_2) \pm t_{\alpha/2}\sqrt{\frac{s_1^2}{n_1} + \frac{s_2^2}{n_2}}$$

(10.6)

Degrees of freedom for the t distribution using two independent random samples

$$df = \frac{\left(\dfrac{s_1^2}{n_1} + \dfrac{s_2^2}{n_2}\right)^2}{\left(\dfrac{1}{n_1 - 1}\right)\left(\dfrac{s_1^2}{n_1}\right)^2 + \left(\dfrac{1}{n_2 - 1}\right)\left(\dfrac{s_2^2}{n_2}\right)^2}$$

(10.7)

Test statistic for hypothesis tests about $\mu_1 - \mu_2$: σ_1 and σ_2 unknown

$$t = \frac{(\bar{x}_1 - \bar{x}_2) - D_0}{\sqrt{\dfrac{s_1^2}{n_1} + \dfrac{s_2^2}{n_2}}}$$

(10.8)

Test statistic for hypothesis test involving matched samples

$$t = \frac{\bar{d} - \mu_d}{s_d/\sqrt{n}}$$

(10.9)

Point estimator of the difference between two population proportions

$$P_1 - P_2$$

(10.10)

Standard error of $P_1 - P_2$

$$\sigma_{P_1 - P_2} = \sqrt{\frac{\pi_1(1 - \pi_1)}{n_1} + \frac{\pi_2(1 - \pi_2)}{n_2}}$$

(10.11)

Interval estimate of the difference between two population proportions

$$(p_1 - p_2) \pm z_{\alpha/2}\sqrt{\frac{p_1(1 - p_1)}{n_1} + \frac{p_2(1 - p_2)}{n_2}}$$

(10.13)

Standard error of $P_1 - P_2$ when $\pi_1 = \pi_2 = \pi$

$$\sigma_{P_1 - P_2} = \sqrt{\frac{\pi(1 - \pi)}{n_1} + \frac{\pi(1 - \pi)}{n_2}} = \sqrt{\pi(1 - \pi)\left(\frac{1}{n_1} + \frac{1}{n_2}\right)}$$

(10.14)

Pooled estimate of π when $\pi_1 = \pi_2 = \pi$

$$p = \frac{n_1 p_1 + n_2 p_2}{n_1 + n_2}$$

(10.15)

Test statistic for hypothesis tests about $\pi_1 - \pi_2$

$$z = \frac{(p_1 - p_2)}{\sqrt{p(1 - p)\left(\dfrac{1}{n_1} + \dfrac{1}{n_2}\right)}}$$

(10.16)

Case problem Par Products

Par Products is a major manufacturer of golf equipment. Management believes that Par's market share could be increased with the introduction of a cut-resistant, longer-lasting golf ball. Therefore, the research group at Par has been investigating a new golf ball coating designed to resist cuts and provide a more durable ball. The tests with the coating have been promising.

One of the researchers voiced concern about the effect of the new coating on driving distances. Par would like the new cut-resistant ball to offer driving distances comparable to those of the current-model golf ball. To compare the driving distances for the two balls, 40 balls of both the new and current

Model		Model		Model		Model	
Current	New	Current	New	Current	New	Current	New
264	277	270	272	263	274	281	283
261	269	287	259	264	266	274	250
267	263	289	264	284	262	273	253
272	266	280	280	263	271	263	260
258	262	272	274	260	260	275	270
283	251	275	281	283	281	267	263
258	262	265	276	255	250	279	261
266	289	260	269	272	263	274	255
259	286	278	268	266	278	276	263
270	264	275	262	268	264	262	279

© Zimmytws.

models were subjected to distance tests. The testing was performed with a mechanical hitting machine so that any difference between the mean distances for the two models could be attributed to a difference in the two models. The results of the tests, with distances measured to the nearest metre, are available on the CD that accompanies the text, in the file 'Golf'.

Managerial report

1 Formulate and present the rationale for a hypothesis test that Par could use to compare the driving distances of the current and new golf balls.

2 Analyze the data to provide the hypothesis test conclusion. What is the p-value for your test? What is your recommendation for Par Products?

3 Provide descriptive statistical summaries of the data for each model.

4 What is the 95 per cent confidence interval for the population mean of each model, and what is the 95 per cent confidence interval for the difference between the means of the two populations?

5 Do you see a need for larger sample sizes and more testing with the golf balls? Discuss.

Software Section
for Chapter 10

We describe the use of MINITAB to construct interval estimates and do hypothesis tests about the difference between two population means and the difference between two population proportions. MINITAB provides both interval estimation and hypothesis test results within the same module, so the MINITAB procedure is the same for both types of inferences. In the examples that follow, we shall demonstrate interval estimation and hypothesis testing for the same two samples. We note that MINITAB does not provide a routine for inferences about the difference between two population means when the population standard deviations σ_1 and σ_2 are known.

Difference between two population means: σ_1 and σ_2 unknown

We shall use the data for the cheque account balances example presented in Section 10.2 (file 'CheqAcct.MTW' on the accompanying CD). The cheque account balances at the Northern Branch are in column C1, and the cheque account balances at the Eastern Branch are in column C2. In this example, we will use the MINITAB 2-Sample t procedure to provide a 95 per cent confidence interval estimate of the difference between population means for the cheque account balances at the two branch banks. The output of the procedure also provides the p-value for the hypothesis test: $H_0: \mu_1 - \mu_2 = 0$ versus $H_1: \mu_1 - \mu_2 \neq 0$.

Step 1   **Stat > Basic Statistics > 2-Sample t**   [Main menu bar]

Step 2   Check **Samples in different columns**
                                    [**2-Sample t (Test and Confidence Interval)** panel]
           Enter **C1** in the **First** box
           Enter **C2** in the **Second** box
           Click **Options**

Step 3   Enter **95** in the **Confidence level** box   [**2-Sample t - Options** panel]
           Enter **0** in the **Test difference** box
           Select **not equal** on the **Alternative** menu
           Click **OK**

Step 4   Click **OK**   [**2-Sample t (Test and Confidence Interval)** panel]

> **Figure 10.2** MINITAB output for the hypothesis test and confidence interval for the cheque account balances
>
> **Results for: CheqAcct.MTW**
>
> **Two-Sample T-Test and CI: Northern, Eastern**
>
> ```
> Two-sample T for Northern vs Eastern
>
>                             SE
>           N   Mean  StDev  Mean
> Northern  28  1025    150    28
> Eastern   22   910    125    27
>
>
> Difference = mu (Northern) – mu (Eastern)
> Estimate for difference:  115.0
> 95% CI for difference:  (36.7, 193.2)
> T-Test of difference = 0 (vs not =): T-Value = 2.95  P-Value = 0.005  DF = 47
> ```

The MINITAB output is shown above in Figure 10.2. The 95 per cent confidence interval estimate is (€37 to €193) as described in Section 10.2. The p-value = 0.005 shows the null hypothesis of equal population means can be rejected at the $\alpha = 0.01$ level of significance. Note that MINITAB used equation (10.7) to compute 47 degrees of freedom for this analysis. In other applications, the Options dialogue panel may be used to provide different confidence levels, different hypothesized values, and different forms of the hypotheses.

Difference between two population means with matched samples

We use the data on production times in Table 10.2 to illustrate the matched-sample procedure (file 'Matched.MTW' on the accompanying CD). The completion times for method 1 are entered into column C1 and the completion times for method 2 are entered into column C2. The MINITAB steps for a matched sample are as follows:

Step 1 **Stat > Basic Statistics > Paired t**   [Main menu bar]

Step 2 Select **Samples in columns**
                              [**Paired t (Test and Confidence Interval)** panel]
          Enter **C1** in the **First sample** box
          Enter **C2** in the **Second sample** box
          Click **Options**

Step 3 Enter **95** in the **Confidence level** box   [**Paired t - Options** panel]
          Enter **0** in the **Test difference** box
          Select **not equal** on the **Alternative** menu
          Click **OK**

Step 4 Click **OK**   [**Paired t (Test and Confidence Interval)** panel]

The **Paired t - Options** dialogue panel may be used to provide different confidence levels, different hypothesized values, and different forms of the hypotheses.

Difference between two population proportions

TAXPREP

We shall use the data on tax return errors presented in Section 10.4 (file 'TaxPrep' on the accompanying CD). The sample results for 250 tax returns prepared at Office 1 are in column C1 and the sample results for 300 tax returns prepared at Office 2 are in column C2. Yes denotes an error was found in the tax return and No indicates no error was found. The procedure we describe provides both a 95 per cent confidence interval for $\pi_1 - \pi_2$ and hypothesis test results for $H_0: \pi_1 - \pi_2 = 0$ versus $H_1: \pi_1 - \pi_2 \neq 0$.

Step 1 **Stat > Basic Statistics > 2 Proportions**   [Main menu bar]

Step 2 Select **Samples in different columns**
                          [**2 Proportions (Test and Confidence Interval)** panel]
            Enter **C1** in the **First** box
            Enter **C2** in the **Second** box
            Click **Options**

Step 3 Enter **95** in the **Confidence level** box   [**2 Proportions - Options** panel]
            Enter **0** in the **Test difference** box
            Select **not equal** on the **Alternative** menu
            Check **Use pooled estimate of p for test**
            Click **OK**

Step 4 Click **OK**   [**2 Proportions (Test and Confidence Interval)** panel]

The Options dialogue panel may be used to provide different confidence levels, different hypothesized values, and different forms of the hypotheses.

   In the tax returns example, the data are qualitative. Yes and No are used to indicate whether an error is present. In modules involving proportions, MINITAB calculates proportions for the response coming second in alphabetic order. In the tax preparation example, MINITAB computes the proportion of Yes responses, which is the proportion we wanted. If MINITAB's alphabetical ordering does not compute the proportion for the response of interest, we can fix it. Select any cell in the data column, go to the MINITAB menu bar, and select **Editor > Column > Value Order**. This sequence will provide the option of entering a user-specified order. Simply make sure that the response of interest is listed second in the **define-an-order** box. MINITAB's **2 Proportion** routine will then provide the confidence interval and hypothesis testing results for the population proportion of interest.

   Finally, we note that MINITAB's **2 Proportion** routine uses a computational procedure different from the procedure described in the text. Consequently, the MINITAB output can be expected to provide slightly different interval estimates and slightly different *p*-values. However, results from the two methods should be close and are expected to provide the same interpretation and conclusion.

Inferences about two populations using EXCEL

EXCEL

We describe the use of EXCEL to conduct hypothesis tests about the difference between two population means. We begin with hypothesis tests for the difference between the means of two populations when the population standard deviations σ_1 and σ_2 are known. (Routines are not available for interval estimation of the difference between two population means, nor for inferences about the difference between two population proportions.)

Difference between two population means: σ_1 and σ_2 known

We shall use the examination scores for the two training centres discussed in Section 10.1 (file 'ExamScores.XLS' on the accompanying CD). The label Centre A is in cell A1 and the label Centre B is in cell B1. The examination scores for Centre A are in cells A2:A31 and examination scores for Centre B are in cells B2:B41. The population standard deviations are assumed known with $\sigma_1 = 10$ and $\sigma_2 = 10$. The EXCEL routine will request the input of variances which are $\sigma_1^2 = 100$ and $\sigma_2^2 = 100$. The following steps can be used to conduct a hypothesis test about the difference between the two population means.

Step 1 Click the **Data** tab on the Ribbon

Step 2 In the **Analysis** group, click **Data Analysis**

Step 3 Choose **z-Test: Two Sample for Means**
Click **OK**

Step 4 Enter **A1:A31** in the **Variable 1 Range** box
[**z-Test: Two Sample for Means** panel]
Enter **B1:B41** in the **Variable 2 Range** box
Enter **0** in the **Hypothesized Mean Difference** box
Enter **100** in the **Variable 1 Variance (known)** box
Enter **100** in the **Variable 2 Variance (known)** box
Select **Labels**
Enter **.05** in the **Alpha** box
Select **Output Range** and enter **C1** in the box
Click **OK**

The two-tailed p-value is denoted 'P($Z <= z$) two-tail'. Its value of 0.0977 does not allow us to reject the null hypothesis at $\alpha = 0.05$.

Difference between two population means: σ_1 and σ_2 unknown

We use the data for the software testing study in Table 10.1 (file 'SoftwareTest.XLS' on the accompanying CD). The data are already entered into an EXCEL worksheet with the label Current in cell A1 and the label New in cell B1. The completion times for the current technology are in cells A2:A13, and the completion times for the new software are in cells B2:B13. The following steps can be used to conduct a hypothesis test about the difference between two population means with σ_1 and σ_2 unknown.

Step 1 Click the **Data** tab on the Ribbon

Step 2 In the **Analysis** group, click **Data Analysis**

Step 3 Choose **t-Test: Two Sample Assuming Unequal Variances**
Click **OK**

Step 4 Enter **A1:A13** in the **Variable 1 Range** box
[**t-Test: Two Sample Assuming Unequal Variances** panel]
Enter **B1:B13** in the **Variable 2 Range** box
Enter **0** in the **Hypothesized Mean Difference** box
Select **Labels**
Enter **.05** in the **Alpha** box
Select **Output Range** and enter **C1** in the box
Click **OK**

The appropriate *p*-value is denoted 'P(T <= t) one-tail'. Its value of 0.017 allows us to reject the null hypothesis at $\alpha = 0.05$.

Difference between two population means with matched samples

We use the matched-sample completion times in Table 10.2 to illustrate (file 'Matched. XLS' on the accompanying CD). The data are entered into a worksheet with the label Method 1 in cell A1 and the label Method 2 in cell B1. The completion times for method 1 are in cells A2:A7 and the completion times for method 2 are in cells B2:B7. The EXCEL procedure uses the steps previously described for the *t*-Test except the user chooses the **t-Test: Paired Two Sample for Means** data analysis tool in step 3. The variable 1 range is A1:A7 and the variable 2 range is B1:B7.

The appropriate *p*-value is denoted 'P(T <= t) two-tail'. Its value of 0.08 does not allow us to reject the null hypothesis at $\alpha = 0.05$.

Inferences about two populations using PASW

We describe the use of PASW to construct interval estimates and do hypothesis tests about the difference between two population means, first for the independent-samples case when the population standard deviations σ_1 and σ_2 are unknown, and second for the matched-samples case. PASW provides both interval estimation and hypothesis test results within the same module, so the PASW procedure is the same for both types of inferences. In the examples that follow, we will demonstrate interval estimation and hypothesis testing for the same two samples. We note that PASW does not provide a routine for inferences about the difference between two population means when the population standard deviations σ_1 and σ_2 are known, nor a routine for inferences about the difference between two population proportions.

Difference between two population means: σ_1 and σ_2 unknown

We shall use the data for the cheque account balances example presented in Section 10.2 (file 'CheqAcct.SAV' on the accompanying CD). The cheque account balances at both the Northern and Eastern branches are in the first column of the PASW data file. The second column indicates whether the account balance is for a Northern Branch account or an Eastern Branch account, i.e. either Northern or Eastern appears in each row of the second column. For an independent-samples analysis in PASW, the data must be set out in this way. In this example, we will use the PASW **Independent-Samples t** procedure to provide a 95 per cent confidence interval estimate of the difference between population means for the cheque account balances at the two branch banks. The output of the procedure also provides the p-value for the hypothesis test: $H_0: \mu_1 - \mu_2 = 0$ versus $H_1: \mu_1 - \mu_2 \neq 0$.

Step 1 **Analyze > Compare Means > Independent-Samples T Test**

[Main menu bar]

Step 2 Transfer the account balance to the **Test Variable(s)** box

[**Independent-Samples T Test** panel]

Transfer the branch variable to the **Grouping Variable** box

Click **Define Groups**

Step 3 Enter **1** (code for Northern branch) in the **Group 1** box   [**Define Groups** panel]

Enter **2** (code for Eastern branch) in the **Group 2** box

Click **Continue**

Step 4 Click **Options**   [**Independent-Samples T Test** panel]

Step 5 Enter **95** in the **Confidence Interval** box

[**Independent-Samples T Test:Options** panel]

Click **Continue**

Step 6 Click **OK**   [**Independent-Samples T Test** panel]

Part of the PASW output is shown below in Figure 10.3. PASW computes two versions of the t procedure; one assuming equal variances for the two populations, the other not making this assumption. Only the latter, more general procedure has been described fully in this text. The results for this procedure are in the second row of the table. The 95 per cent confidence interval estimate is €37 to €193 as described in Section 10.2. The p-value $= 0.005$ shows the null hypothesis of equal population means can be rejected at the $\alpha = 0.01$ level of significance. Note that PASW used equation (10.7) to compute 47.8 degrees of freedom for this analysis.

Figure 10.3 PASW output for the hypothesis test and confidence interval for the cheque account balances

Independent Samples Test

| | | Levene's Test for Equality of Variances | | t-test for Equality of Means | | | | | | |
| | | | | | | | | | 95% Confidence Interval of the Difference | |
		F	Sig.	t	df	Sig. (2-tailed)	Mean Difference	Std. Error Difference	Lower	Upper
Account balance (€)	Equal variances assumed	1.571	.216	2.891	48	.006	115.014	39.777	35.037	194.991
	Equal variances not assumed			2.956	47.805	.005	115.014	38.908	36.775	193.252

Difference between two population means with matched samples

We use the data on production times in Table 10.2 to illustrate the matched-samples procedure (file 'Matched.SAV' on the accompanying CD). The completion times for method 1 are entered into the first column of the PASW data file and the completion times for method 2 are entered into the second column. The PASW steps for matched samples are as follows:

Step 1 **Analyze > Compare Means > Paired-Samples T Test**   [Main menu bar]

Step 2 Transfer Method 1 to **Variable1** in the **Paired variables** area
                                                              [Paired-Samples T Test panel]
Transfer Method 2 to **Variable2** in the **Paired variables** area
Click **Options**

Step 3 Enter **95** in the **Confidence Interval** box
                                                              [Paired-Samples T Test:Options panel]
Click **Continue**

Step 4 Click **OK**   **[Paired-Samples T Test** panel]

Chapter 11

Inferences about Population Variances

In the preceding four chapters we examined methods of statistical inference involving population means and population proportions. In this chapter we extend the discussion to situations involving inferences about population variances.

In many manufacturing applications, controlling the process variance is extremely important in maintaining quality. Consider the production process of filling containers with a liquid detergent product, for example. The filling mechanism for the process is adjusted so that the mean filling weight is 500 grams per container. In addition, the variance of the filling weights is critical. Even with the filling mechanism properly adjusted for the mean of 500 grams, we cannot expect every container to contain exactly 500 grams. By selecting a sample of containers, we can compute a sample variance for the number of grams placed in a container. This value will serve as an estimate of the variance for the population of containers being filled by the production process. If the sample variance is modest, the production process will be continued. However, if the sample variance is excessive, over-filling and under-filling may be occurring, even though the mean is correct at 500 grams. In this case, the filling mechanism will be re-adjusted in an attempt to reduce the filling variance for the containers.

In the first section we consider inferences about the variance of a single population. Subsequently, we will discuss procedures that can be used to make inferences comparing the variances of two populations.

11.1 Inferences about a population variance

Recall that sample variance is calculated as follows:

$$s^2 = \frac{\Sigma(x_i - \bar{x})^2}{n - 1} \tag{11.1}$$

The sample variance (S^2) is a point estimator of the population variance σ^2. To make inferences about σ^2, the sampling distribution of the quantity $(n - 1)S^2/\sigma^2$ can be used, under appropriate circumstances.

Statistics in Practice

Takeovers and mergers in the UK brewing industry

In the last 50 or so years, the structure of the UK brewing industry has changed radically. The *Statistical Handbook* of the British Beer & Pub Association records only 48 brewery companies in the UK in 2003, compared with 362 in 1950. Much of the re-structuring, involving mergers and takeovers, took place during the 1950s and 1960s. By 1969, the number of brewery companies had already fallen to 96. Alison Dean, of the City University Business School, undertook some research to characterize the firms that were the targets of takeovers in the 1945 to 1960 period. She studied samples of firms that were taken over during this period, firms that merged during the period, and firms that remained independent.

Dean's research questioned whether performance criteria were mainly responsible for making firms vulnerable to takeover, or whether non-performance criteria such as size and property holdings were more influential. Her broad conclusion was that performance criteria appeared to be relatively unimportant. Brewery companies typically

Boddington's famous strangeways brewery in Manchester, England.
© Dominic Harrison/Alamy.

own 'tied houses' – pubs leased to tenant managers on condition that the tenant buys beer from the brewery. Dean found that a distinguishing characteristic of taken-over firms was a low average asset value per tied house.

In respect of performance criteria, the research looked at profitability, earnings per share, dividend payments, liquidity and share price/net asset ratio. No statistically significant differences were found between the group of taken-over companies and the group of companies that remained independent. In two respects, though, there were significant differences between the group of merged companies and the other two groups. These two criteria were profitability and liquidity, and in both cases it was the standard deviation (rather than the mean of the performance criterion) that distinguished the group of merged companies. For example, in the case of liquidity, measured as the ratio of current assets to current liabilities, the mean values for the three groups of companies were quite similar: 1.6 for the taken-over companies, 1.5 for the merged companies and 1.7 for the independent companies. However, the standard deviation for this ratio was only 0.2 for the merged companies, compared with 0.6 for the independent companies and 0.8 for the taken-over companies. In other words, the group of merged companies seemed very homogeneous in respect of this performance characteristic.

The researcher investigated differences between standard deviations using a statistical test called the *F* test, which you will learn about in this chapter.

Source: Alison Dean, The characteristics of takeover target firms: the case of the English Brewing Industry, 1945–1960. *Review of Industrial Organization*, 12, 579–591 (1997)

Sampling distribution of $(n - 1)S^2/\sigma^2$

When a simple random sample of size n is selected from a normal population, the sampling distribution of

$$\frac{(n - 1)S^2}{\sigma^2} \tag{11.2}$$

has a chi-squared distribution with $n - 1$ degrees of freedom.

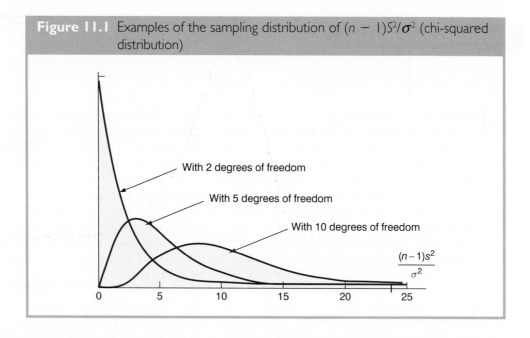

Figure 11.1 Examples of the sampling distribution of $(n - 1)S^2/\sigma^2$ (chi-squared distribution)

Figure 11.1 shows some possible forms of the sampling distribution of $(n - 1)S^2/\sigma^2$. Because the sampling distribution of $(n - 1)S^2/\sigma^2$ is a chi-squared distribution, under the conditions described above, we can use this distribution to construct interval estimates and do hypothesis tests about a population variance. Tables of areas or probabilities are readily available for the chi-squared distribution.

Interval estimation

Suppose we are interested in estimating the population variance for the production filling process mentioned at the beginning of this chapter. A sample of 20 containers is taken and the sample variance for the filling quantities is found to be $s^2 = 2.50$ (in appropriate units). However, we cannot expect the variance of a sample of 20 containers to provide the exact value of the variance for the population of containers filled by the production process. Our interest will be in constructing an interval estimate for the population variance.

The Greek letter chi is χ, so chi-squared is often denoted χ^2. We shall use the notation χ^2_α to denote the value for the chi-squared distribution that provides an area or probability of α to the *right* of the χ^2_α value. For example, in Figure 11.2 the chi-squared distribution with 19 degrees of freedom is shown, with $\chi^2_{0.025} = 32.852$ indicating that 2.5 per cent of the chi-squared values are to the right of 32.852, and $\chi^2_{0.975} = 8.907$ indicating that 97.5 per cent of the chi-squared values are to the right of 8.907. Refer to Table 3 of Appendix B and verify that these chi-squared values with 19 degrees of freedom are correct (19th row of the table).

From Figure 11.2 we see that 0.95, or 95 per cent, of the chi-squared values are between $\chi^2_{0.975}$ and $\chi^2_{0.025}$. That is, there is a 0.95 probability of obtaining a χ^2 value such that

$$\chi^2_{0.975} \leq \chi^2 \leq \chi^2_{0.025}$$

We stated in expression (11.2) that the random variable $(n - 1)S^2/\sigma^2$ follows a chi-squared distribution, therefore we can substitute $(n - 1)s^2/\sigma^2$ for χ^2 and write

$$\chi^2_{0.975} \leq \frac{(n - 1)s^2}{\sigma^2} \leq \chi^2_{0.025} \tag{11.3}$$

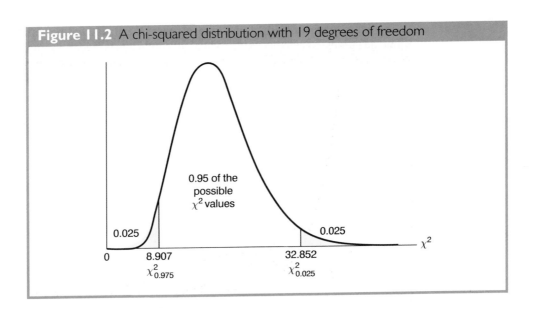

Figure 11.2 A chi-squared distribution with 19 degrees of freedom

0.95 of the possible χ^2 values

0.025

0.025

0   8.907   32.852 χ^2

$\chi^2_{0.975}$ $\chi^2_{0.025}$

Expression (11.3) provides the basis for an interval estimate in that 0.95, or 95 per cent, of all possible values for $(n-1) S^2/\sigma^2$ will be in the interval $\chi^2_{0.975}$ to $\chi^2_{0.025}$. We now need to do some algebraic manipulations with expression (11.3) to develop an interval estimate for the population variance σ^2. Using the leftmost inequality in expression (11.3), we have

$$\chi^2_{0.975} \leq \frac{(n-1)s^2}{\sigma^2}$$

So

$$\chi^2_{0.975}\sigma^2 \leq (n-1)s^2$$

or

$$\sigma^2 \leq \frac{(n-1)s^2}{\chi^2_{0.975}} \tag{11.4}$$

Doing similar algebraic manipulations with the rightmost inequality in expression (11.3) gives

$$\frac{(n-1)s^2}{\chi^2_{0.025}} \leq \sigma^2 \tag{11.5}$$

Expressions (11.4) and (11.5) can be combined to provide

$$\frac{(n-1)s^2}{\chi^2_{0.025}} \leq \sigma^2 \leq \frac{(n-1)s^2}{\chi^2_{0.975}} \tag{11.6}$$

Because expression (11.3) is true for 95 per cent of the $(n-1)s^2/\sigma^2$ values, expression (11.6) provides a 95 per cent confidence interval estimate for the population variance σ^2.

We return to the problem of providing an interval estimate for the population variance of filling quantities. Recall that the sample of 20 containers provided a sample variance of $s^2 = 2.50$. With a sample size of 20, we have 19 degrees of freedom. As shown in Figure 11.2, we have already determined that $\chi^2_{0.975} = 8.907$ and $\chi^2_{0.025} = 32.852$. Using these values in expression (11.6) provides the following interval estimate for the population variance.

$$\frac{(19)(2.50)}{32.825} \leq \sigma^2 \leq \frac{(19)(2.50)}{8.907}$$

or

$$1.45 \leq \sigma^2 \leq 5.33$$

Taking the square root of these values provides the following 95 per cent confidence interval for the population standard deviation.

$$1.20 \leq \sigma \leq 2.31$$

Note that because $\chi^2_{0.975} = 8.907$ and $\chi^2_{0.025} = 32.852$ were used, the interval estimate has a 0.95 confidence coefficient. Extending expression (11.6) to the general case of any confidence coefficient, we have the following interval estimate of a population variance.

Interval estimate of a population variance

$$\frac{(n-1)s^2}{\chi^2_{\alpha/2}} \leq \sigma^2 \leq \frac{(n-1)s^2}{\chi^2_{1-\alpha/2}} \tag{11.7}$$

where the χ^2 values are based on a chi-squared distribution with $n - 1$ degrees of freedom and where $1 - \alpha$ is the confidence coefficient.

Hypothesis testing

Using σ^2_0 to denote the hypothesized value for the population variance, the three forms for a hypothesis test about a population variance are as follows:

$$H_0: \sigma^2 \geq \sigma^2_0 \qquad H_0: \sigma^2 \leq \sigma^2_0 \qquad H_0: \sigma^2 = \sigma^2_0$$
$$H_1: \sigma^2 < \sigma^2_0 \qquad H_1: \sigma^2 > \sigma^2_0 \qquad H_1: \sigma^2 \neq \sigma^2_0$$

These three forms are similar to the three forms we used to conduct one-tailed and two-tailed hypothesis tests about population means and proportions in Chapters 9 and 10.

Hypothesis tests about a population variance use the hypothesized value for the population variance and the sample variance s^2 to compute the value of a χ^2 test statistic. Assuming that the population has a normal distribution, the test statistic is:

Test statistic for hypothesis tests about a population variance

$$\chi^2 = \frac{(n-1)s^2}{\sigma^2_0} \tag{11.8}$$

where χ^2 has a chi-squared distribution with $n - 1$ degrees of freedom.

After computing the value of the χ^2 test statistic, either the p-value approach or the critical value approach may be used to determine whether the null hypothesis can be rejected.

Here is an example. The Newcastle Metro Bus Company wants to promote an image of reliability by encouraging its drivers to maintain consistent schedules. As a standard policy the company would like arrival times at bus stops to have low variability. The company standard specifies an arrival time variance of four or less when arrival times are measured in minutes. The following hypothesis test is formulated to help the company determine whether the arrival time population variance is excessive.

$$H_0: \sigma^2 \le 4$$
$$H_1: \sigma^2 > 4$$

In tentatively assuming H_0 is true, we are assuming that the population variance of arrival times is within the company guideline. We reject H_0 if the sample evidence indicates that the population variance exceeds the guideline. In this case, follow-up steps should be taken to reduce the population variance. We conduct the hypothesis test using a level of significance of $\alpha = 0.05$.

Suppose that a random sample of 24 bus arrivals taken at a city-centre bus stop provides a sample variance of $s^2 = 4.9$. Assuming that the population distribution of arrival times is approximately normal, the value of the test statistic is as follows.

$$\chi^2 = \frac{(n-1)s^2}{\sigma_0^2} = \frac{(24-1)4.9}{4} = 28.18$$

The chi-squared distribution with $n - 1 = 24 - 1 = 23$ degrees of freedom is shown in Figure 11.3. Because this is an upper tail test, the area under the curve to the right of the test statistic $\chi^2 = 28.18$ is the p-value for the test.

Like the t distribution table, the chi-squared distribution table does not contain sufficient detail to enable us to determine the p-value exactly. However, we can use the chi-squared distribution table to obtain a range for the p-value. For example, using Table 3 of Appendix B, we find the following information for a chi-squared distribution with 23 degrees of freedom.

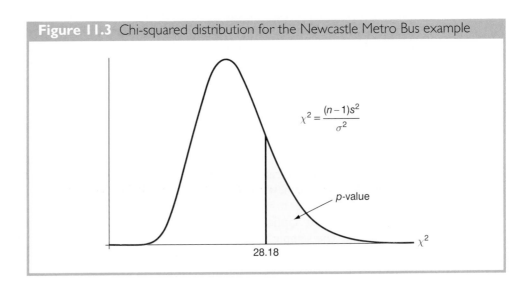

Figure 11.3 Chi-squared distribution for the Newcastle Metro Bus example

Area in upper tail	0.10	0.05	0.025	0.01
χ^2 value (23 df)	32.007	35.172	38.076	41.638

$$\chi^2 = 28.18$$

Because $\chi^2 = 28.18$ is less than 32.007, the area in the upper tail (the p-value) is greater than 0.10. With the p-value $> \alpha = 0.05$, we cannot reject the null hypothesis. The sample does not support the conclusion that the population variance of the arrival times is excessive.

Because of the difficulty of determining the exact p-value directly from the chi-squared distribution table, a computer software package such as MINITAB, PASW or EXCEL is helpful. The sections at the end of the chapter describe the procedures used to show that with 23 degrees of freedom, $\chi^2 = 28.18$ provides a p-value $= 0.2091$.

As with other hypothesis testing procedures, the critical value approach can also be used to draw the conclusion. With $\alpha = 0.05$, $\chi_{0.05}$ provides the critical value for the upper tail hypothesis test. Using Table 3 of Appendix B and 23 degrees of freedom, $\chi_{0.05} = 35.172$. Consequently, the rejection rule for the bus arrival time example is as follows:

$$\text{Reject } H_0 \text{ if } \chi^2 \geq 35.172$$

Because the value of the test statistic is $\chi^2 = 28.18$, we cannot reject the null hypothesis.

In practice, upper tail tests as presented here are the most frequently encountered tests about a population variance. In situations involving arrival times, production times, filling weights, part dimensions and so on, low variances are desirable, whereas large variances are unacceptable. With a statement about the maximum allowable population variance, we can test the null hypothesis that the population variance is less than or equal to the maximum allowable value against the alternative hypothesis that the population variance is greater than the maximum allowable value. With this test structure, corrective action will be taken whenever rejection of the null hypothesis indicates the presence of an excessive population variance.

As we saw with population means and proportions, other forms of hypothesis test can be done. Let us demonstrate a two-tailed test about a population variance by considering a situation faced by a car driver licensing authority. Historically, the variance in test scores for individuals applying for driving licences has been $\sigma^2 = 100$. A new examination with a new style of test questions has been developed. Administrators of the licensing authority would like the variance in the test scores for the new examination to remain at the historical level. To evaluate the variance in the new examination test scores, the following two-tailed hypothesis test has been proposed.

$$H_0: \sigma^2 = 100$$
$$H_1: \sigma^2 \neq 100$$

Rejection of H_0 will indicate that a change in the variance has occurred and suggest that some questions in the new examination may need revision to make the variance of the new test scores similar to the variance of the old test scores.

A sample of 30 applicants for driving licences is given the new version of the examination. The sample provides a sample variance $s^2 = 162$. We shall use a level of significance $\alpha = 0.05$ to do the hypothesis test. The value of the chi-squared test statistic is as follows:

Table 11.1 Summary of hypothesis tests about a population variance

	Lower tail test	Upper tail test	Two-tailed test
Hypotheses	$H_0: \sigma^2 \geq \sigma_0^2$ $H_1: \sigma^2 < \sigma_0^2$	$H_0: \sigma^2 \leq \sigma_0^2$ $H_1: \sigma^2 > \sigma_0^2$	$H_0: \sigma^2 = \sigma_0^2$ $H_1: \sigma^2 \neq \sigma_0^2$
Test statistic	$\chi^2 = \dfrac{(n-1)s^2}{\sigma_0^2}$	$\chi^2 = \dfrac{(n-1)s^2}{\sigma_0^2}$	$\chi^2 = \dfrac{(n-1)s^2}{\sigma_0^2}$
Rejection rule: p-value approach	Reject H_0 if p-value $\leq \alpha$	Reject H_0 if p-value $\leq \alpha$	Reject H_0 if p-value $\leq \alpha$
Rejection rule: critical value approach	Reject H_0 if $\chi^2 \leq \chi_{1-\alpha}^2$	Reject H_0 if $\chi^2 \geq \chi_\alpha^2$	Reject H_0 if $\chi^2 \leq \chi_{1-\alpha/2}^2$ or if $\chi^2 \geq \chi_{\alpha/2}^2$

$$\chi^2 = \frac{(n-1)s^2}{\sigma_0^2} = \frac{(30-1)162}{100} = 46.98$$

Now, let us compute the p-value. Using Table 3 of Appendix B and $n - 1 = 30 - 1 = 29$ degrees of freedom, we find the following.

Area in upper tail	0.10	0.05	0.025	0.01
χ^2 value (29 df)	39.087	42.557	45.722	49.588

$\chi^2 = 46.98$

The value of the test statistic $\chi^2 = 46.98$ gives an area between 0.025 and 0.01 in the upper tail of the chi-squared distribution. Doubling these values shows that the two-tailed p-value is between 0.05 and 0.02. MINITAB, PASW or EXCEL can be used to show the exact p-value $= 0.0374$. With p-value $< \alpha = 0.05$, we reject H_0 and conclude that the new examination test scores have a population variance different from the historical variance of $\sigma^2 = 100$.

A summary of the hypothesis testing procedures for a population variance is shown in Table 11.1.

Exercises

Methods

1 Find the following chi-squared distribution values from Table 3 of Appendix B.

a. $\chi_{0.05}^2$ with df $= 5$
b. $\chi_{0.025}^2$ with df $= 15$
c. $\chi_{0.975}^2$ with df $= 20$
d. $\chi_{0.01}^2$ with df $= 10$
e. $\chi_{0.95}^2$ with df $= 18$

2 A sample of 20 items provides a sample standard deviation of 5.

a. Compute a 90 per cent confidence interval estimate of the population variance.
b. Compute a 95 per cent confidence interval estimate of the population variance.
c. Compute a 95 per cent confidence interval estimate of the population standard deviation.

3 A sample of 16 items provides a sample standard deviation of 9.5. Test the following hypotheses using $\alpha = 0.05$. What is your conclusion? Use both the p-value approach and the critical value approach.

$$H_0: \sigma^2 \leq 50$$
$$H_1: \sigma^2 > 50$$

Applications

4 The variance in drug weights is critical in the pharmaceutical industry. For a specific drug, with weights measured in grams, a sample of 18 units provided a sample variance of $s^2 = 0.36$.

a. Construct a 90 per cent confidence interval estimate of the population variance for the weight of this drug.
b. Construct a 90 per cent confidence interval estimate of the population standard deviation.

5 The table below shows return-on-equity (ROE) figures for 2007, for a sample of six companies listed on the Tel Aviv stock exchange (Source: Datastream, Thomson Financial).

Company	ROE (%)
Bezeq	25.83
Clal Industries	25.35
Harel Insurance	22.60
Koor Industries	28.72
Mizrahi Bank	17.10
Strauss Group	15.40

a. Compute the sample variance and sample standard deviation for these data.
b. What is the 95 per cent confidence interval for the population variance?
c. What is the 95 per cent confidence interval for the population standard deviation?

6 Because of staffing decisions, managers of the Worldview Hotel are interested in the variability in the number of rooms occupied per day during a particular season of the year. A sample of 20 days of operation shows a sample mean of 290 rooms occupied per day and a sample standard deviation of 30 rooms.

a. What is the point estimate of the population variance?
b. Provide a 90 per cent confidence interval estimate of the population variance.
c. Provide a 90 per cent confidence interval estimate of the population standard deviation.

7 The Fidelity Growth & Income mutual fund received a three-star, or neutral, rating from Morningstar. Shown here are the quarterly percentage returns for the five-year period from 2001 to 2005 (*Morningstar Funds 500*, 2006).

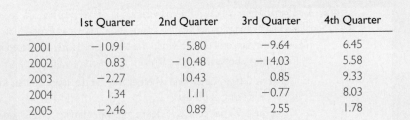

	1st Quarter	2nd Quarter	3rd Quarter	4th Quarter
2001	−10.91	5.80	−9.64	6.45
2002	0.83	−10.48	−14.03	5.58
2003	−2.27	10.43	0.85	9.33
2004	1.34	1.11	−0.77	8.03
2005	−2.46	0.89	2.55	1.78

a. Compute the mean, variance, and standard deviation for the quarterly returns.
b. Financial analysts often use standard deviation of percentage returns as a measure of risk for stocks and mutual funds. Construct a 95 per cent confidence interval for the population standard deviation of quarterly returns for the Fidelity Growth & Income mutual fund.

8   In the file 'Travel' on the accompanying CD, there are estimated daily living costs (in euros) for a businessman travelling to 20 major cities. The estimates include a single room at a four-star hotel, beverages, breakfast, taxi fares, and incidental costs.

a. Compute the sample mean.
b. Compute the sample standard deviation.
c. Compute a 95 per cent confidence interval for the population standard deviation.

City	Daily living cost	City	Daily living cost
Bangkok	242.87	Madrid	283.56
Bogota	260.93	Mexico City	212.00
Bombay	139.16	Milan	284.08
Cairo	194.19	Paris	436.72
Dublin	260.76	Rio de Janeiro	240.87
Frankfurt	355.36	Seoul	310.41
Hong Kong	346.32	Tel Aviv	223.73
Johannesburg	165.37	Toronto	181.25
Lima	250.08	Warsaw	238.20
London	326.76	Washington DC	250.61

9   To analyze the risk, or volatility, associated with investing in Chevron Corporation common stock, a sample of the monthly total percentage return for 12 months was taken. The returns for the 12 months of 2005 are shown here (*Compustat*, February 24, 2006). Total return is price appreciation plus any dividend paid.

Month	Return (%)	Month	Return (%)
January	3.60	July	3.74
February	14.86	August	6.62
March	−6.07	September	5.42
April	−10.82	October	−11.83
May	4.29	November	1.21
June	3.98	December	−0.94

a. Compute the sample variance and sample standard deviation as a measure of volatility of monthly total return for Chevron.
b. Construct a 95 per cent confidence interval for the population variance.
c. Construct a 95 per cent confidence interval for the population standard deviation.

10 Part variability is critical in the manufacturing of ball bearings. Large variances in the size of the ball bearings cause bearing failure and rapid wear. Production standards call for a maximum variance of 0.0025 when the bearing sizes are measured in millimetres. A sample of 15 bearings shows a sample standard deviation of 0.066 mm.

  a. Use $\alpha = 0.10$ to determine whether the sample indicates that the maximum acceptable variance is being exceeded.

  b. Compute a 90 per cent confidence interval estimate for the variance of the ball bearings in the population.

11 The average standard deviation for the annual return of large cap stock mutual funds is 18.2 per cent (*The Top Mutual Funds*, AAII, 2004). The sample standard deviation based on a sample of size 36 for the Vanguard PRIMECAP mutual fund is 22.2 per cent. Construct a hypothesis test that can be used to determine whether the standard deviation for the Vanguard fund is greater than the average standard deviation for large cap mutual funds. With a 0.05 level of significance, what is your conclusion?

12 A sample standard deviation for the number of passengers taking a particular airline flight is 8. A 95 per cent confidence interval estimate of the population standard deviation is 5.86 passengers to 12.62 passengers.

  a. Was a sample size of 10 or 15 used in the statistical analysis?

  b. Suppose the sample standard deviation of $s = 8$ was based on a sample of 25 flights. What change would you expect in the confidence interval for the population standard deviation? Compute a 95 per cent confidence interval estimate of σ with a sample size of 25.

11.2 Inferences about two population variances

In some statistical applications we may want to compare the variances in product quality resulting from two different production processes, the variances in assembly times for two assembly methods, or the variances in temperatures for two heating devices. In making comparisons about the two population variances, we shall be using data collected from two independent random samples, one from population 1 and another from population 2. The two sample variances s_1^2 and s_2^2 will be the basis for making inferences about the two population variances σ_1^2 and σ_2^2. Whenever the variances of two normal populations are equal ($\sigma_1^2 = \sigma_2^2$), the sampling distribution of the ratio of the two sample variances is as follows.

Sampling distribution of S_1^2/S_2^2 when $\sigma_1^2 = \sigma_2^2$

When independent simple random samples of sizes n_1 and n_2 are selected from two normal populations with equal variances, the sampling distribution of

$$\frac{S_1^2}{S_2^2} \qquad\qquad (11.9)$$

has an F distribution with $n_1 - 1$ degrees of freedom for the numerator and $n_2 - 1$ degrees of freedom for the denominator; S_1^2 is the sample variance for the random sample of n_1 items from population 1, and S_2^2 is the sample variance for the random sample of n_2 items from population 2.

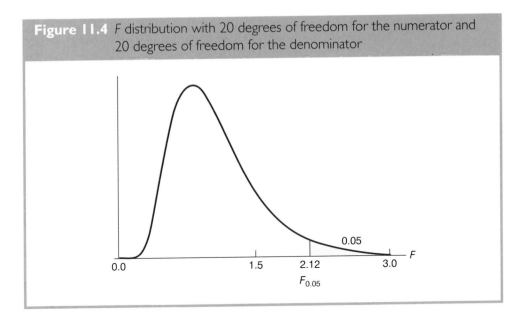

Figure 11.4 *F distribution with 20 degrees of freedom for the numerator and 20 degrees of freedom for the denominator*

Figure 11.4 is a graph of the *F* distribution with 20 degrees of freedom for both the numerator and denominator. As can be seen from this graph, the *F* distribution is not symmetrical, and the *F* values can never be negative. The shape of any particular *F* distribution depends on its numerator and denominator degrees of freedom.

We shall use F_α to denote the value of *F* that provides an area or probability of α in the upper tail of the distribution. For example, as noted in Figure 11.4, $F_{0.05}$ identifies the upper tail area of 0.05 for an *F* distribution with 20 degrees of freedom for both the numerator and for the denominator. The specific value of $F_{0.05}$ can be found by referring to the *F* distribution table, Table 4 of Appendix B. Using 20 degrees of freedom for the numerator, 20 degrees of freedom for the denominator, and the row corresponding to an area of 0.05 in the upper tail, we find $F_{0.05} = 2.12$. Note that the table can be used to find *F* values for upper tail areas of 0.10, 0.05, 0.025 and 0.01.

We now show how the *F* distribution can be used to do a hypothesis test about the equality of two population variances. The hypotheses are stated as follows.

$$H_0: \sigma_1^2 = \sigma_2^2$$
$$H_1: \sigma_1^2 \neq \sigma_2^2$$

We make the tentative assumption that the population variances are equal. If H_0 is rejected, we will draw the conclusion that the population variances are not equal.

The hypothesis test requires two independent random samples, one from each population. The two sample variances are then computed. We refer to the population providing the *larger* sample variance as population 1. A sample size of n_1 and a sample variance of s_1^2 correspond to population 1, and a sample size of n_2 and a sample variance of s_2^2 correspond to population 2. Based on the assumption that both populations have a normal distribution, the ratio of sample variances provides the following *F* test statistic.

Test statistic for hypothesis tests about population variances with $\sigma_1^2 = \sigma_2^2$

$$F = \frac{s_1^2}{s_2^2} \tag{11.10}$$

Denoting the population with the larger sample variance as population 1, the test statistic has an F distribution with $n_1 - 1$ degrees of freedom for the numerator and $n_2 - 1$ degrees of freedom for the denominator.

Because the F test statistic is constructed with the larger sample variance in the numerator, the value of the test statistic will be in the upper tail of the F distribution. Therefore, the F distribution table (Table 4 of Appendix B) need only provide upper tail areas or probabilities.

We now consider an example. Midlands Schools is renewing its school bus service contract for the coming year and must select one of two bus companies, the Red Bus Company or the Route One Company. We shall assume that the two companies have similar performance for average punctuality (i.e. mean arrival time) and use the variance of the arrival times as a primary measure of the quality of the bus service. Low variance values indicate the more consistent and higher quality service. If the variances of arrival times associated with the two services are equal, Midlands Schools' managers will select the company offering the better financial terms. However, if the sample data on bus arrival times for the two companies indicate a significant difference between the variances, the administrators may want to give special consideration to the company with the better or lower variance service. The appropriate hypotheses follow.

SCHOOL BUS

$$H_0: \sigma_1^2 = \sigma_2^2$$
$$H_1: \sigma_1^2 \neq \sigma_2^2$$

If H_0 can be rejected, the conclusion of unequal service quality is appropriate. We shall use a level of significance of $\alpha = 0.10$ to do the hypothesis test. A sample of 26 arrival times for the Red Bus service provides a sample variance of 48 and a sample of 16 arrival times for the Route One service provides a sample variance of 20. Because the Red Bus sample provided the larger sample variance, we shall denote Red Bus as population 1. Using equation (11.10), the value of the test statistic is

$$F = \frac{s_1^2}{s_2^2} = \frac{48}{20} = 2.40$$

The corresponding F distribution has $n_1 - 1 = 26 - 1 = 25$ numerator degrees of freedom and $n_2 - 1 = 16 - 1 = 15$ denominator degrees of freedom. As with other hypothesis testing procedures, we can use the p-value approach or the critical value approach to reach a conclusion. Table 4 of Appendix B shows the following areas in the upper tail and corresponding F values for an F distribution with 25 numerator degrees of freedom and 15 denominator degrees of freedom.

Area in upper tail	0.10	0.05	0.025	0.01
F value (df$_1$ = 25, df$_2$ = 15)	1.89	2.28	2.69	3.28

$F = 2.40$

Because $F = 2.40$ is between 2.28 and 2.69, the area in the upper tail of the distribution is between 0.05 and 0.025. Since this is a two-tailed test, we double the upper tail area, which results in a p-value between 0.10 and 0.05. For this test, we selected $\alpha = 0.10$ as the level of significance, which gives us a p-value $< \alpha = 0.10$. Hence, the null hypothesis is rejected. This finding leads to the conclusion that the two bus services differ in terms of arrival time variances. The recommendation is that the Midlands Schools' managers give special consideration to the better or lower variance service offered by the Route One Company.

We can use MINITAB, PASW or EXCEL to show that the test statistic $F = 2.40$ provides a two-tailed p-value $= 0.0811$. With $0.0811 < \alpha = 0.10$, the null hypothesis of equal population variances is rejected.

To use the critical value approach to do the two-tailed hypothesis test at the $\alpha = 0.10$ level of significance, we select critical values with an area of $\alpha/2 = 0.10/2 = 0.05$ in each tail of the distribution. Because the value of the test statistic computed using equation (11.10) will always be in the upper tail, we only need to determine the upper tail critical value. From Table 4 of Appendix B, we see that $F_{0.05} = 2.28$. So, even though we use a two-tailed test, the rejection rule is stated as follows.

$$\text{Reject } H_0 \text{ if } F \geq 2.28$$

Because the test statistic $F = 2.40$ is greater than 2.28, we reject H_0 and conclude that the two bus services differ in terms of arrival time variances.

One-tailed tests involving two population variances are also possible. In this case, we use the F distribution to determine whether one population variance is significantly greater than the other. If we are using tables of the F distribution to compute the p-value or determine the critical value, a one-tailed hypothesis test about two population variances will always be formulated as an *upper tail* test:

$$H_0: \sigma_1^2 \leq \sigma_2^2$$
$$H_1: \sigma_1^2 > \sigma_2^2$$

This form of the hypothesis test always places the p-value and the critical value in the upper tail of the F distribution. As a result, only upper tail F values will be needed, simplifying both the computations and the table for the F distribution.

As an example of a one-tailed test, consider a public opinion survey. Samples of 31 men and 41 women were used to study attitudes about current political issues. The researcher conducting the study wants to test to see if women show a greater variation in attitude on political issues than men. In the form of the one-tailed hypothesis test given previously, women will be denoted as population 1 and men will be denoted as population 2. The hypothesis test will be stated as follows.

$$H_0: \sigma_{women}^2 \leq \sigma_{men}^2$$
$$H_1: \sigma_{women}^2 > \sigma_{men}^2$$

Rejection of H_0 will give the researcher the statistical support necessary to conclude that women show a greater variation in attitude on political issues.

With the sample variance for women in the numerator and the sample variance for men in the denominator, the F distribution will have $n_1 - 1 = 41 - 1 = 40$ numerator degrees of freedom and $n_2 - 1 = 31 - 1 = 30$ denominator degrees of freedom. We shall use a level of significance $\alpha = 0.05$ for the hypothesis test. The survey results provide a

Table 11.2 Summary of hypothesis tests about two population variances

	Upper tail test	Two-tailed test
Hypotheses	$H_0: \sigma_1^2 \le \sigma_2^2$ $H_1: \sigma_1^2 > \sigma_2^2$	$H_0: \sigma_1^2 = \sigma_2^2$ $H_1: \sigma_1^2 \neq \sigma_2^2$
	Note: Population 1 has the larger sample variance	
Test statistic	$F = \dfrac{s_1^2}{s_2^2}$	$F = \dfrac{s_1^2}{s_2^2}$
Rejection rule: p-value approach	Reject H_0 if p-value $\le \alpha$	Reject H_0 if p-value $\le \alpha$
Rejection rule: critical value approach	Reject H_0 if $F \ge F_\alpha$	Reject H_0 if $F \ge F_{\alpha/2}$

sample variance of $s_1^2 = 120$ for women and a sample variance of $s_2^2 = 80$ for men. The test statistic is as follows.

$$F = \frac{s_1^2}{s_2^2} = \frac{120}{80} = 1.50$$

Referring to Table 4 in Appendix B, we find that an F distribution with 40 numerator degrees of freedom and 30 denominator degrees of freedom has $F_{0.10} = 1.57$. Because the test statistic $F = 1.50$ is less than 1.57, the area in the upper tail must be greater than 0.10. Hence, we can conclude that the p-value is greater than 0.10. Using MINITAB, PASW or EXCEL provides a p-value $= 0.1256$. Because the p-value $> \alpha = 0.05$, H_0 cannot be rejected. Hence, the sample results do not support the conclusion that women show greater variation in attitude on political issues than men.

Table 11.2 provides a summary of hypothesis tests about two population variances. Research confirms that the F distribution is sensitive to the assumption of normal populations. The F distribution should not be used unless it is reasonable to assume that both populations are at least approximately normally distributed.

Exercises

Methods

13   Find the following F distribution values from Table 4 of Appendix B.

    a. $F_{0.05}$ with degrees of freedom 5 and 10
    b. $F_{0.025}$ with degrees of freedom 20 and 15
    c. $F_{0.01}$ with degrees of freedom 8 and 12
    d. $F_{0.10}$ with degrees of freedom 10 and 20

14   A sample of 16 items from population 1 has a sample variance $s^2 = 5.8$ and a sample of 21 items from population 2 has a sample variance $s^2 = 2.4$. Test the following hypotheses at the 0.05 level of significance.

$$H_0: \sigma_1^2 \leq \sigma_2^2$$
$$H_1: \sigma_1^2 > \sigma_2^2$$

a. What is your conclusion using the p-value approach?

b. Repeat the test using the critical value approach.

15 Consider the following hypothesis test.

$$H_0: \sigma_1^2 = \sigma_2^2$$
$$H_1: \sigma_1^2 \neq \sigma_2^2$$

a. What is your conclusion if $n_1 = 21$, $s_1^2 = 8.2$, $n_2 = 26$, $s_2^2 = 4.0$? Use $\alpha = 0.05$ and the p-value approach.

b. Repeat the test using the critical value approach.

Applications

16 Most individuals are aware of the fact that the average annual repair cost for a car depends on its age. A researcher is interested in finding out whether the variance of the annual repair costs also increases with the age of the car. A sample of 26 cars that were eight years old showed a sample standard deviation for annual repair costs of £170 and a sample of 25 cars that were four years old showed a sample standard deviation for annual repair costs of £100.

a. State the null and alternative hypotheses if the research hypothesis is that the variance in annual repair costs is larger for the older cars.

b. At a 0.01 level of significance, what is your conclusion? What is the p-value? Discuss the reasonableness of your findings.

17 On the basis of data provided by a salary survey, the variance in annual salaries for seniors in accounting firms is approximately 2.1 and the variance in annual salaries for managers in accounting firms is approximately 11.1. The salary data were provided in thousands of euros. Assuming that the salary data were based on samples of 25 seniors and 26 managers, test the hypothesis that the population variances in the salaries are equal. At a 0.05 level of significance, what is your conclusion?

18 Fidelity Magellan is a large cap growth mutual fund and Fidelity Small Cap Stock is a small cap growth mutual fund (*Morningstar Funds 500*, 2006). The standard deviation for both funds was computed based on a sample of size 26. For Fidelity Magellan, the sample standard deviation is 8.89 per cent; for Fidelity Small Cap Stock, the sample standard deviation is 13.03 per cent. Financial analysts often use the standard deviation as a measure of risk. Conduct a hypothesis test to determine whether the small cap growth fund is riskier than the large cap growth fund. Use $\alpha = 0.05$ as the level of significance.

19 Two new assembly methods are tested and the variances in assembly times are reported. Use $\alpha = 0.10$ and test for equality of the two population variances.

	Method A	Method B
Sample size	$n_1 = 31$	$n_2 = 25$
Sample variation	$s_1^2 = 25$	$s_2^2 = 12$

20 A research hypothesis is that the variance of stopping distances of cars on wet roads is greater than the variance of stopping distances of cars on dry roads. In the research study, 16 cars travelling at the same speeds are tested for stopping distances on wet roads and then

tested for stopping distances on dry roads. On wet roads, the standard deviation of stopping distances is ten metres. On dry roads, the standard deviation is five metres.

a. At a 0.05 level of significance, do the sample data justify the conclusion that the variance in stopping distances on wet roads is greater than the variance in stopping distances on dry roads? What is the *p*-value?

b. What are the implications of your statistical conclusions in terms of driving safety recommendations?

21 The grade point averages of 352 students who completed a college course in financial accounting have a standard deviation of 0.940. The grade point averages of 73 students who dropped out of the same course have a standard deviation of 0.797. Do the data indicate a difference between the variances of grade point averages for students who completed a financial accounting course and students who dropped out? Use a 0.05 level of significance. *Note:* $F_{0.025}$ with 351 and 72 degrees of freedom is 1.466.

22 The variance in a production process is an important measure of the quality of the process. A large variance often signals an opportunity for improvement in the process by finding ways to reduce the process variance. The file 'Bags' on the accompanying CD contains data for two machines that fill bags with powder. The file has 25 bag weights for Machine 1 and 22 bag weights for Machine 2. Conduct a statistical test to determine whether there is a significant difference between the variances in the bag weights for the two machines. Use a 0.05 level of significance. What is your conclusion? Which machine, if either, provides the greater opportunity for quality improvements?

BAGS

For additional online summary questions and answers go to the companion website at www.cengage.co.uk/aswsbe2

Summary

In this chapter we presented statistical procedures that can be used to make inferences about population variances. In the process we introduced two new probability distributions: the chi-squared distribution and the F distribution. The chi-squared distribution can be used as the basis for interval estimation and hypothesis tests about the variance of a normal population.

We illustrated the use of the F distribution in hypothesis tests about the variances of two normal populations. With independent simple random samples of sizes n_1 and n_2 selected from two normal populations with equal variances, the sampling distribution of the ratio of the two sample variances has an F distribution with $n_1 - 1$ degrees of freedom for the numerator and $n_2 - 1$ degrees of freedom for the denominator.

Key formulae

Interval estimate of a population variance

$$\frac{(n-1)s^2}{\chi^2_{\alpha/2}} \leq \sigma^2 \leq \frac{(n-1)s^2}{\chi^2_{1-\alpha/2}} \tag{11.7}$$

Test statistic for hypothesis tests about a population variance

$$\chi^2 = \frac{(n-1)s^2}{\sigma_0^2} \tag{11.8}$$

Test statistic for hypothesis tests about population variances with $\sigma_1^2 = \sigma_2^2$

$$F = \frac{s_1^2}{s_2^2} \tag{11.10}$$

Case problem Global economic problems in 2008

In 2008, particularly in the latter part of the year, there were global economic problems, including banking crises in a number of countries, clear indications of economic recession, and increased stock market volatility.

One method of measuring volatility in stock markets (a relatively unsophisticated one) is to calculate the standard deviation of percentage changes in stock market prices or share index levels (e.g. daily percentage changes or weekly percentage changes). Indeed, in many texts on finance, this is the first operational definition offered for the concept of 'volatility'.

The data in the file 'Share indices 2007-8' (on the accompanying CD) are samples of daily percentage changes in four well-known stock market indices during each of the years 2007 and 2008. The four indices are the FTSE 100 (London Stock Exchange, UK), the DAX 40 (Frankfurt Stock Exchange, Germany), the FTSE/JSE All Share (Johannesburg Stock Exchange, South Africa) and the ISE National 100 (Istanbul Stock Exchange, Turkey). The samples are of size 50 for each of 2007 and 2008.

The report you are asked to prepare below should be focused particularly on the question of whether the stock markets showed greater volatility in 2008 than in 2007.

SHARE INDICES
2007–8

FTSE 2007	DAX 2007	FTSE/JSE 2007	ISE 2007	FTSE 2008	DAX 2008	FTSE/JSE 2008	ISE 2008
-0.76	0.11	-0.16	-0.40	-1.37	-1.35	-2.62	-2.55
0.96	0.19	1.34	1.27	-0.85	0.08	-1.40	-1.04
0.49	-0.21	0.44	0.69	1.32	0.03	2.19	1.45
1.40	0.13	1.36	1.90	4.75	-0.24	5.28	5.81
-1.22	0.16	-0.89	-1.41	1.64	-0.26	1.44	2.03
1.47	-0.13	2.69	4.40	-0.60	0.01	-3.05	-0.97
-1.06	0.09	0.25	-1.46	-0.20	0.00	-1.17	0.08
0.39	0.06	-1.04	1.17	-1.52	-0.59	-2.13	-2.11
-0.58	0.13	-0.88	-0.25	-0.41	-0.33	-0.75	3.18
0.28	0.19	0.57	0.16	0.74	-0.38	-0.05	2.34

Analyst's report

1   Use appropriate descriptive statistics to summarize the daily percentage change data for each index in 2007 and 2008. What similarities or differences do you observe from the sample data?

2   Use the methods of Chapter 10 to comment on any difference between the population mean daily

percentage change in each index for 2007 versus 2008. Discuss your findings.

3   Compute the standard deviation of the daily percentage changes for each share index, for 2007 and for 2008. For each share index, do a hypothesis test to examine the equality of population variances in 2007 and 2008. Discuss your findings.

4   What conclusions can you reach about any differences between 2007 and 2008?

Sales all year round to try and entice shoppers back to spending. © DBURKE/Alamy.

Software Section for Chapter 11

Population variances using **MINITAB**

Below we describe how to use MINITAB to do the F test to compare the variances of two populations. First we give some guidance on how MINITAB can be used to calculate p-values from either the χ^2 distribution or the F distribution, when the χ^2 statistic or the F statistic has been obtained using a calculator or with aid of a computer. At the end of Chapter 3, we showed how to use MINITAB to calculate sample standard deviations or sample variances.

Calculating p-values

At the end of Chapter 6, we showed how to use MINITAB to compute cumulative probabilities for the normal distribution. Similar steps can be used to obtain cumulative probabilities for the chi-squared or for the F distribution. These can then be used to calculate p-values for the tests described in the current chapter.

For example, in the Newcastle Metro Bus example in Section 11.1 (file 'BusTimes. MTW' on the accompanying CD), the chi-squared test statistic given by equation (11.8) is $\chi^2 = 28.18$. MINITAB can be used to compute an upper tail p-value (appropriate for $H_0: \sigma^2 \leq 4$ versus $H_1: \sigma^2 > 4$).

Step 1   **Calc > Probability Distributions > Chi-Square**   [Main menu bar]

Step 2   Check **Cumulative Probability**   [**Chi-Square** panel]
Enter **23** in the **Degrees of freedom** box
Check **Input constant** and enter **28.18** in the adjacent box
Click **OK**

This gives the cumulative probability 0.7909, which is the area under the curve to the left of $\chi^2 = 28.18$. The p-value is the upper tail area or probability, so p-value $= 1 - 0.7909 = 0.2091$. If the test is a lower tail test, the cumulative probability given by MINITAB is the p-value. For a two-tailed test, the p-value is double the lower or upper tail area, depending on whether the calculated χ^2 value is in the lower or upper tail area of the distribution.

A similar set of steps starting with **Calc > Probability Distributions > F** can be used to obtain p-values for the F distribution. In this case, the degrees of freedom for both the numerator and the denominator are entered at **Step 2**.

F-Test for two populations

We shall use the data for the Midlands Schools bus study in Section 11.2 (file 'SchoolBus.MTW' on the accompanying CD). The arrival times for Red Bus appear in column C1, and the arrival times for Route One appear in column C2. The following MINITAB procedure can be used to do the hypothesis test with hypotheses $H_0: \sigma_1^2 = \sigma_2^2$ and $H_1: \sigma_1^2 \neq \sigma_2^2$.

SCHOOLBUS

Step 1 **Stat > Basic Statistics > 2-Variances**   [Main menu bar]

Step 2 Select **Samples in different columns**   [**2-Variances** panel]
Enter **C1** in the **First** box
Enter **C2** in the **Second** box
Click **OK**

MINITAB produces both textual output in the Session window, and graphical output. The output refers to two tests, the first of which is the *F* test we have discussed in the present chapter. The graphical output is shown in Figure 11.5. This shows the test

Figure 11.5 MINITAB graphical output for the *F* test comparing two variances

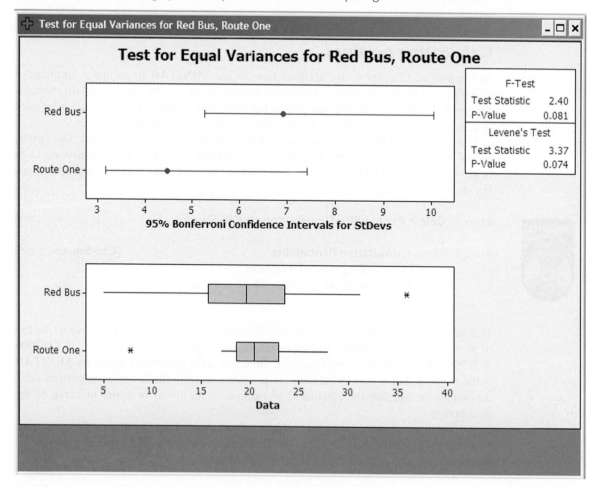

statistic $F = 2.40$ and the p-value $= 0.081$. The p-value is for a two-tailed test. If this MINITAB routine is used for a one-tailed test, the two-tailed p-value should be halved to obtain the appropriate one-tailed p-value.

Population variances using EXCEL

Below we describe how to use EXCEL to do the F test to compare the variances of two populations. First we give some guidance on how EXCEL can be used to calculate p-values from either the χ^2 distribution or the F distribution, when the χ^2 statistic or the F statistic has been obtained using a calculator or with the aid of a computer. At the end of Chapter 3, we showed how to use EXCEL to calculate sample standard deviations or sample variances.

Calculating p-values

At the end of Chapter 6, we showed how to use EXCEL to compute probabilities for the normal distribution. EXCEL also has functions that give probabilities for the χ^2 and the F distributions. In the case of the χ^2 and F distributions, the EXCEL functions return probabilities in the right-hand tail area, so these are directly applicable as p-values for upper-tail tests.

For example, in the Newcastle Metro Bus example in Section 11.1 (file 'BusTimes. XLS' on the accompanying CD), the χ^2 test statistic given by equation (11.8) is $\chi^2 = 28.18$. The EXCEL function **CHIDIST**(χ^2, df) can be used to compute an upper tail p-value (appropriate for H_0: $\sigma^2 \leq 4$ versus H_1: $\sigma^2 > 4$). The first argument for the function is the value of the χ^2 test statistic, the second argument is the number of degrees of freedom. In this case, =CHIDIST(28.18, 23) will return the value 0.2091, which is the p-value for the test. If the test is a lower tail test, the probability given by EXCEL will need to be subtracted from 1 to obtain the p-value. For a two-tailed test, the p-value is double the lower or upper tail area, depending on whether the calculated χ^2 value is in the lower or upper tail area of the distribution.

The EXCEL function **FDIST**(F, df1, df2) returns the probability in the right-hand tail of the F distribution. The first argument for the function is the value of the F statistic, the second argument is the number of degrees of freedom for the numerator, and the third argument is the number of degrees of freedom for the denominator.

F-Test for two populations

We shall use the data for the Midlands Schools bus study in Section 11.2 (file 'SchoolBus. XLS' on the accompanying CD). The EXCEL worksheet has the label Red Bus in cell A1 and the label Route One in cell B1. The times for the Red Bus sample are in cells A2:A27 and the times for the Route One sample are in cells B2:B17. The steps to conduct the hypothesis test H_0: $\sigma_1^2 = \sigma_2^2$ versus H_1: $\sigma_1^2 \neq \sigma_2^2$ are as follows:

Step 1 Click the **Data** tab on the Ribbon

Step 2 In the **Analysis** group, click **Data Analysis**

Figure 11.6 EXCEL output for the *F* test comparing two variances

D	E	F
F-Test Two-Sample for Variances		
	Red Bus	Route One
Mean	20.230769	20.24375
Variance	48.020615	19.999958
Observations	26	16
df	25	15
F	2.4010358	
P(F<=f) one-tail	0.0405271	
F Critical one-tail	2.2797293	

Step 3   Choose **F-Test Two-Sample for Variances**
           Click **OK**

Step 4   Enter **A1:A27** in the **Variable 1 Range** box
                                      [**F-Test Two Sample for Variances** panel]
           Enter **B1:B17** in the **Variable 2 Range** box
           Check **Labels**
           Enter **0.05** in the **Alpha** box
           (*Note:* This EXCEL procedure uses alpha as the area in the upper tail.)
           Select **Output Range** and enter **D1** in the box
           Click **OK**

The EXCEL results are shown in Figure 11.6. The output 'P(F $<=$ f) one-tail' $=$ 0.0405 is the one-tailed area associated with the test statistic $F = 2.40$. So the two-tailed *p*-value is 2(0.0405) $= 0.081$. If the hypothesis test had been a one-tailed test, the one-tailed area in the cell labelled 'P(F $<=$ f) one-tail' directly provides the information necessary to determine the *p*-value for the test.

Population variances using PASW

At the end of Chapter 6, we showed how to use PASW to compute cumulative probabilities for the normal distribution. Similar steps can be used to obtain cumulative probabilities for the chi-squared or for the *F* distribution. These can then be used to calculate *p*-values for the tests described in the current chapter. At the end of Chapter 3, we showed how to use PASW to calculate sample standard deviations or sample variances.

Calculating p-values for the χ^2 distribution

In the Newcastle Metro Bus example in Section 11.1 (file 'BusTimes. SAV' on the accompanying CD), the chi-squared test statistic given by equation (11.8) is $\chi^2 = 28.18$. PASW can be used to compute an upper tail p-value (appropriate for H_0: $\sigma^2 \leq 4$ versus H_1: $\sigma^2 > 4$).

Step 1   Enter **28.18** in the first row of the second column in the Data Editor   [Data Editor]
Name this variable **Chisquared**

Step 2   **Transform > Compute**   [Main menu bar]

Step 3   In the **Target Variable** box, enter the name **Cumprob**
[**Compute Variable** panel]
Select **CDF & Noncentral CDF** in the **Function Group** list
Double click on **Cdf:Chisq** in the **Functions and Special Variables** list, so that
  CDF.CHISQ(?,?) appears in the **Numeric Expression** box
Highlight the first argument (first question mark), then double-click on **Chisquared**
  in the variables list
Enter **23** (degrees of freedom) as the second argument (second question mark) –
  the **Numeric Expression** box should now have **CDF.CHISQ(Chisquared,23)**
Click **OK**

The value 0.7909 will be returned in the first row of column 3 in the Data Editor, now named **Cumprob**. This is a cumulative probability of 0.7909, i.e. the area under the curve to the left of $\chi^2 = 28.18$. The p-value is the upper tail area or probability, so p-value = 1 − 0.7909 = 0.2091. If the test is a lower tail test, the cumulative probability is the p-value. For a two-tailed test, the p-value is double the lower or upper tail area, depending on whether the calculated χ^2 value is in the lower or upper tail area of the distribution.

Calculating p-values for the F distribution

A similar set of steps to the above can be used to obtain p-values for the F distribution. We shall use the data for the Midlands Schools bus study in Section 11.2 (file 'SchoolBus.SAV' on the accompanying CD). The arrival times are in the first column of the PASW file, with codes for the two companies in the second column. Using PASW to calculate the variances yields 48.02 for the Red Bus arrival times and 20.00 for Route 1, giving an F ratio of 2.40.

Step 1   Enter **2.40** in the first row of the third column in the Data Editor   [Data Editor]
Name this variable **Fratio**

Step 2   **Transform > Compute**   [Main menu bar]

Step 3   In the **Target Variable** box, enter the name **Cumprob**
[**Compute Variable** panel]
Select **CDF & Noncentral CDF** in the **Function Group** list
Double click on **Cdf:F** in the **Functions and Special Variables** list, so that
  CDF.CHISQ(?,?,?) appears in the **Numeric Expression** box
Highlight the first argument (first question mark), then double-click on **Fratio** in the
  variables list

Enter **25** and **15** (degrees of freedom) as the second argument and third arguments (second and third question marks) – the **Numeric Expression** box should now have **CDF.CHISQ(Fratio,25,15)**
Click **OK**

The value 0.9594 will be returned in the first row of column 4 in the Data Editor, now named **Cumprob**. This is a cumulative probability, i.e. the area under the curve to the left of $F = 2.40$. The upper tail area is $1 - 0.9594 = 0.0406$. For a two-tailed test, as we did in Section 11.2, the p-value is twice this area, p-value $= 0.081$.

Chapter 12

....

Tests of Goodness of Fit and Independence

Learning objectives

After studying this chapter and doing the exercises, you should be able to construct and interpret the results of goodness of fit tests, using the chi-squared distribution, for several situations:

1 A multinomial population with given probabilities.

2 A test of independence in a two-way contingency table.

3 A Poisson distribution.

4 A normal distribution.

In Chapter 11 we showed how the chi-squared distribution could be used in estimation and in hypothesis tests about a population variance. In the present chapter, we introduce two additional hypothesis testing procedures, both based on the use of the chi-squared distribution. Like other hypothesis testing procedures, these tests compare sample results with those that are expected when the null hypothesis is true.

In the following section we introduce a goodness of fit test for a multinomial population. Later we discuss the test for independence using contingency tables and then show goodness of fit tests for the Poisson and normal distributions.

12.1 Goodness of fit test: a multinomial population

Consider the case in which each element of a population is assigned to one and only one of several classes or categories. Such a population is a **multinomial population**. The multinomial distribution can be thought of as an extension of the binomial distribution to the case of three or more categories of outcomes. On each trial of a multinomial experiment, one and only one of the outcomes occurs. Each trial of the experiment is assumed to be independent of all others, and the probabilities of the outcomes remain the same at each trial.

As an example, consider a market share study being conducted by Scott Market Research. Over the past year market shares stabilized at 30 per cent for company A, 50 per cent for company B and 20 per cent for company C. Recently company C developed a 'new and improved' product to replace its current offering in the market. Company C retained Scott Market Research to assess whether the new product will alter market shares.

In this case, the population of interest is a multinomial population. Each customer is classified as buying from company A, company B or company C. So we have a multinomial population with three possible outcomes. We use the following notation:

$$\pi_A = \text{market share for company A}$$
$$\pi_B = \text{market share for company B}$$
$$\pi_C = \text{market share for company C}$$

Scott Market Research will conduct a sample survey and find the sample proportion preferring each company's product. A hypothesis test will then be done to assess whether the new product will lead to a change in market shares. The null and alternative hypotheses are:

$$H_0: \pi_A = 0.30, \pi_B = 0.50 \text{ and } \pi_C = 0.20$$
$$H_1: \text{The population proportions are not } \pi_A = 0.30, \pi_B = 0.50 \text{ and } \pi_C = 0.20$$

If the sample results lead to the rejection of H_0, Scott Market Research will have evidence that the introduction of the new product may affect market shares.

Statistics in Practice

National lotteries

On 23 December 2008, amid global financial upheavals, the *Guardian* newspaper in the UK had a headline announcing **El Gordo brings £2 bn sparkle to Spain**. The Spanish national Christmas lottery, Lotería de Navidad or El Gordo ('The Fat One'), is traditionally drawn on 22 December. El Gordo claims to be the only lottery in the world that distributes the equivalent of more than one billion US dollars as a result of a single draw.

Many countries have government-sponsored national lotteries, some drawn on a weekly rather than an annual basis. Ireland and the UK, for example, each have a twice-weekly Lotto draw. In the Irish Lotto game, each ticket-holder selects six different numbers in the range 1 to 45. If these numbers match those selected randomly when the

A lottery hawker arranges the traditional El Gordo's tickets as he waits for customers in Madrid, Spain. © Fernando Alvarado/epa/Corbis.

draw takes place, the ticket-holder wins a share in the jackpot prize. The UK Lotto game is similar, but the six numbers are chosen in the range 1 to 49. The chance of winning a jackpot share with a single ticket in the UK Lotto is therefore smaller than in the Irish Lotto: about 1 in 14 million for the UK draw compared to about 1 in 8 million in the Irish draw.

It is important that any national lottery is conducted fairly and transparently. The monitoring of the UK lottery is under the supervision of a public body called the National Lottery Commission (NLC). As part of its monitoring role, the NLC commissions statistical analyses to check that the lottery games are being conducted fairly and that numbers are being drawn randomly. The Centre for the Study of Gambling at the University of Salford, UK has carried out recent commissioned analyses. In its 2004 report on the main Lotto draw, one of the specific objectives was to report on tests for equality of frequency for each Lotto number drawn.

The objective of testing for equality of frequency for each Lotto number drawn was met using a statistical test known as a chi-squared test. This test compares the actual frequency with which the Lotto numbers were drawn with those 'expected' assuming equal probabilities of selection for all 49 numbers, but allowing for possible variation caused by the randomness of the selection process. Other specific aspects of randomness, such as independence between draws, were tested similarly using chi-squared tests, by comparing observed frequencies with those expected assuming randomness to prevail. In no case was any evidence of non-randomness found.

In this chapter you will learn how chi-squared tests like those described here are done.

The market research firm has used a consumer panel of 200 customers for the study, in which each individual is asked to specify a purchase preference for one of three alternatives: company A's product, company B's product and company C's new product. This is equivalent to a multinomial experiment consisting of 200 trials. The 200 responses are summarized here.

Observed frequency		
Company A's product	Company B's product	Company C's new product
48	98	54

We now can do a **goodness of fit test** to assess whether the sample of 200 customer purchase preferences is consistent with the null hypothesis. The goodness of fit test is based on a comparison of the sample of *observed* results with the *expected* results under the assumption that the null hypothesis is true. The next step is therefore to compute expected purchase preferences for the 200 customers under the assumption that $\pi_A = 0.30$, $\pi_B = 0.50$ and $\pi_C = 0.20$. The expected frequency for each category is found by multiplying the sample size of 200 by the hypothesized proportion for the category.

Expected frequency		
Company A's product	Company B's product	Company C's new product
200(0.30) = 60	200(0.50) = 100	200(0.20) = 40

The goodness of fit test now focuses on the differences between the observed frequencies and the expected frequencies. Large differences between observed and expected frequencies cast doubt on the assumption that the hypothesized proportions or market shares are correct. Whether the differences between the observed and expected frequencies are 'large' or 'small' is a question answered with the aid of the following test statistic.

Test statistic for goodness of fit

$$\chi^2 = \sum_{i=1}^{k} \frac{(f_i - e_i)^2}{e_i} \qquad (12.1)$$

where

f_i = observed frequency for category i
e_i = expected frequency for category i
k = the number of categories

Note: The test statistic has a chi-squared distribution with $k - 1$ degrees of freedom provided that the expected frequencies are five or more for all categories.

In the Scott Market Research example we use the sample data to test the hypothesis that the multinomial population has the proportions $\pi_A = 0.30$, $\pi_B = 0.50$ and $\pi_C = 0.20$. We shall use level of significance $\alpha = 0.05$. The computation of the chi-squared test statistic is shown in Table 12.1, giving $\chi^2 = 7.34$.

Table 12.1 Computation of the chi-squared test statistic for the Scott Market Research market share study

	Hypothesized proportion	Observed frequency (f_i)	Expected frequency (e_i)	Difference ($f_i - e_i$)	Squared difference ($f_i - e_i)^2$	Squared difference divided by expected frequency ($f_i - e_i)^2/e_i$
Company A	0.30	48	60	−12	144	2.40
Company B	0.50	98	100	−2	4	0.04
Company C	0.20	54	40	14	196	4.90
Total		200				$\chi^2 = 7.34$

We shall reject the null hypothesis if the differences between the observed and expected frequencies are large, which in turn will result in a large value for the test statistic. Hence the test of goodness of fit will always be an upper tail test. With $k - 1 = 3 - 1 = 2$ degrees of freedom, the chi-squared table (Table 3 of Appendix B) provides the following (an introduction to the chi-squared distribution and the use of the chi-squared table were presented in Section 11.1).

Area in upper tail	0.10	0.05	0.025	0.01
χ^2 value (2 df)	4.605	5.991	7.378	9.210

$$\chi^2 = 7.34$$

The test statistic $\chi^2 = 7.34$ is between 5.991 and 7.378 (very close to 7.378), so the corresponding upper tail area or p-value must be between 0.05 and 0.025 (very close to 0.025). With p-value $< \alpha = 0.05$, we reject H_0 and conclude that the introduction of the new product by company C may alter the current market share structure. MINITAB, PASW or EXCEL can be used to show that $\chi^2 = 7.34$ provides a p-value $= 0.0255$ (see the Software Section at the end of the chapter).

Instead of using the p-value, we could use the critical value approach to draw the same conclusion. With $\alpha = 0.05$ and 2 degrees of freedom, the critical value for the test statistic is $\chi^2 = 5.991$. The upper tail rejection rule becomes

$$\text{Reject } H_0 \text{ if } \chi^2 \geq 5.991$$

With $\chi^2 = 7.34 > 5.991$, we reject H_0. The p-value approach and critical value approach provide the same conclusion.

Although the test itself does not directly tell us about *how* market shares may change, we can compare the observed and expected frequencies descriptively to get an idea of the change in market structure. We see that the observed frequency of 54 for company C is larger than the expected frequency of 40. Because the latter was based on current market shares, the larger observed frequency suggests that the new product will have a positive effect on company C's market share. Similar comparisons for the other two companies suggest that company C's gain in market share will hurt company A more than company B.

Here are the steps for doing a goodness of fit test for a hypothesized multinomial population distribution.

Multinomial distribution goodness of fit test: a summary

1 State the null and alternative hypotheses.

H_0: The population follows a multinomial distribution with specified probabilities for each of the k categories

H_1: The population does not follow a multinomial distribution with the specified probabilities for each of the k categories

2 Select a random sample and record the observed frequencies f_i for each category.

3 Assume the null hypothesis is true and determine the expected frequency e_i in each category by multiplying the category probability by the sample size.

4 Compute the value of the test statistic.

5 Rejection rule:

p-value approach: Reject H_0 if p-value $\leq \alpha$

Critical value approach: Reject H_0 if $\chi^2 \geq \chi_{\alpha}^2$

where α is the level of significance for the test and there are $k - 1$ degrees of freedom.

Exercises

Methods

1   Test the following hypotheses by using the χ^2 goodness of fit test.

H_0: $\pi_A = 0.40$, $\pi_B = 0.40$, $\pi_C = 0.20$

H_1: The population proportions are not $\pi_A = 0.40$, $\pi_B = 0.40$, $\pi_C = 0.20$

A sample of size 200 yielded 60 in category A, 120 in category B, and 20 in category C. Use $\alpha = 0.01$ and test to see whether the proportions are as stated in H_0.

a.   Use the p-value approach.

b.   Repeat the test using the critical value approach.

2   Suppose we have a multinomial population with four categories: A, B, C and D. The null hypothesis is that the proportion of items is the same in every category, i.e.

H_0: $\pi_A = \pi_B = \pi_C = \pi_D = 0.25$

A sample of size 300 yielded the following results.

A: 85   B: 95   C: 50   D: 70

Use $\alpha = 0.05$ to determine whether H_0 should be rejected. What is the p-value?

Applications

3   One of the questions on the *Business Week* Subscriber Study was, 'When making investment purchases, do you use full service or discount brokerage firms?' Survey results showed that 264 respondents use full service brokerage firms only, 255 use discount brokerage firms only and 229 use both full service and discount firms. Use $\alpha = 0.10$ to determine whether there are any differences in preference among the three service choices.

4   How well do airline companies serve their customers? A study by *Business Week* showed the following customer ratings: 3 per cent excellent, 28 per cent good, 45 per cent fair and 24 per cent poor. In a follow-up study of service by telephone companies, assume that a sample of 400 adults found the following customer ratings: 24 excellent, 124 good, 172 fair and 80 poor. Taking the figures from the *Business Week* study as 'population' values, is the distribution of the customer ratings for telephone companies different from the distribution of customer ratings for airline companies? Test with $\alpha = 0.01$. What is your conclusion?

5   In setting sales quotas, the marketing manager of a multinational company makes the assumption that order potentials are the same for each of four sales territories in Africa. A sample of 200 sales follows. Should the manager's assumption be rejected? Use $\alpha = 0.05$.

Sales territories			
1	2	3	4
60	45	59	36

6   A community park will open soon in a large European city. A sample of 210 individuals are asked to state their preference for when they would most like to visit the park. The sample results follow.

Monday	Tuesday	Wednesday	Thursday	Friday	Saturday	Sunday
20	30	30	25	35	20	50

In developing a staffing plan, should the park manager plan on the same number of individuals visiting the park each day? Support your conclusion with a statistical test. Use $\alpha = 0.05$.

7   The results of *ComputerWorld's* Annual Job Satisfaction Survey showed that 28 per cent of information systems (IS) managers are very satisfied with their job, 46 per cent are somewhat satisfied, 12 per cent are neither satisfied or dissatisfied, 10 per cent are somewhat dissatisfied and 4 per cent are very dissatisfied. Suppose that a sample of 500 computer programmers yielded the following results.

Category	Number of respondents
Very satisfied	105
Somewhat satisfied	235
Neither	55
Somewhat dissatisfied	90
Very dissatisfied	15

Taking the *ComputerWorld* figures as 'population' values, use $\alpha = 0.05$ and test to determine whether the job satisfaction for computer programmers is different from the job satisfaction for IS managers.

12.2 Test of independence

Another important application of the chi-squared distribution involves testing for the independence of two variables. Consider a study conducted by the Real Ale Brewery, which manufactures and distributes three types of beer: light ale, lager and best bitter. In an analysis of the market segments for the three beers, the firm's market research group raised the question of whether preferences for the three beers differ between male and female beer drinkers. If beer preference is independent of gender, a single advertising campaign will be initiated for all of the Real Ale beers. However, if beer preference depends on the gender of the beer drinker, the firm will tailor its promotions to different target markets.

A test of independence addresses the question of whether the beer preference (light ale, lager or best bitter) is independent of the gender of the beer drinker (male, female). The hypotheses for this test are:

H_0: Beer preference is independent of the gender of the beer drinker

H_1: Beer preference is not independent of the gender of the beer drinker

Table 12.2 Contingency table for beer preference and gender of beer drinker

Gender	Beer preference		
	Light ale	Lager	Best bitter
Male	cell(1,1)	cell(1,2)	cell(1,3)
Female	cell(2,1)	cell(2,2)	cell(2,3)

Table 12.2 can be used to describe the situation. The population under study is all male and female beer drinkers. A sample can be selected from this population and each individual asked to state his or her preference among the three Real Ale beers. Every individual in the sample will be classified in one of the six cells in the table. For example, an individual may be a male preferring lager (cell (1,2)), a female preferring light ale (cell (2,1)), a female preferring best bitter (cell (2,3)) and so on. Because we have listed all possible combinations of beer preference and gender – in other words, listed all possible contingencies – Table 12.2 is called a **contingency table**. The test of independence is sometimes referred to as a *contingency table test*.

Suppose a simple random sample of 150 beer drinkers is selected. After tasting each beer, the individuals in the sample are asked to state their first-choice preference. The cross-tabulation in Table 12.3 summarizes the responses. The data for the test of independence are collected in terms of counts or frequencies for each cell or category. Of the 150 individuals in the sample, 20 were men who favoured light ale, 40 were men who favoured lager, 20 were men who favoured best bitter and so on. The data in Table 12.3 are the observed frequencies for the six classes or categories.

If we can determine the expected frequencies under the assumption of independence between beer preference and gender of the beer drinker, we can use the chi-squared distribution to determine whether there is a significant difference between observed and expected frequencies.

Expected frequencies for the cells of the contingency table are based on the following rationale. We assume that the null hypothesis of independence between beer preference and gender of the beer drinker is true. Then we note that in the entire sample of 150 beer drinkers, a total of 50 prefer light ale, 70 prefer lager and 30 prefer best bitter. In terms of fractions we conclude that 50/150 of the beer drinkers prefer light ale, 70/150 prefer lager and 30/150 prefer best bitter. If the *independence* assumption is valid, we argue that these fractions must be applicable to both male and female beer drinkers. So we would expect the sample of 80 male beer drinkers to show that (50/150)80 = 26.67 prefer light ale, (70/150)80 = 37.33 prefer lager, and (30/150)80 = 16 prefer best bitter. Application of the same fractions to the 70 female beer drinkers provides the expected frequencies shown in Table 12.4.

Table 12.3 Sample results for beer preferences of male and female beer drinkers (observed frequencies)

Gender	Beer preference			
	Light ale	Lager	Best bitter	Total
Male	20	40	20	80
Female	30	30	10	70
Total	50	70	30	150

Table 12.4 Expected frequencies if beer preference is independent of the gender of the beer drinker

| Gender | Beer preference | | | Total |
	Light ale	Lager	Best bitter	
Male	26.67	37.33	16.00	80
Female	23.33	32.67	14.00	70
Total	50.00	70.00	30.00	150

Let e_{ij} denote the expected frequency for the contingency table category in row i and column j. With this notation, consider the expected frequency calculation for males (row $i = 1$) who prefer lager (column $j = 2$): that is, expected frequency e_{12}. The argument above showed that

$$e_{12} = \left(\frac{70}{150}\right) 80 = 37.33$$

This expression can be written slightly differently as

$$e_{12} = \left(\frac{70}{150}\right) 80 = \frac{(80)(70)}{(150)} = 37.33$$

Note that the 80 in the expression is the total number of males (row 1 total), 70 is the total number of individuals preferring lager (column 2 total) and 150 is the total sample size. Hence, we see that

$$e_{12} = \frac{(\text{Row 1 Total})(\text{Column 2 Total})}{\text{Sample Size}}$$

Generalization of this expression shows that the following formula provides the expected frequencies for a contingency table in the test of independence.

Expected frequencies for contingency tables under the assumption of independence

$$e_{ij} = \frac{(\text{Row } i \text{ Total})(\text{Column } j \text{ Total})}{\text{Sample Size}} \tag{12.2}$$

Using this formula for male beer drinkers who prefer best bitter, we find an expected frequency of $e_{13} = (80)(30)/(150) = 16.00$, as shown in Table 12.4. Use equation (12.2) to verify the other expected frequencies shown in Table 12.4.

The test procedure for comparing the observed frequencies of Table 12.3 with the expected frequencies of Table 12.4 is similar to the goodness of fit calculations made in Section 12.1. Specifically, the χ^2 value based on the observed and expected frequencies is computed as follows.

Test statistic for independence

$$\chi^2 = \sum_i \sum_j \frac{(f_{ij} - e_{ij})^2}{e_{ij}} \tag{12.3}$$

where

f_{ij} = observed frequency for contingency table category in row i and column j

e_{ij} = expected frequency for contingency table category in row i and column j based on the assumption of independence

Note: With n rows and m columns in the contingency table, the test statistic has a chi-squared distribution with $(n - 1)(m - 1)$ degrees of freedom provided that the expected frequencies are five or more for all categories.

The double summation in equation (12.3) is used to indicate that the calculation must be made for all the cells in the contingency table.

The expected frequencies are five or more for each category. We therefore proceed with the computation of the chi-squared test statistic, as shown in Table 12.5. We see that the value of the test statistic is $\chi^2 = 6.12$.

The number of degrees of freedom for the appropriate chi-squared distribution is computed by multiplying the number of rows minus one by the number of columns minus one. With two rows and three columns, we have $(2 - 1)(3 - 1) = 3$ degrees of freedom. Just like the test for goodness of fit, the test for independence rejects H_0 if the differences between observed and expected frequencies provide a large value for the test statistic. So the test for independence is also an upper tail test. Using the chi-squared table (Table 3 of Appendix B), we find that the upper tail area or p-value at $\chi^2 = 6.12$ is between 0.025 and 0.05. At the 0.05 level of significance, p-value $< \alpha = 0.05$. We reject the null hypothesis of independence and conclude that beer preference is not independent of the gender of the beer drinker.

Computer software packages such as PASW, MINITAB and EXCEL can simplify the computations for a test of independence and provide the p-value for the test (see the

Table 12.5 Computation of the chi-squared test statistic for determining whether beer preference is independent of the gender of the beer drinker

Gender	Beer preference	Observed frequency (f_{ij})	Expected frequency (e_{ij})	Difference $(f_{ij} - e_{ij})$	Squared difference $(f_{ij} - e_{ij})^2$	Squared difference divided by expected frequency $(f_{ij} - e_{ij})^2/e_{ij}$
Male	Light ale	20	26.67	−6.67	44.44	1.67
Male	Lager	40	37.33	2.67	7.11	0.19
Male	Best bitter	20	16.00	4.00	16.00	1.00
Female	Light ale	30	23.33	6.67	44.44	1.90
Female	Lager	30	32.67	−2.67	7.11	0.22
Female	Best bitter	10	14.00	−4.00	16.00	1.14
	Total	150				$\chi^2 = 6.12$

Software Section at the end of the chapter). In the Real Ale Brewery example, EXCEL, MINITAB or PASW shows p-value $= 0.0468$.

The test itself does not tell us directly about the nature of the dependence between beer preference and gender, but we can compare the observed and expected frequencies descriptively to get an idea. Refer to Tables 12.3 and 12.4. Male beer drinkers have higher observed than expected frequencies for both lager and best bitter beers, whereas female beer drinkers have a higher observed than expected frequency only for light ale beer. These observations give us insight about the beer preference differences between male and female beer drinkers.

Here are the steps in a contingency table test of independence.

Test of independence: a summary

1 State the null and alternative hypotheses.

H_0: the column variable is independent of the row variable
H_1: the column variable is not independent of the row variable

2 Select a random sample and record the observed frequencies for each cell of the contingency table.

3 Use equation (12.2) to compute the expected frequency for each cell.

4 Use equation (12.3) to compute the value of the test statistic.

5 Rejection rule:

p-value approach: Reject H_0 if p-value $\leq \alpha$
Critical value approach: Reject H_0 if $\chi^2 \geq \chi^2_\alpha$

where α is the level of significance for the test, with n rows and m columns providing $(n - 1) \times (m - 1)$ degrees of freedom.

Note: The test statistic for the chi-squared tests in this chapter requires an expected frequency of five or more for each category. When a category has fewer than five, it is often appropriate to combine two adjacent rows or columns to obtain an expected frequency of five or more in each category.

Exercises

Methods

8 The following 2×3 contingency table contains observed frequencies for a sample of 200. Test for independence of the row and column variables using the χ^2 test with $\alpha = 0.05$.

	Column variable		
Row variable	A	B	C
P	20	44	50
Q	30	26	30

9 The following 3 × 3 contingency table contains observed frequencies for a sample of 240. Test for independence of the row and column variables using the χ^2 test with $\alpha = 0.05$.

Row variable	Column variable		
	A	B	C
P	20	30	20
Q	30	60	25
R	10	15	30

Applications

10 One of the questions on the *Business Week* Subscriber Study was, 'In the past 12 months, when travelling for business, what type of airline ticket did you purchase most often?' The data obtained are shown in the following contingency table.

Type of ticket	Type of flight	
	Domestic flights	International flights
First class	29	22
Business class	95	121
Economy class	518	135

Use $\alpha = 0.05$ and test for the independence of type of flight and type of ticket. What is your conclusion?

11 First-destination jobs for business and engineering graduates are classified by industry as shown in the following table.

Degree major	Industry			
	Oil	Chemical	Electrical	Computer
Business	30	15	15	40
Engineering	30	30	20	20

Use $\alpha = 0.01$ and test for independence of degree major and industry type.

12 Businesses are increasingly placing orders online. The Performance Measurement Group collected data on the rates of correctly filled electronic orders by industry. Assume a sample of 700 electronic orders provided the following results.

Order	Industry			
	Pharmaceutical	Consumer	Computers	Telecommunications
Correct	207	136	151	178
Incorrect	3	4	9	12

a. Test whether order fulfillment is independent of industry. Use $\alpha = 0.05$. What is your conclusion?

b. Which industry has the highest percentage of correctly filled orders?

13 Three suppliers provide the following data on defective parts.

	Part quality		
Supplier	Good	Minor defect	Major defect
A	90	3	7
B	170	18	7
C	135	6	9

Use $\alpha = 0.05$ and test for independence between supplier and part quality. What does the result of your analysis tell the purchasing department?

14 A sample of parts taken in a machine shop in Karachi provided the following contingency table data on part quality by production shift.

Shift	Number good	Number defective
First	368	32
Second	285	15
Third	176	24

Use $\alpha = 0.05$ and test the hypothesis that part quality is independent of the production shift. What is your conclusion?

15 Visa Card studied how frequently consumers of various age groups use plastic cards (debit and credit cards) when making purchases. Sample data for 300 customers shows the use of plastic cards by four age groups.

	Age group			
Payment	18–24	25–34	35–44	45 and over
Plastic	21	27	27	36
Cash or Cheque	21	36	42	90

a. Test for the independence between method of payment and age group. What is the p-value? Using $\alpha = 0.05$, what is your conclusion?
b. If method of payment and age group are not independent, what observation can you make about how different age groups use plastic to make purchases?
c. What implications does this study have for companies such as Visa and MasterCard?

16 The following cross-tabulation shows industry type and P/E ratio for 100 companies in the consumer products and banking industries.

	P/E ratio					
Industry	5–9	10–14	15–19	20–24	25–29	Total
Consumer	4	10	18	10	8	50
Banking	14	14	12	6	4	50
Total	18	24	30	16	12	100

Does there appear to be a relationship between industry type and P/E ratio? Support your conclusion with a statistical test using $\alpha = 0.05$.

12.3 Goodness of fit test: Poisson and normal distributions

In general, the chi-squared goodness of fit test can be used with any hypothesized probability distribution. In this section we illustrate for cases in which the population is hypothesized to have a Poisson or a normal distribution. The goodness of fit test and the use of the chi-squared distribution for the test follow the same general procedure used for the goodness of fit test in Section 12.1.

Poisson distribution

Consider the arrival of customers at the Mediterranean Food Market. Because of recent staffing problems, the Mediterranean's managers asked a local consultancy to assist with the scheduling of checkout assistants. After reviewing the checkout operation, the consultancy will make a recommendation for a scheduling procedure. The procedure, based on a mathematical analysis of waiting times, is applicable only if the number of customers arriving during a specified time period follows the Poisson distribution. Therefore, before the scheduling process is implemented, data on customer arrivals must be collected and a statistical test done to see whether an assumption of a Poisson distribution for arrivals is reasonable.

We define the arrivals at the store in terms of the *number of customers* entering the store during five-minute intervals. The following null and alternative hypotheses are appropriate:

H_0: The number of customers entering the store during five-minute intervals
   has a Poisson probability distribution

H_1: The number of customers entering the store during five-minute intervals
   does not have a Poisson distribution

If a sample of customer arrivals indicates H_0 cannot be rejected, the Mediterranean will proceed with the implementation of the consultancy's scheduling procedure. However, if the sample leads to the rejection of H_0, the assumption of the Poisson distribution for the arrivals cannot be made and other scheduling procedures will be considered.

To test the assumption of a Poisson distribution for the number of arrivals during weekday morning hours, a store assistant randomly selects a sample, $n = 128$, of five-minute intervals during weekday mornings over a three-week period. For each five-minute interval in the sample, the store employee records the number of customer arrivals. In summarizing the data, the store assistant determines the number of five-minute intervals having no arrivals, the number of five-minute intervals having one arrival, the number of five-minute intervals having two arrivals, and so on. These data are summarized in Table 12.6, which gives the observed frequencies for the ten categories.

To do the goodness of fit test, we need to consider the expected frequency for each of the ten categories, under the assumption that the Poisson distribution of arrivals is true. The Poisson probability function, first introduced in Chapter 5, is

$$p(X = x) = \frac{\mu^x e^{-\mu}}{x!} \tag{12.4}$$

In this function, μ represents the mean or expected number of customers arriving per five-minute period, X is a random variable indicating the number of customers arriving during a five-minute period, and $p(X = x)$ is the probability that exactly x customers will arrive in a five-minute interval.

Table 12.6 Observed frequency of the Mediterranean's customer arrivals for a sample of 128 five-minute time periods

Number of customers arriving	Observed frequency
0	2
1	8
2	10
3	12
4	18
5	22
6	22
7	16
8	12
9	6
Total	128

To use (12.4), we must obtain an estimate of μ, the mean number of customer arrivals during a five-minute time period. The sample mean for the data in Table 12.6 provides this estimate. With no customers arriving in two five-minute time periods, one customer arriving in eight five-minute time periods and so on, the total number of customers who arrived during the sample of 128 five-minute time periods is given by $0(2) + 1(8) + 2(10) + \cdots + 9(6) = 640$. The 640 customer arrivals over the sample of 128 periods provide an estimated mean arrival rate of $640/128 = 5$ customers per five-minute period. With this value for the mean of the distribution, an estimate of the Poisson probability function for the Mediterranean Food Market is

$$p(X = x) = \frac{5^x e^{-5}}{x!} \qquad \textbf{(12.5)}$$

This probability function can be evaluated for different values x to determine the probability associated with each category of arrivals. These probabilities, which can also be found in Table 7 of Appendix B, are given in Table 12.7. For example, the probability of zero customers arriving during a five-minute interval is $p(0) = 0.0067$, the probability of one customer arriving during a five-minute interval is $p(1) = 0.0337$ and so on. As we saw in Section 12.1, the expected frequencies for the categories are found by multiplying the probabilities by the sample size. For example, the expected number of periods with zero arrivals is given by $(0.0067)(128) = 0.86$, the expected number of periods with one arrival is given by $(0.0337)(128) = 4.31$, and so on.

Note that in Table 12.7, four of the categories have an expected frequency less than five. This condition violates the requirements for use of the chi-squared distribution. However, adjacent categories can be combined to satisfy the 'at least five' expected frequency requirement. In particular, we shall combine 0 and 1 into a single category, and then combine 9 with '10 or more' into another single category. Table 12.8 shows the observed and expected frequencies after combining categories.

As in Section 12.1, the goodness of fit test focuses on the differences between observed and expected frequencies, $f_i - e_i$. The calculations are shown in Table 12.8. The value of the test statistic is $\chi^2 = 10.96$.

Table 12.7 Expected frequency of Mediterranean's customer arrivals, assuming a Poisson distribution with $\mu = 5$

Number of customers arriving (x)	Poisson probability $p(x)$	Expected number of five-minute time periods with x arrivals, $128p(x)$
0	0.0067	0.86
1	0.0337	4.31
2	0.0842	10.78
3	0.1404	17.97
4	0.1755	22.46
5	0.1755	22.46
6	0.1462	18.71
7	0.1044	13.36
8	0.0653	8.36
9	0.0363	4.65
10 or more	0.0318	4.07
	Total	128.00

In general, the chi-squared distribution for a goodness of fit test has $k - p - 1$ degrees of freedom, where k is the number of categories and p is the number of population parameters estimated from the sample data. Table 12.8 shows $k = 9$ categories. Because the sample data were used to estimate the mean of the Poisson distribution, $p = 1$. Hence, there are $k - p - 1 = 9 - 1 - 1 = 7$ degrees of freedom.

Suppose we test the null hypothesis with a 0.05 level of significance. We need to determine the p-value for the test statistic $\chi^2 = 10.96$ by finding the area in the upper tail of a chi-squared distribution with 7 degrees of freedom. Using Table 3 of Appendix B,

Table 12.8 Observed and expected frequencies for the Mediterranean's customer arrivals after combining categories, and computation of the chi-squared test statistic

Number of customers arriving (x)	Observed frequency (f_i)	Expected frequency (e_i)	Difference ($f_i - e_i$)	Squared difference ($f_i - e_i)^2$	Squared difference divided by expected frequency ($f_i - e_i)^2/e_i$
0 or 1	10	5.17	4.83	23.28	4.50
2	10	10.78	−0.78	0.61	0.06
3	12	17.97	−5.97	35.62	1.98
4	18	22.46	−4.46	19.89	0.89
5	22	22.46	−0.46	0.21	0.01
6	22	18.72	3.28	10.78	0.58
7	16	13.37	2.63	6.92	0.52
8	12	8.36	3.64	13.28	1.59
9 or more	6	8.72	−2.72	7.38	0.85
Total	128	128.00			$\chi^2 = 10.96$

we find that $\chi^2 = 10.96$ provides an area in the upper tail greater than 0.10. So we know that the p-value is greater than 0.10. MINITAB, PASW or EXCEL shows p-value $=$ 0.1403. With p-value $> \alpha = 0.10$, we cannot reject H_0. The assumption of a Poisson probability distribution for weekday morning customer arrivals cannot be rejected. As a result, the Mediterranean's management may proceed with the consulting firm's scheduling procedure for weekday mornings.

Poisson distribution goodness of fit test: a summary

1 State the null and alternative hypotheses.

H_0: The population has a Poisson distribution

H_1: The population does not have a Poisson distribution

2 Select a random sample and

a. Record the observed frequency f_i for each value of the Poisson random variable.

b. Compute the mean number of occurrences.

3 Compute the expected frequency of occurrences e_i for each value of the Poisson random variable. Multiply the sample size by the Poisson probability of occurrence for each value of the Poisson random variable. If there are fewer than five expected occurrences for some values, combine adjacent values and reduce the number of categories as necessary.

4 Compute the value of the test statistic.

$$\chi^2 = \sum_{i=1}^{k} \frac{(f_i - e_i)^2}{e_i}$$

5 Rejection rule:

p-value approach: Reject H_0 if p-value $\leq \alpha$

Critical value approach: Reject H_0 if $\chi^2 \geq \chi_\alpha^2$

where α is the level of significance for the test, and there are $k - 2$ degrees of freedom.

Normal distribution

A goodness of fit test for a normal distribution can also be based on the use of the chi-squared distribution. It is similar to the procedure we discussed for the Poisson distribution. In particular, observed frequencies for several categories of sample data are compared to expected frequencies under the assumption that the population has a normal distribution. Because the normal distribution is continuous, we must modify the way the categories are defined and how the expected frequencies are computed.

Consider the job applicant test data for Pharmaco plc, listed in Table 12.9. Pharmaco hires approximately 400 new employees annually for its four plants located throughout Europe. The personnel director asks whether a normal distribution applies for the population of test scores. If such a distribution can be used, the distribution would be helpful in evaluating specific test scores; that is, scores in the upper 20 per cent, lower 40 per cent and so on, could be identified quickly. Hence, we want to test the null hypothesis that the population of test scores has a normal distribution.

We first use the data in Table 12.9 to calculate estimates of the mean and standard deviation of the normal distribution that will be considered in the null hypothesis. We use

Table 12.9 Pharmaco employee aptitude test scores for 50 randomly chosen job applicants

71	65	54	93	60	86	70	70	73	73
55	63	56	62	76	54	82	79	76	68
53	58	85	80	56	61	64	65	62	90
69	76	79	77	54	64	74	65	65	61
56	63	80	56	71	79	84	66	61	61

the sample mean and the sample standard deviation as point estimators of the mean and standard deviation of the normal distribution. The calculations follow.

$$\bar{x} = \frac{\Sigma x_i}{n} = \frac{3421}{50} = 68.42$$

$$s = \sqrt{\frac{\Sigma(x_i - \bar{x})^2}{n-1}} = \sqrt{\frac{5310.0369}{49}} = 10.41$$

Using these values, we state the following hypotheses about the distribution of the job applicant test scores.

H_0: The population of test scores has a normal distribution with mean 68.42 and standard deviation 10.41.

H_1: The population of test scores does not have a normal distribution with mean 68.42 and standard deviation 10.41.

Now we look at how to define the categories for a goodness of fit test involving a normal distribution. For the discrete probability distribution in the Poisson distribution test, the categories were readily defined in terms of the number of customers arriving, such as 0, 1, 2 and so on. However, with the continuous normal probability distribution, we must use a different procedure for defining the categories. We need to define the categories in terms of *intervals* of test scores.

Recall the rule of thumb for an expected frequency of at least five in each interval or category. We define the categories of test scores such that the expected frequencies will be at least five for each category. With a sample size of 50, one way of establishing categories is to divide the normal distribution into ten equal-probability intervals (see Figure 12.2). With a sample size of 50, we would expect five outcomes in each interval or category and the rule of thumb for expected frequencies would be satisfied.

When the normal probability distribution is assumed, the standard normal distribution tables can be used to determine the category boundaries. First consider the test score cutting off the lowest 10 per cent of the test scores. From Table 1 of Appendix B we find that the z value for this test score is −1.28. Therefore, the test score x = 68.42 − 1.28 (10.41) = 55.10 provides this cut-off value for the lowest 10 per cent of the scores. For the lowest 20 per cent, we find z = −0.84, and so x = 68.42 − 0.84(10.41) = 59.68. Working through the normal distribution in that way provides the following test score values.

Lower 10%: 68.42 − 1.28(10.41) = 55.10

Lower 20%: 68.42 − 0.84(10.41) = 59.68

Lower 30%: $68.42 - 0.52(10.41) = 63.01$
Lower 40%: $68.42 - 0.25(10.41) = 65.82$
Mid-score: $68.42 - 0(10.41)\ \ \ \ \ = 68.42$
Upper 40%: $68.42 + 0.25(10.41) = 71.02$
Upper 30%: $68.42 + 0.52(10.41) = 73.83$
Upper 20%: $68.42 + 0.84(10.41) = 77.16$
Upper 10%: $68.42 + 1.28(10.41) = 81.74$

These cutoff or interval boundary points are identified on the graph in Figure 12.1.

We can now return to the sample data of Table 12.9 and determine the observed frequencies for the categories. The results are in Table 12.10. The goodness of fit calculations now proceed exactly as before. Namely, we compare the observed and expected results by computing a χ^2 value. The computations are also shown in Table 12.10. We see that the value of the test statistic is $\chi^2 = 7.2$.

To determine whether the computed χ^2 value of 7.2 is large enough to reject H_0, we need to refer to the appropriate chi-squared distribution tables. Using the rule for computing the number of degrees of freedom for the goodness of fit test, we have $k - p - 1 = 10 - 2 - 1 = 7$ degrees of freedom based on $k = 10$ categories and $p = 2$ parameters (mean and standard deviation) estimated from the sample data.

Suppose we do the test with a 0.10 level of significance. To test this hypothesis, we need to determine the p-value for the test statistic $\chi^2 = 7.2$ by finding the area in the upper tail of a chi-squared distribution with 7 degrees of freedom. Using Table 3 of Appendix B, we find that $\chi^2 = 7.2$ provides an area in the upper tail greater than 0.10. So we know that the p-value is greater than 0.10. EXCEL, PASW or MINITAB shows p-value $= 0.4084$.

With p-value $> \alpha = 0.10$, the hypothesis that the probability distribution for the Pharmaco job applicant test scores is a normal distribution cannot be rejected. The normal distribution may be applied to assist in the interpretation of test scores.

A summary of the goodness fit test for a normal distribution follows.

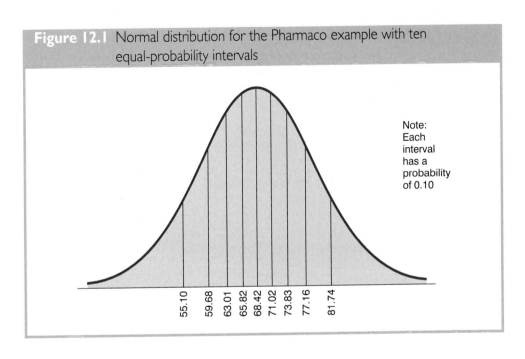

Figure 12.1 Normal distribution for the Pharmaco example with ten equal-probability intervals

Note:
Each
interval
has a
probability
of 0.10

55.10 59.68 63.01 65.82 68.42 71.02 73.83 77.16 81.74

Table 12.10 Observed and expected frequencies for Pharmaco job applicant test scores, and computation of the chi-squared test statistic

Test score interval	Observed frequency (f_i)	Expected frequency (e_i)	Difference $(f_i - e_i)$	Squared difference $(f_i - e_i)^2$	Squared difference divided by expected frequency $(f_i - e_i)^2/e_i$
Less than 55.10	5	5	0	0	0.0
55.10 to 59.67	5	5	0	0	0.0
59.68 to 63.00	9	5	4	16	3.2
63.01 to 65.81	6	5	1	1	0.2
65.82 to 68.41	2	5	3	9	1.8
68.42 to 71.01	5	5	0	0	0.0
71.02 to 73.82	2	5	3	9	1.8
73.83 to 77.15	5	5	0	0	0.0
77.16 to 81.73	5	5	0	0	0.0
81.74 and over	6	5	1	1	0.2
Total	50	50			$\chi^2 = 7.2$

Normal distribution goodness of fit test: a summary

1 State the null and alternative hypotheses.

H_0: The population has a normal distribution
H_1: The population does not have a normal distribution

2 Select a random sample and

a. Compute the sample mean and sample standard deviation.
b. Define intervals of values so that the expected frequency is at least five for each interval. Using equal probability intervals is a good approach.
c. Record the observed frequency of data values f_i in each interval defined.

3 Compute the expected number of occurrences e_i for each interval of values defined in step 2(b). Multiply the sample size by the probability of a normal random variable being in the interval.

4 Compute the value of the test statistic.

$$\chi^2 = \sum_{i=1}^{k} \frac{(f_i - e_i)^2}{e_i}$$

5 Rejection rule:

p-value approach: Reject H_0 if p-value $\leq \alpha$
Critical value approach: Reject H_0 if $\chi^2 \geq \chi^2_\alpha$

where α is the level of significance for the test, and there are $k - 3$ degrees of freedom.

Exercises

Methods

17 The following data are believed to have come from a normal distribution. Use a goodness of fit test and $\alpha = 0.05$ to test this claim.

17	23	22	24	19	23	18	22	20	13	11	21	18	20	21
21	18	15	24	23	23	43	29	27	26	30	28	33	23	29

18 Data on the number of occurrences per time period and observed frequencies follow. Use $\alpha = 0.05$ and a goodness of fit test to see whether the data fit a Poisson distribution.

Number of occurrences	Observed frequency
0	39
1	30
2	30
3	18
4	3

Applications

19 The number of incoming phone calls to a small call centre in Mumbai, during one-minute intervals, is believed to have a Poisson distribution. Use $\alpha = 0.10$ and the following data to test the assumption that the incoming phone calls follow a Poisson distribution.

Number of incoming phone calls during a one-minute interval	Observed frequency
0	15
1	31
2	20
3	15
4	13
5	4
6	2
Total	100

20 The weekly demand for a particular product in a white-goods store is thought to be normally distributed. Use a goodness of fit test and the following data to test this assumption. Use $\alpha = 0.10$. The sample mean is 24.5 and the sample standard deviation is 3.0.

18	20	22	27	22	25	22	27	25	24
26	23	20	24	26	27	25	19	21	25
26	25	31	29	25	25	28	26	28	24

21 A random sample of final examination grades for a college course in Middle-East studies follows.

55	85	72	99	48	71	88	70	59	98
80	74	93	85	74	82	90	71	83	60
95	77	84	73	63	72	95	79	51	85
76	81	78	65	75	87	86	70	80	64

Use $\alpha = 0.05$ and test to determine whether a normal distribution should be rejected as being representative of the population's distribution of grades.

22 The number of car accidents per day in a particular city is believed to have a Poisson distribution. A sample of 80 days during the past year gives the following data. Do these data support the belief that the number of accidents per day has a Poisson distribution? Use $\alpha = 0.05$.

Number of accidents	Observed frequency (days)
0	34
1	25
2	11
3	7
4	3

Summary

The purpose of a goodness of fit test is to determine whether a hypothesized probability distribution can be used as a model for a particular population of interest. The computations for conducting the goodness of fit test involve comparing observed frequencies from a sample with expected frequencies when the hypothesized probability distribution is assumed true. A chi-squared distribution is used to determine whether the differences between observed and expected frequencies are large enough to reject the hypothesized probability distribution.

In this chapter we introduced the goodness of fit test for a multinomial distribution. A test of independence for two variables is an extension of the methodology used in the goodness of fit test for a multinomial population. A contingency table is used to set out the observed and expected frequencies. Then a chi-squared value is computed.

We also illustrated the goodness of fit test for Poisson and normal distributions.

Key terms

Contingency table   Multinomial population
Goodness of fit test

Key formulae

Test statistic for goodness of fit

$$\chi^2 = \sum_{i=1}^{k} \frac{(f_i - e_i)^2}{e_i} \qquad (12.1)$$

Expected frequencies for contingency tables under the assumption of independence

$$e_{ij} = \frac{(\text{Row } i \text{ Total})(\text{Column } j \text{ Total})}{\text{Sample Size}} \qquad (12.2)$$

Test statistic for independence

$$\chi^2 = \sum_i \sum_j \frac{(f_{ij} - e_{ij})^2}{e_{ij}} \qquad (12.3)$$

Case problem I Evaluation of Management School website pages

WEBSITES

A group of MSc students at an international university conducted a survey to assess the students' views regarding the web pages of the university's Management School. Among the questions in the survey were items that asked respondents to express agreement or disagreement with the following statements.

1 The Management School web pages are attractive for prospective students.

2 I find it easy to navigate the Management School web pages.

Students using a computer to research business schools. © John Henley/CORBIS.

3 There is up-to-date information about courses on the Management School web pages.

4 If I were to recommend the university to someone else, I would suggest that he/she goes to the Management School web pages.

Responses were originally given on a five-point scale, but in the data file on CD that accompanies this case problem ('Web Pages'), the responses have been recoded as binary variables. For each questionnaire item, those who agreed or agreed strongly with the statement have been grouped into one category (Agree). Those who disagreed, disagreed strongly, were indifferent or opted for a 'Don't know' response, have been grouped into a second category (Don't Agree). The data file also contains particulars of respondent gender and level of study (undergraduate or postgraduate). A screenshot of the first few rows of the data file is shown below.

Managerial report

1 Use descriptive statistics to summarize the data from this study. What are your preliminary conclusions about the independence of the response (Agree or Don't Agree) and gender for each of the four items? What are your preliminary conclusions about the independence of the response (Agree or Don't Agree) and level of study for each of the four items?

2 With regard to each of the four items, test for the independence of the response (Agree or Don't Agree) and gender. Use $\alpha = 0.05$.

3 With regard to each of the four items, test for the independence of the response (Agree or Don't Agree) and level of study. Use $\alpha = 0.05$.

4 Does it appear that views regarding the web pages are consistent for students of both genders and both levels of study? Explain.

Gender	Study level	Attractiveness	Navigation	Up-to-date	Referrals
Female	Undergraduate	Don't Agree	Agree	Agree	Agree
Female	Undergraduate	Agree	Agree	Agree	Agree
Male	Undergraduate	Don't Agree	Don't Agree	Don't Agree	Don't Agree
Male	Undergraduate	Agree	Agree	Agree	Agree
Male	Undergraduate	Agree	Agree	Agree	Agree
Female	Undergraduate	Don't Agree	Don't Agree	Agree	Agree
Male	Undergraduate	Don't Agree	Agree	Agree	Agree
Male	Undergraduate	Agree	Agree	Agree	Agree
Male	Undergraduate	Don't Agree	Agree	Agree	Agree

Case problem 2 Checking for randomness in Lotto draws

National Lottery play slips. © Positive Image/Alamy.

In the main Lotto game of the UK National Lottery, six balls are randomly selected from a set of balls numbered 1, 2, . . . , 49. The file 'Lotto' on the accompanying CD contains details of the numbers drawn in the main Lotto game from early January 1995 up to early March 2008.

A screen shot of the first few rows of the data file is shown below. In addition to showing the six numbers drawn in the game each time, and the order in which they were drawn, the file also gives details of the day on which the draw took place, the machine that was used to do the draw, and the set of balls that was used. In recent years, Lotto draws have taken place on both Wednesday and Saturday each week. A number of similar machines are used for the draws: Sapphire, Topaz, etc, and eight sets of balls are used.

Analyst's report

1   Use an appropriate hypothesis test to assess whether there is any evidence of non-randomness in the first ball drawn. Similarly, test for non-randomness in the second ball drawn, third ball drawn, . . . , sixth ball drawn.

2   Use an appropriate hypothesis test to assess whether there is any evidence of non-randomness overall in the drawing of the 49 numbers (regardless of the order of selection).

3   Use an appropriate hypothesis test to assess whether there is evidence of any dependence between the numbers drawn and the day on which the draw is made.

4   Use an appropriate hypothesis test to assess whether there is evidence of any dependence between the numbers drawn and the machine on which the draw is made.

5   Use an appropriate hypothesis test to assess whether there is evidence of any dependence between the numbers drawn and the set of balls that is used.

No.	Day	DD	MMM	YYYY	N1	N2	N3	N4	N5	N6	Machine	Set
1378	Sat	7	Mar	2009	30	33	32	44	21	6	Sapphire	3
1377	Wed	4	Mar	2009	46	3	32	37	25	21	Topaz	2
1376	Sat	28	Feb	2009	43	6	33	37	25	42	Sapphire	1
1375	Wed	25	Feb	2009	19	6	39	32	25	37	Topaz	6
1374	Sat	21	Feb	2009	18	42	45	46	26	22	Sapphire	6
1373	Wed	18	Feb	2009	24	44	29	5	23	39	Topaz	3
1372	Sat	14	Feb	2009	17	45	4	2	19	38	Topaz	2
1371	Wed	11	Feb	2009	46	38	14	9	47	16	Sapphire	1
1370	Sat	7	Feb	2009	41	26	7	3	24	25	Sapphire	5

Software Section for Chapter 12

···

Tests of goodness of fit and independence using MINITAB

Goodness of fit test

This MINITAB procedure can be used for goodness of fit tests for the multinomial distribution in Section 12.1 and the Poisson and normal distributions in Section 12.3. The user must obtain the observed frequencies, calculate the expected frequencies and enter both the observed and expected frequencies in a MINITAB worksheet. Using the Scott Market Research example presented in Section 12.1, open a MINITAB worksheet, enter the observed frequencies 48, 98 and 54 in column C1 and enter the expected frequencies 60, 100 and 40 in column C2. The MINITAB steps for the goodness of fit test follow.

Step 1 **Calc > Calculator**   [Main menu bar]

Step 2 Enter **ChiSquared** in the **Store result in variable** box   [**Calculator** panel]
Enter **Sum((C1-C2)**2/C2)** in the **Expression** box
Click **OK**

Step 3 **Calc > Probability Distributions > Chi-square**   [Main menu bar]

Step 4 Select **Cumulative probability**   [**Chi-square Distribution** panel]
Enter **2** in the **Degrees of freedom** box
Select **Input column** and enter **ChiSquared** in the adjacent box
Click **OK**

The MINITAB output provides the chi-squared statistic $\chi^2 = 7.34$, and the cumulative probability 0.9745, which is the area under the curve to the left of $\chi^2 = 7.34$. The area remaining in the upper tail is the p-value. We have p-value $= 1 - 0.9745 = 0.0255$.

Test of independence

We begin with a new MINITAB worksheet and enter the observed frequency data for the Real Ale Brewery example from Section 12.2 into columns 1, 2 and 3, respectively. We enter the observed frequencies corresponding to a light ale preference (20 and 30) in C1, the observed frequencies corresponding to a lager preference (40 and 30) in C2, and the observed frequencies corresponding to a best bitter preference (20 and 10) in C3. The MINITAB steps for the test of independence are given below. The screen shot in Figure 12.2 shows the MINITAB output, with χ^2 (2 df) $= 6.122, p = 0.047$.

Figure 12.2 MINITAB output for the Real Ale Brewery test of independence

Chi-Square Test: Light Ale, Lager, Best Bitter

Expected counts are printed below observed counts
Chi-Square contributions are printed below expected counts

	Light Ale	Lager	Best Bitter	Total
1	20	40	20	80
	26.67	37.33	16.00	
	1.667	0.190	1.000	
2	30	30	10	70
	23.33	32.67	14.00	
	1.905	0.218	1.143	
Total	50	70	30	150

Chi-Sq = 6.122, DF = 2, P-Value = 0.047

Step 1 Stat > Tables > Chi-square Test (Two-Way Table in Worksheet)
[Main menu bar]

Step 2 Enter **C1-C3** in the **Columns containing the table** box
[**Chi-square Test (Table in Worksheet)** panel]
Click **OK**

Tests of goodness of fit and independence using **EXCEL**

FITTEST

Goodness of fit test

This EXCEL procedure can be used for goodness of fit tests for the multinomial distribution in Section 12.1 and the Poisson and normal distributions in Section 12.3. The user must obtain the observed frequencies, calculate the expected frequencies, and enter both the observed and expected frequencies in an EXCEL worksheet.

The observed frequencies and expected frequencies for the Scott Market Research example presented in Section 12.1 are entered in columns A and B as shown in Figure 12.3 (refer to the file 'FitTest.XLS' on the accompanying CD). The test statistic $\chi^2 = 7.34$ is calculated in column D. With $k = 3$ categories, the user enters the degrees of freedom $k - 1 = 3 - 1 = 2$ in cell D11. The CHIDIST function provides the p-value in cell D13. The left-hand panel shows the cell formulae.

INDEPEN-
DENCE

Test of independence

The EXCEL procedure for the test of independence requires the user to obtain the observed frequencies and enter them in the worksheet. The Real Ale Brewery example from Section 12.2 provides the observed frequencies, which are entered in cells B7 to D8 as shown in the worksheet in Figure 12.4 (refer to the file 'Independence.XLS' on the

Figure 12.3 EXCEL worksheet for the Scott Market Research goodness of fit test

	C	D
1		
2		
3		
4		Calculations
5		=(A5-B5)^2/B5
6		=(A6-B6)^2/B6
7		=(A7-B7)^2/B7
8		
9	Test Statistic	=SUM(D5:D7)
10		
11	Degrees of Freedom	2
12		
13	p-Value	=CHIDIST(D9,D11)
14		

	A	B	C	D
1	Goodness of Fit Test			
2				
3	Observed	Expected		
4	Frequency	Frequency		Calculations
5	48	60		2.40
6	98	100		0.04
7	54	40		4.90
8				
9			Test Statistic	7.34
10				
11			Degrees of Freedom	2
12				
13			p-Value	0.0255
14				

accompanying CD). The cell formulae in the left-hand panel show the procedure used to compute the expected frequencies. With two rows and three columns, the user enters the degrees of freedom $(2 - 1)(3 - 1) = 2$ in cell E22. The CHITEST function provides the p-value in cell E24.

Figure 12.4 EXCEL worksheet for the Real Ale Brewery test of independence

	B	C	D	E
1				
2				
3				
4				
5		Beer Peference		
6	Light	Regular	Dark	Total
7	20	40	20	=SUM(B7:D7)
8	30	30	10	=SUM(B8:D8)
9	=SUM(B7:B8)	=SUM(C7:C8)	=SUM(D7:D8)	=SUM(E7:E8)
10				
11				
12				
13				
14		Beer Peference		
15	Light	Regular	Dark	Total
16	=E7*B$9/$E$9	=E7*C$9/$E$9	=E7*D$9/$E$9	=SUM(B16:D16)
17	=E8*B$9/$E$9	=E8*C$9/$E$9	=E8*D$9/$E$9	=SUM(B17:D17)
18	=SUM(B16:B17)	=SUM(C16:C17)	=SUM(D16:D17)	=SUM(E16:E17)
19				
20			Test Statistic	=CHIINV(E24,E22)
21				
22			Degrees of Freedom	2
23				
24			p-value	=CHITEST(B7:D8,B16:D
25				

	A	B	C	D	E
1	Test of Independence				
2					
3	Observed Frequencies				
4					
5			Beer Peference		
6	Gender	Light	Regular	Dark	Total
7	Male	20	40	20	80
8	Female	30	30	10	70
9	Total	50	70	30	150
10					
11					
12	Expected Frequencies				
13					
14			Beer Peference		
15	Gender	Light	Regular	Dark	Total
16	Male	26.67	37.33	16	80
17	Female	23.33	32.67	14	70
18	Total	50	70	30	150
19					
20				Test Statistic	6.12
21					
22				Degrees of Freedom	2
23					
24				p-value	0.0468
25					

Tests of goodness of fit and independence using PASW

A PASW data file suitable for the Real Ale Brewery example is on the accompanying CD ('RealAle.SAV'). There are three columns in the data file, which between them define the contingency table for this example. The first column contains the row indices for the table (1, 2 for Male, Female), the second column contains the column indices (1, 2, 3 for Light Ale, Lager, Best Bitter) and the third column contains the frequency counts. The data file has six rows corresponding to the six cells of the contingency table. The PASW steps for the test of independence are as follows.

Step 1 Data > Weight Cases   [Main menu bar]

Step 2 Select **Weight cases by**   [**Weight Cases** panel]
Transfer **Frequency count** (third column) to the **Frequency Variable** box
Click **OK**

Step 3 Analyze > Descriptive Statistics > Crosstabs   [Main menu bar]

Step 4 Transfer **Gender** (first column) to the **Row(s)** box   [**Crosstabs** panel]
Transfer **Type of beer** (second column) to the Column(s) box
Click on the **Statistics** button

Step 5 Check **Chi-square**   [**Crosstabs: Statistics** panel]
Click **Continue**

Step 6 Click **OK**   [**Crosstabs** panel]

The PASW output is shown below in Figure 12.5. The first table is the contingency table. Expected values, together with row and/or column percentages, can be added to this table using the **Cells** button on the **Crosstabs** dialogue panel. The result of the chi-squared test is shown in the first row of the second table, χ^2 (2 df) = 6.122, p = 0.047.

Figure 12.5 PASW output for the Real Ale Brewery test of independence

Gender * Type of beer Crosstabulation

Count

		Type of beer			Total
		Light Ale	Lager	Best Bitter	
Gender	Male	20	40	20	80
	Female	30	30	10	70
Total		50	70	30	150

Chi-Square Tests

	Value	df	Asymp. Sig. (2-sided)
Pearson Chi-Square	6.122[a]	2	.047
Likelihood Ratio	6.178	2	.046
Linear-by-Linear Association	5.872	1	.015
N of Valid Cases	150		

a. 0 cells (.0%) have expected count less than 5. The minimum expected count is 14.00.

Chapter 13

Analysis of Variance and Experimental Design

Learning objectives

After reading this chapter and doing the exercises, you should be able to:

1 Understand how the analysis of variance procedure can be used to determine if the means of more than two populations are equal.

2 Know the assumptions necessary to use the analysis of variance procedure.

3 Understand the use of the *F* distribution in performing the analysis of variance procedure.

4 Know how to set up an ANOVA table and interpret the entries in the table.

5 Use output from computer software packages to solve analysis of variance problems.

6 Know how to use Fisher's least significant difference (LSD) procedure and Fisher's LSD with the Bonferroni adjustment to conduct

statistical comparisons between pairs of population means.

7 Understand the difference between a completely randomized design, a randomized block design and factorial experiments.

8 Know the definition of the following terms:

comparisonwise Type I error rate
experimentwise Type I error rate
factor
level
treatment
partitioning
blocking
main effect
interaction
replication

In this chapter we introduce a statistical procedure called *analysis of variance* (ANOVA).

First, we show how ANOVA can be used to test for the equality of three or more population means using data obtained from an observational study. Then, we discuss the use of ANOVA for analyzing data obtained from three types of experimental studies: a completely randomized design, a randomized block design and a factorial experiment. In the following chapters we will see that ANOVA plays a key role in analyzing the results of regression analysis involving both experimental and observational data.

13.1 An introduction to analysis of variance

National Computer Products (NCP) manufactures printers and fax machines at plants located in Ayr, Dusseldorf and Stockholm. To measure how much employees at these plants know about total quality management, a random sample of six employees was selected from each plant and given a quality awareness examination. The examination scores obtained for these 18 employees are listed in Table 13.1. The sample means, sample variances and sample standard deviations for each group are also provided. Managers want to use these data to test the hypothesis that the mean examination score is the same for all three plants.

We will define population 1 as all employees at the Ayr plant, population 2 as all employees at the Dusseldorf plant, and population 3 as all employees at the Stockholm plant. Let

μ_1 = mean examination score for population 1
μ_2 = mean examination score for population 2
μ_3 = mean examination score for population 3

Statistics in Practice

Product customization and manufacturing trade-offs

The analysis of variance technique was used recently in a study to investigate trade-offs between product customization and other manufacturing priorities. A total of 102 UK manufacturers from eight industrial sectors were involved in the research. Three levels of customization were

Interior of a car manufacturing plant. © George Clerk.

considered: full customization where customer input was incorporated at the product design or fabrication stages; partial customization with customer input incorporated into product assembly or delivery stages and standard products which did not incorporate any customer input at all.

The impact of customization was considered against four competitive imperatives – cost, quality, delivery and volume flexibility.

It was found that customization had a significant effect on delivery (both in terms of speed and lead times); also on manufacturer's costs (though not design, component, delivery and servicing costs).

The findings suggest that customization is not cost-free and that the advent of mass customization is unlikely to see the end of trade-offs with other key priorities.

Source: Squire, B., Brown, S., Readman, J. and Bessant J. (2005) The impact of mass customization on manufacturing trade-offs. *Production and Operations Management* Journal 15(1) 10–21

Although we will never know the actual values of μ_1, μ_2 and μ_3, we want to use the sample results to test the following hypotheses.

$$H_0: \mu_1 = \mu_2 = \mu_3$$
$$H_1: \text{Not all population means are equal}$$

As we will demonstrate shortly, analysis of variance is a statistical procedure that can be used to determine whether the observed differences in the three sample means are large enough to reject H_0.

Table 13.1 Examination scores for 18 employees

Observation	Plant 1 Ayr	Plant 2 Dusseldorf	Plant 3 Stockholm
1	85	71	59
2	75	75	64
3	82	73	62
4	76	74	69
5	71	69	75
6	85	82	67
Sample mean	79	74	66
Sample variance	34	20	32
Sample standard deviation	5.83	4.47	5.66

In the introduction to this chapter we stated that analysis of variance can be used to analyse data obtained from both an observational study and an experimental study.

To provide a common set of terminology for discussing the use of analysis of variance in both types of studies, we introduce the concepts of a response variable, a factor and a treatment.

The two variables in the NCP example are plant location and score on the quality awareness examination. Because the objective is to determine whether the mean examination score is the same for plants located in Ayr, Dusseldorf and Stockholm, examination score is referred to as the dependent or *response variable* and plant location as the independent variable or *factor*. In general, the values of a factor selected for investigation are referred to as levels of the factor or *treatments*. Thus, in the NCP example the three treatments are Ayr, Dusseldorf and Stockholm. These three treatments define the populations of interest in the NCP example. For each treatment or population, the response variable is the examination score.

Assumptions for analysis of variance

Three assumptions are required to use analysis of variance.

1 **For each population, the response variable is normally distributed**. Implication: In the NCP example, the examination scores (response variable) must be normally distributed at each plant.

2 **The variance of the response variable, denoted σ^2, is the same for all of the populations**. Implication: In the NCP example, the variance of examination scores must be the same for all three plants.

3 **The observations must be independent**. Implication: In the NCP example, the examination score for each employee must be independent of the examination score for any other employee.

A conceptual overview

If the means for the three populations are equal, we would expect the three sample means to be close together. In fact, the closer the three sample means are to one another, the more evidence we have for the conclusion that the population means are equal. Alternatively, the more the sample means differ, the more evidence we have for the conclusion that the population means are not equal. In other words, if the variability among the sample means is 'small', it supports H_0; if the variability among the sample means is 'large', it supports H_1.

If the null hypothesis, $H_0: \mu_1 = \mu_2 = \mu_3$, is true, we can use the variability among the sample means to develop an estimate of σ^2. First, note that if the assumptions for analysis of variance are satisfied, each sample will have come from the same normal distribution with mean μ and variance σ^2. Recall from Chapter 7 that the sampling distribution of the sample mean for a simple random sample of size n from a normal population will be normally distributed with mean μ and variance σ^2/n. Figure 13.1 illustrates such a sampling distribution.

Therefore, if the null hypothesis is true, we can think of each of the three sample means, $\bar{x}_1 = 79$, $\bar{x}_2 = 74$, and $\bar{x}_3 = 66$, from Table 13.1 as values drawn at random from the sampling distribution shown in Figure 13.1. In this case, the mean and variance of the three values can be used to estimate the mean and variance of the sampling distribution. When the sample sizes are equal, as in the NCP example, the best estimate of the mean of the sampling distribution of $\bar{X}$ is the mean or average of the sample means. Thus, in the NCP example, an estimate of the mean of the sampling

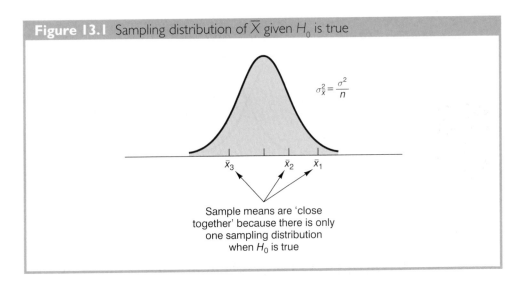

Figure 13.1 Sampling distribution of $\overline{X}$ given H_0 is true

$$\sigma_{\overline{x}}^2 = \frac{\sigma^2}{n}$$

$\overline{X}_3 \quad \overline{X}_2 \quad \overline{X}_1$

Sample means are 'close
together' because there is only
one sampling distribution
when H_0 is true

distribution of $\overline{X}$ is $(79 + 74 + 66)/3 = 73$. We refer to this estimate as the *overall sample mean*. An estimate of the variance of the sampling distribution of $\overline{X}$, $\sigma_{\overline{x}}^2$ is provided by the variance of the three sample means.

$$s_{\overline{x}}^2 = \frac{(79 - 73)^2 + (74 - 73)^2 + (66 - 73)^2}{3 - 1} = \frac{86}{2} = 43$$

Because $\sigma_{\overline{x}}^2 = \sigma^2 / n$, solving for σ^2 gives

$$\sigma^2 = n\sigma_{\overline{x}}^2$$

Hence,

$$\text{Estimate of } \sigma^2 = n \, (\text{Estimate of } \sigma_{\overline{x}}^2) = ns_{\overline{x}}^2 = 6(43) = 258$$

The result, $ns_{\overline{x}}^2 = 258$, is referred to as the *between-treatments* estimate of σ^2.

The between-treatments estimate of σ^2 is based on the assumption that the null hypothesis is true. In this case, each sample comes from the same population, and there is only one sampling distribution of $\overline{X}$. To illustrate what happens when H_0 is false, suppose the population means all differ. Note that because the three samples are from normal populations with different means, they will result in three different sampling distributions. Figure 13.2 shows that in this case, the sample means are not as close together as they were when H_0 was true. Thus, $s_{\overline{x}}^2$ will be larger, causing the between-treatments estimate of σ^2 to be larger. In general, when the population means are not equal, the between-treatments estimate will overestimate the population variance σ^2.

The variation within each of the samples also has an effect on the conclusion we reach in analysis of variance. When a simple random sample is selected from each population, each of the sample variances provides an unbiased estimate of σ^2. Hence, we can combine or pool the individual estimates of σ^2 into one overall estimate. The estimate of σ^2 obtained in this way is called the *pooled* or *within-treatments* estimate of σ^2. Because each sample variance provides an estimate of σ^2 based only on the variation within each sample, the within-treatments estimate of σ^2 is not affected by whether the population means are equal.

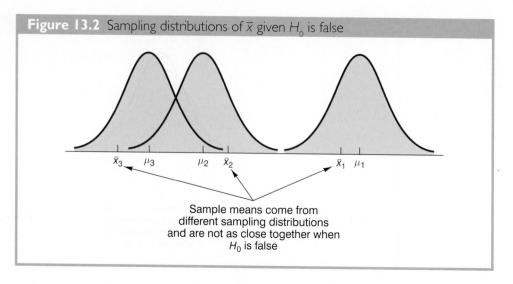

Figure 13.2 Sampling distributions of $\bar{x}$ given H_0 is false

When the sample sizes are equal, the within-treatments estimate of σ^2 can be obtained by computing the average of the individual sample variances. For the NCP example we obtain

$$\text{Within-treatments estimate of } \sigma^2 = \frac{34 + 20 + 32}{3} = \frac{86}{3} = 28.67$$

In the NCP example, the between-treatments estimate of σ^2 (258) is much larger than the within-treatments estimate of σ^2 (28.67). In fact, the ratio of these two estimates is $258/28.67 = 9.00$. Recall, however, that the between-treatments approach provides a good estimate of σ^2 only if the null hypothesis is true; if the null hypothesis is false, the between treatments approach overestimates σ^2. The within-treatments approach provides a good estimate of σ^2 in either case. Thus, if the null hypothesis is true, the two estimates will be similar and their ratio will be close to 1. If the null hypothesis is false, the between-treatments estimate will be larger than the within-treatments estimate, and their ratio will be large. In the next section we will show how large this ratio must be to reject H_0.

In summary, the logic behind ANOVA is based on the development of two independent estimates of the common population variance σ^2. One estimate of σ^2 is based on the variability among the sample means themselves, and the other estimate of σ^2 is based on the variability of the data within each sample. By comparing these two estimates of σ^2, we will be able to determine whether the population means are equal.

13.2 Analysis of variance: testing for the equality of k population means

Analysis of variance can be used to test for the equality of k population means. The general form of the hypotheses tested is

$$H_0: \mu_1 = \mu_2 = \ldots = \mu_k$$
$$H_1: \text{Not all population means are equal}$$

where

$$\mu_j = \text{mean of the } j\text{th population}$$

We assume that a simple random sample of size n_j has been selected from each of the k populations or treatments. For the resulting sample data, let

x_{ij} = value of observation i for treatment j
n_j = number of observations for treatment j
$\bar{x}_j$ = sample mean for treatment j
s_j^2 = sample variance for treatment j
s_j = sample standard deviation for treatment j

The formulae for the sample mean and sample variance for treatment j are as follows.

Testing for the Equality of k Population means sample mean for Treatment j

$$\bar{x}_j = \frac{\sum_{i=1}^{n_j} x_{ij}}{n_j} \tag{13.1}$$

Sample Variance for Treatment j

$$s_j^2 = \frac{\sum_{i=1}^{n_j} (x_{ij} - \bar{x}_j)^2}{n_j - 1} \tag{13.2}$$

The overall sample mean, denoted $\bar{\bar{x}}$, is the sum of all the observations divided by the total number of observations. That is,

Overall Sample Mean

$$\bar{\bar{x}} = \frac{\sum_{j=1}^{k} \sum_{i=1}^{n_j} x_{ij}}{n_T} \tag{13.3}$$

where

$$n_T = n_1 + n_2 + \ldots + n_k \tag{13.4}$$

If the size of each sample is n, $n_T = kn$; in this case equation (13.3) reduces to

$$\bar{\bar{x}} = \frac{\sum_{j=1}^{k} \sum_{i=1}^{n_j} x_{ij}}{kn} = \frac{\sum_{j=1}^{k} \sum_{i=1}^{n_j} x_i / n}{k} = \frac{\sum_{j=1}^{k} \bar{x}_j}{k} \tag{13.5}$$

In other words, whenever the sample sizes are the same, the overall sample mean is just the average of the k sample means.

Because each sample in the NCP example consists of $n = 6$ observations, the overall sample mean can be computed by using equation (13.5). For the data in Table 13.1 we obtained the following result.

$$\bar{\bar{x}} = \frac{79 + 74 + 66}{3} = 73$$

If the null hypothesis is true ($\mu_1 = \mu_2 = \mu_3 = \mu$), the overall sample mean of 73 is the best estimate of the population mean μ.

Between-treatments estimate of population variance

In the preceding section, we introduced the concept of a between-treatments estimate of σ^2 and showed how to compute it when the sample sizes were equal. This estimate of σ^2 is called the *mean square due to treatments* and is denoted MSTR. The general formula for computing MSTR is

$$\text{MSTR} = \frac{\displaystyle\sum_{j=1}^{k} n_j\,(\bar{x}_j - \bar{\bar{x}})^2}{k - 1} \tag{13.6}$$

The numerator in equation (13.6) is called the *sum of squares due to treatments* and is denoted SSTR. The denominator, $k - 1$, represents the degrees of freedom associated with SSTR. Hence, the mean square due to treatments can be computed by the following formula.

Mean square due to treatments

$$\text{MSTR} = \frac{\text{SSTR}}{k - 1} \tag{13.7}$$

where

$$\text{SSTR} = \sum_{j=1}^{k} n_j\,(\bar{x}_j - \bar{\bar{x}})^2 \tag{13.8}$$

If H_0 is true, MSTR provides an unbiased estimate of σ^2. However, if the means of the k populations are not equal, MSTR is not an unbiased estimate of σ^2; in fact, in that case, MSTR should overestimate σ^2.

For the NCP data in Table 13.1, we obtain the following results.

$$\text{SSTR} = \sum_{j=1}^{k} n_j\,(\bar{x}_j - \bar{\bar{x}})^2 = 6(79 - 73)^2 + 6(74 - 73)^2 + 6(66 - 73)^2 = 516$$

$$\text{MSTR} = \frac{\text{SSTR}}{k - 1} = \frac{516}{2} = 258$$

Within-treatments estimate of population variance

Earlier, we introduced the concept of a within-treatments estimate of σ^2 and showed how to compute it when the sample sizes were equal. This estimate of σ^2 is called the *mean square due to error* and is denoted MSE. The general formula for computing MSE is

$$\text{MSE} = \frac{\displaystyle\sum_{j=1}^{k} (n_j - 1)s_j^2}{n_T - k} \tag{13.9}$$

The numerator in equation (13.9) is called the *sum of squares due to error* and is denoted SSE. The denominator of MSE is referred to as the degrees of freedom associated with SSE. Hence, the formula for MSE can also be stated as follows.

Mean square due to error

$$MSE = \frac{SSE}{n_T - k} \tag{13.10}$$

where

$$SSE = \sum_{j=1}^{k} (n_j - 1)s_j^2 \tag{13.11}$$

Note that MSE is based on the variation within each of the treatments; it is not influenced by whether the null hypothesis is true. Thus, MSE always provides an unbiased estimate of σ^2.

For the NCP data in Table 13.1 we obtain the following results.

$$SSE = \sum_{j=1}^{k} (n_j - 1)s_j^2 = (6-1)34 + (6-1)20 + (6-1)32 = 430$$

$$MSE = \frac{SSE}{n_T - k} = \frac{430}{18 - 3} = \frac{430}{15} = 28.67$$

Comparing the variance estimates: the *F* test

If the null hypothesis is true, MSTR and MSE provide two independent, unbiased estimates of σ^2. Based on the material covered in Chapter 11 we know that for normal populations, the sampling distribution of the ratio of two independent estimates of σ^2 follows an *F* distribution. Hence, if the null hypothesis is true and the ANOVA assumptions are valid, the sampling distribution of MSTR/MSE is an *F* distribution with numerator degrees of freedom equal to $k - 1$ and denominator degrees of freedom equal to $n_T - k$. In other words, if the null hypothesis is true, the value of MSTR/MSE should appear to have been selected from this *F* distribution.

However, if the null hypothesis is false the value of MSTR/MSE will be inflated because MSTR overestimates σ^2. Hence, we will reject H_0 if the resulting value of MSTR/MSE appears to be too large to have been selected from an *F* distribution with $k - 1$ numerator degrees of freedom and $n_T - k$ denominator degrees of freedom.

Because the decision to reject H_0 is based on the value of MSTR/MSE, the test statistic used to test for the equality of k population means is as follows.

Test statistic for the equality of k population means

$$F = \frac{MSTR}{MSE} \tag{13.12}$$

The test statistic follows an *F* distribution with $k - 1$ degrees of freedom in the numerator and $n_T - k$ degrees of freedom in the denominator.

Returning to the National Computer Products example we use a level of significance $\alpha = 0.05$ to conduct the hypothesis test. The value of the test statistic is

$$F = \frac{MSTR}{MSE} = \frac{258}{28.67} = 9$$

The numerator degrees of freedom is $k - 1 = 3 - 1 = 2$ and the denominator degrees of freedom is $n_T - k = 18 - 3 = 15$. Because we will only reject the null hypothesis for large values of the test statistic, the p-value is the upper tail area of the F distribution to the right of the test statistic $F = 9$. Figure 13.3 shows the sampling distribution of $F = MSTR/MSE$, the value of the test statistic, and the upper tail area that is the p-value for the hypothesis test.

From Table 4 of Appendix B we find the following areas in the upper tail of an F distribution with two numerator degrees of freedom and 15 denominator degrees of freedom.

Area in upper tail	0.10	0.05	0.025	0.01
F value (df$_1$ = 2, df$_2$ = 15)	2.70	3.68	4.77	6.36

$F = 9$

Because $F = 9$ is greater than 6.36, the area in the upper tail at $F = 9$ is less than 0.01. Thus, the p-value is less than 0.01. With a p-value $\leq \alpha = 0.05$, H_0 is rejected. The test provides sufficient evidence to conclude that the means of the three populations are not equal. In other words, analysis of variance supports the conclusion that the population mean examination scores at the three NCP plants are not equal.

Because the F table only provides values for upper tail areas of 0.10, 0.05, 0.025 and 0.01, we cannot determine the exact p-value directly from the table. MINITAB, EXCEL or PASW provide the p-value as part of the standard ANOVA output. The software section at the end of the chapter shows the procedures that can be used. For the NCP example, the exact p-value corresponding to the test statistic $F = 9$ is 0.003.

As with other hypothesis testing procedures, the critical value approach may also be used. With $\alpha = 0.05$, the critical F value occurs with an area of 0.05 in the upper tail of an F distribution with 2 and 15 degrees of freedom. From the F distribution table, we find $F_{0.05} = 3.68$. Hence, the appropriate upper tail rejection rule for the NCP example is

Reject H_0 if $F \geq 3.68$

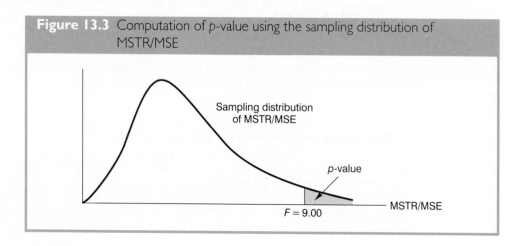

Figure 13.3 Computation of p-value using the sampling distribution of MSTR/MSE

With $F = 9$, we reject H_0 and conclude that the means of the three populations are not equal. A summary of the overall procedure for testing for the equality of k population means follows.

Test for the equality of k population means

$$H_0: \mu_1 = \mu_2 = \ldots = \mu_k$$
$$H_1: \text{Not all population means are equal}$$

Test statistic

$$F = \frac{MSTR}{MSE}$$

Rejection rule

p-value approach: Reject H_0 if p-value $\leq \alpha$

Critical value approach: Reject H_0 if $F \geq F_\alpha$

where the value of F_α is based on an F distribution with $k - 1$ numerator degrees of freedom and $n_T - k$ denominator degrees of freedom

---

ANOVA table

The results of the preceding calculations can be displayed conveniently in a table referred to as the analysis of variance or **ANOVA table**. Table 13.2 is the analysis of variance table for the National Computer Products example. The sum of squares associated with the source of variation referred to as 'total' is called the total sum of squares (SST). Note that the results for the NCP example suggest that SST = SSTR + SSE, and that the degrees of freedom associated with this total sum of squares is the sum of the degrees of freedom associated with the between-treatments estimate of σ^2 and the within-treatments estimate of σ^2.

We point out that SST divided by its degrees of freedom $n_T - 1$ is nothing more than the overall sample variance that would be obtained if we treated the entire set of 18 observations as one data set. With the entire data set as one sample, the formula for computing the total sum of squares, SST, is

---

Total sum of squares

$$SST = \sum_{j=1}^{k} \sum_{i=1}^{n_j} (x_{ij} - \bar{\bar{x}})^2 \qquad (13.13)$$

It can be shown that the results we observed for the analysis of variance table for the NCP example also apply to other problems. That is,

---

Partitioning of sum of squares

$$SST = SSTR + SSE \qquad (13.14)$$

Table 13.2 Analysis of variance table for the NCP example

Source of variation	Degrees of freedom	Sum of squares	Mean square	F
Treatments	2	516	258.00	9.00
Error	15	430	28.67	
Total	17	946		

In other words, SST can be partitioned into two sums of squares: the sum of squares due to treatments and the sum of squares due to error. Note also that the degrees of freedom corresponding to SST, $n_T - 1$, can be partitioned into the degrees of freedom corresponding to SSTR, $k - 1$, and the degrees of freedom corresponding to SSE, $n_T - k$. The analysis of variance can be viewed as the process of **partitioning** the total sum of squares and the degrees of freedom into their corresponding sources: treatments and error. Dividing the sum of squares by the appropriate degrees of freedom provides the variance estimates and the F value used to test the hypothesis of equal population means.

Computer results for analysis of variance

Because of the widespread availability of statistical computer packages, analysis of variance computations with large sample sizes or a large number of populations can be performed easily. In Figure 13.4 we show output for the NCP example obtained from the MINITAB computer package. The first part of the computer output contains the familiar ANOVA table format. Comparing Figure 13.4 with Table 13.2, we see that the same information is available, although some of the headings are slightly different. The heading Source is used for the source of variation column, and Factor identifies the treatments row. A p-value is provided for the F test. Thus, at the $\alpha = 0.05$ level of significance, we reject H_0 because the p value $= 0.003 < \alpha = 0.05$.

Figure 13.4 MINITAB output for the NCP analysis of variance

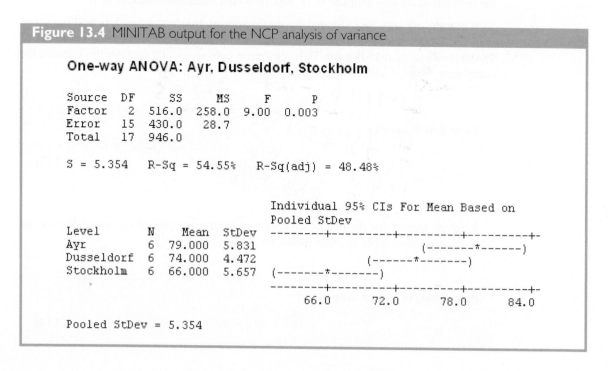

```
One-way ANOVA: Ayr, Dusseldorf, Stockholm

Source   DF     SS      MS     F      P
Factor    2   516.0   258.0  9.00  0.003
Error    15   430.0    28.7
Total    17   946.0

S = 5.354    R-Sq = 54.55%    R-Sq(adj) = 48.48%


                                  Individual 95% CIs For Mean Based on
                                  Pooled StDev
Level         N    Mean   StDev  --------+---------+---------+---------+-
Ayr           6  79.000   5.831                          (-------*------)
Dusseldorf    6  74.000   4.472                  (------*-------)
Stockholm     6  66.000   5.657  (-------*-------)
                                  --------+---------+---------+---------+-
                                      66.0      72.0      78.0      84.0

Pooled StDev = 5.354
```

Note that following the ANOVA table the computer output contains the respective sample sizes, the sample means, and the standard deviations. In addition, MINITAB provides a figure that shows individual 95 per cent confidence interval estimates of each population mean.

In developing these confidence interval estimates, MINITAB uses MSE as the estimate of σ^2. Thus, the square root of MSE provides the best estimate of the population standard deviation σ. This estimate of σ on the computer output is Pooled StDev; it is equal to 5.354. To provide an illustration of how these interval estimates are developed, we will compute a 95 per cent confidence interval estimate of the population mean for the Ayr plant.

From our study of interval estimation in Chapter 8, we know that the general form of an interval estimate of a population mean is

$$\bar{x} \pm t_{\alpha/2} \frac{s}{\sqrt{n}} \qquad (13.15)$$

where s is the estimate of the population standard deviation σ. In the analysis of variance the best estimate of σ is provided by the square root of MSE or the Pooled StDev, therefore we use a value of 5.354 for s in expression (13.15). The degrees of freedom for the t value is 15, the degrees of freedom associated with the within-treatments estimate of σ^2. Hence, with $t_{0.025} = 2.131$ we obtain

$$79 \pm 2.131 \frac{5.354}{\sqrt{6}} = 79 \pm 4.66$$

Therefore, the individual 95 per cent confidence interval for the Ayr plant goes from $79 - 4.66 = 74.34$ to $79 + 4.66 = 83.66$. Because the sample sizes are equal for the NCP example, the individual confidence intervals for the Dusseldorf and Stockholm plants are also constructed by adding and subtracting 4.66 from each sample mean. Thus, in the figure provided by MINITAB we see that the widths of the confidence intervals are the same.

Exercises

Methods

1 Five observations were selected from each of three populations. The data obtained follow.

Observation	Sample 1	Sample 2	Sample 3
1	32	44	33
2	30	43	36
3	30	44	35
4	26	46	36
5	32	48	40
Sample mean	30	45	36
Sample variance	6.00	4.00	6.50

a. Compute the between-treatments estimate of σ^2.
b. Compute the within-treatments estimate of σ^2.

c. At the $\alpha = 0.05$ level of significance, can we reject the null hypothesis that the means of the three populations are equal?

d. Set up the ANOVA table for this problem.

2   Four observations were selected from each of three populations. The data obtained follow.

Observation	Sample 1	Sample 2	Sample 3
1	165	174	169
2	149	164	154
3	156	180	161
4	142	158	148
Sample mean	153	169	158
Sample variance	96.67	97.33	82.00

a. Compute the between-treatments estimate of σ^2.

b. Compute the within-treatments estimate of σ^2.

c. At the $\alpha = 0.05$ level of significance, can we reject the null hypothesis that the three population means are equal? Explain.

d. Set up the ANOVA table for this problem.

3   Samples were selected from three populations. The data obtained follow.

	Sample 1	Sample 2	Sample 3
	93	77	88
	98	87	75
	107	84	73
	102	95	84
	85	75	82
$\bar{x}_j$	100	85	79
s_j^2	35.33	35.60	43.50

a. Compute the between-treatments estimate of σ^2.

b. Compute the within-treatments estimate of σ^2.

c. At the $\alpha = 0.05$ level of significance, can we reject the null hypothesis that the three population means are equal? Explain.

d. Set up the ANOVA table for this problem.

4   A random sample of 16 observations was selected from each of four populations. A portion of the ANOVA table follows.

Source of variation	Degrees of freedom	Sum of squares	Mean square	F
Treatments			400	
Error				
Total		1500		

    a. Provide the missing entries for the ANOVA table.

    b. At the $\alpha = 0.05$ level of significance, can we reject the null hypothesis that the means of the four populations are equal?

5 Random samples of 25 observations were selected from each of three populations. For these data, SSTR = 120 and SSE = 216.

    a. Set up the ANOVA table for this problem.

    b. At the $\alpha = 0.05$ level of significance, can we reject the null hypothesis that the three population means are equal?

Applications

6 To test whether the mean time needed to mix a batch of material is the same for machines produced by three manufacturers, the Jacobs Chemical Company obtained the following data on the time (in minutes) needed to mix the material. Use these data to test whether the population mean times for mixing a batch of material differ for the three manufacturers. Use $\alpha = 0.05$.

Manufacturer		
1	2	3
20	28	20
26	26	19
24	31	23
22	27	22

7 Managers at all levels of an organization need adequate information to perform their respective tasks. One study investigated the effect the source has on the dissemination of information. In this particular study the sources of information were a superior, a peer and a subordinate. In each case, a measure of dissemination was obtained, with higher values indicating greater dissemination of information. Use $\alpha = 0.05$ and the following data to test whether the source of information significantly affects dissemination. What is your conclusion, and what does it suggest about the use and dissemination of information?

Superior	Peer	Subordinate
8	6	6
5	6	5
4	7	7
6	5	4
6	3	3
7	4	5
5	7	7
5	6	5

8 A study investigated the perception of corporate ethical values among individuals specializing in marketing. Use $\alpha = 0.05$ and the following data (higher scores indicate higher ethical values) to test for significant differences in perception among the three groups.

Marketing managers	Marketing research	Advertising
6	5	6
5	5	7
4	4	6
5	4	5
6	5	6
4	4	6

9 A study reported in the *Journal of Small Business Management* concluded that self-employed individuals experience higher job stress than individuals who are not self-employed. In this study job stress was assessed with a 15-item scale designed to measure various aspects of ambiguity and role conflict. Ratings for each of the 15 items were made using a scale with 1–5 response options ranging from strong agreement to strong disagreement. The sum of the ratings for the 15 items for each individual surveyed is between 15 and 75, with higher values indicating a higher degree of job stress. Suppose that a similar approach, using a 20-item scale with 1–5 response options, was used to measure the job stress of individuals for 15 randomly selected property agents, 15 architects and 15 stockbrokers. The results obtained follow.

Property agent	Architect	Stockbroker
81	43	65
48	63	48
68	60	57
69	52	91
54	54	70
62	77	67
76	68	83
56	57	75
61	61	53
65	80	71
64	50	54
69	37	72
83	73	65
85	84	58
75	58	58

Use $\alpha = 0.05$ to test for any significant difference in job stress among the three professions.

10 *Condé Nast Traveler* conducts an annual survey in which readers rate their favourite cruise ships. Ratings are provided for small ships (carrying up to 500 passengers), medium ships (carrying 500 to 1500 passengers) and large ships (carrying a minimum of 1500 passengers). The following data show the service ratings for eight randomly selected small ships, eight randomly selected medium ships and eight randomly selected large ships. All ships are rated on a 100-point scale, with higher values indicating better service (*Condé Nast Traveler*, February 2003).

Small ships		Medium ships		Large ships	
Name	Rating	Name	Rating	Name	Rating
Hanseactic	90.5	Amsterdam	91.1	Century	89.2
Mississippi Queen	78.2	Crystal Symphony	98.9	Disney Wonder	90.2
Philae	92.3	Maasdam	94.2	Enchantment of the Seas	85.9
Royal Clipper	95.7	Noordam	84.3	Grand Princess	84.2
Seabourn Pride	94.1	Royal Princess	84.8	Infinity	90.2
Seabourn Spirit	100.0	Ryndam	89.2	Legend of the Seas	80.6
Silver Cloud	91.8	Statendam	86.4	Paradise	75.8
Silver Wind	95.0	Veendam	88.3	Sun Princess	82.3

Use $\alpha = 0.05$ to test for any significant difference in the mean service ratings among the three sizes of cruise ships.

13.3 Multiple comparison procedures

When we use analysis of variance to test whether the means of k populations are equal, rejection of the null hypothesis allows us to conclude only that the population means are *not all equal.* In some cases we will want to go a step further and determine where the differences among means occur. The purpose of this section is to introduce two **multiple comparison procedures** that can be used to conduct statistical comparisons between pairs of population means.

Fisher's LSD

Suppose that analysis of variance provides statistical evidence to reject the null hypothesis of equal population means. In this case, Fisher's least significant difference (LSD) procedure can be used to determine where the differences occur. To illustrate the use of Fisher's LSD procedure in making pairwise comparisons of population means, recall the NCP example introduced in Section 13.1. Using analysis of variance, we concluded that the population mean examination scores are not the same at the three plants. In this case, the follow-up question is: We believe the plants differ, but where do the differences occur? That is, do the means of populations 1 and 2 differ? Or those of populations 1 and 3? Or those of populations 2 and 3?

In Chapter 10 we presented a statistical procedure for testing the hypothesis that the means of two populations are equal. With a slight modification in how we estimate the population variance, Fisher's LSD procedure is based on the t test statistic presented for the two-population case. The following table summarizes Fisher's LSD procedure.

Fisher's LSD Procedure

$$H_0: \mu_i = \mu_j$$
$$H_1: \mu_i \neq \mu_j$$

Test statistic for Fisher's LSD procedure

$$t = \frac{\bar{x}_i - \bar{x}_j}{\sqrt{MSE\left(\dfrac{1}{n_i} + \dfrac{1}{n_j}\right)}} \qquad (13.16)$$

Rejection rule

p-value approach: Reject H_0 if p-value $\leq \alpha$

Critical value approach: Reject H_0 if $t \leq -t_{\alpha/2}$ or $t \geq t_{\alpha/2}$

where the value of $t_{\alpha/2}$ is based on a t distribution with $n_T - k$ degrees of freedom.

Let us now apply this procedure to determine whether there is a significant difference between the means of population 1 (Ayr) and population 2 (Dusseldorf) at the $\alpha = 0.05$ level of significance. Table 13.1 shows that the sample mean is 79 for the Ayr plant and 74 for the Dusseldorf plant. Table 13.2 shows that the value of MSE is 28.67; it is the estimate of σ^2 and is based on 15 degrees of freedom. For the NCP data the value of the test statistic is

$$t = \frac{79 - 74}{\sqrt{28.67\left(\dfrac{1}{6} + \dfrac{1}{6}\right)}} = 1.62$$

The t distribution table (Table 2 in Appendix B) shows that with 15 degrees of freedom $t = 1.341$ for an area of 0.10 in the upper tail and $t = 1.753$ for an area of 0.05 in the upper tail. Because the test statistic $t = 1.62$ is between 1.341 and 1.753, we know that the area in the upper tail must be between 0.05 and 0.10. Because this test is a two-tailed test, we double these values to conclude that the p-value is between 0.10 and 0.20. MINITAB or EXCEL can be used to show that the p-value corresponding to $t = 1.62$ is 0.1261. Because the p-value is greater than $\alpha = 0.05$, we cannot reject the null hypothesis. Hence, we cannot conclude that the population mean score at the Ayr plant is different from the population mean score at the Dusseldorf plant.

Many practitioners find it easier to determine how large the difference between the sample means must be to reject H_0. In this case the test statistic is $\bar{x}_i - \bar{x}_j$ and the test is conducted by the following procedure.

Fisher's LSD procedure based on the test statistic $\bar{x}_i - \bar{x}_j$

$$H_0: \mu_i - \mu_j$$
$$H_1: \mu_i \neq \mu_j$$

Test statistic

$$\bar{x}_i - \bar{x}_j$$

Rejection rule at a level of significance α

$$\text{Reject } H_0 \text{ if } \quad |\bar{x}_i - \bar{x}_j| > \text{LSD}$$

where

$$\text{LSD} = t_{\alpha/2}\sqrt{MSE\left(\frac{1}{n_i} + \frac{1}{n_j}\right)} \qquad (13.17)$$

For the NCP example the value of LSD is

$$\text{LSD} = 2.131\sqrt{28.67\left(\frac{1}{6} + \frac{1}{6}\right)} = 6.59$$

Note that when the sample sizes are equal, only one value for LSD is computed. In such cases we can simply compare the magnitude of the difference between any two sample means with the value of LSD. For example, the difference between the sample means for population 1 (Ayr) and population 3 (Stockholm) is $79 - 66 = 13$. This difference is greater than 6.59, which means we can reject the null hypothesis that the population mean examination score for the Ayr plant is equal to the population mean score for the Stockholm plant.

Similarly, with the difference between the sample means for populations 2 and 3 of $74 - 66 = 8 > 6.59$, we can also reject the hypothesis that the population mean examination score for the Dusseldorf plant is equal to the population mean examination score for the Stockholm plant. In effect, our conclusion is that the Ayr and Dusseldorf plants both differ from the Stockholm plant.

Fisher's LSD can also be used to develop a confidence interval estimate of the difference between the means of two populations. The general procedure follows.

Confidence interval estimate of the difference between two Population means using Fisher's LSD procedure

$$\bar{x}_i - \bar{x}_j \pm \text{LSD} \qquad (13.18)$$

where

$$\text{LSD} = t_{\alpha/2}\sqrt{MSE\left(\frac{1}{n_i} + \frac{1}{n_j}\right)} \qquad (13.19)$$

and $t_{\alpha/2}$ is based on a t distribution with $n_T - k$ degrees of freedom. If the confidence interval in expression (13.18) includes the value zero, we cannot reject the hypothesis that the two population means are equal. However, if the confidence interval does not include the value zero, we conclude that there is a difference between the population means. For the NCP example, recall that $\text{LSD} = 6.59$ (corresponding to $t_{0.025} = 2.131$). Thus, a 95 per cent confidence interval estimate of the difference between the means of populations 1 and 2 is $79 - 74 \pm 6.59 = 5 \pm 6.59 = -1.59$ to 11.59; because this interval includes zero, we cannot reject the hypothesis that the two population means are equal.

Type I error rates

We began the discussion of Fisher's LSD procedure with the premise that analysis of variance gave us statistical evidence to reject the null hypothesis of equal population means.

We showed how Fisher's LSD procedure can be used in such cases to determine where the differences occur. Technically, it is referred to as a *protected* or *restricted* LSD test because it is employed only if we first find a significant F value by using analysis of variance.

To see why this distinction is important in multiple comparison tests, we need to explain the difference between a *comparisonwise* Type I error rate and an *experimentwise* Type I error rate.

In the NCP example we used Fisher's LSD procedure to make three pairwise comparisons.

Test 1	Test 2	Test 3
$H_0: \mu_1 = \mu_2$	$H_0: \mu_1 = \mu_3$	$H_0: \mu_2 = \mu_3$
$H_1: \mu_1 \neq \mu_2$	$H_1: \mu_1 \neq \mu_3$	$H_1: \mu_2 \neq \mu_3$

In each case, we used a level of significance of $\alpha = 0.05$. Therefore, for each test, if the null hypothesis is true, the probability that we will make a Type I error is $\alpha = 0.05$; hence, the probability that we will not make a Type I error on each test is $1 - 0.05 = 0.95$. In discussing multiple comparison procedures we refer to this probability of a Type I error ($\alpha = 0.05$) as the **comparisonwise Type I error rate**; comparisonwise Type I error rates indicate the level of significance associated with a single pairwise comparison.

Let us now consider a slightly different question. What is the probability that in making three pairwise comparisons, we will commit a Type I error on at least one of the three tests? To answer this question, note that the probability that we will not make a Type I error on any of the three tests is $(0.95)(0.95)(0.95) = 0.8574$.* Therefore, the probability of making at least one Type I error is $1 - 0.8574 = 0.1426$. Thus, when we use Fisher's LSD procedure to make all three pairwise comparisons, the Type I error rate associated with this approach is not 0.05, but actually 0.1426; we refer to this error rate as the *overall* or **experimentwise Type I error rate**. To avoid confusion, we denote the experimentwise Type I error rate as α_{EW}.

The experimentwise Type I error rate gets larger for problems with more populations. For example, a problem with five populations has ten possible pairwise comparisons. If we tested all possible pairwise comparisons by using Fisher's LSD with a comparisonwise error rate of $\alpha = 0.05$, the experimentwise Type I error rate would be $1 - (1 - 0.05)^{10} = 0.40$. In such cases, practitioners look to alternatives that provide better control over the experimentwise error rate.

One alternative for controlling the overall experimentwise error rate, referred to as the *Bonferroni adjustment*, involves using a smaller comparisonwise error rate for each test. For example, if we want to test C pairwise comparisons and want the maximum probability of making a Type I error for the overall experiment to be α_{EW}, we simply use a comparisonwise error rate equal to α_{EW}/C. In the NCP example, if we want to use Fisher's LSD procedure to test all three pairwise comparisons with a maximum experimentwise error rate of $\alpha_{EW} = 0.05$, we set the comparisonwise error rate to be $\alpha = 0.05/3 = 0.017$. For a problem with five populations and ten possible pairwise comparisons, the Bonferroni adjustment would suggest a comparisonwise error rate of $0.05/10 = 0.005$. Recall from our discussion of hypothesis testing in Chapter 9 that for a fixed sample size, any decrease in the probability of making a Type I error will result in an increase in the probability of making a Type II error, which corresponds to accepting the hypothesis that the two population means are

---

*The assumption is that the three tests are independent, and hence the joint probability of the three events can be obtained by simply multiplying the individual probabilities. In fact, the three tests are not independent because MSE is used in each test; therefore, the error involved is even greater than that shown.

equal when in fact they are not equal. As a result, many practitioners are reluctant to perform individual tests with a low comparisonwise Type I error rate because of the increased risk of making a Type II error.

Several other procedures, such as Tukey's procedure and Duncan's multiple range test, have been developed to help in such situations. However, there is considerable controversy in the statistical community as to which procedure is 'best'. The truth is that no one procedure is best for all types of problems.

Exercises

Methods

11 In exercise 1, five observations were selected from each of three populations. For these data, $\bar{x}_1 = 30, \bar{x}_2 = 45, \bar{x}_3 = 36$ and MSE = 5.5. At the $\alpha = 0.05$ level of significance, the null hypothesis of equal population means was rejected. In the following calculations, use $\alpha = 0.05$.

 a. Use Fisher's LSD procedure to test whether there is a significant difference between the means of populations 1 and 2, populations 1 and 3, and populations 2 and 3.
 b. Use Fisher's LSD procedure to develop a 95 per cent confidence interval estimate of the difference between the means of populations 1 and 2.

12 Four observations were selected from each of three populations. The data obtained are shown. In the following calculations, use $\alpha = 0.05$.

	Sample 1	Sample 2	Sample 3
	63	82	69
	47	72	54
	54	88	61
	40	66	48
$\bar{x}_j$	51	77	58
s_j^2	96.67	97.34	81.99

 a. Use analysis of variance to test for a significant difference among the means of the three populations.
 b. Use Fisher's LSD procedure to see which means are different.

Applications

13 Refer to exercise 6. At the $\alpha = 0.05$ level of significance, use Fisher's LSD procedure to test for the equality of the means for manufacturers 1 and 3. What conclusion can you draw after carrying out this test?

14 Refer to exercise 13. Use Fisher's LSD procedure to develop a 95 per cent confidence interval estimate of the difference between the means of population 1 and population 2.

15 Refer to exercise 8. At the $\alpha = 0.05$ level of significance, we can conclude that there are differences in the perceptions for marketing managers, marketing research specialists and advertizing specialists. Use the procedures in this section to determine where the differences occur. Use $\alpha = 0.05$.

16 To test for any significant difference in the number of hours between breakdowns for four machines, the following data were obtained.

Machine 1	Machine 2	Machine 3	Machine 4
6.4	8.7	11.1	9.9
7.8	7.4	10.3	12.8
5.3	9.4	9.7	12.1
7.4	10.1	10.3	10.8
8.4	9.2	9.2	11.3
7.3	9.8	8.8	11.5

a. At the $\alpha = 0.05$ level of significance, what is the difference, if any, in the population mean times among the four machines?

b. Use Fisher's LSD procedure to test for the equality of the means for machines 2 and 4. Use a 0.05 level of significance.

17 Refer to exercise 16. Use the Bonferroni adjustment to test for a significant difference between all pairs of means. Assume that a maximum overall experimentwise error rate of 0.05 is desired.

18 Refer to exercise 10. At the 0.05 level of significance, we can conclude that there are differences between the mean service ratings of small ships, medium ships, and large ships. Use the procedures in this section to determine where the differences occur. Use $\alpha = 0.05$.

13.4 An introduction to experimental design

Statistical studies can be classified as being either experimental or observational. In an experimental study, variables of interest are identified. Then, one or more factors in the study are controlled so that data can be obtained about how the factors influence the variables. In *observational* or *non-experimental* studies, no attempt is made to control the factors. A survey (see Chapter 22) is perhaps the most common type of observational study.

The NCP example that we used to introduce analysis of variance is an illustration of an observational statistical study. To measure how much NCP employees knew about total quality management, a random sample of six employees was selected from each of NCP's three plants and given a quality-awareness examination. The examination scores for these employees were then analyzed by analysis of variance to test the hypothesis that the population mean examination scores were equal for the three plants.

As an example of an experimental statistical study, let us consider the problem facing the Chemietech company. Chemietech developed a new filtration system for municipal water supplies.

The components for the new filtration system will be purchased from several suppliers, and Chemietech will assemble the components at its plant in North Saxony. The industrial engineering group is responsible for determining the best assembly method for the new filtration system. After considering a variety of possible approaches, the group narrows the alternatives to three: method A, method B and method C. These methods differ in the sequence of steps used to assemble the product. Managers at Chemietech want to determine which assembly method can produce the greatest number of filtration systems per week.

In the Chemietech experiment, assembly method is the independent variable or **factor**. Because three assembly methods correspond to this factor, we say that three treatments are associated with this experiment; each **treatment** corresponds to each of the three assembly methods. The Chemietech problem is an example of a **single-factor experiment** involving a qualitative factor (method of assembly). Other experiments may consist of multiple factors; some factors may be qualitative and some may be quantitative.

The three assembly methods or treatments define the three populations of interest for the Chemietech experiment. One population is all Chemietech employees who use assembly method A, another is those who use method B and the third is those who use method C. Note that for each population the dependent or response variable is the number of filtration systems assembled per week, and the primary statistical objective of the experiment is to determine whether the mean number of units produced per week is the same for all three populations.

Suppose a random sample of three employees is selected from all assembly workers at the Chemietech production facility. In experimental design terminology, the three randomly selected workers are the **experimental units**. The experimental design that we will use for the Chemietech problem is called a **completely randomized design**. This type of design requires that each of the three assembly methods or treatments be assigned randomly to one of the experimental units or workers. For example, method A might be randomly assigned to the second worker, method B to the first worker and method C to the third worker. The concept of *randomization*, as illustrated in this example, is an important principle of all experimental designs.

Note that this experiment would result in only one measurement or number of units assembled for each treatment. To obtain additional data for each assembly method, we must repeat or replicate the basic experimental process. Suppose, for example, that instead of selecting just three workers at random we selected 15 workers and then randomly assigned each of the three treatments to five of the workers. Because each method of assembly is assigned to five workers, we say that five replicates have been obtained. The process of *replication* is another important principle of experimental design. Figure 13.5 shows the completely randomized design for the Chemietech experiment.

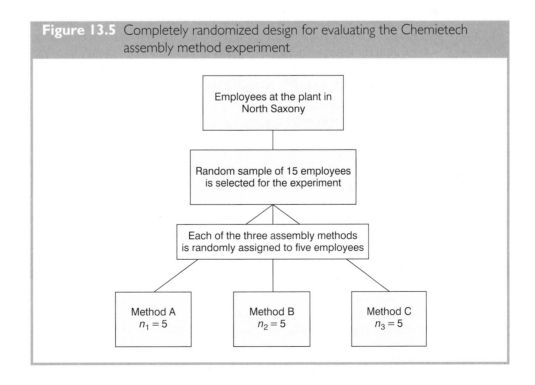

Figure 13.5 Completely randomized design for evaluating the Chemietech assembly method experiment

Data collection

Once we are satisfied with the experimental design, we proceed by collecting and analysing the data. In the Chemietech case, the employees would be instructed in how to perform the assembly method assigned to them and then would begin assembling the new filtration systems using that method. After this assignment and training, the number of units assembled by each employee during one week is as shown in Table 13.3. The sample mean number of units produced with each of the three assembly methods is reported in the following table.

Assembly method	Mean number produced
A	62
B	66
C	52

From these data, method B appears to result in higher production rates than either of the other methods.

The real issue is whether the three sample means observed are different enough for us to conclude that the means of the populations corresponding to the three methods of assembly are different. To write this question in statistical terms, we introduce the following notation.

μ_1 = mean number of units produced per week for method A
μ_2 = mean number of units produced per week for method B
μ_3 = mean number of units produced per week for method C

Although we will never know the actual values of μ_1, μ_2 and μ_3, we want to use the sample means to test the following hypotheses.

$$H_0: \mu_1 = \mu_2 = \mu_3$$
$$H_1: \text{Not all population means are equal}$$

The problem we face in analysing data from a completely randomized experimental design is the same problem we faced when we first introduced analysis of variance as a method for testing whether the means of more than two populations are equal. In the next section we will show how analysis of variance is applied in problem situations such as the Chemietech assembly method experiment.

Table 13.3 Number of units produced by 15 workers

| | | Method | |
Observation	A	B	C
1	58	58	48
2	64	69	57
3	55	71	59
4	66	64	47
5	67	68	49
Sample mean	62	66	52
Sample variance	27.5	26.5	31.0
Sample standard deviation	5.24	5.15	5.57

CHEMIETECH

13.5 Completely randomized designs

The hypotheses we want to test when analysing the data from a completely randomized design are exactly the same as the general form of the hypotheses we presented in Section 13.2.

$$H_0: \mu_1 = \mu_2 = \ldots = \mu_k$$
$$H_1: \text{Not all means are equal}$$

Hence, to test for the equality of means in situations where the data are collected in a completely randomized experimental design, we can use analysis of variance as introduced in Sections 13.1 and 13.2. Recall that analysis of variance requires the calculation of two independent estimates of the population variance σ^2.

Between-treatments estimate of population variance

The between-treatments estimate of σ^2 is referred to as the *mean square due to treatments* and is denoted MSTR. The formula for computing MSTR follows:

Completely randomized designs
Mean square due to treatments

$$MSTR = \frac{\sum_{j=1}^{k} n_j (\bar{x}_j - \bar{\bar{x}})^2}{k-1} \qquad (13.20)$$

The numerator in equation (13.20) is called the *sum of squares between* or *sum of squares due to treatments* and is denoted SSTR. The denominator $k - 1$ represents the degrees of freedom associated with SSTR.

For the Chemietech data in Table 13.3, we obtain the following results (note: $\bar{\bar{x}} = 60$).

$$SSTR = \sum_{j=1}^{k} n_j (\bar{x}_j - \bar{\bar{x}})^2 = 5(62 - 60)^2 + 5(66 - 60)^2 + 5(52 - 60)^2 = 520$$

$$MSTR = \frac{SSTR}{k-1} = \frac{520}{3-1} = 260$$

Within-treatments estimate of population variance

The within-treatments estimate of σ^2 is referred to as the *mean square due to error* and is denoted MSE. The formula for computing MSE follows.

Mean square due to error

$$MSE = \frac{\sum_{j=1}^{k} (n_j - 1)s_j^2}{n_T - k} \qquad (13.21)$$

The numerator in equation (13.21) is called the *sum of squares within* or *sum of squares due to error* and is denoted SSE. The denominator of MSE is referred to as the degrees of freedom associated with SSE.

For the Chemietech data in Table 13.3, we obtain the following results.

$$SSE = \sum_{j=1}^{k} (n_j - 1)s_j^2 = 4(27.5) + 4(26.5) + 4(31) = 340$$

$$MSE = \frac{SSE}{n_T - k} = \frac{340}{15 - 3} = 28.33$$

Comparing the variance estimates: the *F* test

If the null hypothesis is true and the ANOVA assumptions are valid, the sampling distribution of MSTR/MSE is an F distribution with numerator degrees of freedom equal to $k - 1$ and denominator degrees of freedom equal to $n_T - k$. Recall also that if the means of the k populations are not equal, the value of MSTR/MSE will be inflated because MSTR overestimates σ^2. Hence we will reject H_0 if the resulting value of MSTR/MSE appears to be too large to have been selected at random from an F distribution with degrees of freedom $k - 1$ in the numerator and $n_T - k$ in the denominator.

Let us return to the Chemietech problem and use a level of significance $\alpha = 0.05$ to conduct the hypothesis test. The value of the test statistic is

$$F = \frac{MSTR}{MSE} = \frac{260}{28.33} = 9.18$$

The numerator degrees of freedom is $k - 1 - 3 - 1 = 2$ and the denominator degrees of freedom is $n_T - k = 15 - 3 = 12$. Because we will only reject the null hypothesis for large values of the test statistic, the p-value is the area under the F distribution to the right of $F = 9.18$. From Table 4 of Appendix B we find that the F value with an area of 0.01 in the upper tail is 6.93. Because the area in the upper tail for an F value of 9.18 must be less than 0.01, the p-value for the Chemietech hypothesis test is less than 0.01. Alternatively, we can use MINITAB, PASW or EXCEL to show that the exact p-value corresponding to $F = 9.18$ is 0.0038. With p-value $\leq \alpha = 0.05$, H_0 is rejected. The test gives us sufficient evidence to conclude that not all the population means are equal.

ANOVA table

We can now write the result that shows how the total sum of squares, SST, is partitioned.

$$SST = SSTR + SSE \tag{13.22}$$

This result also holds true for the degrees of freedom associated with each of these sums of squares; that is, the total degrees of freedom is the sum of the degrees of freedom associated with SSTR and SSE. The general form of the ANOVA table for a completely randomized design is shown in Table 13.4; Table 13.5 is the corresponding ANOVA table for the Chemietech problem.

Table 13.4 ANOVA table for a completely randomized design

Source of variation	Degrees of freedom	Sum of squares	Mean square	F
Treatments	$k - 1$	SSTR	$MSTR = \dfrac{SSTR}{k - 1}$	$\dfrac{MSTR}{MSE}$
Error	$n_T - k$	SSE	$MSE = \dfrac{SSE}{n_T - k}$	
Total	$n_T - 1$	SST		

Table 13.5 ANOVA table for the chemietech problem

Source of variation	Degrees of freedom	Sum of squares	Mean square	F
Treatments	2	520	260.00	9.18
Error	12	340	28.33	
Total	14	860		

Pairwise comparisons

We can use Fisher's LSD procedure to test all possible pairwise comparisons for the Chemietech problem. At the 5 per cent level of significance, the t distribution table shows that with $n_T - k = 15 - 3 = 12$ degrees of freedom, $t_{0.025} = 2.179$. Using MSE = 28.33 in equation (13.17), we obtain Fisher's least significant difference.

$$LSD = t_{\alpha/2} \sqrt{MSE\left(\frac{1}{n_i} + \frac{1}{n_j}\right)} = 2.179 \sqrt{28.33\left(\frac{1}{5} + \frac{1}{5}\right)} = 7.34$$

If the magnitude of the difference between any two sample means exceeds 7.34, we can reject the hypothesis that the corresponding population means are equal. For the Chemietech data in Table 13.3, we obtain the following results.

Sample differences significant?
Method A − Method B = 62 − 66 = −4 No
Method A − Method C = 62 − 52 = 10 Yes
Method B − Method C = 66 − 52 = 14 Yes

Thus, the difference in the population means is attributable to the difference between the means for method A and method C and the difference between the means for method B and method C. Methods A and B therefore are preferred to method C. However, more testing should be done to compare method A with method B. The current study does not provide sufficient evidence to conclude that these two methods differ.

Exercises

Methods

19 The following data are from a completely randomized design.

	Treatment		
Observation	A	B	C
1	162	142	126
2	142	156	122
3	165	124	138
4	145	142	140
5	148	136	150
6	174	152	128
$\bar{x}_j$	156	142	134
s_j^2	164.4	131.2	110.4

a. Compute the sum of squares between treatments.
b. Compute the mean square between treatments.
c. Compute the sum of squares due to error.
d. Compute the mean square due to error.
e. At the $\alpha = 0.05$ level of significance, test whether the means for the three treatments are equal.

20 Refer to exercise 19.

a. Set up the ANOVA table.
b. At the $\alpha = 0.05$ level of significance, use Fisher's least significant difference procedure to test all possible pairwise comparisons. What conclusion can you draw after carrying out this procedure?

21 In a completely randomized experimental design, seven experimental units were used for each of the five levels of the factor. Complete the following ANOVA table.

Source of variation	Degrees of freedom	Sum of squares	Mean square	F
Treatments		300		
Error				
Total		460		

22 Refer to exercise 21.

a. What hypotheses are implied in this problem?
b. At the $\alpha = 0.05$ level of significance, can we reject the null hypothesis in part (a)? Explain.

23 In an experiment designed to test the output levels of three different treatments, the following results were obtained: SST = 400, SSTR = 150, $n_T = 19$. Set up the ANOVA table and test for any significant difference between the mean output levels of the three treatments. Use $\alpha = 0.05$.

24 In a completely randomized experimental design, 12 experimental units were used for the first treatment, 15 for the second treatment and 20 for the third treatment. Complete the following analysis of variance. At a 0.05 level of significance, is there a significant difference between the treatments?

Source of variation	Degrees of freedom	Sum of squares	Mean square	F
Treatments		1200		
Error				
Total		1800		

EXER25

25 Develop the analysis of variance computations for the following experimental design. At $\alpha = 0.05$, is there a significant difference between the treatment means?

	Treatment		
	A	B	C
	136	107	92
	120	114	82
	113	125	85
	107	104	101
	131	107	89
	114	109	117
	129	97	110
	102	114	120
		104	98
		89	106
$\bar{x}_j$	119	107	100
s_j^2	146.86	96.44	173.78

Applications

26 Three different methods for assembling a product were proposed by an industrial engineer. To investigate the number of units assembled correctly with each method, 30 employees were randomly selected and randomly assigned to the three proposed methods in such a way that each method was used by ten workers. The number of units assembled correctly was recorded, and the analysis of variance procedure was applied to the resulting data set.

The following results were obtained: SST = 10 800; SSTR = 4560.

a. Set up the ANOVA table for this problem.
b. Use $\alpha = 0.05$ to test for any significant difference in the means for the three assembly methods.

27 In an experiment designed to test the breaking strength of four types of cables, the following results were obtained: SST = 85.05, SSTR = 61.64, $n_T = 24$. Set up the ANOVA table and test for any significant difference in the mean breaking strength of the four cables. Use $\alpha = 0.05$.

28 To study the effect of temperature on yield in a chemical process, five batches were produced at each of three temperature levels. The results follow. Construct an analysis of variance table. Use a 0.05 level of significance to test whether the temperature level has an effect on the mean yield of the process.

Temperature		
50°C	60°C	70°C
34	30	23
24	31	28
36	34	28
39	23	30
32	27	31

29 Auditors must make judgments about various aspects of an audit on the basis of their own direct experience, indirect experience, or a combination of the two. In a study, auditors were asked to make judgments about the frequency of errors to be found in an audit. The judgments by the auditors were then compared with the actual results. Suppose the following data were obtained from a similar study; lower scores indicate better judgments.

Direct	Indirect	Combination
17.0	16.6	25.2
18.5	22.2	24.0
15.8	20.5	21.5
18.2	18.3	26.8
20.2	24.2	27.5
16.0	19.8	25.8
13.3	21.2	24.2

Use $\alpha = 0.05$ to test to see whether the basis for the judgment affects the quality of the judgment. What is your conclusion?

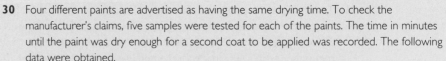

30 Four different paints are advertised as having the same drying time. To check the manufacturer's claims, five samples were tested for each of the paints. The time in minutes until the paint was dry enough for a second coat to be applied was recorded. The following data were obtained.

Paint 1	Paint 2	Paint 3	Paint 4
128	144	133	150
137	133	143	142
135	142	137	135
124	146	136	140
141	130	131	153

At the $\alpha = 0.05$ level of significance, test to see whether the mean drying time is the same for each type of paint.

31 Details of independent random samples of average hourly output for three manufacturing plants are as follows:

PLANT

	Plant	
1	2	3
83	77	81.6
86	82	83
79	82	83.6
79.8	80.6	88
81.6	81	85
83.6	80	
	79.8	

Analyze these data appropriately. Do average outputs differ significantly by plant and if so how?

32 Refer to Exercise 29. Use Fisher's least significant difference procedure to test all possible pairwise comparisons. What conclusion can you draw after carrying out this procedure? Use $\alpha = 0.05$.

33 Refer to the NCP data in Table 13.1. Use Fisher's least significant difference procedure allowing for the Bonferroni adjustment to test all pairwise comparisons. What conclusion can you draw after carrying out this procedure? Use $\alpha = 0.05$.

13.6 Randomized block design

Thus far we have considered the completely randomized experimental design. Recall that to test for a difference among treatment means, we computed an F value by using the ratio

F Test Statistic

$$F = \frac{MSTR}{MSE} \tag{13.23}$$

A problem can arise whenever differences due to extraneous factors (ones not considered in the experiment) cause the MSE term in this ratio to become large. In such cases, the F value in equation (13.23) can become small, signalling no difference among treatment means when in fact such a difference exists.

In this section we present an experimental design known as a **randomized block design**. Its purpose is to control some of the extraneous sources of variation by removing such variation from the MSE term. This design tends to provide a better estimate of the true error variance and leads to a more powerful hypothesis test in terms of the ability to detect differences among treatment means. To illustrate, let us consider a stress study for air traffic controllers.

Air traffic controller stress test

A study measuring the fatigue and stress of air traffic controllers resulted in proposals for modification and redesign of the controller's work station. After consideration of several designs for the work station, three specific alternatives are selected as having the best potential for reducing controller stress. The key question is: to what extent do the three alternatives differ in terms of their effect on controller stress? To answer this question, we need to design an experiment that will provide measurements of air traffic controller stress under each alternative.

In a completely randomized design, a random sample of controllers would be assigned to each work station alternative. However, controllers are believed to differ substantially in their ability to handle stressful situations. What is high stress to one controller might be only moderate or even low stress to another. Hence, when considering the within-group source of variation (MSE), we must realize that this variation includes both random error and error due to individual controller differences. In fact, managers expected controller variability to be a major contributor to the MSE term.

One way to separate the effect of the individual differences is to use a randomized block design. Such a design will identify the variability stemming from individual controller differences and remove it from the MSE term. The randomized block design calls for a single sample of controllers. Each controller in the sample is tested with each of the three work station alternatives. In experimental design terminology, the work station is the *factor of interest* and the controllers are the *blocks*. The three treatments or populations associated with the work station factor correspond to the three work station alternatives. For simplicity, we refer to the work station alternatives as system A, system B and system C.

The *randomized* aspect of the randomized block design is the random order in which the treatments (systems) are assigned to the controllers. If every controller were to test the three systems in the same order, any observed difference in systems might be due to the order of the test rather than to true differences in the systems.

To provide the necessary data, the three work station alternatives were installed at the Berlin control centre. Six controllers were selected at random and assigned to operate each of the systems. A follow-up interview and a medical examination of each controller participating in the study provided a measure of the stress for each controller on each system. The data are reported in Table 13.6.

Table 13.6 A randomized block design for the air traffic controller stress test

	Treatments		
	System A	System B	System C
Controller 1	15	15	18
Controller 2	14	14	14
Blocks Controller 3	10	11	15
Controller 4	13	12	17
Controller 5	16	13	16
Controller 6	13	13	13

Table 13.7 Summary of stress data for the air traffic controller stress test

		Treatments			
		System A	System B	System C	Row or block totals	Block means
	Controller 1	15	15	18	48	$\bar{x}_{1.} = 48/3 = 16.0$
	Controller 2	14	14	14	42	$\bar{x}_{2.} = 42/3 = 14.0$
Blocks	Controller 3	10	11	15	36	$\bar{x}_{3.} = 36/3 = 12.0$
	Controller 4	13	12	17	42	$\bar{x}_{4.} = 42/3 = 14.0$
	Controller 5	16	13	16	45	$\bar{x}_{5.} = 45/3 = 15.0$
	Controller 6	13	13	13	39	$\bar{x}_{6.} = 39/3 = 13.0$
Column or Treatment Totals		81	78	93	252	$\bar{\bar{x}} = \dfrac{252}{18} = 14.0$
Treatment Means		$\bar{x}_{.1} = \dfrac{81}{6}$ $= 13.5$	$\bar{x}_{.2} = \dfrac{78}{6}$ $= 13.0$	$\bar{x}_{.3} = \dfrac{93}{6}$ $= 15.5$		

Table 13.7 is a summary of the stress data collected. In this table we include column totals (treatments) and row totals (blocks) as well as some sample means that will be helpful in making the sum of squares computations for the ANOVA procedure. Because lower stress values are viewed as better, the sample data seem to favour system B with its mean stress rating of 13. However, the usual question remains: do the sample results justify the conclusion that the population mean stress levels for the three systems differ? That is, are the differences statistically significant? An analysis of variance computation similar to the one performed for the completely randomized design can be used to answer this statistical question.

ANOVA procedure

The ANOVA procedure for the randomized block design requires us to partition the sum of squares total (SST) into three groups: sum of squares due to treatments, sum of squares due to blocks and sum of squares due to error. The formula for this partitioning follows.

$$SST = SSTR + SSBL + SSE \qquad (13.24)$$

This sum of squares partition is summarized in the ANOVA table for the randomized block design as shown in Table 13.8. The notation used in the table is

$$k = \text{the number of treatments}$$
$$b = \text{the number of blocks}$$
$$n_T = \text{the total sample size } (n_T = kb)$$

Note that the ANOVA table also shows how the $n_T - 1$ total degrees of freedom are partitioned such that $k - 1$ degrees of freedom go to treatments, $b - 1$ go to blocks, and $(k - 1)(b - 1)$ go to the error term. The mean square column shows the sum of squares

Table 13.8 ANOVA table for the randomized block design with k treatments and b blocks

Source of variation	Degrees of freedom	Sum of squares	Mean square	F
Treatments	$k - 1$	SSTR	$\text{MSTR} = \dfrac{\text{SSTR}}{k - 1}$	$\dfrac{\text{MSTR}}{\text{MSE}}$
Blocks	$b - 1$	SSBL	$\text{MSBL} = \dfrac{\text{SSBL}}{b - 1}$	
Error	$(k - 1)(b - 1)$	SSE	$\text{MSE} = \dfrac{\text{SSE}}{(k - 1)(b - 1)}$	
Total	$n_T - 1$	SST		

divided by the degrees of freedom, and $F = \text{MSTR/MSE}$ is the F ratio used to test for a significant difference among the treatment means. The primary contribution of the randomized block design is that, by including blocks, we remove the individual controller differences from the MSE term and obtain a more powerful test for the stress differences in the three work station alternatives.

Computations and conclusions

To compute the F statistic needed to test for a difference among treatment means with a randomized block design, we need to compute MSTR and MSE. To calculate these two mean squares, we must first compute SSTR and SSE; in doing so, we will also compute SSBL and SST. To simplify the presentation, we perform the calculations in four steps. In addition to k, b and n_T as previously defined, the following notation is used.

x_{ij} = value of the observation corresponding to treatment j in block i
$\bar{x}_j$ = sample mean of the jth treatment
$\bar{x}_i$ = sample mean for the ith block
$\bar{\bar{x}}$ = overall sample mean

Step 1 Compute the total sum of squares (SST).

$$\text{SST} = \sum_{i=1}^{b} \sum_{j=1}^{k} (x_{ij} - \bar{\bar{x}})^2 \tag{13.25}$$

Step 2 Compute the sum of squares due to treatments (SSTR).

$$\text{SSTR} = b \sum_{j=1}^{k} (\bar{x}_j - \bar{\bar{x}})^2 \tag{13.26}$$

Step 3 Compute the sum of squares due to blocks (SSBL).

$$\text{SSBL} = k \sum_{i=1}^{b} (\bar{x}_i - \bar{\bar{x}})^2 \tag{13.27}$$

Step 4 Compute the sum of squares due to error (SSE).

$$\text{SSE} = \text{SST} - \text{SSTR} - \text{SSBL} \tag{13.28}$$

For the air traffic controller data in Table 13.7, these steps lead to the following sums of squares.

Step 1 $SST = (15 - 14)^2 + (15 - 14)^2 + (18 - 14)^2 + \ldots + (13 - 14)^2 = 70$

Step 2 $SSTR = 6[(13.5 - 14)^2 + (13.0 - 14)^2 + (15.5 - 14)^2] = 21$

Step 3 $SSBL = 3[(16 - 14)^2 + (14 - 14)^2 + (12 - 14)^2 + (14 - 14)^2 + (15 - 14)^2$
$+ (13 - 14)^2] = 30$

Step 4 $SSE = 70 - 21 - 30 = 19$

These sums of squares divided by their degrees of freedom provide the corresponding mean square values shown in Table 13.9.

Let us use a level of significance $\alpha = 0.05$ to conduct the hypothesis test. The value of the test statistic is

$$F = \frac{MSTR}{MSE} = \frac{10.5}{1.9} = 5.53$$

The numerator degrees of freedom is $k - 1 = 3 - 1 = 2$ and the denominator degrees of freedom is $(k - 1)(b - 1) = (3 - 1)(6 - 1) = 10$. Because we will only reject the null hypothesis for large values of the test statistic, the p-value is the area under the F distribution to the right of $F = 5.53$. From Table 4 of Appendix B we find that with the degrees of freedom 2 and 10, $F = 5.53$ is between $F_{0.025} = 5.46$ and $F_{0.01} = 7.56$. As a result, the area in the upper tail, or the p-value, is between 0.01 and 0.025. Alternatively, we can use MINITAB, PASW or EXCEL to show that the exact p-value for $F = 5.53$ is 0.0241. With p-value $\le \alpha = 0.05$, we reject the null hypothesis $H_0: \mu_1 = \mu_2 = \mu_3$ and conclude that the population mean stress levels differ for the three work station alternatives.

Some general comments can be made about the randomized block design. The experimental design described in this section is a *complete* block design; the word 'complete' indicates that each block is subjected to all k treatments. That is, all controllers (blocks) were tested with all three systems (treatments). Experimental designs in which some but not all treatments are applied to each block are referred to as *incomplete block* designs. A discussion of incomplete block designs is beyond the scope of this text.

Because each controller in the air traffic controller stress test was required to use all three systems, this approach guarantees a complete block design. In some cases, however, **blocking** is carried out with 'similar' experimental units in each block. For example, assume that in a pretest of air traffic controllers, the population of controllers was divided into groups ranging from extremely high-stress individuals to extremely low-stress individuals.

Table 13.9 ANOVA table for the air traffic controller stress test

F	Source of variation	Degrees of freedom	Sum of squares	Mean square
10.5/1.9 = 5.53	Treatments	2	21	10.5
	Blocks	5	30	6.0
	Error	10	19	1.9
	Total	17	70	

The blocking could still be accomplished by having three controllers from each of the stress classifications participate in the study. Each block would then consist of three controllers in the same stress group. The randomized aspect of the block design would be the random assignment of the three controllers in each block to the three systems.

Finally, note that the ANOVA table shown in Table 13.8 provides an F value to test for treatment effects but *not* for blocks. The reason is that the experiment was designed to test a single factor – work station design. The blocking based on individual stress differences was conducted to remove such variation from the MSE term. However, the study was not designed to test specifically for individual differences in stress.

Some analysts compute $F = \text{MSB/MSE}$ and use that statistic to test for significance of the blocks. Then they use the result as a guide to whether the same type of blocking would be desired in future experiments. However, if individual stress difference is to be a factor in the study, a different experimental design should be used. A test of significance on blocks should not be performed as a basis for a conclusion about a second factor.

Exercises

Methods

34 Consider the experimental results for the following randomized block design. Make the calculations necessary to set up the analysis of variance table.

		Treatments	
Blocks	A	B	C
1	0	9	8
2	2	6	5
3	8	15	14
4	0	18	18
5	8	7	8

Use $\alpha = 0.05$ to test for any significant differences.

35 The following data were obtained for a randomized block design involving five treatments and three blocks: SST = 430, SSTR = 310, SSB = 85. Set up the ANOVA table and test for any significant differences. Use $\alpha = 0.05$.

36 An experiment has been conducted for four treatments with eight blocks. Complete the following analysis of variance table.

Source of variation	Degrees of freedom	Sum of squares	Mean Square	F
Treatments		900		
Blocks		400		
Error				
Total		1800		

Use $\alpha = 0.05$ to test for any significant differences.

Applications

37 A car dealer conducted a test to determine if the time in minutes needed to complete a minor engine tune-up depends on whether a computerized engine analyzer or an electronic analyzer is used. Because tune-up time varies among compact, intermediate and full-sized cars, the three types of cars were used as blocks in the experiment. The data obtained follow.

	Analyzer	
Car	Computerized	Electronic
Compact	50	42
Intermediate	55	44
Full-sized	63	46

Use $\alpha = 0.05$ to test for any significant differences.

38 A textile mill produces a silicone proofed fabric for making into rainwear. The chemist in charge thinks that a silicone solution of about 12 per cent strength should yield a fabric with maximum waterproofing-index. He also suspected there may be some batch to batch variation because of slight differences in the cloth. To allow for this possibility five different strengths of solution were used on each of the three different batches of fabric. The following values of water-proofing index were obtained:

		[Strength of silicone solution (%)]				
		6	9	12	15	18
	1	20.8	20.6	22.0	22.6	20.9
Fabric	2	19.4	21.2	21.8	23.9	22.4
	3	19.9	21.1	22.7	22.7	22.1

Using $\alpha = 0.05$, carry out an appropriate test of these data and comment on the chemist's original beliefs.

39 An important factor in selecting software for word-processing and database management systems is the time required to learn how to use the system. To evaluate three file management systems, a firm designed a test involving five word-processing operators. Because operator variability was believed to be a significant factor, each of the five operators was trained on each of the three file management systems. The data obtained follow.

	System		
Operator	A	B	C
1	6	16	24
2	9	17	22
3	4	13	19
4	3	12	18
5	8	17	22

Use $\alpha = 0.05$ to test for any difference in the mean training time (in hours) for the three systems.

13.7 Factorial experiments

The experimental designs we considered thus far enable us to draw statistical conclusions about one factor. However, in some experiments we want to draw conclusions about more than one variable or factor. **Factorial experiments** and their corresponding ANOVA computations are valuable designs when simultaneous conclusions about two or more factors are required. The term *factorial* is used because the experimental conditions include all possible combinations of the factors. For example, for *a* levels of factor A and *b* levels of factor B, the experiment will involve collecting data on *ab* treatments. In this section we will show the analysis for a two-factor factorial experiment. The basic approach can be extended to experiments involving more than two factors.

As an illustration of a two-factor factorial experiment, we will consider a study involving the Graduate Management Admissions Test (GMAT), a standardized test used by graduate schools of business to evaluate an applicant's ability to pursue a graduate programme in that field. Scores on the GMAT range from 200 to 800, with higher scores implying higher aptitude.

In an attempt to improve students' performance on the GMAT exam, a major Spanish university is considering offering the following three GMAT preparation programmes.

1 A three-hour review session covering the types of questions generally asked on the GMAT.

2 A one-day programme covering relevant exam material, along with the taking and grading of a sample exam.

3 An intensive ten-week course involving the identification of each student's weaknesses and the setting up of individualized programmes for improvement.

Therefore, one factor in this study is the GMAT preparation programme, which has three levels: three-hour review, one-day programme and ten-week course. Before selecting the preparation programme to adopt, further study will be conducted to determine how the proposed programmes affect GMAT scores.

The GMAT is usually taken by students from three colleges: the College of Business, the College of Engineering and the College of Arts and Sciences. Therefore, a second factor of interest in the experiment is whether a student's undergraduate college affects the GMAT score. This second factor, undergraduate college, also has three levels: business, engineering and arts and sciences. The factorial design for this experiment with three levels corresponding to factor A, the preparation programme, and three levels corresponding to factor B, the undergraduate college, will give rise to a total of $3 \times 3 = 9$ treatments. These treatments or combinations of factor levels are summarized in Table 13.10.

Assume that a sample of two students will be selected corresponding to each of the nine treatments shown in Table 13.10: two business students will take the three-hour

Table 13.10 Nine treatments for the two-factor GMAT experiment

		Factor B: College		
		Business	Engineering	Arts and sciences
Factor A:	Three-hour review	1	2	3
Preparation	One-day programme	4	5	6
Programme	Ten-week course	7	8	9

review, two will take the one-day programme and two will take the ten-week course. In addition, two engineering students and two arts and sciences students will take each of the three preparation programmes. In experimental design terminology, the sample size of two for each treatment indicates that we have two **replications**. Additional replications and a larger sample size could easily be used, but we elect to minimize the computational aspects for this illustration.

This experimental design requires that six students who plan to attend graduate school be randomly selected from *each* of the three undergraduate colleges. Then two students from each college should be assigned randomly to each preparation programme, resulting in a total of 18 students being used in the study.

Let us assume that the randomly selected students participated in the preparation programmes and then took the GMAT. The scores obtained are reported in Table 13.11.

The analysis of variance computations with the data in Table 13.11 will provide answers to the following questions.

- **Main effect (factor A):** Do the preparation programmes differ in terms of effect on GMAT scores?
- **Main effect (factor B):** Do the undergraduate colleges differ in terms of effect on GMAT scores?
- **Interaction effect (factors A and B):** Do students in some colleges do better on one type of preparation programme whereas others do better on a different type of preparation programme?

The term **interaction** refers to a new effect that we can now study because we used a factorial experiment. If the interaction effect has a significant impact on the GMAT scores, we can conclude that the effect of the type of preparation programme depends on the undergraduate college.

ANOVA procedure

The ANOVA procedure for the two-factor factorial experiment is similar to the completely randomized experiment and the randomized block experiment in that we again partition the sum of squares and the degrees of freedom into their respective sources. The formula for partitioning the sum of squares for the two-factor factorial experiments follows.

$$SST = SSA + SSB + SSAB + SSE \qquad (13.29)$$

Table 13.11   GMAT scores for the two-factor experiment

		Factor B: College		
		Business	Engineering	Arts and sciences
Factor A: Preparation Programme	Three-hour review	500	540	480
		580	460	400
	One-day programme	460	560	420
		540	620	480
	Ten-week course	560	600	480
		600	580	410

Table 13.12 ANOVA table for the two-factor factorial experiment with r replications

Source of variation	Degrees of freedom	Sum of squares	Mean square	F
Factor A	$a - 1$	SSA	$MSA = \dfrac{SSA}{a - 1}$	$\dfrac{MSA}{MSE}$
Factor B	$b - 1$	SSB	$MSB = \dfrac{SSB}{b - 1}$	$\dfrac{MSB}{MSE}$
Interaction	$(a - 1)(b - 1)$	SSAB	$MSAB = \dfrac{SSAB}{(a - 1)(b - 1)}$	$\dfrac{MSAB}{MSE}$
Error	$ab(r - 1)$	SSE	$MSE = \dfrac{SSE}{ab(r - 1)}$	
Total	$n_T - 1$	SST		

The partitioning of the sum of squares and degrees of freedom is summarized in Table 13.12. The following notation is used.

a = number of levels of factor A
b = number of levels of factor B
r = number of replications
n_T = total number of observations taken in the experiment; $n_T = abr$

Computations and conclusions

To compute the F statistics needed to test for the significance of factor A, factor B, and interaction, we need to compute MSA, MSB, MSAB, and MSE. To calculate these four mean squares, we must first compute SSA, SSB, SSAB, and SSE; in doing so we will also compute SST. To simplify the presentation, we perform the calculations in five steps. In addition to a, b, r and n_T as previously defined, the following notation is used.

x_{ijk} = observation corresponding to the kth replicate taken from treatment i of factor A and treatment j of factor B
$\bar{x}_{i.}$ = sample mean for the observations in treatment i (factor A)
$\bar{x}_{.j}$ = sample mean for the observations in treatment j (factor B)
$\bar{x}_{ij}$ = sample mean for the observations corresponding to the combination of treatment i (factor A) and treatment j (factor B)
$\bar{\bar{x}}$ = overall sample mean of all n_T observations

Step 1 Compute the total sum of squares.

$$SST = \sum_{i=1}^{a} \sum_{j=1}^{b} \sum_{k=1}^{r} (x_{ijk} - \bar{\bar{x}})^2 \qquad (13.30)$$

Step 2 Compute the sum of squares for factor A.

$$SSA = br \sum_{i=1}^{a} (\bar{x}_{i.} - \bar{\bar{x}})^2 \qquad (13.31)$$

Step 3 Compute the sum of squares for factor B.

$$SSBR = ar \sum_{i=1}^{b} (x_{.j} - \bar{\bar{x}})^2 \tag{13.32}$$

Step 4 Compute the sum of squares for interaction.

$$SSAB = r \sum_{i=}^{a} \sum_{j=1}^{b} (x_{ij} - \bar{x}_{i.} - \bar{x}_{.j} + \bar{\bar{x}})^2 \tag{13.33}$$

Step 5 Compute the sum of squares due to error.

$$SSE = SST - SSA - SSB - SSAB \tag{13.34}$$

Table 13.13 reports the data collected in the experiment and the various sums that will help us with the sum of squares computations. Using equations (13.30) through (13.34), the sums of squares for the GMAT two-factor factorial experiment can be calculated as follows.

Step 1 $SST = (500 - 515)^2 + (580 - 515)^2 + (540 - 515)^2 + \cdots$
$\qquad\qquad + (410 - 515)^2 = 82\ 450$

Step 2 $SSA = (3)(2)[(493.33 - 515)^2 + (513.33 - 515)^2 + (538.33 - 515)^2] = 6100$

Step 3 $SSB = (3)(2)[(540 - 515)^2 + (560 - 515)^2 + (445 - 515)^2] = 45\ 300$

Step 4 $SSAB = 2[(540 - 493.33 - 540 - 515)^2 + (500 - 493.33 - 560 + 515)^2 + \cdots$
$\qquad\qquad + (445 - 538.33 - 445 + 515)^2] = 11\ 200$

Step 5 $SSE = 82\ 450 - 6100 - 45\ 300 - 11\ 200 = 19\ 850$

These sums of squares divided by their corresponding degrees of freedom, as shown to prepare students from the different colleges for the GMAT in Table 13.14, provide the appropriate mean square values for testing the two main effects (preparation programme and undergraduate college) and the interaction effect.

Let us use a level of significance $\alpha = 0.05$ to conduct the hypothesis tests for the two-factor GMAT study. Because of the computational effort involved in any modest- to large-size factorial experiment, the computer usually plays an important role in performing the analysis of variance computations and in the calculation of the p-values used to make the hypothesis testing decisions. Figure 13.6 shows the MINITAB output for the analysis of variance for the GMAT two-factor factorial experiment Because the p-value used to test for significant differences among the three preparation programmes (factor A) = 0.299 is greater than $\alpha = 0.05$, we deduce there is no significant difference in the mean GMAT test scores for the three preparation programmes. However, for the undergraduate college effect, the p-value = 0.005 is less than $\alpha = 0.05$; thus, there is a significant difference in the mean GMAT test scores among the three undergraduate colleges. Finally, because the p-value of 0.350 for the interaction effect is greater than $\alpha = 0.05$, there is no significant interaction effect. Therefore, the study provides no reason to believe that the three preparation programmes differ in their ability to prepare students from the different colleges for the GMAT.

Undergraduate college however was found to be a significant factor. Checking the calculations in Table 13.13, we see that the sample means are: business students $\bar{x}_{.1} = 540$, engineering students $\bar{x}_{.2} = 560$ and arts and sciences students $\bar{x}_{.3} = 445$. Tests on individual treatment means can be conducted; yet after reviewing the three sample means, we would

Table 13.13 GMAT summary data for the two-factor experiment

Treatment combination totals	Factor B: College			Row totals	Factor A means
	Business	Engineering	Arts and sciences		
Three-hour review	500	540	480		
	580	460	400		
	1080	1000	880	2960	$\bar{x}_{1\cdot} = \dfrac{2960}{6} = 493.33$
	$\bar{x}_{11} = \dfrac{1080}{2} = 540$	$\bar{x}_{12} = \dfrac{1000}{2} = 500$	$\bar{x}_{13} = \dfrac{880}{2} = 440$		
One-day programme	460	560	420		
	540	620	480		
	1000	1180	900	3080	$\bar{x}_{2\cdot} = \dfrac{3080}{6} = 513.33$
	$\bar{x}_{21} = \dfrac{1000}{2} = 500$	$\bar{x}_{22} = \dfrac{1180}{2} = 590$	$\bar{x}_{23} = \dfrac{900}{2} = 450$		
10-week course	560	600	480		
	600	580	410		
	1160	1180	890	3230	$\bar{x}_{3\cdot} = \dfrac{3230}{6} = 538.33$
	$\bar{x}_{31} = \dfrac{1160}{2} = 580$	$\bar{x}_{32} = \dfrac{1180}{2} = 590$	$\bar{x}_{33} = \dfrac{890}{2} = 445$		
Column totals	3240	3360	2670	9270 ⟶ Overall total	
Factor B means	$\bar{x}_{\cdot 1} = \dfrac{3240}{6} = 540$	$\bar{x}_{\cdot 2} = \dfrac{3360}{6} = 560$	$\bar{x}_{\cdot 3} = \dfrac{2670}{6} = 445$	$\bar{\bar{x}} = \dfrac{9270}{18} = 515$	

Table 13.14 ANOVA table for the two-factor GMAT study

Source of variation	Degrees of freedom	Sum of squares	Mean square	F
Factor A	2	6100	3050	3050/2206 = 1.38
Factor B	2	45 300	22 650	22 650/2206 = 10.27
Interaction	4	11 200	2800	2800/2206 = 1.27
Error	9	19 850	2206	
Total	17	82 450		

Figure 13.6 MINITAB output for the GMAT two-factor design

```
Two-way ANOVA: Score versus Factor A, Factor B

Source       DF    SS       MS       F       P
Factor A     2     6100     3050.0   1.38    0.299
Factor B     2     45300    22650.0  10.27   0.005
Interaction  4     11200    2800.0   1.27    0.350
Error        9     19850    2205.6
Total        17    82450

S = 46.96    R-Sq = 75.92%    R-Sq(adj) = 54.52%
```

anticipate no difference in preparation for business and engineering graduates. However, the arts and sciences students appear to be significantly less prepared for the GMAT than students in the other colleges. Perhaps this observation will lead the university to consider other options for assisting these students in preparing for graduate management admission tests.

Exercises

Methods

40 A factorial experiment involving two levels of factor A and three levels of factor B resulted in the following data.

		Factor B		
		Level 1	Level 2	Level 3
Factor A	Level 1	135	90	75
		165	66	93
	Level 2	125	127	120
		95	105	136

Test for any significant main effects and any interaction. Use $\alpha = 0.05$.

41 The calculations for a factorial experiment involving four levels of factor A, three levels of factor B, and three replications resulted in the following data: SST = 280, SSA = 26, SSB = 23, SSAB = 175. Set up the ANOVA table and test for any significant main effects and any interaction effect. Use $\alpha = 0.05$.

Applications

42 A mail-order catalogue firm designed a factorial experiment to test the effect of the size of a magazine advertisement and the advertisement design on the number of catalogue requests received (data in thousands). Three advertising designs and two different-size advertisements were considered. The data obtained follow.

		Size of advertisement	
		Small	Large
	A	8	12
		12	8
Design	B	22	26
		14	30
	C	10	18
		18	14

Use the ANOVA procedure for factorial designs to test for any significant effects due to type of design, size of advertizement, or interaction. Use $\alpha = 0.05$.

43 A factorial experiment involved measurement of average fuel consumption for 36 long journeys for three different types of vehicle by value and three different types of fuel additive. The data (km / litre) obtained follow:

		Fuel additive	
Vehicle type	1	2	3
A	7	8	8
	7	8	8
	7	7	8
	8	7	8
B	6	8	7
	6	8	7
	6	8	8
	6	8	7
C	6	8	7
	6	7	7
	6	7	7
	6	7	7

Perform an appropriate analysis of these data. Use $\alpha = 0.05$. What are your conclusions?

44 A study reported in *The Accounting Review* examined the separate and joint effects of two levels of time pressure (low and moderate) and three levels of knowledge (naïve, declarative and procedural) on key word selection behaviour in tax research. Subjects were given a tax case containing a set of facts, a tax issue and a key word index consisting of 1336 key words. They were asked to select the key words they believed would refer them to a tax authority relevant to resolving the tax case. Prior to the experiment, a group of tax experts determined that the text contained 19 relevant key words. Subjects in the naïve group had little or no declarative or procedural knowledge, subjects in the declarative group had significant declarative knowledge but little or no procedural knowledge, and subjects in the procedural group had significant declarative knowledge and procedural knowledge. Declarative knowledge consists of knowledge of both the applicable tax rules and the technical terms used to describe such rules. Procedural knowledge is knowledge of the rules that guide the tax researcher's search for relevant key words. Subjects in the low time pressure situation were told they had 25 minutes to complete the problem, an amount of time which should be 'more than adequate' to complete the case; subjects in the moderate time pressure situation were told they would have 'only' 11 minutes to complete the case. Suppose 25 subjects were selected for each of the six treatments and the sample means for each treatment are as follows (standard deviations are in parentheses).

| | | Knowledge | | |
		Naïve	Declarative	Procedural
Time pressure	Low	1.13	1.56	2.00
		(1.12)	(1.33)	(1.54)
	Moderate	0.48	1.68	2.86
		(0.80)	(1.36)	(1.80)

Use the ANOVA procedure to test for any significant differences due to time pressure, knowledge, and interaction. Use a 0.05 level of significance. Assume that the total sum of squares for this experiment is 327.5

Summary

In this chapter we showed how analysis of variance can be used to test for differences among means of several populations or treatments. We introduced the completely randomized design, the randomized block design and the two-factor factorial experiment and confirmed corresponding assumptions. The completely randomized design and the randomized block design are used to draw conclusions about differences in the means of a single factor. The primary purpose of blocking in the randomized block design is to remove extraneous sources of variation from the error term. Such blocking provides a better estimate of the true error variance and a better test for determining whether the population or treatment means of the factor differ significantly. Correspondingly factorial experiments involve conclusions being drawn about two or more factors including their potential interactions.

We showed that the basis for the statistical tests used in analysis of variance and experimental design is the development of two independent estimates of the population variance σ^2. In the single-factor case, one estimator is based on the variation between the treatments; this estimator provides an unbiased estimate of σ^2 only if the treatment means are all equal. A second estimator of σ^2 is based on the variation of the observations within each sample; this estimator will always provide an unbiased estimate of σ^2. By computing the ratio of these two estimators (the F statistic) we developed a rejection rule for determining whether to reject the null hypothesis that the population or treatment means are equal. In all the experimental designs considered, the partitioning of the sum of squares and degrees of freedom into their various sources enabled us to compute the appropriate values for the analysis of variance calculations and tests. We also showed how Fisher's LSD procedure and the Bonferroni adjustment can be used to perform pairwise comparisons to determine which means are different.

Key terms

ANOVA table
Blocking
Comparisonwise Type I error rate
Completely randomized design
Experimental units
Experimentwise Type I error rate
Factor
Factorial experiments

Interaction
Multiple comparison procedures
Partitioning
Randomized block design
Replications
Single-factor experiment
Treatment

Key formulae

Testing for the equality of k population means
Sample mean for treatment j

$$\bar{x}_j = \frac{\sum_{i=1}^{n_j} x_{ij}}{n_j} \tag{13.1}$$

Sample variance for treatment j

$$s_j^2 = \frac{\sum_{i=1}^{n_j} (x_{ij} - \bar{x}_j)^2}{n_j - 1} \tag{13.2}$$

Overall sample mean

$$\bar{\bar{x}} = \frac{\sum_{j=1}^{k} \sum_{i=1}^{n_j} x_{ij}}{n_T} \tag{13.3}$$

$$n_T = n_2 + \cdots + n_k \tag{13.4}$$

Mean square due to treatments

$$\text{MSTR} = \frac{\text{SSTR}}{k-1} \tag{13.7}$$

Sum of squares due to treatments

$$\text{SSTR} = \sum_{j=1}^{k} n_j (\bar{x}_j - \bar{\bar{x}})^2 \tag{13.8}$$

Mean square due to error

$$\text{MSE} = \frac{\text{SSE}}{n_T - k} \tag{13.10}$$

Sum of squares due to error

$$\text{SSE} = \sum_{j=1}^{k} (n_j - 1)s_j^2 \tag{13.11}$$

Test statistic for the equality of k population means

$$F = \frac{\text{MSTR}}{\text{MSE}} \tag{13.12}$$

Total sum of squares

$$\text{SST} = \sum_{j=1}^{k} \sum_{i=1}^{n_j} (x_{ij} - \bar{\bar{x}})^2 \tag{13.13}$$

Partitioning of sum of squares

$$\text{SST} = \text{SSTR} + \text{SSE} \tag{13.14}$$

Multiple comparison procedures Test statistic for Fisher's LSD procedure

$$t = \frac{\bar{x}_i - \bar{x}_j}{\sqrt{\text{MSE}\left(\frac{1}{n_i} + \frac{1}{n_j}\right)}} \tag{13.16}$$

Fisher's LSD

$$LSD = t_{\alpha/2} \sqrt{MSE\left(\frac{1}{n_i} + \frac{1}{n_j}\right)} \qquad (13.17)$$

Completely randomized designs
Mean square due to treatments

$$MSTR = \frac{\sum_{j=1}^{k} n_j (\bar{x}_j - \bar{\bar{x}})^2}{k - 1} \qquad (13.20)$$

Mean square due to error

$$MSE = \frac{\sum_{j=1}^{k} (n_j - 1)s_j^2}{n_T - k} \qquad (13.21)$$

F test statistic

$$F = \frac{MSTR}{MSE} \qquad (13.23)$$

Randomized block designs
Total sum of squares

$$SST = \sum_{i=1}^{b} \sum_{j=1}^{k} (x_{ij} - \bar{\bar{x}})^2 \qquad (13.25)$$

Sum of squares due to treatments

$$SSTR = b \sum_{j=1}^{k} (\bar{x}_j - \bar{\bar{x}})^2 \qquad (13.26)$$

Sum of squares due to blocks

$$SSBL = k \sum_{i=1}^{b} (\bar{x}_{i.} - \bar{\bar{x}})^2 \qquad (13.27)$$

Sum of squares due to error

$$SSE = SST - SSTR - SSBL \qquad (13.28)$$

Factorial experiments
Total sum of squares

$$SST = \sum_{i=1}^{a} \sum_{j=1}^{b} \sum_{k=1}^{r} (x_{ijk} - \bar{\bar{x}})^2 \qquad (13.30)$$

Sum of squares for factor A

$$SSA = br \sum_{i=1}^{a} (\bar{x}_{i.} - \bar{\bar{x}})^2 \qquad (13.31)$$

Sum of squares for factor B

$$SSTR = ar \sum_{i=1}^{b} (\bar{x}_{j} - \bar{\bar{x}})^2 \qquad \text{(13.32)}$$

Sum of squares for interaction

$$SSAB = r \sum_{i=1}^{a} \sum_{j=1}^{b} (x_{ij} - \bar{x}_{i.} - \bar{x}_{j} + \bar{\bar{x}})^2 \qquad \text{(13.33)}$$

Sum of squares for error

$$SSE = SST - SSA - SSB - SSAB \qquad \text{(13.34)}$$

Case problem 1 Wentworth Medical Centre

As part of a long-term study of individuals 65 years of age or older, sociologists and physicians at the Wentworth Medical Centre in Britain investigated the relationship between geographic location and depression. A sample of 60 individuals, all in reasonably good health, was selected; 20 individuals were residents of Scotland, 20 were residents of England, and 20 were residents of Wales. Each of the individuals sampled was given a standardized test to measure depression. The data collected follow; higher test scores indicate higher levels of depression. These data are available on the data disk in the file Medical1.

A second part of the study considered the relationship between geographic location and depression for individuals 65 years of age or older who had a chronic health condition such as arthritis, hypertension, and/or heart ailment. A sample of 60 individuals with such conditions was identified. Again, 20 were residents of Scotland, 20 were residents of England and 20 were residents of Wales. The levels of depression recorded for this study follow.

These data are available on the CD accompanying the text in the file named Medical2.

	Data from Medical1			Data from Medical2		
	Scotland	England	Wales	Scotland	England	Wales
	3	8	10	13	14	10
	7	11	7	12	9	12
	7	9	3	17	15	15
	3	7	5	17	12	18
	8	8	11	20	16	12
	8	7	8	21	24	14
	8	8	4	16	18	17
	5	4	3	14	14	8
	5	13	7	13	15	14
	2	10	8	17	17	16
	6	6	8	12	20	18

Data from Medical1			Data from Medical2		
Scotland	England	Wales	Scotland	England	Wales
2	8	7	9	11	17
6	12	3	12	23	19
6	8	9	15	19	15
9	6	8	16	17	13
7	8	12	15	14	14
5	5	6	13	9	11
4	7	3	10	14	12
7	7	8	11	13	13
3	8	11	17	11	11

An elderly lady taking part in a depression study. © Mark Papas.

Managerial Report

1 Use descriptive statistics to summarize the data from the two studies. What are your preliminary observations about the depression scores?

2 Use analysis of variance on both data sets. State the hypotheses being tested in each case. What are your conclusions?

3 Use inferences about individual treatment means where appropriate. What are your conclusions?

4 Discuss extensions of this study or other analyses that you feel might be helpful.

Case problem 2 Product Design Testing

An engineering manager has been designated the task of evaluating a commercial device subject to marked variations in temperature. Three different types of component are being considered for the device. When the device is manufactured and is shipped to the field, the manager has no control over the temperature extremes that the device will encounter, but knows from experience that temperature is an important factor in relation to the component's life. Notwithstanding this, temperature can be controlled in the laboratory for the purposes of the test.

DEVICE

The engineering manager arranges for all three components to be tested at the temperature levels: −10°C, 20°C, and 50°C – as these temperature levels are consistent with the product end-use environment. Four components are tested for each combination of type and temperature, and all 36 tests are run in random order. The resulting observed component life data are presented in Table 1.

Table I	Component lifetimes (000s of hours)					
	Temperature(°C)					
Type	−10		20		50	
I	3.12	3.70	0.82	0.96	0.48	1.68
	1.80	4.32	1.92	1.80	1.97	1.39
2	3.60	4.51	3.02	2.93	0.60	1.68
	3.82	3.02	2.54	2.76	1.39	1.08
3	3.31	2.64	4.18	2.88	2.30	2.50
	4.03	3.84	3.60	3.34	1.97	1.44

Testing the effects of extreme temperatures on products in a laboratory. © Bartee Inc/Phototake Science.

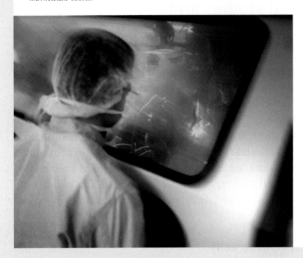

Managerial Report

1 What are the effects of the chosen factors on the life of the component?

2 Do any components have a consistently long life regardless of temperature?

3 What recommendation would you make to the engineering manager?

Software Section for Chapter 13

Analysis of variance and experimental design using MINITAB

Single factor observational studies and completely randomized designs

In Section 13.2 we showed how analysis of variance could be used to test for the equality of k population means using data from an observational study. In Section 13.5 we showed how the same approach could be used to test for the equality of k population means in situations where the data have been collected in a completely randomized design. To illustrate how MINITAB can be used to test for the equality of k population means for both of these cases, we show how to test whether the mean examination score is the same at each plant in the National Computer Products example introduced in Section 13.1. The examination score data are entered into the first three columns of a MINITAB worksheet; column 1 is labelled Ayr, column 2 is labelled Dusseldorf and column 3 is labelled Stockholm. The steps involved in producing the output in Figure 13.4 in MINITAB follow.

Step 1 **Stat > ANOVA > One-way (Unstacked)**   [Main menu bar]

Step 2 Enter C1-C3 in the **Responses (in separate columns)** box
                                                     [**One-way (Unstacked)** panel]

Click **OK**

Randomized block designs

In Section 13.6 we showed how analysis of variance could be used to test for the equality of k population means using data from a randomized block design. To illustrate how MINITAB can be used for this type of experimental design, we show how to test whether the mean stress levels for air traffic controllers is the same for three work stations. The stress level scores shown in Table 13.6 are entered into column 1 of a MINITAB worksheet. Coding the treatments as 1 for System A, 2 for System B and 3 for System C, the coded values for the treatments are entered into column 2 of the worksheet. Finally, the corresponding number of each controller (1, 2, 3, 4, 5, 6) is entered into column 3. Thus, the values in the first row of the worksheet are 15, 1, 1; the values in row 2 are 15, 2, 1; the values in row 3 are 18, 3, 1; the values in row 4 are 14, 1, 2 and so on. In particular, the steps involved in producing the MINITAB output corresponding to the ANOVA table shown in Table 13.9 follow.

Step 1   Select **Stat > ANOVA Two-way**   [Main menu bar]

Step 2   Enter C1 in the **Response** box   [**ANOVA Two-way** panel]
Enter C2 in the **Row factor** box
Enter C3 in the **Column factor** box
Select **Fit additive model**
Click **OK**

Factorial experiments

In Section 13.7 we showed how analysis of variance could be used to test for the equality of k population means using data from a factorial experiment. To illustrate how MINITAB can be used for this type of experimental design, we show how to analyse the data for the two-factor GMAT experiment introduced in that section. The GMAT scores shown in Table 13.11 are entered into column 1 of a MINITAB worksheet; column 1 is labelled Score, column 2 is labelled Factor A, and column 3 is labelled Factor B. Coding the factor A preparation programmes as 1 for the three-hour review, 2 for the one-day programme, and 3 for the ten-week course, the coded values for factor A are entered into column 2 of the worksheet. Coding the factor B colleges as 1 for Business, 2 for Engineering, and 3 for Arts and Sciences, the coded values for factor B are entered into column 3. Thus, the values in the first row of the worksheet are 500, 1, 1; the values in row 2 are 580, 1, 1; the values in row 3 are 540, 1, 2; the values in row 4 are 460, 1, 2 and so on. In particular, the steps involved in producing the MINITAB output corresponding to the ANOVA table shown in Figure 13.6 follow.

Step 1   **Stat > ANOVA > Two-way**   [Main menu bar]

Step 2   [**ANOVA Two-way** panel]
Enter C1 in the **Response** box
Enter C2 in the **Row factor** box
Enter C3 in the **Column factor** box
Click **OK**

Analysis of variance and experimental design using EXCEL

Single-factor observational studies and completely randomized designs

In Section 13.2 we showed how analysis of variance could be used to test for the equality of k population means using data from an observational study. In Section 13.5 we showed how the same approach could be used to test for equality of k population means in situations where the data are collected in a completely randomized design. To illustrate how EXCEL can be used to test for the equality of k population means for both of these cases, we show how to test whether the mean examination score is the same at each plant in the National Computer Products example introduced in Section 13.1. The examination

score data are entered into worksheet rows 2 to 7 of columns B, C and D as shown in Figure 13.7. Note that cells A2:A7 are used to identify the observations at each of the plants. The steps involved in using EXCEL to produce the output shown in cells A9:G23 follow; the ANOVA portion of this output corresponds to the ANOVA table shown in Table 13.2.

Step 1   Select **Data > Data Analysis > Anova: Single-Factor**   [Main menu bar]
           Click **OK**

Step 2   **[Anova: Single-Factor** panel]
           Enter B1:D7 in **Input Range** box
           Select **Columns**
           Select **Labels in First Row**
           Select **Output Range** and enter A9 in the box
           Click **OK**

Randomized block designs

In Section 13.6 we showed how analysis of variance could be used to test for the equality of k population means using data from a randomized block design. To illustrate how EXCEL can be used for this type of experimental design, we show

Figure 13.7 EXCEL solution for the NCP analysis of variance example

	A	B	C	D	E	F	G
1	**Observation**	**Ayr**	**Dusseldorf**	**Stockholm**			
2	1	85	71	59			
3	2	75	75	64			
4	3	82	73	62			
5	4	76	74	69			
6	5	71	69	75			
7	6	85	82	67			
8							
9	Anova: Single Factor						
10							
11	SUMMARY						
12	*Groups*	*Count*	*Sum*	*Average*	*Variance*		
13	Ayr	6	474	79	34		
14	Dusseldorf	6	444	74	20		
15	Stockholm	6	396	66	32		
16							
17							
18	ANOVA						
19	*Source of Variation*	*SS*	*df*	*MS*	*F*	*P-value*	*F crit*
20	Between Groups	516	2	258	9	0.0027	3.68
21	Within Groups	430	15	28.666667			
22							
23	Total	946	17				
24							

how to test whether the mean stress levels for air traffic controllers are the same for three work stations. The stress level scores shown in Table 13.6 are entered into worksheet rows 2 to 7 of columns B, C and D as shown in Figure 13.8. The cells in rows 2 to 7 of column A contain the number of each controller (1, 2, 3, 4, 5, 6). The steps involved in using EXCEL to produce output corresponding to the ANOVA table shown in Table 13.9, follow.

NCP

Step 1 **Data > Data Analysis > Anova: Two-Factor Without Replication**

[Main menu bar]

Click **OK**

Step 2 Enter A1:D7 in **Input Range** box

[**Anova: Two-Factor Without Replication** panel]

Select **Labels**.
Select **Output Range** and enter A9 in the box
Click **OK**

Figure 13.8 EXCEL solution for the air traffic controller stress test

	A	B	C	D	E	F	G
1	Controller	System A	System B	System C			
2	1	15	15	18			
3	2	14	14	14			
4	3	10	11	15			
5	4	13	12	17			
6	5	16	13	16			
7	6	13	13	13			
8							
9	Anova: Two-Factor Without Replication						
10							
11	SUMMARY	Count	Sum	Average	Variance		
12	1	3	48	16	3		
13	2	3	42	14	0		
14	3	3	36	12	7		
15	4	3	42	14	7		
16	5	3	45	15	3		
17	6	3	39	13	0		
18							
19	System A	6	81	13.5	4.3		
20	System B	6	78	13	2		
21	System C	6	93	15.5	3.5		
22							
23							
24	ANOVA						
25	urce of Variat	SS	df	MS	F	P-value	F crit
26	Rows	30	5	6	3.16	0.0574	3.33
27	Columns	21	2	10.5	5.53	0.0242	4.10
28	Error	19	10	1.9			
29							
30	Total	70	17				
31							

Factorial experiments

In Section 13.7 we showed how analysis of variance could be used to test for the equality of k population means using data from a factorial experiment. To illustrate how EXCEL can be used for this type of experimental design, we show how to analyse the data for the two-factor GMAT experiment introduced in that section. The GMAT scores shown in Table 13.11 are entered into worksheet rows 2 to 7 of columns B, C, and D as shown in Figure 13.9. The steps involved in using EXCEL to produce output shown in cells A10:G45 follows; the ANOVA portion of this output corresponds to the ANOVA table shown in Table 13.14.

GMAT

Step 1 **Data > Data Analysis > Anova: Two-Factor With Replication**

[Main menu bar]

Click **OK**

Figure 13.9 EXCEL solution for the two-factor GMAT experiment

	A	B	C	D	E	F	G
1		**Business**	**Engineering**	**Arts and Sciences**			
2	**3-hour review**	500	540	480			
3		580	460	400			
4	**1-day program**	460	560	420			
5		540	620	480			
6	**10-week course**	560	600	480			
7		600	580	410			
8							
9							
10	Anova: Two-Factor With Replication						
11							
12	SUMMARY	Business	Engineering	Arts and Sciences	Total		
13	*3-hour review*						
14	Count	2	2	2	6		
15	Sum	1080	1000	880	2960		
16	Average	540	500	440	493.33333		
17	Variance	3200	3200	3200	3946.6667		
18							
19	*1-day program*						
20	Count	2	2	2	6		
21	Sum	1000	1180	900	3080		
22	Average	500	590	450	513.33333		
23	Variance	3200	1800	1800	5386.6667		
24							
25	*10-week course*						
26	Count	2	2	2	6		
27	Sum	1160	1180	890	3230		
28	Average	580	590	445	538.33333		
29	Variance	800	200	2450	5936.6667		
30							
31	*Total*						
32	Count	6	6	6			
33	Sum	3240	3360	2670			
34	Average	540	560	445			
35	Variance	2720	3200	1510			
36							
37							
38	ANOVA						
39	*Source of Variation*	*SS*	*df*	*MS*	*F*	*P-value*	*F crit*
40	Sample	6100	2	3050	1.38	0.2994	4.26
41	Columns	45300	2	22650	10.27	0.0048	4.26
42	Interaction	11200	4	2800	1.27	0.3503	3.63
43	Within	19850	9	2205.5556			
44							
45	Total	82450	17				
46							

Step 2   Enter A1:D7 in **Input Range** box

[**Anova: Two-Factor With Replication** panel]

Enter 2 in **Rows per sample** box
Select **Labels**
Select **Output Range** and enter A10 in the box
Click **OK**

Analysis of variance and experimental design using PASW

Single-factor observational studies and completely randomized designs

To illustrate how PASW can be used to test for the equality of k population means, we show how to test whether the mean examination score is the same at each plant in the National Computer Products example introduced in Section 13.1. First, the data must be entered in a PASW worksheet. In 'Data View' mode, the examination score data are entered into the leftmost column of a PASW worksheet; the six values for Ayr, followed by the six for Dusseldorf and then the six for Stockholm. This is automatically labelled by the system V1. In the adjacent column to the right the code 1 (corresponding to the Ayr plant) is entered six times followed by the code 2 (corresponding to the Dusseldorf plant) six times and the code 3 (corresponding to the Stockholm plant) six times. Thus, the values in the first row of the worksheet are 85, 1; the values in row 2 are 75, 1; the values in row 3 are 82, 1; the values in row 4 are 76, 1; the values in row 5 are 71, 1; the values in row 6 are 85, 1; the values in row 7 are 71, 2 and so on.

The latter variable names can then be changed to score and plant in 'Variable View' mode. The codes used for the plant variable can also be relabelled by following the steps below.

Step 1   **Data > Define Variable Properties**   [Main menu bar]

Step 2   Select plant   [**Define Variable Properties** panel]
Click on **Continue**
Select plant
Attach the **Value Labels** Ayr to code 1, Dusseldorf to code 2 and Stockholm to code 3.
Click **OK**

The following steps show how PASW generates the ANOVA results shown in Figure 13.4.

Step 1   **Analyze > Compare Means > One-Way ANOVA**   [Main menu bar]

Step 2   Enter score in the **Dependent List** box   [**One-Way ANOVA** panel]
Enter plant in the **Factor** box
Click on **Options**
Select **Descriptive Statistics**
Click **Continue**
Click **OK**

Randomized block designs

In Section 13.6 we showed how analysis of variance could be used to test for the equality of k population means using data from a randomized block design. To illustrate how PASW can be used for this type of experimental design, we show how to test whether the mean stress levels for air traffic controllers is the same for three work stations. The stress level scores shown in Table 13.6 are entered into the leftmost column of an PASW worksheet. Coding the treatments as 1 for System A, 2 for System B, and 3 for System C, the coded values for the treatments are entered into the adjacent column to the right in the worksheet. Finally, the corresponding number of each controller (1, 2, 3, 4, 5, 6) is entered into the next adjacent column to the right. The columns are automatically labelled by the system V1, V2 and V3 but can be relabelled in Variable View mode as stress, system and controller respectively. Thus, the values in the first row of the worksheet are 15, 1, 1; the values in row 2 are 15, 2, 1; the values in row 3 are 18, 3, 1; the values in row 4 are 14, 1, 2 and so on. The following steps show how PASW generates the ANOVA results shown in Table 13.9.

AIRTRAF

Step 1 Analyze > General Linear Model > Univariate   [Main menu bar]

Step 2 Enter stress in the **Dependent Variable** box   **[Univariate** panel]
Enter system and controller in the **Fixed Factors** box
Click on **Model**
Click on **Custom**
Select system and controller
Select Main Effects in **Build Terms** box
Click on **Continue**
Click **OK**

(Note that the F test provided in the output for the controller (blocking) factor can be effectively ignored.)

Factorial experiments

GMAT

In Section 13.7 we showed how analysis of variance could be used to test for the equality of k population means using data from a factorial experiment. To illustrate how PASW can be used for this type of experimental design, we show how to analyse the data for the two-factor GMAT experiment introduced in that section. The GMAT scores shown in Table 13.11 are entered into leftmost column of a PASW worksheet in Data View mode. Coding the factor A preparation programmes as 1 for the three-hour review, 2 for the one-day programme, and 3 for the ten-week course, the coded values for factor A are entered in Data View mode into the next adjacent column to the right in the worksheet. Coding the factor B colleges as 1 for Business, 2 for Engineering and 3 for Arts and Sciences, the coded values for factor B are entered into the next rightmost column of the worksheet. The columns are automatically labelled by the system, V1, V2 and V3 but can be relabelled in Variable View mode as score, factorA and factorB respectively. (Note that variable names in PASW are not allowed to contain spaces.) Thus, the values in the first row of the worksheet are 500, 1, 1; the values in row 2 are 580, 1, 1; the values in row 3 are 540, 2, 1; the values in row 4 are 460, 2, 1 and so on. The following steps show how to produce the PASW output corresponding to the ANOVA table shown in Figure 13.6.

Step 1   [Main menu bar]

Analyze > General Linear Model > Univariate

Step 2   [**Univariate** panel]

Enter stress in the **Dependent Variable** box
Enter system and controller in the **Fixed Factors** box
Click on **Model**
Click on **Full factorial**
Click on **Continue**
Click **OK**

Chapter 14

Simple Linear Regression

Learning objectives

After reading this chapter and doing the exercises, you should be able to:

1 Understand how regression analysis can be used to develop an equation that estimates mathematically how two variables are related.

2 Understand the differences between the regression model, the regression equation, and the estimated regression equation.

3 Know how to fit an estimated regression equation to a set of sample data based upon the least-squares method.

4 Determine how good a fit is provided by the estimated regression equation and compute the sample correlation coefficient from the regression analysis output.

5 Understand the assumptions necessary for statistical inference and be able to test for a significant relationship.

6 Know how to develop confidence interval estimates of the mean value of Y and an individual value of Y for a given value of X.

7 Learn how to use a residual plot to make a judgment as to the validity of the regression assumptions, recognise outliers and identify influential observations.

8 Use the Durbin-Watson test to test for autocorrelation.

9 Know the definition of the following terms:
independent and dependent variable
simple linear regression
regression model
regression equation and estimated regression equation
scatter diagram
coefficient of determination
standard error of the estimate
confidence interval
prediction interval
residual plot
standardized residual plot
outlier
influential observation
leverage

Managerial decisions are often based on the relationship between two or more variables. For example, after considering the relationship between advertising expenditures and sales, a marketing manager might attempt to predict sales for a given level of advertising expenditure. In another case, a public utility might use the relationship between the daily high temperature and the demand for electricity to predict electricity usage on the basis of next month's anticipated daily high temperatures. Sometimes a manager will rely on intuition to judge how two variables are related. However, if data can be obtained, a statistical procedure called *regression analysis* can be used to develop an equation showing how the variables are related.

In regression terminology, the variable being predicted is called the **dependent variable**. The variable or variables being used to predict the value of the dependent variable are called the **independent variables**. For example, in analyzing the effect of advertising expenditures on sales, a marketing manager's desire to predict sales would suggest making sales the dependent variable. Advertising expenditure would be the independent variable used to help predict sales. In statistical notation, Y denotes the dependent variable and X denotes the independent variable.

Statistics in Practice

Foreign direct investment (FDI) in China

In a recent study by Kingston Business School, regression modelling was used to investigate patterns of FDI in China as well as to assess the particular potential of the autonomous region of Guangxi in SW China as an FDI attractor. A variety of simple models were developed based on positive correlations between GDP and FDI

American coffee shop Starbucks in Shanghai, China. Keren Su/China Span/Alamy.

in provinces using data collected from official statistical sources.

Estimated regression equations obtained were as follows:

$$\hat{y} = 1.1m + 21.7x \qquad 1990\text{--}1993$$
$$\hat{y} = 2.1m + 8.9x \qquad 1995\text{--}1998$$
$$\hat{y} = 3.3m + 14.6x \qquad 2000\text{--}2003$$

where $\quad \hat{y}$ = estimated GDP

x = FDI

across all provinces.

In terms of FDI *per capita*, Guangxi has been ranked around 27 of 31 over the last ten years or so. FDI is a key driver of economic growth in modern China. But clearly Guangxi needs to improve its ranking if it is to be able to compete effectively with the more successful eastern coastal provinces and great municipalities.

Source: Foster MJ (2002) 'On evaluation of FDI's: Principles, Actualities and Possibilities' *International Journal of Management and Decision-Making* 3(1) 67–82

In this chapter we consider the simplest type of regression analysis involving one independent variable and one dependent variable in which the relationship between the variables is approximated by a straight line. It is called **simple linear regression**. Regression analysis involving two or more independent variables is called *multiple regression analysis*; multiple regression and cases involving curvilinear relationships are covered in Chapters 15 and 16.

14.1 Simple linear regression model

Armand's Pizza Parlours is a chain of Italian-food restaurants located in northern Italy. Armand's most successful locations are near college campuses. The managers believe that quarterly sales for these restaurants (denoted by Y) are related positively to the size of the student population (denoted by X); that is, restaurants near campuses with a large student population tend to generate more sales than those located near campuses with a small student population. Using regression analysis, we can develop an equation showing how the dependent variable Y is related to the independent variable X.

Regression model and regression equation

In the Armand's Pizza Parlours example, the population consists of all the Armand's restaurants.

For every restaurant in the population, there is a value x of X (student population) and a corresponding value y of Y (quarterly sales). The equation that describes how Y is related to x and an error term is called the **regression model**. The regression model used in simple linear regression follows.

Simple linear regression model

$$Y = \beta_0 + \beta_1 x + \varepsilon \qquad (14.1)$$

β_0 and β_1 are referred to as the parameters of the model, and ε (the Greek letter epsilon) is a random variable referred to as the *error term*. The error term ε accounts for the variability in Y that cannot be explained by the linear relationship between X and Y.

The population of all Armand's restaurants can also be viewed as a collection of sub-populations, one for each distinct value of X. For example, one subpopulation consists of all Armand's restaurants located near college campuses with 8000 students; another subpopulation consists of all Armand's restaurants located near college campuses with 9000 students and so on. Each subpopulation has a corresponding distribution of Y values. Thus, a distribution of Y values is associated with restaurants located near campuses with 8000 students a distribution of Y values is associated with restaurants located near campuses with 9000 students and so on. Each distribution of Y values has its own mean or expected value. The equation that describes how the expected value of Y – denoted by $E(Y)$ or equivalently $E(Y|X = x)$ – is related to x is called the **regression equation**. The regression equation for simple linear regression follows.

Simple linear regression equation

$$E(Y) = \beta_0 + \beta_1 x \qquad (14.2)$$

The graph of the simple linear regression equation is a straight line; β_0 is the y-intercept of the regression line, β_1 is the slope and $E(Y)$ is the mean or expected value of Y for a given value of X.

Examples of possible regression lines are shown in Figure 14.1. The regression line in Panel A shows that the mean value of Y is related positively to X, with larger values of

---

Figure 14.1 Possible regression lines in simple linear regression

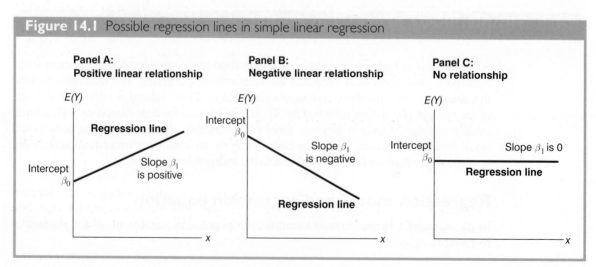

Figure 14.2 The estimation process in simple linear regression

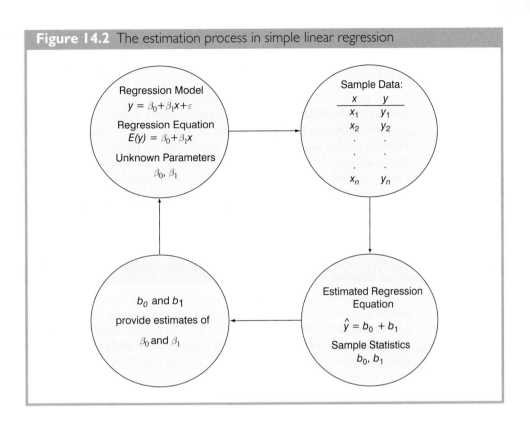

$E(Y)$ associated with larger values of X. The regression line in Panel B shows the mean value of Y is related negatively to X, with smaller values of $E(Y)$ associated with larger values of X. The regression line in Panel C shows the case in which the mean value of Y is not related to X; that is, the mean value of Y is the same for every value of X.

Estimated regression equation

If the values of the population parameters β_0 and β_1 were known, we could use equation (14.2) to compute the mean value of Y for a given value of X. In practice, the parameter values are not known, and must be estimated using sample data. Sample statistics (denoted b_0 and b_1) are computed as estimates of the population parameters β_0 and β_1. Substituting the values of the sample statistics b_0 and b_1 for β_0 and β_1 in the regression equation, we obtain the **estimated regression equation**. The estimated regression equation for simple linear regression follows.

Estimated simple linear regression equation

$$\hat{y} = b_0 + b_1 x \tag{14.3}$$

The graph of the estimated simple linear regression equation is called the *estimated regression line*; b_0 is the y intercept and b_1 is the slope. In the next section, we show how the least squares method can be used to compute the values of b_0 and b_1 in the estimated regression equation.

In general, $\hat{y}$ is the point estimator of $E(Y)$, the mean value of Y for a given value of X. Thus, to estimate the mean or expected value of quarterly sales for all restaurants located near campuses with 10 000 students, Armand's would substitute the value of 10 000 for X in equation (14.3). In some cases, however, Armand's may be more interested in predicting sales for one particular restaurant. For example, suppose Armand's would like to predict quarterly sales for the restaurant located near Cabot College, a school with 10 000 students.

As it turns out, the best estimate of Y for a given value of X is also provided by $\hat{y}$. Thus, to predict quarterly sales for the restaurant located near Cabot College, Armand's would also substitute the value of 10 000 for X in equation (14.3). Because the value of $\hat{y}$ provides both a point estimate of $E(Y)$ and an individual value of Y for a given value of X, we will refer to $\hat{y}$ simply as the *estimated value of* y.

Figure 14.2 provides a summary of the estimation process for simple linear regression.

14.2 Least squares method

The **least squares method** is a procedure for using sample data to find the estimated regression equation. To illustrate the least squares method, suppose data were collected from a sample of ten Armand's Pizza Parlour restaurants located near college campuses. For the ith observation or restaurant in the sample, x_i is the size of the student population (in thousands) and y_i is the quarterly sales (in thousands of euros). The values of x_i and y_i for the ten restaurants in the sample are summarized in Table 14.1. We see that restaurant 1, with $x_1 = 2$ and $y_1 = 58$, is near a campus with 2000 students and has quarterly sales of €58 000. Restaurant 2, with $x_2 = 6$ and $y_2 = 105$, is near a campus with 6000 students and has quarterly sales of €105 000. The largest sales value is for restaurant 10, which is near a campus with 26 000 students and has quarterly sales of €202 000.

Figure 14.3 is a scatter diagram of the data in Table 14.1. Student population is shown on the horizontal axis and quarterly sales are shown on the vertical axis. **Scatter diagrams** for regression analysis are constructed with the independent variable X on the horizontal axis and the dependent variable Y on the vertical axis. The scatter diagram enables us to observe the data graphically and to draw preliminary conclusions about the possible relationship between the variables.

Table 14.1 Student population and quarterly sales data for ten Armand's Pizza Parlours

Restaurant i	Student population (000s) x_i	Quarterly sales (€000s) y_i
1	2	58
2	6	105
3	8	88
4	8	118
5	12	117
6	16	137
7	20	157
8	20	169
9	22	149
10	26	202

ARMANDS

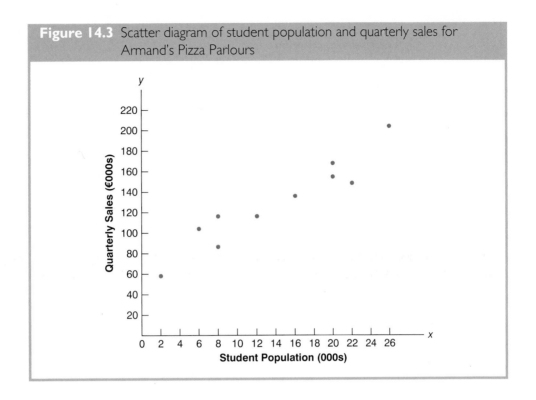

Figure 14.3 Scatter diagram of student population and quarterly sales for Armand's Pizza Parlours

What preliminary conclusions can be drawn from Figure 14.3? Quarterly sales appear to be higher at campuses with larger student populations. In addition, for these data the relationship between the size of the student population and quarterly sales appears to be approximated by a straight line; indeed, a positive linear relationship is indicated between X and Y. We therefore choose the simple linear regression model to represent the relationship between quarterly sales and student population. Given that choice, our next task is to use the sample data in Table 14.1 to determine the values of b_0 and b_1 in the estimated simple linear regression equation. For the ith restaurant, the estimated regression equation provides

$$\hat{y}_i = b_0 + b_1 x_i \tag{14.4}$$

where

$\hat{y}_i$ = estimated value of quarterly sales (€000s) for the ith restaurant

b_0 = the y intercept of the estimated regression line

b_1 = the slope of the estimated regression line

x_i = size of the student population (000s) for the ith restaurant

Every restaurant in the sample will have an observed value of sales y_i and an estimated value of sales $\hat{y}_i$. For the estimated regression line to provide a good fit to the data, we want the differences between the observed sales values and the estimated sales values to be small.

The least squares method uses the sample data to provide the values of b_0 and b_1 that minimize the *sum of the squares of the deviations* between the observed values of the dependent variable y_i and the estimated values of the dependent variable. The criterion for the least squares method is given by expression (14.5).

Least squares criterion

$$\text{Min } \Sigma(y_i - \hat{y}_i)^2 \tag{14.5}$$

where

y_i = observed value of the dependent variable for the ith observation
$\hat{y}_i$ = estimated value of the dependent variable for the ith observation

Differential calculus can be used to show that the values of b_0 and b_1 that minimize expression (14.5) can be found by using equations (14.6) and (14.7).

Slope and y-intercept for the estimated regression equation*

$$b_1 = \frac{\Sigma(x_i - \bar{x})(y_i - y)}{\Sigma(x_i - \bar{x})^2} \tag{14.6}$$

$$b_0 = \bar{y} - b_1\bar{x} \tag{14.7}$$

where

x_i = value of the independent variable for the ith observation
Y_i = value of the dependent variable for the ith observation
$\bar{x}$ = mean value for the independent variable
$\bar{y}$ = mean value for the dependent variable
n = total number of observations

Some of the calculations necessary to develop the least squares estimated regression equation for Armand's Pizza Parlours are shown in Table 14.2. With the sample of ten restaurants, we have $n = 10$ observations. Because equations (14.6) and (14.7) require $\bar{x}$ and $\bar{y}$ we begin the calculations by computing $\bar{x}$ and $\bar{y}$.

$$\bar{x} = \frac{\Sigma x_i}{n} = \frac{140}{10} = 14$$

$$\bar{y} = \frac{\Sigma y_i}{n} = \frac{1300}{10} = 130$$

Using equations (14.6) and (14.7) and the information in Table 14.2, we can compute the slope and intercept of the estimated regression equation for Armand's Pizza Parlours. The calculation of the slope (b_1) proceeds as follows.

$$b_1 = \frac{\Sigma(x_i - \bar{x})(y_i - \bar{y})}{\Sigma(x_i - \bar{x})^2}$$

$$= \frac{2840}{568} = 5$$

---

*An alternative formula for b_1 is

$$b_1 = \frac{\Sigma x_i y_i - (\Sigma x_i \Sigma y_i)n}{\Sigma x^2 - (\Sigma x_i)^2/n}$$

This form of equation (14.6) is often recommended when using a calculator to compute b_1.

Table 14.2 Calculations for the least squares estimated regression equation for Armand's Pizza Parlours

Restaurant i	x_i	y_i	$x_i - \bar{x}$	$y_i - \bar{y}$	$(x_i - \bar{x})(y_i - \bar{y})$	$(x_i - \bar{x})^2$
1	2	58	−12	−72	864	144
2	6	105	−8	−25	200	64
3	8	88	−6	−42	252	36
4	8	118	−6	−12	72	36
5	12	117	−2	−13	26	4
6	16	137	2	7	14	4
7	20	157	6	27	162	36
8	20	169	6	39	234	36
9	22	149	8	19	152	64
10	26	202	12	72	864	144
Totals	140	1300			2840	568
	Σx_i	Σy_i			$\Sigma(x_i - \bar{x})(y_i - \bar{y})$	$\Sigma(x_i - \bar{x})^2$

The calculation of the y intercept (b_0) follows.

$$b_0 = \bar{y} - b_1 \bar{x}$$
$$= 130 - 5(14)$$
$$= 60$$

Thus, the estimated regression equation is

$$\hat{y} = 60 + 5x$$

Figure 14.4 shows the graph of this equation on the scatter diagram.

The slope of the estimated regression equation ($b_1 = 5$) is positive, implying that as student population increases, sales increase. In fact, we can conclude (based on sales measured in €000s and student population in 000s) that an increase in the student population of 1000 is associated with an increase of €5000 in expected sales; that is, quarterly sales are expected to increase by €5 per student.

If we believe the least squares estimated regression equation adequately describes the relationship between X and Y, it would seem reasonable to use the estimated regression equation to predict the value of Y for a given value of X. For example, if we wanted to predict quarterly sales for a restaurant to be located near a campus with 16 000 students, we would compute

$$\hat{y} = 60 + 5(16) = 140$$

Therefore, we would predict quarterly sales of €140 000 for this restaurant. In the following sections we will discuss methods for assessing the appropriateness of using the estimated regression equation for estimation and prediction.

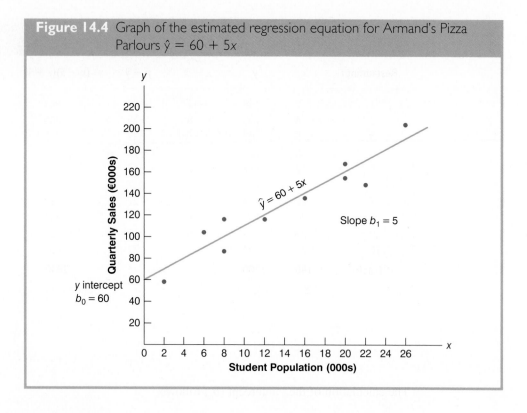

Figure 14.4 Graph of the estimated regression equation for Armand's Pizza Parlours $\hat{y} = 60 + 5x$

Exercises

Methods

1   Given are five observations for two variables, X and Y

x_i	1	2	3	4	5
y_i	3	7	5	11	14

a. Develop a scatter diagram for these data.
b. What does the scatter diagram developed in part (a) indicate about the relationship between the two variables?
c. Try to approximate the relationship between X and Y by drawing a straight line through the data.
d. Develop the estimated regression equation by computing the values of b_0 and b_1 using equations (14.6) and (14.7).
e. Use the estimated regression equation to predict the value of Y when $X = 4$.

2   Given are five observations for two variables, X and Y.

x_i	2	3	5	1	8
y_i	25	25	20	30	16

a. Develop a scatter diagram for these data.
b. What does the scatter diagram developed in part (a) indicate about the relationship between the two variables?

c. Try to approximate the relationship between X and Y by drawing a straight line through the data.

d. Develop the estimated regression equation by computing the values of b_0 and b_1 using equations (14.6) and (14.7).

e. Use the estimated regression equation to predict the value of Y when $X = 6$.

3 Given are five observations collected in a regression study on two variables.

$$\begin{array}{cccccc} x_i & 2 & 4 & 5 & 7 & 8 \\ y_i & 2 & 3 & 2 & 6 & 4 \end{array}$$

a. Develop a scatter diagram for these data.

b. Develop the estimated regression equation for these data.

c. Use the estimated regression equation to predict the value of Y when $X = 4$.

Applications

4 The following data were collected on the height (cm) and weight (kg) of women swimmers.

$$\begin{array}{cccccc} \text{Height} & 173 & 163 & 157 & 165 & 168 \\ \text{Weight} & 60 & 49 & 46 & 52 & 58 \end{array}$$

a. Develop a scatter diagram for these data with height as the independent variable.

b. What does the scatter diagram developed in part (a) indicate about the relationship between the two variables?

c. Try to approximate the relationship between height and weight by drawing a straight line through the data.

d. Develop the estimated regression equation by computing the values of b_0 and b_1.

e. If a swimmer's height is 160 cm, what would you estimate her weight to be?

5 The Dow Jones Industrial Average (DJIA) and the Standard & Poor's 500 (S&P) indexes are both used as measures of overall movement in the stock market. The DJIA is based on the price movements of 30 large companies; the S&P 500 is an index composed of 500 stocks. Some say the S&P 500 is a better measure of stock market performance because it is broader based. The closing prices for the DJIA and the S&P 500 for ten weeks, beginning with 11 February 2009, follow (*uk.finance.yahoo.com*, 21 April 2009).

DOWS&P

Date	DJIA	S&P
11 Feb 09	7939.53	833.74
18 Feb 09	7555.63	788.42
25 Feb 09	7270.89	764.90
03 Mar 09	6726.02	696.33
10 Mar 09	6926.49	719.60
17 Mar 09	7395.70	778.12
24 Mar 09	7660.21	806.12
31 Mar 09	7608.92	797.87
07 Apr 09	7789.56	815.55
14 Apr 09	7920.18	841.50

a. Develop a scatter diagram for these data with DJIA as the independent variable.

b. Develop the least squares estimated regression equation.

c. Suppose the closing price for the DJIA is 8000. Estimate the closing price for the S&P 500.

6 A sales manager collected the following data on annual sales and years of experience.

Salesperson	Years of experience	Annual sales (€000s)
1	1	80
2	3	97
3	4	92
4	4	102
5	6	103
6	8	111
7	10	119
8	10	123
9	11	117
10	13	136

a. Develop a scatter diagram for these data with years of experience as the independent variable.
b. Develop an estimated regression equation that can be used to predict annual sales given the years of experience.
c. Use the estimated regression equation to predict annual sales for a salesperson with nine years of experience.

14.3 Coefficient of determination

For the Armand's Pizza Parlours example, we developed the estimated regression equation $\hat{y} = 60 + 5x$ to approximate the linear relationship between the size of student population X and quarterly sales Y. A question now is: How well does estimated regression equation fit the data? In this section, we show that **coefficient of determination** provides a measure of the goodness of fit for the estimated regression equation.

For the ith observation, the difference between the observed value of the dependent variable, y_i, and the estimated value of the dependent variable, $\hat{y}_i$, is called the **ith residual**. The ith residual represents the error in using y_i to estimate $\hat{y}_i$. Thus, for the ith observation, the residual is $y_i - \hat{y}_i$. The sum of squares of these residuals or errors is the quantity that is minimized by the least squares method. This quantity, also known as the *sum of squares due to error*, is denoted by SSE.

Sum of squares due to error

$$SSE = \Sigma(y_i - \hat{y}_i)^2 \qquad (14.8)$$

The value of SSE is a measure of the error in using the least squares regression equation to estimate the values of the dependent variable in the sample.

In Table 14.3 we show the calculations required to compute the sum of squares due to error for the Armand's Pizza Parlours example. For instance, for restaurant 1 the values of the independent and dependent variables are $x_1 = 2$ and $y_1 = 58$. Using the estimated regression equation, we find that the estimated value of quarterly sales for restaurant 1 is

Table 14.3 Calculation of SSE for Armand's Pizza Parlours

Restaurant i	x_i = Student population (000s)	y_i = Quarterly sales (€000s)	Predicted sales $\hat{y}_i = 60 + 5x_i$	Error $y_i - \hat{y}_i$	Squared error $(y_i - \hat{y}_i)^2$
1	2	58	70	−12	144
2	6	105	90	15	225
3	8	88	100	−12	144
4	8	118	100	18	324
5	12	117	120	−3	9
6	16	137	140	−3	9
7	20	157	160	−3	9
8	20	169	160	9	81
9	22	149	170	−21	441
10	26	202	190	12	144
					SSE = 1530

$\hat{y}_1 = 60 + 5(2) = 70$. Thus, the error in using $\hat{y}_1$ to estimate y_1 for restaurant 1 is $y_1 - \hat{y}_1 = 58 - 70 = -12$. The squared error, $(-12)^2 = 144$, is shown in the last column of Table 14.3. After computing and squaring the residuals for each restaurant in the sample, we sum them to obtain SSE = 1530. Thus, SSE = 1530 measures the error in using the estimated regression equation $\hat{y}_1 = 60 + 5x$ to predict sales.

Now suppose we are asked to develop an estimate of quarterly sales without knowledge of the size of the student population. Without knowledge of any related variables, we would use the sample mean as an estimate of quarterly sales at any given restaurant. Table 14.2 shows that for the sales data, $\Sigma y_i = 1300$. Hence, the mean value of quarterly sales for the sample of ten Armand's restaurants is $\bar{y} = \Sigma y/n = 1300/10 = 130$. In Table 14.4 we show the sum of squared deviations obtained by using the sample mean $\bar{y} = 130$ to estimate the value of quarterly sales for each restaurant in the sample. For the ith restaurant in the sample, the difference $y_i - \bar{y}$ provides a measure of the error involved in using $\bar{y}$ to estimate sales. The corresponding sum of squares, called the *total sum of squares*, is denoted SST.

Total sum of squares

$$SST = \Sigma(y_i - \bar{y})^2 \qquad (14.9)$$

The sum at the bottom of the last column in Table 14.4 is the total sum of squares for Armand's Pizza Parlours; it is SST = 15 730.

In Figure 14.5 we show the estimated regression line $\hat{y}_i = 60 + 5x$ and the line corresponding to $\bar{y} = 130$. Note that the points cluster more closely around the estimated regression line than they do about the line $\bar{y} = 130$. For example, for the tenth restaurant in the sample we see that the error is much larger when $\bar{y} = 130$ is used as an estimate of y_{10} than when $\hat{y}_i = 60 + 5(26) = 190$ is used. We can think of SST as a measure of how well the observations cluster about the y line and SSE as a measure of how well the observations cluster about the $\hat{y}$ line.

Table 14.4 Computation of the total sum of squares for Armand's Pizza Parlours

Restaurant i	x_i = Student population (000s)	y_i = Quarterly sales (€000s)	Deviation $y_i - \bar{y}$	Squared deviation $(y_i - \bar{y})^2$
1	2	58	−72	5 184
2	6	105	−25	625
3	8	88	−42	1 764
4	8	118	−12	144
5	12	117	−13	169
6	16	137	7	49
7	20	157	27	729
8	20	169	39	1 521
9	22	149	19	361
10	26	202	72	5 184
				SST = 15 730

To measure how much the $\hat{y}$ values on the estimated regression line deviate from $\bar{y}$, another sum of squares is computed. This sum of squares, called the *sum of squares due to regression*, is denoted SSR.

Sum of squares due to regression

$$SSR = \Sigma(\hat{y}_i - \bar{y})^2 \qquad\qquad (14.10)$$

Figure 14.5 Deviations about the estimated regression line and the line $y = \bar{y}$ for Armand's Pizza Parlours

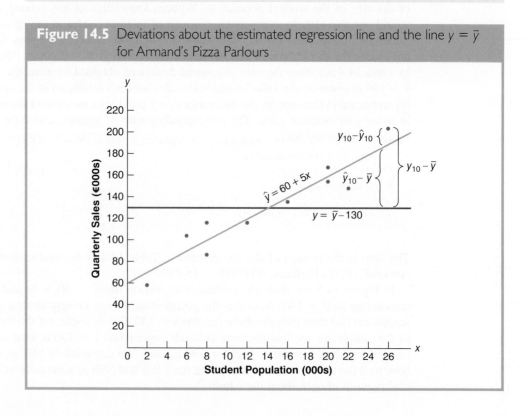

From the preceding discussion, we should expect that SST, SSR and SSE are related. Indeed, the relationship among these three sums of squares provides one of the most important results in statistics.

Relationship among SST, SSR and SSE

$$SST = SSR + SSE \tag{14.11}$$

where

SST = total sum of squares
SSR = sum of squares due to regression
SSE = sum of squares due to error

Equation (14.11) shows that the total sum of squares can be partitioned into two components, the regression sum of squares and the sum of squares due to error. Hence, if the values of any two of these sum of squares are known, the third sum of squares can be computed easily. For instance, in the Armand's Pizza Parlours example, we already know that SSE = 1530 and SST = 15 730; therefore, solving for SSR in equation (14.11), we find that the sum of squares due to regression is

$$SSR = SST - SSE = 15\ 730 - 1530 = 14\ 200$$

Now let us see how the three sums of squares, SST, SSR and SSE, can be used to provide a measure of the goodness of fit for the estimated regression equation. The estimated regression equation would provide a perfect fit if every value of the dependent variable y_i happened to lie on the estimated regression line. In this case, $y_i - \hat{y}_i$ would be zero for each observation, resulting in SSE = 0. Because SST = SSR + SSE, we see that for a perfect fit SSR must equal SST and the ratio (SSR/SST) must equal one. Poorer fits will result in larger values for SSE. Solving for SSE in equation (14.11), we see that SSE = SST - SSR. Hence, the largest value for SSE (and hence the poorest fit) occurs when SSR = 0 and SSE = SST. The ratio SSR/SST, which will take values between zero and one, is used to evaluate the goodness of fit for the estimated regression equation. This ratio is called the *coefficient of determination* and is denoted by r^2.

Coefficient of determination

$$r^2 = \frac{SSR}{SST} \tag{14.12}$$

For the Armand's Pizza Parlours example, the value of the coefficient of determination is

$$r^2 = \frac{SSR}{SST} = \frac{14\ 200}{15\ 730} = 0.9027$$

When we express the coefficient of determination as a percentage, r^2 can be interpreted as the percentage of the total sum of squares that can be explained by using the estimated regression equation. For Armand's Pizza Parlours, we can conclude that 90.27 per cent of the total sum of squares can be explained by using the estimated regression equation $\hat{y} = 60 + 5x$ to predict quarterly sales. In other words, 90.27 per cent of the variability in sales can be explained by the linear relationship between the size of the student population and sales. We should be pleased to find such a good fit for the estimated regression equation.

Correlation coefficient

In Chapter 3 we introduced the **correlation coefficient** as a descriptive measure of the strength of linear association between two variables, X and Y. Values of the correlation coefficient are always between -1 and $+1$. A value of $+1$ indicates that the two variables X and Y are perfectly related in a positive linear sense. That is, all data points are on a straight line that has a positive slope. A value of -1 indicates that X and Y are perfectly related in a negative linear sense, with all data points on a straight line that has a negative slope. Values of the correlation coefficient close to zero indicate that X and Y are not linearly related.

In Section 3.5 we presented the equation for computing the sample correlation coefficient. If a regression analysis has already been performed and the coefficient of determination r^2 computed, the sample correlation coefficient can be computed as follows.

Sample correlation coefficient

$$r_{XY} = (\text{sign of } b_1) \sqrt{\text{Coefficient of determination}}$$
$$= (\text{sign of } b_1) \sqrt{r^2} \qquad (14.13)$$

where

$$b_1 = \text{the slope of the estimated regression equation } \hat{y} = b_0 + b_1 x$$

The sign for the sample correlation coefficient is positive if the estimated regression equation has a positive slope ($b_1 > 0$) and negative if the estimated regression equation has a negative slope ($b_1 < 0$).

For the Armand's Pizza Parlour example, the value of the co efficient of determination corresponding to the estimated regression equation $\hat{y} = 60 + 5x$ is 0.9027. Because the slope of the estimated regression equation is positive, equation (14.13) shows that the sample correlation coefficient is $= \sqrt{0.9027} = +0.9501$.

With a sample correlation coefficient of $r_{XY} = +0.9501$, we would conclude that a strong positive linear association exists between X and Y.

In the case of a linear relationship between two variables, both the coefficient of determination and the sample correlation coefficient provide measures of the strength of the relationship. The coefficient of determination provides a measure between zero and one whereas the sample correlation coefficient provides a measure between -1 and $+1$. Although the sample correlation coefficient is restricted to a linear relationship between two variables, the coefficient of determination can be used for nonlinear relationships and for relationships that have two or more independent variables. Thus, the coefficient of determination provides a wider range of applicability.

Exercises

Methods

7   The data from exercise 1 follow.

x_i	1	2	3	4	5
y_i	3	7	5	11	14

The estimated regression equation for these data is $\hat{y} = 0.20 + 2.60x$.

a. Compute SSE, SST and SSR using equations (14.8), (14.9) and (14.10).
b. Compute the coefficient of determination r^2. Comment on the goodness of fit.
c. Compute the sample correlation coefficient.

8   The data from exercise 2 follow.

x_i	2	3	5	1	8
y_i	25	25	20	30	16

The estimated regression equation for these data is $\hat{y} = 30.33 - 1.88x$.

a. Compute SSE, SST and SSR.
b. Compute the coefficient of determination r^2. Comment on the goodness of fit.
c. Compute the sample correlation coefficient.

9   The data from exercise 3 follow.

x_i	2	4	5	7	8
y_i	2	3	2	6	4

The estimated regression equation for these data is $\hat{y} = 0.75 + 0.51x$. What percentage of the total sum of squares can be accounted for by the estimated regression equation? What is the value of the sample correlation coefficient?

Applications

10   The estimated regression equation for the data in exercise 5 can be shown to be $\hat{y} = -75.586 + 0.115x$. What percentage of the total sum of squares can be accounted for by the estimated regression equation?

Comment on the goodness of fit. What is the sample correlation coefficient?

11   An important application of regression analysis in accounting is in the estimation of cost. By collecting data on volume and cost and using the least squares method to develop an estimated regression equation relating volume and cost, an accountant can estimate the cost associated with a particular manufacturing volume. Consider the following sample of production volumes and total cost data for a manufacturing operation.

Production volume (units)	Total cost (€)
400	4000
450	5000
550	5400
600	5900
700	6400
750	7000

a. Use these data to develop an estimated regression equation that could be used to predict the total cost for a given production volume.

b. What is the variable cost per unit produced?

c. Compute the coefficient of determination. What percentage of the variation in total cost can be explained by production volume?

d. The company's production schedule shows 500 units must be produced next month. What is the estimated total cost for this operation?

12 *PCWorld* provided details for ten of the most economical laser printers (*PCWorld*, April 2009). The following data show the maximum printing speed in pages per minute (ppm) and the price (in euros including 15 per cent value added tax) for each printer.

PRINTERS-2009

Name	Speed (ppm)	Price (€)
Brother HL 2035	18	61.35
HP Laserjet P1005	15	70.13
Samsung ML-1640	16	77.39
HP Laserjet P1006	17	82.93
Brother HL-2140	22	92.34
Brother DCP7030	22	96.04
HP Laserjet P1009	16	99.52
HP Laserjet P1505	24	119.10
Samsung 4300	18	121.64
Epson EPL-6200 Mono	20	133.53

a. Develop the estimated regression equation with speed as the independent variable.

b. Compute r^2. What percentage of the variation in cost can be explained by the printing speed?

c. What is the sample correlation coefficient between speed and price? Does it reflect a strong or weak relationship between printing speed and cost?

14.4 Model assumptions

We saw in the previous section that the value of the coefficient of determination (r^2) is a measure of the goodness of fit of the estimated regression equation. However, even with a large value of r^2, the estimated regression equation should not be used until further analysis of the appropriateness of the assumed model has been conducted. An important step in determining whether the assumed model is appropriate involves testing for the significance of the relationship. The tests of significance in regression analysis are based on the following assumptions about the error term ε.

Assumptions about the error term ε in the regression model

$$Y = \beta_0 + \beta_1 x + \varepsilon$$

1 The error term ε is a random variable with a mean or expected value of zero; that is, $E(\varepsilon) = 0$.

Implication: β_0 and β_1 are constants, therefore $E(\beta_0) = \beta_0$ and $E(\beta_1) = \beta_1$; thus, for a given value x of X, the expected value of Y is

$$E(Y) = \beta_0 + \beta_1 x \qquad (14.14)$$

As we indicated previously, equation (14.14) is referred to as the regression equation.

2 The variance of ε, denoted by σ^2, is the same for all values of X
   Implication: The variance of Y about the regression line equals σ^2 and is the same for all values of X.

3 The values of ε are independent.
   Implication: The value of ε for a particular value of X is not related to the value of ε for any other value of X; thus, the value of Y for a particular value of X is not related to the value of Y for any other value of X.

4 The error term ε is a normally distributed random variable.
   Implication: Because Y is a linear function of ε, Y is also a normally distributed random variable.

Figure 14.6 illustrates the model assumptions and their implications; note that in this graphical interpretation, the value of $E(Y)$ changes according to the specific value of X considered. However, regardless of the X value, the probability distribution of ε and hence the probability distributions of Y are normally distributed, each with the same

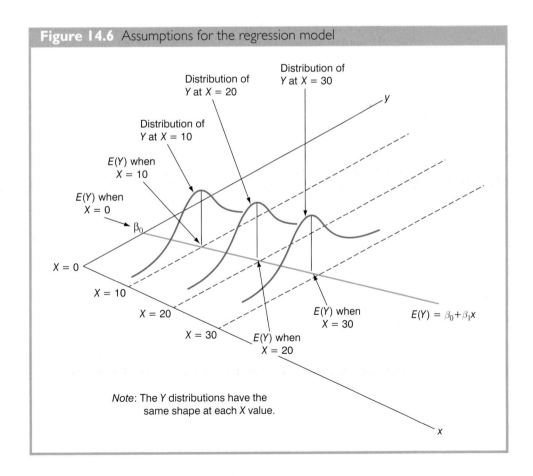

Figure 14.6 Assumptions for the regression model

Note: The Y distributions have the same shape at each X value.

variance. The specific value of the error ε at any particular point depends on whether the actual value of Y is greater than or less than $E(Y)$.

At this point, we must keep in mind that we are also making an assumption or hypothesis about the form of the relationship between X and Y. That is, we assume that a straight line represented by $\beta_0 + \beta_1 x$ is the basis for the relationship between the variables. We must not lose sight of the fact that some other model, for instance $Y = \beta_0 + \beta_1 x^2 + \varepsilon$ may turn out to be a better model for the underlying relationship.

14.5 Testing for significance

In a simple linear regression equation, the mean or expected value of Y is a linear function of x: $E(Y) = \beta_0 + \beta_1 x$. If the value of β_1 is zero, $E(Y) = \beta_0 + (0) x = \beta_0$. In this case, the mean value of Y does not depend on the value of X and hence we would conclude that X and Y are not linearly related. Alternatively, if the value of β_1 is not equal to zero, we would conclude that the two variables are related. Thus, to test for a significant regression relationship, we must conduct a hypothesis test to determine whether the value of β_1 is zero. Two tests are commonly used. Both require an estimate of σ^2, the variance of ε in the regression model.

Estimate of σ^2

From the regression model and its assumptions we can conclude that σ^2, the variance of ε, also represents the variance of the Y values about the regression line. Recall that the deviations of the Y values about the estimated regression line are called residuals. Thus, SSE, the sum of squared residuals, is a measure of the variability of the actual observations about the estimated regression line. The **mean square error** (MSE) provides the estimate of σ^2; it is SSE divided by its degrees of freedom.

With $\hat{y}_i = b_0 + b_1 x_i$, SSE can be written as

$$\text{SSE} = \Sigma(y_1 - \hat{y}_i)^2 = \Sigma(y_1 - b_0 - b_1 x_i)^2$$

Every sum of squares is associated with a number called its degrees of freedom. Statisticians have shown that SSE has $n - 2$ degrees of freedom because two parameters (β_0 and β_1) must be estimated to compute SSE. Thus, the mean square is computed by dividing SSE by $n - 2$. MSE provides an unbiased estimator of σ^2. Because the value of MSE provides an estimate of σ^2, the notation s^2 is also used.

Mean square error (estimate of σ^2)

$$s^2 = \text{MSE} = \frac{\text{SSE}}{n - 2} \tag{14.15}$$

In Section 14.3 we showed that for the Armand's Pizza Parlours example, SSE = 1530; hence,

$$s^2 = \text{MSE} = \frac{1530}{8} = 191.25$$

provides an unbiased estimate of σ^2.

To estimate σ we take the square root of s^2. The resulting value, s, is referred to as the **standard error of the estimate**.

Standard error of estimate

$$s = \sqrt{\text{MSE}} = \sqrt{\frac{\text{SSE}}{n-2}}$$ (14.16)

For the Armand's Pizza Parlours example, $s = \sqrt{\text{MSE}} = \sqrt{191.25} = 13.829$. In the following discussion, we use the standard error of the estimate in the tests for a significant relationship between X and Y.

t test

The simple linear regression model is $Y = \beta_0 + \beta_1 x + \varepsilon$. If X and Y are linearly related, we must have $\beta_1 \neq 0$. The purpose of the t test is to see whether we can conclude that $\beta_1 \neq 0$.

We will use the sample data to test the following hypotheses about the parameter β_1.

$$H_0: \beta_1 = 0$$
$$H_1: \beta_1 \neq 0$$

If H_0 is rejected, we will conclude that $\beta_1 \neq 0$ and that a statistically significant relationship exists between the two variables. However, if H_0 cannot be rejected, we will have insufficient evidence to conclude that a significant relationship exists. The properties of the sampling distribution of b_1, the least squares estimator of β_1, provide the basis for the hypothesis test.

First, let us consider what would happen if we used a different random sample for the same regression study. For example, suppose that Armand's Pizza Parlours used the sales records of a different sample of ten restaurants. A regression analysis of this new sample might result in an estimated regression equation similar to our previous estimated regression equation $\hat{y} = 60 + 5x$. However, it is doubtful that we would obtain exactly the same equation (with an intercept of exactly 60 and a slope of exactly 5). Indeed, b_0 and b_1, the least squares estimators, are sample statistics with their own sampling distributions. The properties of the sampling distribution of b_1 follow.

Sampling distribution of b_1

Expected value $E(b_1) = \beta_1$
Standard deviation

$$\sigma_{b_1} = \frac{\sigma}{\sqrt{\Sigma(x_i - \bar{x})^2}}$$ (14.17)

Distribution form
Normal

Note that the expected value of b_1 is equal to β_1, so b_1 is an unbiased estimator of β_1. As we do not know the value of σ, so we estimate σ_{b_1} by s_{b_1} where s_{b_1} is derived by substituting s for σ in equation (14.17):

Estimated standard deviation of b_1

$$s_{b_1} = \frac{s}{\sqrt{\Sigma(x_i - \bar{x})^2}} \qquad \text{(14.18)}$$

For Armand's Pizza Parlours, $s = 13.829$. Hence, using $\Sigma(x_i - \bar{x})^2 = 568$ as shown in Table 14.2, we have

$$s_{b_1} = \frac{13.829}{\sqrt{568}} = 0.5803$$

as the estimated standard deviation of b_1.

The t test for a significant relationship is based on the fact that the test statistic

$$\frac{b_1 - \beta_1}{s_{b_1}}$$

follows a t distribution with $n - 2$ degrees of freedom. If the null hypothesis is true, then $\beta_1 = 0$ and $t = b_1/s_{b_1}$.

Let us conduct this test of significance for Armand's Pizza Parlours at the $\alpha = 0.01$ level of significance. The test statistic is

$$t = \frac{b_1}{s_{b_1}} = \frac{5}{0.5803} = 8.62$$

The t distribution table shows that with $n - 2 = 10 - 2 = 8$ degrees of freedom, $t = 3.355$ provides an area of 0.005 in the upper tail. Thus, the area in the upper tail of the t distribution corresponding to the test statistic $t = 8.62$ must be less than 0.005. Because this test is a two-tailed test, we double this value to conclude that the p-value associated with $t = 8.62$ must be less than $2(0.005) = 0.01$. MINITAB, PASW or EXCEL show the p-value $= 0.000$. Because the p-value is less than $\alpha = 0.01$, we reject H_0 and conclude that β_1 is not equal to zero. This evidence is sufficient to conclude that a significant relationship exists between student population and quarterly sales. A summary of the t test for significance in simple linear regression follows.

t test for significance in simple linear regression

$$H_0: \beta_1 = 0$$
$$H_1: \beta_1 \neq 0$$

Test statistic

$$t = \frac{b_1}{s_{b_1}} \qquad \text{(14.19)}$$

Rejection rule

p-value approach: Reject H_0 if p-value $\leq \alpha$

Critical value approach: Reject H_0 if $t \leq -t_{\alpha/2}$ or if $t \geq t_{\alpha/2}$

where $t_{\alpha/2}$ is based on a t distribution with $n - 2$ degrees of freedom.

Confidence interval for β_1

The form of a confidence interval for β_1 is as follows:

$$b_1 \pm t_{\alpha/2} s_{b_1}$$

The point estimator is b_1 and the margin of error is $t_{\alpha/2} s_{b_1}$. The confidence coefficient associated with this interval is $1 - \alpha$, and $t_{\alpha/2}$ is the t value providing an area of $\alpha/2$ in the upper tail of a t distribution with $n - 2$ degrees of freedom. For example, suppose that we wanted to develop a 99 per cent confidence interval estimate of β_1 for Armand's Pizza Parlours. From Table 2 of Appendix B we find that the t value corresponding to $\alpha = 0.01$ and $n - 2 = 10 - 2 = 8$ degrees of freedom is $t_{0.005} = 3.355$. Thus, the 99 per cent confidence interval estimate of β_1 is

$$b_1 \pm t_{\alpha/2} s_{b_1} = 5 \pm 3.355(0.5803) = 5 \pm 1.95$$

or 3.05 to 6.95.

In using the t test for significance, the hypotheses tested were

$$H_0: \beta_1 = 0$$
$$H_1: \beta_1 \neq 0$$

At the $\alpha = 0.01$ level of significance, we can use the 99 per cent confidence interval as an alternative for drawing the hypothesis testing conclusion for the Armand's data. Because 0, the hypothesized value of β_1, is not included in the confidence interval (3.05 to 6.95), we can reject H_0 and conclude that a significant statistical relationship exists between the size of the student population and quarterly sales. In general, a confidence interval can be used to test any two-sided hypothesis about β_1. If the hypothesized value of β_1 is contained in the confidence interval, do not reject H_0. Otherwise, reject H_0.

F test

An F test, based on the F probability distribution, can also be used to test for significance in regression. With only one independent variable, the F test will provide the same conclusion as the t test; that is, if the t test indicates $\beta_1 \neq 0$ and hence a significant relationship, the F test will also indicate a significant relationship*. But with more than one independent variable, only the F test can be used to test for an overall significant relationship.

The logic behind the use of the F test for determining whether the regression relationship is statistically significant is based on the development of two independent estimates of σ^2. We explained how MSE provides an estimate of σ^2. If the null hypothesis $H_0: \beta_1 = 0$ is true, the sum of squares due to regression, SSR, divided by its degrees of freedom provides another independent estimate of σ^2. This estimate is called the *mean square due to regression*, or simply the *mean square regression*, and is denoted MSR. In general,

$$MSR = \frac{SSR}{\text{Regression degrees of freedom}}$$

---

*In fact $F = t^2$ for a simple regression model.

For the models we consider in this text, the regression degrees of freedom is always equal to the number of independent variables in the model:

Mean square regression

$$MSR = \frac{SSR}{\text{Number of independent variables}}$$

(14.20)

Because we consider only regression models with one independent variable in this chapter, we have $MSR = SSR/1 = SSR$. Hence, for Armand's Pizza Parlours, $MSR = SSR = 14\ 200$.

If the null hypothesis (H_0: $\beta_1 = 0$) is true, MSR and MSE are two independent estimates of σ^2 and the sampling distribution of MSR/MSE follows an F distribution with numerator degrees of freedom equal to one and denominator degrees of freedom equal to $n - 2$. Therefore, when $\beta_1 = 0$, the value of MSR/MSE should be close to one. However, if the null hypothesis is false ($\beta_1 \neq 0$), MSR will overestimate σ^2 and the value of MSR/MSE will be inflated; thus, large values of MSR/MSE lead to the rejection of H_0 and the conclusion that the relationship between X and Y is statistically significant.

Let us conduct the F test for the Armand's Pizza Parlours example. The test statistic is

$$F = \frac{MSR}{MSE} = \frac{14\ 200}{191.25} = 74.25$$

The F distribution table (Table 4 of Appendix B) shows that with one degree of freedom in the denominator and $n - 2 = 10 - 2 = 8$ degrees of freedom in the denominator, $F = 11.26$ provides an area of 0.01 in the upper tail. Thus, the area in the upper tail of the F distribution corresponding to the test statistic $F = 74.25$ must be less than 0.01. Thus, we conclude that the p-value must be less than 0.01. MINITAB, PASW or EXCEL show the p-value $= 0.000$. Because the p-value is less than $\alpha = 0.01$, we reject H_0 and conclude that a significant relationship exists between the size of the student population and quarterly sales. A summary of the F test for significance in simple linear regression follows.

F test for significance in simple linear regression

$$H_0: \beta_1 = 0$$
$$H_1: \beta_1 \neq 0$$

Test statistic

$$F = \frac{MSR}{MSE}$$

(14.21)

Rejection rule

p-value approach: Reject H_0 if p-value $\leq \alpha$
Critical value approach: Reject H_0 if $F \geq F_\alpha$

where F_α is based on an F distribution with 1 degree of freedom in the numerator and $n - 2$ degrees of freedom in the denominator.

Table 14.5 General form of the ANOVA table for simple linear regression

Source of variation	Degrees of freedom	Sum of squares	Mean square	F
Regression	1	SSR	$MSR = \dfrac{SSR}{1}$	$\dfrac{MSR}{MSE}$
Error	$n - 2$	SSE	$MSE = \dfrac{SSE}{n - 2}$	
Total	$n - 1$	SST		

In Chapter 13 we covered analysis of variance (ANOVA) and showed how an **ANOVA table** could be used to provide a convenient summary of the computational aspects of analysis of variance. A similar ANOVA table can be used to summarize the results of the F test for significance in regression. Table 14.5 is the general form of the ANOVA table for simple linear regression. Table 14.6 is the ANOVA table with the F test computations performed for Armand's Pizza Parlours. Regression, Error and Total are the labels for the three sources of variation, with SSR, SSE and SST appearing as the corresponding sum of squares in column 3. The degrees of freedom, 1 for SSR, $n - 2$ for SSE and $n - 1$ for SST, are shown in column 2. Column 4 contains the values of MSR and MSE and column 5 contains the value of F = MSR/MSE. Almost all computer printouts of regression analysis include an ANOVA table summary and the F test for significance.

Some cautions about the interpretation of significance tests

Rejecting the null hypothesis H_0: $\beta_1 = 0$ and concluding that the relationship between X and Y is significant does not enable us to conclude that a cause-and-effect relationship is present between X and Y. Concluding a cause-and-effect relationship is warranted only if the analyst can provide some type of theoretical justification that the relationship is in fact causal. In the Armand's Pizza Parlours example, we can conclude that there is a significant relationship between the size of the student population X and quarterly sales Y; moreover, the estimated regression equation $\hat{y} = 60 + 5x$ provides the least squares estimate of the relationship. We cannot, however, conclude that changes in student population X cause changes in quarterly sales Y just because we identified a statistically significant relationship. The appropriateness of such a cause-and-effect conclusion is left to supporting theoretical justification and to good judgment on the part of the analyst.

Table 14.6 ANOVA table for the Armand's Pizza Parlours problem

Source of variation	Degrees of freedom	Sum of squares	Mean square	F
Regression	1	14 200	$\dfrac{14\,200}{1} = 14\,200$	$\dfrac{14\,200}{191.25} = 74.25$
Error	8	1 530	$\dfrac{1530}{8} = 191.25$	
Total	9	15 730		

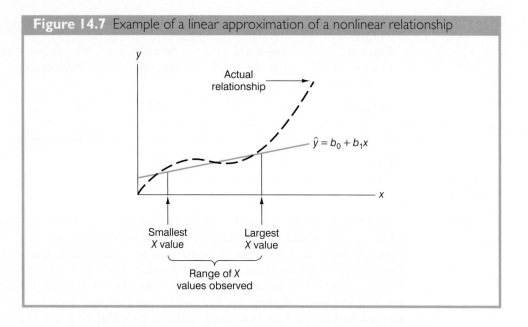

Figure 14.7 Example of a linear approximation of a nonlinear relationship

Armand's managers felt that increases in the student population were a likely cause of increased quarterly sales. Thus, the result of the significance test enabled them to conclude that a cause-and-effect relationship was present.

In addition, just because we are able to reject H_0: $\beta_1 = 0$ and demonstrate statistical significance does not enable us to conclude that the relationship between X and Y is linear. We can state only that X and Y are related and that a linear relationship explains a significant portion of the variability in Y over the range of values for X observed in the sample. Figure 14.7 illustrates this situation. The test for significance calls for the rejection of the null hypothesis H_0: $\beta_1 = 0$ and leads to the conclusion that X and Y are significantly related, but the figure shows that the actual relationship between X and Y is not linear. Although the linear approximation provided by $\hat{y} = b_0 + b_1 x$ is good over the range of X values observed in the sample, it becomes poor for X values outside that range.

Given a significant relationship, we should feel confident in using the estimated regression equation for predictions corresponding to X values within the range of the X values observed in the sample. For Armand's Pizza Parlours, this range corresponds to values of X between 2 and 26. Unless other reasons indicate that the model is valid beyond this range, predictions outside the range of the independent variable should be made with caution. For Armand's Pizza Parlours, because the regression relationship has been found significant at the 0.01 level, we should feel confident using it to predict sales for restaurants where the associated student population is between 2000 and 26 000.

Exercises

Methods

13   The data from exercise 1 follow.

x_i	1	2	3	4	5
y_i	3	7	5	11	14

a. Compute the mean square error using equation (14.15).
b. Compute the standard error of the estimate using equation (14.16).
c. Compute the estimated standard deviation of b_1 using equation (14.18). d. Use the t test to test the following hypotheses ($\alpha = 0.05$):

$$H_0: \beta_1 = 0$$
$$H_1: \beta_1 \neq 0$$

e. Use the F test to test the hypotheses in part (d) at a 0.05 level of significance. Present the results in the analysis of variance table format.

14 The data from exercise 2 follow.

x_i	2	3	5	1	8
y_i	25	25	20	30	16

a. Compute the mean square error using equation (14.15).
b. Compute the standard error of the estimate using equation (14.16).
c. Compute the estimated standard deviation of b_1 using equation (14.18).
d. Use the t test to test the following hypotheses ($\alpha = 0.05$):

$$H_0: \beta_1 = 0$$
$$H_1: \beta_1 \neq 0$$

e. Use the F test to test the hypotheses in part (d) at a 0.05 level of significance. Present the results in the analysis of variance table format.

15 The data from exercise 3 follow.

x_i	2	4	5	7	8
y_i	2	3	2	6	4

a. What is the value of the standard error of the estimate?
b. Test for a significant relationship by using the t test. Use $\alpha = 0.05$.
c. Use the F test to test for a significant relationship. Use $\alpha = 0.05$. What is your conclusion?

Applications

16 Refer to exercise 11, where data on production volume and cost were used to develop an estimated regression equation relating production volume and cost for a particular manufacturing operation. Use $\alpha = 0.05$ to test whether the production volume is significantly related to the total cost. Show the ANOVA table. What is your conclusion?

17 Refer to exercise 12 where the data were used to determine whether the price of a printer is related to the speed for plain text printing (*PC World*, April 2009). Does the evidence indicate a significant relationship between printing speed and price? Conduct the appropriate statistical test and state your conclusion. Use $\alpha = 0.05$.

14.6 Using the estimated regression equation for estimation and prediction

When using the simple linear regression model we are making an assumption about the relationship between X and Y. We then use the least squares method to obtain the estimated simple linear regression equation. If a significant relationship exists between X and

Y, and the coefficient of determination shows that the fit is good, the estimated regression equation should be useful for estimation and prediction.

Point estimation

In the Armand's Pizza Parlours example, the estimated regression equation $\hat{y} = 60 + 5x$ provides an estimate of the relationship between the size of the student population X and quarterly sales Y. We can use the estimated regression equation to develop a point estimate of either the mean value of Y or an individual value of Y corresponding to a given value of X. For instance, suppose Armand's managers want a point estimate of the mean quarterly sales for all restaurants located near college campuses with 10 000 students. Using the estimated regression equation $\hat{y} = 60 + 5x$, we see that for $X = 10$ (or 10 000 students), $\hat{y} = 60 + 5(10) = 110$. Thus, a point estimate of the mean quarterly sales for all restaurants located near campuses with 10 000 students is €110 000.

Now suppose Armand's managers want to predict sales for an individual restaurant located near Cabot College, a school with 10 000 students. Then, as the point estimate for an individual value of Y is the same as the point estimate for the mean value of Y we would predict quarterly sales of $\hat{y} = 60 + 5(10) = 110$ or €110 000 for this one restaurant.

Interval estimation

Point estimates do not provide any information about the precision associated with an estimate. For that we must develop interval estimates much like those in Chapters 8, 10 and 11. The first type of interval estimate, a **confidence interval**, is an interval estimate of the *mean value of Y* for a given value of X. The second type of interval estimate, a **prediction interval**, is used whenever we want an interval estimate of an *individual value* of Y for a given value of X. The point estimate of the mean value of Y is the same as the point estimate of an individual value of Y. But, the interval estimates we obtain for the two cases are different. The margin of error is larger for a prediction interval.

Confidence interval for the mean value of Y

The estimated regression equation provides a point estimate of the mean value of Y for a given value of X. In developing the confidence interval, we will use the following notation.

$$x_p = \text{the particular or given value of the independent variable } X$$
$$Y_p = \text{the dependent variable } Y \text{ corresponding to the given } x_p$$
$$E(Y_p) = \text{the mean or expected value of the dependent variable } Y_p \text{ corresponding to the given } x_p$$
$$\hat{y}_p = b_0 + b_1 x_p = \text{the point estimate of } E(Y_p) \text{ when } X = x_p$$

Using this notation to estimate the mean sales for all Armand's restaurants located near a campus with 10 000 students, we have $x_p = 10$, and $E(Y_p)$ denotes the unknown mean value of sales for all restaurants where $x_p = 10$. The point estimate of $E(Y_p)$ is given by $\hat{y}_p = 60 + 5(10) = 110$.

In general, we cannot expect $\hat{y}_p$ to equal $E(Y_p)$ exactly. If we want to make an inference about how close $\hat{y}_p$ is to the true mean value $E(Y_p)$, we will have to estimate the variance of $\hat{y}_p$. The formula for estimating the variance of $\hat{y}_p$ given x_p, denoted by $s^2_{\hat{y}_p}$ is

$$s^2_{\hat{y}_p} = s^2 \left[\frac{1}{n} + \frac{(x_p - \bar{x})^2}{\Sigma(x_i - \bar{x})^2} \right]$$

The general expression for a confidence interval follows.

Confidence interval for $E(Y_p)$

$$\hat{y}_p \pm t_{\alpha/2}s\sqrt{\frac{1}{n} + \frac{(x_p - \bar{x})^2}{\Sigma(x_i - \bar{x})^2}} \qquad (14.22)$$

where the confidence coefficient is $1 - \alpha$ and $t_{\alpha/2}$ is based on a t distribution with $n - 2$ degrees of freedom.

Using expression (14.22) to develop a 95 per cent confidence interval of the mean quarterly sales for all Armand's restaurants located near campuses with 10 000 students, we need the value of t for $\alpha/2 = 0.025$ and $n - 2 = 10 - 2 = 8$ degrees of freedom. Using Table 2 of Appendix B, we have $t_{0.025} = 2.306$. Thus, with $\hat{y}_p = 110$, the 95 per cent confidence interval estimate is

$$\hat{y}_p \pm t_{\alpha/2}s\sqrt{\frac{1}{n} + \frac{(x_p - \bar{x})^2}{\Sigma(x_i - \bar{x})^2}}$$

$$110 \pm 2.306 \times 13.829\sqrt{\frac{1}{10} + \frac{(10 - 14)^2}{568}}$$

$$= 110 \pm 11.415$$

In euros, the 95 per cent confidence interval for the mean quarterly sales of all restaurants near campuses with 10 000 students is €110 000 ± €11 415. Therefore, the 95 per cent confidence interval for the mean quarterly sales when the student population is 10 000 is €98 585 to €121 415.

Note that the estimated standard deviation of $\hat{y}_p$ is smallest when $x_p = \bar{x}$ so that the quantity $x_p - \bar{x} = 0$. In this case, the estimated standard deviation of $\hat{y}_p$ becomes

$$s\sqrt{\frac{1}{n} + \frac{(x_p - \bar{x})^2}{\Sigma(x_i - \bar{x})^2}} = s\sqrt{\frac{1}{n}}$$

This result implies that the best or most precise estimate of the mean value of Y occurs when $x_p = \bar{x}$. But, the further x_p is from $\bar{x}$ the larger $x_p - \bar{x}$ becomes and thus the wider confidence intervals will be for the mean value of Y. This pattern is shown graphically in Figure 14.8.

Prediction interval for an individual value of Y

Suppose that instead of estimating the mean value of sales for all Armand's restaurants located near campuses with 10 000 students, we want to estimate the sales for an individual restaurant located near Cabot College, a school with 10 000 students. As noted previously,

Figure 14.8 Confidence intervals for the mean sales Y at given values of student population x

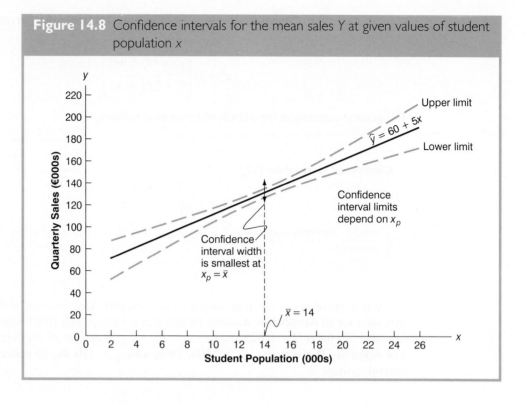

the point estimate of y_p, the value of Y corresponding to the given x_p, is provided by the estimated regression equation $\hat{y}_p = b_0 + b_1 x_p$. For the restaurant at Cabot College, we have $x_p = 10$ and a corresponding predicted quarterly sales of $\hat{y}_p = 60 + 5(10) = 110$, or €110 000.

Note that this value is the same as the point estimate of the mean sales for all restaurants located near campuses with 10 000 students.

To develop a prediction interval, we must first determine the variance associated with using $\hat{y}_p$ as an estimate of an individual value of Y when $X = x_p$. This variance is made up of the sum of the following two components.

1   The variance of individual Y values about the mean $E(Y_p)$, an estimate of which is given by s^2.

2   The variance associated with using $\hat{y}_p$ to estimate $E(Y_p)$, an estimate of which is given by

$$s_{\hat{y}_p}^2 = s^2 \left[\frac{1}{n} + \frac{(x_p - \bar{x})^2}{\Sigma(x_i - \bar{x})^2} \right]$$

Thus the formula for estimating the variance of an individual value of Y_p, is

$$s^2 + s_{\hat{y}_p}^2 = s^2 + s^2 \left[\frac{1}{n} + \frac{(x_p - \bar{x})^2}{\Sigma(x_i - \bar{x})^2} \right] = s^2 \left[1 + \frac{1}{n} + \frac{(x_p - \bar{x})^2}{\Sigma(x_i - \bar{x})^2} \right]$$

The general expression for a prediction interval follows.

Prediction interval for y_p

$$\hat{y}_p \pm t_{\alpha/2}\, s\sqrt{1 + \frac{1}{n} + \frac{(x_p - \bar{x})^2}{\Sigma(x_i - \bar{x})^2}} \qquad\qquad (14.23)$$

where the confidence coefficient is $1 - \alpha$ and $t_{\alpha/2}$ is based on a t distribution with $n - 2$ degrees of freedom.

Thus the 95 per cent prediction interval of sales for one specific restaurant located near a campus with 10 000 students is

$$\hat{y}_p \pm t_{\alpha/2}\, s\sqrt{1 + \frac{1}{n} + \frac{(x_p - \bar{x})^2}{\Sigma(x_i - \bar{x})^2}}$$

$$= 110 \pm 2.306 \times 13.829\sqrt{1 + \frac{1}{10} + \frac{(10 - 14)^2}{568}}$$

$$= 110 \pm 33.875$$

In euros, this prediction interval is €110 000 ± €33 875 or €76 125 to €143 875. Note that the prediction interval for an individual restaurant located near a campus with 10 000 students is wider than the confidence interval for the mean sales of all restaurants located near campuses with 10 000 students. The difference reflects the fact that we are able to estimate the mean value of Y more precisely than we can an individual value of Y.

Both confidence interval estimates and prediction interval estimates are most precise when the value of the independent variable is $x_p = \bar{x}$. The general shapes of confidence intervals and the wider prediction intervals are shown together in Figure 14.9.

Figure 14.9 Confidence and prediction intervals for sales Y at given values of student population X

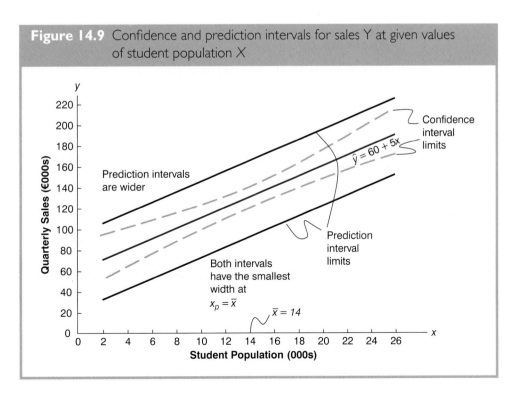

Exercises

Methods

18 The data from exercise 1 follow.

x_i	1	2	3	4	5
y_i	3	7	5	11	14

a. Use expression (14.22) to develop a 95 per cent confidence interval for the expected value of Y when $X = 4$.
b. Use expression (14.23) to develop a 95 per cent prediction interval for Y when $X = 4$.

19 The data from exercise 2 follow.

x_i	2	3	5	1	8
y_i	25	25	20	30	16

a. Estimate the standard deviation of $\hat{y}_p$ when $X = 3$.
b. Develop a 95 per cent confidence interval for the expected value of Y when $X = 3$.
c. Estimate the standard deviation of an individual value of Y when $X = 3$.
d. Develop a 95 per cent prediction interval for Y when $X = 3$.

20 The data from exercise 3 follow.

x_i	2	4	5	7	8
y_i	2	3	2	6	4

Develop the 95 per cent confidence and prediction intervals when $X = 3$. Explain why these two intervals are different.

Applications

21 Refer to Exercise 11, where data on the production volume X and total cost Y for a particular manufacturing operation were used to develop the estimated regression equation $\hat{y} = 1246.67 + 7.6x$.

a. The company's production schedule shows that 500 units must be produced next month. What is the point estimate of the total cost for next month?
b. Develop a 99 per cent prediction interval for the total cost for next month.
c. If an accounting cost report at the end of next month shows that the actual production cost during the month was €6000, should managers be concerned about incurring such a high total cost for the month? Discuss.

14.7 Computer solution

Performing the regression analysis computations without the help of a computer can be quite time consuming. In this section we discuss how the computational burden can be minimized by using a computer software package such as MINITAB.

We entered Armand's student population and sales data into a MINITAB worksheet. The independent variable was named Pop and the dependent variable was named Sales to assist with interpretation of the computer output. Using MINITAB, we obtained the

Figure 14.10 MINITAB output for the Armand's Pizza Parlours problem

```
Regression Analysis: Sales versus Pop

The regression equation is
Sales = 60.0 + 5.00 Pop


Predictor    Coef   SE Coef     T      P
Constant   60.000    9.226   6.50  0.000
Pop        5.0000   0.5803   8.62  0.000


S = 13.8293   R-Sq = 90.3%   R-Sq(adj) = 89.1%


Analysis of Variance

Source          DF     SS     MS      F      P
Regression       1  14200  14200  74.25  0.000
Residual Error   8   1530    191
Total            9  15730


Predicted Values for New Observations

New
Obs     Fit  SE Fit        95% CI            95% PI
  1  110.00    4.95  (98.58, 121.42)  (76.13, 143.87)


Values of Predictors for New Observations

New
Obs    Pop
  1   10.0
```

printout for Armand's Pizza Parlours shown in Figure 14.10.* The interpretation of this printout follows.

1 MINITAB prints the estimated regression equation as Sales = 60.0 + 5.00Pop.

2 A table is printed that shows the values of the coefficients b_0 and b_1, the standard deviation of each coefficient, the t value obtained by dividing each coefficient value by its standard deviation, and the p-value associated with the t test. Because the p-value is zero (to three decimal places), the sample results indicate that the null hypothesis ($H_0: \beta_1 = 0$) should be rejected. Alternatively, we could compare 8.62 (located in the t-ratio column) to the appropriate critical value. This procedure for the t test was described in Section 14.5.

3 MINITAB prints the standard error of the estimate, $s = 13.83$, as well as information about the goodness of fit. Note that 'R-sq = 90.3 per cent' is the coefficient of determination expressed as a percentage. The value 'R-Sq (adj) = 89.1 per cent' is discussed in Chapter 15.

*The MINITAB steps necessary to generate the output are given in the software section at the end of the chapter.

4 The ANOVA table is printed below the heading Analysis of Variance. MINITAB uses the label Residual Error for the error source of variation. Note that DF is an abbreviation for degrees of freedom and that MSR is given as 14 200 and MSE as 191.

The ratio of these two values provides the F value of 74.25 and the corresponding p-value of 0.000. Because the p-value is zero (to three decimal places), the relationship between Sales and Pop is judged statistically significant.

5 The 95 per cent confidence interval estimate of the expected sales and the 95 per cent prediction interval estimate of sales for an individual restaurant located near a campus with 10 000 students are printed below the ANOVA table. The confidence interval is (98.58, 121.42) and the prediction interval is (76.12, 143.88) as we showed in Section 14.6.

Exercises

Applications

22 The commercial division of a real estate firm is conducting a regression analysis of the relationship between X, annual gross rents (in thousands of euros), and Y, selling price (in thousands of euros) for apartment buildings. Data were collected on several properties recently sold and the following computer selective output was obtained.

```
The regression equation is
Y = 20.0 + 7.21 X

Predictor      Coef     SE Coef       T
Constant     20.000      3.2213    6.21
X             7.210      1.3626    5.29

Analysis of Variance

SOURCE             DF          SS
Regression          1     41587.3
Residual Error      7
Total               8     51984.1
```

a. How many apartment buildings were in the sample?
b. Write the estimated regression equation.
c. What is the value of s_{b_1}?
d. Use the F statistic to test the significance of the relationship at a 0.05 level of significance.
e. Estimate the selling price of an apartment building with gross annual rents of €50 000.

23 Following is a portion of the computer output for a regression analysis relating Y = maintenance expense (euros per month) to X = usage (hours per week) of a particular brand of computer terminal.

```
The regression equation is
Y = 6.1092 + .8951 X

Predictor      Coef     SE Coef
Constant     6.1092      0.9361
X            0.8951      0.1490

Analysis of Variance

SOURCE             DF          SS         MS
Regression          1     1575.76    1575.76
Residual Error      8      349.14      43.64
Total               9     1924.90
```

a. Write the estimated regression equation.

b. Use a *t* test to determine whether monthly maintenance expense is related to usage at the 0.05 level of significance.

c. Use the estimated regression equation to predict mean monthly maintenance expense for any terminal that is used 25 hours per week.

24 A regression model relating *X*, number of salespersons at a branch office, to *Y*, annual sales at the office (in thousands of euros) provided the following computer output from a regression analysis of the data.

```
The regression equation is
Y = 80.0 + 50.00 X

Predictor        Coef     SE Coef        T
Constant         80.0      11.333     7.06
X                50.0       5.482     9.12

Analysis of Variance

SOURCE             DF          SS        MS
Regression          1      6828.6    6828.6
Residual Error     28      2298.8      82.1
Total              29      9127.4
```

a. Write the estimated regression equation.

b. How many branch offices were involved in the study?

c. Compute the *F* statistic and test the significance of the relationship at a 0.05 level of significance.

d. Predict the annual sales at the Marseilles branch office. This branch employs 12 salespersons.

14.8 Residual analysis: validating model assumptions

As we noted previously, the *residual* for observation *i* is the difference between the observed value of the dependent variable (y_i) and the estimated value of the dependent variable $(\hat{y}_i)$.

Residual for observation *i*

$$y_i - \hat{y}_i \tag{14.24}$$

where

y_i is the observed value of the dependent variable

$\hat{y}_i$ is the estimated value of the dependent variable

In other words, the *i*th residual is the error resulting from using the estimated regression equation to predict the value of the dependent variable. The residuals for the Armand's Pizza Parlours example are computed in Table 14.7. The observed values of the dependent variable are in the second column and the estimated values of the dependent variable, obtained using the estimated regression equation $\hat{y} = 60 + 5x$, are in the third column. An analysis of the corresponding residuals in the fourth column

Table 14.7 Residuals for Armand's Pizza Parlours

Student population x_i	Sales y_i	Estimated sales $\hat{y}_i = 60 - 5x_i$	Residuals $y_i - \hat{y}_i$
2	58	70	−12
6	105	90	15
8	88	100	−12
8	118	100	18
12	117	120	−3
16	137	140	−3
20	157	160	−3
20	169	160	9
22	149	170	−21
26	202	190	12

will help determine whether the assumptions made about the regression model are appropriate.

Recall that for the Armand's Pizza Parlours example it was assumed the simple linear regression model took the form:

$$Y = \beta_0 + \beta_1 x + \varepsilon \tag{14.25}$$

In other words we assumed quarterly sales (Y) to be a linear function of the size of the student population (X) plus an error term ε. In Section 14.4 we made the following assumptions about the error term ε.

1 $E(\varepsilon) = 0$.

2 The variance of ε, denoted by σ^2, is the same for all values of X.

3 The values of ε are independent.

4 The error term ε has a normal distribution.

These assumptions provide the theoretical basis for the t test and the F test used to determine whether the relationship between X and Y is significant, and for the confidence and prediction interval estimates presented in Section 14.6. If the assumptions about the error term ε appear questionable, the hypothesis tests about the significance of the regression relationship and the interval estimation results may not be valid.

The residuals provide the best information about ε; hence an analysis of the residuals is an important step in determining whether the assumptions for ε are appropriate. Much of **residual analysis** is based on an examination of graphical plots. In this section, we discuss the following residual plots.

1 A plot of the residuals against values of the independent variable X.

2 A plot of residuals against the predicted values $\hat{y}$ of the dependent variable.

3 A standardized residual plot.

4 A normal probability plot.

Residual plot against *X*

A **residual plot** against the independent variable *X* is a graph in which the values of the independent variable are represented by the horizontal axis and the corresponding residual values are represented by the vertical axis. A point is plotted for each residual. The first coordinate for each point is given by the value of x_i and the second coordinate is given by the corresponding value of the residual $y_i - \hat{y}_i$. For a residual plot against *X* with the Armand's Pizza Parlours data from Table 14.7, the coordinates of the first point are $(2, -12)$, corresponding to $x_1 = 2$ and $y_1 - \hat{y}_1 = -12$; the coordinates of the second point are $(6, 15)$, corresponding to $x_2 = 6$ and $y_2 - \hat{y}_2 = 15$ and so on. Figure 14.11 shows the resulting residual plot.

Before interpreting the results for this residual plot, let us consider some general patterns that might be observed in any residual plot. Three examples appear in Figure 14.12.

If the assumption that the variance of ε is the same for all values of *X* and the assumed regression model is an adequate representation of the relationship between the variables, the residual plot should give an overall impression of a horizontal band of points such as the one in Panel A of Figure 14.12. However, if the variance of ε is not the same for all values of *X* – for example, if variability about the regression line is greater for larger values of *X* – a pattern such as the one in Panel B of Figure 14.12 could be observed. In this case, the assumption of a constant variance of ε is violated. Another possible residual plot is shown in Panel C. In this case, we would conclude that the assumed regression model is not an adequate representation of the relationship between the variables. A curvilinear regression model or multiple regression model should be considered.

Now let us return to the residual plot for Armand's Pizza Parlours shown in Figure 14.11. The residuals appear to approximate the horizontal pattern in Panel A of Figure 14.12. Hence, we conclude that the residual plot does not provide evidence that the assumptions made for Armand's regression model should be challenged. At

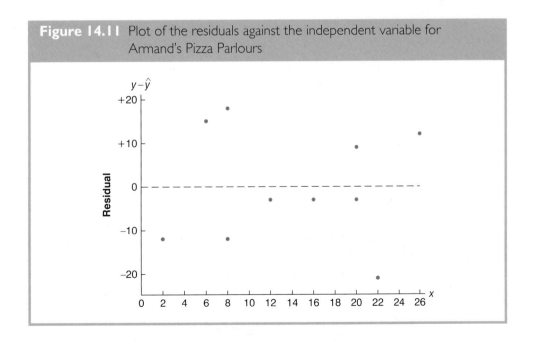

Figure 14.11 Plot of the residuals against the independent variable for Armand's Pizza Parlours

Figure 14.12 Residual plots from three regression studies

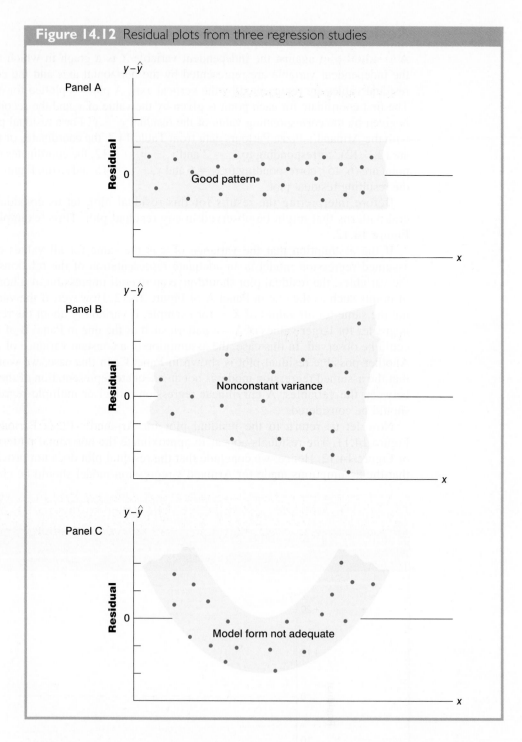

this point, we are confident in the conclusion that Armand's simple linear regression model is valid.

Experience and good judgment are always factors in the effective interpretation of residual plots. Seldom does a residual plot conform precisely to one of the patterns in Figure 14.12. Yet analysts who frequently conduct regression studies and frequently review residual plots become adept at understanding the differences between patterns that are reasonable and patterns that indicate the assumptions of the model should be questioned. A residual plot provides one technique to assess the validity of the assumptions for a regression model.

Residual plot against $\hat{y}$

Another residual plot represents the predicted value of the dependent variable $\hat{y}$ on the horizontal axis and the residual values on the vertical axis. A point is plotted for each residual. The first coordinate for each point is given by $\hat{y}_i$ and the second coordinate is given by the corresponding value of the ith residual $y_i - \hat{y}_i$. With the Armand's data from Table 14.7, the coordinates of the first point are $(70, -12)$, corresponding to $\hat{y}_1 = 70$ and $y_1 - \hat{y}_1 = -12$; the coordinates of the second point are $(90, 15)$, and so on. Figure 14.13 provides the residual plot. Note that the pattern of this residual plot is the same as the pattern of the residual plot against the independent variable X. It is not a pattern that would lead us to question the model assumptions. For simple linear regression, both the residual plot against X and the residual plot against $\hat{y}$ provide the same pattern. For multiple regression analysis, the residual plot against $\hat{y}$ is more widely used because of the presence of more than one independent variable.

Standardized residuals

Many of the residual plots provided by computer software packages use a standardized version of the residuals. As demonstrated in preceding chapters, a random variable is standardized by subtracting its mean and dividing the result by its standard deviation. With the least squares method, the mean of the residuals is zero. Thus, simply dividing each residual by its standard deviation provides the **standardized residual**.

It can be shown that the standard deviation of residual i depends on the standard error of the estimate s and the corresponding value of the independent variable x_i.

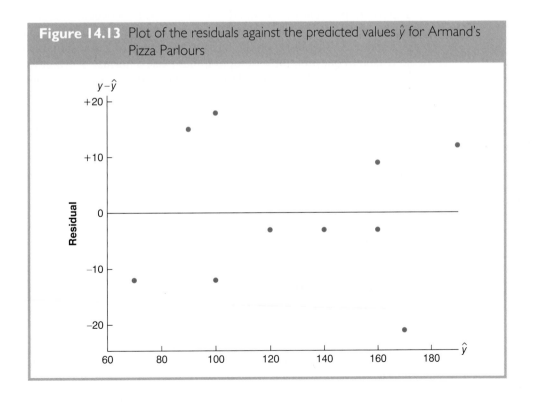

Figure 14.13 Plot of the residuals against the predicted values $\hat{y}$ for Armand's Pizza Parlours

Note that equation (14.26) shows that the standard deviation of the ith residual depends on x_i because of the presence of h_i in the formula.[†] Once the standard deviation of each residual is calculated, we can compute the standardized residual by dividing each residual by its corresponding standard deviation.

Standard deviation of the ith residual*

$$s_{y_i - \hat{y}_i} = s\sqrt{1 - h_i} \tag{14.26}$$

where

$$s_{y_i - \hat{y}_i} = \text{the standard deviation of residual } i$$
$$s = \text{the standard error of the estimate}$$

$$h_i = \frac{1}{n} + \frac{(x_i - \bar{x})^2}{\Sigma(x_i - \bar{x})^2} \tag{14.27}$$

Standardized residual for observation i

$$\frac{y_i - \hat{y}_i}{s_{y_i - \hat{y}_i}} \tag{14.28}$$

Table 14.8 shows the calculation of the standardized residuals for Armand's Pizza Parlours. Recall that previous calculations showed $s = 13.829$. Figure 14.14 is the plot of the standardized residuals against the independent variable X.

The standardized residual plot can provide insight about the assumption that the error term ε has a normal distribution. If this assumption is satisfied, the distribution of the standardized residuals should appear to come from a standard normal probability distribution.[‡]

Thus, when looking at a standardized residual plot, we should expect to see approximately 95 per cent of the standardized residuals between -2 and $+2$. We see in Figure 14.14 that for the Armand's example all standardized residuals are between -2 and $+2$. Therefore, on the basis of the standardized residuals, this plot gives us no reason to question the assumption that ε has a normal distribution.

Because of the effort required to compute the estimated values $\hat{y}$, the residuals, and the standardized residuals, most statistical packages provide these values as optional regression output. Hence, residual plots can be easily obtained. For large problems computer packages are the only practical means for developing the residual plots discussed in this section.

Normal probability plot

Another approach for determining the validity of the assumption that the error term has a normal distribution is the **normal probability plot**. To show how a normal probability plot is developed, we introduce the concept of *normal scores*.

[†] h_i is referred to as the *leverage* of observation i. Leverage will be discussed further when we consider influential observations in Section 14.9.

[*] This equation actually provides an estimate of the standard deviation of the ith residual, because s is used instead of σ.

[‡] Because s is used instead of σ in equation (14.26), the probability distribution of the standardized residuals is not technically normal. However, in most regression studies, the sample size is large enough that a normal approximation is very good.

Table 14.8 Computation of standardized residuals for Armand's Pizza Parlours

Restaurant i	x_i	$x_i - \bar{x}$	$(x_i - \bar{x})^2$	$\dfrac{(x_i - \bar{x})^2}{\Sigma(x_i - \bar{x})^2}$	h_i	$s_{y_i - \hat{y}_i}$	$y_i - \hat{y}_i$	Standardized residual
1	2	−12	144	0.2535	0.3535	11.1193	−12	−1.0792
2	6	−8	64	0.1127	0.2127	12.2709	15	1.2224
3	8	−6	36	0.0634	0.1634	12.6493	−12	−0.9487
4	8	−6	36	0.0634	0.1634	12.6493	18	1.4230
5	12	−2	4	0.0070	0.1070	13.0682	−3	−0.2296
6	16	2	4	0.0070	0.1070	13.0682	−3	−0.2296
7	20	6	36	0.0634	0.1634	12.6493	−3	−0.2372
8	20	6	36	0.0634	0.1634	12.6493	9	0.7115
9	22	8	64	0.1127	0.2127	12.2709	−21	−1.7114
10	26	12	144	0.2535	0.3535	11.1193	12	1.0792
		Total	568					

Note: The values of the residuals were computed in Table 14.7.

Suppose ten values are selected randomly from a normal probability distribution with a mean of zero and a standard deviation of one, and that the sampling process is repeated over and over with the values in each sample of ten ordered from smallest to largest. For now, let us consider only the smallest value in each sample. The random variable representing the smallest value obtained in repeated sampling is called the first order statistic.

Figure 14.14 Plot of the standardized residuals against the independent variable X for Armand's Pizza Parlours

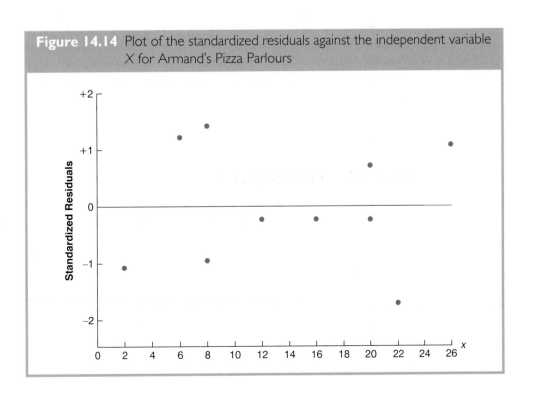

Table 14.9 Normal scores for $n = 10$

Order Statistic	Normal score
1	−1.55
2	−1.00
3	−0.65
4	−0.37
5	−0.12
6	0.12
7	0.37
8	0.65
9	1.00
10	1.55

Statisticians show that for samples of size ten from a standard normal probability distribution, the expected value of the first-order statistic is − 1.55. This expected value is called a normal score. For the case with a sample of size $n = 10$, there are ten order statistics and ten normal scores (see Table 14.9). In general, a data set consisting of n observations will have n order statistics and hence n normal scores.

Let us now show how the ten normal scores can be used to determine whether the standardized residuals for Armand's Pizza Parlours appear to come from a standard normal probability distribution. We begin by ordering the ten standardized residuals from Table 14.8. The ten normal scores and the ordered standardized residuals are shown together in Table 14.10. If the normality assumption is satisfied, the smallest standardized residual should be close to the smallest normal score, the next smallest standardized residual should be close to the next smallest normal score and so on. If we were to develop a plot with the normal scores on the horizontal axis and the corresponding standardized residuals on the vertical axis, the plotted points should cluster closely around a 45-degree line passing through the origin if the standardized residuals are approximately normally distributed. Such a plot is referred to as a *normal probability plot*.

Figure 14.15 is the normal probability plot for the Armand's Pizza Parlours example. Judgment is used to determine whether the pattern observed deviates from

Table 14.10 Normal scores and ordered standardized residuals for Armand's Pizza Parlours

Ordered normal scores	Standardized residuals
−1.55	−1.7114
−1.00	−1.0792
−0.65	−0.9487
−0.37	−0.2372
−0.12	−0.2296
0.12	−0.2296
0.37	0.7115
0.65	1.0792
1.00	1.2224
1.55	1.4230

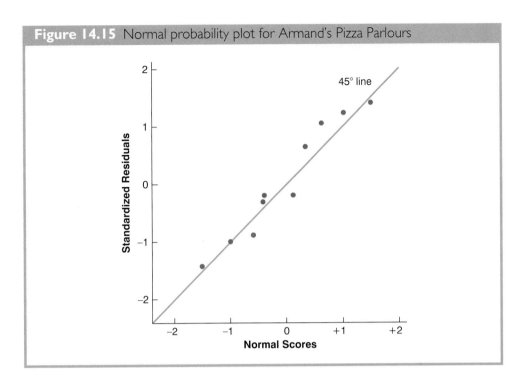

Figure 14.15 Normal probability plot for Armand's Pizza Parlours

the line enough to conclude that the standardized residuals are not from a standard normal probability distribution. In Figure 14.15, we see that the points are grouped closely about the line. We therefore conclude that the assumption of the error term having a normal probability distribution is reasonable. In general, the more closely the points are clustered about the 45-degree line, the stronger the evidence supporting the normality assumption. Any substantial curvature in the normal probability plot is evidence that the residuals have not come from a normal distribution. Normal scores and the associated normal probability plot can be obtained easily from statistical packages such as MINITAB.

14.9 Residual analysis: autocorrelation

In the last section we showed how residual plots can be used to detect violations of assumptions about the error term ε in the regression model. In many regression studies, particularly involving data collected over time, a special type of correlation among the error terms can cause problems; it is called **serial correlation** or **autocorrelation**. In this section we show how the **Durbin-Watson test** can be used to detect significant autocorrelation.

Autocorrelation and the Durbin-Watson test

Often, the data used for regression studies in business and economics are collected over time. It is not uncommon for the value of Y at time t, denoted by y_t, to be related to the value of Y at previous time periods. In such cases, we say autocorrelation (also called serial correlation) is present in the data. If the value of Y in time period t is related to its value in time period $t - 1$, first-order autocorrelation is present. If the value of Y in time period t is related to the value of Y in time period $t - 2$, second-order autocorrelation is present and so on.

When autocorrelation is present, one of the assumptions of the regression model is violated: the error terms are not independent. In the case of first-order autocorrelation, the error at time t, denoted ε_t, will be related to the error at time period $t - 1$, denoted ε_{t-1}. Two cases of first-order autocorrelation are illustrated in Figure 14.16. Panel A is the case of positive autocorrelation; panel B is the case of negative autocorrelation. With positive autocorrelation we expect a positive residual in one period to be followed by a positive residual in the next period, a negative residual in one period to be followed by a negative residual in the next period and so on. With negative autocorrelation, we expect a positive residual in one period to be followed by a negative residual in the next period, then a positive residual and so on. When autocorrelation is present, serious errors can be made in performing tests of statistical significance based upon the assumed regression model. It is therefore important to be able to detect autocorrelation and take corrective action. We will show how the Durbin-Watson statistic can be used to detect first-order autocorrelation.

Suppose the values of ε are not independent but are related in the following manner:

First order autocorrelation

$$\varepsilon_t = \rho\varepsilon_{t-1} + z_t \tag{14.29}$$

where ρ is a parameter with an absolute value less than one and z_t is a normally and independently distributed random variable with a mean of zero and a variance of σ^2. From equation (16.16) we see that if $\rho = 0$, the error terms are not related, and each has a mean of zero and a variance of σ^2. In this case, there is no autocorrelation and the regression assumptions are satisfied. If $\rho > 0$, we have positive autocorrelation; if $\rho < 0$, we have negative autocorrelation. In either of these cases, the regression assumptions about the error term are violated.

The Durbin-Watson test for autocorrelation uses the residuals to determine whether $\rho = 0$. To simplify the notation for the Durbin-Watson statistic, we denote the ith residual by $e_t = y_t - \hat{y}_t$. The Durbin-Watson test statistic is computed as follows.

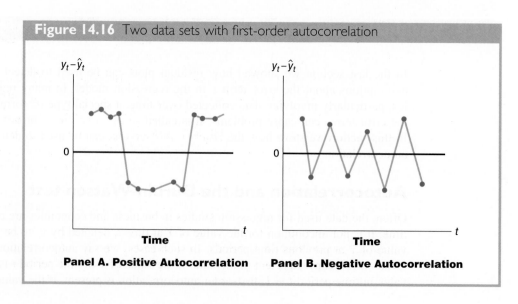

Figure 14.16 Two data sets with first-order autocorrelation

Panel A. Positive Autocorrelation

Panel B. Negative Autocorrelation

Durbin-Watson test statistic

$$d = \frac{\sum_{t=2}^{n}(e_t - e_{t-1})^2}{\sum_{t=1}^{n}e_t^2}$$

(14.30)

If successive values of the residuals are close together (positive autocorrelation), the value of the Durbin-Watson test statistic will be small. If successive values of the residuals are far apart (negative autocorrelation), the value of the Durbin-Watson statistic will be large.

The Durbin-Watson test statistic ranges in value from zero to four, with a value of two indicating no autocorrelation is present. Durbin and Watson developed tables that can be used to determine when their test statistic indicates the presence of autocorrelation. Table 8 in Appendix B shows lower and upper bounds (d_L and d_U) for hypothesis tests using $\alpha = 0.05$, $\alpha = 0.025$, and $\alpha = 0.01$; n denotes the number of observations.

The null hypothesis to be tested is always that there is no autocorrelation.

$$H_0: \rho = 0$$

The alternative hypothesis to test for positive autocorrelation is

$$H_1: \rho > 0$$

The alternative hypothesis to test for negative autocorrelation is

$$H_1: \rho < 0$$

A two-sided test is also possible. In this case the alternative hypothesis is

$$H_1: \rho \neq 0$$

Figure 14.17 shows how the values of d_L and d_U in Table 8 are used to test for autocorrelation.

Panel A illustrates the test for positive autocorrelation. If $d < d_L$, we conclude that positive autocorrelation is present. If $d_L \leq d \leq d_U$, we say the test is inconclusive. If $d > d_U$, we conclude that there is no evidence of positive autocorrelation.

Panel B illustrates the test for negative autocorrelation. If $d > 4 - d_L$, we conclude that negative autocorrelation is present. If $4 - d_U \leq d \leq 4 - d_L$, we say the test is inconclusive. If $d < 4 - d_U$, we conclude that there is no evidence of negative autocorrelation.

Note: Entries in Table 8 are the critical values for a one-tailed Durbin-Watson test for autocorrelation. For a two-tailed test, the level of significance is doubled.

Panel C illustrates the two-sided test. If $d < d_L$ or $d > 4 - d_L$, we reject H_0 and conclude that autocorrelation is present. If $d_L \leq d \leq d_U$ or $4 - d_U \leq d \leq 4 - d_L$, we say the test is inconclusive. If $d_U \leq d \leq 4 - d_U$, we conclude that there is no evidence of autocorrelation.

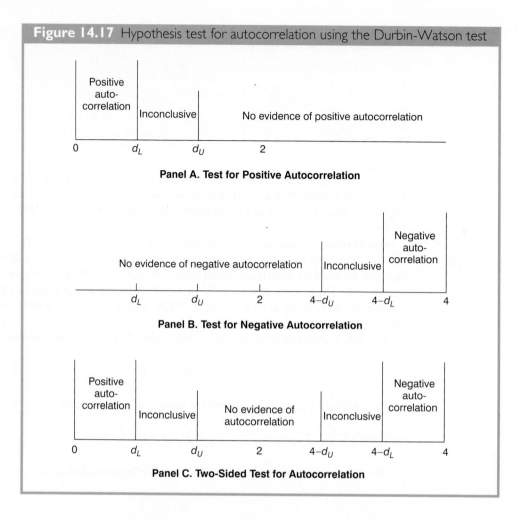

Figure 14.17 Hypothesis test for autocorrelation using the Durbin-Watson test

Panel A. Test for Positive Autocorrelation

Panel B. Test for Negative Autocorrelation

Panel C. Two-Sided Test for Autocorrelation

If significant autocorrelation is identified, we should investigate whether we omitted one or more key independent variables that have time-ordered effects on the dependent variable. If no such variables can be identified, including an independent variable that measures the time of the observation (for instance, the value of this variable could be one for the first observation, two for the second observation and so on) will sometimes eliminate or reduce the autocorrelation. When these attempts to reduce or remove auto-correlation do not work, transformations on the dependent or independent variables can prove helpful; a discussion of such transformations can be found in more advanced texts on regression analysis.

Note that the Durbin-Watson tables list the smallest sample size as 15. The reason is that the test is generally inconclusive for smaller sample sizes; in fact, many statisticians believe the sample size should be at least 50 for the test to produce worthwhile results.

Exercises

Methods

25 Given are data for two variables, X and Y.

x_i	6	11	15	18	20
y_i	6	8	12	20	30

a. Develop an estimated regression equation for these data.
b. Compute the residuals.
c. Develop a plot of the residuals against the independent variable X. Do the assumptions about the error terms seem to be satisfied?
d. Compute the standardized residuals.
e. Develop a plot of the standardized residuals against $\hat{y}$. What conclusions can you draw from this plot?

26 The following data were used in a regression study.

Observation	x_i	y_i	Observation	x_i	y_i
1	2	4	6	7	6
2	3	5	7	7	9
3	4	4	8	8	5
4	5	6	9	9	11
5	7	4			

a. Develop an estimated regression equation for these data.
b. Construct a plot of the residuals. Do the assumptions about the error term seem to be satisfied?

Applications

27 Data on advertising expenditures and revenue (in thousands of euros) for the Four Seasons Restaurant follow.

Advertising expenditures	Revenue
1	19
2	32
4	44
6	40
10	52
14	53
20	54

a. Let X equal advertising expenditures and Y equal revenue. Use the method of least squares to develop a straight line approximation of the relationship between the two variables.
b. Test whether revenue and advertising expenditures are related at a 0.05 level of significance.
c. Prepare a residual plot of $y - \hat{y}$ versus $\hat{y}$. Use the result from part (a) to obtain the values of $\hat{y}$.
d. What conclusions can you draw from residual analysis? Should this model be used, or should we look for a better one?

28 Refer to exercise 6, where an estimated regression equation relating years of experience and annual sales was developed.

a. Compute the residuals and construct a residual plot for this problem.
b. Do the assumptions about the error terms seem reasonable in light of the residual plot?

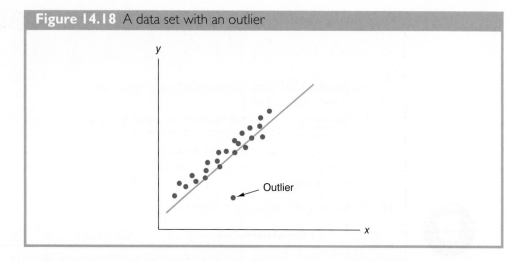

Figure 14.18 A data set with an outlier

14.10 Residual analysis: outliers and influential observations

In Section 14.8 we showed how residual analysis could be used to determine when violations of assumptions about the regression model occur. In this section, we discuss how residual analysis can be used to identify observations that can be classified as outliers or as being especially influential in determining the estimated regression equation. Some steps that should be taken when such observations occur are discussed.

Detecting outliers

Figure 14.18 is a scatter diagram for a data set that contains an **outlier**, a data point (observation) that does not fit the trend shown by the remaining data. Outliers represent observations that are suspect and warrant careful examination. They may represent erroneous data; if so, the data should be corrected. They may signal a violation of model assumptions; if so, another model should be considered. Finally, they may simply be unusual values that occurred by chance. In this case, they should be retained.

To illustrate the process of detecting outliers, consider the data set in Table 14.11; Figure 14.19 is a scatter diagram. Except for observation 4 ($x_4 = 3$, $y_4 = 75$), a pattern

Table 14.11 Data set illustrating the effect of an outlier

x_i	y_i
1	45
1	55
2	50
3	75
3	40
3	45
4	30
4	35
5	25
6	15

Figure 14.19 Scatter diagram for outlier data set

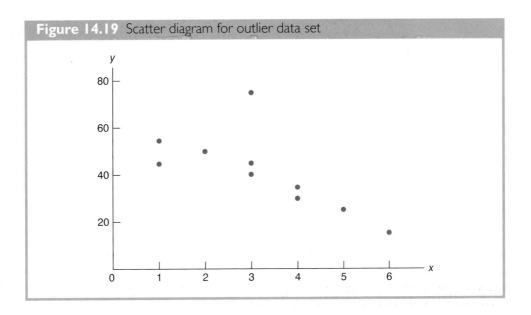

suggesting a negative linear relationship is apparent. Indeed, given the pattern of the rest of the data, we would expect y_4 to be much smaller and hence would identify the corresponding observation as an outlier. For the case of simple linear regression, one can often detect outliers by simply examining the scatter diagram.

The standardized residuals can also be used to identify outliers. If an observation deviates greatly from the pattern of the rest of the data (e.g. the outlier in Figure 14.18), the corresponding standardized residual will be large in absolute value. Many computer packages automatically identify observations with standardized residuals that are large in absolute value. In Figure 14.20 we show the MINITAB output from a regression analysis of the data in Table 14.11. The next to last line of the output shows that the standardized residual for observation 4 is 2.67. MINITAB identifies any observation with a standardized residual of less than -2 or greater than $+2$ as an unusual observation; in such cases, the observation is printed on a separate line with an R next to the standardized residual, as shown in Figure 14.20. With normally distributed errors, standardized residuals should be outside these limits approximately 5 per cent of the time.

In deciding how to handle an outlier, we should first check to see whether it is a valid observation. Perhaps an error was made in initially recording the data or in entering the data into the computer file. For example, suppose that in checking the data for the outlier in Table 14.11, we find an error; the correct value for observation 4 is $x_4 = 3$, $y_4 = 30$. Figure 14.21 is the MINITAB output obtained after correction of the value of y_4. We see that using the incorrect data value substantially affected the goodness of fit. With the correct data, the value of R-sq increased from 49.7 per cent to 83.8 per cent and the value of b_0 decreased from 64.958 to 59.237. The slope of the line changed from -7.331 to -6.949. The identification of the outlier enabled us to correct the data error and improve the regression results.

Detecting influential observations

Sometimes one or more observations exert a strong influence on the results obtained. Figure 14.22 shows an example of an **influential observation** in simple linear regression. The estimated regression line has a negative slope. However, if the influential observation were dropped from the data set, the slope of the estimated regression line

Figure 14.20 MINITAB output for regression analysis of the outlier data set

```
Regression Analysis: y versus x

The regression equation is
y = 65.0 - 7.33 x


Predictor     Coef   SE Coef       T      P
Constant    64.958     9.258    7.02  0.000
x           -7.331     2.608   -2.81  0.023


S = 12.6704   R-Sq = 49.7%   R-Sq(adj) = 43.4%


Analysis of Variance

Source         DF       SS      MS      F      P
Regression      1   1268.2  1268.2   7.90  0.023
Residual Error  8   1284.3   160.5
Total           9   2552.5


Unusual Observations

Obs    x       y    Fit  SE Fit  Residual  St Resid
  4  3.00  75.00  42.97    4.04     32.03     2.67R

R denotes an observation with a large standardized residual.
```

Figure 14.21 MINITAB output for the revised outlier data set

```
Regression Analysis: y versus x

The regression equation is
y = 59.2 - 6.95 x


Predictor     Coef   SE Coef       T      P
Constant    59.237     3.835   15.45  0.000
x           -6.949     1.080   -6.43  0.000


S = 5.24808   R-Sq = 83.8%   R-Sq(adj) = 81.8%


Analysis of Variance

Source         DF       SS      MS      F      P
Regression      1   1139.7  1139.7   41.38  0.000
Residual Error  8    220.3    27.5
Total           9   1360.0
```

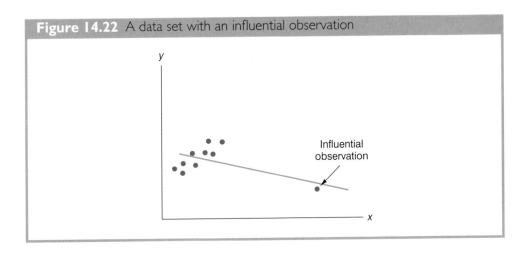

Figure 14.22 A data set with an influential observation

would change from negative to positive and the y-intercept would be smaller. Clearly, this one observation is much more influential in determining the estimated regression line than any of the others; dropping one of the other observations from the data set would have little effect on the estimated regression equation.

Influential observations can be identified from a scatter diagram when only one independent variable is present. An influential observation may be an outlier (an observation with a Y value that deviates substantially from the trend), it may correspond to an X value far away from its mean (e.g. see Figure 14.22), or it may be caused by a combination of the two (a somewhat off-trend Y value and a somewhat extreme X value).

Because influential observations may have such a dramatic effect on the estimated regression equation, they must be examined carefully. We should first check to make sure that no error was made in collecting or recording the data. If an error occurred, it can be corrected and a new estimated regression equation can be developed. If the observation is valid, we might consider ourselves fortunate to have it. Such a point, if valid, can contribute to a better understanding of the appropriate model and can lead to a better estimated regression equation. The presence of the influential observation in Figure 14.22, if valid, would suggest trying to obtain data on intermediate values of X to understand better the relationship between X and Y.

Observations with extreme values for the independent variables are called **high leverage points**. The influential observation in Figure 14.22 is a point with high leverage. The leverage of an observation is determined by how far the values of the independent variables are from their mean values. For the single-independent-variable case, the leverage of the ith observation, denoted h_i, can be computed by using equation (14.31).

Leverage of observation i

$$h_i = \frac{1}{n} + \frac{(x_i - \bar{x})^2}{\Sigma(x_i - \bar{x})^2}$$

(14.31)

From the formula, it is clear that the farther x_i is from its mean $\bar{x}$, the higher the leverage of observation i.

Many statistical packages automatically identify observations with high leverage as part of the standard regression output. As an illustration of how the MINITAB statistical package identifies points with high leverage, let us consider the data set in Table 14.12.

Table 14.12 Data set with a high leverage observation

x_i	y_i
10	125
10	130
15	120
20	115
20	120
25	110
70	100

From Figure 14.23, a scatter diagram for the data set in Table 14.12, it is clear that observation 7 ($X = 70$, $Y = 100$) is an observation with an extreme value of X. Hence, we would expect it to be identified as a point with high leverage. For this observation, the leverage is computed by using equation (14.31) as follows.

$$h_7 = \frac{1}{n} + \frac{(x_7 - \bar{x})^2}{\Sigma (x_i - \bar{x})^2} = \frac{1}{7} + \frac{(70 - 24.286)^2}{2621.43} = 0.94$$

For the case of simple linear regression, MINITAB identifies observations as having high leverage if $h_i > 6/n$; for the data set in Table 14.12, $6/n = 6/7 = 0.86$. Because $h_7 = 0.94 > 0.86$, MINITAB will identify observation 7 as an observation whose X value gives it large influence. Figure 14.24 shows the MINITAB output for a regression analysis of this data set. Observation 7 ($X = 70$, $Y = 100$) is identified as having large influence; it is printed on a separate line at the bottom, with an X in the right margin.

Influential observations that are caused by an interaction of large residuals and high leverage can be difficult to detect. Diagnostic procedures are available that take both into account in determining when an observation is influential. One such measure, called Cook's D statistic, will be discussed in Chapter 15.

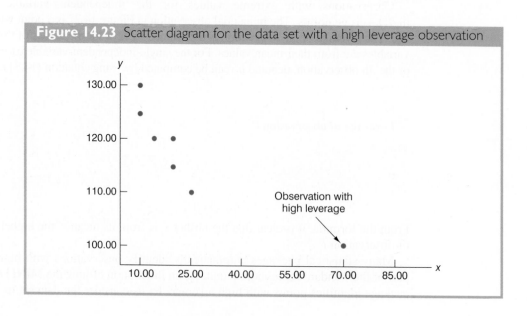

Figure 14.23 Scatter diagram for the data set with a high leverage observation

Figure 14.24 MINITAB output for the data set with a high leverage observation

```
Regression Analysis: y versus x

The regression equation is
y = 127 - 0.425 x


Predictor      Coef   SE Coef      T      P
Constant     127.466    2.961  43.04  0.000
x           -0.42507  0.09537  -4.46  0.007


S = 4.88282   R-Sq = 79.9%   R-Sq(adj) = 75.9%


Analysis of Variance

Source          DF      SS      MS      F      P
Regression       1  473.65  473.65  19.87  0.007
Residual Error   5  119.21   23.84
Total            6  592.86


Unusual Observations

Obs    x       y     Fit  SE Fit  Residual  St Resid
  7  70.0  100.00  97.71    4.73      2.29     1.91 X

X denotes an observation whose X value gives it large leverage.
```

Exercises

Methods

29 Consider the following data for two variables, X and Y.

x_i	135	110	130	145	175	160	120
y_i	145	100	120	120	130	130	110

a. Compute the standardized residuals for these data. Do there appear to be any outliers in the data? Explain.
b. Plot the standardized residuals against $\hat{y}$. Does this plot reveal any outliers?
c. Develop a scatter diagram for these data. Does the scatter diagram indicate any outliers in the data? In general, what implications does this finding have for simple linear regression?

30 Consider the following data for two variables, X and Y.

x_i	4	5	7	8	10	12	12	22
y_i	12	14	16	15	18	20	24	19

a. Compute the standardized residuals for these data. Do there appear to be any outliers in the data? Explain.

b. Compute the leverage values for these data. Do there appear to be any influential observations in these data? Explain.

c. Develop a scatter diagram for these data. Does the scatter diagram indicate any influential observations? Explain.

For additional online summary questions and answers go to the companion website at www.cengage.co.uk/aswsbe2

Summary

In this chapter we showed how regression analysis can be used to determine how a dependent variable Y is related to an independent variable X. In simple linear regression, the regression model is $Y = \beta_0 + \beta_1 x + \varepsilon$. The simple linear regression equation $E(\hat{Y}) = \beta_0 + \beta_1 x$ describes how the mean or expected value of Y is related to X. We used sample data and the least squares method to develop the estimated regression equation $\hat{y} = b_0 + b_1 x$ for a given value x of X. In effect, b_0 and b_1 are the sample statistics used to estimate the unknown model parameters β_0 and β_1.

The coefficient of determination was presented as a measure of the goodness of fit for the estimated regression equation; it can be interpreted as the proportion of the variation in the dependent variable Y that can be explained by the estimated regression equation. We reviewed correlation as a descriptive measure of the strength of a linear relationship between two variables.

The assumptions about the regression model and its associated error term ε were discussed, and t and F tests, based on those assumptions, were presented as a means for determining whether the relationship between two variables is statistically significant. We showed how to use the estimated regression equation to develop confidence interval estimates of the mean value of Y and prediction interval estimates of individual values of Y.

The chapter concluded with a section on the computer solution of regression problems and two sections on the use of residual analysis to validate the model assumptions and to identify outliers and influential observations.

Key terms

ANOVA table
Autocorrelation
Coefficient of determination
Confidence interval
Correlation coefficient
Dependent variable
Durbin-Watson test
Estimated regression equation
High leverage points
Independent variable
Influential observation
ith residual
Least squares method

Mean square error
Normal probability plot
Outlier
Prediction interval
Regression equation
Regression model
Residual analysis
Residual plot
Scatter diagram
Serial correlation
Simple linear regression
Standard error of the estimate
Standardized residual

Key formulae

Simple linear regression model

$$Y = \beta_0 + \beta_1 x + \varepsilon \qquad (14.1)$$

Simple linear regression equation

$$E(Y) = \beta_0 + \beta_1 x \qquad (14.2)$$

Estimated simple linear regression equation

$$\hat{y} = b_0 + b_1 x \qquad (14.3)$$

Least squares criterion

$$\text{Min } \Sigma (y_i - \hat{y}_i)^2 \qquad (14.5)$$

Slope and y-intercept for the estimated regression equation

$$b_1 = \frac{\Sigma (x_i - \bar{x})(y_i - \bar{y})}{\Sigma (x_i - \bar{x})^2} \qquad (14.6)$$

$$b_0 = \bar{y} - b_1 \bar{x} \qquad (14.7)$$

Sum of squares due to error

$$SSE = \Sigma (y_i - \hat{y}_i)^2 \qquad (14.8)$$

Total sum of squares

$$SST = \Sigma (y_i - \bar{y})^2 \qquad (14.9)$$

Sum of squares due to regression

$$SSR = \Sigma (\hat{y}_i - \bar{y})^2 \qquad (14.10)$$

Relationship among SST, SSR, and SSE

$$SST = SSR + SSE \qquad (14.11)$$

Coefficient of determination

$$r^2 = \frac{SSR}{SST} \qquad (14.12)$$

Sample correlation coefficient

$$r_{XY} = (\text{sign of } b_1) \sqrt{\text{Coefficient of determination}}$$
$$= (\text{sign of } b_1) \sqrt{r^2} \qquad (14.13)$$

Mean square error (estimate of s^2)

$$s^2 = MSE = \frac{SSE}{n - 2} \qquad (14.15)$$

Standard error of the estimate

$$s = \sqrt{\frac{SSE}{(n - 2)}} \qquad (14.16)$$

Standard deviation of b_1

$$\sigma_{b_1} = \frac{\sigma}{\sqrt{\Sigma(x_i - \overline{x})^2}}$$

(14.17)

Estimated standard deviation of b_1

$$s_{b_1} = \frac{s}{\sqrt{\Sigma(x_i - \overline{x})^2}}$$

(14.18)

t test statistic

$$t = \frac{b_1}{s_{b_1}}$$

(14.19)

Mean square regression

$$MSR = \frac{SSR}{\text{Number of independent variables}}$$

(14.20)

F test statistic

$$F = \frac{MSR}{MSE}$$

(14.21)

Confidence interval for $E(Y_p)$

$$\hat{y}_p \pm t_{\alpha/2}s\sqrt{\frac{1}{n} + \frac{(x_p - \overline{x})^2}{\Sigma(x_i - \overline{x})^2}}$$

(14.22)

Prediction interval for Y_p

$$\hat{y}_p \pm t_{\alpha/2}s\sqrt{1 + \frac{1}{n} + \frac{(x_p - \overline{x})^2}{\Sigma(x_i - \overline{x})^2}}$$

(14.23)

Residual for observation i

$$y_i - \hat{y}_i$$

(14.24)

Standard deviation of the ith residual

$$s_{y_i - \hat{y}_i} = s\sqrt{1 - h_i}$$

(14.26)

Standardized residual for observation i

$$\frac{y_i - \hat{y}_i}{s_{y_i - \hat{y}_i}}$$

(14.28)

First order autocorrelation

$$\varepsilon_t = \rho\,\varepsilon_{t-1} + z_t$$

(14.29)

Durbin-Watson test statistic

$$d = \frac{\sum_{t=2}^{n}(e_t - e_{t-1})^2}{\sum_{t=1}^{n}e_t^2}$$

(14.30)

Leverage of observation i

$$h_i = \frac{1}{n} + \frac{(x_i - \bar{x})^2}{\Sigma(x_i - \bar{x})^2}$$

(14.31)

Case problem 1   Investigating the relationship between weight loss and triglyceride level reduction[†]

Epidemiological studies have shown that there is a relationship between raised blood levels of triglyceride and coronary heart disease but it is not certain how important a risk factor triglycerides are. It is believed that exercise and lower consumption of fatty acids can help to reduce triglyceride levels.[*]

In 1998 Knoll Pharmaceuticals received authorization to market sibutramine for the treatment of obesity in the US. One of their suite of studies involved 35 obese patients who followed a treatment regime comprising a combination of diet, exercise and drug treatment.

Each patient's weight and triglyceride level were recorded at the start (known as *baseline*) and at week eight. The information recorded for each patient was:

- Patient ID.
- Weight at baseline (kg).
- Weight at week 8 (kg).
- Triglyceride level at baseline (mg/dl).
- Triglyceride level at week 8 (mg/dl).

TRIGLYCERID

Doctor checking an overweight patient's blood pressure. Digital readout indicates high blood pressure and pulse rate. © Eliza Snow.

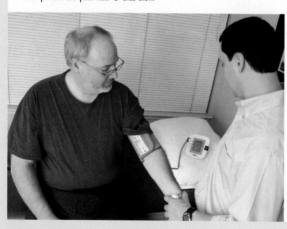

The results are shown below.

Patient ID	Weight at baseline	Weight at week 8	Triglyceride level at baseline	Triglyceride level at week 8
201	84.0	82.4	90	131
202	88.8	87.0	137	82
203	87.0	81.8	182	152
204	84.5	80.4	72	72
205	69.4	69.0	143	126
206	104.7	102.0	96	157
207	90.0	87.6	115	88
208	89.4	86.8	124	123
209	95.2	92.8	188	255
210	108.1	100.9	167	87
211	93.9	90.2	143	213
212	83.4	75.0	143	102
213	104.4	102.9	276	313
214	103.7	95.7	84	84
215	99.2	99.2	142	135
216	95.6	88.5	64	114
217	126.0	123.2	226	152
218	103.7	95.5	199	120
219	133.1	130.8	212	156
220	85.0	80.0	268	250
221	83.8	77.9	111	107
222	104.5	98.3	132	117
223	76.8	73.2	165	96
224	90.5	88.9	57	63
225	106.9	103.7	163	131
226	81.5	78.9	111	54
227	96.5	94.9	300	241
228	103.0	97.2	192	124
229	127.5	124.7	176	215
230	103.2	102.0	146	138

(continued)

Patient ID	Weight at baseline	Weight at week 8	Triglyceride level at baseline	Triglyceride level at week 8
231	113.5	115.0	446	795
232	107.0	99.2	232	63
233	106.0	103.5	255	204
234	114.9	105.3	187	144
235	103.4	96.0	154	96

Managerial report

1 Are weight loss and triglyceride level reduction (linearly) correlated?

2 Is there a linear relationship between weight loss and triglyceride level reduction?

3 How can a more detailed regression analysis be undertaken?

[Data in this case study reproduced with permission from STARS (www.stars.ac.uk).
*Triglycerides are lipids (fats) which are formed from glycerol and fatty acids. They can be absorbed into the body from food intake, particularly from fatty food, or produced in the body itself when the uptake of energy (food) exceeds the expenditure (exercise). Triglycerides provide the principal energy store for the body. Compared with carbohydrates or proteins, triglycerides produce a substantially higher number of calories per gram.

Case Problem 2 US Department of Transportation

As part of a study on transportation safety, the US Department of Transportation collected data on the number of fatal accidents per 1000 licences and the percentage of licensed drivers under the age of 21 in a sample of 42 cities. Data collected over a one-year period follow. These data are available on the CD accompanying the text in the file named Safety.

Percentage under 21	Fatal accidents per 1000 licences	Percentage under 21	Fatal accidents per 1000 licences
13	2.962	17	4.100
12	0.708	8	2.190
8	0.885	16	3.623
12	1.652	15	2.623
11	2.091	9	0.835
17	2.627	8	0.820
18	3.830	14	2.890
8	0.368	8	1.267
13	1.142	15	3.224
8	0.645	10	1.014
9	1.028	10	0.493
16	2.801	14	1.443
12	1.405	18	3.614
9	1.433	10	1.926
10	0.039	14	1.643
9	0.338	16	2.943
11	1.849	12	1.913
12	2.246	15	2.814
14	2.855	13	2.634
14	2.352	9	0.926
11	1.294	17	3.256

A fatal car accident. © Celso Pupo.

Managerial report

1   Develop numerical and graphical summaries of the data.

2   Use regression analysis to investigate the relationship between the number of fatal accidents and the percentage of drivers under the age of 21. Discuss your findings.

3   What conclusion and recommendations can you derive from your analysis?

Case Problem 3 Can we detect dyslexia?*

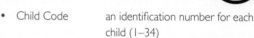

DYSLEXIA

Data were collected on 34 pre-school children and then in follow-up tests (on the same children) three years later when they were seven years old.

Scores were obtained from a variety of tests on all the children at age four when they were at nursery school. The tests were:

- Knowledge of vocabulary, measured by the British Picture Vocabulary Test (BPVT) in three versions – as raw scores, standardized scores and percentile norms.
- Another vocabulary test – non-word repetition.
- Motor skills, where the children were scored on the time in seconds to complete five different peg board tests.
- Knowledge of prepositions, scored as the number correct out of ten.
- Three tests on the use of rhyming, scored as the number correct out of ten.

Three years later the same children were given a reading test, from which a reading deficiency was calculated as Reading Age – Chronological Age (in months), this being known as Reading Age Deficiency (RAD). The children were then classified into 'poor' or 'normal' readers, depending on their RAD scores. Poor reading ability is taken as an indication of potential dyslexia.

One purpose of this study is to identify which of the tests at age four might be used as predictors of poor reading ability, which in turn is a possible indication of dyslexia.

*Data in this case study reproduced with permission from STARS (www.stars.ac.uk)

Data

The data set *Dyslexia* contains 18 variables:

- Child Code   an identification number for each child (1–34)
- Sex   m for male, f for female

The BPVT scores:

- BPVT raw   the raw score
- BPVT std   the standardized score
- BPVT % norm   cumulative percentage scores
- Non-wd repn   score for non-word repetition

Scores in motor skills:

- Pegboard set1 to   the time taken to complete
  Pegboard set5   each test
- Mean   child's average over the pegboard tests
- Preps Score   knowledge of prepositions (6–10)

Scores in rhyming tests (2–10):

- Rhyme set1
- Rhyme set2
- Rhyme set3
- RAD
- Poor/Normal   RAD scores, categorized as 1 = normal, 2 = poor

Details for ten records from the dataset are shown below.

Child code	Sex	BPVT raw	BPVT std	BPVT % norm	Non-wd repn	Pegboard set1	Pegboard set2	Pegboard set3	Pegboard set4	Pegboard set5
1	m	29	88	22	15	20.21	28.78	28.04	20.00	24.37
2	m	21	77	6	11	26.34	26.20	20.35	28.25	20.87
3	m	50	107	68	17	21.13	19.88	17.63	16.25	19.76
4	m	23	80	9	5	16.46	16.47	16.63	14.16	17.25
5	f	35	91	28	13	17.88	15.13	17.81	18.41	15.99
6	m	36	97	42	16	20.41	18.64	17.03	16.69	14.47
7	f	47	109	72	25	21.31	18.06	28.00	21.88	18.03
8	m	32	92	30	12	14.57	14.22	13.47	12.29	18.38
9	f	38	101	52	14	22.07	22.69	21.19	22.72	20.62
10	f	44	105	63	15	16.40	14.48	13.83	17.59	34.68

Child code	Mean	Preps score	Rhyme set1	Rhyme set2	Rhyme set3	RAD	Poor/ normal
1	24.3	6	5	5	5	−6.50	P
2	24.4	9	3	3	4	−7.33	P
3	18.9	10	9	8	*	49.33	N
4	16.2	7	4	6	4	−11.00	P
5	17.0	10	10	6	6	−2.67	N
6	17.5	10	6	5	5	−8.33	P
7	21.5	8	9	10	10	26.33	N
8	14.6	10	8	6	3	9.00	N
9	21.9	9	10	10	7	2.67	N
10	19.4	10	7	8	4	9.67	N

A young boy with dyslexia reads a book. © karen squires.

Managerial report

1 Is there a (linear) relationship between scores in tests at ages four and seven?

2 Can we predict RAD from scores at age four?

Software Section for Chapter 14

Regression analysis using MINITAB

In Section 14.7 we discussed the computer solution of regression problems by showing MINITAB's output for the Armand's Pizza Parlours problem. In this section, we describe the steps required to generate the MINITAB computer solution. First, the data must be entered in a MINITAB worksheet. Student population data are entered in column C1 and quarterly sales data are entered in column C2. The variable names Pop and Sales are entered as the column headings on the worksheet. In subsequent steps, we refer to the data by using the variable names Pop and Sales or the column indicators C1 and C2. The steps involved in using MINITAB to produce the regression results shown in Figure 14.10 follow.

Step 1 **Stat > Regression > Regression**   [Main menu bar]

Step 2 Enter Sales in the **Response** box   [**Regression** panel]
Enter Pop in the **Predictors** box
Click the **Options** button
Enter 10 in the **Prediction intervals for new observations** box
Click **OK**
(The MINITAB regression panel provides additional capabilities that can be obtained by selecting the desired options. For instance, to obtain a residual plot that shows the predicted value of the dependent variable on the horizontal axis and the standardized residual values on the vertical axis, click the **Graphs** button. Select **Standardized** under Residuals for Plots. Select **Residuals versus fits** under Residual Plots). Click **OK.** When the Regression panel reappears: Click **OK**.

Regression analysis using EXCEL

In this section we will illustrate how EXCEL's Regression tool can be used to perform the regression analysis computations for the Armand's Pizza Parlours problem. Refer to Figure 14.25 as we describe the steps involved. The labels Restaurant, Population and Sales are entered into cells A1:C1 of the worksheet. To identify each of the ten observations, we entered the numbers 1 through 10 into cells A2:A11. The sample data are entered into cells B2:C11. The steps involved in using the Regression tool for regression analysis follow.

Figure 14.25 EXCEL solution to the Armand's Pizza Parlours problem

	A	B	C	D	E	F	G	H	I
1	Restaurant	Population	Sales						
2	1	2	58						
3	2	6	105						
4	3	8	88						
5	4	8	118						
6	5	12	117						
7	6	16	137						
8	7	20	157						
9	8	20	169						
10	9	22	149						
11	10	26	202						
12									
13	SUMMARY OUTPUT								
14									
15	*Regression Statistics*								
16	Multiple R	0.9501							
17	R Square	0.9027							
18	Adjusted R Sq	0.8906							
19	Standard Error	13.8293							
20	Observations	10							
21									
22	ANOVA								
23		*df*	*SS*	*MS*	*F*	*Significance F*			
24	Regression	1	14200	14200	74.2484	0.0000			
25	Residual	8	1530	191.25					
26	Total	9	15730						
27									
28		*Coefficients*	*Standard Error*	*t Stat*	*P-value*	*Lower 95%*	*Upper 95%*	*Lower 99.0%*	*Upper 99.0%*
29	Intercept	60	9.2260	6.5033	0.0002	38.7247	81.2753	29.0431	90.9569
30	Population	5	0.5803	8.6167	0.0000	3.6619	6.3381	3.0530	6.9470
31									

Step 1 Data > Data Analysis > Regression [Main menu bar]

ARMAND'S

Step 2 Enter C1:C11 in the **Input Y Range** box [**Regression** panel]
Enter B1:B11 in the **Input X Range** box
Select **Labels**
Select Confidence **Level.** Enter 99 in the **Confidence Level** box
Select **Output Range**
Enter A13 in the **Output Range** box (to identify the upper left corner of the section of the worksheet where the output will appear)
Click **OK**

The first section of the output, titled *Regression Statistics*, contains summary statistics such as the coefficient of determination (R Square). The second section of the output, titled ANOVA, contains the analysis of variance table. The last section of the output, which is not titled, contains the estimated regression coefficients and related information. We will begin our discussion of the interpretation of the regression output with the information contained in cells A28:I30.

Interpretation of estimated regression equation output

The y intercept of the estimated regression line, $b_0 = 60$, is shown in cell B29, and the slope of the estimated regression line, $b_1 = 5$, is shown in cell B30. The label Intercept in cell A29 and the label Population in cell A30 are used to identify these two values. In Section 14.5 we showed that the estimated standard deviation of b_1 is $s_{b_1} = 0.5803$.

Note that the value in cell C30 is the standard error, or standard deviation, s_{b_1} of b_1. Recall that the t test for a significant relationship required the computation of the t statistic, $t = b_1/s_{b_1}$. For the Armand's data, the value of t that we computed was $t = 5/0.5803 = 8.62$. The label in cell D28, t Stat, reminds us that cell D30 contains the value of the t test statistic.

The value in cell E30 is the p-value associated with the t test for significance. EXCEL has displayed the p-value in cell E30 using scientific notation. To obtain the decimal value, we move the decimal point five places to the left, obtaining a value of 0.0000255. Because the p-value $= 0.0000255 < \alpha = 0.01$, we can reject H_0 and conclude that we have a significant relationship between student population and quarterly sales.

Cells F28:I30 refer to confidence interval estimates of the y intercept and slope of the estimated regression equation. EXCEL always provides the lower and upper limits for a 95 per cent confidence interval. Recall that in step 4 we selected Confidence Level and entered 99 in the Confidence Level box. As a result, EXCEL's Regression tool also provides the lower and upper limits for a 99 per cent confidence interval. The value in cell H30 is the lower limit for the 99 per cent confidence interval estimate of β_1 and the value in cell I30 is the upper limit. Thus, after rounding, the 99 per cent confidence interval estimate of β_1 is 3.05 to 6.95. The values in cells F30 and G30 provide the lower and upper limits for the 95 per cent confidence interval. Thus, the 95 per cent confidence interval is 3.66 to 6.34.

Interpretation of ANOVA output

The information in cells A22:F26 is a summary of the analysis of variance computations. The three sources of variation are labelled Regression, Residual and Total. The label df in cell B23 stands for degrees of freedom, the label SS in cell C23 stands for sum of squares, and the label MS in cell D23 stands for mean square.

In Section 14.5 we stated that the mean square error, obtained by dividing the error or residual sum of squares by its degrees of freedom, provides an estimate of σ^2. The value s^2 in cell D25, 191.25, is the mean square error for the Armand's regression output. In Section 14.5 we showed that an F test could also be used to test for significance in regression.

The value in cell F24, 0.0000, is the p-value associated with the F test for significance. Because the p-value $= 0.0000 < \alpha = 0.01$, we can reject H_0 and conclude that we have a significant relationship between student population and quarterly sales. The label EXCEL uses to identify the p-value for the F test for significance, shown in cell F23, is *Significance F*.

Interpretation of regression statistics output

The coefficient of determination, 0.9027, appears in cell B17; the corresponding label, R Square, is shown in cell A17. The square root of the coefficient of determination provides the sample correlation coefficient (though EXCEL always shows the positive square root of R^2) of 0.9501 shown in cell B16. Note that EXCEL uses the label Multiple R (cell A16) to identify this value. In cell A19, the label Standard Error is used to identify the value of the standard error of the estimate shown in cell B19. Thus, the standard error of the estimate is 13.8293. We caution the reader to keep in mind that in the EXCEL output, the label Standard Error appears in two different places. In the Regression Statistics section of the output, the label Standard Error refers to the estimate of σ. In the Estimated Regression Equation section of the output, the label *Standard Error* refers to s_{b_1}, the standard deviation of the sampling distribution of b_1.

Regression analysis using PASW

First, the data must be entered in a PASW worksheet. In 'Data View' mode, restaurants are entered in rows 1 to 10 of the leftmost column. This is automatically labelled by the system V1. Similarly population and sales details are entered in the two immediately adjacent columns to the right and are labelled V2 and V3 respectively. The latter variable names can then be changed to Restaurants, Pop and Sales in 'Variable View' mode. The following command sequence describes how PASW generates the regression results shown in Figure 14.26.

Step 1 **Analyze > Regression > Linear**   [Main menu bar]

Step 2 Enter Sales in the **Dependent** box   [**Linear** panel]
Enter Pop in the **Independent(s)** box
(In an analogous way to MINITAB, by clicking on the **Plots** button, a variety of residual plots can also be obtained.)
Click **OK**

Figure 14.26 PASW solution to the Armand's Pizza Parlours problem

Model Summary

Model	R	R Square	Adjusted R Square	Std. Error of the Estimate
1	.950[a]	.903	.891	13.82932

a. Predictors: (Constant), Population

ANOVA[b]

Model		Sum of Squares	df	Mean Square	F	Sig.
1	Regression	14200.000	1	14200.000	74.248	.000[a]
	Residual	1530.000	8	191.250		
	Total	15730.000	9			

a. Predictors: (Constant), Population

b. Dependent Variable: Sales

Coefficients[a]

Model		Unstandardized Coefficients		Standardized Coefficients	t	Sig.
		B	Std. Error	Beta		
1	(Constant)	60.000	9.226		6.503	.000
	Population	5.000	.580	.950	8.617	.000

a. Dependent Variable: Sales

Chapter 15

Multiple Regression

After reading this chapter and doing the exercises you should be able to:

1 Understand how multiple regression analysis can be used to develop relationships involving one dependent variable and several independent variables.

2 Interpret the coefficients in a multiple regression analysis.

3 Appreciate the background assumptions necessary to conduct statistical tests involving the hypothesized regression model.

4 Understand the role of computer packages in performing multiple regression analysis.

5 Interpret and use computer output to develop the estimated regression equation.

6 Determine how good a fit is provided by the estimated regression equation.

7 Test the significance of the regression equation.

8 Understand how multicollinearity affects multiple regression analysis.

9 Understand how residual analysis can be used to make a judgment as to the appropriateness of the model, identify outliers and determine which observations are influential.

10 Understand how logistic regression is used for regression analyses involving a binary dependent variable.

In Chapter 14 we presented simple linear regression and demonstrated its use in developing an estimated regression equation that describes the relationship between two variables. Recall that the variable being predicted or explained is called the dependent variable and the variable being used to predict or explain the dependent variable is called the independent variable. In this chapter we continue our study of regression analysis by considering situations involving two or more independent variables. This subject area, called **multiple regression analysis**, enables us to consider more than one potential predictor and thus obtain better estimates than are possible with simple linear regression.

15.1 Multiple regression model

Multiple regression analysis is the study of how a dependent variable Y is related to two or more independent variables. In the general case, we will use p to denote the number of independent variables.

Regression model and regression equation

The concepts of a regression model and a regression equation introduced in the preceding chapter are applicable in the multiple regression case. The equation that describes how the dependent variable Y is related to the independent variables $X_1, X_2, \ldots X_p$ and an error term is called the **multiple regression model**. We begin with the assumption that the multiple regression model takes the following form.

Stastics in Practice

Jura

Jura is a large island (380 sq km) off the South West of Scotland famous for its malt whisky and the large deer population that wander the quartz mountains ('the Paps') that dominate the landscape. With a population of a mere 461 it has one of the lowest population densities of any place in the UK. Currently Jura is only accessible via the adjoining island, Islay, which has three ferry services a day – crossings taking about two hours. However, because Jura is only four miles from the mainland it has been suggested that a direct car ferry taking less than half an hour would be preferable and more economical than existing provisions.

In exploring the case for an alternative service, Riddington (1994) arrives at a number of alternative mathematical formulations that essentially reduce to multiple regression analysis. In particular using historical data that also encompasses other inner Hebridean islands of Arran, Bute, Mull and Skye he obtains the estimated binary logistic regression model:

$$\text{Log}_e \frac{Q_{1it}}{Q_{2it}} = 6.48 - 0.89 \frac{P_{1it}}{P_{2it}} + 0.129 \frac{F_{1it}}{F_{2it}} - 6.18 \frac{J_{1it}}{J_{2it}}$$

where

Q_{1it}/Q_{2it} it is the number of cars travelling by route 1 relative to the number travelling by route 2 to island i in year t

P_{1it}/P_{2it} is the relative price between route 1 and route 2 to i in year t

F_{1it}/F_{2it} is the relative frequency between route 1 and route 2 to i in year t

J_{1it}/J_{2it} is the relative journey time between route 1 and route 2 to i in year t

Based on appropriate economic assumptions he estimates from this that some 132 000 passengers and 38 000 cars would use the new service each year rising over time. Initially this would yield a revenue of £426 000. Allowing for annual running costs of £322 000, the resultant gross profit would therefore be of the order of £100 000.

Source: Riddington, Geoff (1996) How many for the ferry boat? OR Insight Vol 9 Issue 2 pp 26–32.

Isle of Jura off the west coast of Scotland. The mountains in the distance are the distinctive Paps of Jura. © Martin McCarthy.

Multiple regression model

$$Y = \beta_0 + \beta_1 x_1 + \beta_2 x_2 + \cdots + \beta_p x_p + \varepsilon \qquad (15.1)$$

where $X_1 = x_1, X_2 = x_2, \ldots, X_p = x_p$.

In the multiple regression model, $\beta_0, \beta_1, \ldots, \beta_p$ are the parameters and ε (the Greek letter epsilon) is a random variable. A close examination of this model reveals that Y is a linear function of $x_1, x_2, \ldots, x_p$ (the $\beta_0 + \beta_1 x_1 + \beta_2 x_2 + \cdots\cdots + \beta_p x_p$ part) plus an error term ε. The error term accounts for the variability in Y that cannot be explained by the linear effect of the p independent variables.

In Section 15.4 we will discuss the assumptions for the multiple regression model and ε. One of the assumptions is that the mean or expected value of ε is zero. A consequence of this assumption is that the mean or expected value of Y, denoted $E(Y)$, is equal to $\beta_0 + \beta_1 x_1 + \beta_2 x_2 + \cdots\cdots + \beta_p x_p$. The equation that describes how the mean value of Y is related to $x_1, x_2, \ldots x_p$ is called the **multiple regression equation**.

Multiple regression equation

$$E(Y) = \beta_0 + \beta_1 x_1 + \beta_2 x_2 + \cdots + \beta_p x_p \tag{15.2}$$

Estimated multiple regression equation

If the values of $\beta_0, \beta_1, \ldots, \beta_p$ were known, equation (15.2) could be used to compute the mean value of Y at given values of $x_1, x_2, \ldots x_p$. Unfortunately, these parameter values will not, in general, be known and must be estimated from sample data. A simple random sample is used to compute sample statistics $b_0, b_1, \ldots, b_p$ that are used as the point estimators of the parameters $\beta_0, \beta_1, \ldots, \beta_p$. These sample statistics provide the following **estimated multiple regression equation**.

Estimated multiple regression equation

$$\hat{y} = b_0 + b_1 x_1 + b_2 x_2 + \cdots + b_p x_p \tag{15.3}$$

where

$b_0, b_1, \ldots, b_p$ are the estimates of $\beta_0, \beta_1, \ldots, \beta_p$
$\hat{y} =$ estimated value of the dependent variable

The estimation process for multiple regression is shown in Figure 15.1.

15.2 Least squares method

In Chapter 14, we used the **least squares method** to develop the estimated regression equation that best approximated the straight-line relationship between the dependent and independent variables. This same approach is used to develop the estimated multiple regression equation. The least squares criterion is restated as follows.

Least squares criterion

$$\min \Sigma (y_i - \hat{y}_i)^2 \tag{15.4}$$

where

$y_i =$ observed value of the dependent variable for the i th observation
$\hat{y}_i =$ estimated value of the dependent variable for the i th observation

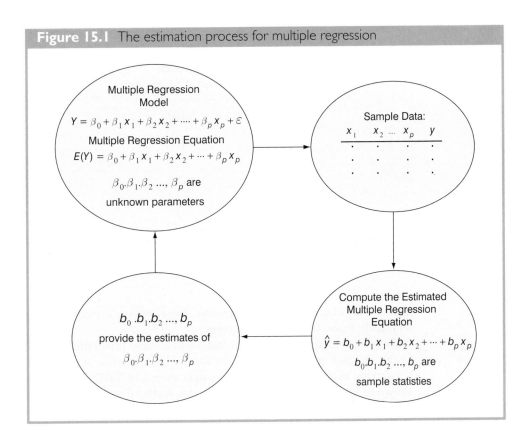

Figure 15.1 The estimation process for multiple regression

The estimated values of the dependent variable are computed by using the estimated multiple regression equation,

$$\hat{y} = b_0 + b_1 x_1 + b_2 x_2 + \cdots + b_p x_p$$

As expression (15.4) shows, the least squares method uses sample data to provide the values of $b_0, b_1, \ldots, b_p$ that make the sum of squared residuals {the deviations between the observed values of the dependent variable (y_i) and the estimated values of the dependent variable $\hat{y}_i$} a minimum.

In Chapter 14 we presented formulae for computing the lea st squares estimators b_0 and b_1 for the estimated simple linear regression equation $\hat{y} = b_0 + b_1 x$. With relatively small data sets, we were able to use those formulae to compute b_0 and b_1 by manual calculations. In multiple regression, however, the presentation of the formulae for the regression coefficients $b_0, b_1, \ldots, b_p$ involves the use of matrix algebra and is beyond the scope of this text. Therefore, in presenting multiple regression, we focus on how computer software packages can be used to obtain the estimated regression equation and other information. The emphasis will be on how to interpret the computer output rather than on how to make the multiple regression computations.

An example: Eurodistributor Company

As an illustration of multiple regression analysis, we will consider a problem faced by the Eurodistributor Company, an independent distribution company in the Netherlands. A major portion of Eurodistributor's business involves deliveries throughout its local

Table 15.1 Preliminary data for Eurodistributor

Driving assignment	X_1 = Distance travelled (kilometres)	Y = Travel time (hours)
1	100	9.3
2	50	4.8
3	100	8.9
4	100	6.5
5	50	4.2
6	80	6.2
7	75	7.4
8	65	6.0
9	90	7.6
10	90	6.1

area. To develop better work schedules, the company's managers want to estimate the total daily travel time for their drivers.

Initially the managers believed that the total daily travel time would be closely related to the distance travelled in making the daily deliveries. A simple random sample of ten driving assignments provided the data shown in Table 15.1 and the scatter diagram shown in Figure 15.2. After reviewing this scatter diagram, the managers hypothesized that the simple linear regression model $Y = \beta_0 + \beta_1 x_1 + \varepsilon$ could be used to describe the relationship between the total travel time (Y) and the distance travelled (X_1). To estimate the parameters β_0 and β_1, the least squares method was used to develop the estimated regression equation.

$$\hat{y} = b_0 + b_1 x_1 \tag{15.5}$$

In Figure 15.3, we show the MINITAB computer output from applying simple linear regression to the data in Table 15.1. The estimated regression equation is

$$\hat{y} = 1.27 + 0.0678 x_1$$

Figure 15.2 Scatter diagram of preliminary data for Eurodistributor

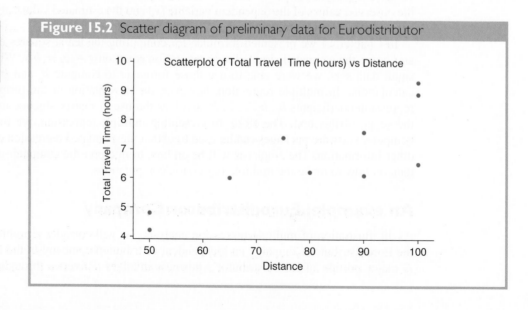

Figure 15.3 MINITAB output for Eurodistributor with one independent variable

```
Regression Analysis: Time versus Distance

The regression equation is
Time = 1.27 + 0.0678 Distance


Predictor        Coef    SE Coef      T       P
Constant        1.274      1.401    0.91   0.390
Distance      0.06783    0.01706    3.98   0.004


S = 1.00179    R-Sq = 66.4%    R-Sq(adj) = 62.2%


Analysis of Variance

Source             DF       SS       MS       F       P
Regression          1   15.871   15.871   15.81   0.004
Residual Error      8    8.029    1.004
Total               9   23.900
```

At the 0.05 level of significance, the F value of 15.81 and its corresponding p-value of 0.004 indicate that the relationship is significant; that is, we can reject H_0: $\beta_1 = 0$ because the p-value is less than $\alpha = 0.05$. Thus, we can conclude that the relationship between the total travel time and the distance travelled is significant; longer travel times are associated with more distance. With a coefficient of determination (expressed as a percentage) of R-sq = 66.4 per cent, we see that 66.4 per cent of the variability in travel time can be explained by the linear effect of the distance travelled. This finding is fairly good, but the managers might want to consider adding a second independent variable to explain some of the remaining variability in the dependent variable.

In attempting to identify another independent variable, the managers felt that the number of deliveries could also contribute to the total travel time. The Eurodistributor data, with the number of deliveries added, are shown in Table 15.2. The MINITAB computer solution with both distance (X_1) and number of deliveries (X_2) as independent variables is shown in Figure 15.4. The estimated regression equation is

$$\hat{y} = -0.869 + 0.0611x_1 + 0.923x_2 \tag{15.6}$$

In the next section we will discuss the use of the coefficient of multiple determination in measuring how good a fit is provided by this estimated regression equation. Before doing so, let us examine more carefully the values of $b_1 = 0.0611$ and $b_2 = 0.923$ in equation (15.6).

Note on interpretation of coefficients

One observation can be made at this point about the relationship between the estimated regression equation with only the distance as an independent variable and the equation that includes the number of deliveries as a second independent variable. The value of b_1

Table 15.2 Data for Eurodistributor with distance (X_1) and number of deliveries (X_2) as the independent variables

Driving assignment	X_1 = Distance travelled (kilometres)	X_2 = Number of deliveries	Y = Travel time (hours)
1	100	4	9.3
2	50	3	4.8
3	100	4	8.9
4	100	2	6.5
5	50	2	4.2
6	80	2	6.2
7	75	3	7.4
8	65	4	6.0
9	90	3	7.6
10	90	2	6.1

is not the same in both cases. In simple linear regression, we interpret b_1 as an estimate of the change in Y for a one-unit change in the independent variable. In multiple regression analysis, this interpretation must be modified somewhat. That is, in multiple regression analysis, we interpret each regression coefficient as follows: b_i represents an estimate

Figure 15.4 MINITAB output for Eurodistributor with two independent variables

```
Regression Analysis: Time versus Distance, Deliveries

The regression equation is
Time = - 0.869 + 0.0611 Distance + 0.923 Deliveries


Predictor        Coef     SE Coef        T       P
Constant      -0.8687      0.9515    -0.91   0.392
Distance     0.061135    0.009888     6.18   0.000
Deliveries     0.9234      0.2211     4.18   0.004


S = 0.573142    R-Sq = 90.4%    R-Sq(adj) = 87.6%


Analysis of Variance

Source           DF       SS       MS       F       P
Regression        2   21.601   10.800   32.88   0.000
Residual Error    7    2.299    0.328
Total             9   23.900


Source        DF   Seq SS
Distance       1   15.871
Deliveries     1    5.729
```

of the change in Y corresponding to a one-unit change in X_i when all other independent variables are held constant.

In the Eurodistributor example involving two independent variables, $b_1 = 0.0611$. Thus, 0.0611 hours is an estimate of the expected increase in travel time corresponding to an increase of one kilometre in the distance travelled when the number of deliveries is held constant. Similarly, because $b_2 = 0.923$, an estimate of the expected increase in travel time corresponding to an increase of one delivery when the distance travelled is held constant is 0.923 hours.

Exercises

Note to student: The exercises involving data in this and subsequent sections were designed to be solved using a computer software package.

Methods

1 The estimated regression equation for a model involving two independent variables and ten observations follows.

$$\hat{y} = 29.1270 + 0.5906x_1 + 0.4980x_2$$

a. Interpret b_1 and b_2 in this estimated regression equation.
b. Estimate Y when $X_1 = 180$ and $X_2 = 310$.

2 Consider the following data for a dependent variable Y and two independent variables, X_1 and X_2.

x_1	x_2	y
30	12	94
47	10	108
25	17	112
51	16	178
40	5	94
51	19	175
74	7	170
36	12	117
59	13	142
76	16	211

a. Develop an estimated regression equation relating Y to X_1. Estimate Y if $X_1 = 45$.
b. Develop an estimated regression equation relating Y to X_2. Estimate Y if $X_2 = 15$.
c. Develop an estimated regression equation relating Y to X_1 and X_2. Estimate Y if $X_1 = 45$ and $X_2 = 15$.

3 In a regression analysis involving 30 observations, the following estimated regression equation was obtained.

$$\hat{y} = 17.6 + 0\ 3.8x_1 - 2.3x_2 + 7.6x_3 + 2.7x_4$$

a. Interpret b_1, b_2, b_3, and b_4 in this estimated regression equation.
b. Estimate Y when $X_1 = 10$, $X_2 = 5$, $X_3 = 1$, and $X_4 = 2$.

Applications

4   A shoe store developed the following estimated regression equation relating sales to inventory investment and advertising expenditures.

$$\hat{y} = 25 + 10x_1 + 8x_2$$

where

$$X_1 = \text{inventory investment (€000s)}$$
$$X_2 = \text{advertising expenditures (€000s)}$$
$$Y = \text{sales (€000s)}$$

a.   Estimate sales resulting from a €15 000 investment in inventory and an advertising budget of €10 000.

b.   Interpret b_1 and b_2 in this estimated regression equation.

5   The owner of Toulon Theatres would like to estimate weekly gross revenue as a function of advertising expenditures. Historical data for a sample of eight weeks follow.

TOULON

Weekly gross revenue (€000s)	Television advertising (€000s)	Newspaper advertising (€000s)
96	5.0	1.5
90	2.0	2.0
95	4.0	1.5
92	2.5	2.5
95	3.0	3.3
94	3.5	2.3
94	2.5	4.2
94	3.0	2.5

a.   Develop an estimated regression equation with the amount of television advertising as the independent variable.

b.   Develop an estimated regression equation with both television advertizing and newspaper advertising as the independent variables.

c.   Is the estimated regression equation coefficient for television advertising expenditures the same in part (a) and in part (b)? Interpret the coefficient in each case.

d.   What is the estimate of the weekly gross revenue for a week when €3500 is spent on television advertising and €1800 is spent on newspaper advertising?

6   The following table gives the annual return, the safety rating (0 = riskiest, 10 = safest), and the annual expense ratio for 20 foreign funds.

	Annual safety rating	Expense ratio (%)	Annual return (%)
Accessor Int'l Equity 'Adv'	7.1	1.59	49
Aetna 'I' International	7.2	1.35	52
Amer Century Int'l Discovery 'Inv'	6.8	1.68	89
Columbia International Stock	7.1	1.56	58
Concert Inv 'A' Int'l Equity	6.2	2.16	131
Dreyfus Founders Int'l Equity 'F'	7.4	1.80	59
Driehaus International Growth	6.5	1.88	99

FORFUNDS

	Annual safety rating	Expense ratio (%)	Annual return (%)
Excelsior 'Inst' Int'l Equity	7.0	0.90	53
Julius Baer International Equity	6.9	1.79	77
Marshall International Stock 'Y'	7.2	1.49	54
MassMutual Int'l Equity 'S'	7.1	1.05	57
Morgan Grenfell Int'l Sm Cap 'Inst'	7.7	1.25	61
New England 'A' Int'l Equity	7.0	1.83	88
Pilgrim Int'l Small Cap 'A'	7.0	1.94	122
Republic International Equity	7.2	1.09	71
Sit International Growth	6.9	1.50	51
Smith Barney 'A' Int'l Equity	7.0	1.28	60
State St Research 'S' Int'l Equity	7.1	1.65	50
Strong International Stock	6.5	1.61	93
Vontobel International Equity	7.0	1.50	47

a. Develop an estimated regression equation relating the annual return to the safety rating and the annual expense ratio.
b. Estimate the annual return for a firm that has a safety rating of 7.5 and annual expense ratio of 2.

15.3 Multiple coefficient of determination

In simple linear regression we showed that the total sum of squares can be partitioned into two components: the sum of squares due to regression and the sum of squares due to error. The same procedure applies to the sum of squares in multiple regression.

Relationship among SST, SSR and SSE

$$SST = SSR + SSE \qquad (15.7)$$

where

$$SST = \text{total sum of squares} = \Sigma(y_i - \bar{y})^2$$
$$SSR = \text{sum of squares due to regression} = \Sigma(\hat{y}_i - \bar{y})^2$$
$$SSE = \text{sum of squares due to error} = \Sigma(y_i - \hat{y}_i)^2$$

Because of the computational difficulty in computing the three sums of squares, we rely on computer packages to determine those values. The analysis of variance part of the MINITAB output in Figure 15.4 shows the three values for the Eurodistributor problem with two independent variables: SST = 23.900, = SSR 21.601 and SSE = 2.299. With only one independent variable (distance travelled), the MINITAB output in Figure 15.3 shows that SST = 23.900, SSR = 15.871 and SSE = 8.029. The value of SST is the same in both cases because it does not depend on $\hat{y}$ but SSR increases and SSE decreases when a second independent variable (number of deliveries) is added. The implication is that the estimated multiple regression equation provides a better fit for the observed data.

In Chapter 14, we used the coefficient of determination, $R^2 = $ SSR/SST, to measure the goodness of fit for the estimated regression equation. The same concept applies to multiple regression. The term **multiple coefficient of determination** indicates that we are measuring the goodness of fit for the estimated multiple regression equation. The multiple coefficient of determination, denoted R^2, is computed as follows.

Multiple coefficient of determination

$$R^2 = \frac{SSR}{SST}$$

(15.8)

The multiple coefficient of determination can be interpreted as the proportion of the variability in the dependent variable that can be explained by the estimated multiple regression equation. Hence, when multiplied by 100, it can be interpreted as the percentage of the variability in Y that can be explained by the estimated regression equation.

In the two-independent-variable Eurodistributor example, with SSR = 21.601 and SST = 23.900, we have

$$R^2 = \frac{21.601}{23.900} = 0.904$$

Therefore, 90.4 per cent of the variability in travel time Y is explained by the estimated multiple regression equation with distance and number of deliveries as the independent variables. In Figure 15.4, we see that the multiple coefficient of determination is also provided by the MINITAB output; it is denoted by R-sq = 90.4 per cent.

Figure 15.3 shows that the R-sq value for the estimated regression equation with only one independent variable, distance travelled (X_1), is 66.4 per cent. Thus, the percentage of the variability in travel times that is explained by the estimated regression equation increases from 66.4 per cent to 90.4 per cent when number of deliveries is added as a second independent variable. In general, R^2 increases as independent variables are added to the model.

Many analysts prefer adjusting R^2 for the number of independent variables to avoid overestimating the impact of adding an independent variable on the amount of variability explained by the estimated regression equation. With n denoting the number of observations and p denoting the number of independent variables, the **adjusted multiple coefficient of determination** is computed as follows.

Adjusted multiple coefficient of determination

$$\text{adj } R^2 = 1 - (1 - R^2)\frac{n - 1}{n - p - 1}$$

(15.9)

For the Eurodistributor example with $n = 10$ and $p = 2$, we have

$$\text{adj } R^2 = 1 - (1 - 0.904)\frac{10 - 1}{10 - 2 - 1} = 0.88$$

Therefore, after adjusting for the two independent variables, we have an adjusted multiple coefficient of determination of 0.88. This value, allowing for rounding, corresponds with the value in the MINITAB output in Figure 15.4 of R-sq(adj) = 87.6 per cent.

Exercises

Methods

7 In exercise 1, the following estimated regression equation based on ten observations was presented.

$$\hat{y} = 29.1270 + 0.5906x_1 + 0.4980x_2$$

The values of SST and SSR are 6724.125 and 6216.375, respectively.

a. Find SSE.
b. Compute R^2.
c. Compute $Adj\ R^2$.
d. Comment on the goodness of fit.

8 In exercise 2, ten observations were provided for a dependent variable Y and two independent variables X_1 and X_2; for these data SST = 15 182.9, and SSR = 14 052.2.

a. Compute R^2.
b. Compute $Adj\ R^2$.
c. Does the estimated regression equation explain a large amount of the variability in the data? Explain.

9 In exercise 3, the following estimated regression equation based on 30 observations was presented.

$$\hat{y} = 17.6 + 3.8x_1 - 2.3x_2 + 7.6x_3 + 2.7x_4$$

The values of SST and SSR are 1805 and 1760, respectively.

a. Compute R^2.
b. Compute $Adj\ R^2$.
c. Comment on the goodness of fit.

Applications

10 In exercise 4, the following estimated regression equation relating sales to inventory investment and advertising expenditures was given.

$$\hat{y} = 25 + 10x_1 + 8x_2$$

The data used to develop the model came from a survey of ten stores; for those data, SST = 16 000 and SSR = 12 000.

a. For the estimated regression equation given, compute R^2.
b. Compute $Adj\ R^2$.
c. Does the model appear to explain a large amount of variability in the data? Explain.

11 In exercise 5, the owner of Toulon Theatres used multiple regression analysis to predict gross revenue (Y) as a function of television advertising (X_1) and newspaper advertising (X_2). The estimated regression equation was

$$\hat{y} = 83.2 + 2.29x_1 + 1.30x_2$$

The computer solution provided SST = 25.5 and SSR = 23.435.

a. Compute and interpret R^2 and $Adj\ R^2$.
b. When television advertising was the only independent variable, $R^2 = 0.653$ and $Adj\ R^2 = 0.595$. Do you prefer the multiple regression results? Explain.

15.4 Model assumptions

In Section 15.1 we introduced the following multiple regression model.

Multiple regression model

$$Y = \beta_0 + \beta_1 x_1 + \beta_2 x_2 + \cdots + \beta_p x_p + \varepsilon \qquad (15.10)$$

The assumptions about the error term ε in the multiple regression model parallel those for the simple linear regression model.

Assumptions about the error term in the multiple regression model

$$Y = \beta_0 + \beta_1 x_1 + \beta x_2 + \cdots + \beta_p x_p + \varepsilon$$

1. The error ε is a random variable with mean or expected value of zero; that is, $E(\varepsilon) = 0$.
   Implication: For given values of $X_1, X_2, \ldots X_p$, the expected, or average, value of Y is given by

   $$E(Y) = \beta_0 + \beta_1 x_1 + \beta_2 x_2 + \cdots + \beta_p x_p \qquad (15.11)$$

   Equation (15.11) is the multiple regression equation we introduced in Section 15.1. In this equation, $E(Y)$ represents the average of all possible values of Y that might occur for the given values of $X_1, X_2, \ldots, X_p$.
2. The variance of ε is denoted by σ^2 and is the same for all values of the independent variables $X_1, X_2, \ldots, X_p$.
   Implication: The variance of Y about the regression line equals σ^2 and is the same for all values of $X_1, X_2, \ldots, X_p$.
3. The values of ε are independent.
   Implication: The size of the error for a particular set of values for the independent variables is not related to the size of the error for any other set of values.
4. The error ε is a normally distributed random variable reflecting the deviation between the Y value and the expected value of Y given by $\beta_0 + \beta_1 x_1 + \beta_2 x_2 + \cdots + \beta_p x_p$.
   Implication: Because $\beta_0, \beta_1, \ldots, \beta_p$ are constants for the given values of $x_1, x_2, \ldots x_p$, the dependent variable Y is also a normally distributed random variable.

To obtain more insight about the form of the relationship given by equation (15.11), consider the following two-independent-variable multiple regression equation.

$$E(\text{Y}) = \beta_0 + \beta_1 x_1 + \beta_2 x_2$$

The graph of this equation is a plane in three-dimensional space. Figure 15.5 provides an example of such a graph. Note that the value of ε shown is the difference between the actual Y value and the expected value of y, $E(\text{Y})$, when $X_1 = x_1{}^*$ and $X_2 = x_2{}^*$.

In regression analysis, the term *response variable* is often used in place of the term *dependent variable*. Furthermore, since the multiple regression equation generates a plane or surface, its graph is called a *response surface*.

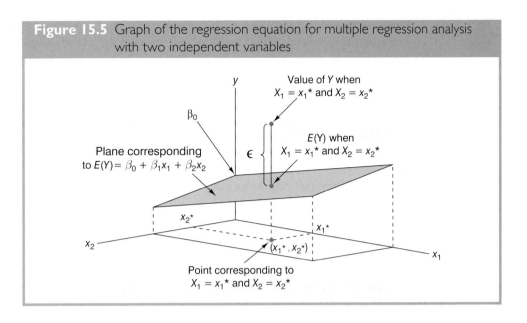

Figure 15.5 Graph of the regression equation for multiple regression analysis with two independent variables

15.5 Testing for significance

In this section we show how to conduct significance tests for a multiple regression relationship.

The significance tests we used in simple linear regression were a t test and an F test. In simple linear regression, both tests provide the same conclusion; that is, if the null hypothesis is rejected, we conclude that the slope parameter $\beta_1 \neq 0$. In multiple regression, the t test and the F test have different purposes.

1 The F test is used to determine whether a significant relationship exists between the dependent variable and the set of all the independent variables; we will refer to the F test as the test for *overall significance*.

2 If the F test shows an overall significance, the t test is used to determine whether each of the individual independent variables is significant. A separate t test is conducted for each of the independent variables in the model; we refer to each of these t tests as a test for *individual significance*.

In the material that follows, we will explain the F test and the t test and apply each to the Eurodistributor Company example.

F test

Given the multiple regression model defined in (15.1)

$$Y = \beta_0 + \beta_1 x_1 + \beta_2 x_2 + \cdots\cdots + \beta_p x_p + \varepsilon$$

the hypotheses for the F test can be written as follows:

$$H_0: \beta_1 = \beta_2 = \ldots\ldots = \beta_p = 0$$
$$H_1: \text{One or more of the parameters is not equal to zero}$$

If H_0 is rejected, the test gives us sufficient statistical evidence to conclude that one or more of the parameters is not equal to zero and that the overall relationship between Y and the set of independent variables $X_1, X_2, \ldots X_p$ is significant. However, if H_0 cannot be rejected, we deduce there is not sufficient evidence to conclude that a significant relationship is present.

Before confirming the steps involved in performing the F test, it might be helpful if we first review the concept of *mean square*. A mean square is a sum of squares divided by its corresponding degrees of freedom. In the multiple regression case, the total sum of squares has $n - 1$ degrees of freedom, the sum of squares due to regression (SSR) has p degrees of freedom, and the sum of squares due to error has $n - p - 1$ degrees of freedom. Hence, the mean square due to regression (MSR) is

Mean square regression

$$MSR = \frac{SSR}{p} \tag{15.12}$$

and

Mean square error

$$MSE = s^2 = \frac{SSE}{n - p - 1} \tag{15.13}$$

As has already been acknowledged in Chapter 14, MSE provides an unbiased estimate of σ^2, the variance of the error term ε. If $H_0: \beta_1 = \beta_2 = \ldots\ldots = \beta_p = 0$ is true, MSR also provides an unbiased estimate of σ^2, and the value of MSR/MSE should be close to 1. However, if H_0 is false, MSR overestimates σ^2 and the value of MSR/MSE becomes larger. To determine how large the value of MSR/MSE must be to reject H_0, we make use of the fact that if H_0 is true and the assumptions about the multiple regression model are valid, the sampling distribution of MSR/MSE is an F distribution with p degrees of freedom in the numerator and $n - p - 1$ in the denominator. A summary of the F test for significance in multiple regression follows.

F test for overall significance

$H_0: \beta_1 = \beta_2 = \ldots = \beta_p = 0$
$H_1:$ One or more of the parameters is not equal to zero

Test statistic

$$F = \frac{MSR}{MSE} \tag{15.14}$$

Rejection rule

p-value approach: Reject H_0 if p-value $\leq \alpha$
Critical value approach: Reject H_0 if $F \geq F_\alpha$

where F_α is based on an F distribution with p degrees of freedom in the numerator and $n - p - 1$ degrees of freedom in the denominator.

Applying the F test to the Eurodistributor Company multiple regression problem with two independent variables, the hypotheses can be written as follows.

$$H_0: \beta_1 = \beta_2 = 0$$
$$H_1: \beta_1 \text{ and/or } \beta_2 \text{ is not equal to zero}$$

Figure 15.6 shows the MINITAB output for the multiple regression model with distance (X_1) and number of deliveries (X_2) as the two independent variables. In the analysis of variance part of the output, we see that MSR = 10.8 and MSE = 0.328. Using equation (15.14), we obtain the test statistic.

$$F = \frac{10.8}{0.328} = 32.9$$

Note that the F value on the MINITAB output is $F = 32.88$; the value we calculated differs because we used rounded values for MSR and MSE in the calculation. Using $\alpha = 0.01$, the p-value = 0.000 in the last column of the analysis of variance table (Figure 15.6) indicates that we can reject $H_0: \beta_1 = \beta_2 = 0$ because the p-value is less than $\alpha = 0.01$. Alternatively, Table 4 of Appendix B shows that with two degrees of freedom in the numerator and seven degrees of freedom in the denominator, $F_{0.01} = 9.55$. With $32.9 > 9.55$, we reject $H_0: \beta_1 = \beta_2 = 0$ and conclude that a significant relationship is present between travel time Y and the two independent variables, distance and number of deliveries.

As noted previously, the mean square error provides an unbiased estimate of σ^2, the variance of the error term ε. Referring to Figure 15.6, we see that the estimate of σ^2 is MSE = 0.328. The square root of MSE is the estimate of the standard deviation of the error term. As defined in Section 14.5, this standard deviation is called the standard error

Figure 15.6 MINITAB output for Eurodistributor with two independent variables, distance (X_1) and number of deliveries (X_2)

```
Regression Analysis: Time versus Distance, Deliveries

The regression equation is
Time = - 0.869 + 0.0611 Distance + 0.923 Deliveries


Predictor        Coef     SE Coef       T       P
Constant      -0.8687      0.9515   -0.91   0.392
Distance     0.061135    0.009888    6.18   0.000
Deliveries     0.9234      0.2211    4.18   0.004


S = 0.573142    R-Sq = 90.4%    R-Sq(adj) = 87.6%


Analysis of Variance

Source            DF      SS       MS       F       P
Regression         2  21.601   10.800   32.88   0.000
Residual Error     7   2.299    0.328
Total              9  23.900
```

Table 15.3 ANOVA table for a multiple regression model with p independent variables

Source	Degrees of freedom	Sum of squares	Mean squares	F
Regression	p	SSR	$MSR = \dfrac{SSR}{p}$	$F = \dfrac{MSR}{MSE}$
Error	$n - p - 1$	SSE	$MSE = \dfrac{SSE}{n - p - 1}$	
Total	$n - 1$	SST		

of the estimate and is denoted s. Hence, we have $s = \sqrt{MSE} = \sqrt{0.328} = 0.573$. Note that the value of the standard error of the estimate appears in the MINITAB output in Figure 15.6.

Table 15.3 is the general analysis of variance (ANOVA) table that provides the F test results for a multiple regression model. The value of the F test statistic appears in the last column and can be compared to F_α with p degrees of freedom in the numerator and $n - p - 1$ degrees of freedom in the denominator to make the hypothesis test conclusion.

By reviewing the MINITAB output for Eurodistributor Company in Figure 15.6, we see that MINITAB's analysis of variance table contains this information. In addition, MINITAB provides the p-value corresponding to the F test statistic.

t test

If the F test shows that the multiple regression relationship is significant, a t test can be conducted to determine the significance of each of the individual parameters. The t test for individual significance follows.

t test for individual significance

For any parameter β_i

$$H_0: \beta_i = 0$$
$$H_1: \beta_i \neq 0$$

Test statistic

$$t = \frac{b_i}{s_{b_i}}$$

Rejection rule

p-value approach: Reject H_0 if p-value $\leq \alpha$.
Critical value approach: Reject H_0 if $t \leq -t_{\alpha/2}$ or if $t \geq t_{\alpha/2}$
where $t_{\alpha/2}$ is based on a t distribution with $n - p - 1$ degrees of freedom.

In the test statistic, s_{b_i} is the estimate of the standard deviation of b_i. The value of s_{b_i} will be provided by the computer software package.

Let us conduct the t test for the Eurodistributor regression problem. Refer to the section of Figure 15.6 that shows the MINITAB output for the t-ratio calculations. Values of b_1, b_2, s_{b_1} and s_{b_2} are as follows.

$$b_1 = 0.061135 \qquad s_{b_1} = 0.009888$$
$$b_2 = 0.9234 \qquad s_{b_2} = 0.2211$$

Using equation (15.15), we obtain the test statistic for the hypotheses involving parameters β_1 and β_2.

$$t = 0.061135/0.009888 = 6.18$$
$$t = 0.9234/0.2211 = 4.18$$

Note that both of these t-ratio values and the corresponding p-values are provided by the MINITAB output in Figure 15.6. Using $\alpha = 0.01$, the p-values of 0.000 and 0.004 from the MINITAB output indicate that we can reject $H_0: \beta_1 = 0$ and $H_0: \beta_2 = 0$. Hence, both parameters are statistically significant. Alternatively, Table 2 of Appendix B shows that with $n - p - 1 = 10 - 2 - 1 = 7$ degrees of freedom, $t_{0.005} = 3.499$. With $6.18 > 3.499$, we reject $H_0: \beta_1 = 0$. Similarly, with $4.18 > 3.499$, we reject $H_0: \beta_2 = 0$.

Multicollinearity

In multiple regression analysis, **multicollinearity** refers to the correlation among the independent variables. We used the term independent variable in regression analysis to refer to any variable being used to predict or explain the value of the dependent variable. The term does not mean, however, that the independent variables themselves are independent in any statistical sense. On the contrary, most independent variables in a multiple regression problem are correlated to some degree with one another. For example, in the Eurodistributor example involving the two independent variables X_1 (distance) and X_2 (number of deliveries), we could treat the distance as the dependent variable and the number of deliveries as the independent variable to determine whether those two variables are themselves related. We could then compute the sample correlation coefficient to determine the extent to which the variables are related. Doing so yields

Pearson correlation of Distance and Deliveries $= 0.162$

which suggests only a small degree of linear association exists between the two variables. The implication from this would be that multicollinearity is not a problem for the data. If however the association had been more pronounced the resultant multicollinearity might seriously have jeopardized the estimation of the model.

To provide a better perspective of the potential problems of multicollinearity, let us consider a modification of the Eurodistributor example. Instead of X_2 being the number of deliveries, let X_2 denote the number of litres of petrol consumed. Clearly, X_1 (the distance) and X_2 are related; that is, we know that the number of litres of petrol used depends on the distance travelled. Hence, we would conclude logically that X_1 and X_2 are highly correlated independent variables.

Assume that we obtain the equation $\hat{y} = b_0 + b_1 x_1 + b_2 x_2$ and find that the F test shows the relationship to be significant. Then suppose we conduct a t test on β_1 to determine whether $\beta_1 = 0$, and we cannot reject $H_0: \beta_1 = 0$. Does this result mean that travel time is not related to distance? Not necessarily. What it probably means is that with X_2 already in the model, X_1 does not make a significant contribution to determining the value of Y. This interpretation makes sense in our example; if we know the amount of petrol consumed, we do not gain much additional information useful in predicting Y by knowing the distance. Similarly, a t test might lead us to conclude $\beta_2 = 0$ on the grounds that, with X_1 in the model, knowledge of the amount of petrol consumed does not add much.

One useful way of detecting multicollinearity is to calculate the **variance inflation factor** (VIF) for each independent variable (X_j) in the model. The VIF is defined as

Variance inflation factor

$$VIF(X_j) = \frac{1}{1 - R_j^2} \qquad (15.16)$$

where R_j^2 is the coefficient of determination obtained when X_j ($j = 1, 2, \ldots, p$) is regressed on all remaining independent variables in the model. If X_j is not correlated with other predictors $R_j^2 \approx 0$ and VIF ≈ 1. Correspondingly if R_j^2 is close to 1 the VIF will be very large. Typically VIF values of ten or more are regarded as problematic.

For the Eurodistributor data, the VIF for X_1 (and also X_2 by symmetry) would be

$$VIF(X_j) = \frac{1}{1 - 0.162^2} = 1.027$$

signifying as before there is no problem with multicollinearity.

To summarize, for t tests associated with testing for the significance of individual parameters, the difficulty caused by multicollinearity is that it is possible to conclude that none of the individual parameters are significantly different from zero when an F test on the overall multiple regression equation indicates there is a significant relationship. This problem is avoided however when little correlation among the independent variables exists.

If possible, every attempt should be made to avoid including independent variables that are highly correlated. In practice, however, strict adherence to this policy is not always possible. When decision-makers have reason to believe substantial multicollinearity is present, they must realize that separating the effects of the individual independent variables on the dependent variable is difficult.

Exercises

Methods

12 In exercise 1, the following estimated regression equation based on ten observations was presented.

$$\hat{y} = 29.1270 + 0.5906x_1 + 0.4980x_2$$

Here SST = 6724.125, SSR = 6216.375, s_{b_1} = 0.0813 and s_{b_2} = 0.0567.

a. Compute MSR and MSE.
b. Compute F and perform the appropriate F test. Use $\alpha = 0.05$.
c. Perform a t test for the significance of β_1. Use $\alpha = 0.05$.
d. Perform a t test for the significance of β_2. Use $\alpha = 0.05$.

13 Refer to the data presented in exercise 2. The estimated regression equation for these data is

$$\hat{y} = -18.4 + 2.01x_1 + 4.74x_2$$

Here SST = 15 182.9, SSR = 14 052.2, s_{b_1} = 0.2471 and s_{b_2} = 0.9484.

a. Test for a significant relationship among X_1, X_2 and Y. Use α = 0.05.
b. Is β_1 significant? Use α = 0.05.
c. Is β_2 significant? Use α = 0.05.

14 The following estimated regression equation was developed for a model involving two independent variables.

$$\hat{y} = 40.7 + 8.63x_1 + 2.71x_2$$

After X_2 was dropped from the model, the least squares method was used to obtain an estimated regression equation involving only X_1 as an independent variable.

$$\hat{y} = 42.0 + 9.01x_1$$

a. Give an interpretation of the coefficient of X_1 in both models.
b. Could multicollinearity explain why the coefficient of X_1 differs in the two models? If so, how?

Applications

15 In exercise 4 the following estimated regression equation relating sales to inventory investment and advertizing expenditures was given.

$$\hat{y} = 25 + 10x_1 + 8x_2$$

The data used to develop the model came from a survey of ten stores; for these data SST = 16 000 and SSR = 12 000.

a. Compute SSE, MSE and MSR.
b. Use an F test and a 0.05 level of significance to determine whether there is a relationship among the variables.

16 Refer to exercise 5.

a. Use α = 0.01 to test the hypotheses

$$H_0: \beta_1 = \beta_2 = 0$$
$$H_1: \beta_1 \text{ and/or } \beta_2 \text{ is not equal to zero}$$

for the model $Y = \beta_0 + \beta_1 x_1 + \beta_2 x_2 + \varepsilon$, where

$$X_1 = \text{television advertising (€1000s)}$$
$$X_2 = \text{newspaper advertising (€1000s)}$$

b. Use α = 0.05 to test the significance of β_1. Should X_1 be dropped from the model?
c. Use α = 0.05 to test the significance of β_2. Should X_2 be dropped from the model?

15.6 Using the estimated regression equation for estimation and prediction

The procedures for estimating the mean value of Y and predicting an individual value of Y in multiple regression are similar to those in regression analysis involving one independent variable. First, recall that in Chapter 14 we showed that the point estimate of the expected value of Y for a given value of X was the same as the point estimate of an individual value of Y. In both cases, we used $\hat{y} = b_0 + b_1 x$ as the point estimate.

Table 15.4 The 95% confidence and prediction intervals for Eurodistributor

		Confidence Interval		Prediction Interval	
Value of X_1	Value of X_2	Lower Limit	Upper Limit	Lower Limit	Upper Limit
50	2	3.146	4.924	2.414	5.656
50	3	4.127	5.789	3.368	6.548
50	4	4.815	6.948	4.157	7.607
100	2	6.258	7.926	5.500	8.683
100	3	7.385	8.645	6.520	9.510
100	4	8.135	9.742	7.362	10.515

In multiple regression we use the same procedure. That is, we substitute the given values of $X_1, X_2, \ldots X_p$ into the estimated regression equation and use the corresponding value of $\hat{y}$ as the point estimate. Suppose that for the Eurodistributor example we want to use the estimated regression equation involving X_1 (distance) and X_2 (number of deliveries) to develop two interval estimates:

1 A *confidence interval* of the mean travel time for all trucks that travel 100 kilometres and make two deliveries.

2 A *prediction interval* of the travel time for *one specific* truck that travels 100 kilometres and makes two deliveries.

Using the estimated regression equation $\hat{y} = -0.869 + 0.0611x_1 + 0.923x_2$ with $X_1 = 100$ and $X_2 = -2$, we obtain the following value of $\hat{y}$.

$$\hat{y} = -0.869 + 0.0611(100) + 0.923(2) = 7.09$$

Hence, the point estimate of travel time in both cases is approximately seven hours.

To develop interval estimates for the mean value of Y and for an individual value of Y, we use a procedure similar to that for regression analysis involving one independent variable.

The formulae required are beyond the scope of the text, but computer packages for multiple regression analysis will often provide confidence intervals once the values of $X_1, X_2, \ldots X_p$ are specified by the user. In Table 15.4 we show the 95 per cent confidence and prediction intervals for the Eurodistributor example for selected values of X_1 and X_2; these values were obtained using MINITAB. Note that the interval estimate for an individual value of Y is wider than the interval estimate for the expected value of Y. This difference simply reflects the fact that for given values of X_1 and X_2 we can estimate the mean travel time for all trucks with more precision than we can predict the travel time for one specific truck.

Exercises

Methods

17 In exercise 1, the following estimated regression equation based on ten observations was presented.

$$\hat{y} = 29.1270 + 0.5906x_1 + 0.4980x_2$$

a. Develop a point estimate of the mean value of Y when $X_1 = 180$ and $X_2 = 310$.
b. Develop a point estimate for an individual value of Y when $X_1 = 180$ and $X_2 = 310$.

18   Refer to the data in exercise 2. The estimated regression equation for those data is

$$\hat{y} = -18.4 + 2.01x_1 + 4.74x_2$$

a. Develop a 95 per cent confidence interval for the mean value of Y when $X_1 = 45$ and $X_2 = 15$.
b. Develop a 95 per cent prediction interval for Y when $X_1 = 45$ and $X_2 = 15$.

Applications

19   In exercise 5, the owner of Toulon Theatres used multiple regression analysis to predict gross revenue (Y) as a function of television advertising (X_1) and newspaper advertising (X_2). The estimated regression equation was

$$\hat{y} = 83.2 + 2.29x_1 + 1.30x_2$$

a. What is the gross revenue expected for a week when €3500 is spent on television advertising ($X_1 = 3.5$) and €1800 is spent on newspaper advertising ($X_2 = 1.8$)?
b. Provide a 95 per cent confidence interval for the mean revenue of all weeks with the expenditures listed in part (a).
c. Provide a 95 per cent prediction interval for next week's revenue, assuming that the advertising expenditures will be allocated as in part (a).

15.7 Qualitative independent variables

Thus far, the examples we considered involved quantitative independent variables such as distance travelled and number of deliveries. In many situations, however, we must work with **qualitative independent variables** such as gender (male, female), method of payment (cash, credit card, cheque) and so on. The purpose of this section is to show how qualitative variables are handled in regression analysis. To illustrate the use and interpretation of a qualitative independent variable, we will consider a problem facing the managers of Johansson Filtration.

An example: Johansson Filtration

Johansson Filtration provides maintenance service for water-filtration systems throughout southern Denmark. Customers contact Johansson with requests for maintenance service on their water-filtration systems. To estimate the service time and the service cost, Johansson's managers wish to predict the repair time necessary for each maintenance request. Hence, repair time in hours is the dependent variable. Repair time is believed to be related to two factors, the number of months since the last maintenance service and the type of repair problem (mechanical or electrical). Data for a sample of ten service calls are reported in Table 15.5.

Let Y denote the repair time in hours and X_1 denote the number of months since the last maintenance service. The regression model that uses only X_1 to predict Y is

$$Y = \beta_0 + \beta_1 x_1 + \varepsilon$$

JOHANNSON

Table 15.5 Data for the Johansson Filtration example

Service call	Months since last service	Type of repair	Repair time in hours
1	2	electrical	2.9
2	6	mechanical	3.0
3	8	electrical	4.8
4	3	mechanical	1.8
5	2	electrical	2.9
6	7	electrical	4.9
7	9	mechanical	4.2
8	8	mechanical	4.8
9	4	electrical	4.4
10	6	electrical	4.5

Using MINITAB to develop the estimated regression equation, we obtained the output shown in Figure 15.7. The estimated regression equation is

$$\hat{y} = 2.15 + 0.304x_1 \tag{15.17}$$

At the 0.05 level of significance, the p-value of 0.016 for the t (or F) test indicates that the number of months since the last service is significantly related to repair time. R-sq = 53.4 per cent indicates that X_1 alone explains 53.4 per cent of the variability in repair time.

Figure 15.7 MINITAB output for Johansson Filtration with months since last service (X_1) as the independent variable

```
Regression Analysis: Time versus Months

The regression equation is
Time = 2.15 + 0.304 Months


Predictor     Coef    SE Coef      T       P
Constant    2.1473    0.6050    3.55   0.008
Months      0.3041    0.1004    3.03   0.016


S = 0.781022   R-Sq = 53.4%   R-Sq(adj) = 47.6%


Analysis of Variance

Source           DF       SS       MS      F       P
Regression        1   5.5960   5.5960   9.17   0.016
Residual Error    8   4.8800   0.6100
Total             9  10.4760
```

To incorporate the type of failure into the regression model, we define the following variable.

$$X_2 = 0 \text{ if the type of repair is mechanical}$$
$$X_2 = 1 \text{ if the type of repair is electrical}$$

In regression analysis X_2 is called a **dummy** or *indicator* **variable**. Using this dummy variable, we can write the multiple regression model as

$$Y = \beta_0 + \beta_1 x_1 + \beta_2 x_2 + \varepsilon$$

Table 15.6 is the revised data set that includes the values of the **dummy variable**. Using MINITAB and the data in Table 15.6, we can develop estimates of the model parameters. The MINITAB output in Figure 15.8 shows that the estimated multiple regression equation is

$$\hat{y} = 0.93 + 0.388 x_1 + 1.26 x_2 \qquad \text{(15.18)}$$

At the 0.05 level of significance, the *p*-value of 0.001 associated with the *F* test ($F = 21.36$) indicates that the regression relationship is significant. The *t* test part of the printout in Figure 15.8 shows that both months since last service (*p*-value = 0.000) and type of repair (*p*-value = 0.005) are statistically significant. In addition, *R*-sq = 85.9 per cent and *R*-sq(adj) = 81.9 per cent indicate that the estimated regression equation does a good job of explaining the variability in repair times. Thus, equation (15.18) should prove helpful in estimating the repair time necessary for the various service calls.

Interpreting the parameters

The multiple regression equation for the Johansson Filtration example is

$$E(Y) = \beta_0 + \beta_1 x_1 + \beta_2 x_2 \qquad \text{(15.19)}$$

Table 15.6 Data for the Johansson Filtration example with type of repair indicated by a dummy variable ($X_2 = 0$ for mechanical; $X_2 = 1$ for electrical)

Customer	Months since last service (X_1)	Type of repair (X_2)	Repair time in hours (Y)
1	2	1	2.9
2	6	0	3.0
3	8	1	4.8
4	3	0	1.8
5	2	1	2.9
6	7	1	4.9
7	9	0	4.2
8	8	0	4.8
9	4	1	4.4
10	6	1	4.5

Figure 15.8 MINITAB output for Johansson Filtration with months since last service (X_1) and type of repair (X_2) as the independent variables

```
Regression Analysis: Time versus Months, Type

The regression equation is
Time = 0.930 + 0.388 Months + 1.26 Type


Predictor       Coef   SE Coef      T      P
Constant      0.9305    0.4670   1.99  0.087
Months       0.38762   0.06257   6.20  0.000
Type          1.2627    0.3141   4.02  0.005


S = 0.459048    R-Sq = 85.9%    R-Sq(adj) = 81.9%


Analysis of Variance

Source             DF       SS      MS      F      P
Regression          2   9.0009  4.5005  21.36  0.001
Residual Error      7   1.4751  0.2107
Total               9  10.4760


Source  DF  Seq SS
Months   1  5.5960
Type     1  3.4049
```

To understand how to interpret the parameters β_0, β_1, and β_2 when a qualitative variable is present, consider the case when $X_2 = 0$ (mechanical repair). Using $E(Y|\text{mechanical})$ to denote the mean or expected value of repair time *given* a mechanical repair, we have

$$E(Y|\text{mechanical}) = \beta_0 + \beta_1 x_1 + \beta_2(0) = \beta_0 + \beta_1 x_1 \qquad (15.20)$$

Similarly, for an electrical repair ($X_2 = 1$), we have

$$E(Y|\text{electrical}) = \beta_0 + \beta_1 x_1 + \beta_2(1) = \beta_0 + \beta_1 x_1 + \beta_2 \qquad (15.21)$$
$$= (\beta_0 + \beta_2) + \beta_1 x_1$$

Comparing equations (15.20) and (15.21), we see that the mean repair time is a linear function of X_1 for both mechanical and electrical repairs. The slope of both equations is β_1, but the y-intercept differs. The y-intercept is β_0 in equation (15.20) for mechanical repairs and $(\beta_0 + \beta_2)$ in equation (15.21) for electrical repairs. The interpretation of β_2 is that it indicates the difference between the mean repair time for an electrical repair and the mean repair time for a mechanical repair.

If β_2 is positive, the mean repair time for an electrical repair will be greater than that for a mechanical repair; if β_2 is negative, the mean repair time for an electrical repair will be less than that for a mechanical repair. Finally, if $\beta_2 = 0$, there is no difference in the mean repair time between electrical and mechanical repairs and the type of repair is not related to the repair time.

Using the estimated multiple regression equation $\hat{y} = 0.93 + 0.388x_1 + 1.26x_2$, we see that 0.93 is the estimate of β_0 and 1.26 is the estimate of β_2. Thus, when $X_2 = 0$ (mechanical repair)

$$\hat{y} = 0.93 + 0.388x_1 \qquad \textbf{(15.22)}$$

and when $X_2 = 1$ (electrical repair)

$$\begin{aligned} \hat{y} &= 0.93 + 0.388x_1 + 1.26(1) \\ &= 2.19 + 0.388x_1 \end{aligned} \qquad \textbf{(15.23)}$$

In effect, the use of a dummy variable for type of repair provides two equations that can be used to predict the repair time, one corresponding to mechanical repairs and one corresponding to electrical repairs. In addition, with $b_2 = 1.26$, we learn that, on average, electrical repairs require 1.26 hours longer than mechanical repairs.

Figure 15.9 is the plot of the Johansson data from Table 15.6. Repair time in hours (Y) is represented by the vertical axis and months since last service (X_1) is represented by the horizontal axis. A data point for a mechanical repair is indicated by an M and a data point for an electrical repair is indicated by an E. Equations (15.22) and (15.23) are plotted on the graph to show graphically the two equations that can be used to predict the repair time, one corresponding to mechanical repairs and one corresponding to electrical repairs.

Figure 15.9 Scatter diagram for the Johansson Filtration repair data from Table 15.6

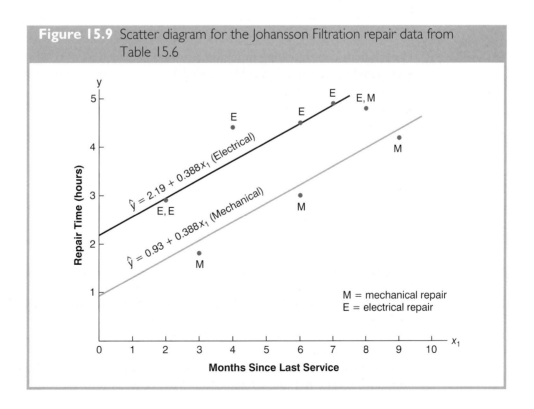

More complex qualitative variables

Because the qualitative variable for the Johansson Filtration example had two levels (mechanical and electrical), defining a dummy variable with zero indicating a mechanical repair and one indicating an electrical repair was easy. However, when a qualitative variable has more than two levels, care must be taken in both defining and interpreting the dummy variables. As we will show, if a qualitative variable has k levels, $k - 1$ dummy variables are required, with each dummy variable being coded as 0 or 1.

For example, suppose a manufacturer of copy machines organized the sales territories for a particular area into three regions: A, B and C. The managers want to use regression analysis to help predict the number of copiers sold per week. With the number of units sold as the dependent variable, they are considering several independent variables (the number of sales personnel, advertising expenditures and so on). Suppose the managers believe sales region is also an important factor in predicting the number of copiers sold. Because sales region is a qualitative variable with three levels, A, B and C, we will need $3 - 1 = 2$ dummy variables to represent the sales region. Each variable can be coded 0 or 1 as follows.

$$X_1 = \begin{cases} 1 \text{ if sales region B} \\ 0 \text{ otherwise} \end{cases}$$

$$X_2 = \begin{cases} 1 \text{ if sales region C} \\ 0 \text{ otherwise} \end{cases}$$

With this definition, we have the following values of X_1 and X_2.

Region	X_1	X_2
A	0	0
B	1	0
C	0	1

Observations corresponding to region A would be coded $X_1 = 0$, $X_2 = 0$; observations corresponding to region B would be coded $X_1 = 1$, $X_2 = 0$; and observations corresponding to region C would be coded $X_1 = 0$, $X_2 = 1$.

The regression equation relating the expected value of the number of units sold, $E(Y)$, to the dummy variables would be written as

$$E(Y) = \beta_0 + \beta_1 x_1 + \beta_2 x_2$$

To help us interpret the parameters β_0, β_1 and β_2, consider the following three variations of the regression equation.

$$E(Y \mid \text{region A}) = \beta_0 + \beta_1(0) + \beta_2(0) = \beta_0$$
$$E(Y \mid \text{region B}) = \beta_0 + \beta_1(1) + \beta_2(0) = \beta_0 + \beta_1$$
$$E(Y \mid \text{region C}) = \beta_0 + \beta_1(0) + \beta_2(1) = \beta_0 + \beta_2$$

Therefore, β_0 is the mean or expected value of sales for region A; β_1 is the difference between the mean number of units sold in region B and the mean number of units sold in region A; and β_2 is the difference between the mean number of units sold in region C and the mean number of units sold in region A.

Two dummy variables were required because sales region is a qualitative variable with three levels. But the assignment of $X_1 = 0$, $X_2 = 0$ to indicate region A, $X_1 = 1$,

$X_2 = 0$ to indicate region B, and $X_1 = 0$, $X_2 = 1$ to indicate region C was arbitrary. For example, we could have chosen $X_1 = 1$, $X_2 = 0$ to indicate region A, $X_1 = 0$, $X_2 = 0$ to indicate region B, and $X_1 = 0$, $X_2 = 1$ to indicate region C. In that case, β_1 would have been interpreted as the mean difference between regions A and B and β_2 as the mean difference between regions C and B.

Exercises

Methods

20 Consider a regression study involving a dependent variable Y, a quantitative independent variable X_1 and a qualitative variable with two levels (level 1 and level 2).

   a. Write a multiple regression equation relating X_1 and the qualitative variable to Y.
   b. What is the expected value of Y corresponding to level 1 of the qualitative variable?
   c. What is the expected value of Y corresponding to level 2 of the qualitative variable?
   d. Interpret the parameters in your regression equation.

21 Consider a regression study involving a dependent variable Y, a quantitative independent variable X_1, and a qualitative independent variable with three possible levels (level 1, level 2 and level 3).

   a. How many dummy variables are required to represent the qualitative variable?
   b. Write a multiple regression equation relating X_1 and the qualitative variable to Y.
   c. Interpret the parameters in your regression equation.

Applications

22 Management proposed the following regression model to predict sales at a fast-food outlet.

$$Y = \beta_0 + \beta_1 x_1 + \beta_2 x_2 + \beta_3 x_3 + \varepsilon,$$

where

$$X_1 = \text{number of competitors within one kilometre}$$
$$X_2 = \text{population within one kilometre (000s)}$$
$$X_3 = 1 \text{ if drive-up window present}$$
$$0 \text{ otherwise}$$
$$Y = \text{sales (€000s)}$$

The following estimated regression equation was developed after 20 outlets were surveyed.

$$\hat{y} = 10.1 - 4.2x_1 + 6.8x_2 + 15.3x_3$$

   a. What is the expected amount of sales attributable to the drive-up window?
   b. Predict sales for a store with two competitors, a population of 8000 within one kilometre and no drive-up window.
   c. Predict sales for a store with one competitor, a population of 3000 within one kilometre and a drive-up window.

23 Refer to the Johansson Filtration problem introduced in this section. Suppose that in addition to information on the number of months since the machine was serviced and whether a mechanical or an electrical failure had occurred, the managers obtained a list showing which engineer performed the service. The revised data follow.

Repair time in hours	Months since last service	Type of repair	Engineer
2.9	2	Electrical	Heinz Kolb
3.0	6	Mechanical	Heinz Kolb
4.8	8	Electrical	Wolfgang Linz
1.8	3	Mechanical	Heinz Kolb
2.9	2	Electrical	Heinz Kolb
4.9	7	Electrical	Wolfgang Linz
4.2	9	Mechanical	Wolfgang Linz
4.8	8	Mechanical	Wolfgang Linz
4.4	4	Electrical	Wolfgang Linz
4.5	6	Electrical	Heinz Kolb

REPAIR

a. Ignore for now the months since the last maintenance service (X_1) and the engineer who performed the service. Develop the estimated simple linear regression equation to predict the repair time (Y) given the type of repair (X_2). Recall that $X_2 = 0$ if the type of repair is mechanical and 1 if the type of repair is electrical.

b. Does the equation that you developed in part (a) provide a good fit for the observed data? Explain.

c. Ignore for now the months since the last maintenance service and the type of repair associated with the machine. Develop the estimated simple linear regression equation to predict the repair time given the engineer who performed the service. Let $X_3 = 0$ if Heinz Kolb performed the service and $X_3 = 1$ if Wolfgang Linz performed the service.

d. Does the equation that you developed in part (c) provide a good fit for the observed data? Explain.

24 In a multiple regression analysis by McIntyre (1994), Tar, Nicotine and Weight are considered as possible predictors of Carbon Monoxide (CO) content for 25 different brands of cigarette. Details of variables and data follow.

Brand	The cigarette brand
Tar	The tar content (in mg)
Nicotine	The nicotine content (in mg)
Weight	The weight (in g)
CO	The carbon monoxide (CO) content (in mg)

Brand	Tar	Nicotine	Weight	CO
Alpine	14.1	0.86	.9853	13.6
Benson&Hedges	16.0	1.06	1.0938	16.6
BullDurham	29.8	2.03	1.1650	23.5
CamelLights	8.0	0.67	0.9280	10.2
Carlton	4.1	0.40	0.9462	5.4
Chesterfield	15.0	1.04	0.8885	15.0
GoldenLights	8.8	0.76	1.0267	9.0
Kent	12.4	0.95	0.9225	12.3
Kool	16.6	1.12	0.9372	16.3

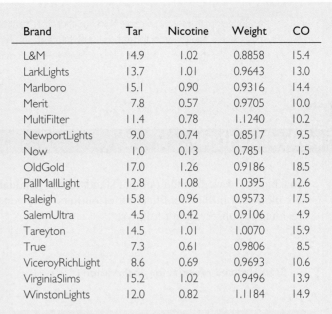

Brand	Tar	Nicotine	Weight	CO
L&M	14.9	1.02	0.8858	15.4
LarkLights	13.7	1.01	0.9643	13.0
Marlboro	15.1	0.90	0.9316	14.4
Merit	7.8	0.57	0.9705	10.0
MultiFilter	11.4	0.78	1.1240	10.2
NewportLights	9.0	0.74	0.8517	9.5
Now	1.0	0.13	0.7851	1.5
OldGold	17.0	1.26	0.9186	18.5
PallMallLight	12.8	1.08	1.0395	12.6
Raleigh	15.8	0.96	0.9573	17.5
SalemUltra	4.5	0.42	0.9106	4.9
Tareyton	14.5	1.01	1.0070	15.9
True	7.3	0.61	0.9806	8.5
ViceroyRichLight	8.6	0.69	0.9693	10.6
VirginiaSlims	15.2	1.02	0.9496	13.9
WinstonLights	12.0	0.82	1.1184	14.9

GIGARETTES

a. Examine correlations between variables in the study and hence assess the possibility of problems of multicollinearity affecting any subsequent regression model involving independent variables Tar and Nicotine.

b. Thus develop an estimated multiple regression equation using an appropriate number of the independent variables featured in the study.

c. Are your predictors statistically significant? Use $\alpha = 0.05$. What explanation can you give for the results observed?

25 The data below (Dunn, 2007) come from a study investigating a new method of measuring body composition. Body fat percentage, age and gender is given for 18 adults aged between 23 and 61.

Age	Percent.Fat	Gender
23	9.5	M
23	27.9	F
27	7.8	M
27	17.8	M
39	31.4	F
41	25.9	F
45	27.4	M
49	25.2	F
50	31.1	F
53	34.7	F
53	42	F
54	29.1	F
56	32.5	F
57	30.3	F
58	33	F
58	33.8	F
60	41.1	F
61	34.5	F

BODYFAT

a. Develop an estimated regression equation that relates Age and Gender to Percent.Fat
b. Is Age a significant factor in predicting Percent.Fat? Explain. Use $\alpha = 0.05$.
c. What is the estimated body fat percentage for a female aged 45?

15.8 Residual analysis

In Chapter 14 we pointed out that standardized residuals were frequently used in residuals plots and in the identification of outliers. The general formula for the standardized residual for observation i follows.

Standardized residual for observation i

$$\frac{y_i - \hat{y}_i}{s_{y_i - \hat{y}_i}}$$

where

$s_{y_i - \hat{y}_i}$ = the standard deviation of residual i

The general formula for the standard deviation of residual i is defined as follows.

Standard deviation of residual i

$$s_{y_i - \hat{y}_i} = s \sqrt{1 - h_i} \qquad (15.25)$$

where

s = standard error of the estimate
h_i = leverage of observation i

The **leverage** of an observation is determined by how far the values of the independent variables are from their means. The computation of h_i, $s_{y_i - \hat{y}_i}$ and hence the standardized residual for observation i in multiple regression analysis is too complex to be done by hand. However, the standardized residuals can be easily obtained as part of the output from statistical software packages. Table 15.7 lists the predicted values, the residuals, and the standardized residuals for the Eurodistributor example presented previously in this chapter; we obtained these values by using the MINITAB statistical software package. The predicted values in the table are based on the estimated regression equation

$$\hat{y} = -0.869 + 0.0611x_1 + 0.923x_2$$

The standardized residuals and the predicted values of Y from Table 15.7 are used in the standardized residual plot in Figure 15.10.

This standardized residual plot does not indicate any unusual abnormalities. Also, all of the standardized residuals are between -2 and $+2$; hence, we have no reason

Table 15.7 Residuals and standardized residuals for the Eurodistributor regression analysis

Distance travelled (X_1)	Deliveries (X_2)	Travel time (Y)	Predicted value ($\hat{y}$)	Residual ($y - \hat{y}$)	Standardized residual
100	4	9.3	8.93846	0.361540	0.78344
50	3	4.8	4.95830	−0.158305	−0.34962
100	4	8.9	8.93846	−0.038460	−0.08334
100	2	6.5	7.09161	−0.591609	−1.30929
50	2	4.2	4.03488	0.165121	0.38167
80	2	6.2	5.86892	0.331083	0.65431
75	3	7.4	6.48667	0.913330	1.68917
65	4	6.0	6.79875	−0.798749	−1.77372
90	3	7.6	7.40369	0.196311	0.36703
90	2	6.1	6.48026	−0.380263	−0.77639

to question the assumption that the error term ε is normally distributed. We conclude that the model assumptions are reasonable.

A normal probability plot also can be used to determine whether the distribution of ε appears to be normal. The procedure and interpretation for a normal probability plot were discussed in Section 14.8. The same procedure is appropriate for multiple regression. Again, we would use a statistical software package to perform the computations and provide the normal probability plot.

Figure 15.10 Standardized residual plot for the Eurodistributor multiple regression analysis

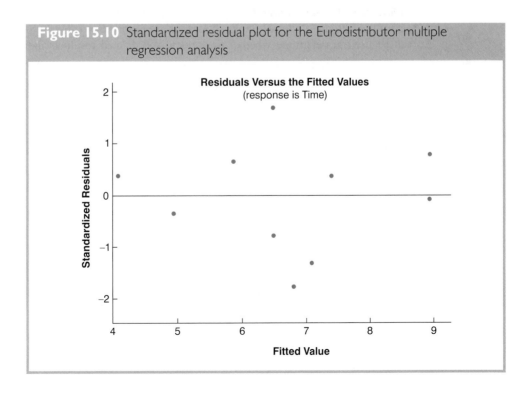

Detecting outliers

An **outlier** is an observation that is unusual in comparison with the other data; in other words, an outlier does not fit the pattern of the other data. In Chapter 14 we showed an example of an outlier and discussed how standardized residuals can be used to detect outliers.

MINITAB classifies an observation as an outlier if the value of its standardized residual is less than -2 or greater than $+2$. Applying this rule to the standardized residuals for the Eurodistributor example (see Table 15.7), we do not detect any outliers in the data set.

In general, the presence of one or more outliers in a data set tends to increase s, the standard error of the estimate, and hence increase $s_{y_i - \hat{y}_i}$, the standard deviation of residual i. Because $s_{y_i - \hat{y}_i}$ appears in the denominator of the formula for the standardized residual (15.24), the size of the standardized residual will decrease as s increases.

As a result, even though a residual may be unusually large, the large denominator in expression (15.24) may cause the standardized residual rule to fail to identify the observation as being an outlier. We can circumvent this difficulty by using a form of standardized residuals called **studentized deleted residuals**.

Studentized deleted residuals and outliers

Suppose the ith observation is deleted from the data set and a new estimated regression equation is developed with the remaining $n - 1$ observations. Let $s_{(i)}$ denote the standard error of the estimate based on the data set with the ith observation deleted. If we compute the standard deviation of residual i (15.25) using $s_{(i)}$ instead of s, and then compute the standardized residual for observation i (15.24) using the revised value, the resulting standardized residual is called a studentized deleted residual.

If the ith observation is an outlier, $s_{(i)}$ will be less than s. The absolute value of the ith studentized deleted residual therefore will be larger than the absolute value of the standardized residual. In this sense, studentized deleted residuals may detect outliers that standardized residuals do not detect. Many statistical software packages provide an option for obtaining studentized deleted residuals. Using MINITAB, we obtained the studentized deleted residuals for the Eurodistributor example; the results are reported in Table 15.8. The t distribution can be used to determine whether the studentized deleted residuals indicate the presence

Table 15.8 Studentized deleted residuals for Eurodistributor

Distance travelled (X_1)	Deliveries (X_2)	Travel time (Y)	Standardized residual	Studentized deleted residual
100	4	9.3	0.78344	0.75938
50	3	4.8	−0.34962	−0.32654
100	4	8.9	−0.08334	−0.0772
100	2	6.5	−1.30929	−1.39494
50	2	4.2	0.38167	0.35709
80	2	6.2	0.65431	0.62519
75	3	7.4	1.68917	2.03187
65	4	6.0	−1.77372	−2.21314
90	3	7.6	0.36703	0.34312
90	2	6.1	−0.77639	−0.7519

of outliers. Recall that p denotes the number of independent variables and n denotes the number of observations. Hence, if we delete the ith observation, the number of observations in the reduced data set is $n - 1$; in this case the error sum of squares has $(n - 1) - p - 1$ degrees of freedom. For the Eurodistributor example with $n = 10$ and $p = 2$, the degrees of freedom for the error sum of squares with the ith observation deleted is $9 - 2 - 1 = 6$. At a 0.05 level of significance, the t distribution (Table 2 of Appendix B) shows that with six degrees of freedom, $t_{0.025} = 2.447$. If the value of the ith studentized deleted residual is less than -2.447 or greater than $+2.447$, we can conclude that the ith observation is an outlier. The studentized deleted residuals in Table 15.8 do not exceed those limits; therefore, we conclude that outliers are not present in the data set.

Influential observations

In Section 14.9 we discussed how the leverage of an observation can be used to identify observations for which the value of the independent variable may have a strong influence on the regression results. As we acknowledged, the leverage (h_i) of an observation, measures how far the values of the independent variables are from their mean values. The leverage values are easily obtained as part of the output from statistical software packages. MINITAB computes the leverage values and uses the rule of thumb

$$h_i > 3(p + 1)/n$$

to identify **influential observations**. For the Eurodistributor example with $p = 2$ independent variables and $n = 10$ observations, the critical value for leverage is $3(2 + 1)/10 = 0.9$. The leverage values for the Eurodistributor example obtained by using MINITAB are reported in Table 15.9. As h_i does not exceed 0.9, no influential observations in the data set are detected.

Using Cook's distance measure to identify influential observations

A problem that can arise in using leverage to identify influential observations is that an observation can be identified as having high leverage and not necessarily be influential in terms of the resulting estimated regression equation. For example, Table 15.10 shows a

Table 15.9 Leverage and Cook's distance measures for Eurodistributor

Distance travelled (X_1)	Deliveries (X_2)	Travel time (Y)	Leverage (h_i)	Cook's D (D_i)
100	4	9.3	0.351704	0.110994
50	3	4.8	0.375863	0.024536
100	4	8.9	0.351704	0.001256
100	2	6.5	0.378451	0.347923
50	2	4.2	0.430220	0.036663
80	2	6.2	0.220557	0.040381
75	3	7.4	0.110009	0.117561
65	4	6.0	0.382657	0.650029
90	3	7.6	0.129098	0.006656
90	2	6.1	0.269737	0.074217

Table 15.10 Data set illustrating potential problem using the leverage criterion

x_i	y_i	Leverage h_i
1	18	0.204170
1	21	0.204170
2	22	0.164205
3	21	0.138141
4	23	0.125977
4	24	0.125977
5	26	0.127715
15	39	0.909644

data set consisting of eight observations and their corresponding leverage values (obtained by using MINITAB). Because the leverage for the eighth observation is $0.91 > 0.75$ (the critical leverage value), this observation is identified as influential. Before reaching any final conclusions, however, let us consider the situation from a different perspective.

Figure 15.11 shows the scatter diagram and the estimated regression equation corresponding to the data set in Table 15.10. We used MINITAB to develop the following estimated regression equation for these data.

$$\hat{y} = 18.2 + 1.39\,x$$

The straight line in Figure 15.11 is the graph of this equation. Now, let us delete the observation $X = 15$, $Y = 39$ from the data set and fit a new estimated regression equation to the remaining seven observations; the new estimated regression equation is

$$\hat{y} = 18.1 + 1.42\,x$$

We note that the y-intercept and slope of the new estimated regression equation are not fundamentally different from the values obtained by using all the data. Although the leverage criterion identified the eighth observation as influential, this observation clearly had little influence on the results obtained. Thus, in some situations using only leverage to identify influential observations can lead to wrong conclusions.

Cook's distance measure uses both the leverage of observation i, h_i, and the residual for observation i, $(y_i - \hat{y}_i)$, to determine whether the observation is influential.

Cook's distance measure

$$D_i = \frac{(y_i - \hat{y}_i)^2\, h_i}{(p - 1)s^2\,(1 - h_i)^2}$$

(15.26)

where

D_i = Cook's distance measure for observation i

$y_i - \hat{y}_i$ = the residual for observation i

h_i = the leverage for observation i

p = the number of independent variables

s = the standard error of the estimate

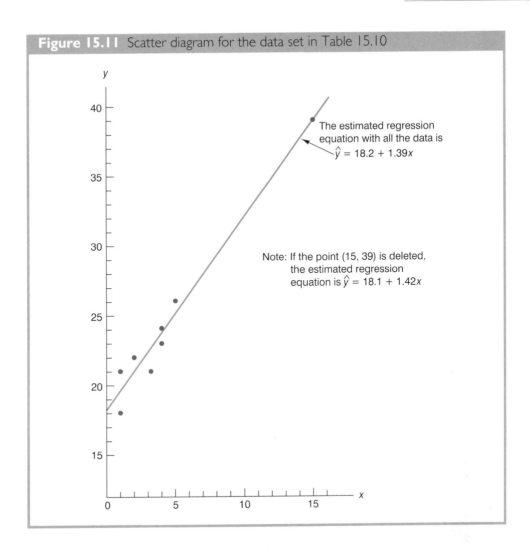

Figure 15.11 Scatter diagram for the data set in Table 15.10

The estimated regression equation with all the data is $\hat{y} = 18.2 + 1.39x$

Note: If the point (15, 39) is deleted, the estimated regression equation is $\hat{y} = 18.1 + 1.42x$

The value of Cook's distance measure will be large and indicate an influential observation if the residual or the leverage is large. As a rule of thumb, values of $D_i > 1$ indicate that the ith observation is influential and should be studied further. The last column of Table 15.9 provides Cook's distance measure for the Eurodistributor problem as given by MINITAB. Observation 8 with $D_i = 0.650029$ has the most influence. However, applying the rule $D_i > 1$, we should not be concerned about the presence of influential observations in the Eurodistributor data set.

Exercises

Methods

26 Data for two variables, X and Y, follow.

x_i	1	2	3	4	5
y_i	3	7	5	11	14

a. Develop the estimated regression equation for these data.
b. Plot the standardized residuals versus $\hat{y}$. Do there appear to be any outliers in these data? Explain.
c. Compute the studentized deleted residuals for these data. At the 0.05 level of significance, can any of these observations be classified as an outlier? Explain.

27 Data for two variables, X and Y, follow.

x_i	22	24	26	28	40
y_i	12	21	31	35	70

a. Develop the estimated regression equation for these data.
b. Compute the studentized deleted residuals for these data. At the 0.05 level of significance, can any of these observations be classified as an outlier? Explain.
c. Compute the leverage values for these data. Do there appear to be any influential observations in these data? Explain.
d. Compute Cook's distance measure for these data. Are any observations influential? Explain.

Applications

GIGARETTES

28 Exercise 5 gave data on weekly gross revenue, television advertising, and newspaper advertising for Toulon theatres.

a. Find an estimated regression equation relating weekly gross revenue to television and newspaper advertising.
b. Plot the standardized residuals against $\hat{y}$. Does the residual plot support the assumptions about ε? Explain.
c. Check for any outliers in these data. What are your conclusions?
d. Are there any influential observations? Explain.

29 Data (Tufte, 1974) on male deaths per million in 1950 for lung cancer (Y) and *per capita* cigarette consumption in 1930 (X) are given below:

Country	y	x	Country	y	x
Ireland	58	220	Norway	90	250
Sweden	115	310	Canada	150	510
Denmark	165	380	Australia	170	455
USA	190	1280	Holland	245	460
Switzerland	250	530	Finland	350	1115
GB	465	1145			

Results from a simple regression analysis of this information are as follows:

```
Regression Analysis: y versus x

The regression equation is
y = 65.7 + 0.229 x


Predictor      Coef  SE Coef     T      P
Constant      65.75    48.96  1.34  0.212
x           0.22912  0.06921  3.31  0.009


S = 84.1296   R-Sq = 54.9%   R-Sq(adj) = 49.9%

Analysis of Variance

Source           DF      SS      MS      F      P
Regression        1   77554   77554  10.96  0.009
Residual Error    9   63700    7078
Total            10  141255

Unusual Observations

Obs    x      y     Fit  SE Fit  Residual  St Resid
 4  1280  190.0  359.0    53.2    -169.0    -2.59R

R denotes an observation with a large standardized residual.

Durbin-Watson statistic = 2.07188
```

Corresponding lererage and cook distance details are as follows.

HI1	COOK1
0.191237	0.06985
0.149813	0.00694
0.125175	0.00172
0.399306	2.23320
0.094716	0.03222
0.288283	0.75365
0.176211	0.02001
0.097018	0.00893
0.106139	0.00000
0.105140	0.05060
0.266962	0.02909

Carry out any further statistical tests you deem appropriate, otherwise comment on the effectiveness of the linear modes.

15.9 Logistic regression

In many regression applications the dependent variable may only assume two discrete values. For instance, a bank might like to develop an estimated regression equation for predicting whether a person will be approved for a credit card. The dependent variable can be coded as $Y = 1$ if the bank approves the request for a credit card and $Y = 0$ if the bank rejects the request for a credit card. Using logistic regression we can estimate the probability that the bank will approve the request for a credit card given a particular set of values for the chosen independent variables.

Consider an application of logistic regression involving a direct mail promotion being used by Stamm Stores. Stamm owns and operates a national chain of women's fashion stores. Five thousand copies of an expensive four-colour sales catalogue have been printed, and each catalogue includes a coupon that provides a €50 discount on purchases of €200 or more.

The catalogues are expensive and Stamm would like to send them to only those customers who have the highest probability of making a €200 purchase using the discount coupons.

Management thinks that annual spending at Stamm Stores and whether a customer has a Stamm credit card are two variables that might be helpful in predicting whether a customer who receives the catalogue will use the coupon to make a €200 purchase. Stamm conducted a pilot study using a random sample of 50 Stamm credit card customers and 50 other customers who do not have a Stamm credit card. Stamm sent the catalogue to each of the 100 customers selected. At the end of a test period, Stamm noted whether the customer made a purchase (coded 1 if the customer made a purchase and 0 if not). The sample data for the first ten catalogue recipients are shown in Table 15.11. The amount each customer spent last year at Stamm is shown in thousands of euros and the credit card information has been coded as 1 if the customer has a Stamm credit card and 0 if not. In the Purchase column, a 1 is recorded if the sampled customer used the €50 discount coupon to make a purchase of €200 or more.

We might think of building a multiple regression model using the data in Table 15.11 to help Stamm predict whether a catalogue recipient will make a purchase. We would use Annual Spending and Stamm Card as independent variables and Purchase as the dependent variable.

Because the dependent variable may only assume the values of 0 or 1, however, the ordinary multiple regression model is not applicable. This example shows the type of situation for which logistic regression was developed. Let us see how logistic regression can be used to help Stamm predict which type of customer is most likely to take advantage of their promotion.

Table 15.11 Sample data for Stamm Stores

Customer	Annual spending (€000s)	Stamm card	Purchase
1	2.291	1	0
2	3.215	1	0
3	2.135	1	0
4	3.924	0	0
5	2.528	1	0
6	2.473	0	1
7	2.384	0	0
8	7.076	0	0
9	1.182	1	1
10	3.345	0	0

Logistic regression equation

In many ways logistic regression is like ordinary regression. It requires a dependent variable, Y, and one or more independent variables. In multiple regression analysis, the mean or expected value of Y, is referred to as the multiple regression equation.

$$E(Y) = \beta_0 + \beta_1 x_1 + \beta_2 x_2 + \cdots\cdots + \beta_p x_p \qquad (15.27)$$

In logistic regression, statistical theory as well as practice has shown that the relationship between $E(Y)$ and $X_1, X_2, \ldots X_p$ is better described by the following nonlinear equation.

Logistic regression equation

$$E(Y) = \frac{e^{\beta_0 + \beta_1 x_1 + \beta_2 x_2 + \cdots + \beta_p x_p}}{1 + e^{\beta_0 + \beta_1 x_1 + \beta_2 x_2 + \cdots + \beta_p x_p}} \qquad (15.28)$$

If the two values of the dependent variable Y are coded as 0 or 1, the value of $E(Y)$ in equation (15.28) provides the *probability* that $Y = 1$ given a particular set of values for the independent variables $X_1, X_2, \ldots X_p$. Because of the interpretation of $E(Y)$ as a probability, the **logistic regression equation** is often written as follows.

Interpretation of $E(Y)$ as a probability in logistic regression

$$E(Y) = P(y = 1 | x_1, x_2, \ldots x_p) \qquad (15.29)$$

To provide a better understanding of the characteristics of the logistic regression equation, suppose the model involves only one independent variable X and the values of the model parameters are $\beta_0 = -7$ and $\beta_1 = 3$. The logistic regression equation corresponding to these parameter values is

$$E(Y) = P(Y = 1 | x) = \frac{e^{\beta_0 + \beta_1 x}}{1 + e^{\beta_0 + \beta_1 x}} = \frac{e^{-7 + 3x}}{1 + e^{-7 + 3x}} \qquad (15.30)$$

Figure 15.12 shows a graph of equation (15.30). Note that the graph is S-shaped. The value of $E(Y)$ ranges from 0 to 1, with the value of $E(Y)$ gradually approaching 1 as the value of X becomes larger and the value of $E(Y)$ approaching 0 as the value of X becomes smaller. Note also that the values of $E(Y)$, representing probability, increase fairly rapidly as X increases from 2 to 3. The fact that the values of $E(Y)$ range from 0 to 1 and that the curve is S-shaped makes equation (15.30) ideally suited to model the probability the dependent variable is equal to 1.

Estimating the logistic regression equation

In simple linear and multiple regression the least squares method is used to compute $b_0, b_1, \ldots, b_p$ as estimates of the model parameters ($\beta_0, \beta_1, \ldots, \beta_p$). The nonlinear form of the logistic regression equation makes the method of computing estimates

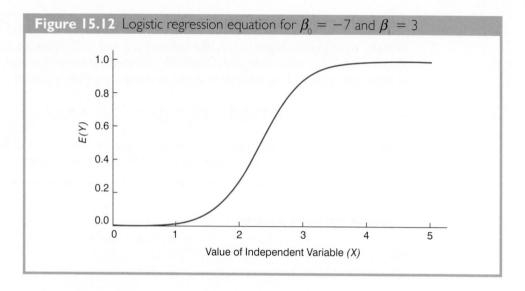

Figure 15.12 Logistic regression equation for $\beta_0 = -7$ and $\beta_1 = 3$

more complex and beyond the scope of this text. We will use computer software to provide the estimates. The **estimated logistic regression equation** is

Estimated logistic regression equation

$$\hat{y} = \text{estimate of } P(Y = 1 \mid x_1, x_2, \ldots x_p) = \frac{e^{b_0 + b_1x_1 + b_2x_2 + \cdots + b_px_p}}{1 + e^{b_0 + b_1x_1 + b_2x_2 + \cdots + b_px_p}}$$

(15.31)

Here $\hat{y}$ provides an estimate of the probability that $Y = 1$, given a particular set of values for the independent variables.

Let us now return to the Stamm Stores example. The variables in the study are defined as follows:

$$Y = \begin{cases} 0 \text{ if the customer made no purchase during the test period} \\ 1 \text{ if the customer made a purchase during the test period} \end{cases}$$

$$X_1 = \text{annual spending at Stamm Stores (€000s)}$$

$$X_2 = \begin{cases} 0 \text{ if the customer does not have a Stamm credit card} \\ 1 \text{ if the customer has a Stamm credit card} \end{cases}$$

Therefore, we choose a logistic regression equation with two independent variables.

$$E(Y) = \frac{e^{\beta_0 + \beta_1x_1 + \cdots + \beta_px_p}}{1 + e^{\beta_0 + \beta_1x_1 + \cdots + \beta_px_p}}$$

(15.32)

Using the sample data (see Table 15.11), MINITAB's binary logistic regression procedure was used to compute estimates of the model parameters β_0, β_1, and β_2. A portion of the output obtained is shown in Figure 15.13. We see that $b_0 = -2.1464$, $b_1 = 0.3416$, and $b_2 = 1.0987$. Thus, the estimated logistic regression equation is

$$\hat{y} = \frac{e^{b_0 + b_1x_1 + \cdots + b_px_p}}{1 + e^{b_0 + b_1x_1 + \cdots + b_px_p}} = \frac{e^{-2.1464 + 0.3416x_1 + 1.0987x_2}}{1 + e^{-2.1464 + 0.3416x_1 + 1.0987x_2}}$$

(15.33)

Figure 15.13 Partial logistic regression output for the Stamm Stores example

```
Logistic Regression Table

                                               Odds      95% CI
Predictor     Coef    SE Coef      Z      P   Ratio   Lower  Upper
Constant   -2.14637  0.577245  -3.72  0.000
Spending    0.341643 0.128672   2.66  0.008   1.41    1.09   1.81
Card        1.09873  0.444696   2.47  0.013   3.00    1.25   7.17


Log-Likelihood = -60.487
Test that all slopes are zero: G = 13.628, DF = 2, P-Value = 0.001
```

We can now use equation (15.33) to estimate the probability of making a purchase for a particular type of customer. For example, to estimate the probability of making a purchase for customers that spend €2000 annually and do not have a Stamm credit card, we substitute $X_1 = 2$ and $X_2 = 0$ into equation (15.33).

$$\hat{y} = \frac{e^{-2.1464 + 0.3416(2) + 1.0987(0)}}{1 + e^{-2.1464 + 0.3416(2) + 1.0987(0)}} = \frac{e^{-1.4632}}{1 + e^{-1.4632}} = \frac{0.2315}{1.2315} = 0.1880$$

Thus, an estimate of the probability of making a purchase for this particular group of customers is approximately 0.19. Similarly, to estimate the probability of making a purchase for customers that spent €2000 last year and have a Stamm credit card, we substitute $X_1 = 2$ and $X_2 = 1$ into equation (15.33).

$$\hat{y} = \frac{e^{-2.1464 + 0.3416(2) + 1.0987(1)}}{1 + e^{-2.1464 + 0.3416(2) + 1.0987(1)}} = \frac{e^{-0.3645}}{1 + e^{-0.3645}} = \frac{0.6945}{1.6945} = 0.4099$$

Thus, for this group of customers, the probability of making a purchase is approximately 0.41. It appears that the probability of making a purchase is much higher for customers with a Stamm credit card. Before reaching any conclusions, however, we need to assess the statistical significance of our model.

Testing for significance

Testing for significance in logistic regression is similar to testing for significance in multiple regression. First we conduct a test for overall significance. For the Stamm Stores example, the hypotheses for the test of overall significance follow:

$$H_0: \beta_1 = \beta_2 = 0$$
$$H_1: \beta_1 \text{ and/or } \beta_2 \text{ is not equal to zero}$$

The test for overall significance is based upon the value of a G test statistic. This is commonly referred to as the 'Deviance Statistic'. If the null hypothesis is true, the sampling distribution of G follows a chi-square distribution with degrees of freedom equal to the number of independent variables in the model. Although the computation of G is beyond the scope of the book, the value of G and its corresponding p-value are provided as part of MINITAB's binary logistic regression output. Referring to the last line in Figure 15.13, we see that the value of G is 13.628, its degrees of freedom are 2, and its p-value is 0.001. Thus, at any level of significance $\alpha \geq 0.001$, we would reject the null hypothesis and conclude that the overall model is significant.

If the G test shows an overall significance, a z test can be used to determine whether each of the individual independent variables is making a significant contribution to the overall model. For the independent variables X_i, the hypotheses are

$$H_0: \beta_i = 0$$
$$H_1: \beta_i \neq 0$$

If the null hypothesis is true, the value of the estimated coefficient divided by its standard error follows a standard normal probability distribution. The column labelled Z in the MINITAB output contains the values of $z_i = b_i / s_{b_i}$ for each of the estimated coefficients and the column labelled p contains the corresponding p-values. The z_i ratio is also known as a 'Wald Statistic'. Suppose we use $\alpha = 0.05$ to test for the significance of the independent variables in the Stamm model. For the independent variable X_1 the z value is 2.66 and the corresponding p-value is 0.008. Thus, at the 0.05 level of significance we can reject $H_0: \beta_1 = 0$. In a similar fashion we can also reject $H_0: \beta_2 = 0$ because the p-value corresponding to $z = 2.47$ is 0.013. Hence, at the 0.05 level of significance, both independent variables are statistically significant.

Managerial use

We now use the estimated logistic regression equation to make a decision recommendation concerning the Stamm Stores catalogue promotion. For Stamm Stores, we already computed

$$P(Y = 1 | X_1 = 2, X_2 = 1) = 0.4099 \quad \text{and} \quad P(Y = 1 | X_1 = 2, X_2 = 0) = 0.1880$$

These probabilities indicate that for customers with annual spending of €2000 the presence of a Stamm credit card increases the probability of making a purchase using the discount coupon. In Table 15.12 we show estimated probabilities for values of annual spending ranging from €1000 to €7000 for both customers who have a Stamm credit card and customers who do not have a Stamm credit card. How can Stamm use this information to better target customers for the new promotion? Suppose Stamm wants to send the promotional catalogue only to customers who have a 0.40 or higher probability of making a purchase. Using the estimated probabilities in Table 15.12, Stamm promotion strategy would be:

Customers who have a Stamm credit card: Send the catalogue to every customer that spent €2000 or more last year.

Customers who do not have a Stamm credit card: Send the catalogue to every customer that spent €6000 or more last year.

Looking at the estimated probabilities further, we see that the probability of making a purchase for customers who do not have a Stamm credit card, but spend €5000 annually is 0.3921. Thus, Stamm may want to consider revising this strategy by including those customers who do not have a credit card as long as they spent €5000 or more last year.

Table 15.12 Estimated probabilities for Stamm Stores

		€1000	€2000	€3000	€4000	€5000	€6000	€7000
				Annual spending				
Credit card	Yes	0.3305	0.4099	0.4943	0.5790	0.6593	0.7314	0.7931
	No	0.1413	0.1880	0.2457	0.3143	0.3921	0.4758	0.5609

Interpreting the logistic regression equation

Interpreting a regression equation involves relating the independent variables to the business question that the equation was developed to answer. With logistic regression, it is difficult to interpret the relation between the independent variables and the probability that $Y = 1$ directly because the logistic regression equation is nonlinear. However, statisticians have shown that the relationship can be interpreted indirectly using a concept called the odds ratio.

The **odds in favour of an event occurring** is defined as the probability the event will occur divided by the probability the event will not occur. In logistic regression the event of interest is always $Y = 1$. Given a particular set of values for the independent variables, the odds in favour of $Y = 1$ can be calculated as follows:

$$\text{Odds} = \frac{P(Y = 1 \mid X_1, X_2, \ldots X_y)}{P(Y = 0 \mid X_1, X_2, \ldots X_y)} = \frac{P(Y = 1 \mid X_1, X_2, \ldots X_y)}{1 - P(Y = 1 \mid X_1, X_2, \ldots X_y)} \qquad (15.34)$$

$$\ldots$$

The **odds ratio** measures the impact on the odds of a one-unit increase in only one of the independent variables. The odds ratio is the odds that $Y = 1$ given that one of the independent variables has been increased by one unit (odds_1) divided by the odds that $Y = 1$ given no change in the values for the independent variables (odds_0).

Odds ratio

$$\text{Odds ratio} = \frac{\text{odds}_1}{\text{odds}_0} \qquad (15.35)$$

For example, suppose we want to compare the odds of making a purchase for customers who spend 2000 annually and have a Stamm credit card ($X_1 = 2$ and $X_2 = 1$) to the odds of making a purchase for customers who spend €2000 annually and do not have a Stamm credit card ($X_1 = 2$ and $X_2 = 0$). We are interested in interpreting the effect of a one-unit increase in the independent variable X_2. In this case

$$\text{Odds}_1 = \frac{P(Y = 1 \mid X_1 = 2, X_2 = 1)}{1 - P(Y = 1 \mid X_1 = 2, X_2 = 1)}$$

and

$$\text{Odds}_0 = \frac{P(Y = 1 \mid X_1 = 2, X_2 = 0)}{1 - P(Y = 1 \mid X_1 = 2, X_2 = 0)}$$

Previously we showed that an estimate of the probability that $Y = 1$ given $X_1 = 2$ and $X_2 = 1$ is 0.4099, and an estimate of the probability that $Y = 1$ given $X_1 = 2$ and $X_2 = 0$ is 0.1880. Thus,

$$\text{Estimate of odds}_1 = \frac{0.4099}{1 - 0.4099} = 0.6946$$

and

$$\text{Estimate of odds}_0 = \frac{0.1880}{1 - 0.1880} = 0.2315$$

The estimated odds ratio is

$$\text{Estimated odds ratio} = \frac{0.6946}{0.2315} = 3.00$$

Thus, we can conclude that the estimated odds in favour of making a purchase for customers who spent €2000 last year and have a Stamm credit card are three times greater than the estimated odds in favour of making a purchase for customers who spent €2000 last year and do not have a Stamm credit card.

The odds ratio for each independent variable is computed while holding all the other independent variables constant. But it does not matter what constant values are used for the other independent variables. For instance, if we computed the odds ratio for the Stamm credit card variable (X_2) using €3000, instead of €2000, as the value for the annual spending variable (X_1), we would still obtain the same value for the estimated odds ratio (3.00). Thus, we can conclude that the estimated odds of making a purchase for customers who have a Stamm credit card are three times greater than the estimated odds of making a purchase for customers who do not have a Stamm credit card.

The odds ratio is standard output for logistic regression software packages. Refer to the MINITAB output in Figure 15.13. The column with the heading Odds Ratio contains the estimated odds ratios for each of the independent variables. The estimated odds ratio for X_1 is 1.41 and the estimated odds ratio for X_2 is 3.00. We already showed how to interpret the estimated odds ratio for the binary independent variable X_2. Let us now consider the interpretation of the estimated odds ratio for the continuous independent variable X_1.

The value of 1.41 in the Odds Ratio column of the MINITAB output tells us that the estimated odds in favour of making a purchase for customers who spent €3000 last year is 1.41 times greater than the estimated odds in favour of making a purchase for customers who spent €2000 last year. Moreover, this interpretation is true for any one-unit change in X_1.

For instance, the estimated odds in favour of making a purchase for someone who spent €5000 last year is 1.41 times greater than the odds in favour of making a purchase for a customer who spent €4000 last year. But suppose we are interested in the change in the odds for an increase of more than one unit for an independent variable. Note that X_1 can range from 1 to 7. The odds ratio as printed by the MINITAB output does not answer this question.

To answer this question we must explore the relationship between the odds ratio and the regression coefficients.

A unique relationship exists between the odds ratio for a variable and its corresponding regression coefficient. For each independent variable in a logistic regression equation it can be shown that

$$\text{Odds ratio} = e^{b_i}$$

To illustrate this relationship, consider the independent variable X_1 in the Stamm example. The estimated odds ratio for X_1 is

$$\text{Estimated odds ratio} = e^{b_1} = e^{0.3416} = 1.41$$

Similarly, the estimated odds ratio for X_2 is

$$\text{Estimated odds ratio} = e^{b_2} = e^{1.0987} = 3.00$$

This relationship between the odds ratio and the coefficients of the independent variables makes it easy to compute estimates of the odds ratios once we develop estimates of the model parameters. Moreover, it also provides us with the ability to investigate changes in the odds ratio of more than or less than one unit for a continuous independent variable.

The odds ratio for an independent variable represents the change in the odds for a one unit change in the independent variable holding all the other independent variables constant. Suppose that we want to consider the effect of a change of more than one unit, say c units. For instance, suppose in the Stamm example that we want to compare the odds of making a purchase for customers who spend €5000 annually ($X_1 = 5$) to the odds of making a purchase for customers who spend €2000 annually ($X_1 = 2$). In this case $c = 5 - 2 = 3$ and the corresponding estimated odds ratio is

$$e^{cb_i} = e^{3(0.3416)} = e^{1.0248} = 2.79$$

This result indicates that the estimated odds of making a purchase for customers who spend €5000 annually is 2.79 times greater than the estimated odds of making a purchase for customers who spend €2000 annually. In other words, the estimated odds ratio for an increase of €3000 in annual spending is 2.79.

In general, the odds ratio enables us to compare the odds for two different events. If the value of the odds ratio is 1, the odds for both events are the same. Thus, if the independent variable we are considering (such as Stamm credit card status) has a positive impact on the probability of the event occurring, the corresponding odds ratio will be greater than 1. Most logistic regression software packages provide a confidence interval for the odds ratio. The MINITAB output in Figure 15.13 provides a 95 per cent confidence interval for each of the odds ratios. For example, the point estimate of the odds ratio for X_1 is 1.41 and the 95 per cent confidence interval is 1.09 to 1.81. Because the confidence interval does not contain the value of 1, we can conclude that X_1, has a significant effect on the odds ratio. Similarly, the 95 per cent confidence interval for the odds ratio for X_2 is 1.25 to 7.17. Because this interval does not contain the value of 1, we can also conclude that X_2 has a significant effect on the odds ratio.

Logit transformation

An interesting relationship can be observed between the odds in favour of $Y = 1$ and the exponent for e in the logistic regression equation. It can be shown that

$$\ln(\text{odds}) = \beta_0 + \beta_1 x_1 + \beta_2 x_2 + \cdots + \beta_p x_p$$

This equation shows that the natural logarithm of the odds in favour of $Y = 1$ is a linear function of the independent variables. This linear function is called the **logit**. We will use the notation $g(x_1, x_2, \ldots x_p)$ to denote the logit.

Logit	
$g(x_1, x_2, \ldots, x_p) = \beta_0 + \beta_1 x_1 + \beta_2 x_2 + \cdots + \beta_p x_p$	(15.36)

Substituting $g(x_1, x_2, \ldots x_p)$ for $\beta_0 + \beta_1 x_1 + \beta_2 x_2 + \cdots + \beta_p x_p$ in equation (15.28), we can write the logistic regression equation as

$$E(Y) = \frac{e^{g(x_1, x_2, \ldots, x_p)}}{1 + e^{g(x_1, x_2, \ldots, x_p)}} \tag{15.37}$$

Once we estimate the parameters in the logistic regression equation, we can compute an estimate of the logit. Using $\hat{g}\,(x_1, x_2 \ldots x_p)$ to denote the **estimated logit**, we obtain

Estimated logit

$$\hat{g}\,(x_1, x_2, \ldots x_p) = b_0 + b_1 x_1 + b_2 x_2 + \cdots + b_p x_p \tag{15.38}$$

Therefore, in terms of the estimated logit, the estimated regression equation is

$$\hat{y} = \frac{e^{b_0 + b_1 x_1 + b_2 x_2 + \cdots + b_p x_p}}{1 + e^{b_0 + b_1 x_1 + b_2 x_2 + \cdots + b_p x_p}} = \frac{e^{\hat{g}\,(x_1, x_2, \ldots, x_p)}}{1 + e^{\hat{g}(x_1, x_2, \ldots, x_p)}} \tag{15.39}$$

For the Stamm Stores example, the estimated logit is

$$\hat{g}\,(x_1, x_2) = -2.1464 + 0.3416 x_1 + 1.0987 x_2$$

and the estimated regression equation is

$$y = \frac{e^{\hat{g}(x_1, x_2)}}{1 + e^{\hat{g}(x_1, x_2)}} = \frac{e^{-2.1464 + 0.3416 x_1 + 1.0987 x_2}}{1 + e^{-2.1464 + 0.3416 x_1 + 1.0987 x_2}}$$

Therefore, because of the unique relationship between the estimated logit and the estimated logistic regression equation, we can compute the estimated probabilities for Stamm Stores by dividing $e^{\hat{g}\,(x_1,\,x_2)}$ by $1 + e^{\hat{g}\,(x_1,\,x_2)}$.

Exercises

Applications

30 Refer to the Stamm Stores example introduced in this section. The dependent variable is coded as $Y = 1$ if the customer makes a purchase and 0 if not.

 Suppose that the only information available to help predict whether the customer will make a purchase is the customer's credit card status, coded as $X = 1$ if the customer has a Stamm credit card and $X = 0$ if not.

a. Write the logistic regression equation relating X to Y.
b. What is the interpretation of $E(Y)$ when $X = 0$?
c. For the Stamm data in Table 15.11, use MINITAB to compute the estimated logit.

d. Use the estimated logit computed in part (c) to compute an estimate of the probability of making a purchase for customers who do not have a Stamm credit card and an estimate of the probability of making a purchase for customers who have a Stamm credit card.

e. What is the estimate of the odds ratio? What is its interpretation?

31 In Table 15.12 we provided estimates of the probability of a purchase in the Stamm Stores catalogue promotion. A different value is obtained for each combination of values for the independent variables.

a. Compute the odds in favour of a purchase for a customer with annual spending of €4000 who does not have a Stamm credit card ($X_1 = 4, X_2 = 0$).

b. Use the information in Table 15.12 and part (a) to compute the odds ratio for the Stamm credit card variable X_2 holding annual spending constant at $X_1 = 4$.

c. In the text, the odds ratio for the credit card variable was computed using the information in the €2000 column of Table 15.12. Did you get the same value for the odds ratio in part (b)?

32 Community Bank would like to increase the number of customers who use payroll direct deposit. Management is considering a new sales campaign that will require each branch manager to call each customer who does not currently use payroll direct deposit. As an incentive to sign up for payroll direct deposit, each customer contacted will be offered free banking for two years. Because of the time and cost associated with the new campaign, management would like to focus their efforts on customers who have the highest probability of signing up for payroll direct deposit. Management believes that the average monthly balance in a customer's current account may be a useful predictor of whether the customer will sign up for direct payroll deposit. To investigate the relationship between these two variables, Community Bank tried the new campaign using a sample of 50 current account customers that do not currently use payroll direct deposit. The sample data show the average monthly current account balance (in hundreds of euros) and whether the customer contacted signed up for payroll direct deposit (coded 1 if the customer signed up for payroll direct deposit and 0 if not). The data are contained in the data set named Bank; a portion of the data follows.

BANK

Customer	X Monthly balance	Y Direct deposit
1	1.22	0
2	1.56	0
3	2.10	0
4	2.25	0
5	2.89	0
6	3.55	0
7	3.56	0
8	3.65	1
.	.	.
.	.	.
.	.	.
48	18.45	1
49	24.98	0
50	26.05	1

a. Write the logistic regression equation relating X to Y.

b. For the Community Bank data, use MINITAB to compute the estimated logistic regression equation.

c. Conduct a test of significance using the G test statistic. Use $\alpha = 0.05$.
d. Estimate the probability that customers with an average monthly balance of €1000 will sign up for direct payroll deposit.
e. Suppose Community Bank only wants to contact customers who have a 0.50 or higher probability of signing up for direct payroll deposit. What is the average monthly balance required to achieve this level of probability?
f. What is the estimate of the odds ratio? What is its interpretation?

For additional online summary questions and answers go to the companion website at www.cengage.co.uk/aswsbe2

Summary

In this chapter, we introduced multiple regression analysis as an extension of simple linear regression analysis presented in Chapter 14. Multiple regression analysis enables us to understand how a dependent variable is related to two or more independent variables. The regression equation $E(Y) = \beta_0 + \beta_1 x_1 + \beta_2 x_2 + \cdots + \beta_p x_p$ shows that the expected value or mean value of the dependent variable Y is related to the values of the independent variables $X_1, X_2, \ldots, X_p$. Sample data and the least squares method are used to develop the estimated regression equation $\hat{y} = b_0 + b_1 x_1 + b_2 x_2 + \cdots + b_p x_p$. In effect $b_0, b_1, b_2, \ldots, b_p$ are sample statistics used to estimate the unknown model parameters $\beta_0, \beta_1, \beta_2, \ldots, \beta_p$. Computer printouts were used throughout the chapter to emphasize the fact that statistical software packages are the only realistic means of performing the numerous computations required in multiple regression analysis.

The multiple coefficient of determination was presented as a measure of the goodness of fit of the estimated regression equation. It determines the proportion of the variation of Y that can be explained by the estimated regression equation. The adjusted multiple coefficient of determination is a similar measure of goodness of fit that adjusts for the number of independent variables and thus avoids overestimating the impact of adding more independent variables. Model assumptions for multiple regression are shown to parallel those for simple regression analysis.

An F test and a t test were presented as ways of determining statistically whether the relationship among the variables is significant. The F test is used to determine whether there is a significant overall relationship between the dependent variable and the set of all independent variables. The t test is used to determine whether there is a significant relationship between the dependent variable and an individual independent variable given the other independent variables in the regression model. Correlation among the independent variables, known as multicollinearity, was discussed.

The section on qualitative independent variables showed how dummy variables can be used to incorporate qualitative data into multiple regression analysis. The section on residual analysis showed how residual analysis can be used to validate the model assumptions, detect outliers and identify influential observations. Standardized residuals, leverage, studentized deleted residuals and Cook's distance measure were discussed. The chapter concluded with a section on how logistic regression can be used to model situations in which the dependent variable may only assume two values.

Key terms

Adjusted multiple coefficient of determination
Cook's distance measure
Dummy variable
Estimated logistic regression equation
Estimated logit
Estimated multiple regression equation
Influential observation
Least squares method
Leverage
Logistic regression equation

Logit
Multicollinearity
Multiple coefficient of determination
Multiple regression analysis
Multiple regression equation
Multiple regression model
Odds in favour of an event occurring
Odds ratio
Outlier
Qualitative independent variable
Studentized deleted residuals
Variance inflation factor

Key formulae

Multiple regression model

$$Y = \beta_0 + \beta_1 x_1 + \beta_2 x_2 + \cdots + \beta_p x_p + \varepsilon \qquad (15.1)$$

Multiple regression equation

$$E(Y) = \beta_0 + \beta_1 x_1 + \beta_2 x_2 + \cdots + \beta_p x_p \qquad (15.2)$$

Estimated multiple regression equation

$$\hat{y} = b_0 + b_1 x_1 + b_2 x_2 + \cdots + b_p x_p \qquad (15.3)$$

Least squares criterion

$$\min \Sigma (y_i - \hat{y}_i)^2 \qquad (15.4)$$

Relationship among SST, SSR and SSE

$$SST = SSR + SSE \qquad (15.7)$$

Multiple coefficient of determination

$$R^2 = \frac{SSR}{SST} \qquad (15.8)$$

Adjusted multiple coefficient of determination

$$\text{adj } R^2 = 1 - (1 - R^2)\,\frac{n-1}{n-p-1} \qquad (15.9)$$

Mean square regression

$$MSR = \frac{SSR}{P} \qquad (15.12)$$

Mean square error

$$MSE = s^2 = \frac{SSE}{n-p-1} \qquad (15.13)$$

F test statistic

$$F = \frac{MSR}{MSE} \qquad (15.14)$$

t test statistic

$$t = \frac{b_i}{s_{b_i}} \qquad (15.15)$$

Variance Inflation Factor

$$VIF(X_j) = \frac{1}{1 - R_j^2} \qquad \text{(15.16)}$$

Standardized residual for observation i

$$\frac{y_i - \hat{y}_i}{s_{y_i - \hat{y}_i}} \qquad \text{(15.24)}$$

Standard deviation of residual i

$$s_{y_i - \hat{y}_i} = s\sqrt{1 - h_i} \qquad \text{(15.25)}$$

Cook's distance measure

$$D_i = \frac{(y_i - \hat{y}_i)^2 \, h_i}{(p - 1)s^2 \, (1 - h_i)^2} \qquad \text{(15.26)}$$

Logistic regression equation

$$E(Y) = \frac{e^{\beta_0 + \beta_1 x_1 + \beta_2 x_2 + \cdots + \beta_p x_p}}{1 + e^{\beta_0 + \beta_1 x_1 + \beta_2 x_2 + \cdots + \beta_p x_p}} \qquad \text{(15.28)}$$

Interpretation of $E(Y)$ as a probability in logistic regression

$$E(Y) = P(Y = 1 | x_1, x_2, \ldots x_p) \qquad \text{(15.29)}$$

Estimated logistic regression equation

$$\hat{y} = \text{estimate of } P(Y = 1 \mid x_1, x_2, \ldots x_p) = \frac{e^{b_0 + b_1 x_1 + b_2 x_2 + \cdots + b_p x_p}}{1 + e^{b_0 + b_1 x_1 + b_2 x_2 + \cdots + b_p x_p}} \qquad \text{(15.31)}$$

Odds ratio

$$\text{Odds ratio} = \frac{\text{odds}_1}{\text{odds}_0} \qquad \text{(15.35)}$$

Logit

$$g(x_1, x_2, \ldots, x_p) = \beta_0 + \beta_1 x_1 + \beta_2 x_2 + \cdots + \beta_p x_p \qquad \text{(15.36)}$$

Estimated logits

$$\hat{g}(x_1, x_2, \ldots x_p) = b_0 + b_1 x_1 + b_2 x_2 + \cdots + b_p x_p \qquad \text{(15.38)}$$

Case problem Consumer Research

Consumer Research is an independent agency that conducts research on consumer attitudes and behaviours for a variety of firms. In one study, a client asked for an investigation of consumer characteristics that tend to be used to predict the amount charged by credit card users. Data were collected on annual income, household size and annual credit card charges for a sample of 50 consumers. The following data are on the CD accompanying the text in the data set named Consumer.

Managerial report

1 Use methods of descriptive statistics to summarize the data. Comment on the findings.

2 Develop estimated regression equations, first using annual income as the independent variable and then using household size as the independent variable. Which variable is the better predictor of annual credit card charges? Discuss your findings.

3 Develop an estimated regression equation with annual income and household size as the independent variables. Discuss your findings.

4 What is the predicted annual credit card charge for a three-person household with an annual income of €40 000?

5 Discuss the need for other independent variables that could be added to the model.

What additional variables might be helpful?

CONSUMER

Income (€000s)	Household size	Amount charged (€)	Income (€000s)	Household size	Amount charged (€)
54	3	4016	54	6	5573
30	2	3159	30	1	2583
32	4	5100	48	2	3866
50	5	4742	34	5	3586
.	.	.	.	.	.
.	.	.	.	.	.
.	.	.	.	.	.
42	2	3020	46	5	4820
41	7	4828	66	4	5149

Shopper paying for purchase with a credit card. © Marcus Clackson.

Software Section
for Chapter 15

Multiple regression using MINITAB

In this section we show how MINITAB can be used to model multiple regression problems using data for the Eurodistributor Company. First, the data must be entered in a MINITAB worksheet. The distances are entered in column C1, the number of deliveries are entered in column C2, and the travel times (hours) are entered in column C3. The variable names Distance, Deliveries and Time are entered as the column headings on the worksheet. In subsequent steps, we refer to the data by using the variable names Distance, Deliveries and Time. The steps involved in using MINITAB to produce the regression results shown in Figure 15.4 follow.

EURODIS-TRIBUTOR

Step 1 **Stat > Regression > Regression**   [Main menu bar]

Step 2 Enter Time in the **Response** box   [**Regression** panel]
Enter Distance and Deliveries in the **Predictors** box
Click **OK**

Step 3 Click **OK**   [**Regression** panel]

Logistic regression using MINITAB

MINITAB calls logistic regression with a dependent variable that can only assume the values 0 and 1 Binary Logistic Regression. In this section we describe the steps required to use MINITAB's Binary Logistic Regression procedure to generate the computer output for the Stamm Stores problem shown in Figure 15.13. First, the data must be entered in a MINITAB worksheet. The amounts customers spent last year at Stamm (in thousands of euros) are entered into column C2, the credit card data (1 if a Stamm card; 0 otherwise) are entered into column C3, and the purchase data (1 if the customer made a purchase; 0 otherwise) are entered in column C4. The variable names Spending, Card and Purchase are entered as the column headings on the worksheet. In subsequent steps, we refer to the data by using the variable names Spending, Card and Purchase. The steps involved in using MINITAB to generate the logistic regression output follow.

STAMM

Step 1 **Regression > Binary Logistic Regression**   [Main menu bar]

Step 2 Enter Purchase in the **Response** box   [**Binary Logistic Regression** panel]
Enter Spending and Card in the **Model** box
Click **OK**

Step 3 Click **OK**   [**Regression** panel]

Figure 15.4 PASW output for Eurodistributor with two independent variables

Model Summary

Model	R	R Square	Adjusted R Square	Std. Error of the Estimate
1	.951[a]	.904	.876	.5731

a. Predictors: (Constant), Deliveries, Distance

ANOVA[b]

Model		Sum of Squares	df	Mean Square	F	Sig.
1	Regression	21.601	2	10.800	32.878	.000[a]
	Residual	2.299	7	.328		
	Total	23.900	9			

a. Predictors: (Constant), Deliveries, Distance

b. Dependent Variable: Time

Coefficients[a]

Model		Unstandardized Coefficients		Standardized Coefficients	t	Sig.
		B	Std. Error	Beta		
1	(Constant)	-.869	.952		-.913	.392
	Distance	.061	.010	.735	6.182	.000
	Deliveries	.923	.221	.496	4.176	.004

a. Dependent Variable: Time

Multiple regression using EXCEL

EXCEL

In Section 15.2 we discussed the computer solution of multiple regression problems by showing MINITAB's output for the Eurodistributor Company problem. In this section we describe how to use EXCEL's Regression tool to develop the estimated multiple regression equation for the Eurodistributor problem. Refer to Figure 15.14 as we describe the tasks involved. First, the labels Assignment, Distance, Deliveries and Time are entered into cells A1:D1 of the worksheet, and the sample data into cells B2:D11. The numbers 1–10 in cells A2:A11 identify each observation. The steps involved in using the Regression tool for multiple regression analysis follow.

Figure 15.15 EXCEL output for Eurodistributor with two independent variables

	A	B	C	D	E	F	G	H	I
1	Assignment	Distance	Deliveries	Time					
2	1	100	4	9.3					
3	2	50	3	4.8					
4	3	100	4	8.9					
5	4	100	2	6.5					
6	5	50	2	4.2					
7	6	80	2	6.2					
8	7	75	3	7.4					
9	8	65	4	6.0					
10	9	90	3	7.6					
11	10	90	2	6.1					
12									
13	SUMMARY OUTPUT								
14									
15	*Regression Statistics*								
16	Multiple R	0.9507							
17	R Square	0.9038							
18	Adjusted R Sq	0.8763							
19	Standard Error	0.5731							
20	Observations	10							
21									
22	ANOVA								
23		df	SS	MS	F	Significance F			
24	Regression	2	21.6006	10.8003	32.8784	0.0003			
25	Residual	7	2.2994	0.3285					
26	Total	9	23.9						
27									
28		Coefficients	Standard Error	t Stat	P-value	Lower 95%	Upper 95%	Lower 99.0%	Upper 99.0%
29	Intercept	-0.8687	0.9515	-0.9129	0.3916	-3.1188	1.3814	-4.1986	2.4612
30	Distance	0.0611	0.0099	6.1824	0.0005	0.0378	0.0845	0.0265	0.0957
31	Deliveries	0.9234	0.2211	4.1763	0.0042	0.4006	1.4463	0.1496	1.6972
32									

EURODIS-
TRIBUTOR

Step 1 Select **Data > Data Analysis > Regression**   [Main menu bar]
Click **OK**

Step 2 Enter D1:D11 in the **Input Y Range** box   [**Regression** panel]
Enter B1:C11 in the **Input X Range** box
Select **Labels**
Select **Confidence Level.** Enter 99 in the **Confidence Level** box
Select **Output Range**
Enter A13 in the **Output Range** box (to identify the upper left corner of the section of the worksheet where the output will appear)
Click **OK**

In the EXCEL output shown in Figure 15.14 the label for the independent variable X_1 is Distance (see cell A30), and the label for the independent variable X_2 is Deliveries (see cell A31). The estimated regression equation is

$$\hat{y} = -0.8687 + 0.0611x_1 + 0.9234x_2$$

Note that using EXCEL's Regression tool for multiple regression is almost the same as using it for simple linear regression. The major difference is that in the multiple regression case a larger range of cells is required in order to identify the independent variables.

Note that Logistic Regression is not a standard analysis feature of EXCEL.

Multiple regression using PASW

PASW

First, the data must be entered in a PASW worksheet. In 'Data View' mode, distances are entered in rows 1 to 10 of the leftmost column. This is automatically labelled by the system V1. Similarly the number of deliveries and travel times are entered in the two immediately adjacent columns to the right and are labelled V2 and V3 respectively. The latter variable names can then be changed to Distance, Deliveries and Time in 'Variable View' mode. The steps involved in using PASW to produce the regression results shown in Figure 15.4 follow.

Step 1 Analyze > Regression > Linear   [Main menu bar]

Step 2 Enter Time in the **Dependent** box   [**Linear** panel]
Enter Distance and Deliveries in the **Independent(s)** box
Click **OK**

Logistic regression using PASW

First, the data must be entered in a PASW worksheet in Data View mode. The amounts customers spent last year at Stamm (in thousands of euros) are entered into rows 1 to 100 of the leftmost column. Corresponding credit card details (1 if a Stamm card; 0 otherwise) and purchase data (1 if the customer made a purchase; 0 otherwise) are entered into the immediately adjacent columns to the right. The system automatically assigns headings V1, V2 and V3 to these columns but these can be easily changed to Spending, Card and Purchase in Variable View mode. The following command sequence will generate the logistic regression output.

Step 1 Analyze > Regression > Binary Logistic   Select the menubar item

Step 2 Enter Purchase in the **Dependent** box   [**Logistic Regression** panel]
Enter Spending and Card in the **Covariates** box
Click **OK**
Click **OK**

Selective output is as follows:

Classification Table[a]

			Predicted		
			Purchase		Percentage
Observed			0	1	Correct
Step 1	Purchase	0	52	8	86.7
		1	20	20	50.0
	Overall Percentage				72.0

a. The cut value is .500

Variables in the Equation

		B	S.E.	Wald	df	Sig.	Exp(B)
Step 1[a]	Spending	.342	.129	7.050	1	.008	1.407
	Card	1.099	.445	6.105	1	.013	3.000
	Constant	-2.146	.577	13.826	1	.000	.117

a. Variable(s) entered on step 1: Spending, Card.

Chapter 16

Regression Analysis: Model Building

Learning objectives

After reading this chapter and doing the exercises, you should be able to:

1 Appreciate how the general linear model can be used to model problems involving curvilinear relationships.

2 Understand the concept of interaction and how it can be accounted for in the general linear model.

3 Understand how an F test can be used to determine when to add or delete one or more variables.

4 Appreciate the complexities involved in solving larger regression analysis problems.

5 Understand how variable selection procedures can be used to choose a set of independent variables for an estimated regression equation.

Model building in regression analysis is the process of developing an estimated regression equation that describes the relationship between a dependent variable and one or more independent variables. The major issues in model building are finding an effective functional form of the relationship and selecting the independent variables to be included in the model. In Section 16.1 we establish the framework for model building by introducing the concept of a general linear model. Section 16.2, which provides the foundation for the more sophisticated computer-based procedures, introduces a general approach for determining when to add or delete independent variables. In Section 16.3 we consider a larger regression problem involving eight independent variables and 25 observations; this problem is used to illustrate the variable selection procedures presented in Section 16.4, including stepwise regression, the forward selection procedure, the backward elimination procedure, and best-subsets regression.

16.1 General linear model

Suppose we collected data for one dependent variable Y and k independent variables $X_1, X_2, \ldots X_k$. Our objective is to use these data to develop an estimated regression equation that provides the best relationship between the dependent and independent variables. As a general framework for developing more complex relationships among the independent variables we introduce the concept of the **general linear model** involving p independent variables.

General linear model

$$Y = \beta_0 + \beta_1 z_1 + \beta_2 z_2 + \cdots + \beta_p z_p + \varepsilon \tag{16.1}$$

In equation (16.1), each of the independent variables Z_j (where $j = 1, 2, \ldots, p$) is a function of $X_1, X_2, \ldots, X_k$ (the variables for which data are collected). In some cases, each Z_j may be a function of only one X variable. The simplest case is when we collect data for just one variable X_1 and want to estimate Y by using a straight-line relationship. In this case $Z_1 = X_1$ and equation (16.1) becomes

$$Y = \beta_0 + \beta_1 x_1 + \varepsilon \tag{16.2}$$

Statistics in Practice

Selecting a university

To demonstrate an application of their new decision analysis methodology, Philip and Green (2002) consider a school-leaver, Jenny, who is hoping to go to university. Before applying, however, she wishes to prioritize alternatives from the 97 choices available. She undertakes an analysis based on the nine classifications used in the *Times Good University Guide (2000)* to construct the 2000 League Table, a section of which is summarized below.

Jenny's priorities vary by each of these variables but on degree quality she is not so impressed by how students perform at university overall so much as the amount of 'gain' that takes place for a given intake standard. Having noted that degree quality (Deg) is related to entry

University	T	R	As	St	L	Fac	Deg	Des	Com
Aberdeen	86	66	72	53	70	78	76	95	86
Abertay Dundee	76	23	30	50	72	62	59	86	90
Aberystwyth	81	61	60	38	68	78	66	88	93
Anglia	76	20	42	47	61	67	66	86	80
Aston	87	58	70	42	69	79	74	96	93
Bangor	81	53	55	44	72	71	59	90	93
Bath	82	81	82	50	84	94	80	93	96
Birmingham	89	73	82	57	68	79	78	94	94
Bournemouth	63	19	43	50	62	62	52	89	87
Bradford	68	67	58	40	67	84	55	90	89

where T denotes teaching quality, R, research assessment, As, 'A' level examination points (required for entry), St, student-staff ratio, L, library and computing facilities spending and Fac, student facilities spending.

Aberystwyth University building. © Paul Littler.

standards, she therefore estimated the effect of this using a polynomial regression analysis and then removed it to create a new variable 'degree quality gain' which was used instead of Deg. The relevant equation was as follows:

Degree quality gain,
$$DQG = Deg - 0.0047As^2 - 0.0088As - 43.8$$

Similarly she developed new variables in place of Des (subsequent employment), and Com (course completion), 'removing' the effects of entry standards from Com, and degree quality from Des, as follows:

Destination gain, $DG = Des - 0.246Deg - 74.1$
Completion gain, $CG = Com - 0.292As - 69.3$

These decision variables were then later combined with the other original variables, to create the function

Relative Value $= 0.00218T + 0.00182/(St + DQG + L + Fac) + 0.0011(R + As + CG + DG)$

which could then be used to represent Jenny's distinctive outlook.

Sources: Sutton, P. P. & Green, R.H. (2002) A data envelope approach to decision analysis. J. Opl. Res. Soc. 53 1215–1224. The Times, the Good University Guide 14 April 2000.

Equation (16.2) is the simple linear regression model introduced in Chapter 14 with the exception that the independent variable is labelled X_1 instead of X. In the statistical modelling literature, this model is called a *simple first-order model with one predictor variable*.

Modelling curvilinear relationships

More complex types of relationships can be modelled with equation (16.1). To illustrate, let us consider the problem facing Reynard Ltd, a manufacturer of industrial scales and laboratory equipment. Managers at Reynard want to investigate the relationship between length of employment of their salespeople and the number of electronic laboratory scales sold. Table 16.1 gives the number of scales sold by 15 randomly selected salespeople for the most recent sales period and the number of months each salesperson has been employed by the firm. Figure 16.1 is the scatter diagram for these data. The scatter diagram indicates a possible curvilinear relationship between the length of time employed and the number of units sold. Before considering how to develop a curvilinear relationship for Reynard, let us consider the MINITAB output in Figure 16.2 corresponding to a simple first-order model; the estimated regression is

$$\text{Sales} = 111 + 2.38 \text{ Months}$$

where

$$\text{Sales} = \text{number of electronic laboratory scales sold}$$
$$\text{Months} = \text{the number of months the salesperson has been employed}$$

Figure 16.3 is the corresponding standardized residual plot. Although the computer output shows that the relationship is significant (*p*-value = 0.000) and that a linear relationship explains a high percentage of the variability in sales (R-sq = 78.1 per cent), the standardized residual plot suggests that a curvilinear relationship is needed.

Table 16.1 Data for the Reynard example

Months employed	Scales sold
41	275
106	296
76	317
104	376
22	162
12	150
85	367
111	308
40	189
51	235
9	83
12	112
6	67
56	325
19	189

REYNARD

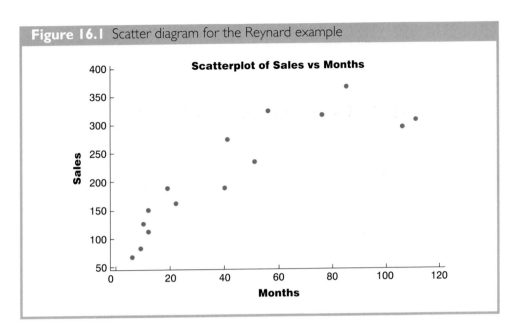

Figure 16.1 Scatter diagram for the Reynard example

Figure 16.2 MINITAB output for the Reynard example: first-order model

Regression Analysis: Sales versus Months

```
The regression equation is
Sales = 111 + 2.38 Months


Predictor      Coef   SE Coef      T       P
Constant     111.23     21.63   5.14   0.000
Months       2.3768    0.3489   6.81   0.000


S = 49.5158    R-Sq = 78.1%    R-Sq(adj) = 76.4%


Analysis of Variance

Source           DF        SS       MS      F       P
Regression        1    113783   113783  46.41   0.000
Residual Error   13     31874     2452
Total            14    145657
```

To account for the curvilinear relationship, we set $Z_1 = X_1$ and $Z_2 = X_1^2$ in equation (16.1) to obtain the model

$$Y = \beta_0 + \beta_1 x_1 + \beta_2 x_1^2 + \varepsilon \qquad (16.3)$$

This model is called a *second-order model with one predictor variable.* To develop an estimated regression equation corresponding to this second-order model, the statistical

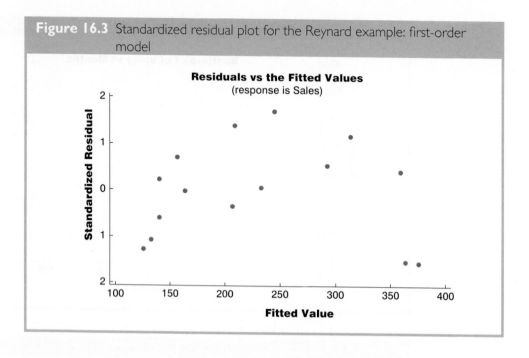

Figure 16.3 Standardized residual plot for the Reynard example: first-order model

software package we are using needs the original data in Table 16.1, as well as the data corresponding to adding a second independent variable that is the square of the number of months the employee has been with the firm. In Figure 16.4 we show the MINITAB output corresponding to the second-order model; the estimated regression equation is

$$\text{Sales} = 45.3 + 6.34 \text{ Months} - 0.0345 \text{ MonthsSq}$$

where

$\quad\quad$ MonthsSq = the square of the number of months the salesperson has been employed

Figure 16.5 is the corresponding standardized residual plot. It shows that the previous curvilinear pattern has been removed. At the 0.05 level of significance, the computer output shows that the overall model is significant (p-value for the F test is 0.000); note also that the p-value corresponding to the t-ratio for MonthsSq (p-value = 0.002) is less than 0.05, and hence we can conclude that adding MonthsSq to the model involving Months is significant. With an R-sq(adj) value of 88.6 per cent, we should be pleased with the fit provided by this estimated regression equation. More important, however, is seeing how easy it is to handle curvilinear relationships in regression analysis.

$\quad\quad$ Clearly, many types of relationships can be modelled by using equation (16.1). The regression techniques with which we have been working are definitely not limited to linear, or straight-line, relationships. In multiple regression analysis the word *linear* in the term 'general linear model' refers only to the fact that $\beta_0, \beta_1, \dots, \beta_p$ all have exponents of 1; it does not imply that the relationship between Y and the X_i's is linear. Indeed, in this section we have seen one example of how equation (16.1) can be used to model a curvilinear relationship.

Figure 16.4 MINITAB output for the Reynard example: second-order model

```
Regression Analysis: Sales versus Months, MonthsSq

The regression equation is
Sales = 45.3 + 6.34 Months - 0.0345 MonthsSq


Predictor        Coef    SE Coef       T      P
Constant        45.35      22.77    1.99  0.070
Months          6.345      1.058    6.00  0.000
MonthsSq    -0.034486   0.008948   -3.85  0.002


S = 34.4528    R-Sq = 90.2%    R-Sq(adj) = 88.6%


Analysis of Variance

Source            DF       SS      MS      F      P
Regression         2   131413   65707  55.36  0.000
Residual Error    12    14244    1187
Total             14   145657


Source      DF   Seq SS
Months       1   113783
MonthsSq     1    17630
```

Figure 16.5 Standardized residual plot for the Reynard example: second-order model

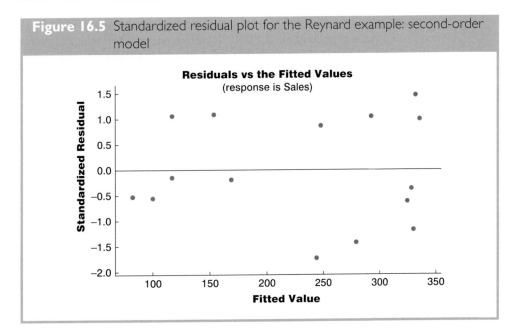

Interaction

To provide an illustration of **interaction** and what it means, let us review the regression study conducted by Veneto Care for one of its new shampoo products. Two factors believed to have the most influence on sales are unit selling price and advertizing expenditure.

Table 16.2 Data for the Veneto Care example

VENETO

Price	Advertising expenditure (€000s)	Sales (000s)	Price	Advertising expenditure (€000s)	Sales (000s)
€2.00	50	478	€2.00	100	810
€2.50	50	373	€2.50	100	653
€3.00	50	335	€3.00	100	345
€2.00	50	473	€2.00	100	832
€2.50	50	358	€2.50	100	641
€3.00	50	329	€3.00	100	372
€2.00	50	456	€2.00	100	800
€2.50	50	360	€2.50	100	620
€3.00	50	322	€3.00	100	390
€2.00	50	437	€2.00	100	790
€2.50	50	365	€2.50	100	670
€3.00	50	342	€3.00	100	393

To investigate the effects of these two variables on sales, prices of €2.00, €2.50 and €3.00 were paired with advertising expenditures of €50 000 and €100 000 in 24 test markets.

The unit sales (in thousands) that were observed are reported in Table 16.2.

Table 16.3 is a summary of these data. Note that the mean sales corresponding to a price of €2.00 and an advertising expenditure of €50 000 is 461 000, and the mean sales corresponding to a price of €2.00 and an advertising expenditure of €100 000 is 808 000. Hence, with price held constant at €2.00, the difference in mean sales between advertising expenditures of €50 000 and €100 000 is 808 000 − 461 000 = 347 000 units. When the price of the product is €2.50, the difference in mean sales is 646 000 − 364 000 = 282 000 units. Finally, when the price is €3.00, the difference in mean sales is 375 000 − 332 000 = 43 000 units. Clearly, the difference in mean sales between advertising expenditures of €50 000 and €100 000 depends on the price of the product. In other words, at higher selling prices, the effect of increased advertising expenditure diminishes. These observations provide evidence of interaction between the price and advertising expenditure variables.

When interaction between two variables is present, we cannot study the effect of one variable on the response Y independently of the other variable. In other words, meaningful conclusions can be developed only if we consider the joint effect that both variables have on the response.

Table 16.3 Mean unit sales (1000s) for the Veneto Care example

		Price		
		€2.00	€2.50	€3.00
Advertising	€50 000	461	364	332
Expenditure	€100 000	808	646	375

Mean sales of 808 000 units when price = €2.00 and advertizing expenditure = €100 000

To account for the effect of interaction, we will use the following regression model.

$$Y = \beta_0 + \beta_1 x_1 + \beta_2 x_2 + \beta_3 x_1 x_2 + \varepsilon \tag{16.4}$$

where

$$Y = \text{unit sales (000s)}$$
$$X_1 = \text{price (€)}$$
$$X_2 = \text{advertizing expenditure (€000s)}$$

Note that equation (16.4) reflects Veneto's belief that the number of units sold depends linearly on selling price and advertising expenditure (accounted for by the $\beta_1 x_1$ and $\beta_2 x_2$ terms), and that there is interaction between the two variables (accounted for by the $\beta_3 x_1 x_2$ term).

To develop an estimated regression equation, a general linear model involving three independent variables (Z_1, Z_2, and Z_3) was used.

$$Y = \beta_0 + \beta_1 z_1 + \beta_2 z_2 + \beta_3 z_3 + \varepsilon \tag{16.5}$$

where

$$z_1 = x_1$$
$$z_2 = x_2$$
$$z_3 = x_1 x_2$$

Figure 16.6 is the MINITAB output corresponding to the interaction model for the Veneto Care example. The resulting estimated regression equation is

$$\text{Sales} = -276 + 175\,\text{Price} + 19.7\,\text{AdvExp} - 6.08\,\text{PriceAdv}$$

where

$$\text{Sales} = \text{unit sales (000s)}$$
$$\text{Price} = \text{price of the product (€)}$$
$$\text{AdvExp} = \text{advertising expenditure (€000s)}$$
$$\text{PriceAdv} = \text{interaction term (Price times AdvExp)}$$

Because the model is significant (p-value for the F test is 0.000) and the p-value corresponding to the t test for PriceAdv is 0.000, we conclude that interaction is significant given the linear effect of the price of the product and the advertising expenditure. Thus, the regression results show that the effect of advertising expenditure on sales depends on the price.

Transformations involving the dependent variable

In showing how the general linear model can be used to model a variety of possible relationships between the independent variables and the dependent variable, we have focused attention on transformations involving one or more of the independent variables. Often it is worthwhile to consider transformations involving the dependent

> **Figure 16.6** MINITAB output for the Veneto Care example
>
> ```
> Regression Analysis: Sales versus Price, AdvExpen, PriceAdv
>
> The regression equation is
> Sales = - 276 + 175 Price + 19.7 AdvExpen - 6.08 PriceAdv
>
>
> Predictor      Coef   SE Coef       T      P
> Constant     -275.8     112.8    -2.44  0.024
> Price        175.00     44.55     3.93  0.001
> AdvExpen     19.680     1.427    13.79  0.000
> PriceAdv    -6.0800    0.5635   -10.79  0.000
>
>
> S = 28.1739    R-Sq = 97.8%    R-Sq(adj) = 97.5%
>
>
> Analysis of Variance
>
> Source          DF       SS       MS       F      P
> Regression       3   709316   236439  297.87  0.000
> Residual Error  20    15875      794
> Total           23   725191
>
>
> Source    DF   Seq SS
> Price      1   315844
> AdvExpen   1   301056
> PriceAdv   1    92416
>
>
> Unusual Observations
>
> Obs  Price   Sales      Fit  SE Fit  Residual  St Resid
>  23   2.50  670.00   609.67    8.13     60.33      2.24R
>
> R denotes an observation with a large standardized residual.
> ```

variable *y*. As an illustration of when we might want to transform the dependent variable, consider the data in Table 16.4, which shows the kilometres-per-litre ratings and weights (kg) for 12 cars.

The scatter diagram in Figure 16.7 indicates a negative linear relationship between these two variables. Therefore, we use a simple first-order model to relate the two variables. The MINITAB output is shown in Figure 16.8; the resulting estimated regression equation is

$$KPL = 19.8 - 0.00907 \text{ Weight}$$

where

$$KPL = \text{kilometres-per-litre rating}$$
$$Weight = \text{weight of the car in kilograms}$$

The model is significant (*p*-value for the *F* test is 0.000) and the fit is very good (R-sq = 93.6 per cent). However, we note in Figure 16.8 that observation 3 is identified as having a large standardized residual.

Table 16.4 Kilometres-per-litre ratings and weights for 12 cars

Weight	Kilometres per litre
1038	10.2
958	10.3
989	12.1
1110	9.9
919	11.8
1226	9.3
1205	8.5
955	10.8
1463	6.4
1457	6.9
1636	5.1
1310	7.4

Figure 16.7 Scatter diagram for the kilometres-per-litre problem

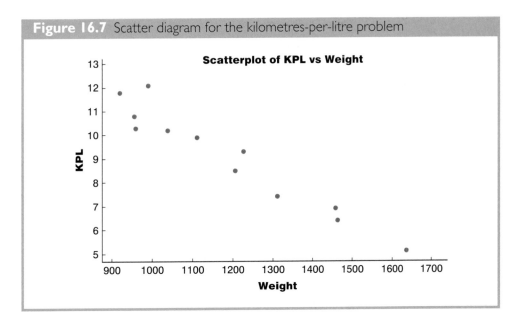

Figure 16.9 is the standardized residual plot corresponding to the first-order model. The pattern we observe does not look like the horizontal band we should expect to find if the assumptions about the error term are valid. Instead, the variability in the residuals appears to increase as the value of increases. In other words, we see the wedge-shaped pattern referred to in Chapters 14 and 15 as being indicative of a nonconstant variance. We are not justified in reaching any conclusions about the statistical significance of the resulting estimated regression equation when the underlying assumptions for the tests of significance do not appear to be satisfied.

Often the problem of non-constant variance can be corrected by transforming the dependent variable to a different scale. For instance, if we work with the logarithm of the dependent variable instead of the original dependent variable, the effect will be to compress the values of the dependent variable and thus diminish the effects of non-constant variance.

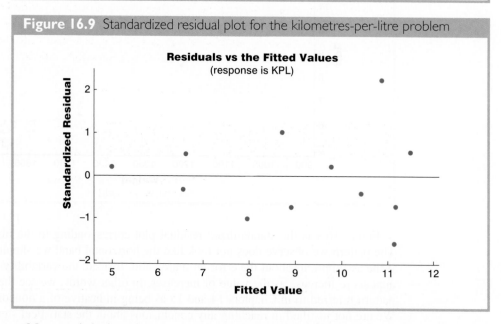

Figure 16.8 MINITAB output for the kilometres-per-litre problem

```
Regression Analysis: KPL versus Weight

The regression equation is
KPL = 19.8 - 0.00907 Weight


Predictor        Coef     SE Coef         T      P
Constant      19.8381      0.9099     21.80  0.000
Weight      -0.0090675   0.0007519    -12.06  0.000


S = 0.588710   R-Sq = 93.6%   R-Sq(adj) = 92.9%


Analysis of Variance

Source          DF       SS       MS       F      P
Regression       1   50.403   50.403  145.43  0.000
Residual Error  10    3.466    0.347
Total           11   53.869


Unusual Observations

Obs   Weight      KPL     Fit  SE Fit  Residual  St Resid
  3      989   12.100  10.870   0.227     1.230     2.26R

R denotes an observation with a large standardized residual.
```

Figure 16.9 Standardized residual plot for the kilometres-per-litre problem

Residuals vs the Fitted Values
(response is KPL)

Most statistical packages provide the ability to apply logarithmic transformations using either the base 10 (common logarithm) or the base $e = 2.71828 \ldots$ (natural logarithm). We applied a natural logarithmic transformation to the kilometres-per-litre data and developed the estimated regression equation relating weight to the natural logarithm of kilometres-per-litre. The regression results obtained by using the natural logarithm of kilometres-per-litre as the dependent variable, labelled LogeKPL in the output, are shown in Figure 16.10; Figure 16.11 is the corresponding standardized residual plot.

Figure 16.10 MINITAB output for the kilometres-per-litre problem: logarithmic transformation

Regression Analysis: LogeKPL versus Weight

The regression equation is
LogeKPL = 3.48 - 0.00110 Weight

Predictor	Coef	SE Coef	T	P
Constant	3.48115	0.09780	35.60	0.000
Weight	-0.00110033	0.00008082	-13.62	0.000

S = 0.0632759   R-Sq = 94.9%   R-Sq(adj) = 94.4%

Analysis of Variance

Source	DF	SS	MS	F	P
Regression	1	0.74221	0.74221	185.37	0.000
Residual Error	10	0.04004	0.00400		
Total	11	0.78225			

Looking at the residual plot in Figure 16.11, we see that the wedge-shaped pattern has now disappeared. Moreover, none of the observations are identified as having a large standardized residual. The model with the logarithm of kilometres per litre as the dependent variable is statistically significant and provides an excellent fit to the observed data. Hence, we would recommend using the estimated regression equation

$$\text{Log}_e \text{ KPL} = 3.49 - 0.00110 \text{ Weight}$$

Figure 16.11 Standardized residual plot for the kilometres-per-litre problem: logarithmic transformation

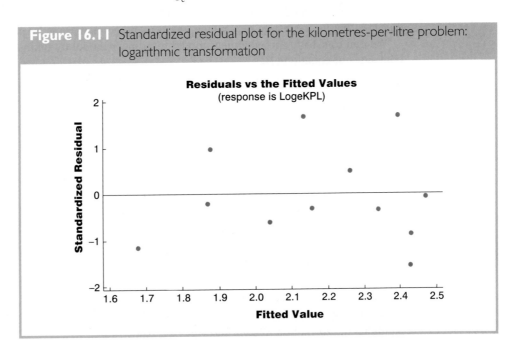

To estimate the kilometres-per-litre rating for an car that weighs 1500 kilograms, we first develop an estimate of the logarithm of the kilometres-per-litre rating.

$$\text{Log}_e \text{ KPL} = 3.49 - 0.00110 \,(1500) = 1.84$$

The kilometres-per-litre estimate is obtained by finding the number whose natural logarithm is 1.84. Using a calculator with an exponential function, or raising e to the power 1.84, we obtain 6.2 kilometres per litre.

Another approach to problems of non-constant variance is to use $1/Y$ as the dependent variable instead of Y. This type of transformation is called a *reciprocal transformation*. For instance, if the dependent variable is measured in kilometres per litre, the reciprocal transformation would result in a new dependent variable whose units would be 1/(kilometres per litre) or litres per kilometre. In general, there is no way to determine whether a logarithmic transformation or a reciprocal transformation will perform better without actually trying each of them.

Nonlinear models that are intrinsically linear

Models in which the parameters $(\beta_0, \beta_1, \ldots, \beta_p)$ have exponents other than 1 are called nonlinear models. The exponential model involves the following regression equation.

$$E(Y) = \beta_0 \beta_1^x \qquad (16.6)$$

This model is appropriate when the dependent variable Y increases or decreases by a constant percentage, instead of by a fixed amount, as X increases.

As an example, suppose sales for a product Y are related to advertizing expenditure X (in thousands of euros) according to the following exponential model.

$$E(Y) = 500(1.2)^x$$

Thus, for $X = 1$, $E(Y) = 500(1.2)^1 = 600$; for $X = 2$, $E(Y) = 500(1.2)^2 = 720$; and for $X = 3$, $E(Y) = 500(1.2)^3 = 864$. Note that $E(Y)$ is not increasing by a constant amount in this case, but by a constant percentage; the percentage increase is 20 per cent.

We can transform this nonlinear model to a linear model by taking the logarithm of both sides of equation (16.6).

$$\log E(Y) = \log \beta_0 + x \log \beta_1 \qquad (16.7)$$

Now if we let $y' = \log E(Y)$, $\beta_0' = \log \beta_0$, and $\beta_1' = \log \beta_1$, we can rewrite equation (16.7) as:

$$y' = \beta_0' + \beta_1' x \qquad (16.8)$$

It is clear that the formulae for simple linear regression can now be used to develop estimates of β'_0 and β'_1. Denoting the estimates as b_0' and b_1' leads to the following estimated regression equation.

$$\hat{y}' = b_0' + b_1' x \qquad (16.9)$$

To obtain predictions of the original dependent variable Y given a value of X, we would first substitute the value of X into equation (16.8) and compute $\hat{y}$. The antilog of $\hat{y}$ would be the prediction of Y, or the expected value of Y.

Many nonlinear models cannot be transformed into an equivalent linear model. However, such models have had limited use in business and economic applications. Furthermore, the mathematical background needed for study of such models is beyond the scope of this text.

Exercises

Methods

1 Consider the following data for two variables, X and Y.

x	22	24	26	30	35	40
y	12	21	33	35	40	36

a. Develop an estimated regression equation for the data of the form $\hat{y} = b_0 + b_1 x$.
b. Use the results from part (a) to test for a significant relationship between X and Y. Use $\alpha = 0.05$.
c. Develop a scatter diagram for the data. Does the scatter diagram suggest an estimated regression equation of the form $\hat{y} = b_0 + b_1 x + b_2 x^2$? Explain.
d. Develop an estimated regression equation for the data of the form $\hat{y} = b_0 + b_1 x + b_2 x^2$.
e. Refer to part (d). Is the relationship between X, X^2, and Y significant? Use $\alpha = 0.05$.
f. Predict the value of Y when X = 25.

2 Consider the following data for two variables, X and Y.

x	9	32	18	15	26
y	10	20	21	16	22

a. Develop an estimated regression equation for the data of the form $\hat{y} = b_0 + b_1 x$. Comment on the adequacy of this equation for predicting Y.
b. Develop an estimated regression equation for the data of the form $\hat{y} = b_0 + b_1 x + b_2 x^2$. Comment on the adequacy of this equation for predicting Y.
c. Predict the value of Y when X = 20.

3 Consider the following data for two variables, X and Y.

x	2	3	4	5	7	7	7	8	9
y	4	5	4	6	4	6	9	5	11

a. Does there appear to be a linear relationship between X and Y? Explain.
b. Develop the estimated regression equation relating X and Y.
c. Plot the standardized residuals versus for the estimated regression equation developed in part (b). Do the model assumptions appear to be satisfied? Explain.
d. Perform a logarithmic transformation on the dependent variable Y. Develop an estimated regression equation using the transformed dependent variable. Do the model assumptions appear to be satisfied by using the transformed dependent variable? Does a reciprocal transformation work better in this case? Explain.

Applications

4 A highway department is studying the relationship between traffic flow and speed. The following model has been hypothesized.

$$Y = \beta_0 + \beta_1 x + \varepsilon$$

where

$$Y = \text{traffic flow in vehicles per hour}$$
$$X = \text{vehicle speed in kilometres per hour}$$

The following data were collected during rush hour for six main roads leading out of the city.

Traffic flow (y)	Vehicle speed (x)
1256	56
1329	64
1226	48
1335	72
1349	80
1124	40

a. Develop an estimated regression equation for the data.
b. Use $\alpha = 0.01$ to test for a significant relationship.

5 In working further with the problem of exercise 4, statisticians suggested the use of the following curvilinear estimated regression equation.

$$\hat{y} = b_0 + b_1 x + b_2 x^2$$

a. Use the data of exercise 4 to determine estimated regression equation.
b. Use $\alpha = 0.01$ to test for a significant relationship.
c. Estimate the traffic flow in vehicles per hour at a speed of 60 kilometres per hour.

6 An international study of life expectancy by Ross (1994) covers variables

LifeExp	Life expectancy in years
People.per.TV	Average number of people per TV
People.per.Dr	Average number of people per physician
LifeExp.Male	Male life expectancy in years
LifeExp.Female	Female life expectancy in years

with data details as follows:

	LifeExp	People. per.TV	People. per.Dr	LifeExp. Male	LifeExp. Female
Argentina	70.5	4	370	74	67
Bangladesh	53.5	315	6 166	53	54
Brazil	65	4	684	68	62
Canada	76.5	1.7	449	80	73
China	70	8	643	72	68
Colombia	71	5.6	1 551	74	68
Egypt	60.5	15	616	61	60

	LifeExp	People. per.TV	People. per.Dr	LifeExp. Male	LifeExp. Female
Ethiopia	51.5	503	36 660	53	50
France	78	2.6	403	82	74
Germany	76	2.6	346	79	73
India	57.5	44	2 471	58	57
Indonesia	61	24	7 427	63	59
Iran	64.5	23	2 992	65	64
Italy	78.5	3.8	233	82	75
Japan	79	1.8	609	82	76
Kenya	61	96	7 615	63	59
Korea.North	70	90	370	73	67
Korea.South	70	4.9	1 066	73	67
Mexico	72	6.6	600	76	68
Morocco	64.5	21	4 873	66	63
Burma	54.5	592	3 485	56	53
Pakistan	56.5	73	2 364	57	56
Peru	64.5	14	1 016	67	62
Philippines	64.5	8.8	1 062	67	62
Poland	73	3.9	480	77	69
Romania	72	6	559	75	69
Russia	69	3.2	259	74	64
South.Africa	64	11	1 340	67	61
Spain	78.5	2.6	275	82	75
Sudan	53	23	12 550	54	52
Taiwan	75	3.2	965	78	72
Tanzania	52.5	NA	25 229	55	50
Thailand	68.5	11	4 883	71	66
Turkey	70	5	1 189	72	68
Ukraine	70.5	3	226	75	66
UK	76	3	611	79	73
USA	75.5	1.3	404	79	72
Venezuela	74.5	5.6	576	78	71
Vietnam	65	29	3 096	67	63
Zaire	54	NA	23 193	56	52

(Note that the average number of people per TV is not given for Tanzania and Zaire.)

LIFE EXPECTANCY

a. Develop scatter diagrams for these data, treating LifeExp as the dependent variable.
b. Does a simple linear model appear to be appropriate? Explain.
c. Estimate simple regression equations for the data accordingly. Which do you prefer and why?

7 To assess the reliability of computer media, *Choice* magazine (www.choice.com.au) has obtained data by:

price (A$)	Paid in April 2005
pack	the number of disks in the pack
media	one of CD (CD), DVD (DVD-R) or DVDRW (DVD+/-RW)

with details as follows:

MEDIA

Price	Pack	Media	Price	Pack	Media
0.48	50	CD	1.85	10	DVD
0.60	25	CD	0.72	25	DVD
0.64	25	CD	2.28	10	DVD
0.50	50	CD	2.34	5	DVD
0.89	10	CD	2.40	10	DVD
0.89	10	CD	1.49	5	DVD
1.20	10	CD	3.60	5	DVDRW
1.30	10	CD	5.00	10	DVDRW
1.29	10	CD	2.79	5	DVDRW
0.50	10	CD	2.79	10	DVDRW
0.57	50	DVD	4.37	5	DVDRW
2.60	10	DVD	1.50	10	DVDRW
1.59	10	DVD	2.50	5	DVDRW
1.85	10	DVD	3.90	10	DVDRW

a. Develop scatter diagrams for these data with pack and media as potential independent variables.
b. Does a simple or multiple linear regression model appear to be appropriate?
c. Develop an estimated regression equation for the data you believe will best explain the relationship between these variables.

8 In Europe the number of Internet users varies widely from country to country. In 1999, 44.3 per cent of all Swedes used the Internet, while in France the audience was less than 10 per cent. The disparities are expected to persist even though Internet usage is expected to grow dramatically over the next several years. The following table shows the number of Internet users in 1999 and the projected number of users in 2005 for European countries.

INTERNET

	1999 Internet users (%)	2005 projected users (%)
Austria	12.6	53.4
Belgium	24.2	60.2
Denmark	40.4	71.2
Finland	40.9	71.4
France	9.7	53.1
Germany	15.0	59.6
Greece	3.4	15.8
Ireland	12.1	46.7
Italy	8.4	34.4
Netherlands	18.6	61.4
Norway	38.0	71.7
Portugal	4.63	36.6
Spain	7.4	39.9
Sweden	44.3	71.9
Switzerland	28.1	66.7
UK	23.6	66.8

a. Develop a scatter diagram of the data using the 1999 Internet user percentage as the independent variable. Does a simple linear regression model appear to be appropriate? Discuss.

b. Develop an estimated multiple regression equation with X = the number of 1999 Internet users and X^2 as the two independent variables.

c. Consider the nonlinear relationship shown by equation (16.6). Use logarithms to develop an estimated regression equation for this model.

d. Do you prefer the estimated regression equation developed in part (b) or part (c)? Explain.

16.2 Determining when to add or delete variables

General case

Consider the following multiple regression model involving q independent variables, where $q < p$.

$$Y = \beta_0 + \beta_1 x_1 + \beta_2 x_2 + \cdots + \beta_q x_q + \varepsilon \tag{16.10}$$

If we add variables $X_{q+1}, X_{q+2}, \ldots, X_p$ to this model, we obtain a model involving p independent variables.

$$\begin{aligned} Y = \beta_0 &+ \beta_1 x_1 + \beta_2 x_2 + \cdots + \beta_q x_q \\ &+ \beta_{q+1} x_{q+1} + \beta_{q+2} x_{q+2} + \cdots + \beta_p x_p + \varepsilon \end{aligned} \tag{16.11}$$

To test whether the addition of $X_{q+1}, X_{q+2}, \ldots, X_p$ is statistically significant, the null and alternative hypotheses can be stated as follows.

$H_0. \ \beta_{q+1} = \beta_{q+2} = \cdots = \beta_p = 0$
H_1: One or more of the parameters is not equal to zero

The following F statistic provides the basis for testing whether the additional independent variables are statistically significant.

F test statistic for adding or deleting $p-q$ variables

$$F = \frac{\dfrac{\text{SSE}(x_1, x_2, \ldots, x_q) - \text{SSE}(x_1, x_2, x_q, x_{q+1}, x_p)}{p - q}}{\dfrac{\text{SSE}(x_1, x_2, x_q, x_{q+1}, x_p)}{n - p - 1}} \tag{16.12}$$

This computed F value is then compared with F_α, the table value with $p - q$ numerator degrees of freedom and $n - p - 1$ denominator degrees of freedom. If $F > F$ we reject H_0 and conclude that the set of additional independent variables is statistically significant.

Many students find equation (16.12) somewhat complex. To provide a simpler description of this F ratio, we can refer to the model with the smaller number of independent variables as the reduced model and the model with the larger number of independent variables as the full model. If we let SSE (reduced) denote the error sum of squares for the reduced model and SSE (full) denote the error sum of squares for the full model, we can write the numerator of (16.12) as

$$\frac{\text{SSE(reduced)} - \text{SSE(full)}}{\text{number of extra terms}} \qquad (16.13)$$

Note that 'number of extra terms' denotes the difference between the number of independent variables in the full model and the number of independent variables in the reduced model. The denominator of equation (16.12) is the error sum of squares for the full model divided by the corresponding degrees of freedom; in other words, the denominator is the mean square error for the full model. Denoting the mean square error for the full model as MSE(full) enables us to write it as

$$F = \frac{\dfrac{\text{SSE(reduced)} - \text{SSE(full)}}{\text{number of extra terms}}}{\text{MSE(full)}} \qquad (16.14)$$

To illustrate the use of this F statistic, let us return to the Eurodistributor data introduced in Chapter 15. Recall that the managers were trying to develop a regression model to predict total daily travel time for trucks using two independent variables: distance travelled (X_1) and number of deliveries (X_2). With one model using only X_1 as an independent variable the error sum of squares was found to be 8.029. For the second however, using both X_1 and X_2, the error sum of squares was 2.299. The question is did the addition of the second independent variable X_2 result in a significant reduction in the error sum of squares?

Using formula 16.14 with $n = 10$, $q = 1$ and $p = 2$ it is easily shown the test statistic is:

$$F = \frac{\dfrac{8.029 - 2.299}{1}}{\dfrac{2.299}{7}} = 17.47$$

which is statistically significant since $17.47 > F_{0.05}(1, 7) = 5.59$.

Use of p-values

Note also that the p-value associated with $F(1, 7) = 17.47$ is 0.004. As this is less than $\alpha = 0.05$ we can conclude once again that the addition of the second independent variable is statistically significant. In general p-values cannot be looked up directly from tables of the F distribution, but can be straightforwardly obtained using computer software packages, such as MINITAB, PASW or EXCEL.

Exercises

Methods

9   In a regression analysis involving 27 observations, the following estimated regression equation was developed.

$$\hat{y} = 25.2 + 5.5x_1$$

For this estimated regression equation SST = 1550 and SSE = 520.

a. At $\alpha = 0.05$, test whether X_1 is significant.

Suppose that variables X_2 and X_3 are added to the model and the following regression equation is obtained.

$$\hat{y} = 16.3 + 2.3x_1 + 12.1x_2 - 5.8x_3$$

For this estimated regression equation SST = 1550 and SSE = 100.

b. Use an F test and a 0.05 level of significance to determine whether X_2 and X_3 contribute significantly to the model.

10 In a regression analysis involving 30 observations, the following estimated regression equation was obtained.

$$\hat{y} = 17.6 + 3.8x_1 - 2.3x_2 + 7.6x_3 + 2.7x_4$$

For this estimated regression equation SST = 1805 and SSR = 1760.

a. At $\alpha = 0.05$, test the significance of the relationship among the variables.

Suppose variables X_1 and X_4 are dropped from the model and the following estimated regression equation is obtained.

$$\hat{y} = 11.1 - 3.6x_2 + 8.1x_3$$

For this model SST = 1805 and SSR = 1705.

b. Compute $SSE(x_1, x_2, x_3, x_4)$.

c. Compute $SSE(x_2, x_3)$.

d. Use an F test and a 0.05 level of significance to determine whether X_1 and X_4 contribute significantly to the model.

Applications

11 A ten-year study conducted by the American Heart Association provided data on how age, blood pressure and smoking relate to the risk of strokes. Data from a portion of this study follow. Risk is interpreted as the probability (times 100) that a person will have a stroke over the next ten-year period. For the smoker variable, 1 indicates a smoker and 0 indicates a non-smoker.

Risk	Age	Blood pressure	Smoker
12	57	152	0
24	67	163	0
13	58	155	0
56	86	177	1
28	59	196	0
51	76	189	1

Risk	Age	Blood pressure	Smoker
18	56	155	1
31	78	120	0
37	80	135	1
15	78	98	0
22	71	152	0
36	70	173	1
15	67	135	1
48	77	209	1
15	60	199	0
36	82	119	1
8	66	166	0
34	80	125	1
3	62	117	0
37	59	207	1

a. Develop an estimated regression equation that can be used to predict the risk of stroke given the age and blood-pressure level.

b. Consider adding two independent variables to the model developed in part (a), one for the interaction between age and blood-pressure level and the other for whether the person is a smoker. Develop an estimated regression equation using these four independent variables.

c. At a 0.05 level of significance, test to see whether the addition of the interaction term and the smoker variable contribute significantly to the estimated regression equation developed in part (a).

12 Failure data obtained in the course of the development of a silver-zinc battery for a NASA programme were analyzed by Sidik, Leibecki and Bozek in 1980. Relevant variables were as follows:

$x1$ charge rate (amps):
$x2$ discharge rate (amps)
$x3$ depth of discharge (% of rated ampere – hours)
$x4$ temperature (°C)
$x5$ end of charge voltage (volts)
y cycles to failure

Adopting $\ln(y)$ as the response variable, a number of regression models were estimated for the data using MINITAB:

Regression Analysis: lny versus x1, x2, x3, x4, x5

```
The regression equation is
lny = 63.7 - 0.459 x1 - 0.327 x2 - 0.0111 x3 + 0.116 x4 + 33.8 x5

Predictor       Coef   SE Coef        T       P   VIF
Constant      -63.68     51.18    -1.24   0.234
x1           -0.4593    0.5493    -0.84   0.417   1.1
x2           -0.3267    0.1761    -1.85   0.085   1.0
x3          -0.01113   0.01699    -0.66   0.523   1.1
x4           0.11577   0.02499     4.63   0.000   1.0
x5             33.81     25.59     1.32   0.208   1.0

S = 1.070   R-Sq = 66.3%   R-Sq(adj) = 54.3%
```

```
Analysis of Variance

Source            DF      SS      MS      F      P
Regression         5   31.578   6.316   5.52   0.005
Residual Error    14   16.032   1.145
Total             19   47.610


Source  DF   Seq SS
x1       1    1.464
x2       1    4.512
x3       1    0.291
x4       1   23.311
x5       1    1.999


Unusual Observations

Obs    x1     lny     Fit   StDev Fit   Residual   St Resid
  1   0.38   4.615   6.708     0.651     -2.093     -2.46R

R denotes an observation with a large standardized residual


Durbin-Watson statistic = 1.72
```

Regression Analysis: lny versus x4

```
The regression equation is lny = 1.78 + 0.114 x4


Predictor     Coef   SE Coef      T      P
Constant    1.7777    0.5660   3.14   0.006
x4          0.11395  0.02597   4.39   0.000


S = 1.130   R-Sq = 51.7%   R-Sq(adj) = 49.0%


Analysis of Variance

Source            DF      SS      MS      F      P
Regression         1   24.607  24.607  19.26  0.000
Residual Error    18   23.002   1.278
Total             19   47.610


Unusual Observations

Obs    x4     lny     Fit   StDev Fit   Residual   St Resid
 12   10.0   0.693   2.917     0.353     -2.224     -2.07R

R denotes an observation with a large standardized residual
```

a. Explain this computer output, carrying out any additional tests you think necessary or appropriate.
b. Is the first model significantly better than the second?
c. Which model do you prefer and why?

13 A section of MINITAB output from an analysis of data relating to truck exhaust emissions under different atmospheric conditions (Hare and Bradow, 1977) is as follows:

```
Regression Analysis: nox versus humi, temp, HT

The regression equation is
nox = 1.61 - 0.0146 humi - 0.00681 temp + 0.000150 HT


Predictor          Coef      SE Coef        T       P
Constant         1.6104       0.2287     7.04   0.000
humi           -0.014572     0.003091    -4.71   0.000
temp           -0.006806     0.002889    -2.36   0.023
HT            0.00014985    0.00003733    4.01   0.000


S = 0.0595096   R-Sq = 71.5%   R-Sq(adj) = 69.4%


Analysis of Variance

Source            DF        SS        MS       F       P
Regression         3   0.35544   0.11848   33.46   0.000
Residual Error    40   0.14166   0.00354
Total             43   0.49710


Source   DF   Seq SS
humi      1   0.28446
temp      1   0.01392
HT        1   0.05706


Unusual Observations

Obs  humi      nox      Fit   SE Fit   Residual   St Resid
  6    13  1.11000  1.09407  0.03316    0.01593     0.32 X
 14    11  1.10000  0.99555  0.03105    0.10445     2.06R

R denotes an observation with a large standardized residual.
X denotes an observation whose X value gives it large leverage.


Durbin-Watson statistic = 1.63335
```

Variables used in this analysis are defined as follows:

nox	Nitrous oxides, NO and NO_2, (grams/km)
humi	Humidity (grains H_2O/lbm dry air)
temp	Temperature (°F)
HT	humi × temp

EMISSIONS

a. Provide a descriptive summary of this information, carrying out any further calculations or statistical tests you think relevant or necessary.

b. It has been argued that the inclusion of quadratic terms

$$HH = humi \times humi$$
$$TT = temp \times temp$$

on the right hand side of the model will lead to a significantly improved *R*-square outcome. Details of the revised analysis are shown below. Is the claim justified?

```
Regression Analysis: nox versus humi, temp, HT, HH, TT

The regression equation is
nox = 2.69 - 0.0102 humi - 0.0371 temp + 0.000057 HT + 0.000022 HH
      + 0.000222 TT


Predictor          Coef     SE Coef        T      P
Constant          2.685       1.306     2.06  0.047
humi          -0.010167    0.003015    -3.37  0.002
temp           -0.03714     0.03414    -1.09  0.284
HT           0.00005662  0.00004073     1.39  0.173
HH           0.00002209  0.00000592     3.73  0.001
TT            0.0002221   0.0002224     1.00  0.324


S = 0.0515260   R-Sq = 79.7%   R-Sq(adj) = 77.0%


Analysis of Variance

Source            DF         SS         MS       F      P
Regression         5   0.396213   0.079243   29.85  0.000
Residual Error    38   0.100887   0.002655
Total             43   0.497100


Source  DF    Seq SS
humi     1   0.284462
temp     1   0.013924
HT       1   0.057058
HH       1   0.038121
TT       1   0.002648


Unusual Observations

Obs   humi      nox      Fit   SE Fit   Residual   St Resid
  5     10  0.99000  1.08738  0.02324   -0.09738     -2.12R
 14     11  1.10000  1.08224  0.03654    0.01776      0.49 X
 40    139  0.70000  0.82314  0.01759   -0.12314     -2.54R

R denotes an observation with a large standardized residual.
X denotes an observation whose X value gives it large leverage.


Durbin-Watson statistic = 1.77873
```

16.3 Analysis of a larger problem

In introducing multiple regression analysis, we used the Eurodistributor example extensively. The small size of this problem was an advantage in exploring introductory concepts, but would make it difficult to illustrate some of the variable selection issues involved in model building. To provide an illustration of the variable selection procedures discussed in the next section, we introduce a data set

CRAVENS

consisting of 25 observations on eight independent variables. Permission to use these data was provided by Dr David W. Cravens of the Department of Marketing at Texas Christian University. Consequently, we refer to the data set as the Cravens data.*

The Cravens data are for a company that sells products in several sales territories, each of which is assigned to a single sales representative. A regression analysis was conducted to determine whether a variety of predictor (independent) variables could explain sales in each territory. A random sample of 25 sales territories resulted in the data in Table 16.5; the variable definitions are given in Table 16.6.

As a preliminary step, let us consider the sample correlation coefficients between each pair of variables. Figure 16.12 is the correlation matrix obtained using MINITAB. Note that the sample correlation coefficient between Sales and Time is 0.623, between Sales and Poten is 0.598 and so on.

Looking at the sample correlation coefficients between the independent variables, we see that the correlation between Time and Accounts is 0.758 and significant; hence, if Accounts were used as an independent variable, Time would not add much

Table 16.5 Cravens data

Sales	Time	Poten	AdvExp	Share	Change	Accounts	Work	Rating
3669.88	43.10	74065.1	4582.9	2.51	0.34	74.86	15.05	4.9
3473.95	108.13	58117.3	5539.8	5.51	0.15	107.32	19.97	5.1
2295.10	13.82	21118.5	2950.4	10.91	−0.72	96.75	17.34	2.9
4675.56	186.18	68521.3	2243.1	8.27	0.17	195.12	13.40	3.4
6125.96	161.79	57805.1	7747.1	9.15	0.50	180.44	17.64	4.6
2134.94	8.94	37806.9	402.4	5.51	0.15	104.88	16.22	4.5
5031.66	365.04	50935.3	3140.6	8.54	0.55	256.10	18.80	4.6
3367.45	220.32	35602.1	2086.2	7.07	−0.49	126.83	19.86	2.3
6519.45	127.64	46176.8	8846.2	12.54	1.24	203.25	17.42	4.9
4876.37	105.69	42053.2	5673.1	8.85	0.31	119.51	21.41	2.8
2468.27	57.72	36829.7	2761.8	5.38	0.37	116.26	16.32	3.1
2533.31	23.58	33612.7	1991.8	5.43	−0.65	142.28	14.51	4.2
2408.11	13.82	21412.8	1971.5	8.48	0.64	89.43	19.35	4.3
2337.38	13.82	20416.9	1737.4	7.80	1.01	84.55	20.02	4.2
4586.95	86.99	36272.0	10694.2	10.34	0.11	119.51	15.26	5.5
2729.24	165.85	23093.3	8618.6	5.15	0.04	80.49	15.87	3.6
3289.40	116.26	26878.6	7747.9	6.64	0.68	136.58	7.81	3.4
2800.78	42.28	39572.0	4565.8	5.45	0.66	78.86	16.00	4.2
3264.20	52.84	51866.1	6022.7	6.31	−0.10	136.58	17.44	3.6
3453.62	165.04	58749.8	3721.1	6.35	−0.03	138.21	17.98	3.1
1741.45	10.57	23990.8	861.0	7.37	−1.63	75.61	20.99	1.6
2035.75	13.82	25694.9	3571.5	8.39	−0.43	102.44	21.66	3.4
1578.00	8.13	23736.3	2845.5	5.15	0.04	76.42	21.46	2.7
4167.44	58.44	34314.3	5060.1	12.88	0.22	136.58	24.78	2.8
2799.97	21.14	22809.5	3552.0	9.14	−0.74	88.62	24.96	3.9

*For details see David W. Cravens, Robert B. Woodruff, and Joe C. Stamper, 'Analytical Approach for Evaluating Sales Territory Performance', *Journal of Marketing,* 36 (January 1972): 31–37. Copyright © 1972 Americal Marketing Association.

Table 16.6 Variable definitions for the Cravens data

Variable	Definition
Sales	Total sales credited to the sales representative
Time	Length of time employed in months
Poten	Market potential; total industry sales in units for the sales territory*
AdvExp	Advertising expenditure in the sales territory
Share	Market share; weighted average for the past four years
Change	Change in the market share over the previous four years
Accounts	Number of accounts assigned to the sales representative*
Work	Workload; a weighted index based on annual purchases and concentrations of accounts
Rating	Sales representative overall rating on eight performance dimensions; an aggregate rating on a 1−7 scale

*These data were coded to preserve confidentiality.

more explanatory power to the model. Recall that inclusion of highly correlated independent variables, as discussed in the Section 15.4 on multicollinearity, can cause problems for the model. If possible, then, we should avoid including both Time and Accounts in the same regression model. The sample correlation coefficient of 0.549 between Change and Rating is also significant (p-value < 0.05) and this may also prove problematic.

Looking at the sample correlation coefficients between Sales and each of the independent variables can give us a quick indication of which independent variables are, by

Figure 16.12 Sample correlation coefficients for the Cravens data

```
Correlations: Sales, Time, Poten, AdvExp, Share, Change, Accounts, Work, Rating

            Sales     Time    Poten   AdvExp    Share   Change  Accounts

Time        0.623
            0.001

Poten       0.598    0.454
            0.002    0.023

AdvExp      0.596    0.249    0.174
            0.002    0.230    0.405

Share       0.484    0.106   -0.211    0.264
            0.014    0.614    0.312    0.201

Change      0.489    0.251    0.268    0.377    0.085
            0.013    0.225    0.195    0.064    0.685

Accounts    0.754    0.758    0.479    0.200    0.403    0.327
            0.000    0.000    0.016    0.338    0.046    0.110

Work       -0.117   -0.179   -0.259   -0.272    0.349   -0.288   -0.199
            0.577    0.391    0.212    0.188    0.087    0.163    0.341

Rating      0.402    0.101    0.359    0.411   -0.024    0.549    0.229
            0.046    0.630    0.078    0.041    0.911    0.004    0.272

Cell Contents: Pearson correlation
               P-Value
```

themselves, good predictors. We see that the single best predictor of Sales is Accounts, because it has the highest sample correlation coefficient (0.754). Recall that for the case of one independent variable, the square of the sample correlation coefficient is the coefficient of determination.

Thus, Accounts can explain $(0.754)^2(100)$, or 56.85 per cent, of the variability in Sales. The next most important independent variables are Time, Poten and AdvExp, each with a sample correlation coefficient of approximately 0.6.

Although there are potential multicollinearity problems, let us consider developing an estimated regression equation using all eight independent variables. The MINITAB computer package provided the results in Figure 16.13. The eight-variable multiple regression model has an adjusted coefficient of determination of 88.3 per cent. Note, however, that the p-values for the t tests of individual parameters show that only Poten, AdvExp and Share are significant at the $\alpha = 0.05$ level, given the effect of all the other

Figure 16.13 MINITAB output for the model involving all eight independent variables

```
Regression Analysis: Sales versus Time, Poten, ...

The regression equation is
Sales = - 1508 + 2.01 Time + 0.0372 Poten + 0.151 AdvExp + 199 Share
        + 291 Change + 5.55 Accounts + 19.8 Work + 8 Rating


Predictor       Coef    SE Coef       T       P
Constant     -1507.8      778.6   -1.94   0.071
Time           2.010      1.931    1.04   0.313
Poten       0.037206   0.008202    4.54   0.000
AdvExp       0.15098    0.04711    3.21   0.006
Share         199.04      67.03    2.97   0.009
Change        290.9       186.8    1.56   0.139
Accounts       5.550      4.775    1.16   0.262
Work          19.79       33.68    0.59   0.565
Rating          8.2       128.5    0.06   0.950


S = 449.015   R-Sq = 92.2%   R-Sq(adj) = 88.3%


Analysis of Variance

Source             DF         SS        MS       F       P
Regression          8   38153712   4769214   23.66   0.000
Residual Error     16    3225837    201615
Total              24   41379549


Source      DF     Seq SS
Time         1   16054463
Poten        1    5173018
AdvExp       1    7701943
Share        1    8145442
Change       1     788073
Accounts     1     219388
Work         1      70567
Rating       1        819
```

Figure 16.14 MINITAB output for the model involving Poten, AdvExp and Share

```
Regression Analysis: Sales versus Poten, AdvExp, Share

The regression equation is
Sales = - 1604 + 0.0543 Poten + 0.167 AdvExp + 283 Share


Predictor        Coef    SE Coef        T      P
Constant      -1603.6      505.6    -3.17  0.005
Poten        0.054286   0.007474     7.26  0.000
AdvExp        0.16748    0.04427     3.78  0.001
Share          282.75      48.76     5.80  0.000


S = 545.515   R-Sq = 84.9%   R-Sq(adj) = 82.7%


Analysis of Variance

Source          DF        SS        MS      F      P
Regression       3  35130228  11710076  39.35  0.000
Residual Error  21   6249321    297587
Total           24  41379549


Source  DF    Seq SS
Poten    1  14788203
AdvExp   1  10333728
Share    1  10008297
```

variables. Hence, we might be inclined to investigate the results that would be obtained if we used just those three variables. Figure 16.14 shows the MINITAB results obtained for the estimated regression equation with those three variables. We see that the estimated regression equation has an adjusted coefficient of determination of 82.7 per cent, which, although not quite as good as that for the eight-independent-variable estimated regression equation, is high.

How can we find an estimated regression equation that will do the best job given the data available? One approach is to compute all possible regressions. That is, we could develop eight one-variable estimated regression equations (each of which corresponds to one of the independent variables), 28 two-variable estimated regression equations (the number of combinations of eight variables taken two at a time), and so on. In all, for the Cravens data, 255 different estimated regression equations involving one or more independent variables would have to be fitted to the data.

With the excellent computer packages available today, it is possible to compute all possible regressions. But doing so involves a great amount of computation and requires the model builder to review a large volume of computer output, much of which is associated with obviously poor models. Statisticians prefer a more systematic approach to selecting the subset of independent variables that provide the best estimated regression equation. In the next section, we introduce some of the more popular approaches.

16.4 Variable selection procedures

In this section we discuss four **variable selection procedures**: stepwise regression, forward selection, backward elimination, and best-subsets regression. Given a data set with several possible independent variables, we can use these procedures to identify which independent variables provide the best model. The first three procedures are iterative; at each step of the procedure a single independent variable is added or deleted and the new model is evaluated. The process continues until a stopping criterion indicates that the procedure cannot find a better model. The last procedure (best subsets) is not a one-variable-at-a-time procedure; it evaluates regression models involving different subsets of the independent variables.

In the stepwise regression, forward selection, and backward elimination procedures, the criterion for selecting an independent variable to add or delete from the model at each step is based on the F statistic introduced in Section 16.2.

Stepwise regression

Based on this statistic, the stepwise regression procedure begins each step by determining whether any of the variables *already in the model* should be removed. If none of the independent variables can be removed from the model, the procedure checks to see whether any of the independent variables that are not currently in the model can be entered.

Because of the nature of the stepwise regression procedure, an independent variable can enter the model at one step, be removed at a subsequent step, and then enter the model at a later step. The procedure stops when no independent variables can be removed from or entered into the model.

Figure 16.15 shows the results obtained by using the MINITAB stepwise regression procedure for the Cravens data using values of 0.05 for *Alpha to remove* and 0.05 for *Alpha to enter*. (These are the technical settings used by the software for deciding whether an independent variable should be removed or entered into the model.) The stepwise procedure terminated after four steps. The estimated regression equation identified by the MINITAB stepwise regression procedure is

$$\hat{y} = -1441.93 + 9.2 \text{ Accounts} + 0.175 \text{ AdvExp} + 0.0382 \text{ Poten} + 190 \text{ Share}$$

Note also in Figure 16.15 that $s = \sqrt{\text{MSE}}$ has been reduced from 881 with the best one variable model (using Accounts) to 454 after four steps. The value of R-sq has been increased from 56.85 per cent to 90.04 per cent, and the recommended estimated regression equation has an R-Sq(adj) value of 88.05 per cent.

Forward selection

The forward selection procedure starts with no independent variables. It adds variables one at a time using the same procedure as stepwise regression for determining whether an independent variable should be entered into the model. However, the forward selection procedure does not permit a variable to be removed from the model once it has been entered.

The estimated regression equation obtained using MINITAB's forward selection procedure is

$$\hat{y} = -1441.93 + 9.2 \text{ Accounts} + 0.175 \text{ AdvExp} + 0.0382 \text{ Poten} + 190 \text{ Share}$$

Figure 16.15 MINITAB stepwise regression output for the Cravens data

```
Stepwise Regression: Sales versus Time, Poten, ...

   Alpha-to-Enter: 0.05   Alpha-to-Remove: 0.05


Response is Sales on 8 predictors, with N = 25


Step                  1        2        3        4
Constant         709.32    50.29  -327.24  -1441.93

Accounts           21.7     19.0     15.6       9.2
T-Value            5.50     6.41     5.19      3.22
P-Value           0.000    0.000    0.000     0.004

AdvExp                     0.227    0.216     0.175
T-Value                     4.50     4.77      4.74
P-Value                    0.000    0.000     0.000

Poten                              0.0219    0.0382
T-Value                              2.53      4.79
P-Value                             0.019     0.000

Share                                          190
T-Value                                       3.82
P-Value                                      0.001

S                   881      650      583       454
R-Sq              56.85    77.51    82.77     90.04
R-Sq(adj)         54.97    75.47    80.31     88.05
Mallows Cp         67.6     27.2     18.4       5.4
```

Thus, for the Cravens data, the forward selection procedure leads to the same estimated regression equation as the stepwise procedure.

Backward elimination

The backward elimination procedure begins with a model that includes all the independent variables. It then deletes one independent variable at a time using the same procedure as stepwise regression. However, the backward elimination procedure does not permit an independent variable to be re-entered once it has been removed.

The estimated regression equation obtained using MINITAB's backward elimination procedure for the Cravens data is

$$\hat{y} = -1312 + 3.8 \text{ Time} + 0.0444 \text{ Poten} + 0.152 \text{ AdvExp} + 259 \text{ Share}$$

Comparing the estimated regression equation identified using the backward elimination procedure to the estimated regression equation identified using the forward selection procedure, we see that three independent variables – AdvExp, Poten and Share – are common to both. However, the backward elimination procedure has included Time instead of Accounts.

Forward selection and backward elimination are the two extremes of model building; the forward selection procedure starts with no independent variables in the model and adds independent variables one at a time, whereas the backward elimination procedure starts with all independent variables in the model and deletes variables one at a time. The two procedures may lead to the same estimated regression equation. It is possible, however, for them to lead to two different estimated regression equations, as we saw with the Cravens data. Deciding which estimated regression equation to use remains a topic for discussion. Ultimately, the analyst's judgment must be applied. The best-subsets model-building procedure we discuss next provides additional model-building information to be considered before a final decision is made.

Best-subsets regression

Stepwise regression, forward selection and backward elimination are approaches to choosing the regression model by adding or deleting independent variables one at a time. None of them guarantees that the best model for a given number of variables will be found. Hence, these one-variable-at-a-time methods are properly viewed as heuristics for selecting a good regression model.

Some software packages use a procedure called best-subsets regression that enables the user to find, given a specified number of independent variables, the best regression model. MINITAB has such a procedure. Figure 16.16 is a portion of the computer output obtained by using the best-subsets procedure for the Cravens data set.

Figure 16.16 Portion of MINITAB best-subsets regression output

Best Subsets Regression: Sales versus Time, Poten, ...

Response is Sales

Vars	R-Sq	R-Sq(adj)	Mallows Cp	S	Time	Poten	AdvExp	Share	AcctChange	Accounts	WkLoad	Rating
1	56.8	55.0	67.6	881.09						X		
1	38.8	36.1	104.6	1049.3	X							
2	77.5	75.5	27.2	650.39			X		X			
2	74.6	72.3	33.1	691.11	X	X						
3	84.9	82.7	14.0	545.52	X	X	X					
3	82.8	80.3	18.4	582.64	X	X			X			
4	90.0	88.1	5.4	453.84	X	X	X		X			
4	89.6	87.5	6.4	463.93	X	X	X	X				
5	91.5	89.3	4.4	430.21	X	X	X	X	X			
5	91.2	88.9	5.0	436.75		X	X	X	X	X		
6	92.0	89.4	5.4	427.99	X	X	X	X	X	X		
6	91.6	88.9	6.1	438.20		X	X	X	X	X	X	
7	92.2	89.0	7.0	435.66	X	X	X	X	X	X	X	
7	92.0	88.8	7.3	440.29	X	X	X	X	X	X		X
8	92.2	88.3	9.0	449.02	X	X	X	X	X	X	X	X

This output identifies the two best one-variable estimated regression equations, the two best two-variable equations, the two best three-variable equations and so on. The criterion used in determining which estimated regression equations are best for any number of predictors is the value of the coefficient of determination (R-sq). For instance, Accounts, with an R-sq = 56.8 per cent, provides the best estimated regression equation using only one independent variable; AdvExp and Accounts, with an R-sq = 77.5 per cent, provides the best estimated regression equation using two independent variables; and Poten, AdvExp, and Share, with an R-sq = 84.9 per cent, provides the best estimated regression equation with three independent variables. For the Cravens data, the adjusted coefficient of determination (R-sq(adj)) = 89.4 per cent is largest for the model with six independent variables: Time, Poten, AdvExp, Share, Change and Accounts. However, the best model with four independent variables (Poten, AdvExp, Share, Accounts) has an adjusted coefficient of determination almost as high (88.1 per cent). All other things being equal, a simpler model with fewer variables is usually preferred.

Making the final choice

The analysis performed on the Cravens data to this point is good preparation for choosing a final model, but more analysis should be conducted before the final choice is made. As we noted in Chapters 14 and 15, a careful analysis of the residuals should be undertaken. We want the residual plot for the chosen model to resemble approximately a horizontal band. Let us assume the residuals are not a problem and that we want to use the results of the best-subsets procedure to help choose the model.

The best-subsets procedure shows us that the best four-variable model contains the independent variables Poten, AdvExp, Share and Accounts. This result also happens to be the four-variable model identified with the stepwise regression procedure. Note also that the S and R-Sq(adj) results are virtually identical between the two models. Also there is very little difference between the corresponding R-Sq values.

Exercises

Applications

14 A sample of 16 companies taken from the *Stock Investor Pro* database was used to obtain the following data on the price/earnings (P/E) ratio, the gross profit margin, and the sales growth for each company (*Stock Investor Pro*, American Association of Individual Investors, 21 August 1997). The data in the Industry column are codes used to define the industry for each company: 1 = energy-international oil; 2 = health-drugs; and 3 = other.

SKTDATA

Firm	P/E ratio	Gross profit margin (%)	Sales growth (%)	Industry
Abbott Laboratories	22.3	23.7	10.0	2
American Home Products	22.6	21.1	5.3	2
Amoco	16.7	11.0	16.5	1
Bristol Meyers Squibb Co.	25.9	26.6	9.4	2
Chevron	18.3	11.6	18.4	1
Exxon	18.7	9.8	8.3	1
General Electric Company	13.1	13.4	13.1	3

Firm	P/E ratio	Gross profit margin (%)	Sales growth (%)	Industry
Hewlett-Packard	23.3	9.7	21.9	3
IBM	17.3	11.5	5.6	3
Merck & Co. Inc.	26.2	25.6	18.9	2
Mobil	18.7	8.2	8.1	1
Pfizer	34.6	25.1	12.8	2
Pharmacia & Upjohn, Inc.	22.3	15.0	2.7	2
Procter & Gamble Co.	5.4	14.9	5.4	3
Texaco	12.3	7.3	23.7	1
Travelers Group Inc.	28.7	17.8	28.7	3

Develop an estimated regression equation that can be used to predict price/earnings ratio. Briefly discuss the process you used to develop a recommended estimated regression equation for these data.

15 A sales executive is interested in predicting sales of a newly released record (Field, 2005). Details are available for 200 individual past recordings as follows:

airplay = *number of times a record is played on Radio 1*
sales = *record sales (thousands)*
advert = *advertizing budget (£000s)*
attract = *attractiveness rating (1–10) of recording act*

Selective modelling details using MINITAB are given below:

```
Correlations: adverts, sales, airplay, attract

          adverts    sales   airplay
sales       0.578
            0.000

airplay     0.102    0.599
            0.151    0.000

attract     0.081    0.326    0.182
            0.256    0.000    0.010


Cell Contents: Pearson correlation
               P-Value


Stepwise Regression: sales versus adverts, airplay, attract

  Alpha-to-Enter: 0.15   Alpha-to-Remove: 0.15


Response is sales on 3 predictors, with N = 200


Step            1        2        3
Constant    84.87    41.12   -26.61

airplay      3.94     3.59     3.37
T-Value     10.52    12.51    12.12
P-Value      0.000    0.000    0.000

adverts              0.0869   0.0849
T-Value              11.99    12.26
P-Value               0.000    0.000
```

```
attract                        11.1
T-Value                        4.55
P-Value                        0.000


S            64.8    49.4    47.1
R-Sq         35.87   62.93   66.47
R-Sq(adj)    35.55   62.55   65.95
Mallows Cp   178.8   22.7     4.0
```

Regression Analysis: sales versus adverts, airplay, attract

```
The regression equation is
sales = - 26.6 + 0.0849 adverts + 3.37 airplay + 11.1 attract


Predictor       Coef    SE Coef       T      P     VIF
Constant      -26.61      17.35   -1.53  0.127
adverts     0.084885   0.006923   12.26  0.000   1.015
airplay       3.3674     0.2778   12.12  0.000   1.043
attract       11.086      2.438    4.55  0.000   1.038


S = 47.0873   R-Sq = 66.5%   R-Sq(adj) = 66.0%


Analysis of Variance

Source             DF        SS       MS       F      P
Regression          3    861377   287126  129.50  0.000
Residual Error    196    434575     2217
Total             199   1295952


Source   DF   Seq SS
adverts   1   433688
airplay   1   381836
attract   1    45853


Unusual Observations

Obs   adverts   sales     Fit   SE Fit   Residual   St Resid
  1        10  330.00  229.92    10.23     100.08      2.18R
  2       986  120.00  228.95     4.21    -108.95     -2.32R
  7       472   70.00   91.87    14.21     -21.87     -0.49 X
 10       174  300.00  200.47     5.85      99.53      2.13R
 12       611   70.00  114.81    11.92     -44.81     -0.98 X
 47       103   40.00  154.97     5.90    -114.97     -2.46R
 52       406  190.00   92.60     8.05      97.40      2.10R
 55      1542  190.00  304.12     7.61    -114.12     -2.46R
 61       579  300.00  201.19     3.44      98.81      2.10R
 68        57   70.00  180.42     5.90    -110.42     -2.36R
100      1000  250.00  152.71     7.85      97.29      2.10R
138        30   60.00   81.34    14.79     -21.34     -0.48 X
164         9  120.00  241.32     9.34    -121.32     -2.63R
169       146  360.00  215.87     6.79     144.13      3.09R
181       179   70.00   63.65    14.33       6.35      0.14 X
184      2272  320.00  326.06    12.97      -6.06     -0.13 X
200       786  110.00  207.21     7.07     -97.21     -2.09R

R denotes an observation with a large standardized residual.
X denotes an observation whose X value gives it large leverage.

Durbin-Watson statistic = 1.94982
```

```
Best Subsets Regression: sales versus adverts, airplay, attract

Response is sales
```

Vars	R-Sq	R-Sq(adj)	Mallows Cp	S	adverts	airplay	attract
1	35.9	35.5	178.8	64.787			X
1	33.5	33.1	192.9	65.991	X		
2	62.9	62.6	22.7	49.383	X	X	
2	41.3	40.7	149.0	62.129	X		X
3	66.5	66.0	4.0	47.087	X	X	X

a. How would you interpret this information?

b. Which of the various models shown here do you favour and why?

16 In a study of car ownership in 24 countries, data (OECD, 1982) have been collected on the following variables:

OECDCARS

ao	cars per person
pop	population (millions)
den	population density
gdp	*per capita* income ($)
pr	petrol price (cents per litre)
con	petrol consumption (tonnes per car per year)
tr	bus and rail use (passenger km per person)

Selective results from a linear modelling analysis (ao is the dependent variable) are as follows:

Best Subsets Regression: ao versus pop, den, gdp, pr, con, tr

```
Response is ao
```

Vars	R-Sq	R-Sq(adj)	Mallows Cp	S	pop	den	gdp	pr	con	tr
1	53.0	50.9	41.2	0.085534	X					
1	10.7	6.7	96.4	0.11791		X				
2	67.8	64.7	24.0	0.072526	X	X				
2	67.3	64.2	24.6	0.073035	X				X	
3	72.5	68.4	19.8	0.068579		X	X		X	
3	72.1	68.0	20.3	0.069090	X		X		X	
4	83.0	79.5	8.1	0.055298			X	X	X	X
4	77.1	72.3	15.8	0.064197	X		X		X	X
5	86.2	82.4	6.0	0.051208	X		X	X	X	X
5	83.2	78.5	9.9	0.056611		X	X	X	X	X
6	87.0	82.4	7.0	0.051270	X	X	X	X	X	X

Correlations: ao, pop, den, gdp, pr, con, tr

	ao	pop	den	gdp	pr	con
pop	0.278					
	0.188					
den	-0.042	0.109				
	0.846	0.612				
gdp	0.728	0.057	0.193			
	0.000	0.791	0.365			
pr	-0.327	-0.437	0.338	0.076		
	0.118	0.033	0.106	0.724		

```
con    0.076    0.342   -0.357   -0.085   -0.723
       0.723    0.101    0.087    0.694    0.000

tr    -0.119   -0.025    0.397    0.328    0.483   -0.602
       0.581    0.906    0.055    0.118    0.017    0.002


Cell Contents: Pearson correlation
              P-Value
```

Regression Analysis: ao versus pop, gdp, pr, con, tr

```
The regression equation is
ao = 0.472 + 0.000521 pop + 0.0319 gdp - 0.00429 pr - 0.104 con - 0.0735 tr


Predictor         Coef    SE Coef       T       P
Constant       0.47190    0.09081    5.20   0.000
pop          0.0005211  0.0002556    2.04   0.056
gdp           0.031889   0.003423    9.32   0.000
pr           -0.004289   0.001245   -3.44   0.003
con           -0.10449    0.02626   -3.98   0.001
tr            -0.07354    0.01733   -4.24   0.000


S = 0.0512085    R-Sq = 86.2%    R-Sq(adj) = 82.4%
```

```
Analysis of Variance

Source             DF         SS         MS       F       P
Regression          5   0.295364   0.059073   22.53   0.000
Residual Error     18   0.047202   0.002622
Total              23   0.342565


Source  DF    Seq SS
pop      1   0.026535
gdp      1   0.174352
pr       1   0.033131
con      1   0.014144
tr       1   0.047202


Unusual Observations

Obs  pop      ao     Fit   SE Fit   Residual   St Resid
 11   57  0.3000  0.1914   0.0235     0.1086      2.39R
 23  218  0.5300  0.5178   0.0454     0.0122      0.52 X
```

```
R denotes an observation with a large standardized residual.
X denotes an observation whose X value gives it large leverage.
```

a. Which of the various model options considered here do you prefer and why?

b. Corresponding stepwise output from MINITAB terminates after two stages, gdp being the first independent variable selected and pr the second. How does this latest information reconcile with that summarized earlier?

c. Does it alter in any way, your inferences for a.? If so, why and if not, why not?

17 In an analysis of the effects of rainfall, temperature and time of exposure on the ret loss of flax, the following MINITAB output has been obtained:

(Note: X_1 = Mean daily rainfall (0.01 inches per day)

X_2 = Retting period (days)

X_3 = Mean maximum daily temperature (°F)

Y = per cent ret loss of flax

```
Regression Analysis: y versus x1, x2, x3

The regression equation is
y = 10.8 + 1.81 x1 + 0.109 x2 + 0.0926 x3

Predictor       Coef    SE Coef      T       P     VIF
Constant      10.819      7.258    1.49   0.150
x1             1.8101     0.5451   3.32   0.003     1.2
x2             0.10887    0.05858  1.86   0.076     1.5
x3             0.09263    0.09296  1.00   0.329     1.7

S = 2.197        R-Sq = 42.3%      R-Sq(adj) = 34.7%

Analysis of Variance

SOURCE         DF        SS        MS       F       P
Regression      3     81.285    27.095    5.61   0.005
Error          23    111.045     4.828
Total          26    192.330
```

```
SOURCE    DF    SEQ SS
x1         1    37.060
x2         1    39.430
x3         1     4.795

Unusual Observations
Obs.    x1         y      Fit    SE Fit   Residual   St. Resid
21     4.80    29.500   34.004   1.013     -4.504      -2.31R
24     5.40    38.900   34.050   0.890      4.850       2.41R

R denotes an obs. with a large st. resid.

Durbin-Watson statistic = 1.64
```

```
Stepwise Regression: y versus x1, x2, x3

Stepwise regression of y on 3 predictors, with N    27
      STEP      1      2
CONSTANT     27.39  16.42

x1            1.36   1.59
T-RATIO       2.44   3.20

x2                   0.141
T-RATIO              2.86
S             2.49   2.20
R-SQ         19.27  39.77
```

```
Best Subsets Regression: y versus x1, x2, x3

Best Subsets
Regression of y
                Adj.                   x   x   x
Vars   R-sq    R-sq    C-p       s     1   2   3
  1    19.3    16.0    9.2    2.4921   X
  1    14.1    10.7   11.2    2.5700           X
  2    39.8    34.8    3.0    2.1970   X   X
  2    33.6    28.1    5.5    2.3069   X       X
  3    42.3    34.7    4.0    2.1973   X   X   X
```

a. How would you interpret this information?
b. Confirm details of any tests you carry out to support your inferences.
c. Which is your preferred model of those covered here?

18 A senior police manager is reviewing manpower allocation of police officers to a number of geographical districts which fall under her responsibility (Wisniewski, 2002). Detailed regression analysis results have been obtained involving the following variables:

POLICE

Crimes	number of reported crimes
Officers	number of full-time equivalent police officers
Support	number of civilian support staff
Unemployment	unemployment rate (%) for the area
Retired	percentage of the local population who are retired

Selected MINITAB output is given below:

```
Correlations: Crimes, Officers, Support, Unemployment, Retired

                 Crimes     Officers      Support   Unemployment
Officers         -0.735
                  0.000

Support           0.259       -0.345
                  0.202        0.085

Unemployment      0.760       -0.434        0.128
                  0.000        0.027        0.535

Retired          -0.867        0.655       -0.138       -0.661
                  0.000        0.000        0.501        0.000


Cell Contents: Pearson correlation
               P-Value


Best Subsets Regression: Crimes versus Officers, Support, ...

Response is Crimes

                                            U
                                            n
                                            e
                                            m
                                        O   p
                                        f S l R
                                        f u o e
                                        i p y t
                                        c p m i
                                        e o e r
                             Mallows     r r n e
Vars  R-Sq  R-Sq(adj)     Cp        S    s t t d
  1   75.1      74.0    16.8   100.55            X
  1   57.7      55.9    43.9   131.05        X
  2   81.3      79.7     9.2    89.041       X X
  2   80.0      78.3    11.1    91.970   X       X
  3   86.2      84.3     3.5    78.170   X   X X
  3   82.9      80.6     8.6    86.946       X X X
  4   86.5      83.9     5.0    79.085   X X X X


Stepwise Regression: Crimes versus Officers, Support, ...

Backward elimination.  Alpha-to-Remove: 0.1


Response is Crimes on 4 predictors, with N = 26
```

```
Step                    1     2
Constant             1344  1411

Officers            -14.1 -15.5
T-Value             -2.36 -2.80
P-Value             0.028 0.010

Support                10
T-Value              0.70
P-Value             0.490
```

```
Unemployment   17.0   17.1
T-Value         3.06   3.14
P-Value        0.006  0.005

Retired       -20.6  -20.0
T-Value       -3.64  -3.62
P-Value        0.002  0.002

S              79.1   78.2
R-Sq          86.52  86.20
R-Sq(adj)     83.95  84.32
Mallows Cp      5.0    3.5
```

If a new variable *Total staff* = *Officers* + *Support* is created and a further analysis undertaken, the following results are obtained.

Regression Analysis: Crimes versus Unemployment, Retired, Total staff

```
The regression equation is
Crimes = 1433 + 17.4 Unemployment - 21.5 Retired - 13.4 Total staff


Predictor        Coef   SE Coef      T      P
Constant       1433.0     164.6   8.71  0.000
Unemployment   17.412     5.786   3.01  0.006
Retired       -21.511     5.883  -3.66  0.001
Total staff   -13.398     6.205  -2.16  0.042


S = 82.7009   R-Sq = 84.6%   R-Sq(adj) = 82.4%


Analysis of Variance

Source           DF      SS      MS      F      P
Regression        3  823591  274530  40.14  0.000
Residual Error   22  150468    6839
Total            25  974059


Source          DF  Seq SS
Unemployment     1  561901
Retired          1  229809
Total staff      1   31881


Durbin-Watson statistic = 2.22341
```

a. Explain this computer output, carrying out any additional tests you think necessary or appropriate.
b. Is the last model a significant improvement on the corresponding two predictor model (best subsets option with $R^2 = 81.3$ per cent) for which details were summarized earlier?
c. Which of the various models shown do you prefer and why?

For additional online summary questions and answers go to the companion website at www.cengage.co.uk/aswsbe2

Summary

In this chapter we discussed several concepts used by model builders in identifying the best estimated regression equation. First, we introduced the concept of a general linear model to show how the methods discussed in Chapters 14 and 15 could be extended to handle curvilinear relationships and interaction effects. Then we discussed how transformations involving the dependent variable could be used to account for problems such as non-constant variance in the error term.

In many applications of regression analysis, a large number of independent variables are considered. We presented a general approach based on an F statistic for adding or deleting variables from a regression model. We then introduced a larger problem involving 25 observations and eight independent variables. We saw that one issue encountered in solving larger problems is finding the best subset of the independent variables. To help in that task, we discussed several variable selection procedures: stepwise regression, forward selection, backward elimination and best-subsets regression.

Key terms

General linear model
Interaction

Variable selection procedures

Key formulae

General linear model

$$Y = \beta_0 + \beta_1 z_1 + \beta_2 z_2 + \cdots + \beta_p z_p + \varepsilon \tag{16.1}$$

F test statistic for adding or deleting $p-q$ variables

$$F = \frac{\dfrac{SSE(x_1, x_2, \ldots, x_q) - SSE(x_1, x_2, , x_q, x_{q+1}, \ldots, x_p)}{p - q}}{\dfrac{SSE(x_1, x_2, \ldots, x_q, x_{q+1}, \ldots, x_p)}{n - p - 1}} \tag{16.12}$$

Case problem I Unemployment study

LAYOFFS

Weeks	Age	Educ	Married	Head	Tenure	Manager	Sales
37	30	14	1	1	1	0	0
62	27	14	1	0	6	0	0
49	32	10	0	1	11	0	0
73	44	11	1	0	2	0	0
8	21	14	1	1	2	0	0
15	26	13	1	0	7	1	0
52	26	15	1	0	6	0	0
72	33	13	0	1	6	0	0
11	27	12	1	1	8	0	0
13	33	12	0	1	2	0	0

A study provided data on variables that may be related to the number of weeks a manufacturing worker has been jobless. The dependent variable in the study (Weeks) was defined as the number of weeks a worker has been jobless due to a layoff. The following independent variables were used in the study.

Age	The age of the worker
Educ	The number of years of education
Married	A dummy variable: 1 if married, 0 otherwise
Head	A dummy variable: 1 if the head of household, 0 otherwise
Tenure	The number of years on the previous job
Manager	A dummy variable: 1 if management occupation, 0 otherwise
Sales	A dummy variable: 1 if sales occupation, 0 otherwise.

Looking for a job. © RESO.

Altogether, details were collected for 50 displaced workers and these are available on the CD accompanying the text in the file named Layoffs.

Managerial report

Use the methods presented in this and previous chapters to analyze this data set. Present a summary of your analysis, including key statistical results, conclusions, and recommendations, in a managerial report. Include any appropriate technical material (computer output, residual plots, etc.) in an appendix.

Case problem 2 Treating obesity*

Obesity is a major health risk throughout Europe and the USA, leading to a number of possibly life-threatening diseases. Developing a successful treatment for obesity is therefore important, as a reduction in weight can greatly reduce the risk of illness. A sustained weight loss of 5–10 per cent of initial body weight reduces the health risks associated with obesity. Diet and exercise are useful in weight control but may not always be successful in the long term. An integrated programme of diet, exercise and drug treatment may be beneficial for obese patients.

The study

In 1998 Knoll Pharmaceuticals received authorization to market sibutramine for the treatment of obesity in the USA. One of their suite of studies involved 37 obese patients who followed a treatment regime comprising a combination of diet, exercise and drug treatment. Patients taking part in this study were healthy adults (aged 18 to 65 years) and were between 30 per cent and 80 per cent above their ideal body

OBESITY

weight. Rigorous criteria were defined to ensure that only otherwise-healthy individuals took part.

Patients received either the new drug or placebo for an eight-week period and body weight was recorded at the start (week 0, also known as baseline) and at week eight. The information recorded for each patient was:

- Age (years)
- Gender (F: female, M: male)
- Height (cm)
- Family history of obesity? (N: no, Y: yes) Missing for patient number 134
- Motivation rating (1: some, 2: moderate, 3: great)
- Number of previous weight loss attempts
- Age of onset of obesity (1: 11 years, 2: 12–17 years, 3: 18–65 years)
- Weight at week 0 (kg)
- Weight at week 8 (kg)
- Treatment group (1 = placebo, 2 = new drug)

Results are shown below for a selection of ten of the 37 patients that took part in the study:

Age	Gender	Height	Family history?	Motivation rating	Previous weight loss attempts	Age of onset	Weight at week 0	Weight at week 8	Treatment group
40	F	170	N	2	1	3	83.4	75.0	2
50	F	164	Y	2	5	2	102.2	96.3	1
39	F	154	Y	2	1	3	84.0	82.6	1
40	F	169	Y	1	7	3	103.7	95.7	2
44	F	169	N	2	1	1	99.2	99.2	2
44	M	177	Y	2	2	2	126.0	123.2	2
38	M	171	Y	1	1	1	103.7	95.5	2
42	M	175	N	2	4	3	117.9	117.0	1
53	M	177	Y	2	3	3	112.4	111.8	1
52	F	166	Y	1	3	3	85.0	80.0	2

Clinical trials

The study is an example of a clinical trial commonly used to assess the effectiveness of a new treatment. Clinical trials are subject to rigorous controls to ensure that individuals are not unnecessarily put at risk and that they

are fully informed and give their consent to take part in the study. As giving any patient a treatment may have a psychological effect, many studies compare a new drug with a dummy treatment (placebo) where, to avoid bias, neither the patient nor the doctor recording information knows whether the patient is on the new treatment

*Data in this case study reproduced with permission from STARS (www.stars.ac.uk)

or placebo as the tablets/capsules look identical; this approach is known as double-blinding. Bias could also occur if the treatment given to a patient was based on their characteristics; for example, if the more-overweight patients were given the new treatment rather than the placebo they would have a greater chance of weight loss. To avoid such bias the decision as to which individuals will receive the new treatment or placebo must be made using a process known as randomization. Using this approach each individual has the same chance of being given either the new treatment or the placebo.

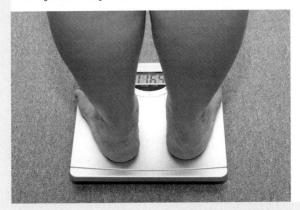

Overweight woman standing on scales at doctor's office. © Kokhanchikov.

Managerial report

1 Use the methods presented in this and previous chapters to analyze this data set. The priority is to use regression modelling to help determine which variables most influence weight loss. The treatment group variable is a particular concern in this respect.

2 Present a summary of your analysis, including key statistical results, conclusions and recommendations, in a managerial report. Include any appropriate technical material (computer output, residual plots, etc.) in an appendix.

Chapter 17

···

Index Numbers

Learning objectives

After studying this chapter and doing the exercises, you should be able to:

1 Calculate price relatives.

2 Calculate Laspeyres and Paasche weighted aggregate price index numbers.

3 Give an outline account of some important index numbers such as consumer price indices, producer price indices and share price indices.

4 Use a price index number to deflate a time series to constant prices.

5 Give an outline account of some of the issues involved in constructing a price index number.

6 Calculate Laspeyres and Paasche weighted aggregate quantity index numbers.

Each month or quarter, governments in most developed and developing countries publish a variety of index numbers designed to provide indications of current business and economic conditions. In many countries, the most widely known and cited of these index numbers is the CPI, the Consumer Price (or Prices) Index (see Statistics in Practice).

As its name implies, the CPI is an indicator of what is happening to prices people (consumers) pay for items purchased. More specifically, the CPI measures changes in average prices consumers pay over a period of time. It has a nominated starting point or *base period*, usually assigned the index value of 100, and can be used to compare current period prices with those in the base period. For example, a CPI value of 125 implies that prices as a whole are running approximately 25 per cent above the base period prices for the same items. Although relatively few individuals know exactly what this number means, they do know enough about the CPI to understand that an increase in the CPI means higher prices.

In addition to the CPI, many other government and private-sector index numbers are available to help us measure and understand how economic conditions in one period compare with economic conditions in other periods. The purpose of this chapter is to describe the most widely used types of index numbers. We shall begin by constructing some simple index numbers to gain a better understanding of how they are computed.

17.1 Price relatives

The simplest form of a price index shows how the current price per unit for a given item compares to the base period price per unit for the same item. For example, Table 17.1 reports the cost of one litre of diesel fuel in Norway for the years 2000 to 2009 (the figures are for January each year, in Norwegian Krone). To facilitate comparisons with other years, the actual cost-per-litre figure can be converted to a **price relative**, which expresses the unit price in each period as a percentage of the unit price in the nominated base period.

Price relative in period t

$$\text{Price relative in period } t = \frac{\text{Price in period } t}{\text{Base period price}} (100) \qquad (17.1)$$

Statistics in Practice

Index numbers in the headlines

In the turbulent global financial climate of late 2008 and early 2009, index numbers of various kinds were rarely out of the news headlines. During this period, stock markets across the world experienced substantial falls in value, and these were most frequently reported in terms of the day-to-day changes in well-known share price indices. On 6 October 2008, for example, it was reported that the Dow Jones index (New York Stock Exchange) had fallen below the 10 000 level for the first time since October 2004. On 2 March 2009, the Dow Jones fell below 7000 for the first time since October 1977. On the same day, it was reported that the FTSE 100 (London Stock Exchange) had fallen to a six-year low, and that the Nikkei Index (Tokyo Stock Exchange) had fallen to its lowest level since 6 October 1982.

Central banks in many countries reacted to the financial problems by lowering interest rates, and in turn this had an effect on price inflation, as reflected in Consumer Price Indices (CPIs). For instance, in December 2008, the Central Statistics Office Ireland reported that average prices, as measured by the CPI, had fallen by 0.9 per cent during November, bringing the annual rate of inflation sharply down to 2.5 per cent. In January 2009, it was reported that the 2008 fourth-quarter New Zealand CPI had registered its first negative move for two years. On 5 March 2009, India's rate of inflation, based on the official wholesale price index, was reported to have fallen for the fifth consecutive week.

In January 2009, the UK annual inflation rate, as measured by the Retail Prices Index, fell to a 49-year low of just 0.1 per cent. The UK has two commonly quoted price indices: the Consumer Price Index (CPI) and the Retail Prices Index (RPI). It was widely expected that annual inflation in February 2009 as measured by the RPI would become negative. In the event, it fell to zero, but not into negative territory. At the same time, annual inflation as measured by the CPI, which had been expected to fall, rose slightly to 3.2 per cent. Annual inflation as measured by the RPI moved into negative territory in March 2009.

The purpose of this chapter is to describe the most widely used types of index numbers. We shall begin by constructing some simple index numbers to gain a better understanding of how they are computed. Later in the chapter, the differences between the UK's CPI and RPI are described.

Newspaper headline regarding economic downturn. © JustASC.

For the diesel fuel prices in Table 17.1, and with January 2000 as the nominated base period, the price relatives for one litre of diesel in January in the years 2000 to 2009 can be calculated. These price relatives are listed in Table 17.2. Note how easily the price in any given year can be compared with the price in the base year by knowing the price relative. For example, the price relative of 121.0 in 2008 indicates that the diesel price in January 2008 was 21.0 per cent above the January 2000 base-period price. In the years 2001–2006, the January price relatives are below 100.0, indicating that the diesel price was below the 2000 base-period price. Price relatives, such as the ones shown for diesel fuel, are extremely helpful in terms of understanding and interpreting changing economic and business conditions over time.

Table 17.1   Diesel fuel prices in Norway

Year	Price per litre (Norwegian Krone)
2000	9.85
2001	8.81
2002	8.29
2003	8.35
2004	8.30
2005	9.00
2006	9.50
2007	10.70
2008	11.92
2009	10.99

Source: Figures published by The Automobile Association (UK)

Table 17.2   Price relatives for one litre of diesel fuel (2000–2009)

Year	Price relative (base 2000)
2000	(9.85/9.85)100 = 100.0
2001	(8.81/9.85)100 = 89.4
2002	(8.29/9.85)100 = 84.2
2003	(8.35/9.85)100 = 84.8
2004	(8.30/9.85)100 = 84.3
2005	(9.00/9.85)100 = 91.4
2006	(9.50/9.85)100 = 96.4
2007	(10.70/9.85)100 = 108.6
2008	(11.92/9.85)100 = 121.0
2009	(10.99/9.85)100 = 111.6

17.2 Aggregate price index numbers

Although price relatives can be used to identify price changes over time for individual items, we are often more interested in the general price change for a group of items taken as a whole. For example, if we want an index that measures the change in the overall cost of living over time, we will want the index to be based on the price changes for a variety of items, including food, housing, clothing, transportation, entertainment and so on. An **aggregate price index** has the specific purpose of measuring the combined change in price of a group of items.

Consider the development of an aggregate price index for a group of items categorized as motoring costs in the UK (i.e. the normal costs of running a car). For illustration, we limit the items included in the group to petrol, tyres, labour costs for service and insurance.

Table 17.3 gives the data (in £) for the four components of our motoring costs index for the years 2000 and 2008. With 2000 as the base period, an aggregate price index for the four components will give us a measure of the change in motoring costs over the 2000–2008 period.

Table 17.3 Data for motoring costs index

	Unit price (£)	
Item	2000	2008
Litre of petrol	0.85	1.05
One tyre	75.00	50.00
One hour of service	55.00	60.00
Insurance policy	450.00	475.00

We can construct an unweighted aggregate index by simply summing the unit prices in the year of interest (e.g. 2008) and dividing that sum by the sum of the unit prices in the base year (2000). Let

$$P_{it} = \text{unit price for item } i \text{ in period } t$$
$$P_{i0} = \text{unit price for item } i \text{ in the base period}$$

An unweighted aggregate price index in period t, denoted by I_t, is given by

Unweighted aggregate price index in period t

$$I_t = \frac{\Sigma P_{it}}{\Sigma P_{i0}}(100) \tag{17.2}$$

where the sums are for all items in the group.

An unweighted aggregate index for motoring costs in 2008 ($t = 2008$) is given by

$$I_{2008} = \frac{1.05 + 50.00 + 60.00 + 475.00}{0.85 + 75.00 + 55.00 + 450.00}(100)$$

$$= \frac{586.05}{580.85}(100) = 100.9$$

From the unweighted aggregate price index, we might conclude that motoring costs increased by just 0.9 per cent over the period from 2000 to 2008. But note that the unweighted aggregate approach to establishing a composite price index is heavily influenced by the items with large per-unit prices, which dominate items with relatively low unit prices. The unweighted aggregate index for motoring costs is too heavily influenced by the price changes in tyres, labour costs for servicing and insurance. Petrol, with a low unit price (but a price increase of 23.5 per cent over the period) has very little influence on the aggregate index. Because of the sensitivity of an unweighted index to one or more high-priced items, this form of aggregate index is not widely used. A weighted aggregate price index provides a better comparison.

The philosophy behind the **weighted aggregate price index** is that each item in the group should be weighted according to its importance. In most cases, the quantity used is the best measure of importance. Therefore, we must obtain a measure of the quantity used for the various items in the group. Table 17.4 gives annual usage information for

Table 17.4 Annual usage information for motoring costs index

Item	Quantity weights
Litres of petrol	1200
Tyres	2
Service hours	5
Insurance policy for one year	1

each item of motoring costs based on typical running of a small car for approximately 16 000 kilometres (10 000 miles) per year. The quantity weights listed show the expected annual usage for this type of driving situation.

Let Q_i be the quantity used of item i. The weighted aggregate price index in period t is given by

Weighted aggregate price index in period t

$$I_t = \frac{\Sigma P_{it} Q_i}{\Sigma P_{i0} Q_i} (100)$$

(17.3)

where the sums are for all items in the group. Applied to our motoring costs, the weighted aggregate price index is based on dividing total running costs in 2008 by total running costs in 2008.

Let $t = 2008$, and use the quantity weights in Table 17.4. We obtain the following weighted aggregate price index for motoring costs in 2008.

$$I_{2008} = \frac{1.05(1200) + 50.00(2) + 60.00(5) + 475.00(1)}{0.85(1200) + 75.00(2) + 55.00(5) + 450.00(1)} (100)$$

$$= \frac{2135.00}{1895.00} (100) = 112.7$$

From this weighted aggregate price index, we would conclude that motoring costs have increased by 12.7 per cent over the period from 2000 to 2008.

Compared with the unweighted aggregate index, the weighted index provides a more realistic indication of the price change for overall motoring costs over the 2000–2008 period. The weighted index indicates a larger increase in motoring costs than the unweighted index, because taking into account the quantity of petrol used helps to reflect the relatively large increase in petrol price. If the quantity used is the same for each item, an unweighted index gives the same value as a weighted index. In practice, however, quantities of usage are rarely the same, or indeed even comparable between different items in the index. In general, the weighted aggregate index with quantities used as weights is the preferred method for establishing a price index for a group of items.

In the weighted aggregate price index formula (17.3), note that the quantity term Q_i does not have a second subscript to indicate the time period. The reason is that the quantities Q_i are considered fixed and do not vary with time as the prices do. The fixed weights or quantities are specified by the designer of the index at levels believed to be representative of typical usage.

In a particular case of the fixed-weight aggregate index, the quantities are determined from base-period usages. In this case we write $Q_i = Q_{i0}$, with the zero subscript indicating base-period quantity weights; formula (17.3) becomes

$$I_t = \frac{\Sigma P_{it} Q_{i0}}{\Sigma P_{i0} Q_{i0}} (100) \qquad (17.4)$$

When the fixed quantity weights are determined from base-period usage, the weighted aggregate index is known as the **Laspeyres price index**.

Another option for determining quantity weights is to revise the quantities each period. A quantity Q_{it} is determined for each period that the index is computed. The weighted aggregate index in period t with these quantity weights is given by

$$I_t = \frac{\Sigma P_{it} Q_{it}}{\Sigma P_{i0} Q_{it}} (100)$$

Note that the same quantity weights are used in the denominator (base period 0) and in the numerator (period t). However, the weights are based on usage in period t, not the base period. This weighted aggregate index is known as the **Paasche price index**. It has the advantage of being based on current usage patterns. However, this method of computing a weighted aggregate index presents a serious disadvantage: the normal usage quantities Q_{it} must be revised each period, thereby adding to the time and cost of data collection. Because of this, the Laspeyres index is more widely used. The motoring costs index above was computed with base-period quantities. Therefore, it is a Laspeyres index. If usage figures for 2008 were used, it would be a Paasche index.

Exercises

Methods

1 The following table reports prices and quantities used for two items in 2007 and 2009.

	Quantity		Unit price (€)	
Item	2007	2009	2007	2009
A	1500	1800	7.50	7.75
B	2	1	630.00	1500.00

a. Compute price relatives for each item in 2009 using 2007 as the base period.
b. Compute an unweighted aggregate price index for the two items in 2009 using 2007 as the base period.
c. Compute a weighted aggregate price index for the two items using the Laspeyres method.
d. Compute a weighted aggregate price index for the two items using the Paasche method.

2 An item with a price relative of 132 cost €10.75 in 2009. The base year was 2000.

a. What was the percentage increase or decrease in cost of the item over the 9-year period?
b. What did the item cost in 2000?

Applications

3 The average selling prices for an apartment in Yorkshire, England for the years 2007 and 2008, by quarter, were as follows (Source: Nationwide Building Society).

Quarter & Year	Price (£)
Q1 2007	132 383
Q2 2007	133 063
Q3 2007	133 762
Q4 2007	131 401
Q1 2008	121 205
Q2 2008	121 647
Q3 2008	116 224
Q4 2008	121 916

a. Use Q1 2007 as the base period and construct a price index for apartments in Yorkshire over this two-year period.

b. Use Q1 2008 as the base year and construct a price index for apartments in Yorkshire over this two-year period.

4 A large manufacturer purchases an identical component from three independent suppliers that differ in respect of unit price and quantity supplied. The relevant data for 2007 and 2009 are given here.

Supplier	Quantity (2007)	Unit price (€)	
		2007	2009
A	150	5.45	6.00
B	200	5.60	5.95
C	120	5.50	6.20

a. Compute the 2009 price relatives for each of the component suppliers separately, using 2007 as the base year. Compare the price increases by the suppliers over the two-year period.

b. Compute an unweighted aggregate price index for the component part in 2009, using 2007 as the base year.

c. Compute a 2009 weighted aggregate price index for the component part, using 2007 as the base year. What is the interpretation of this index value for the manufacturing firm?

5 Tipple Limited provides a complete range of beer, wine and soft drink products for distribution through retail outlets in Ireland. Below are unit price data for 2005 and 2009, and quantities sold (cases of 12 one-litre bottles) for 2005.

Product	2005 quantity (cases)	Unit price (€)	
		2005	2009
Beer	35 000	21.50	23.25
Wine	5 000	85.70	91.50
Soft drink	60 000	14.00	14.30

Compute a weighted aggregate price index for Tipple Limited products in 2009, with 2005 as the base period.

6   The data below refer to four products produced by a company. Under the LIFO (Last In, First Out) inventory valuation method, a price index for inventory must be established for tax purposes with quantity weights based on year-ending inventory levels. Using the beginning-of-the-year price per unit as the base-period price, construct a weighted aggregate price index for total inventory at the end of the year. What type of index number is this?

Product	Year-end inventory	Unit price (€) Beginning	Ending
A	500	0.15	0.19
B	50	1.60	1.80
C	100	4.50	4.20
D	40	12.00	12.20

7   Exports Europe Limited has the following data on quantities exported and unit costs of shipping for each of its four products:

Products	Quantities exported in 2005	Mean freight cost per unit (€) 2005	2009
A	2000	0.50	15.90
B	5000	6.25	32.00
C	6500	2.20	17.40
D	2500	20.00	35.50

Using 2005 as the base year, compute a Laspeyres price index that reflects the freight cost change over the four-year period.

8   With 2005 as the base year, use the price data in exercise 7 to compute a Paasche index for the freight cost in 2009, if 2009 quantities are 4000, 3000, 7500 and 3000 for each of the four products.

17.3 Computing an aggregate price index from price relatives

In Section 17.1 we defined the concept of a price relative and showed how a price relative can be computed with knowledge of the current-period unit price and the base-period unit price. We now want to show how aggregate price indexes like the ones developed in Section 17.2 can be computed directly from information about the price relative of each item in the group. Because of the limited use of unweighted indexes, we restrict our attention to weighted aggregate price indexes. Let us return to the motoring costs index of the preceding section. The relevant information for the four items is given in Table 17.5.

Let w_i be the weight applied to the price relative for item i. The general expression for a weighted average of price relatives is given by

Table 17.5 Price relatives for motoring costs index

	Unit Price (€)			
	2000	2008	Price relative	Annual usage
Item	(P_0)	(P_t)	$(P_t/P_0)100$	
Litre of petrol	0.85	1.05	123.5	1200
One tyre	75.00	50.00	66.7	2
One hour labour	55.00	60.00	109.1	5
Insurance policy for one year	450.00	475.00	105.6	1

Weighted average of price relatives

$$I_t = \frac{\sum \frac{P_i}{P_0}(100)w_i}{\sum w_i} \tag{17.6}$$

The proper choice of weights in equation (17.6) will enable us to compute a weighted aggregate price index from the price relatives. The proper choice of weights is given by multiplying the base-period price by the quantity of usage.

Weighting factor for equation (17.6)

$$w_i = P_{i0}Q_i \tag{17.7}$$

We must be sure that prices and quantities are in corresponding units. For example, if prices are per case, quantity must be the number of cases and not, for instance, the number of individual units.

Substituting $w_i = P_{i0}Q_i$ into equation (17.6) provides the following expression for a weighted price relatives index.

$$I_t = \frac{\sum \frac{P_{it}}{P_{i0}}(100)(P_{i0}Q_i)}{\sum P_{i0}Q_i} \tag{17.8}$$

With the cancelling of the P_{i0} terms in the numerator, an equivalent expression for the weighted price relatives index is

$$I_t = \frac{\sum P_{it}Q_i}{\sum P_{i0}Q_i}(100)$$

We see that the weighted price relatives index with $w = P_{i0}Q_i$ provides a price index identical to the weighted aggregate index presented by equation (17.3) in Section 17.2. Use of base-period quantities (i.e. $Q_i = Q_{i0}$) in equation (17.7) leads to a Laspeyres index.

Table 17.6 Motoring costs index (2000–2008) based on weighted price relatives

Item	Price relatives $(P_t/P_0)100$	Base price (€) P_{i0}	Quantity Q_i	Weight $w_i = P_{i0}Q_i$	Weighted price relatives $(P_{it}/P_{i0})(100)w_i$
Petrol	123.5	0.85	1200	1020	125 970
Tyres	66.7	75.00	2	150	10 005
Road tax	109.1	55.00	5	275	30 002.5
Insurance	105.6	450.00	1	450	47 520
			Totals	1895	213 497.5

$$I_{2008} = \frac{213\ 497.5}{1895} = 112.7$$

We return to the motoring costs data, and use the price relatives in Table 17.5 and equation (17.6) to compute a weighted average of price relatives. The results obtained by using the weights specified by equation (17.7) are reported in Table 17.6. The index number 112.7 represents a 12.7 per cent increase in overall motoring costs, the same increase as identified by the weighted aggregate index computation in Section 17.2.

Exercises

Methods

9   Price relatives for three items, along with base-period prices and usage, are shown in the following table. Compute a weighted aggregate price index for the current period.

Item	Current period price relative	Base period Price (€ per unit)	Usage (number of units)
A	150	22.00	20
B	90	5.00	50
C	120	14.00	40

Applications

10 The Europa Chemical Company produces a special industrial chemical that is a blend of three chemical ingredients. The beginning-of-year cost per pound, the end-of-year cost per kilogram and the blend proportions follow.

Ingredient	Cost per kilogram (€) Beginning	Ending	Quantity (kg) per 100 kg of product
A	2.50	3.95	25
B	8.75	9.90	15
C	0.99	0.95	60

a. Compute a price relative for end-of-year, using beginning-of-year as base period, for each of the three ingredients.

b. Compute a weighted average of the price relatives to develop an end-of-year cost index for raw materials used in the product. What is your interpretation of this index value?

11 An investment portfolio consists of shares in four companies. The purchase price, current price, and number of shares are reported in the following table.

Company	Purchase price per share (£)	Current price per share (£)	Number of shares
Euro Leisure	1.55	1.70	5000
UK Utilities	1.85	2.05	2000
Scand Gas	2.65	2.60	5000
PQ Domestic	4.25	4.55	3000

Construct a weighted average of price relatives as an index of the performance of the portfolio to date. Interpret this price index.

12 Compute the price relatives for the Tipple Limited products in exercise 5. Use a weighted average of price relatives to show that this method provides the same index as the weighted aggregate method.

13 Using 2005 as the base year, compute the 2009 price relatives for the four products making up the index in exercise 7. Use the weighted aggregates of price relatives method to compute a value for the 2009 index.

17.4 Some important price index numbers

In this section, we consider some price indexes that are important measures of business and economic conditions.

Consumer price index and retail prices index

The Statistics In Practice section of this chapter has already made reference to the UK **Consumer Price Index** (CPI) and Retail Prices Index (RPI), both of which are used as measures of inflation. The group of items used to develop the two index numbers consists of a *market basket* of about 650 items including food, clothing, transportation and entertainment. The RPI includes housing costs, but the CPI does not. They are weighted aggregate price indices with weights that are fixed for a calendar year and revised annually. The CPI is the UK's implementation of the Harmonised Index of Consumer Prices (HICP), a set of consumer price index numbers produced to common standards by countries in the European Union.

Producer price index numbers

The UK Office for National Statistics also produces, on a monthly basis, a set of **Producer Price Index (PPI)** numbers, which measure the price changes of goods bought and sold by UK manufacturers. Input price indices measure changes in the prices of materials bought by manufacturers for processing. Output price indices have a similar aim for the prices of goods produced by UK manufacturers, often known as 'factory gate' prices. PPIs are a further important tool in the monitoring of trends in price inflation, because increases in factory gate prices are usually, in due course, passed on to consumers in the form of increases in retail prices.

Share price index numbers

In free-market economies, stock market activity is an important component of economic activity, and the movements of share price indices are frequently reported in the media, particularly when there are sudden rises or falls in the market. In the UK the two most commonly quoted share price indices are the *FTSE 100 index* and the *FTSE All-Share index.*

The FTSE 100 index gives an indication of the average change in price of shares of the top 100 companies (judged by market value) quoted on the London Stock Exchange. The FTSE All-Share Index aims to track the average movement of the UK stock market as a whole. Both are Laspeyres-type index numbers, with the weighting patterns (including the issue of which companies to include in the index) reviewed regularly.

Because of the dependence of markets globally on the US economy, there is also frequent reporting of movements in the Dow Jones Industrial Average (DJIA), an index number designed to show price trends and movements on the New York Stock Exchange. The DJIA is based on stock prices of 30 large US companies. The DJIA is unusual in being an unweighted price index (share price indices of this type are often referred to as 'price-weighted' index numbers). You will also see and hear references in the media to share indices that track the Japanese stock market (Nikkei indices) and the Hong Kong stock market (Hang Seng indices).

17.5 Deflating a series using a price index number

Many business and economic series reported over time, such as company sales, industry sales and inventories, are measured in monetary amounts. These time series often show an increasing growth pattern over time, which is generally interpreted as indicating an increase in the physical volume associated with the activities. For example, if sales in a high street store increase from one year to the next by 10 per cent, measured in euros or in pounds sterling, this might be interpreted to mean that the amount of stock sold was 10 per cent larger.

Such interpretations can be misleading if a time series is measured in terms of euros or pounds, and the total monetary amount is a combination of both price and quantity changes. Hence, in periods when price changes are significant, the changes in monetary amounts may not be indicative of quantity changes unless we are able to adjust the time series to eliminate the price change effect.

For example, from 1980 to 2000, the total amount spent on beer in the UK increased by about 220 per cent. That figure suggests a staggering increase (no pun intended) in beer consumption. However, beer prices were increasing even faster: by a total of about 270 per cent over the period. To make a correct interpretation for beer consumption over the 1980–2000 period, we must adjust the total spending series by a price index to remove the price increase effect. When we remove the price increase effect from a time series, we are *deflating the series*. The deflated series is expressed in *constant prices*.

In relation to personal income and wages, we often hear discussions about issues such as 'real wages' or the 'purchasing power' of wages. These concepts are based on the notion of deflating an hourly wage or annual salary. For example, Figure 17.1 shows

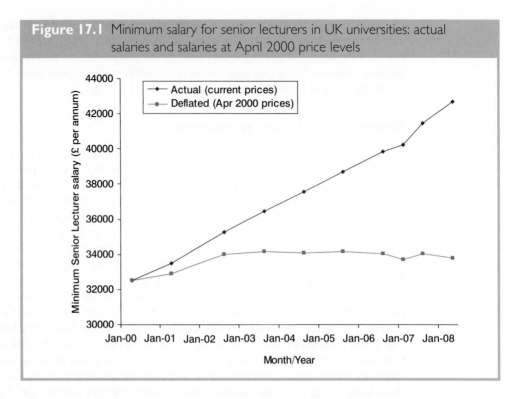

Figure 17.1 Minimum salary for senior lecturers in UK universities: actual salaries and salaries at April 2000 price levels

the minimum point on the senior lecturer salary scale in UK universities for the period 2000–2008 (these figures are for staff in the 'old universities' which had university status pre-1992). The upper line in the chart shows the actual salaries (£ per annum), at current prices. We see a quite steady trend of salary increases from around £32 500 per annum to over £42 500 per annum. Should academic staff on the lowest point of the senior lecturer scale be pleased with this growth in annual salary? The answer depends on what happened to the purchasing power of their salary. If we can compare the purchasing power of the £32 500 annual salary in 2000 with the purchasing power of the £42 700 annual salary in 2008, we shall be better able to judge the relative improvement in salary.

Table 17.7 reports both the annual salaries and the RPI for the period 2000–2008. With these data, we will show how the RPI can be used to deflate the annual salaries. We calculate the deflated series by dividing the annual salary in each year by the corresponding

Table 17.7 UK senior lecturer salaries and retail prices index: 2000–2008

Month/year	Minimum SL annual salary (£)	RPI (Jan 1987 = 100)
April 2000	32 510	170.1
April 2001	33 485	173.1
August 2002	35 251	176.4
August 2003	36 464	181.6
August 2004	37 558	187.4
August 2005	38 685	192.6
August 2006	39 846	199.2
February 2007	40 244	203.1
August 2007	41 451	207.3
May 2008	42 695	215.1

Table 17.8 Deflated series of UK senior lecturer salaries: 2000–2008

Month/year	Deflated annual salary (April 2000 £)
April 2000	(32 510/170.1)(170.1) = 32 510
April 2001	(33 485/173.1)(170.1) = 32 905
August 2002	(35 251/176.4)(170.1) = 33 992
August 2003	(36 464/181.6)(170.1) = 34 155
August 2004	(37 558/187.4)(170.1) = 34 091
August 2005	(38 685/192.6)(170.1) = 34 166
August 2006	(39 846/199.2)(170.1) = 34 025
February 2007	(40 244/203.1)(170.1) = 33 705
August 2007	(41 451/207.3)(170.1) = 34 013
May 2008	(42 695/215.1)(170.1) = 33 763

value of the RPI, and multiplying the result by 170.1, the value of the RPI in April 2000. The deflated annual salaries, at April 2000 prices, are given in Table 17.8.

What does the deflated series of annual salaries tell us about the real salaries or purchasing power of academic staff on the lowest point of the senior lecturer scale during the 2000–2008 period? The deflated salary values are also plotted in Figure 17.1 (lower line on the chart). In terms of April 2000 pounds, the annual salary increased by 3.85 per cent over the period, compared with a rise of 31.3 per cent in the actual salaries expressed at current prices. The increases in constant-price salaries over the period are substantially smaller than the increases in current-price salaries. The salary in constant prices has, in effect, shown no increase since 2002. The advantage of using price index numbers to deflate a series is that they give us a clearer picture of the real purchasing power changes that are occurring.

This process of deflating a series measured over time has an important application in the computation of the gross domestic product (GDP). The GDP is the total value of all goods and services produced in a given country. Obviously, over time the GDP will show gains that are in part due to price increases if the GDP is not deflated by a price index. Therefore, to adjust the total value of goods and services to reflect actual changes in the volume of goods and services produced and sold, the GDP must be computed with a price index deflator. The process is similar to that discussed in the real wages computation.

Exercises

Applications

14 On 1 April 1999 when the UK National Minimum Wage was introduced, the minimum rate for workers aged 22 and over was £3.60 per hour. In October 2008, the rate was £5.73 per hour. The RPI in April 1999 was 165.2; in October 2008 it was 217.7 (1 January 1987 = 100).

a. Deflate the hourly wage rates in 1999 and 2008 to find the real wage rates at January 1987 price levels.

b. What is the percentage change from 1999 to 2008 in actual minimum hourly wage rates?

c. What is the percentage change from 1999 to 2008 in real wage rates at January 1987 price levels?

15 Statistics Sweden report the following exports of motor vehicles from Sweden over the period 2000 to 2008, and a relevant Export Price Index (EPI) for motor vehicles.

Year	Exports (Swedish Krone, millions)	EPI (1990 = 100)
2000	97.69	123.7
2001	100.26	128.5
2002	100.14	128.5
2003	113.77	121.7
2004	128.95	115.1
2005	133.65	115.0
2006	146.55	116.4
2007	154.27	116.9
2008	143.02	119.1

Use the Export Price Index figures to deflate the export values and comment on the pattern shown by the 'real' export values.

16 Marks & Spencer UK retail sales figures for 2000 to 2008, in millions of pounds sterling, are shown in the following table. Also shown are annual values for the RPI, based at 1 January 1987. Deflate the Marks & Spencer sales figures on the basis of 1987 constant pounds sterling, and comment on the firm's sales volumes in terms of deflated pounds.

Year	Retail sales (£ million)	RPI (1 Jan 1987 = 100)
2000	6483	170.3
2001	6293	173.4
2002	6575	176.2
2003	7027	181.3
2004	7160	186.7
2005	7035	192.0
2006	7275	198.1
2007	7978	206.6
2008	8309	214.8

17 Annual salaries in England and Wales for school teachers on the bottom point of the 'leadership' scale were as shown in the following table. Use the CPI to deflate the salary data to constant values. Comment on the trend in school teachers' salaries in England and Wales as indicated by these data.

Year	Salary (£)	CPI (base year 2005)
2000	28 158	93.1
2001	29 499	94.2
2002	30 531	95.4
2003	31 416	96.7
2004	32 202	98.0
2005	33 249	100.0
2006	34 083	102.3
2007	34 938	104.7
2008	35 794	108.5

17.6 Price index numbers: other considerations

In the preceding sections we described several methods used to compute price index numbers, discussed the use of some important indices, and presented a procedure for using price index numbers to deflate a time series. Several other issues must be considered to enhance our understanding of how price index numbers are constructed and how they are used.

Selection of items

The primary purpose of a price index is to measure the price change over time for a specified class of items, products and so on. Whenever the class of items is very large, the index cannot be based on all items in the class. Rather, a sample of representative items must be used. By collecting price information for the sampled items, we hope to obtain a good idea of the price behaviour of all items that the index is representing.

For example, in the UK Retail Price Index (RPI) and Consumer Price Index (CPI) the total number of items that might be considered in the population of items bought by consumers could run into the thousands. However, the RPI and CPI are based on the price movements of about 650 representative items. The selection of these representative items is not a trivial task. Surveys of user purchase patterns as well as good judgment go into the selection process. A simple random sample is not used to select the 650 items. The group of representative items is regularly reviewed and revised to take account of changing consumer purchase patterns.

Selection of a base period

Most index numbers are established with a base-period value of 100 either at some specific time point or over a specific time period. All future values of the index are then related to the base-period value. For example, the current base period for the UK RPI is 1 January 1987. The base period for the UK Index of Retail Sales Volume is the year 2000. Although as a general guideline the base period should not be too far from the current period, choosing a base period is not normally a critical issue, because interest in an index number is usually focused on percentage changes in the index value from period to period, rather than on its actual level.

Quality changes

The purpose of a price index is to measure changes in prices over time. Ideally, price data are collected for the same set of items at different times and then the index is computed. A basic assumption is that the prices are identified for the same items in each period. A problem is encountered when a product changes in quality from one period to the next. For example, a manufacturer may alter the quality of a product by using less expensive materials, fewer features and so on, from year to year. The price may stay constant, but the price is for a lower-quality product and ideally the price index would take this into account. Similarly, a substantial quality improvement may cause an increase in the price of a product, and ideally the portion of the price related to the quality improvement should be excluded from the index computation. In some situations, however, a substantial improvement in quality is followed by a decrease in the price. This less typical situation has been the case with personal computers and other electronic equipment during the 1990s and the first decade of the 2000s.

It is difficult to adjust an index for changes in the quality of an item. Although common practice is to ignore minor quality changes in developing a price index, major quality changes must be addressed, and that is one of the aims of periodic review. The item might be deleted from the index, and replaced by a suitable alternative. There is also a technique known as 'hedonic regression' that is used to take account of quality changes.

17.7 Quantity index numbers

In addition to the price index numbers described in the preceding sections, other types of index numbers are useful. In particular, one other application of index numbers is to measure changes in quantity levels over time. This type of index is called a **quantity index**.

Recall that in the development of the weighted aggregate price index in Section 17.2, to compute an index number for period t we needed data on unit prices at a base period (P_0) and period t (P_t). Equation (17.3) provided the weighted aggregate price index as

$$I_t = \frac{\Sigma P_{it} Q_i}{\Sigma P_{i0} Q_i} (100)$$

The numerator, $\Sigma P_{it} Q_i$, represents the total value of fixed quantities of the index items in period t. The denominator, $\Sigma P_{i0} Q_i$, represents the total value of the same fixed quantities of the index items in period 0.

Computation of a weighted aggregate quantity index is similar to that of a weighted aggregate price index. Quantities for each item are measured in the base period and period t, with Q_{i0} and Q_{it}, respectively, representing those quantities for item i. The quantities are then weighted by a fixed price, the value added, or some other factor. The 'value added' to a product is the sales value minus the cost of purchased inputs. The formula for computing a weighted aggregate quantity index for period t is

Weighted aggregate quantity index

$$I_t = \frac{\Sigma Q_{it} w_i}{\Sigma Q_{i0} w^i} (100) \qquad \text{(17.9)}$$

In some quantity index numbers the weight for item i is taken to be the base-period price (P_{i0}), In which case the weighted aggregate quantity index is

$$I_t = \frac{\Sigma Q_{it} P_{i0}}{\Sigma Q_{i0} P_{i0}} (100) \qquad \text{(17.10)}$$

Quantity index numbers can also be computed on the basis of weighted quantity relatives. one formula for this version of a quantity index follows.

$$I_t = \frac{\Sigma \dfrac{Q_{it}}{Q_{i0}} (Q_{i0} P_i)}{\Sigma Q_{i0} P_i} (100) \qquad \text{(17.11)}$$

This formula is the quantity version of the weighted price relatives formula developed in Section 17.3, as in equation (17.8).

A number of quantity index numbers feature regularly in news bulletins about the UK economy. For example, the Index of Production is a quantity index designed to measure changes in the overall volume of production in manufacturing industries. The base period is the year 2001. The Index of Retail Sales Volume is an index number, base period 2000, that aims to track changes in the level of aggregate retail sales, corrected for any price changes that may occur.

Exercises

Methods

18 Data on quantities of three items sold in 2005 and 2009 are given here along with the sales prices of the items in 2005. Compute a weighted aggregate quantity index for 2009.

	Quantity sold		
Item	2005	2009	Price per unit 2005 (€)
A	350	300	18.00
B	220	400	4.90
C	730	850	15.00

Applications

19 A freight company handles four commodities for a particular distributor. Total shipments for the commodities in 2000 and 2009, measured in terms of standard containers, as well as the 2000 prices, are reported in the following table.

	Containers shipped		
Commodity	2000	2009	Price per shipment 2000 (€)
A	120	95	1200
B	86	75	1800
C	35	50	2000
D	60	70	1500

Calculate a weighted aggregate quantity index number for 2009 with a 2000 base. Comment on the growth or decline in shipments over the 2000–2009 period.

20 A car dealer reports the 2005 and 2008 sales for three models in the following table. Compute quantity relatives and use them to develop a weighted aggregate quantity index for 2008 using the two years of data.

	Sales		
Model	2005	2008	Mean price per sale 2005 (£)
Two-door hatchback	200	170	15 200
Two-door cabriolet	100	80	17 000
Four-door saloon	75	60	16 800

21 A major manufacturing company reports the quantity and product value information for 2005 and 2009 in the table that follows. Compute a weighted aggregate quantity index for the data. Comment on what this quantity index means.

	Quantities		
Product	2005	2009	Value (£)
A	800	1200	30.00
B	600	500	20.00
C	200	500	25.00

For additional online summary questions and answers go to the companion website at www.cengage.co.uk/aswsbe2

Summary

Price and quantity index numbers are important measures of changes in price and quantity levels within the business and economic environment. Price relatives are simply the ratio of the current unit price of an item to a base-period unit price, multiplied by 100, with a value of 100 indicating no difference in the current and base-period prices.

Aggregate price index numbers are created as a composite measure of the overall change in prices for a given group of items or products. Usually the items in an aggregate price index are weighted by their quantity of usage. A weighted aggregate price index can also be computed by weighting the price relatives by the usage quantities for the items in the index. The most common type of weighted price index is a Laspeyres index, which uses base period weighting. A Paasche index uses current period weighting.

The Consumer Price Index and the Retail Prices Index are both widely used index numbers in the UK. The FTSE 100 share index is another widely quoted price index, along with share price indices for stock markets elsewhere, like the Dow Jones average (US), the Nikkei index (Japan) and the Hang Seng index (Hong Kong).

Price index numbers are used to deflate other economic series reported over time. We saw how the RPI could be used to deflate hourly wages to obtain an index of real wages.

Selection of the items to be included in the index, selection of a base period for the index, and adjustment for changes in quality are important additional considerations in the development of an index number.

Quantity index numbers were briefly discussed, and the Index of Production and Index of Retail Sales Volume were mentioned as important quantity indices.

Key terms

Aggregate price index

Consumer Price Index

Laspeyres price index

Paasche price index

Price relative

Producer Price Index

Quantity index

Weighted aggregate price index

Key formulae

Price relative in period t

$$\text{Price relative in period } t = \frac{\text{Price in period } t}{\text{Base period price}}(100) \tag{17.1}$$

Unweighted aggregate price index in period t

$$I_t = \frac{\Sigma P_{it}}{\Sigma P_{i0}}(100) \tag{17.2}$$

Weighted aggregate price index in period t

$$I_t = \frac{\Sigma P_{it} Q_i}{\Sigma P_{i0} Q_i}(100)$$

(17.3)

Weighted average of price relatives

$$I_t = \frac{\Sigma \frac{P_i}{P_0}(100)w_i}{\Sigma w_i}$$

(17.6)

Weighting factor for equation (17.6)

$$w_i = P_{i0} Q_i$$

(17.7)

Weighted aggregate quantity index

$$I_t = \frac{\Sigma Q_{it} w_i}{\Sigma Q_{i0} w_i}(100)$$

(17.9)

Case problem Indices

The file 'Indices', on the CD that accompanies the text, has monthly values of consumer price indices for five countries over the mid-1998 to mid-2008 period. The five countries are the UK, US, Germany, Turkey and South Africa. The file also contains monthly values for leading share price indices for the stock markets in each of these five countries. The share price indices are: the FTSE All-Share index (UK), the Dow Jones Wilshire 5000 (US), the DAX 200 (Germany), the ISE National 100 (Turkey) and the FTSE/JSE (South Africa). A screenshot of the first few rows from the file is shown below.

INDICES

(Source: Datastream, Thomson Financial)

Analyst's report

1 The consumer price indices for the five countries have different base periods. Re-scale the five CPIs to a common base period, July 1998. Represent the five re-scaled CPI series graphically on a single time-series chart using MINITAB, PASW or EXCEL. Comment on the changes in each of the five CPIs over the ten-year period.

2 Similarly, the share price indices for the five stock markets have different base periods. As with the CPIs, re-scale the five share price indices to a common base period, July 1998. Represent the five re-scaled share price index series on a single time-series chart

| Date | Consumer Price Indices | | | | | Share Price Indices | | | | |
	Germany	South Africa	US	UK	Turkey	DAX 200	FTSE/ JSE	DJ Wilshire 5000	FTSE All Share	ISE National 100
Jul-98	98.50	90.7	94.8	97.58	38.66	4731.3	6564.9	11014.6	2873.6	4435.4
Aug-98	98.28	91.7	94.9	98.01	40.19	4909.6	5653.6	9913.8	2562.2	3351.7
Sep-98	98.07	93.3	95.0	98.44	42.89	5050.3	4483.7	9453.7	2444.9	2199.0
Oct-98	97.96	93.7	95.2	98.44	45.49	5083.0	4926.7	9434.9	2328.4	1985.4
Nov-98	97.96	93.7	95.2	98.55	47.44	5121.4	5270.2	10307.3	2518.4	2115.4
Dec-98	98.07	93.7	95.2	98.76	49.00	5144.9	4814.2	10591.3	2536.8	2436.0
Jan-99	97.85	94.4	95.4	98.23	51.35	5158.4	5233.3	11450.7	2697.4	2482.3
Feb-99	98.07	94.4	95.5	98.33	52.98	5136.7	5533.3	11242.8	2761.6	3095.0
Mar-99	98.07	94.4	95.8	98.87	55.13	5081.0	5989.5	11865.4	2863.5	4047.2

© age fotostock.

using MINITAB, PASW or EXCEL. Comment on the changes in each of the five share price indices over the ten-year period.

3 Deflate each of the five re-scaled share price indices, to July 1998 values, using the CPI for the appropriate country. Represent the five deflated share price index series graphically on a single time-series chart, using MINITAB, PASW or EXCEL. Comment on the changes in the five deflated share price indices over the ten-year period, and draw comparisons with your comments in (2) above.

Chapter 18

Forecasting

Learning objectives

After reading this chapter and doing the exercises you should be able to:

1 Understand that the long-run success of an organization is often closely related to how well management is able to predict future aspects of the operation.

2 Know the various components of a time series.

3 Use smoothing techniques such as moving averages and exponential smoothing.

4 Use the least squares method to identify the trend component of a time series.

5 Understand how the classical time series model can be used to explain the pattern or behaviour of the data in a time series and to develop a forecast for the time series.

6 Be able to determine and use seasonal indices for a time series.

7 Know how regression models can be used in forecasting.

8 Know the definition of the following terms:

time series
forecast
trend component
cyclical component
seasonal component
irregular component
mean squared error
moving averages
weighted moving averages
smoothing constant
seasonal constant

An essential aspect of managing any organization is planning for the future. Indeed, the long-run success of an organization is closely related to how well management is able to anticipate the future and develop appropriate strategies. Good judgment, intuition and an awareness of the state of the economy may give a manager a rough idea or 'feeling' of what is likely to happen in the future. However, converting that feeling into a number that can be used for next quarter's sales volume or next year's raw material cost is difficult. The purpose of this chapter is to introduce several forecasting methods.

Suppose we are asked to provide quarterly forecasts of the sales volume for a particular product during the coming one-year period. Production schedules, raw material purchasing, inventory policies and sales quotas will all be affected by the quarterly forecasts we provide. Consequently, poor forecasts may result in poor planning and hence increased costs for the firm. How should we go about providing the quarterly sales volume forecasts?

We will certainly want to review the actual sales data for the product in past periods. Using these historical data, we can identify the general level of sales and any trend such as an increase or decrease in sales volume over time. A further review of the data might reveal a seasonal pattern such as peak sales occurring in the third quarter of each year and sales volume bottoming out during the first quarter. By reviewing historical data, we can often develop a better understanding of the pattern of past sales, leading to better predictions of future sales for the product.

Historical sales form a time series. A **time series** is a set of observations on a variable measured at successive points in time or over successive periods of time. In this chapter, we will introduce several procedures for analyzing time series. The objective of such analyses is to provide good **forecasts** or predictions of future values of the time series.

Statistics in Practice

Asylum applications

Asylum applications to the UK have been a major concern for the authorities for a number of years (Langham, 2005). In the autumn of 2002 the monthly rate of applicants seeking political asylum in the UK exceeded 7500 for the first time in history. Responding to charges that immigration was running out of control, the Labour

Afghan refugees protesting outside Downing Street at forced removal back to Afghanistan. Photo fusion Picture Library/Alamy.

government of the time introduced a series of initiatives with the aim of drastically reducing the numbers of asylum seekers coming into the country. The effect of these was dramatic, the number of asylum applications halving between October 2002 and September 2003. In a report[1] commissioned by the Home Office subsequently, relevant datasets were checked and analyzed using regression (trend) and correlation analysis to see if the reduction in the number of asylum applications had had a significant impact on other forms of migration. Though no clear connection was found it was accepted that reasons for migration were extremely complex. The report also recognized that government measures to manage down the intake of asylum seekers had played a part in reducing the number of asylum applications.

Source: Langham, Alison (2005) Asylum and migration: A review of Home Office statistics. *Significance*, Vol 2 Issue 2 pp 78–80.

[1] Can be obtained from: http://www.nao.org.uk

Forecasting methods can be classified as quantitative or qualitative. Quantitative forecasting methods can be used when (1) past information about the variable being forecast is available, (2) the information can be quantified and (3) a reasonable assumption is that the pattern of the past will continue into the future. In such cases, a forecast can be developed using a time series method or a causal method.

If the historical data are restricted to past values of the variable, the forecasting procedure is called a *time series method*. The objective of time series methods is to discover a pattern in the historical data and then extrapolate the pattern into the future; the forecast is based solely on past values of the variable and/or on past forecast errors. In this chapter we discuss three time series methods: smoothing (moving averages, weighted moving averages and exponential smoothing), trend projection and trend projection adjusted for seasonal influence.

Causal forecasting methods are based on the assumption that the variable we are forecasting has a cause-effect relationship with one or more other variables. In this chapter we discuss the use of regression analysis as a causal forecasting method. For instance, the sales volume for many products is influenced by advertising expenditures, so regression analysis may be used to develop an equation showing how these two variables are related. Then, once the advertising budget is set for the next period, we could substitute this value into the equation to develop a prediction or forecast of the sales volume for that period. Note that if a time series method was used to develop the forecast, advertising expenditures would not be considered; that is, a time series method would base the forecast solely on past sales.

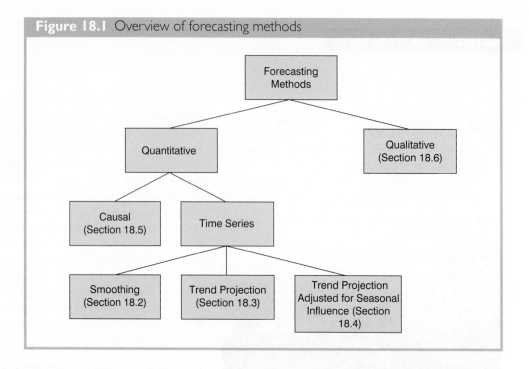

Figure 18.1 Overview of forecasting methods

Qualitative methods generally involve the use of expert judgement to develop forecasts. For instance, a panel of experts might develop a consensus forecast of the base rate for a year from now. An advantage of qualitative procedures is that they can be applied when the information on the variable being forecast cannot be quantified and when historical data are either not applicable or unavailable. Figure 18.1 provides an overview of the types of forecasting methods.

18.1 Components of a time series

The pattern or behaviour of the data in a time series typically involves four separate components – trend, cyclical, seasonal and irregular – which combine to provide specific values for the time series. Let us look more closely at each of these components.

Trend component

Time series analysis measurements may be taken every hour, day, week, month or year, or at any other regular interval.* Although time series data generally exhibit random fluctuations, the time series may still show gradual shifts or movements to relatively higher or lower values over a longer period of time. The gradual shifting of the time series is referred to as the **trend** in the time series; this shifting or trend is usually the result of long-term factors such as changes in the population, demographic characteristics of the population, technology and/or consumer preferences.

---

*We limit our discussion to time series in which the values of the series are recorded at equal intervals. Cases in which the observations are not made at equal intervals are beyond the scope of this text.

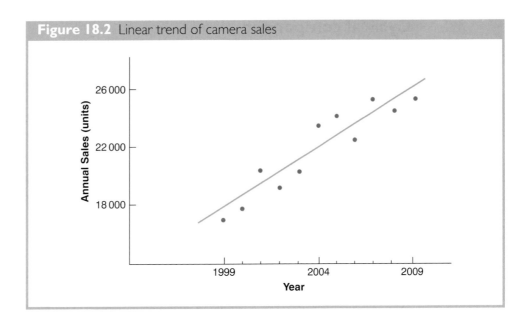

Figure 18.2 Linear trend of camera sales

For example, a manufacturer of photographic equipment may see substantial month-to-month variability in the number of cameras sold. However, in reviewing the sales over the past ten to 15 years, the manufacturer may find a gradual increase in the annual sales volume.

Suppose the sales volume was approximately 17 000 cameras in 1999, 23 000 cameras in 2004 and 25 000 cameras in 2009. This gradual growth in sales over time shows an upward trend for the time series. Figure 18.2 shows a straight line that may be a good approximation of the trend in camera sales. Although the trend for camera sales appears to be linear and increasing over time, sometimes the trend in a time series can be described better by some other patterns.

Figure 18.3 shows some other possible time series trend patterns. Panel A shows a nonlinear trend; in this case, the time series indicates little growth initially, then a period of rapid growth, and finally a levelling off. This trend might be a good approximation of sales for a product from introduction through a growth period and into a period of market saturation.

The linear decreasing trend in panel B is useful for a time series displaying a steady decrease over time. The horizontal line in panel C represents a time series with no consistent increase or decrease over time and thus no trend.

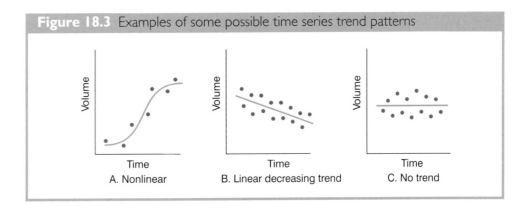

Figure 18.3 Examples of some possible time series trend patterns

Cyclical component

Although a time series may exhibit a trend over long periods of time, all future values of the time series will not necessarily fall on the trend line. In fact, time series often show alternating sequences of points below and above the trend line. Any recurring sequence of points above and below the trend line lasting more than one year can be attributed to the **cyclical component** of the time series. Figure 18.4 shows the graph of a time series with an obvious cyclical component. The observations are taken at intervals one year apart.

Generally, this component of the time series is due to multi-year cyclical movements in the economy. For example, periods of moderate inflation followed by periods of rapid inflation can lead to time series that alternate below and above a generally increasing trend line (e.g. a time series for housing costs).

Seasonal component

Whereas the trend and cyclical components of a time series are identified by analysing multi-year movements in historical data, many time series show a regular pattern over one-year periods. For example, a manufacturer of swimming pools expects low sales activity in the autumn and winter months, with peak sales in the spring and summer months. Manufacturers of snow removal equipment and heavy clothing, however, expect just the opposite yearly pattern. Not surprisingly, the component of the time series that represents the variability in the data due to seasonal influences is called the **seasonal component**. Although we generally think of seasonal movement in a time series as occurring within one year, the seasonal component can also be used to represent any regularly repeating pattern that is less than one year in duration. For example, daily traffic volume data show within-the-day 'seasonal' behaviour, with peak levels occurring during rush hours, moderate flow during the rest of the day and early evening, and light flow from midnight to early morning.

Irregular component

The **irregular component** of the time series is the residual, or 'catch-all', factor that accounts for the deviations of the actual time series values from those expected given the effects of the trend, cyclical and seasonal components. The irregular component is

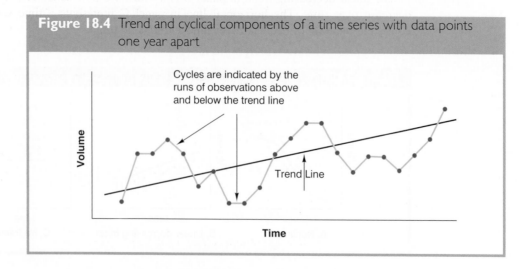

Figure 18.4 Trend and cyclical components of a time series with data points one year apart

caused by the short-term, unanticipated and non-recurring factors that affect the time series. Because this component accounts for the random variability in the time series, it is unpredictable. We cannot attempt to predict its impact on the time series.

18.2 Smoothing methods

In this section we discuss three forecasting methods: moving averages, weighted moving averages, and exponential smoothing. The objective of each of these methods is to 'smooth out' the random fluctuations caused by the irregular component of the time series, therefore they are referred to as smoothing methods. Smoothing methods are appropriate for a stable time series – that is, one that exhibits no significant trend, cyclical or seasonal effects – because they adapt well to changes in the level of the time series. However, without modification, they do not work as well when significant trend, cyclical or seasonal variations are present.

Smoothing methods are easy to use and generally provide a high level of accuracy for short-range forecasts, such as a forecast for the next time period. One of the methods, exponential smoothing, has minimal data requirements and thus is a good method to use when forecasts are required for large numbers of items.

Moving averages

The **moving averages** method uses the average of the most recent n data values in the time series as the forecast for the next period. Mathematically, the moving average calculation is made as follows.

> **Moving average**
>
> $$\text{Moving average} = \frac{\Sigma(\text{most recent } n \text{ data values})}{n} \qquad (18.1)$$

The term *moving* is used because every time a new observation becomes available for the time series, it replaces the oldest observation in equation (18.1) and a new average is computed.

As a result, the average will change, or move, as new observations become available. To illustrate the moving averages method, consider the 12 weeks of data in Table 18.1 and Figure 18.5. These data show the number of litres of petrol sold by a petrol distributor in Sitges, Spain, over the past 12 weeks. Figure 18.5 indicates that, although random variability is present, the time series appears to be stable over time. Hence, the smoothing methods of this section are applicable.

To use moving averages to forecast petrol sales, we must first select the number of data values to be included in the moving average. As an example, let us compute forecasts using a three-week moving average. The moving average calculation for the first three weeks of the petrol sales time series is

$$\text{Moving average (weeks 1–3)} = \frac{17 + 21 + 19}{3} = 19$$

Table 18.1 Petrol sales time series	
Week	Sales (1000s of litres)
1	17
2	21
3	19
4	23
5	18
6	16
7	20
8	18
9	22
10	20
11	15
12	22

We then use this moving average as the forecast for week 4. Because the actual value observed in week 4 is 23, the forecast error in week 4 is $23 - 19 = 4$. In general, the error associated with any forecast is the difference between the observed value of the time series and the forecast.

The calculation for the second three-week moving average is

$$\text{Moving average (weeks 2–4)} = \frac{21 + 19 + 23}{3} = 21$$

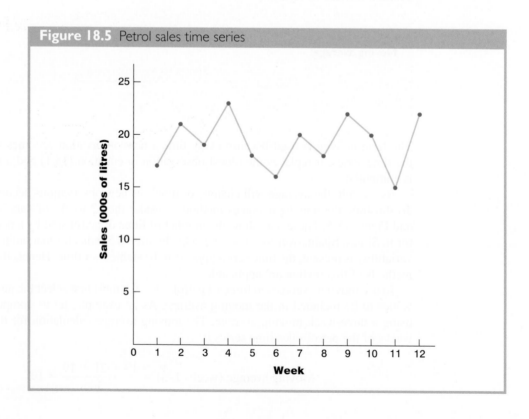

Figure 18.5 Petrol sales time series

Table 18.2 Summary of three-week moving average calculations

Week	Time series value	Moving average forecast	Forecast error	Squared forecast error
1	17			
2	21			
3	19			
4	23	19	4	16
5	18	21	−3	9
6	16	20	−4	16
7	20	19	1	1
8	18	18	0	0
9	22	18	4	16
10	20	20	0	0
11	15	20	−5	25
12	22	19	3	9
			Totals 0	92

Hence, the forecast for week 5 is 21. The error associated with this forecast is $18 - 21 = -3$. Thus, the forecast error may be positive or negative depending on whether the forecast is too low or too high. A complete summary of the three-week moving average calculations for the petrol sales time series is provided in Table 18.2 and Figure 18.6.

Figure 18.6 Petrol sales time series and three-week moving average forecasts

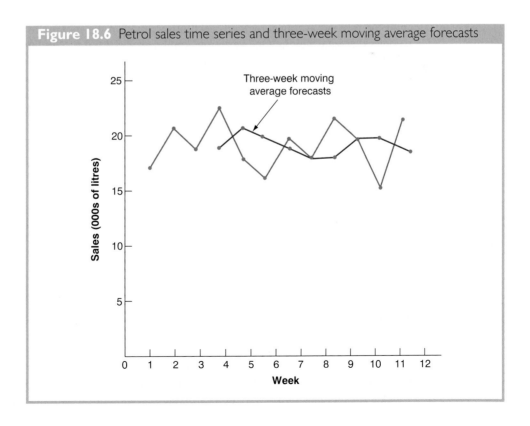

Forecast accuracy

An important consideration in selecting a forecasting method is the accuracy of the forecast. Clearly, we want forecast errors to be small. The last two columns of Table 18.2, which contain the forecast errors and the squared forecast errors, can be used to develop a measure of forecast accuracy.

For the petrol sales time series, we can use the last column of Table 18.2 to compute the average of the sum of the squared errors. Doing so we obtain

$$\text{Average of the sum of squared errors} = \frac{92}{9} = 10.22$$

This average of the sum of squared errors is commonly referred to as the **mean squared error (MSE)**. The MSE is an often-used measure of the accuracy of a forecasting method and is the one we use in this chapter.

As we indicated previously, to use the moving averages method, we must first select the number of data values to be included in the moving average. Not surprisingly, for a particular time series, moving averages of different lengths will differ in their ability to forecast the time series accurately. One possible approach to choosing the number of values to be included in the moving average is to use trial and error to identify the length that minimizes the MSE. Then, if we are willing to assume that the length that is best for the past will also be best for the future, we would could use this to forecast the next value in the time series. Exercise 2 at the end of the section will ask you to consider four-week and five-week moving averages for the petrol sales data. A comparison of the MSEs will indicate the number of weeks of data you may want to include in the moving average calculation.

Weighted moving averages

With the moving averages method, each observation in the moving average calculation receives the same weight. One variation, known as **weighted moving averages**, involves selecting a different weight for each data value and then computing a weighted average of the most recent n values as the forecast. In most cases, the most recent observation receives the most weight, and the weight decreases for older data values. For example, we can use the petrol sales time series to illustrate the computation of a weighted three-week moving average, with the most recent observation receiving a weight three times as great as that given the oldest observation, and the next oldest observation receiving a weight twice as great as the oldest. For week 4 the computation is:

$$\text{Forecast for week 4} = 1/6\,(17) + 2/6\,(21) + 3/6\,(19) = 19.33$$

Note that for the weighted moving average the sum of the weights is equal to 1. Actually the sum of the weights for the simple moving average also equalled 1: each weight was 1/3. However, recall that the simple or unweighted moving average provided a forecast of 19.

Forecast accuracy

To use the weighted moving averages method we must first select the number of data values to be included in the weighted moving average and then choose weights for each of the data values. In general, if we believe that the recent past is a better predictor of the future than the distant past, larger weights should be given to the more recent observations. However, when the time series is highly variable, selecting approximately equal weights for the data values may be best. Note that the only requirement in selecting the weights is that their sum must equal 1. To determine whether one particular combination of number of data values and weights provides a more accurate forecast than another combination, we will continue to use the MSE criterion as the measure of forecast accuracy. That is, if we assume that the combination that is best for the past will also be best for the future, we would use the combination that minimized MSE for the historical time series to forecast the next value in the time series.

Exponential smoothing

Exponential smoothing uses a weighted average of past time series values as the forecast; it is a special case of the weighted moving averages method in which we select only one weight – the weight for the most recent observation. The weights for the other data values are computed automatically and become smaller as the observations move farther into the past. The basic exponential smoothing model follows.

Exponential smoothing model

$$F_{t+1} = \alpha Y_t + (1 - \alpha)F_t \qquad (18.2)$$

where

$$F_{t+1} = \text{forecast of the time series for period } t + 1$$
$$Y_t = \text{actual value of the time series in period } t$$
$$F_t = \text{forecast of the time series for period } t$$
$$\alpha = \text{smoothing constant } (0 \leq \alpha \leq 1)$$

Equation (18.2) shows that the forecast for period $t + 1$ is a weighted average of the actual value in period t and the forecast for period t; note in particular that the weight given to the actual value in period t is α and that the weight given to the forecast in period t is $1 - \alpha$. We can demonstrate that the exponential smoothing forecast for any period is also a weighted average of *all the previous actual values* for the time series with a time series consisting of three periods of data: Y_1, Y_2, and Y_3. To start the calculations, we let F_1 equal the actual value of the time series in period 1; that is, $F_1 = Y_1$. Hence, the forecast for period 2 is

$$F_2 = \alpha Y_1 - (1 - \alpha)F_1$$
$$= \alpha Y_1 - (1 - \alpha)Y_1$$
$$= Y_1$$

Thus, the exponential smoothing forecast for period 2 is equal to the actual value of the time series in period 1.

The forecast for period 3 is

$$F_3 = \alpha Y_2 + (1 - \alpha)F_2 = \alpha Y_2 + (1 - \alpha)Y_1$$

Finally, substituting this expression for F_3 in the expression for F_4, we obtain

$$F_4 = \alpha Y_3 + (1 - \alpha)F_3$$
$$= \alpha Y_3 + (1 - \alpha)[\alpha Y_2 + (1 - \alpha)Y_1]$$
$$= \alpha Y_3 + \alpha(1 - \alpha)Y_2 + (1 - \alpha)^2 Y_1$$

Hence, F_4 is a weighted average of the first three time series values. The sum of the coefficients, or weights, for Y_1, Y_2 and Y_3 equals one. A similar argument can be made to show that, in general, any forecast F_{t+1} is a weighted average of all the previous time series values.

Despite the fact that exponential smoothing provides a forecast that is a weighted average of all past observations, all past data do not need to be saved to compute the forecast for the next period. In fact, once the **smoothing constant** α is selected, only two pieces of information are needed to compute the forecast. Equation (18.2) shows that with a given α we can compute the forecast for period $t + 1$ simply by knowing the actual and forecast time series values for period t – that is, Y_t and F_t.

To illustrate the exponential smoothing approach to forecasting, consider the petrol sales time series in Table 18.1 and Figure 18.5. As indicated, the exponential smoothing forecast for period 2 is equal to the actual value of the time series in period 1. Thus, with $Y_1 = 17$, we will set $F_2 = 17$ to start the exponential smoothing computations. Referring to the time series data in Table 18.1, we find an actual time series value in period 2 of $Y_2 = 21$. Thus, period 2 has a forecast error of $21 - 17 = 4$.

Continuing with the exponential smoothing computations using a smoothing constant of $\alpha = 0.2$, we obtain the following forecast for period 3.

$$F_3 = 0.2Y_2 + 0.8F_2 = 0.2(21) + 0.8(17) = 17.8$$

Once the actual time series value in period 3, $Y_3 = 19$, is known, we can generate a forecast for period 4 as follows.

$$F_4 = 0.2Y_3 + 0.8F_3 = 0.2(19) + 0.8(17.8) = 18.04$$

By continuing the exponential smoothing calculations, we can determine the weekly forecast values and the corresponding weekly forecast errors, as shown in Table 18.3. Note that we have not shown an exponential smoothing forecast or the forecast error for period 1 because no forecast was made. For week 12, we have $Y_{12} = 22$ and $F_{12} = 18.48$. Can we use this information to generate a forecast for week 13 before the actual value of week 13 becomes known? Using the exponential smoothing model, we have

$$F_{13} = 0.2Y_{12} + 0.8F_{12} = 0.2(22) + 0.8(18.48) = 19.18$$

Table 18.3 Summary of the exponential smoothing forecasts and forecast errors for petrol sales with smoothing constant $\alpha = 0.2$

Week (t)	Time series value (Y_t)	Exponential smoothing forecast (F_t)	Forecast error (Y_t, F_t)
1	17		
2	21	17.00	4.00
3	19	17.80	1.20
4	23	18.04	4.96
5	18	19.03	−1.03
6	16	18.83	−2.83
7	20	18.26	1.74
8	18	18.61	−0.61
9	22	18.49	3.51
10	20	19.19	0.81
11	15	19.35	−4.35
12	22	18.48	3.52

Figure 18.7 Actual and forecast petrol sales time series with smoothing constant $\alpha = 0.2$

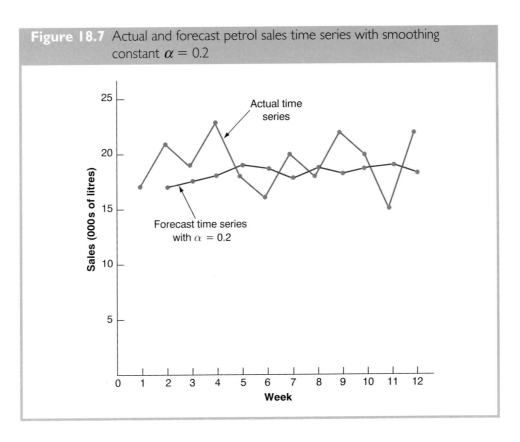

Therefore, the exponential smoothing forecast of the amount sold in week 13 is 19.18, or 19 180 litres of petrol. With this forecast, the firm can make plans and decisions accordingly. The accuracy of the forecast will not be known until the end of week 13.

Figure 18.7 is the plot of the actual and forecast time series values. Note in particular how the forecasts 'smooth out' the irregular fluctuations in the time series.

Forecast accuracy

In the preceding exponential smoothing calculations, we used a smoothing constant of $\alpha = 0.2$. Although any value of α between 0 and 1 is acceptable, some values will yield better forecasts than others. Insight into choosing a good value for α can be obtained by rewriting the basic exponential smoothing model as follows.

$$F_{t+1} = \alpha Y_t + (1 - \alpha)F_t$$
$$F_{t+1} = \alpha Y_t + F_t - \alpha F_t$$
$$F_{t+1} = \underset{\substack{\uparrow \\ \text{Forecast} \\ \text{in Period } t}}{F_t} + \alpha \underset{\substack{\uparrow \\ \text{Forecast error} \\ \text{in period } t}}{(Y_t - F_t)} \tag{18.3}$$

Thus, the new forecast F_{t+1} is equal to the previous forecast F_t plus an adjustment, which is α times the most recent forecast error, $Y_t - F_t$. That is, the forecast in period $t + 1$ is obtained by adjusting the forecast in period t by a fraction of the forecast error. If the time series contains substantial random variability, a small value of the smoothing constant is preferred. The reason for this choice is that, because much of the forecast error is due to random variability, we do not want to over-react and adjust the forecasts too quickly. For a time series with relatively little random variability, larger values of the smoothing

Table 18.4 MSE computations for forecasting petrol sales with $\alpha = 0.2$

Week (t)	Time series value (Y_t)	Forecast (F_t)	Forecast error (Y_t – F_t)	Squared forecast error (Y_t – F_t)²
1	17			
2	21	17.00	4.00	16.00
3	19	17.80	1.20	1.44
4	23	18.04	4.96	24.60
5	18	19.03	–1.03	1.06
6	16	18.83	–2.83	8.01
7	20	18.26	1.74	3.03
8	18	18.61	–0.61	0.37
9	22	18.49	3.51	12.32
10	20	19.19	0.81	0.66
11	15	19.35	–4.35	18.92
12	22	18.48	3.52	12.39
			Total	98.80

$$MSE = \frac{98.80}{11} = 8.98$$

constant provide the advantage of quickly adjusting the forecasts when forecasting errors occur and thus allowing the forecasts to react faster to changing conditions.

The criterion we will use to determine a desirable value for the smoothing constant α is the same as the criterion we proposed for determining the number of periods of data to include in the moving averages calculation. That is, we choose the value of α that minimizes the mean squared error (MSE). A summary of the MSE calculations for the exponential smoothing forecast of petrol sales with $\alpha = 0.2$ is shown in Table 18.4. Note that there is one less squared error term than the number of time periods, because we had no past values with which to make a forecast for period 1. Would a different value of α provide better results in terms of a lower MSE value? Perhaps the most straightforward way to answer this question is simply to try another value for α. We can then compare its mean squared error with the MSE value of 8.98 obtained by using a smoothing constant of $\alpha = 0.2$.

When $\alpha = 0.3$ (see Exercise 4 below) the corresponding MSE obtained can be shown to be 9.35. Thus, we would be inclined to prefer the original smoothing constant of $\alpha = 0.2$. Using a trial-and-error calculation with other values of α, we can find a 'good' value for the smoothing constant. This value can be used in the exponential smoothing model to provide forecasts for the future. At a later date, after new time series observations are obtained, we analyse the newly collected time series data to determine whether the smoothing constant should be revised to provide better forecasting results.

Exercises

Methods

1   Consider the following time series data.

Week	1	2	3	4	5	6
Value	8	13	15	17	16	9

a.   Develop a three-week moving average for this time series. What is the forecast for week 7?

b.   Compute the MSE for the three-week moving average.

c.   Use $\alpha = 0.2$ to compute the exponential smoothing values for the time series. What is the forecast for week 7?

d.   Compare the three-week moving average forecast with the exponential smoothing forecast using $\alpha = 0.2$. Which appears to provide the better forecast?

e.   Use a smoothing constant of 0.4 to compute the exponential smoothing values. Does a smoothing constant of 0.2 or 0.4 appear to provide the better forecast? Explain.

2   Refer to the petrol sales time series data in Table 18.1.

a.   Compute four-week and five-week moving averages for the time series.

b.   Compute the MSE for the four-week and five-week moving average forecasts.

c.   What appears to be the best number of weeks of past data to use in the moving average computation? Remember that the MSE for the three-week moving average is 10.22.

3   Refer again to the petrol sales time series data in Table 18.1.

a.   Using a weight of 1/2 for the most recent observation, 1/3 for the second most recent and 1/6 for third most recent, compute a three-week weighted moving average for the time series.

b.   Compute the MSE for the weighted moving average in part (a). Do you prefer this weighted moving average to the unweighted moving average? Remember that the MSE for the unweighted moving average is 10.22.

c.   Suppose you are allowed to choose any weights as long as they sum to one. Could you always find a set of weights that would make the MSE smaller for a weighted moving average than for an unweighted moving average? Why or why not?

4   Using a smoothing parameter of $\alpha = 0.3$ derive the corresponding results for the petrol time series data to those summarized in Table 18.4. Hence show that the MSE in this case is 9.35.

5   With the petrol time series data from Table 18.1, show the exponential smoothing forecasts using $\alpha = 0.1$. Applying the MSE criterion, would you prefer a smoothing constant of $\alpha = 0.1$ or $\alpha = 0.2$ for the petrol sales time series?

6   With a smoothing constant of $\alpha = 0.2$, equation (18.2) shows that the forecast for week 13 of the petrol sales data from Table 18.1 is given by $F_{13} = 0.2Y_{12} + 0.8F_{12}$. However, the forecast for week 12 is given by $F_{12} = 0.2Y_{11} + 0.8F_{11}$. Thus, we could combine these two results to show that the forecast for week 13 can be written

$$F_{13} = 0.2Y_{12} + 0.8(0.2Y_{11} + 0.8F_{11}) = 0.2Y_{12} + 0.16Y_{11} + 0.64F_{11}$$

a.   Making use of the fact that $F_{11} = 0.2Y_{10} + 0.8F_{10}$ (and similarly for F_{10} and F_9), continue to expand the expression for F_{13} until it is written in terms of the past data values $Y_{12}, Y_{11}, Y_{10}, Y_9, Y_8$, and the forecast for period 8.

b.   Refer to the coefficients or weights for the past values $Y_{12}, Y_{11}, Y_{10}, Y_9, Y_8$; what observation can you make about how exponential smoothing weights past data values in arriving at new forecasts? Compare this weighting pattern with the weighting pattern of the moving averages method.

Applications

7   For the Humfeld Company, the monthly percentages of all shipments received on time over the past 12 months are 80, 82, 84, 83, 83, 84, 85, 84, 82, 83, 84 and 83.

a.   Compare a three-month moving average forecast with an exponential smoothing forecast for $\alpha = 0.2$. Which provides the better forecasts?

b.   What is the forecast for next month?

8 The values of Austrian building contracts (in millions of euros) for a 12-month period follow.

240   350   230   260   280   320   220   310   240   310   240   230

a. Compare a three-month moving averages forecast with an exponential smoothing forecast. Use $\alpha = 0.2$. Which provides the better forecasts?
b. What is the forecast for the next month?

9 The following data represent indices for the seasonally adjusted merchandise trade volumes for New Zealand from 2005–2008.

Year	Quarter	Index	Year	Quarter	Index
2005	Mar	999	2007	Mar	1046
	Jun	998		Jun	1057
	Sep	981		Sep	1052
	Dec	1007		Dec	1157
2006	Mar	993	2008	Mar	1111
	Jun	1004		Jun	1068
	Sep	1062		Sep	1043
	Dec	1005			

a. Compute three- and four-quarter moving averages for this time series. Which moving average provides the better forecast for the fourth quarter of 2008?
b. Plot the data. Do you think the exponential smoothing model would be appropriate for forecasting in this case?

18.3 Trend projection

In this section we show how to forecast a time series that has a long-term linear trend. The type of time series for which the trend projection method is applicable shows a consistent increase or decrease over time; because it is not stable, the smoothing methods described in the preceding section are not applicable.

Consider the time series for bicycle sales of a particular manufacturer over the past ten years, as shown in Table 18.5 and Figure 18.8. Note that 21 600 bicycles were

Table 18.5 Bicycle sales time series

Year (t)	Sales (000s) (Y_t)
1	21.6
2	22.9
3	25.5
4	21.9
5	23.9
6	27.5
7	31.5
8	29.7
9	28.6
10	31.4

BICYCLE

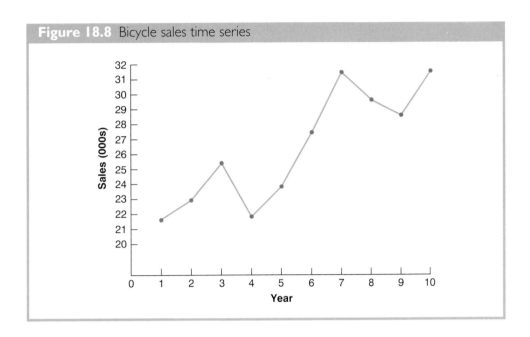

Figure 18.8 Bicycle sales time series

sold in year 1, 22 900 were sold in year 2 and so on. In year ten, the most recent year, 31 400 bicycles were sold. Although Figure 18.8 shows some up and down movement over the past ten years, the time series seems to have an overall increasing or upward trend.

We do not want the trend component of a time series to follow each and every up and down movement. Rather, the trend component should reflect the gradual shifting – in this case, growth – of the time series values. After viewing the time series graph in Figure 18.8, we might agree that a linear trend as shown in Figure 18.9 provides a reasonable description of the long-run movement in the series.

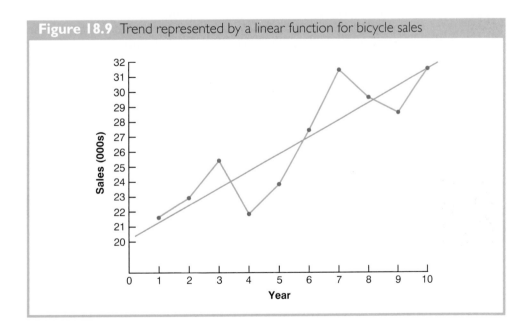

Figure 18.9 Trend represented by a linear function for bicycle sales

We use the bicycle sales data to illustrate the calculations involved in applying regression analysis to identify a linear trend. Recall that in the discussion of simple linear regression in Chapter 14, we described how the least squares method is used to find the best straight-line relationship between two variables. We will use that same methodology to develop the trend line for the bicycle sales time series. Specifically, we will be using regression analysis to estimate the relationship between time and sales volume.

In Chapter 14 the estimated regression equation describing a straight-line relationship between an independent variable x and a dependent variable y was written

$$\hat{y} = b_0 + b_1 x \tag{18.4}$$

To emphasize the fact that, in forecasting, the independent variable is time, we will use t in equation (18.4) instead of x; in addition, we will use T_t in place of $\hat{y}$. Thus, for a linear trend, the estimated sales volume expressed as a function of time can be written as follows.

Equation for linear trend

$$T_t = b_0 + b_1 t \tag{18.5}$$

where

T_t = trend value of the time series in period t
b_0 = intercept of the trend line
b_1 = slope of the trend line
t = time

In equation (18.5), we will let $t = 1$ for the time of the first observation on the time series data, $t = 2$ for the time of the second observation, and so on. Note that for the time series on bicycle sales, $t = 1$ corresponds to the oldest time series value and $t = 10$ corresponds to the most recent year's data. Formulae for computing the estimated regression coefficients (b_1 and b_0) in equation (18.5) follow.

Computing the slope (b_1) and intercept (b_0)

$$b_1 = \frac{\Sigma t Y_t - (\Sigma t \, \Sigma Y_t)/n}{\Sigma t^2 - (\Sigma t)^2/n} \tag{18.6}$$

$$b_0 = \overline{Y} - b_1 \bar{t} \tag{18.7}$$

where

Y_t = value of the time series in period t
n = number of periods
$\overline{Y}$ = average value of the time series; that is, $\overline{Y} = \Sigma Y_t/n$
$\bar{t}$ = average value of t; that is, $\bar{t} = \Sigma t/n$

Using equations (18.6) and (18.7) and the bicycle sales data of Table 18.5, we can compute b_0 and b_1 as follows:

t	Y_t	tY_t	t^2
1	21.6	21.6	1
2	22.9	45.8	4
3	25.5	76.5	9
4	21.9	87.6	16
5	23.9	119.5	25
6	27.5	165.0	36
7	31.5	220.5	49
8	29.7	237.6	64
9	28.6	257.4	81
10	31.4	314.0	100
Totals 55	264.5	1545.5	385

$$\bar{t} = \frac{55}{10} = 5.5$$

$$\bar{Y} = \frac{264.5}{10} = 26.45$$

$$b_1 = \frac{1545.5 - (55)(264.5)/10}{385 - (55)^2/10} = 1.10$$

$$b_0 = 26.45 - 1.10(5.5) = 20.4$$

Therefore,

$$T_t = 20.4 + 1.1t \tag{18.8}$$

is the expression for the linear trend component for the bicycle sales time series.

The slope of 1.1 here indicates that over the past ten years the firm experienced average growth in sales of about 1100 units per year. If we assume that the past ten-year trend in sales is a good indicator of the future, equation (18.8) can be used to project the trend component of the time series. For example, substituting $t = 11$ into equation (18.8) yields next year's trend projection, T_{11}.

$$T_{11} = 20.4 + 1.1(11) = 32.5$$

Thus, using the trend component only, we would forecast sales of 32 500 bicycles next year.

The use of a linear function to model the trend is common. However, as we discussed previously, sometimes time series have a curvilinear, or nonlinear, trend similar to those in Figure 18.10. In Chapter 16 we discussed how regression analysis can be used to model curvilinear relationships of the type shown in panel A of Figure 18.10. More advanced texts discuss in detail how to develop regression models for more complex relationships, such as the one shown in panel B of Figure 18.10.

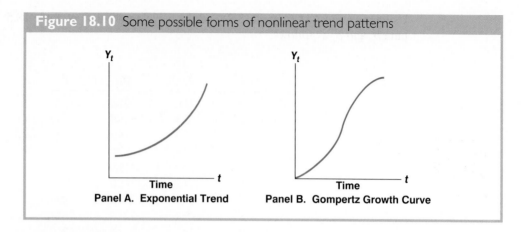

Figure 18.10 Some possible forms of nonlinear trend patterns

Panel A. Exponential Trend

Panel B. Gompertz Growth Curve

Exercises

Methods

10 Consider the following time series.

t	1	2	3	4	5
Y_t	6	11	9	14	15

Develop an equation for the linear trend component of this time series. What is the forecast for $t = 6$?

11 Consider the following time series.

t	1	2	3	4	5	6
Y_t	205	202	195	190	191	188

Develop an equation for the linear trend component for this time series. What is the forecast for $t = 7$?

Applications

12 Car sales at Perez Motors provided the following ten-year time series.

Year	Sales	Year	Sales
1	400	6	260
2	390	7	300
3	320	8	320
4	340	9	340
5	270	10	370

Plot the time series and comment on the appropriateness of a linear trend. What type of functional form do you believe would be most appropriate for the trend pattern of this time series?

13 Numbers of overseas visitors to Ireland (000s) estimated by the Central Statistics Office for the years 2001–2007 are as follows:

2001	2002	2003	2004	2005	2006	2007
5990	6065	6369	6574	6977	7709	8012

a. Graph the data and assess its suitability for linear trend projection.

b. Use a linear trend projection to forecast this time series for 2008–2009.

GDP

14 GDP (Singapore $) for 1990–2007 are tabulated below (*Statistics Singapore, 2009*).

Year	S$	Year	S$
1990	66778	1999	140022
1991	74570	2000	159840
1992	80984	2001	153398
1993	93971	2002	158047
1994	107957	2003	162288
1995	119470	2004	184508
1996	130502	2005	199375
1997	142341	2006	216995
1998	137902	2007	243169

a. Graph this time series. Does a linear trend appear to be present?

b. Develop a linear trend equation for this time series.

c. Use the trend equation to estimate the GDP for the years 2008–2010.

15 Gross revenue data (in millions of euros) for Hispanic Airlines for a ten-year period follow.

Year	Revenue	Year	Revenue
1	2428	6	4264
2	2951	7	4738
3	3533	8	4460
4	3618	9	5318
5	3616	10	6915

a. Develop a linear trend equation for this time series. Comment on what the equation tells about the gross revenue for Hispanic airlines for the ten-year period.

b. Provide the forecasts for gross revenue for years 11 and 12.

18.4 Trend and seasonal components

In this section we extend the discussion by showing how to forecast a time series that has both trend and seasonal components.

Many situations in business and economics involve period-to-period comparisons. We might be interested to learn that unemployment is up 2 per cent compared with last month, that steel production is up 5 per cent over last month or that the production of electric power is down 3 per cent from the previous month. Care must be exercised in using such information, however, because whenever a seasonal influence is present, such comparisons may be misleading. For instance, the fact that electric power consumption is down 3 per cent from August to September might be only the seasonal effect associated with a decrease in the use of air conditioning and not because of a long-term decline in the use of electric power. Indeed, after adjusting for the seasonal effect, we might find that the use of electric power has increased.

Removing the seasonal effect from a time series is known as deseasonalizing the time series. After we do so, period-to-period comparisons are more meaningful and can help identify whether a trend exists. The approach we take in this section is appropriate in situations when only seasonal effects are present or in situations when both seasonal and trend components are present. The first step is to compute seasonal indices and use them to deseasonalize the data. Then, if a trend is apparent in the deseasonalized data, we use regression analysis on the deseasonalized data to estimate the trend component.

Multiplicative model

In addition to a trend component (T) and a seasonal component (S), we will assume that the time series involves an irregular component (I). The irregular component accounts for any random effects in the time series that cannot be explained by the trend and seasonal components. Using T_t, S_t, and I_t to identify the trend, seasonal, and irregular components at time t, we will assume that the time series value, denoted Y_t, can be described by the following **multiplicative time series model**.

Multiplicative time series model with trend, seasonal and irregular components

$$Y_t = T_t \times S_t \times I_t \qquad\qquad (18.9)$$

In this model, T_t is the trend measured in units of the item being forecast. However, the S_t and I_t components are measured in relative terms, with values above 1.00 indicating effects above the trend and values below 1.00 indicating effects below the trend.

We will illustrate the use of the multiplicative model (18.9) by working with the quarterly data in Table 18.6 and Figure 18.11. These data show television set sales (in thousands of units) for a particular manufacturer over the past four years. We begin by showing how to identify the seasonal component of the time series.

Calculating the seasonal indices

Figure 18.11 indicates that sales are lowest in the second quarter of each year and increase in quarters 3 and 4. Thus, we conclude that a seasonal pattern exists for television set sales. The computational procedure used to identify each quarter's seasonal influence begins by computing a moving average to separate the combined seasonal and irregular components, S_t and I_t, from the trend component T_t.

To do so, we use one year of data in each calculation. Because we are working with a quarterly series, we will use four data values in each moving average. The moving average calculation for the first four quarters of the television set sales data is

$$\text{First moving average} = \frac{4.8 + 4.1 + 6.0 + 6.5}{4} = \frac{21.4}{4} = 5.35$$

Table 18.6 Quarterly data for television set sales

Year	Quarter	Sales (000s)
1	1	4.8
	2	4.1
	3	6.0
	4	6.5
2	1	5.8
	2	5.2
	3	6.8
	4	7.4
3	1	6.0
	2	5.6
	3	7.5
	4	7.8
4	1	6.3
	2	5.9
	3	8.0
	4	8.4

Note that the moving average calculation for the first four quarters yields the average quarterly sales over year 1 of the time series. Continuing the moving average calculations, we next add the 5.8 value for the first quarter of year 2 and drop the 4.8 for the first quarter of year 1. Thus, the second moving average is

$$\text{Second moving average} = \frac{4.1 + 6.0 + 6.5 + 5.8}{4} = \frac{22.4}{4} = 5.60$$

Similarly, the third moving average calculation is $(6.0 + 6.5 + 5.8 + 5.2)/4 = 5.875$.

Figure 18.11 Quarterly television set sales time series

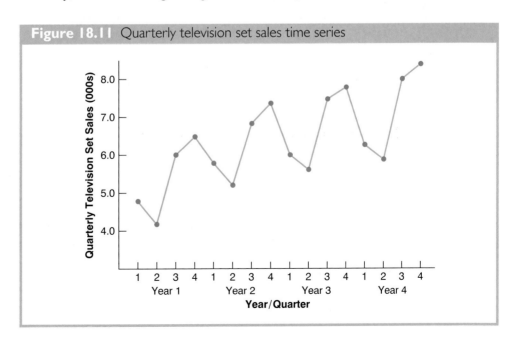

Before we proceed with the moving average calculations for the entire time series, we return to the first moving average calculation, which resulted in a value of 5.35. The 5.35 value represents an average quarterly sales volume (across all seasons) for year 1. As we look back at the calculation of the 5.35 value, associating 5.35 with the 'middle' quarter of the moving average group makes sense. Note, however, that we encounter some difficulty in identifying the middle quarter; with four quarters in the moving average there is no middle quarter. The 5.35 value corresponds to the last half of quarter 2 and the first half of quarter 3. Similarly, if we go to the next moving average value of 5.60, the middle corresponds to the last half of quarter 3 and the first half of quarter 4.

Recall that the reason for computing moving averages is to isolate the combined seasonal and irregular components. However, the moving average values we computed do not correspond directly to the original quarters of the time series. We can resolve this difficulty by using the midpoints between successive moving average values. For example, if 5.35 corresponds to the first half of quarter 3 and 5.60 corresponds to the last half of quarter 3, we can use (5.35 + 5.60)/2 = 5.475 as the moving average value for quarter 3. Similarly, we associate a moving average value of (5.60 + 5.875)/2 = 5.738 with quarter 4. The result is a *centred moving average*. Table 18.7 shows a complete summary of the moving average calculations for the television set sales data.

If the number of data points in a moving average calculation is an odd number, the middle point will correspond to one of the periods in the time series. In such cases, we would not have to centre the moving average values to correspond to a particular time period as we have done in the calculations in Table 18.7.

What do the centred moving averages in Table 18.7 tell us about this time series? Figure 18.12 is a plot of the actual time series values and the centred moving average values. Note particularly how the centred moving average values tend to 'smooth out' both the seasonal and irregular fluctuations in the time series. The moving average

Table 18.7 Centred moving average calculations for the television set sales time series

Year	Quarter	Sales (000s)	Four-quarter moving average	Centred moving average
1	1	4.8		
	2	4.1		
	3	6.0	5.350	5.475
	4	6.5	5.600	5.738
2	1	5.8	5.875	5.975
	2	5.2	6.075	6.188
	3	6.8	6.300	6.325
	4	7.4	6.350	6.400
3	1	6.0	6.450	6.538
	2	5.6	6.625	6.675
	3	7.5	6.725	6.763
	4	7.8	6.800	6.838
4	1	6.3	6.875	6.938
	2	5.9	7.000	7.075
	3	8.0	7.150	
	4	8.4		

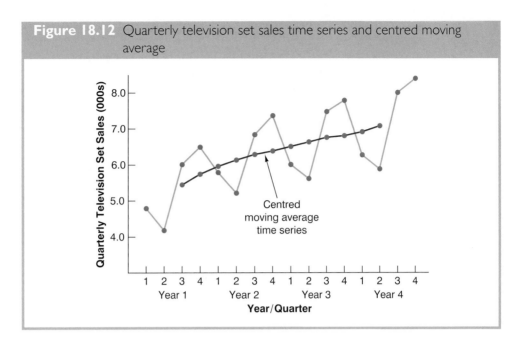

Figure 18.12 Quarterly television set sales time series and centred moving average

values computed for four quarters of data do not include the fluctuations due to seasonal influences because the seasonal effect has been averaged out. Each point in the centred moving average represents the value of the time series as though there were no seasonal or irregular influence.

By dividing each time series observation by the corresponding centred moving average, we can identify the seasonal irregular effect in the time series. For example, the third quarter of year 1 shows 6.0/5.475 = 1.096 as the combined seasonal irregular value. Table 18.8 summarizes the seasonal irregular values for the entire time series.

Consider the third quarter. The results from years 1, 2 and 3 show third-quarter values of 1.096, 1.075 and 1.109, respectively. Thus, in all cases, the seasonal irregular value appears to have an above-average influence in the third quarter. With the year-to-year fluctuations in the seasonal irregular value attributable primarily to the irregular component, we can average the computed values to eliminate the irregular influence and obtain an estimate of the third-quarter seasonal influence.

$$\text{Seasonal effect of third quarter} = \frac{1.096 + 1.075 + 1.109}{3} = 1.09$$

We refer to 1.09 as the *seasonal index* for the third quarter. In Table 18.9 we summarize the calculations involved in computing the seasonal indices for the television set sales time series. Thus, the seasonal indices for the four quarters are: quarter 1, 0.93; quarter 2, 0.84; quarter 3, 1.09; and quarter 4, 1.14.

Interpretation of the values in Table 18.9 provides some observations about the seasonal component in television set sales. The best sales quarter is the fourth quarter, with sales averaging 14 per cent above the average quarterly value. The worst, or slowest, sales quarter is the second quarter; its seasonal index of 0.84 shows that the sales average is 16 per cent below the average quarterly sales. The seasonal component corresponds clearly to the intuitive expectation that television viewing interest and thus television purchase patterns tend to peak in the fourth quarter because of the coming winter season and reduction in outdoor activities. The low second-quarter sales reflect the reduced interest in television viewing due to the spring and pre-summer activities of potential customers.

Table 18.8 Seasonal irregular values for the television set sales time series

Year	Quarter	Sales (000s)	Centred moving average	Seasonal irregular value
1	1	4.8		
	2	4.1		
	3	6.0	5.475	1.096
	4	6.5	5.738	1.133
2	1	5.8	5.975	0.971
	2	5.2	6.188	0.840
	3	6.8	6.325	1.075
	4	7.4	6.400	1.156
3	1	6.0	6.538	0.918
	2	5.6	6.675	0.839
	3	7.5	6.763	1.109
	4	7.8	6.838	1.141
4	1	6.3	6.938	0.908
	2	5.9	7.075	0.834
	3	8.0		
	4	8.4		

One final adjustment is sometimes necessary in obtaining the seasonal indices. The multiplicative model requires that the average seasonal index equal 1.00, so the sum of the four seasonal indices in Table 18.9 must equal 4.00. In other words the seasonal effects must even out over the year. The average of the seasonal indices in our example is equal to 1.00, and hence this type of adjustment is not necessary. In other cases, a slight adjustment may be necessary. To make the adjustment, multiply each seasonal index by the number of seasons divided by the sum of the unadjusted seasonal indices.

For instance, for quarterly data multiply each seasonal index by 4/(sum of the unadjusted seasonal indices). Some of the later exercises will require this adjustment to obtain the appropriate seasonal indices.

Deseasonalizing the time series

The purpose of finding seasonal indices is to remove the seasonal effects from a time series. This process is referred to as *deseasonalizing* the time series. Economic time series adjusted for seasonal variations (**deseasonalized time series**) are often reported

Table 18.9 Seasonal index calculations for the television set sales time series

Quarter	Seasonal irregular component values ($S_t I_t$)	Seasonal index (S_t)
1	0.971, 0.918, 0.908	0.93
2	0.840, 0.839, 0.834	0.84
3	1.096, 1.075, 1.109	1.09
4	1.133, 1.156, 1.141	1.14

Table 18.10 Deseasonalized values for the television set sales time series

Year	Quarter	Sales (000s) (Y_t)	Seasonal index (S_t)	Deseasonalized sales $(Y_t/S_t = T_tI_t)$
1	1	4.8	0.93	5.16
	2	4.1	0.84	4.88
	3	6.0	1.09	5.50
	4	6.5	1.14	5.70
2	1	5.8	0.93	6.24
	2	5.2	0.84	6.19
	3	6.8	1.09	6.24
	4	7.4	1.14	6.49
3	1	6.0	0.93	6.45
	2	5.6	0.84	6.67
	3	7.5	1.09	6.88
	4	7.8	1.14	6.84
4	1	6.3	0.93	6.77
	2	5.9	0.84	7.02
	3	8.0	1.09	7.34
	4	8.4	1.14	7.37

in publications such as the *Financial Times, The Wall Street Journal* and *Euromonitor International.* Using the notation of the multiplicative model, we have

$$Y_t = T_t \times S_t \times I_t$$

By dividing each time series observation by the corresponding seasonal index, we remove the effect of season from the time series. The deseasonalized time series for television set sales is summarized in Table 18.10. A graph of the deseasonalized television set sales time series is shown in Figure 18.13.

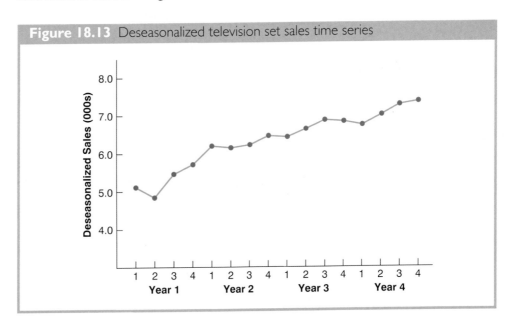

Figure 18.13 Deseasonalized television set sales time series

Using the deseasonalized time series to identify trend

Although the graph in Figure 18.13 shows some random up and down movement over the past 16 quarters, the time series seems to have an upward linear trend. To identify this trend, we will use the same procedure as in the preceding section; in this case, the data are the quarterly deseasonalized sales values. Thus, for a linear trend, the estimated sales volume expressed as a function of time is

$$T_t = b_0 + b_1 t$$

where

T_t = trend value for television set sales in period t
b_0 = intercept of the trend line
b_1 = slope of the trend line

As before, $t = 1$ corresponds to the time of the first observation for the time series, $t = 2$ corresponds to the time of the second observation and so on. Thus, for the deseasonalized television set sales time series, $t = 1$ corresponds to the first deseasonalized quarterly sales value and $t = 16$ corresponds to the most recent deseasonalized quarterly sales value. The formulae for computing the value of b_0 and the value of b_1 follow.

$$b_1 = \frac{\Sigma t Y_t - (\Sigma t \, \Sigma Y_t)/n}{\Sigma t^2 - (\Sigma t)^2/n}$$

$$b_0 = \bar{Y} - b_1 \bar{t}$$

Note, however, that Y_t now refers to the deseasonalized time series value at time t and not to the actual value of the time series. Using the given relationships for b_0 and b_1 and the deseasonalized sales data of Table 18.10, we have the following calculations.

t	Y_t (deseasonalized)	tY_t	t^2
1	5.16	5.16	1
2	4.88	9.76	4
3	5.50	16.50	9
4	5.70	22.80	16
5	6.24	31.20	25
6	6.19	37.14	36
7	6.24	43.68	49
8	6.49	51.92	64
9	6.45	58.05	81
10	6.67	66.70	100
11	6.88	75.68	121
12	6.84	82.08	144
13	6.77	88.01	169
14	7.02	98.28	196
15	7.34	110.10	225
16	7.37	117.92	256
Totals 136	101.74	914.98	1496

where

$$\bar{t} = \frac{136}{16} = 8.5$$

$$\bar{Y} = \frac{101.74}{16} = 6.359$$

$$b_1 = \frac{914.98 - (136)(101.74)/16}{1496 - (136)^2/16} = 0.148$$

$$b_0 = 6.359 - 0.148(8.5) = 5.101$$

Therefore,

$$T_t = 5.101 + 0.148t$$

is the expression for the linear trend component of the time series.

The slope of 0.148 indicates that over the past 16 quarters, the firm averaged deseasonalized growth in sales of about 148 sets per quarter. If we assume that the past 16-quarter trend in sales data is a reasonably good indicator of the future, this equation can be used to project the trend component of the time series for future quarters. For example, substituting $t = 17$ into the equation yields next quarter's trend projection, T_{17}.

$$T_{17} = 5.101 + 0.148(17) = 7.617$$

Thus, the trend component yields a sales forecast of 7617 television sets for the next quarter. Similarly, the trend component produces sales forecasts of 7765, 7913 and 8061 television sets in quarters 18, 19 and 20, respectively.

Seasonal adjustments

The final step in developing the forecast when both trend and seasonal components are present is to use the seasonal index to adjust the trend projection. Returning to the television set sales example, we have a trend projection for the next four quarters. Now we must adjust the forecast for the seasonal effect. The seasonal index for the first quarter of year 5 ($t = 17$) is 0.93, so we obtain the quarterly forecast by multiplying the forecast based on trend ($T_{17} = 7617$) by the seasonal index (0.93). Thus, the forecast for the next quarter is $7617(0.93) = 7084$. Table 18.11 gives the quarterly forecast for quarters 17 through 20. The high-volume fourth quarter has a 9190-unit forecast, and the low-volume second quarter has a 6523-unit forecast.

Table 18.11 Quarterly forecasts for the television set sales time series

Year	Quarter	Trend forecast	Seasonal index (see Table 18.10)	Quarterly forecast
5	1	7617	0.93	(7617)(0.93) = 7084
	2	7765	0.84	(7765)(0.84) = 6523
	3	7913	1.09	(7913)(1.09) = 8625
	4	8061	1.14	(8061)(1.14) = 9190

Models based on monthly data

In the preceding television set sales example, we used quarterly data to illustrate the computation of seasonal indices. However, many businesses use monthly rather than quarterly forecasts. In such cases, the procedures introduced in this section can be applied with minor modifications. First, a 12-month moving average replaces the four-quarter moving average; second, 12 monthly seasonal indices, rather than four quarterly seasonal indices, must be computed. Other than these changes, the computational and forecasting procedures are identical.

Cyclical component

Mathematically, the multiplicative model of equation (18.9) can be expanded to include a cyclical component.

Multiplicative time series model with trend, cyclical, seasonal and irregular components

$$Y_t = T_t \times C_t \times S_t \times I_t$$

(18.10)

The cyclical component, like the seasonal component, is expressed as a percentage of trend. As mentioned in Section 18.1, this component is attributable to multi-year cycles in the time series. It is analogous to the seasonal component, but over a longer period of time. However, because of the length of time involved, obtaining enough relevant data to estimate the cyclical component is often difficult. Another difficulty is that cycles usually vary in length. We leave further discussion of the cyclical component to texts on forecasting methods.

Exercises

Methods

16 Consider the following time series data.

		Year		
Quarter		1	2	3
1		4	6	7
2		2	3	6
3		3	5	6
4		5	7	8

a. Show the four-quarter and centred moving average values for this time series.
b. Compute seasonal indices for the four quarters.

Applications

17 The quarterly sales data (number of copies sold) for a college textbook over the past three years follow.

Quarter	Year 1	Year 2	Year 3
1	1690	1800	1850
2	940	900	1100
3	2625	2900	2930
4	2500	2360	2615

a. Show the four-quarter and centred moving average values for this time series.
b. Compute seasonal indices for the four quarters.
c. When does the textbook publisher have the largest seasonal index? Does this result appear reasonable? Explain.

18 Electric power consumption is measured in kilowatt-hours (kWh). The local utility company offers an interrupt programme whereby commercial customers that participate receive favourable rates but must agree to cut back consumption if the utility requests them to do so. Timko Products cut back consumption at 12:00 noon Thursday. To assess the savings, the utility must estimate Timko's usage without the interrupt. The period of interrupted service was from noon to 8:00 p.m. Data on electric power consumption for the previous 72 hours are available.

Time Period	Monday	Tuesday	Wednesday	Thursday
12–4 a.m.	–	19 281	31 209	27 330
4–8 a.m.	–	33 195	37 014	32 715
8–12 noon	–	99 516	119 968	152 465
12–4 p.m.	124 299	123 666	156 033	
4–8 p.m.	113 545	111 717	128 889	
8–12 midnight	41 300	48 112	73 923	

a. Is there a seasonal effect over the 24-hour period? Compute seasonal indices for the six four-hour periods.
b. Use trend adjusted for seasonal indices to estimate Timko's normal usage over the period of interrupted service.

For additional online summary questions and answers go to the companion website at www.cengage.co.uk/aswsbe2

18.5 Regression analysis

In the discussion of regression analysis in Chapters 14, 15 and 16, we showed how one or more independent variables could be used to predict the value of a single dependent variable. Looking at regression analysis as a forecasting tool, we can view the time series value that we want to forecast as the dependent variable. Hence, if we can identify a good set of related independent, or predictor, variables we may be able to develop an estimated regression equation for predicting or forecasting the time series.

The approach we used in Section 18.3 to fit a linear trend line to the bicycle sales time series is a special case of regression analysis. In that example, two variables – bicycle sales and time – were shown to be linearly related.* The inherent complexity of most real-world problems necessitates the consideration of more than one variable to predict the variable of interest. The statistical technique known as multiple regression analysis can be used in such situations.

Recall that to develop an estimated multiple regression equation, we need a sample of observations for the dependent variable and all independent variables. In time series analysis the n periods of time series data provide a sample of n observations on each variable that can be used in the analysis. For a function involving k independent variables, we use the following notation.

$$Y_t = \text{value of the time series in period } t$$
$$x_{1t} = \text{value of independent variable 1 in period } t$$
$$x_{2t} = \text{value of independent variable 2 in period } t$$

$$x_{kt} = \text{value of independent variable } k \text{ in period } t$$

The n periods of data necessary to develop the estimated regression equation would appear as shown in the following table.

Period	Time series (Y_t)	Value of independent variables						
		x_{1t}	x_{2t}	x_{3t}	.	.	.	x_{kt}
1	Y_1	x_{11}	x_{21}	x_{31}	.	.	.	x_{k1}
2	Y_2	x_{12}	x_{22}	x_{32}	.	.	.	x_{k2}
.	.	.	.	.	.	.	.	.
.	.	.	.	.	.	.	.	.
.	.	.	.	.	.	.	.	.
n	Y_n	x_{1n}	x_{2n}	x_{3n}	.	.	.	x_{kn}

As you might imagine, several choices are possible for the independent variables in a forecasting model. One possible choice for an independent variable is simply time. It was the choice made in Section 18.3 when we estimated the trend of the time series using a linear function of the independent variable time. Letting $x_{1t} = t$, we obtain an estimated regression equation of the form

$$\hat{Y}_t = b_0 + b_1 t$$

where $\hat{Y}_t$ is the estimate of the time series value Y_t and where b_0 and b_1 are the estimated regression coefficients. In a more complex model, additional terms could be added corresponding to time raised to other powers. For example, if $x_{2t} = t^2$ and $x_{3t} = t^3$, the estimated regression equation would become

$$\hat{Y} = b_0 + b_1 x_{1t} + b_2 x_{2t} + b_3 x_{3t}$$
$$= b_0 + b_1 t + b_2 t^2 + b_3 t^3$$

*In a purely technical sense, the number of bicycles sold is not viewed as being related to time; instead, time is used as a surrogate for variables to which the number of bicycles sold is actually related but which are either unknown or too difficult or too costly to measure.

Note that this model provides a forecast of a time series with curvilinear characteristics over time.

Other regression-based forecasting models have a mixture of economic and demographic independent variables. For example, in forecasting sales of refrigerators, we might select the following independent variables.

x_{1t} = price in period t
x_{2t} = total industry sales in period $t - 1$
x_{3t} = number of building permits for new houses in period $t - 1$
x_{4t} = population forecast for period t
x_{5t} = advertising budget for period t

According to the usual multiple regression procedure, an estimated regression equation with five independent variables would be used to develop forecasts.

Whether a regression approach provides a good forecast depends largely on how well we are able to identify and obtain data for independent variables that are closely related to the time series. Generally, during the development of an estimated regression equation, we will want to consider many possible sets of independent variables.

Thus, part of the regression analysis procedure should be the selection of the set of independent variables that provides the best forecasting model.

In the chapter introduction we stated that the **causal forecasting methods** use other time series related to the one being forecast in an effort to explain the cause of a time series' behaviour.

Regression analysis is the tool most often used in developing such causal methods. The related time series become the independent variables, and the time series being forecast is the dependent variable.

Another popular type of regression-based forecasting model is one where the independent variables are all previous values of the same time series. For example, if the time series values are denoted $Y_1, Y_2, \ldots, Y_n$, then with a dependent variable Y_t, we might try to find an estimated regression equation relating Y_t to the most recent times series values Y_{t-1}, Y_{t-2} and so on. With the three most recent periods as independent variables, the estimated regression equation would be

$$\hat{Y}_t = b_0 + b_1 Y_{t-1} + b_2 Y_{t-2} + b_3 Y_{t-3}$$

Regression models in which the independent variables are previous values of the time series are referred to as **autoregressive models**.

More generally, the regression-based forecasting approach incorporates a mixture of the independent variables previously discussed. For example, we might select a combination of time variables, some economic/demographic variables, and some previous values of the time series variable itself.

18.6 Qualitative approaches

In the preceding sections we discussed sexual types of quantitative forecasting methods. Most of those quantitative techniques previously discussed, require historical data on the variable of interest, so they cannot be applied when historical data are not available. Furthermore, even when such data are available, a significant change in environmental conditions can affect the time series making the use of past data questionable for prediction purposes. For example, a government-imposed petrol rationing programme would raise questions about the validity of a petrol sales forecast based on historical data. Qualitative forecasting techniques afford an alternative in these and other cases.

Delphi method

One of the most commonly used qualitative forecasting techniques is the **Delphi method**, originally developed by a research group at the Rand Corporation. It is an attempt to develop forecasts through 'group consensus'. In its usual application, the members of a panel of experts – all of whom are physically separated from and unknown to each other – are asked to respond to a series of questionnaires. The responses from the first questionnaire are tabulated and used to prepare a second questionnaire that contains information and opinions of the entire group. Each respondent is then asked to reconsider and possibly revise his or her previous response in light of the group information provided. This process continues until the coordinator feels that some degree of consensus has been reached. The goal of the Delphi method is not to produce a single answer as output, but instead to produce a relatively narrow spread of opinions within which the majority of experts concur.

Expert judgment

Qualitative forecasts often are based on the judgment of a single expert or represent the consensus of a group of experts. For example, each year groups of experts at Schroders gather to forecast the level of the FTSE Index and the base rate for the next year.

In doing so, the experts individually consider information that they believe will influence the stock market and interest rates; then they combine their conclusions into a forecast. No formal model is used, and no two experts are likely to consider the same information in the same way. Expert judgment is a forecasting method that is commonly recommended when conditions in the past are not likely to hold in the future. Even though no formal quantitative model is used, expert judgment has provided good forecasts in many situations.

Scenario writing

The qualitative procedure known as **scenario writing** consists of developing a conceptual scenario of the future based on a well-defined set of assumptions. Different sets of assumptions lead to different scenarios. The job of the decision-maker is to decide how likely each scenario is and then to make decisions accordingly.

Intuitive approaches

Subjective or *intuitive qualitative* approaches are based on the ability of the human mind to process a variety of information that, in most cases, is difficult to quantify. These techniques are often used in group work, wherein a committee or panel seeks to develop new ideas or solve complex problems through a series of 'brainstorming sessions'. In such sessions, individuals are freed from the usual group restrictions of peer pressure and criticism because they can present any idea or opinion without regard to its relevancy and, even more important, without fear of criticism.

Summary

This chapter provided an introduction to the basic methods of time series analysis and forecasting. First, we acknowledged that the long-run success of an organization is often closely related to how well management is able to predict future aspects of the operation. We then showed that to explain the behaviour of a time series, it is often helpful to think of the time series as consisting of four separate components: trend, cyclical, seasonal and irregular. By isolating these components and measuring their apparent effect, one can forecast future values of the time series.

We discussed how smoothing methods can be used to forecast a time series that exhibits no significant trend, seasonal or cyclical effect. The moving averages approach consists of computing an average of past data values and then using that average as the forecast for the next period. With the exponential smoothing method, a weighted average of past time series values is used to compute a forecast.

For time series that have only a long-term trend, we showed how regression analysis could be used to make trend projections. For time series in which both trend and seasonal influences are significant, we showed how to isolate the effects of the two factors and prepare better forecasts. Specifically we showed how seasonal indices can be determined. Finally, regression analysis was described as a procedure for developing causal forecasting models. A causal forecasting model is one that relates the time series value (dependent variable) to other independent variables that are believed to explain (cause) the time series behaviour.

Qualitative forecasting methods were discussed as approaches that could be used when little or no historical data are available. These methods are also considered most appropriate when the past pattern of the time series is not expected to continue into the future.

Key terms

Autoregressive model
Cyclical component
Causal forecasting methods
Delphi method
Deseasonalized time series
Exponential smoothing
Forecast
Irregular component
Mean squared error (MSE)

Moving averages
Multiplicative time series model
Seasonal component
Scenario writing
Smoothing constant
Time series
Trend
Weighted moving averages

Key formulae

Moving average

$$\text{Moving average} = \frac{\Sigma(\text{most recent } n \text{ data values})}{n}$$

(18.1)

Exponential smoothing model

$$F_{t+1} = \alpha Y_t + (1 - \alpha)F_t \tag{18.2}$$

Equation for linear trend

$$T_t = b_0 + b_1 t \tag{18.5}$$

Multiplicative time series model with trend, seasonal and irregular components

$$Y_t = T_t \times S_t \times I_t \tag{18.9}$$

Multiplicative time series model with trend, cyclical, seasonal and irregular components

$$Y_t = T_t \times C_t \times S_t \times I_t \tag{18.10}$$

Case problem 1 Forecasting food and beverage sales

The Vesuvius Restaurant near Naples, Italy, is owned and operated by Luigi Marconi. The restaurant just completed its third year of operation. During that time, Luigi sought to establish a reputation for the restaurant as a high-quality dining establishment that specializes in fresh seafood. Through the efforts of Luigi and his staff, his restaurant has become one of the best and fastest-growing restaurants in the area.

Luigi believes that to plan for the growth of the restaurant in the future, he needs to develop a system that will enable him to forecast food and beverage sales by month for up to one year in advance. Luigi compiled the following data (in thousands of euros) on total food and beverage sales for the three years of operation.

Managerial report

Perform an analysis of the sales data for the Vesuvius Restaurant. Prepare a report for Luigi that summarizes your findings, forecasts and recommendations. Include the following:

1 A graph of the time series.

2 An analysis of the seasonality of the data. Indicate the seasonal indices for each month, and comment on the high and low seasonal sales months. Do the seasonal indices make intuitive sense? Discuss.

3 A forecast of sales for January through December of the fourth year.

VESUVIUS

Month	First year	Second year	Third year
January	242	263	282
February	235	238	255
March	232	247	265
April	178	193	205
May	184	193	210
June	140	149	160
July	145	157	166
August	152	161	174
September	110	122	126
October	130	130	148
November	152	167	173
December	206	230	235

Outside dinning area of restaurant outside Naples, Italy. © Paul Piebinga.

4 Recommendations as to when the system that you develop should be updated to account for new sales data.

5 Any detailed calculations of your analysis in the appendix of your report.

Assume that January sales for the fourth year turn out to be €295 000. What was your forecast error? If this error is large, Luigi may be puzzled about the difference between your forecast and the actual sales value. What can you do to resolve his uncertainty in the forecasting procedure?

Case problem 2 Allocating patrols to meet future demand for vehicle rescue

The data below summarize actual monthly demands for RAC rescue services over a five-year time period. (The Royal Automobile Club is one of the major motoring organizations that offer emergency breakdown cover in the UK.)

To meet the national demand for its services in the coming year, the RAC's human resources planning department forecasts the number of members expected, using historical data and market forecasts. It then predicts the average number of breakdowns and number of rescue calls expected, by referring to the probability of a member's vehicle breaking down each year. In 2003, an establishment of approximately 1400 patrols was available to deal with the expected workload. Note that this figure

Monthly demand for RAC rescue services 1999–2003

| | Year | | | | |
Month	2003	2002	2001	2000	1999
January	270 093	248 658	253 702	220 332	241 489
February	216 050	210 591	216 575	189 223	193 794
March	211 154	208 969	220 903	188 950	206 068
April	194 909	191 840	191 415	196 343	191 359
May	200 148	194 654	190 436	189 627	179 592
June	195 608	189 892	175 512	177 653	183 712
July	208 493	203 275	193 900	182 219	193 306
August	215 145	213 357	197 628	190 538	199 947
September	200 477	196 811	183 912	183 481	191 231
October	216 821	225 182	213 909	214 009	198 514
November	222 128	244 498	219 336	239 104	202 219
December	250 866	257 704	246 780	254 041	254 217

RAC rescue van repairing a broken down vehicle at the roadside.
© Paul Thompson Images/Alamy.

had to be reviewed monthly since it was an average for the year and did not take into account, fluctuations in demand 'in different seasons'.

Managerial report

1 By undertaking an appropriate statistical analysis of the information provided, describe how you would advise the RAC on its patrol allocation in 2004.

2 State your assumptions.

3 Comment on the validity of your results or otherwise.

Software Section for Chapter 18

18.7 Forecasting using MINITAB

In this section we show how MINITAB can be used to develop forecasts using three forecasting methods: moving averages, exponential smoothing and trend projection.

Moving averages

To show how MINITAB can be used to develop forecasts using the moving averages method, we will develop a forecast for the petrol sales time series in Table 18.1 and Figure 18.5. The sales data for the 12 weeks are entered into column C2 of the worksheet. The steps involved in using MINITAB to compute a three-week moving average forecast for week 13 follow.

Step 1   **Time Series > Moving Average**   [Main menu bar]
Enter 3 in the **MA length** box

Step 2   Enter C2 in the **Variable** box   [**Moving Average** panel]
Select **Generate forecasts**
Enter 1 in the **Number of forecasts** box
Enter 12 in the **Starting from origin** box
Click **OK**

Step 3   Click **OK**   [**Time Series** panel]

The three-week moving average forecast for week 13 is shown in the session window. The mean square error of 10.22 is labelled as MSD in the MINITAB output. Many other output options are available, including a summary table similar to Table 18.2 and graphical output similar to Figure 18.6.

Exponential smoothing

To show how MINITAB can be used to develop an exponential smoothing forecast, we will again develop a forecast of sales in week 13 for the petrol sales time series in Table 18.1 and Figure 18.5. The sales data for the 12 weeks are entered into column 2 of the worksheet. The steps involved in using MINITAB to produce a forecast for week 13 using a smoothing constant of $\alpha = 0.2$ follow.

721

Step 1 **Stat > Time Series > Single Exp Smoothing**   [Main menu bar]

Step 2 Enter C2 in the **Variable** box   [**Single Exp Smoothing** panel]
Select **Use**
Enter 0.2 in the **Use** box
Select **Generate forecasts**
Enter 1 in the **Number of forecasts** box
Enter 12 in the **Starting from origin** box
Select **Options**
Enter 1 in the **Use average of first** box
Click **OK**

Step 3 Click **OK**   [**Time Series** panel]

The exponential smoothing forecast for week 13 is shown in the session window. The mean square error is labelled as MSD in the MINITAB output.* Many other output options are available, including a summary table similar to Table 18.3 and graphical output similar to Figure 18.7.

Trend projection

To show how MINITAB can be used for trend projection, we develop a forecast for the bicycle sales time series in Table 18.5 and Figure 18.8. The year numbers are entered into column C1 and the sales data are entered into column C2 of the worksheet. The steps involved in using MINITAB to produce a forecast for week 13 using trend projection follow.

Step 1 **Stat > Time Series > Trend Analysis**   [Main menu bar]

Step 2 Enter C2 in the **Variable** box   [**Trend Analysis** panel]
Choose **Linear** for the Model Type
Select **Generate forecasts**
Enter 1 in the **Number of forecasts** box
Enter 12 in the **Starting from origin** box
Click **OK**

Step 3 Click **OK**   [**Time Series** panel]

The equation for linear trend and the forecast for the next period are shown in the session window.

Multiplicative model

To show how MINITAB can be used to develop forecasts based on a multiplicative model we will develop forecasts for quarters 17, 18, 19 and 20 for the television set sales time series in Table 18.6. The sales data for the 16 weeks are entered into column C3 of the worksheet.

*The value of MSD computed by MINITAB is not the same as the value of MSE that appears in Table 18.4. MINITAB uses a forecast of 17 for week 1 and computes MSD using all 12 time periods of data. In Section 18.2 we compute MSE using only the data for weeks 2 through 12, because we had no past values with which to make a forecast for period 1.

Step 1 **Stat > Time Series > Decomposition**   [Main menu bar]

Step 4 Enter C3 in the **Variable** box   [**Decomposition** panel]
Enter 4 in the **Seasonal Length** box
Select **Generate forecasts**
Enter 4 in the **Number of forecasts** box
Enter 16 in the **Starting from origin** box
Click **OK**

Step 3 Click **OK**   [**Time Series** panel]

The forecasts for quarters 17,18, 19 and 20 are shown in the session window and plotted on the **Times Series Decomposition Plot** along with past data values. Correspondingly a **Component Analysis** window showing four separate plots respectively of the original data, detrended data, seasonally adjusted data and seasonally adjusted and detrended data. (The latter values are obtained by taking the ratio of seasonally adjusted values with corresponding trend values estimated from the linear regression.) Note that the MINITAB results, though consistent with the ones derived in section 18.4, are slightly different from them because of the particular algorithm employed by the software.

18.8 Forecasting using EXCEL

In this section we show how EXCEL can be used to develop forecasts using three forecasting methods: moving averages, exponential smoothing and trend projection.

Moving averages

To show how EXCEL can be used to develop forecasts using the moving averages method, we will develop a forecast for the petrol sales time series in Table 18.1 and Figure 18.5.
The sales data for the 12 weeks are entered into worksheet rows 2 through 13 of column B.
The steps involved in using EXCEL to produce a three-week moving average, follow.

Step 1 **Data > Data Analysis > Moving Average**   [Main menu bar]

Step 2 Enter B2:B13 in the **Input Range** box   [**Moving Average** panel]
Enter 3 in the **Interval** box
Enter C2 in the **Output Range** box
Click **OK**

The three-week moving average forecasts will appear in column C of the worksheet. Forecasts for periods of other length can be computed easily by entering a different value in the **Interval** box.

Exponential smoothing

To show how EXCEL can be used for exponential smoothing, we again develop a forecast for the petrol sales time series in Table 18.1 and Figure 18.5. The sales data for the 12 weeks are entered into worksheet rows 2 through 13 of column B. The following steps can be used to produce a forecast using a smoothing constant of $\alpha = 0.2$.

Step 1 **Data > Data Analysis > Exponential Smoothing**   [Main menu bar]

Step 2 Enter B2:B13 in the **Input Range** box   [**Exponential Smoothing** panel]
Enter 0.8 in the **Damping factor** box
Enter C2 in the **Output Range** box
Click **OK**

The exponential smoothing forecasts will appear in column C of the worksheet. Note that the value we entered in the Damping factor box is $1 - \alpha$; forecasts for other smoothing constants can be computed easily by entering a different value for $1 - \alpha$ in the Damping factor box.

Trend projection

BICYCLE

To show how EXCEL can be used for trend projection, we develop a forecast for the bicycle sales time series in Table 18.5 and Figure 18.8. The data, with appropriate labels in row 1, are entered into worksheet rows 1 through 11 of columns A and B. Select an empty cell in the worksheet. The following steps can be used to produce a forecast for year 11 by trend projection.

Step 1 **Insert > Function > Statistical**   [Main menu bar]

Step 2 Choose **Forecast**   [**Statistical** panel]

Step 3 Enter 11 in the **x** box. Enter B2:B11 in the
Known y's box.   [**Exponential Smoothing** panel]
Enter A2:A11 in the **Known x's** box
The forecast for year 11, in this case 32.5, will appear in the cell selected in step 1
Click **OK**

Step 4 Click **OK**   [**Statistical** panel]

18.9 Forecasting using PASW

PASW

PETROL

To show how PASW can be used to develop forecasts using the moving average method, we again develop a forecast for the petrol sales time series in Table 18.1 and Figure 18.5.

First, the data must be entered in a PASW worksheet. In 'Data View' mode, weeks are entered in rows 1 to 12 of the leftmost column. This is automatically labelled by the system var00001. Similarly the number of sales are entered in the immediately adjacent column to the right and this is labelled var00002. The latter variable names can be changed to week and sales in 'Variable View' mode. The following steps can be used to generate three point centred moving average results for the data.

Step 1 **Transform > Create Time Series**   [Main menu bar]

Step 2 Select **sales** for the **New Variable**   [**Create Time Series** panel]
Click on **Function**
Select **Centered Moving Average**
Change the value in the **Span** box from 1 to 3

Click on the **Change** button
Click **OK**

The worksheet will now contain a third column headed **sales_1** showing three point centred moving average values for the data. The forecast for week 13 (and indeed any other week after week 11) is simply 19 – the last value in the **sales_1** column, corresponding to week 11.

Exponential smoothing

To show how PASW can be used for exponential smoothing, we again develop a forecast for the petrol sales time series in Table 18.1 and Figure 18.5. As above the data for the 12 weeks are entered into rows 1 to 12 of the two leftmost columns of worksheet. The following steps can be used to produce a forecast using a smoothing constant of $\alpha = 0.2$.

Step 1 **Analyze > Forecasting > Create Models** [Main menu bar]

Step 2 Select **sales** [**Time Series Modeler** panel]
Click on the **Method** and select **Exponential Smoothing**
Click on the **Statistics** and select **Parameter Estimates** under **Statistics for Individual Models**
Click on **Options**
Click on **First case after end of estimation through last case in active datset**
Click on **Plots**
Click on **Forecasts under Each Plot Displays**
Click **OK**

The exponential smoothing forecast will appear in the PASW output.

Trend projection

To show how PASW can be used for trend projection, we develop a forecast for the bicycle sales time series in Table 18.5 and Figure 18.8. First, the data must be entered in an PASW worksheet. In 'Data View' mode, years are entered in rows 1 to 10 of the leftmost column. The column is automatically labelled by the system V1. Similarly the number of sales are entered in the immediately adjacent column to the right and this is labelled V2. Finally the value 11 is entered into row 11 of the V1 column. The latter variable names can then be changed to year and sales in 'Variable View' mode. The following command sequence can be used to produce a forecast. The following steps can be used to produce a forecast for year 11 by trend projection.

Step 1 **Analyze > Regression > Linear** [Main menu bar]

Step 2 Select sales as Dependent variable and year as Independent variable
[**Linear Regression** panel]
Click on **Save**
Under **Predicted Values**, click in the box alongside **Unstandardised**
Click on **Continue**
Click **OK**

The forecast for year 11, in this case 32.5, will appear at the bottom of the third left-most column headed **pre_1**.

Multiplicative model

To show how PASW can be used to develop forecasts based on a multiplicative model we will analyse the television set sales time series in Table 18.6. First, the data must be entered in an PASW worksheet. In 'Data View' mode, sales values are entered in rows 1 to 16 of the leftmost column. The column is automatically labelled by the system V1. The latter variable name can then be changed to sales in 'Variable View' mode. Next the system must be advised that the data constitute a quarterly time series by the following steps.

TVSETS

Step 1 **Analyze > Forecasting > Apply Time Series Models** [Main menu bar]

Step 2 Select **Define Dates** [**Apply Time Series Models** panel]

Step 3 Click on Years, quarters in the **Cases Are** box [**Define Dates** panel]
In the **First Case Is** boxes, enter 1 for **Year** and 1 for **Quarter**
Click **OK**

Three new columns labelled **YEAR_ QUARTER_** and **DATE_** are added to the worksheet. Following the steps:

Step 4 **Analyze > Forecasting > Seasonal Decomposition** [Main menu bar]

Step 5 Enter Sales in the **Variable** box [**Seasonal Decomposition** panel]
Click **OK**

Four new columns are added to the worksheet labelled **ERR_1**, **SAS_1**, **SAF_1** and **STC_1** representing respectively, estimated irregular component values, seasonally adjusted data, estimated seasonal components and estimated trend cycle values for the series. (Note that y = **ERR_1** + **STC_1** so **STC_1** is akin to $\hat{Y}_t$ using our earlier notation.) Unlike MINITAB, which assumes that seasonally adjusted data can be automatically estimated using linear regression, PASW leaves the modelling and extrapolation of seasonally adjusted data for the user to decide on. In support, a plot of the seasonally adjusted data can be obtained by following the steps:

Step 1 **Analyze > Forecasting Sequence Charts** [Main menu bar]

Step 2 Enter **SAS_1** into the **Variable** box [**Sequence Charts panel**]
Click **OK**

Chapter 19

Non-parametric Methods

Statistics in practice: Coffee lovers' preferences: Costa, Starbucks and Caffé Nero

Learning objectives

After studying this chapter and doing the exercises, you should be able to:

1   Explain the essential differences between parametric and non-parametric methods of inference.

2   Recognize the circumstances when it is appropriate to apply the following non-parametric statistical procedures; calculate the appropriate sample statistics; use these statistics to carry out a hypothesis test; interpret the results.

2.1   Sign test.

2.2   Wilcoxon signed-rank test.

2.3   Mann-Whitney-Wilcoxon test.

2.4   Kruskal-Wallis test.

2.5   Spearman rank correlation.

The inferential statistical methods presented so far are generally known as *parametric methods*. In this chapter we introduce several **non-parametric methods**. These are sometimes applicable in situations where parametric methods are not. Non-parametric methods typically require less restrictive assumptions about the level of data measurement and fewer assumptions about the form of the probability distributions generating the sample data.

One consideration in determining whether a parametric or a non-parametric method is appropriate is the *scale of measurement* used to generate the data. As explained in Chapter 1, there are four generally recognized scales of measurement: nominal, ordinal, interval and ratio. Here are definitions and examples of the four scales of measurement.

1   *Nominal scale.* The scale of measurement is nominal if the data are labels or categories used to define an attribute. Nominal data may be numeric or non-numeric.

   Examples. The exchange where a company's shares are listed (LSE, NYSE, NASDAQ, EuroNext) is non-numeric nominal data. The country dialling code for an individual's telephone number is numeric nominal data.

2   *Ordinal scale.* The scale of measurement is ordinal if the data can be used to rank, or order, the observations. Ordinal data may be numeric or non-numeric.

   Examples. The measures small, regular and large for the size of an item (e.g. cups of coffee) are non-numeric ordinal data. The class ranks of individuals measured as 1, 2, 3, … are numeric ordinal data.

3   *Interval scale.* The scale of measurement is interval if the data have the properties of ordinal data and the interval between observations is expressed in terms of a fixed unit of measurement. Interval data must be numeric.

   Examples. Measures of temperature are interval data. Suppose it is 30 degrees in one location and 15 degrees in another. We can rank the locations with respect to warmth: the first location is warmer than the second. The fixed unit of measurement, a degree, enables us to say how much warmer it is at the first location: 15 degrees.

4   *Ratio scale.* The scale of measurement is ratio if the data have the properties of interval data and the ratio of measures is meaningful because the scale has a fundamental zero point. Ratio data must be numeric.

   Examples. Variables such as distance, height and weight are measured on ratio scales. Temperature measures are not ratio data because there is no inherently defined zero point. For instance, the melting point of ice is 32 degrees on a

Statistics in Practice

Coffee lovers' preferences: Costa, Starbucks and Caffé Nero

In spring 2009, Costa Coffee ran a vigorous newspaper advertising and promotional campaign in the UK under the headline **SORRY STARBUCKS THE PEOPLE HAVE VOTED**, with the by-line 'In head-to-head tests, seven out of ten coffee lovers preferred Costa cappuccino to Starbucks'. Some of the advertisements were near full-page in broadsheet newspapers. At the bottom of the advertisements, the small print noted that the research on which the claim was based had been carried out by an independent market research organization, Tangible Branding Limited, that 70 per cent of respondents who identified themselves as coffee lovers preferred Costa cappuccino, and that the total sample size of coffee lovers was 174.

Further details were available on the Costa website (www.costa.co.uk). The research was carried out in three UK towns, High Wycombe, Glasgow and Sheffield, in late November and early December 2008. Each

Costa coffee with advertising poster in the window based on market research. © Authors own image.

respondent was asked to undertake a two-way blind tasting test: these were either Costa versus Starbucks or Costa versus Caffé Nero. 'Runners' transported the coffees to the tasting venue from nearby coffee houses. The order of tasting was rotated. Over the three tasting venues, the total Costa versus Starbucks sample was of size 166, and the Costa versus Caffé Nero sample was 168.

In the Costa versus Caffé Nero comparisons, 64 per cent of tasters preferred Costa. In the Costa versus Starbucks tests, 66 per cent preferred Costa. The website also gave the splits for those who identified themselves as 'coffee lovers', and for those who considered Caffé Nero or Starbucks to be their preferred coffee house. Among self-identified coffee lovers, 69 per cent preferred the Costa coffee to Caffé Nero coffee, and 72 per cent preferred Costa to Starbucks. Among Caffé Nero regulars, 72 per cent expressed a preference for Costa's cappuccino, while 67 per cent of Starbucks regulars preferred Costa's cappuccino. Some of the Costa advertisements had the headline **STARBUCKS DRINKERS PREFER COSTA.**

The website notes that 'All results are significant at the 95 per cent confidence level'. The data on which the results are based are qualitative data: a simple expression of preference between two options. The kind of statistical test needed for data such as these is known as a non-parametric test. Non-parametric tests are the subject of the present chapter. The chapter begins with a discussion of the sign test, a test particularly appropriate for the research situation described by Costa in its advertising.

Fahrenheit scale and zero degrees on a Celsius scale. Ratios are not meaningful with temperature data. For instance, it makes no sense to say that 30 degrees is twice as warm as 15 degrees.

Most parametric statistical methods require the use of interval- or ratio-scaled data. With these levels of measurement, arithmetic operations are meaningful: means, variances, standard deviations, and so on can be computed, interpreted and used in the analysis. With nominal or ordinal data, it is inappropriate to compute means, variances and standard deviations. Non-parametric methods are often the only way to analyze such data and draw statistical conclusions.

In general, for a statistical method to be classified as non-parametric, it must satisfy at least one of the following conditions.*

1 The method can be used with nominal data.

2 The method can be used with ordinal data.

3 The method can be used with interval or ratio data when no assumption can be made about the population distribution.

If the level of data measurement is interval or ratio, and if the necessary distributional assumptions are appropriate, parametric methods provide more powerful or more discerning statistical procedures. However, in some cases where a non-parametric method as well as a parametric method can be applied, the non-parametric method is almost as powerful as the parametric method. In cases where the data are nominal or ordinal, or in cases where the assumptions required by parametric methods are inappropriate, only non-parametric methods are available. The sign test, the Wilcoxon signed-rank test, the Mann-Whitney-Wilcoxon test, the Kruskal-Wallis test, and the Spearman rank correlation are the non-parametric methods presented in this chapter.

19.1 Sign test

A common market-research application of the **sign test** involves using a sample of n potential customers to identify a preference for one of two brands of a product such as coffee, soft drinks or detergents. The n expressions of preference are nominal data because the consumer simply names, or labels, a preference. Given these data, our objective is to determine whether a difference in preference exists between the two items being compared. The sign test is a non-parametric statistical procedure for addressing this question.

Small-sample case

The small-sample case for the sign test is appropriate whenever $n \leq 20$. Consider a study conducted for Sunny Vale Farms, who produce an orange juice product marketed under the name Citrus Delight. A competitor produces an orange juice product known as Tropical Orange. In a study of consumer preferences for the two brands, 12 individuals were given unmarked samples of each product. The brand each individual tasted first was selected randomly. After tasting the two products, the individuals were asked to state a preference for one of the two brands. The purpose of the study is to determine whether consumers in general prefer one product over the other. Letting π indicate the proportion of the population of consumers who favour Citrus Delight, we want to test the following hypotheses.

$$H_0: \pi = 0.50$$
$$H_1: \pi \neq 0.50$$

If H_0 cannot be rejected, we shall have no evidence indicating a difference in preference for the two brands of orange juice. However, if H_0 can be rejected, we can conclude that

---

*See W. J. Conover, *Practical Non-parametric Statistics,* 3rd ed. (John Wiley & Sons, 1998).

the consumer preferences are different for the two brands. In that case, the brand selected by the greater number of consumers can be considered the preferred brand. We shall use a 0.05 level of significance.

To record the preference data for the 12 individuals participating in the study, we use a plus sign if the individual expresses a preference for Citrus Delight and a minus sign if the individual expresses a preference for Tropical Orange. Because the data can be recorded in terms of plus or minus signs, this non-parametric test is called the sign test. The number of plus signs is the test statistic. Under the assumption that H_0 is true ($\pi = 0.50$), its sampling distribution is a binomial distribution with $\pi = 0.50$. Table 5 in Appendix B shows the probabilities for the binomial distribution with a sample size of $n = 12$ and $\pi = 0.50$. Figure 19.1 is a graphical representation of this binomial sampling distribution.

The preference data obtained are shown in Table 19.1. The two plus signs indicate two consumers preferred Citrus Delight. We can now use the binomial probabilities to determine the p-value for the test. With a two-tailed test, the p-value is found by doubling the probability in the tail of the binomial sampling distribution. For Sunny Vale Farms, the number of plus signs (2) is in the lower tail of the distribution. So the probability in the tail is the combined probability of 2, 1 or 0 plus signs. Adding these probabilities, we obtain $0.0161 + 0.0029 + 0.0002 = 0.0192$. Doubling this value, we obtain the p-value $= 2(0.0192) = 0.0384$. With p-value <0.05, we reject H_0. The taste test provides evidence that consumer preference differs significantly for the two brands of orange juice. We would advise Sunny Vale Farms that consumers prefer Tropical Orange.

The Sunny Vale Farms hypothesis test was a two-tailed test, in which the p-value was found by doubling the probability in the tail of the binomial distribution. One-tailed sign tests are also possible. If the test is a lower tail test, the p-value is the probability that the number of plus signs is less than or equal to the observed number. If the test is an upper tail test, the p-value is the probability that the number of plus signs is greater than or equal to the observed number. The binomial probabilities shown in Table 5 of Appendix B can be used for sign tests up to a sample size of $n = 20$.

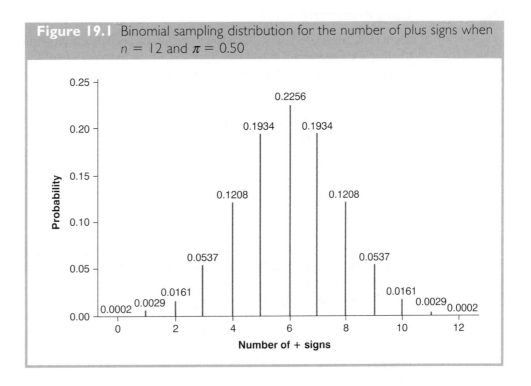

Figure 19.1 Binomial sampling distribution for the number of plus signs when $n = 12$ and $\pi = 0.50$

Table 19.1 Preference data for the Sunny Vale Farms taste test

Individual	Brand preference	Recorded data
1	Tropical Orange	−
2	Tropical Orange	−
3	Citrus Delight	+
4	Tropical Orange	−
5	Tropical Orange	−
6	Tropical Orange	−
7	Tropical Orange	−
8	Tropical Orange	−
9	Citrus Delight	+
10	Tropical Orange	−
11	Tropical Orange	−
12	Tropical Orange	−

In the Sunny Vale Farms taste test, all 12 individuals were able to state a preference for one of the two brands of orange juice. In other applications of the sign test, one or more individuals in the sample may not be able to state a preference. The usual procedure in such cases is to discard the no-preference responses from the sample and base the sign test on a reduced sample size (though this can be criticized on the grounds that it discards data favourable to the null hypothesis).

Large-sample case

The large-sample sign test is equivalent to the test of a population proportion as presented in Chapter 9. Using the null hypothesis H_0: $\pi = 0.50$ and a sample size of $n > 20$, the sampling distribution for the number of plus signs can be approximated by a normal distribution.

Normal approximation of the sampling distribution of the number of plus signs when H_0: $\pi = 0.5$ is true

$$\text{Mean: } \mu = 0.50n \tag{19.1}$$

$$\text{Standard deviation: } \sigma = \sqrt{0.25n} \tag{19.2}$$

Distribution form: approximately normal provided $n > 20$

Consider an application of the sign test to political polling. Take the case of a poll conducted in the UK, in which 200 voters are asked to rate the Conservative and Labour parties in terms of which party has the better policy on European integration. Suppose 103 respondents rate the Conservative policy better, 72 rate the Labour policy better and 25 indicate no preference between the two party policies. Does the poll indicate a significant difference between the two parties in respect of public opinion about their European policies?

Using equations (19.1) and (19.2), and $n = 200 - 25 = 175$, we find that the sampling distribution of the number of plus signs has the following properties.

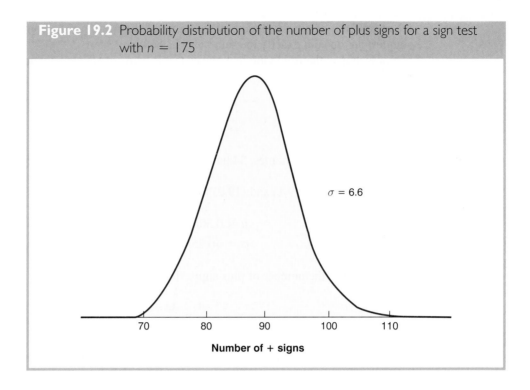

Figure 19.2 Probability distribution of the number of plus signs for a sign test with $n = 175$

$\sigma = 6.6$

70 80 90 100 110

Number of + signs

$$\mu = 0.50n = 0.50(175) = 87.5$$
$$\sigma = \sqrt{0.25n} = \sqrt{0.25(175)} = 6.6$$

With $n = 175$ we can assume that the sampling distribution is approximately normal. This distribution is shown in Figure 19.2.

Let us use a 0.05 level of significance. Based on the number of times the Labour party received the higher European policy rating (number of plus signs $x = 72$), we can calculate the following value for the test statistic.

$$z = \frac{x - \mu}{\sigma} = \frac{72 - 87.5}{6.6} = -2.35$$

The standard normal distribution table shows that the area in the tail to the left of $z = -2.35$ is 0.0094. With a two-tailed test, the p-value $= 2(0.0094) = 0.0188$. With p-value $< \alpha = 0.05$, we reject H_0. The poll indicates that the parties are perceived to differ in terms of public opinion about their European policies. If the analysis used the number of times the Conservative policy was rated higher, $z = 2.35$ would lead us to the same conclusion.

Hypothesis test about a median

In Chapter 9 we described how hypothesis tests can be used to make an inference about a population mean. We now show how the sign test can be used to conduct hypothesis tests about a population median. Recall that the median splits a population in such a way that 50 per cent of the values are at the median or above and 50 per cent are at the median or below. We can apply the sign test by using a plus sign whenever the data in the sample are above the hypothesized value of the median and a minus sign whenever the data in the sample are below the hypothesized value of the median. The computations for the sign test are done in exactly the same way as before.

For example, the following hypothesis test is being conducted about the median price of new homes in Ireland.

$$H_0: \text{Median} = \text{€}260\ 000$$
$$H_1: \text{Median} \neq \text{€}260\ 000$$

In a sample of 60 new homes, 34 have prices above €260 000, and 26 have prices below €260 000.

Using equations (19.1) and (19.2) for the $n = 60$ homes, we obtain

$$\mu = 0.50n = 0.50(60) = 30$$
$$\sigma = \sqrt{0.25n} = \sqrt{0.25(60)} = 3.87$$

With $x = 34$ as the number of plus signs, the test statistic becomes

$$z = \frac{x - \mu}{\sigma} = \frac{34 - 30}{3.87} = 1.03$$

Using the standard normal distribution table and $z = 1.03$, we find the two-tailed p-value $= 2(1 - 0.8485) = 0.303$. With p-value > 0.05, we cannot reject H_0. Based on the sample data, we are unable to reject the null hypothesis that the median selling price of a new home in Ireland is €260 000.

Exercises

Methods

1   The following table lists the preferences indicated by ten individuals in taste tests involving two brands of a product.

Individual	Brand A versus Brand B	Individual	Brand A versus Brand B
1	+	6	+
2	+	7	−
3	+	8	+
4	−	9	−
5	+	10	+

With $\alpha = 0.05$, test for a significant difference in the preferences for the two brands. A plus indicates a preference for brand A over brand B.

2   The following hypothesis test is to be conducted.

$$H_0: \text{Median} \leq 150$$
$$H_1: \text{Median} > 150$$

A sample of size 30 yields 22 cases in which a value greater than 150 is obtained, three cases in which a value of exactly 150 is obtained, and five cases in which a value less than 150 is obtained. Use $\alpha = 0.01$ and conduct the hypothesis test.

Applications

3 A poll asked 1253 adults a series of questions about the state of the economy and their children's future. One question was, 'Do you expect your children to have a better life than you have had, a worse life, or a life about as good as yours?' The responses were 34 per cent better, 29 per cent worse, 33 per cent about the same and 4 per cent not sure. Use the sign test and a 0.05 level of significance to determine whether more adults feel their children will have a better future than feel their children will have a worse future. What is your conclusion?

4 Are stock splits beneficial to stockholders? SNL Securities studied stock splits in the banking industry over an 18-month period and found that stock splits tended to increase the value of an individual's stock holding. Assume that of a sample of 20 recent stock splits, 14 led to an increase in value, four led to a decrease in value, and two resulted in no change. Suppose a sign test is to be used to determine whether stock splits continue to be beneficial for holders of bank stocks.

   a. What are the null and alternative hypotheses?
   b. With $\alpha = 0.05$, what is your conclusion?

5 An opinion survey asked the following question regarding a proposed educational policy. 'Do you favour or oppose providing tax-funded vouchers or tax deductions to parents who send their children to private fee-paying schools?' Of the 2010 individuals surveyed, 905 favoured the support, 1045 opposed the support, and 60 offered no opinion. Do the data indicate a significant tendency towards favouring or opposing the proposed policy? Use a 0.05 level of significance.

6 Suppose a national survey in France has shown that the median annual income adults say would make their dreams come true is €152 000. Suppose further that of a sample of 225 individuals in Calais, 122 individuals report that the amount of income needed to make their dreams come true is less than €152 000 and 103 report that the amount needed is more than €152 000. Test the null hypothesis that the median amount of annual income needed to make dreams come true in Calais is €152 000. Use $\alpha = 0.05$. What is your conclusion?

7 The median number of part-time employees at fast-food restaurants in a particular city was known to be 15 last year. The city council thinks the use of part-time employees may have increased this year. A sample of nine fast-food restaurants showed that more than 15 part-time employees worked at seven of the restaurants, one restaurant had exactly 15 part-time employees, and one had fewer than 15 part-time employees. Test at $\alpha = 0.05$ to see whether the median number of part-time employees has increased.

8 Land Registry figures for January 2009 show the median selling price of houses in England and Wales as £157 000. Assume that the following data were obtained for sales of houses in Greater Manchester and in Oxfordshire.

	Greater than £157 000	Equal to £157 000	Less than £157 000
Greater Manchester	11	2	32
Oxfordshire	27	1	13

a. Is the median selling price in Greater Manchester lower than the national median of £157 000? Use a statistical test with $\alpha = 0.05$ to support your conclusion.

b. Is the median selling price in Oxfordshire higher than the national median of £157 000? Use a statistical test with $\alpha = 0.05$ to support your conclusion.

19.2 Wilcoxon signed-rank test

The **Wilcoxon signed-rank test** is the non-parametric alternative to the parametric matched-sample test presented in Chapter 10. In the matched-sample situation, each experimental unit generates two paired or matched observations, one from population 1 and one from population 2. The differences between the matched observations provide insight about the differences between the two populations.

A manufacturing firm is attempting to determine whether two production methods differ in task completion time. A sample of 11 workers was selected, and each worker completed a production task using each of the two production methods. The production method that each worker used first was selected randomly. Each worker in the sample therefore provided a pair of observations, as shown in the first three columns of Table 19.2. A positive difference in task completion times (column 4 of Table 19.2) indicates that method 1 required more time, and a negative difference in times indicates that method 2 required more time. Do the data indicate that the methods are significantly different in terms of task completion times?

In effect, we have two populations of task completion times, one population associated with each method. The following hypotheses will be tested.

$$H_0: \text{The populations are identical}$$
$$H_1: \text{The populations are not identical}$$

If H_0 cannot be rejected, we will not have evidence to conclude that the task completion times differ for the two methods. However, if H_0 can be rejected, we will conclude that the two methods differ in task completion time.

The first step of the Wilcoxon signed-rank test requires a ranking of the *absolute values* of the differences between the two methods. We discard any differences of zero and then rank the remaining absolute differences from lowest to highest. Tied differences are assigned the average ranking of their positions. The ranking of the absolute values of differences is shown in the sixth column of Table 19.2. Note that the difference of zero for worker 8 is discarded from the rankings. Then the smallest absolute difference of 0.1 is assigned the rank of 1. This ranking of absolute differences continues with the largest absolute difference of 0.9 assigned the rank of 10. The tied absolute differences for workers 3 and 5 are assigned the average rank of 3.5 and the tied absolute differences for workers 4 and 10 are assigned the average rank of 5.5.

Once the ranks of the absolute differences have been determined, the ranks are given the sign of the original difference in the data. For example, the 0.1 difference for worker 7, which was assigned the rank of 1, is given the value of +1 because the observed difference between the two methods was positive. The 0.2 difference (worker 2), which was assigned the rank of 2, is given the value of −2 because the observed difference between the two methods was negative for worker 2. The complete list of signed ranks, as well as their sum, is shown in the last column of Table 19.2.

Table 19.2 Production task completion times (minutes) and ranking of absolute differences

Worker	Method 1	Method 2	Difference	Absolute value of difference	Rank	Signed rank
1	10.2	9.5	0.7	0.7	8.0	+8.0
2	9.6	9.8	-0.2	0.2	2.0	-2.0
3	9.2	8.8	0.4	0.4	3.5	+3.5
4	10.6	10.1	0.5	0.5	5.5	+5.5
5	9.9	10.3	-0.4	0.4	3.5	-3.5
6	10.2	9.3	0.9	0.9	10.0	+10.0
7	10.6	10.5	0.1	0.1	1.0	+1.0
8	10.0	10.0	0.0	0.0	–	–
9	11.2	10.6	0.6	0.6	7.0	+7.0
10	10.7	10.2	0.5	0.5	5.5	+5.5
11	10.6	9.8	0.8	0.8	9.0	+9.0
				Sum of signed ranks		**+44.0**

The null hypothesis is identical population distributions of task completion times for the two methods. In that case, we would expect the positive ranks and the negative ranks to cancel each other, so that the sum of the signed rank values would be approximately zero. Hence, the test for significance in the Wilcoxon signed-rank test involves determining whether the computed sum of signed ranks (+44 in our example) is significantly different from zero. Let T denote the sum of the signed-rank values. The procedure assumes that the distribution of differences between matched pairs is symmetrical, but not necessarily normal in shape. It can be shown that if the two populations are identical and the number of matched pairs of data is ten or more, the sampling distribution of T can be approximated by a normal distribution as follows.

Sampling distribution of T for identical populations

$$\text{Mean: } \mu_T = 0 \tag{19.3}$$

$$\text{Standard deviation: } \sigma_T = \sqrt{\frac{n(n+1)(2n+1)}{6}} \tag{19.4}$$

Distribution form: approximately normal provided $n \geq 10$.

For the example, we have $n = 10$ after discarding the observation with the difference of zero (worker 8). Using equation (19.4), we have

$$\sigma_T = \sqrt{\frac{10(11)(21)}{6}} = 19.62$$

Figure 19.3 is the sampling distribution of T under the null hypothesis.

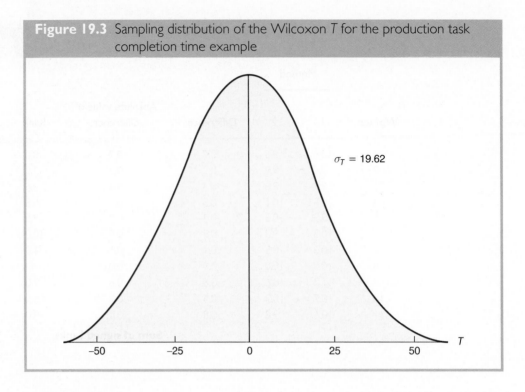

Figure 19.3 Sampling distribution of the Wilcoxon *T* for the production task completion time example

We shall use a 0.05 level of significance to draw a conclusion. With the sum of the signed-rank values $T = 44$, we calculate the following value for the test statistic.

$$z = \frac{T - \mu_T}{\sigma_T} = \frac{44 - 0}{19.62} = 2.24$$

Using the standard normal distribution table and $z = 2.24$, we find the two-tailed p-value $= 2(1 - 0.9875) = 0.025$. With p-value $< \alpha = 0.05$, we reject H_0 and conclude that the two populations are not identical and that the methods differ in task completion time. Method 2's shorter completion times for eight of the workers lead us to conclude that method 2 is the preferred production method.

Exercises

Applications

9   Two fuel additives are tested to determine their effect on litres of fuel consumed per 100 kilometres travelled, for passenger cars. Test results for 12 cars follow. Each car was tested with both fuel additives. Use $\alpha = 0.05$ and the Wilcoxon signed-rank test to see whether there is a significant difference in the additives.

Car	Additive 1	Additive 2	Car	Additive 1	Additive 2
1	7.02	7.82	7	8.74	8.21
2	6.00	6.49	8	7.62	9.43
3	6.41	6.26	9	6.46	7.05
4	7.37	8.28	10	5.83	6.68
5	6.65	6.65	11	6.09	6.20
6	5.70	5.93	12	5.65	5.96

10 A sample of ten men was used in a study to test the effects of a relaxant on the time required to fall asleep for male adults. Data for ten subjects showing the number of minutes required to fall asleep with and without the relaxant follow. Use a 0.05 level of significance to determine whether the relaxant reduces the time required to fall asleep. What is your conclusion?

Participant	Without relaxant	With relaxant	Participant	Without relaxant	With relaxant
1	15	10	6	7	5
2	12	10	7	8	10
3	22	12	8	10	7
4	8	11	9	14	11
5	10	9	10	9	6

11 A test was conducted of two overnight mail delivery services. Two samples of identical deliveries were set up so that both delivery services were notified of the need for a delivery at the same time. The hours required to make each delivery follow. Do the data shown suggest a difference in the delivery times for the two services? Use a 0.05 level of significance for the test.

Delivery	Service 1	Service 2
1	24.5	18.0
2	26.0	25.5
3	28.0	32.0
4	21.0	20.0
5	18.0	19.5
6	36.0	28.0
7	25.0	29.0
8	21.0	22.0
9	24.0	23.5
10	26.0	29.5
11	31.0	30.0

12 Ten test-market cities in France were selected as part of a market research study designed to evaluate the effectiveness of a particular advertising campaign. The sales in euros for each city

were recorded for the week prior to the promotional programme. Then the campaign was conducted for two weeks and new sales data were collected for the week immediately after the campaign. The two sets of sales data (in thousands of euros) follow.

City	Pre-campaign sales	Post-campaign sales
Bordeaux	130	160
Strasbourg	100	105
Nantes	120	140
St Etienne	95	90
Lyon	140	130
Rennes	80	82
Le Havre	65	55
Amiens	90	105
Toulouse	140	152
Marseilles	125	140

Use $\alpha = 0.05$. What conclusion would you draw about the value of the advertising programme?

19.3 Mann-Whitney-Wilcoxon test

In this section we present another non-parametric method that can be used to determine whether a difference exists between two populations. This test, unlike the signed-rank test, is not based on matched samples. Two independent samples are used, one from each population. The test was developed jointly by Mann and Whitney and by Wilcoxon. It is sometimes called the *Mann-Whitney test* and sometimes the *Wilcoxon rank-sum test*. The Mann-Whitney and Wilcoxon versions of this test are equivalent. We refer to it as the **Mann-Whitney-Wilcoxon (MWW) test**.

The non-parametric MWW test does not require interval data nor the assumption that the populations are normally distributed. The only requirement of the MWW test is that the measurement scale for the data is at least ordinal. The MWW test examines whether the two populations are identical. The hypotheses for the MWW test are as follows.

H_0: The two populations are identical

H_1: The two populations are not identical

We first show an application of the MWW test for the small-sample case.

Small-sample case

The small-sample case for the MWW test should be used whenever the sample sizes for both populations are less than or equal to ten. As an example, consider the academic potential of students attending Johnston Secondary School. The majority of students attending Johnston Secondary School previously attended either Garfield Primary School or Mulberry Primary School. The question raised by the secondary school management was whether the population of students who had attended Garfield was identical to the

Table 19.3 Secondary school data: ranking among year-group

Garfield students		Mulberry students	
Student	Year-group ranking	Students	Year-group ranking
Fields	8	Hart	70
Clark	52	Phipps	202
Jones	112	Kirkwood	144
Tibbs	21	Abbott	175
		Guest	146

population of students who had attended Mulberry in terms of academic potential. The following hypotheses were considered.

H_0: The two populations are identical in terms of academic potential

H_1: The two populations are not identical in terms of academic potential

Using school records, Johnston Secondary School management selected a random sample of four students who attended Garfield Primary and another random sample of five students who attended Mulberry Primary. The current secondary school class year-rankings were recorded for each of the nine students used in the study. They are listed in Table 19.3.

The first step in the MWW procedure is to rank the *combined* data from the two samples from low to high. The lowest value (year-ranking 8) receives a rank of 1 and the highest value (year-ranking 202) receives a rank of 9. The ranking of the nine students is given in Table 19.4.

The next step is to sum the ranks for each sample separately. This calculation is shown in Table 19.5. The MWW procedure can use the sum of the ranks for either sample. We use the sum of the ranks for the sample of four students from Garfield, and denote this sum by the symbol T. For our example, $T = 11$.

Let us consider the properties of T. With four students in the sample, Garfield could have the top four students in the study. If this were the case, $T = 1 + 2 + 3 + 4 = 10$ would be the smallest value possible for the rank sum T. Conversely, Garfield could have the bottom four students, in which case $T = 6 + 7 + 8 + 9 = 30$ would be the largest value possible for T. Therefore, T for the Garfield sample must take a value between 10 and 30.

Table 19.4 Ranking of secondary school students

Student	Year-group ranking	Combined sample rank	Student	Year-group ranking	Combined sample rank
Fields	8	1	Kirkwood	144	6
Tibbs	21	2	Guest	146	7
Clark	52	3	Abbott	175	8
Hart	70	4	Phipps	202	9
Jones	112	5			

Note that values of T near 10 imply that Garfield has the significantly better-ranked students, whereas values of T near 30 imply that Garfield has the significantly weaker-ranked students. Hence, if the two populations of students were identical in terms of academic potential, we would expect the value of T to be near the average of the two values, or $(10 + 30)/2 = 20$.

Critical values of the MWW T statistic are provided in Table 9 of Appendix B for cases in which both sample sizes are less than or equal to ten. In that table, n_1 refers to the sample size corresponding to the sample whose rank sum is being used in the test. The value of T_L is read directly from the table and the value of T_U is computed from equation (19.5).

$$T_U = n_1(n_1 + n_2 + 1) - T_L \qquad (19.5)$$

The null hypothesis of identical populations should be rejected only if T is strictly less than T_L or strictly greater than T_U.

Using Table 9 of Appendix B with a 0.05 level of significance, we see that the lower tail critical value for the MWW statistic with $n_1 = 4$ (Garfield) and $n_2 = 5$ (Mulberry) is $T_L = 12$. The upper tail critical value for the MWW statistic computed by using equation (19.5) is

$$T_U = 4(4 + 5 + 1) - 12 = 28$$

The MWW decision rule indicates that the null hypothesis of identical populations can be rejected if the sum of the ranks for the first sample (Garfield) is less than 12 or greater than 28. The rejection rule can be written as

Reject H_0 if $T < 12$ or if $T > 28$

Referring to Table 19.5, we see that $T = 11$. Hence, the null hypothesis H_0 is rejected. We can conclude that the population of students at Garfield differs from the population of students at Mulberry in terms of academic potential, and that there is a tendency for the Garfield students to get better year-rankings at the secondary school than Mulberry students. If we conducted the test with the rank sum of the Mulberry students, we would have $n_1 = 5$, $n_2 = 4$, $T_L = 17$, $T_U = 33$, and $T = 34$. With $T > T_U$, we would reach the same conclusion to reject H_0.

Table 19.5 Rank sums for secondary school students from each primary school

| | Garfield students | | | Mulberry students | |
Student	Year-group ranking	Combined sample rank	Student	Year-group ranking	Combined sample rank
Fields	8	1	Hart	70	4
Clark	52	3	Phipps	202	9
Jones	112	5	Kirkwood	144	6
Tibbs	21	2	Abbott	175	8
			Guest	146	7
Sum of Ranks		11			34

Large-sample case

When both sample sizes are at least ten, a normal approximation of the distribution of T can be used to conduct the analysis for the MWW test. We illustrate the large sample case by considering a situation at People's Bank.

People's Bank has two branch offices. Data collected from two independent simple random samples, one from each branch, are given in Table 19.6 (the rankings in this table are explained below). What do the data indicate regarding the hypothesis that the populations of cheque account balances at the two branch banks are identical?

The first step in the MWW test is to rank the *combined* data from the lowest to the highest values. Using the combined set of 22 observations in Table 19.6, we find the lowest data value of €750 (sixth item of sample 2) and assign to it a rank of 1. Continuing the ranking gives us the following list.

Balance (€)	Item	Assigned rank
750	6th of sample 2	1
800	5th of sample 2	2
805	7th of sample 1	3
850	2nd of sample 2	4
:		
1195	4th of sample 1	21
1200	3rd of sample 1	22

In ranking the combined data, we may find that two or more data values are the same. In that case, the tied values are given the *average* ranking of their positions in the combined data set. For example, the balance of €945 (eighth item of sample 1) will be assigned the rank of 11. However, the next two values in the data set are tied with values of €950 (see the sixth item of sample 1 and the fourth item of sample 2). These two values would be assigned ranks of 12 and 13 if they were distinct, so they are both assigned

Table 19.6 Cheque account balances for two branches of People's Bank, and combined ranking of the data

	Branch 1			Branch 2	
Account	Balance (€)	Rank	Account	Balance (€)	Rank
1	1095	20	1	885	7
2	955	14	2	850	4
3	1200	22	3	915	8
4	1195	21	4	950	12.5
5	925	9	5	800	2
6	950	12.5	6	750	1
7	805	3	7	865	5
8	945	11	8	1000	16
9	875	6	9	1050	18
10	1055	19	10	935	10
11	1025	17	Sum of ranks		83.5
12	975	15			
Sum of ranks		169.5			

the rank of 12.5. The next data value of €955 is then assigned the rank of 14. Table 19.6 shows the assigned rank of each observation.

The next step in the MWW test is to sum the ranks for each sample. The sums are given in Table 19.6. The test procedure can be based on the sum of the ranks for either sample. We use the sum of the ranks for the sample from branch 1. So, for this example, $T = 169.5$.

Given that the sample sizes are $n_1 = 12$ and $n_2 = 10$, we can use the normal approximation to the sampling distribution of the rank sum T. The appropriate sampling distribution is given by the following expressions.

Sampling distribution of T for identical populations

$$\text{Mean: } \mu_T = \frac{1}{2}n_1(n_1 + n_2 + 1) \tag{19.6}$$

$$\text{Standard deviation: } \sigma_T = \sqrt{\frac{1}{12}n_1 n_2(n_1 + n_2 + 1)} \tag{19.7}$$

Distribution form: approximately normal provided $n_1 \geq 10$ and $n_2 \geq 10$.

For branch 1, we have

$$\mu_T = \frac{1}{2}(12)(12 + 10 + 1) = 138$$

$$\sigma_T = \sqrt{\frac{1}{12}(12)(10)(12 + 10 + 1)} = 15.17$$

Figure 19.4 is the sampling distribution of T. We shall use a 0.05 level of significance to draw a conclusion. With the sum of the ranks for branch 1, $T = 169.5$, we calculate the following value for the test statistic.

$$z = \frac{T - \mu_T}{\sigma_T} = \frac{169.5 - 138}{15.17} = 2.08$$

Using the standard normal distribution table and $z = 2.08$, we find the two-tailed p-value $= 2(1 - 0.9812) = 0.0376$. With p-value $< \alpha = 0.05$, we reject H_0 and conclude that the two

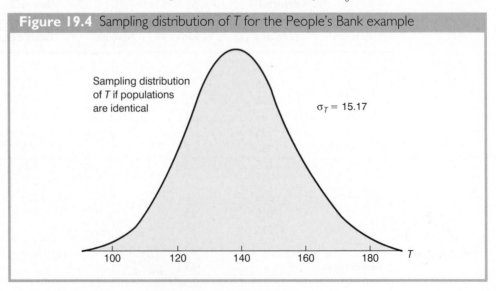

Figure 19.4 Sampling distribution of T for the People's Bank example

Sampling distribution of T if populations are identical

$\sigma_T = 15.17$

100   120   140   160   180 T

populations are not identical; that is, the populations of cheque account balances at the branch banks are not the same. The evidence suggests that the balances at branch 1 tend to be higher (and therefore be assigned higher ranks) than the balances at branch 2.

In summary, the Mann-Whitney-Wilcoxon rank-sum test consists of the following steps to determine whether two independent random samples are selected from identical populations.

1 Rank the combined sample observations from lowest to highest, with tied values being assigned the average of the tied rankings.

2 Compute T, the sum of the ranks for the first sample.

3 In the large-sample case, make the test for significant differences between the two populations by using the observed value of T and comparing it with the sampling distribution of T for identical populations using equations (19.6) and (19.7). The value of the standardized test statistic z and the p-value provide the basis for deciding whether to reject H_0. In the small-sample case, use Table 9 in Appendix B to find the critical values for the test.

The parametric statistical tests described in Chapter 10 test the equality of two population means. When we reject the hypothesis that the means are equal, we conclude that the populations differ in their means. When we reject the hypothesis that the populations are identical by using the MWW test, we cannot state how they differ. The populations could have different means, different medians, different variances, or different forms. Nonetheless, if we believe that the populations are the same in every aspect but the means, a rejection of H_0 by the non-parametric method implies that the means differ.

Exercises

Applications

13 Two fuel additives are being tested to determine their effect on petrol consumption. Seven cars were tested with additive 1 and nine cars were tested with additive 2. The following data show the litres of fuel used per 100 kilometres with the two additives. Use $\alpha = 0.05$ and the MWW test to see whether there is a significant difference in petrol consumption for the two additives.

Additive 1	Additive 2
8.20	7.52
7.69	7.94
7.41	6.62
8.47	6.71
7.75	6.41
7.58	7.52
8.06	7.14
	6.80
	6.99

14 *Business Week* annually publishes statistics on the world's 1000 largest companies. A company's price/earnings (P/E) ratio is the company's current stock price divided by the latest 12 months' earnings per share. Listed below are the P/E ratios for a sample of ten Japanese

and 12 US companies. Is the difference in P/E ratios between the two countries significant? Use the MWW test and $\alpha = 0.01$ to support your conclusion.

Japan		US	
Company	P/E ratio	Company	P/E ratio
Sumitomo Corp.	153	Gannet	19
Kinden	21	Motorola	24
Heiwa	18	Schlumberger	24
NCR Japan	125	Oracle Systems	43
Suzuki Motor	31	Gap	22
Fuji Bank	213	Winn-Dixie	14
Sumitomo Chemical	64	Ingersoll-Rand	21
Seibu Railway	666	American Electric Power	14
Shiseido	33	Hercules	21
Toho Gas	68	Times Mirror	38
		WellPoint Health	15
		Northern States Power	14

15 Samples of starting annual salaries for individuals entering the public accounting and financial planning professions follow. Annual salaries are shown in thousands of euros.

Public Accountant	Public Accountant	Financial Planner	Financial Planner
45.2	50.0	44.0	48.6
53.8	45.9	44.2	44.7
51.3	54.5	48.1	48.9
53.2	52.0	50.9	46.8
49.2	46.9	46.9	43.9

a. Use $\alpha = 0.05$ level of significance and test the hypothesis that there is no difference between the starting annual salaries of public accountants and financial planners. What is your conclusion?
b. What are the sample mean annual salaries for the two professions?

16 A confederation of house builders provided data on the cost (in £) of the most popular home re-modelling projects. Use the Mann-Whitney-Wilcoxon test to see whether it can be concluded that the cost of kitchen re-modelling differs from the cost of master bedroom re-modelling. Use a 0.05 level of significance.

Kitchen	Master Bedroom
13 200	6 000
5 400	10 900
10 800	14 400
9 900	12 800
7 700	14 900
11 000	5 800
7 700	12 600
4 900	9 000
9 800	
11 600	

17 The gap between the earnings of men and women with equal education is narrowing but has not closed. Sample data for seven men and seven women with bachelor's degrees are as follows. Data of earnings are shown in thousands of euros.

Men	30.6	75.5	45.2	62.2	38.2	49.9	55.3
Women	44.5	35.4	27.9	40.5	25.8	47.5	24.8

a. What is the median salary for men? For women?
b. Use $\alpha = 0.05$ and conduct the hypothesis test for equal populations. What is your conclusion?

19.4 Kruskal-Wallis test

The MWW test in Section 19.3 can be used to test whether two populations are identical. Kruskal and Wallis extended the test to the case of three or more populations. The hypotheses for the **Kruskal-Wallis test** with $k \geq 3$ populations can be written as follows.

H_0: All k populations are identical
H_1: Not all k populations are identical

The Kruskal-Wallis test is based on the analysis of independent random samples from each of the k populations.

In Chapter 13 we showed that analysis of variance (ANOVA) can be used to test for the equality of means among three or more populations. The ANOVA procedure requires interval- or ratio-level data and the assumption that the k populations are normally distributed. The non-parametric Kruskal-Wallis test can be used with ordinal data as well as with interval or ratio data. In addition, the Kruskal-Wallis test does not require the assumption of normally distributed populations. We demonstrate the Kruskal-Wallis test by using it in an employee selection application.

Williams Manufacturing Company Limited hires employees for its management staff from three local colleges. Recently, the company's personnel department began collecting and reviewing annual performance ratings in an attempt to determine whether there are differences in performance among the managers hired from these colleges. Performance rating data are available from independent samples of seven employees from college A, six employees from college B and seven employees from college C. These data are summarized in Table 19.7; the overall performance rating of each manager is given on a 0–100 scale, with 100 being the highest possible performance rating (the rankings are explained below).

Suppose we want to test whether the three populations are identical in terms of performance evaluations. We shall use a 0.05 level of significance. The Kruskal-Wallis test statistic, which is based on the sum of ranks for each of the samples, can be computed as follows.

Kruskal-Wallis test statistic

$$W = \left[\frac{12}{n_T(n_T + 1)} \sum_{i=1}^{k} \frac{R_i^2}{n_i} \right] - 3(n_T + 1) \qquad (19.8)$$

where

$$k = \text{the number of populations}$$
$$n_i = \text{the number of items in sample } i$$
$$n_T = \Sigma n_i = \text{total number of items in all samples}$$
$$R_i = \text{sum of the ranks for sample } i$$

Kruskal and Wallis were able to show that, under the null hypothesis that the populations are identical, the sampling distribution of W can be approximated by a chi-squared distribution with $k - 1$ degrees of freedom. This approximation is generally acceptable if each of the sample sizes is greater than or equal to five. The null hypothesis of identical populations will be rejected if the test statistic is large. As a result, the procedure uses an upper tail test.

To compute the W statistic for our example, we must first rank all 20 data items. The lowest data value of 15 from college B sample receives a rank of 1, whereas the highest data value of 95 from college A sample receives a rank of 20. The ranks and the sums of the ranks for the three samples are given in Table 19.7. Note that we assign the average rank to tied items;* for example, the data values of 60, 70, 80 and 90 had ties.

The sample sizes are

$$n_1 = 7 \qquad n_2 = 6 \qquad n_3 = 7$$

and

$$n_T = \Sigma n_i = 7 + 6 + 7 = 20$$

We compute the W statistic by using equation (19.8).

$$W = \frac{12}{20(21)} \left[\frac{(95)^2}{7} + \frac{(27)^2}{6} + \frac{(88)^2}{7} \right] - 3(20 + 1) = 8.92$$

Table 19.7 Performance evaluation ratings for 20 Williams employees

College A	Rank	College B	Rank	College C	Rank
25	3	60	9	50	7
70	12	20	2	70	12
60	9	30	4	60	9
85	17	15	1	80	15.5
95	20	40	6	90	18.5
90	18.5	35	5	70	12
80	15.5			75	14
Sum of ranks	**95**		27		88

*If numerous tied ranks are observed, equation (19.8) must be modified. The modified formula is given in *Practical Non-parametric Statistics* by W. J. Conover.

We can now use the chi-squared distribution table (Table 3 of Appendix B) to determine the p-value for the test. Using $k - 1 = 3 - 1 = 2$ degrees of freedom, we find $\chi^2 = 7.378$ has an area of 0.025 in the upper tail of the distribution and $\chi^2 = 9.21$ has an area of 0.01 in the upper tail. For $W = 8.92$, between 7.378 and 9.21, the area in the upper tail is between 0.025 and 0.01. Because it is an upper tail test, we can conclude that the p-value is between 0.025 and 0.01. (A calculation in MINITAB or EXCEL shows p-value = 0.0116.) Because p-value $< \alpha = 0.05$, we reject H_0 and conclude that the three populations are not identical. Manager performance differs significantly depending on the college attended. Furthermore, because the performance ratings are lowest for college B, it would be reasonable for the company to either cut back recruiting from college B or at least evaluate its graduates more thoroughly.

Exercises

Applications

18 Three college admission test preparation programmes are being evaluated. The scores obtained by a sample of 20 people who used the test preparation programmes provided the following data. Use the Kruskal-Wallis test to determine whether there is a significant difference among the three test preparation programmes. Use $\alpha = 0.01$.

Programme		
A	B	C
540	450	600
400	540	630
490	400	580
530	410	490
490	480	590
610	370	620
	550	570

19 Forty-minute workouts of one of the following activities three days a week will lead to a loss of weight. The following sample data show the number of calories burned during 40-minute workouts for three different activities. Do these data indicate differences in the amount of calories burned for the three activities? Use a 0.05 level of significance. What is your conclusion?

Swimming	Tennis	Cycling
408	415	385
380	485	250
425	450	295
400	420	402
427	530	268

20 *Condé Nast Traveler* magazine conducts an annual survey of its readers in order to rate the top 80 cruise ships in the world. With 100 the highest possible rating, the overall ratings for

a sample of ships from the Holland America, Princess, and Royal Caribbean cruise lines are shown here. Use the Kruskal-Wallis test with $\alpha = 0.05$ to determine whether the overall ratings among the three cruise lines differ significantly.

Holland America		Princess		Royal Caribbean	
Ship	Rating	Ship	Rating	Ship	Rating
Amsterdam	84.5	Coral	85.1	Adventure	84.8
Maasdam	81.4	Dawn	79.0	Jewel	81.8
Ooterdam	84.0	Island	83.9	Mariner	84.0
Volendam	78.5	Princess	81.1	Navigator	85.9
Westerdam	80.9	Star	83.7	Serenade	87.4

21 Course-evaluation ratings for four instructors follow. Use $\alpha = 0.05$ and the Kruskal-Wallis procedure to test for a significant difference in teaching abilities.

Instructor	Course-evaluation rating								
Black	88	80	79	68	96	69			
Jennings	87	78	82	85	99	99	85	94	81
Swanson	88	76	68	82	85	82	84	83	
Wilson	80	85	56	71	89	87			

19.5 Rank correlation

The Pearson product-moment correlation coefficient (see sections 3.5 and 14.3) is a measure of the linear association between two variables for which interval or ratio data are available. In this section, we consider the **Spearman rank correlation coefficient** r_S, which is a measure of association between two variables applicable when only ordinal data are available.

Spearman rank-correlation coefficient

$$r_S = 1 - \frac{6\Sigma d_i^2}{n(n^2 - 1)} \tag{19.9}$$

where

n = the number of items or individuals being ranked
x_i = the rank of item i with respect to one variable
y_i = the rank of item i with respect to the second variable
$d_i = x_i - y_i$

Suppose a company wants to determine whether individuals who were expected at the time of employment to be better salespersons actually turn out to have better sales records. To investigate this question, the personnel manager carefully reviewed the original job interview summaries, academic records and letters of recommendation for

Table 19.8 Sales potential and actual two-year sales data for ten salespeople, and computation of the Spearman rank-correlation coefficient

Salesperson	x_i = Ranking of potential	Two-year sales (units)	y_i = Ranking of sales performance	$d_i = x_i - y_i$	d_i^2
A	9	400	10	−1	1
B	7	360	8	−1	1
C	4	300	6	−2	4
D	10	295	5	5	25
E	5	280	4	1	1
F	8	350	7	1	1
G	1	200	1	0	0
H	2	260	3	−1	1
I	3	220	2	1	1
J	6	385	9	−3	9
					$\Sigma d_i^2 = 44$

$$r_S = 1 - \frac{6\Sigma d_i^2}{n(n^2 - 1)} = 1 - \frac{6(44)}{10(100 - 1)} = 0.73$$

ten current members of the firm's sales force. After the review, the personnel manager ranked the ten individuals in terms of their potential for success, basing the assessment solely on the information available at the time of employment. Then a list was obtained of the number of units sold by each salesperson over the first two years. On the basis of actual sales performance, a second ranking of the ten salespersons was carried out. Table 19.8 gives the relevant data and the two rankings. In the ranking of potential, rank 1 means lowest potential, rank 2 next lowest and so on. The statistical question is whether there is agreement between the ranking of potential at the time of employment and the ranking based on the actual sales performance over the first two years.

The computations for the Spearman rank-correlation coefficient are summarized in Table 19.8. We see that the rank-correlation coefficient is a positive 0.73. The Spearman rank-correlation coefficient ranges from −1.0 to +1.0 and its interpretation is similar to that of the Pearson correlation coefficient, in that positive values near 1.0 indicate a strong association between the rankings; as one rank increases, the other rank increases. Rank correlations near −1.0 indicate a strong negative association between the rankings; as one rank increases, the other rank decreases. The value $r_S = 0.73$ indicates a positive correlation between potential and actual performance. Individuals ranked high on potential tend to rank high on performance.

Test for significant rank correlation

As with many other statistical procedures, we may want to use the sample results to make an inference about the population rank correlation ρ_S. To make an inference about the population rank correlation, we test the following hypotheses.

$$H_0: \rho_S = 0$$
$$H_1: \rho_S \neq 0$$

Under the null hypothesis of no rank correlation ($\rho_S = 0$), the rankings are independent, and the sampling distribution of r_S is as follows.

Sampling distribution of r_S

$$\text{Mean: } \mu_{r_S} = 0 \qquad \text{(19.10)}$$

$$\text{Standard deviation: } \sigma_{r_S} = \sqrt{\frac{1}{n-1}} \qquad \text{(19.11)}$$

Distribution form: approximately normal provided $n \geq 10$.

The sample rank-correlation coefficient for sales potential and sales performance is $r_S = 0.73$. From equation (19.10) we have $\mu_{r_S} = 0$ and from (19.11) we have

$$\sigma_{r_S} = \sqrt{1/(10-1)} = 0.33$$

Using the test statistic, we have

$$z = \frac{r_S - \mu_{r_S}}{\sigma_{r_S}} = \frac{0.73 - 0}{0.33} = 2.20$$

Using the standard normal distribution table and $z = 2.20$, we find the p-value $= 2(1 - 0.9861) = 0.0278$. With a 0.05 level of significance, p-value $< \alpha = 0.05$ leads to the rejection of the hypothesis that the rank correlation is zero. We can conclude that there is a positive rank correlation between sales potential and sales performance.

Exercises

Methods

22 Consider the following set of rankings for a sample of ten elements.

Element	x_i	y_i	Element	x_i	y_i
1	10	8	6	2	7
2	6	4	7	8	6
3	7	10	8	5	3
4	3	2	9	1	1
5	4	5	10	9	9

a. Compute the Spearman rank-correlation coefficient for the data.
b. Use $\alpha = 0.05$ and test for significant rank correlation. What is your conclusion?

23 Consider the following two sets of rankings for six items.

	Case One			Case Two	
Item	First ranking	Second ranking	Item	First ranking	Second ranking
A	1	1	A	1	6
B	2	2	B	2	5
C	3	3	C	3	4
D	4	4	D	4	3
E	5	5	E	5	2
F	6	6	F	6	1

Note that in the first case the rankings are identical, whereas in the second case the rankings are exactly opposite. What value should you expect for the Spearman rank-correlation coefficient for each of these cases? Explain. Calculate the rank-correlation coefficient for each case.

Applications

24 The following two lists show how ten IT companies ranked in a national survey, in terms of reputation and percentage of respondents who said they would purchase the company's shares. A positive rank correlation is anticipated because it seems reasonable to expect that a company with a higher reputation would be a more desirable purchase.

Company	Reputation	Probable purchase
Microsoft	1	3
Intel	2	4
Dell	3	1
Lucent	4	2
Texas Instruments	5	9
Cisco Systems	6	5
Hewlett-Packard	7	10
IBM	8	6
Motorola	9	7
Yahoo	10	8

a. Compute the rank correlation between reputation and probable purchase.
b. Test for a significant positive rank correlation. What is the p-value?
c. At $\alpha = 0.05$, what is your conclusion?

25 A student organization surveyed both recent graduates and current students to obtain information on the quality of teaching at a particular university. An analysis of the responses provided the following teaching-ability rankings. Do the rankings given by the current students agree with the rankings given by the recent graduates? Use $\alpha = 0.10$ and test for a significant rank correlation.

	Ranking by	
Professor	Current students	Recent graduates
A	4	6
B	6	8
C	8	5
D	3	1
E	1	2
F	2	3
G	5	7
H	10	9
J	7	4
K	9	10

26 A sample of 15 students received the following rankings on mid-term and final examinations in a statistics course.

Rank		Rank		Rank	
Mid-term	Final	Mid-term	Final	Mid-term	Final
1	4	6	2	11	14
2	7	7	5	12	15
3	1	8	12	13	11
4	3	9	6	14	10
5	8	10	9	15	13

Compute the Spearman rank-correlation coefficient for the data and test for a significant correlation, with $\alpha = 0.10$.

For additional online summary questions and answers go to the companion website at www.cengage.co.uk/aswsbe2

Summary

In this chapter we presented several statistical procedures that are classified as non-parametric methods. Because non-parametric methods can be applied to ordinal and in some cases nominal data, as well as interval and ratio data, and because they require less restrictive population distribution assumptions, they expand the class of problems that can be subjected to statistical analysis.

The sign test is a non-parametric procedure for identifying differences between two populations when the only data available are nominal data. In the small-sample case, the binomial probability distribution can be used to determine the critical values for the sign test; in the large-sample case, a normal approximation can be used. The Wilcoxon signed-rank test is a procedure for analysing matched-sample data whenever interval- or ratio-scaled data are available for each matched pair. The procedure tests the hypothesis that the two populations being considered are identical. The procedure assumes that the distribution of differences between matched pairs is symmetrical, but not necessarily normal in shape.

The Mann-Whitney-Wilcoxon test is a non-parametric method for testing for a difference between two populations based on two independent random samples. Tables were presented for the small-sample case, and a normal approximation was provided for the large-sample case. The Kruskal-Wallis test extends the Mann-Whitney-Wilcoxon test to the case of three or more populations. The Kruskal-Wallis test is the non-parametric analogue of the parametric ANOVA for differences among population means.

We introduced the Spearman rank-correlation coefficient as a measure of association for two ordinal or rank-ordered sets of items.

Key terms

Kruskal-Wallis test
Mann-Whitney-Wilcoxon (MWW)
  test
Non-parametric methods

Sign test
Spearman rank-correlation
  coefficient
Wilcoxon signed-rank test

Key formulae

Sign test (large-sample case)

$$\text{Mean: } \mu = 0.50n \tag{19.1}$$

$$\text{Standard deviation: } \sigma = \sqrt{0.25n} \tag{19.2}$$

Wilcoxon signed-rank test

$$\text{Mean: } \mu_T = 0 \tag{19.3}$$

$$\text{Standard deviation: } \sigma_T = \sqrt{\frac{n(n+1)(2n+1)}{6}} \tag{19.4}$$

Mann-Whitney-Wilcoxon test (large-sample)

$$\text{Mean: } \mu_T = \frac{1}{2}n_1(n_1 + n_2 + 1) \tag{19.6}$$

$$\text{Standard deviation: } \sigma_T = \sqrt{\frac{1}{12}n_1 n_2(n_1 + n_2 + 1)} \tag{19.7}$$

Kruskal-Wallis test statistic

$$W = \left[\frac{12}{n_T(n_T + 1)} \sum_{i=1}^{k} \frac{R_i^2}{n_i} \right] - 3(n_T + 1)$$ (19.8)

Spearman Rank-correlation coefficient

$$r_S = 1 - \frac{6\Sigma d_i^2}{n(n^2 - 1)}$$ (19.9)

Sampling distribution of r_S in test for significant rank correlation

Mean: $\mu_{r_S} = 0$ (19.10)

Standard deviation: $\sigma_{r_S} = \sqrt{\frac{1}{n - 1}}$ (19.11)

Case problem Company Profiles II

The file 'Companies' on the CD accompanying the text contains a data set compiled in late March 2006. It comprises figures relating to samples of companies whose shares are traded on the stock exchanges in Germany, France, Ireland, South Africa and Israel. The data contained in the file are:

Name of company
Country of stock exchange where the shares are traded
Code for industrial/commercial sector in which the company operates
Market value of company (expressed in £ million) in early 2006
Price change (%) in a recent 12-month period
Dividend yield (%), from the most recently available company accounts
Market to book value (%), from the most recently available company accounts
Return on investment (%), from the most recently available company accounts

BMW - World Munich, Germany. © imagebroker/Alamy.

Percentage of company share in 'free float' (i.e. available for trading)
'Volatility' (standard deviation of share price over last 12 months, divided by mean over last 12 months, multiplied by 40)

A screenshot of the first few rows of data is shown below.

COMPANIES

Company name	Country	Sector	Market value (£ million)	12-month price change (%)	Dividend yield (%)	Market to book value	Return on investment (%)	% of shares in free float	Volatility
Adidas-Salomon	BD	PERSONAL GOODS	5750.63	36.7	0.79	3.02	23.50	100	5
Allianz	BD	NON-LIFE INSURANCE	38271.64	38.8	1.29	1.49	9.04	100	6
Altana	BD	PHARMACEUTICALS	4838.53	2.4	1.91	3.93	27.03	50	3
BASF	BD	CHEMICALS	24746.79	15.7	2.65	1.92	12.16	95	4
Bayer	BD	CHEMICALS	17050.88	32.4	2.82	1.48	4.94	95	5
BMW	BD	AUTOMB	19462.39	26.9	1.37	1.28	13.76	53	3
Commerzbank	BD	BANKS	14454.39	90.6	0.79	0.93	4.32	92	9
Continental	BD	AUTOMOBILES	9037.81	55.1	0.89	2.23	33.98	84	6
Daimler Chrysler	BD	AUTOMOBILES	32925.26	37.5	3.19	1.20	8.49	93	7

Source: *Datastream*, Thomson Financial

Analyst's report

Using non-parametric methods of testing, investigate the following:

1 Is there any evidence of differences between the companies traded on the five different stock exchanges in respect of the company market values, in respect of the 12-month price changes, and in respect of the percentage of company shares in free float?

2 Is there any evidence of differences between the companies traded on the French and German stock markets in respect of the distribution of dividend yields and of the return on investment figures?

3 Is there any evidence of a relationship between 12-month price change and volatility? Would you expect a relationship, given the way that volatility has been measured?

4 Is there any evidence of a relationship between company market value and volatility? Would you expect a relationship?

Software Section for Chapter 19

Non-parametric methods using MINITAB

MINITAB has inbuilt routines for all the non-parametric methods described in this chapter, with the exception of the Spearman rank-correlation coefficient. Routines for the sign test, the Wilcoxon signed-rank test, the Mann-Whitney-Wilcoxon test and the Kruskal-Wallis test are all accessed by opening the **Stat** menu, then choosing appropriately from the **Non-parametrics** sub-menu. For each of the tests, the dialogue panel is very simple, with few options. The dialogues are described briefly below, along with examples of the output that MINITAB produces.

Sign test

The MINITAB sign test routine does a test for the median value of the population, using numerical data from a single column of the worksheet. Assume that the data from the Sunny Vale Farms taste test example of Section 19.1 (see Table 19.1) are in column C1 (named Preference), with the negative signs entered as -1 (preference for Tropical Orange) and the positive signs entered as $(+)1$ (preference for Citrus Delight).

Step 1   **Stat > Non-parametrics > 1-Sample Sign**   [Main menu bar]

Step 2   Transfer **Preference** to the **Variables** box   [**1-Sample Sign** panel]
               Check **Test median,** enter **0** in the adjacent box (population median value under H_0)
               Choose **not equal** from the **Alternative** drop-down menu (two-sided H_1)
               Click **OK**

The MINITAB output is shown in Figure 19.5. MINITAB shows the p-value as 0.0386, almost identical to the value calculated using tables in Section 19.1. H_0 is rejected at the 5 per cent level of significance.

Wilcoxon signed-rank test

The routine in MINITAB assumes that a column of difference scores d_i is available for analysis. If this is not the case, the d_i first need to be calculated. Let us assume that the production task completion time data from Section 19.2 (see Table 19.2) are in columns C1 and C2 of the MINITAB worksheet. In the instructions below, the differences d_i are calculated first and stored in column C3 (named Method1 – Method2):

Figure 19.5 Example of MINITAB output for the sign test

Results for: TASTETEST.MTW

Sign Test for Median: Preference

Sign test of median = 0.00000 versus not = 0.00000

	N	Below	Equal	Above	P	Median
Preference	12	10	0	2	0.0386	-1.000

Step 1 **Calc > Calculator**   [Main menu bar]

Step 2 Transfer **Method1 – Method2** to the **Store result in variable** box
[**Calculator** panel]

Enter **C1 – C2** in the **Expression** box
Click **OK**

Step 3 **Stat > Non-parametrics > 1-sample Wilcoxon**   [Main menu bar]

Step 4 Transfer **Method1 – Method2** to the **Variables** box
[**1-sample Wilcoxon** panel]

Check **Test median,** enter **0** in the adjacent box (population median value under H_0).
Choose **not equal** from the **Alternative** drop-down menu (two-sided H_1)
Click **OK**

The MINITAB output is shown in Figure 19.6. MINITAB shows the *p*-value as 0.028, almost identical to the value we calculated in Section 19.2 with the help of the standard normal table. H_0 is rejected at the 5 per cent level of significance.

Mann-Whitney-Wilcoxon test

To do the MWW test, the data for the two samples must be in separate columns of the MINITAB worksheet. Let us assume that the Secondary School data from section 19.3 (see Table 19.3) are in columns C1 and C2 (named Garfield and Mulberry respectively).

Figure 19.6 Example of MINITAB output for the Wilcoxon signed-rank test

Results for: ProdTask.MTW

Wilcoxon Signed Rank Test: Method1 - Method2

Test of median = 0.000000 versus median not = 0.000000

	N	N for Test	Wilcoxon Statistic	P	Estimated Median
Method1 - Method2	11	10	49.5	0.028	0.3750

Figure 19.7 Example of MINITAB output for the Mann-Whitney test

```
Results for: SecSchool.MTW

Mann-Whitney Test and CI: Garfield, Mulberry

            N  Median
Garfield   4    36.5
Mulberry   5   146.0


Point estimate for ETA1-ETA2 is -108.5
96.3 Percent CI for ETA1-ETA2 is (-181.0,-18.0)
W = 11.0
Test of ETA1 = ETA2 vs ETA1 not = ETA2 is significant at 0.0373
```

Step 1 **Stat > Non-parametrics > Mann-Whitney**   [Main menu bar]

Step 2 Transfer **Garfield** to the **First Sample** box   [**Mann-Whitney** panel]
Transfer **Mulberry** to the **Second Sample** box
Choose **not equal** from the **Alternative** drop-down menu (two-sided H_1)
Click **OK**

The MINITAB output is shown in Figure 19.7. MINITAB shows the p-value as 0.0373. H_0 is rejected at the 5 per cent level of significance, the result we found in Section 19.3 using critical values taken from tables for the MWW test. In the output below, MINITAB refers to the population medians as 'ETA1' and 'ETA2'. The output includes a confidence interval for the difference between the medians of the two populations.

Kruskal-Wallis test

For the Kruskal-Wallis test, the data for the variable to be analyzed must be in a single column of the MINITAB worksheet. A column of codes is needed to denote the groups into which the data is divided. Let us assume that the performance evaluation ratings data from section 19.4 (see Table 19.7) are in column C1 (named Ratings), with a set of codes (A, B, C) in C2 (named College) denoting the three colleges.

Step 1 **Stat > Non-parametrics > Kruskal-Wallis**   [Main menu bar]

Step 2 Transfer **Ratings** to the **Response** box   [**Kruskal-Wallis** panel]
Transfer **College** to the **Factor** box
Click **OK**

The MINITAB output is shown below in Figure 19.8. MINITAB shows the W statistic as 8.92 (MINITAB refers to the statistic as H), as calculated in Section 19.4, with a p-value of 0.012. MINITAB also shows a value for W (or H) = 8.98, with an accompanying p-value = 0.011, that is corrected for the effect of ties in the data. The two p-values are very similar. They confirm the conclusion of Section 19.4 that H_0 is rejected at the 5 per cent level of significance.

Figure 19.8 Example of MINITAB output for the Kruskal-Wallis test

Results for: CollegeRatings.MTW

Kruskal-Wallis Test: Ratings versus College

```
Kruskal-Wallis Test on Ratings

College    N   Median   Ave Rank       Z
A          7    80.00       13.6    1.70
B          6    32.50        4.5   -2.97
C          7    70.00       12.6    1.15
Overall   20                10.5

H = 8.92  DF = 2  P = 0.012
H = 8.98  DF = 2  P = 0.011   (adjusted for ties)
```

Non-parametric methods using PASW

PASW has inbuilt routines for all the non-parametric methods described in this chapter. Routines for the sign test, the Wilcoxon signed-rank test, the Mann-Whitney-Wilcoxon test and the Kruskal-Wallis test are all accessed by opening the **Analyze** menu, then choosing appropriately from the **Non-parametric Tests** sub-menu. The Spearman rank correlation procedure is found by choosing **Analyze > Correlate > Bivariate**. The dialogues are described briefly below, along with examples of the output that PASW produces.

Sign test

The sign test can be found in PASW in two guises: as a test for the median value of the population, and as an alternative to the Wilcoxon signed-ranks test for matched data (see below). As a test of a population median, it is referred to in PASW as the binomial test. Assume that the data from the Sunny Vale Farms taste test example of Section 19.1 (see Table 19.1) are in a column of the PASW Data Editor (variable named and labelled Preference), with the negative signs entered as -1 (preference for Tropical Orange) and the positive signs entered as $(+)1$ (preference for Citrus Delight).

Step 1 **Analyze > Non-parametric Tests > Binomial**   [Main menu bar]

Step 2 Transfer **Preference** to the **Test Variables List** box   [**Binomial Test** panel]
Check **Get from data** under **Define Dichotomy**
Enter **0.5** in **Test Proportion** box
Click **Exact**

Step 3 Check **Exact**   [**Exact Tests** panel]
Click **Continue**

Step 4 Click **OK**   [**Binomial Test** panel]

Figure 19.9 Example of PASW output for the sign test (binomial test in PASW)

NPar Tests

Binomial Test

		Category	N	Observed Prop.	Test Prop.	Exact Sig. (2-tailed)	Point Probability
Preference	Group 1	Tropical Orange	10	.83	.50	.039	.016
	Group 2	Citrus Delight	2	.17			
	Total		12	1.00			

The PASW output is shown above in Figure 19.9. PASW shows the p-value as 0.039, almost identical to the value calculated using tables in Section 19.1. H_0 is rejected at the 5 per cent level of significance.

Wilcoxon signed-rank test

Let us assume that the production task completion time data from Section 19.2 (see Table 19.2) are in two columns of the PASW Data Editor, with variable labels Method 1 and Method 2.

Step 1   Analyze > Non-parametric Tests > 2 Related Samples   [Main menu bar]

Step 2   Transfer **Method 1** to the **Variable1** box (Pair 1)

[Two-Related-Samples Test panel]
Transfer **Method 2** to the **Variable2** box (Pair 1)
Check **Wilcoxon**
Click **Exact**

Step 3   Check **Exact**   **[Exact Tests** panel]
Click **Continue**

Step 4   Click **OK**   **[Two-Related-Samples Test** panel]

The PASW output is shown in Figure 19.10. PASW shows the (exact) p-value as 0.023. H_0 is rejected at the 5 per cent level of significance. The p-value of 0.025 calculated in Section 19.2 is the asymptotic p-value shown in the PASW output.

(The Two-Related-Samples Tests dialogue panel has the sign test as an analysis option. For the sign test, only the signs of the paired differences d_i are taken into account, not their magnitudes. The test examines the null hypothesis that the probabilities of positive and negative differences d_i are equal.)

Mann-Whitney-Wilcoxon test

To do the MWW test, the data for the two samples must be set out in the customary PASW layout for independent samples, with the data to be analyzed in one column of the PASW data editor, and a set of codes distinguishing the samples in a second column. Let us assume that the secondary school data from section 19.3 (see Table 19.3) are set out in this

Figure 19.10 Example of PASW output for the Wilcoxon signed-rank test

NPar Tests

Wilcoxon Signed Ranks Test

Ranks

		N	Mean Rank	Sum of Ranks
Method 2 - Method 1	Negative Ranks	8[a]	6.19	49.50
	Positive Ranks	2[b]	2.75	5.50
	Ties	1[c]		
	Total	11		

a. Method 2 < Method 1

b. Method 2 > Method 1

c. Method 2 = Method 1

Test Statistics[b]

	Method 2 - Method 1
Z	-2.245[a]
Asymp. Sig. (2-tailed)	.025
Exact Sig. (2-tailed)	.023
Exact Sig. (1-tailed)	.012
Point Probability	.004

a. Based on positive ranks.

b. Wilcoxon Signed Ranks Test

way, with the ranking in a variable labelled Year-group ranking and the column of codes (1,2) in a variable labelled Primary School (with value labels Garfield and Mulberry).

Step 1 **Analyze > Non-parametric Tests > 2 Independent Samples**
[Main menu bar]

Step 2 Transfer **Year-group ranking** to the **Test Variable List** box
[**Two-Independent-Samples Tests** panel]
Transfer **Primary School** to the **Grouping Variable** box
Click on **Define Groups**

Step 3 Enter **1** in the **Group 1** box
[**Two Independent Samples: Define Groups** panel]
Enter **2** in the **Group 2** box
Click **Continue**

Step 4 Check **Mann-Whitney** [**Two-Independent-Samples Tests** panel]
Click **Exact**

Figure 19.11 Example of PASW output for the Mann-Whitney-Wilcoxon test

NPar Tests

Mann-Whitney Test

Ranks

	Primary School	N	Mean Rank	Sum of Ranks
Year-group ranking	Garfield	4	2.75	11.00
	Mulberry	5	6.80	34.00
	Total	9		

Test Statistics[b]

	Year-group ranking
Mann-Whitney U	1.000
Wilcoxon W	11.000
Z	-2.205
Asymp. Sig. (2-tailed)	.027
Exact Sig. [2*(1-tailed Sig.)]	.032[a]
Exact Sig. (2-tailed)	.032
Exact Sig. (1-tailed)	.016
Point Probability	.008

a. Not corrected for ties.

b. Grouping Variable: Primary School

Step 5	Check **Exact**	[**Exact Tests** panel]
	Click **Continue**	

Step 6	Click **OK**	[**Two-Independent-Samples Tests** panel]

The PASW output is shown above in Figure 19.11. PASW shows the (exact) p-value as 0.032. H_0 is rejected at the 5 per cent level of significance, the result we found in Section 19.3 using critical values taken from tables for the MWW test.

Kruskal-Wallis test

For the Kruskal-Wallis test, the data must be set out in the customary PASW layout for independent samples, with the data to be analyzed in one column of the PASW data editor, and a set of codes distinguishing the groups in a second column. Let us assume that the performance evaluation ratings data from section 19.4 (see Table 19.7) are entered in a variable in the PASW data editor (labelled Performance ratings), with a set of codes (1, 2, 3, with value labels A, B, C) in a second variable labelled College.

Step 1	**Analyze > Non-parametric Tests > K Independent Samples**

[Main menu bar]

Step 2 Transfer **Performance ratings** to the **Test Variable List** box

[**Test for Several Independent Samples** panel]

Transfer **College** to the **Grouping Variable** box

Click on **Define Range**

Step 3 Enter **1** in the **Minimum** box   [**Define Range** panel]

Enter **3** in the **Maximum** box

Click **Continue**

Step 4 Check **Kruskal-Wallis H**   [**Test for Several Independent Samples** panel]

Click **Exact**

Step 5 Check **Asymptotic only**   [**Exact Tests** panel]

Click **Continue**

Step 6 Click **OK**   [**Test for Several Independent Samples** panel]

The PASW output is shown in Figure 19.12. PASW shows the W statistic as 8.98 (PASW labels the figure as Chi-Square, in the second table below), with a p-value

Figure 19.12 Example of PASW output for the Kruskal-Wallis test

NPar Tests

Kruskal-Wallis Test

Ranks

	College	N	Mean Rank
Performance rating	A	7	13.57
	B	6	4.50
	C	7	12.57
	Total	20	

Test Statistics[a,b]

	Performance rating
Chi-Square	8.984
df	2
Asymp. Sig.	.011

a. Kruskal Wallis Test

b. Grouping Variable: College

of 0.011. These figures differ very slightly from those calculated in Section 19.4, because PASW applies a correction for ties in the data. The p-value confirms the conclusion of Section 19.4 that H_0 is rejected at the 5 per cent level of significance. The Chi-Squared p-value for the Kruskal-Wallis test is based on an 'asymptotic' (i.e. large-sample) approximation. PASW will also calculate an exact p-value for this test if **Exact** is checked on the **Exact Tests** dialogue panel.

Spearman rank correlation

We shall illustrate Spearman rank correlation in PASW using the personnel ranking example of Section 19.5. Let us assume that the sales potential rankings and 2-year sales figures (see Table 19.8) are entered in variables in the PASW data editor with variable labels Ranking of Potential and 2-year Sales respectively.

Step 1 **Analyze > Correlate > Bivariate** [Main menu bar]

Step 2 Transfer **Ranking of Potential** and **2-year Sales** to the **Variables** box

[**Bivariate Correlations** panel]

Uncheck **Pearson**
Check **Spearman**
Check **Two-tailed**
Check **Flag significant correlations**
Click **OK**

The PASW output is shown below in Figure 19.13. PASW shows the Spearman rank correlation r_S as 0.733, with a p-value of 0.016, confirming the conclusion of Section 19.5 that H_0 (no correlation in the population) is rejected at the 5 per cent level of significance. The PASW p-value is slightly different from that given by the normal approximation in Section 19.5.

Figure 19.13 Example of PASW output for Spearman rank-correlation

Nonparametric Correlations

Correlations

			Ranking of Potential	2-year Sales
Spearman's rho	Ranking of Potential	Correlation Coefficient	1.000	.733[*]
		Sig. (2-tailed)	.	.016
		N	10	10
	2-year Sales	Correlation Coefficient	.733[*]	1.000
		Sig. (2-tailed)	.016	.
		N	10	10

*. Correlation is significant at the 0.05 level (2-tailed).

Chapter 20

Statistical Methods for Quality Control

Learning objectives

After reading this chapter and doing the exercises, you should be able to:

1 Understand the importance of quality control and how statistical methods can assist in the quality control process.

2 Understand acceptance sampling procedures.

3 Know the difference between consumer's risk and producer's risk.

4 Use the binomial probability distribution to develop acceptance sampling plans.

5 Know what is meant by multiple sampling plans.

6 Construct quality control charts and understand how they are used for statistical process control.

7 Know the definition of the following terms:

producer's risk
consumer's risk
acceptance sampling
acceptable criterion
operating characteristic curve
assignable causes
common causes
control charts
upper control limit
lower control limit

For its ISO 9001 quality standard, the International Organization for Standardization (ISO) defines quality as a characteristic that a product or service must have. A quality product or service is therefore one that meets the needs and expectations of customers. Organizations recognize that to be competitive in today's global economy, they must strive for high levels of quality.

As a result, an increased emphasis falls on methods for monitoring and maintaining quality. *Quality assurance* refers to the entire system of policies, procedures, and guidelines established by an organization to achieve and maintain quality. Quality assurance consists of two principal functions: quality engineering and quality control. The objective of *quality engineering* is to include quality in the design of products and processes and to identify potential quality problems prior to production. **Quality control** consists of making a series of inspections and measurements to determine whether quality standards are being met. If quality standards are not being met, corrective and/or preventive action can be taken to achieve and maintain conformance. As we will show in this chapter, statistical techniques are extremely useful in quality control.

Traditional manufacturing approaches to quality control have been found to be less than satisfactory and are being replaced by improved managerial tools and techniques. In particular the total quality movement which began in Japan following World War II owed much to the contributions of the two great figures of the 'quality age', Dr W. Edwards Deming and Dr Joseph M. Juran.

Although quality is everybody's job, Deming stressed that quality must be led by managers. He developed a list of 14 points that he believed are the key responsibilities of managers. For instance, Deming stated that managers must cease dependence on mass inspection; must end the practice of awarding business solely on the basis of price; must seek continual improvement in all production processes and services; must foster a team-oriented environment; and must eliminate numerical goals, slogans and work standards that prescribe numerical quotas. Perhaps most importantly, managers must create a work environment in which a commitment to quality and productivity is maintained at all times.

Statistics in Practice

ABC Aerospace

ABC Aerospace specializes in repairing and overhauling aeronautical equipment and one of its operations involves the heat treatment of aluminium sheets to 920°F (≈493°C) so that they are able to withstand pressures of up to 60 pounds per square inch (≈4.2 kg per sq cm). In the past, the operation has taken place in a furnace measuring approximately 5 m (length) by 2 m (height) and 1.5 m (depth) with 13 thermocouples monitoring temperatures within it. Each time the furnace commenced operation, readings from these – recorded by operators at five-minute intervals over a 40-minute period – were analyzed using statistical process control

Aluminium is a key component of aeronautical equipment.
© Ferenc Szelepcsenyi.

methods to ensure the 920°F temperature standard was being maintained.

However, the same charts revealed that temperatures within the furnace often varied slightly – some locations, or areas, being marginally hotter or cooler than others. The reasons for this were thought to be two-fold: because of its counter-clockwise circular flow around the furnace, airflow was not uniform. In addition, sliding doors at the entrance to the furnace did not seal completely, causing temperatures to be lower in this area. Mostafa (2003) describes various countermeasures investigated by the company for dealing with this problem. (A particular concern was that temperature variability at one stage was causing the system to operate at 2°F above its 920°F setting.) Unfortunately none of these countermeasures proved to be productive. Given the well-known quality maxim that adjusting a stable process only makes things worse, perhaps this was unsurprising. But undeterred by their results, the quality improvement team went on to propose further countermeasures which even on a small scale, they were convinced would lead to overall performance improvement. However, these were overruled by ABC's management on cost grounds.

Source: Mostafa, Mohamed M (2003) Deming's Funnel Experiment in Quality Improvement – a Computer Simulation. *OR Insight* Vol 16 4 25–31.

In this chapter we present two statistical methods used in quality control. The first method, *statistical process control*, uses graphical displays known as *control charts* to monitor a production process; the goal is to determine whether the process can be continued or whether it should be adjusted to achieve a desired quality level. The second method, *acceptance sampling*, is used in situations where a decision to accept or reject a group of items must be based on the quality found in a sample.

20.1 Statistical process control

We consider quality control procedures for a production process whereby goods are manufactured continuously. On the basis of sampling and inspection of production output, a decision will be made to either continue the production process or adjust it to bring the items or goods being produced up to acceptable quality standards.

Despite high standards of quality in manufacturing and production operations, machine tools will invariably wear out, vibrations will throw machine settings out of adjustment, purchased materials will be defective and human operators will make mistakes. Any or all of these factors can result in poor quality output. Fortunately, procedures are available for monitoring production output so that poor quality can be detected early and the production process can be adjusted or corrected.

If the variation in the quality of the production output is due to **assignable causes** such as tools wearing out, incorrect machine settings, poor quality raw materials or operator error, the process should be adjusted or corrected as soon as possible. Alternatively, if the variation is due to what are called **common causes** – that is, randomly occurring variations in materials, temperature, humidity, and so on, which the manufacturer cannot possibly control – the process does not need to be adjusted. The main objective of statistical process control is to determine whether variations in output are due to assignable causes or common causes.

Whenever assignable causes are detected, we conclude that the process is *out of control*. In that case, corrective action will be taken to bring the process back to an acceptable level of quality. However, if the variation in the output of a production process is due only to common causes, we conclude that the process is *in statistical control*, or simply *in control*; in such cases, no changes or adjustments are necessary.

The statistical procedures for process control are based on the hypothesis testing methodology presented in Chapter 9. The null hypothesis H_0 is formulated in terms of the production process being in control. The alternative hypothesis H_1 is formulated in terms of the production process being out of control. Table 20.1 shows that correct decisions to continue an in-control process and adjust an out-of-control process are possible. However, as with other hypothesis testing procedures, both a Type I error (adjusting an in-control process) and a Type II error (allowing an out-of-control process to continue) are also possible.

Control charts

A **control chart** provides a basis for deciding whether the variation in the output is due to common causes (in control) or assignable causes (out of control). Whenever an out-of-control situation is detected, adjustments or other corrective action will be taken to bring the process back into control.

Control charts can be classified by the type of data they contain. An $\bar{x}$ **chart** is used if the quality of the output of the process is measured in terms of a variable such as length, weight, temperature and so on. In that case, the decision to continue or to adjust the production process will be based on the mean value found in a sample of the output. To introduce some of the concepts common to all control charts, consider Figure 20.1 which shows the general structure of an $\bar{x}$ chart. The centre line of the chart corresponds

	State of production process	
Table 20.1 The outcomes of statistical process control		
Decision	H_0 True Process in control	H_0 False Process out of control
Continue process	Correct decision	Type II error (allowing an out-of-control process to continue)
Adjust process	Type I error (adjusting an in-control process)	Correct decision

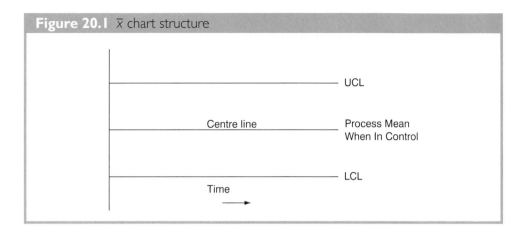

Figure 20.1 $\bar{x}$ chart structure

UCL

Centre line

Process Mean
When In Control

LCL

Time

to the mean of the process when the process is *in control*. The vertical line identifies the scale of measurement for the variable of interest. Each time a sample is taken from the production process, a value of the sample mean $\bar{x}$ is computed and a data point showing the value of $\bar{x}$ is plotted on the control chart.

The two lines labelled UCL and LCL are important in determining whether the process is in control or out of control. The lines are called the *upper control limit* and the *lower control limit*, respectively. They are chosen so that when the process is in control, there will be a high probability that the value of $\bar{x}$ will be between the two control limits. Values outside the control limits provide strong statistical evidence that the process is out of control and corrective action should be taken.

Over time, more and more data points will be added to the control chart. The order of the data points will be from left to right as the process is sampled. In essence, every time a point is plotted on the control chart, we are carrying out a hypothesis test to determine whether the process is in control.

In addition to the $\bar{x}$ chart, other control charts can be used to monitor the range of the measurements in the sample (**R chart**), the proportion defective in the sample (**p chart**), and the number of defective items in the sample (**np chart**). In each case, the control chart has an LCL, a centre line, and a UCL similar to the $\bar{x}$ chart in Figure 20.1. The major difference among the charts is what the vertical axis measures; for instance, in a p chart the measurement scale denotes the proportion of defective items in the sample instead of the sample mean. In the following discussion, we will illustrate the construction and use of the $\bar{x}$ chart, R chart, p chart, and np chart.

$\bar{x}$ chart: process mean and standard deviation known

To illustrate the construction of an $\bar{x}$ chart, let us consider the situation at KJW Packaging. This company operates a production line where cartons of cereal are filled. Suppose KJW knows that when the process is operating correctly – and hence the system is in control – the mean filling weight is $\mu = 450$ grams, and the process standard deviation is $\sigma = 3$ grams. In addition, assume the filling weights are normally distributed. This distribution is shown in Figure 20.2.

The sampling distribution of $\bar{X}$ is as presented in Chapter 7, can be used to determine the variation that can be expected in $\bar{X}$ values for a process that is in control. Let us first briefly review the properties of the sampling distribution of $\bar{X}$. First, recall that the expected value or mean of $\bar{X}$ is equal to μ, the mean filling weight when the production line is in control. For samples of size n, the equation for the standard deviation of $\bar{X}$, called the standard error of the mean, is

Standard error of the mean

$$\sigma_{\bar{x}} = \frac{\sigma}{\sqrt{n}}$$

(20.1)

In addition, because the filling weights are normally distributed, the sampling distribution of $\bar{X}$ is normally distributed for any sample size. Thus, the sampling distribution of $\bar{X}$ is a normal distribution with mean μ and standard deviation $\sigma_{\bar{x}}$. This distribution is shown in Figure 20.3.

The sampling distribution of $\bar{X}$ is used to determine what values of $\bar{X}$ are reasonable if the process is in control. The general practice in quality control is to define as reasonable any value of $\bar{X}$ that is within three standard deviations, or standard errors, above or below the mean value. Recall from the study of the normal probability distribution that

Figure 20.2 Normal distribution of cereal carton filling weights

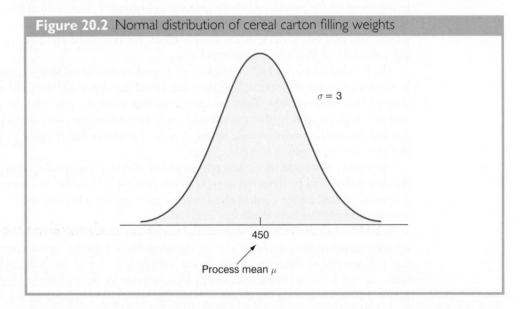

Figure 20.3 Sampling distribution of $\bar{X}$ for a sample of n filling weights

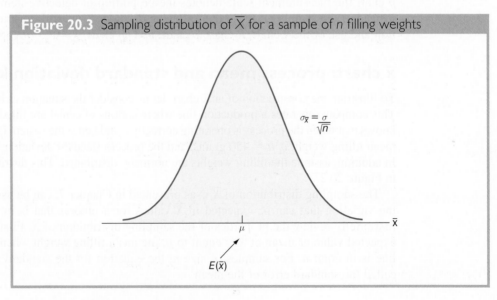

approximately 99.7 per cent of the values of a normally distributed random variable are within ±3 standard deviations of its mean value. Thus, if the value of $\overline{X}$ is within the interval $\mu - 3\sigma_{\overline{x}}$ to $\mu + 3\sigma_{\overline{x}}$, we will assume that the process is in control. In summary, then, the control limits for an x chart are as follows.

Control limits for an $\overline{x}$ chart: process mean and standard deviation known

$$UCL = \mu + 3\sigma_{\overline{x}} \qquad \qquad \textbf{(20.2)}$$
$$LCL = \mu - 3\sigma_{\overline{x}} \qquad \qquad \textbf{(20.3)}$$

Reconsider the KJW Packaging example with the process distribution of filling weights shown in Figure 20.2 and the sampling distribution of $\overline{X}$ shown in Figure 20.3. Assume that a quality control inspector periodically samples six cartons and uses the sample mean filling weight to determine whether the process is in control or out of control. Using equation (20.1), we find that the standard error of the mean is $\sigma_{\overline{x}} = 3/\sqrt{n} = 3/\sqrt{6} = 1.22$. Thus, with the process mean at 450, the control limits are UCL $= 450 + 3(1.22) = 453.66$ and LCL $= 450 - 3(1.22) = 446.34$. Figure 20.4 is the control chart with the results of ten samples taken over a ten-hour period. For ease of reading, the sample numbers 1 through 10 are listed below the chart.

Note that the mean for the fifth sample in Figure 20.4 shows that the process is out of control. The fifth sample mean is below the LCL indicating that assignable causes of output variation are present and that under-filling is occurring. As a result, corrective action was taken at this point to bring the process back into control. The fact that the remaining points on the $\overline{x}$ chart are within the upper and lower control limits indicates that the corrective action was successful.

$\overline{x}$ chart: process mean and standard deviation unknown

In the KJW Packaging example, we showed how an $\overline{x}$ chart can be developed when the mean and standard deviation of the process are known. In most situations, the process mean and standard deviation must be estimated by using samples that are selected from the process when it is in control. For instance, KJW might select a random sample of five boxes

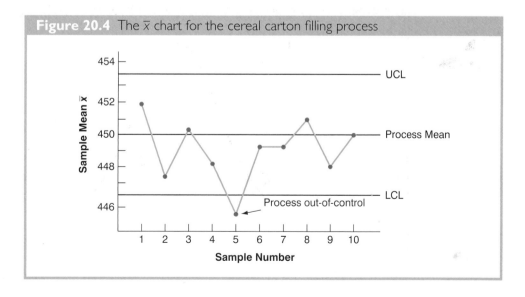

Figure 20.4 The $\overline{x}$ chart for the cereal carton filling process

each morning and five boxes each afternoon for ten days of in-control operation. For each subgroup, or sample, the mean and standard deviation of the sample are computed. The overall averages of both the sample means and the sample standard deviations are used to construct control charts for both the process mean and the process standard deviation.

In practice, it is more common to monitor the variability of the process by using the range instead of the standard deviation because the range is easier to compute. The range can be used to provide good estimates of the process standard deviation; thus it can be used to construct upper and lower control limits for the $\bar{x}$ chart with little computational effort. To illustrate, let us consider the problem facing Jensen Computer Supplies.

Jensen Computer Supplies (JCS) manufactures 12 centimetre-diameter computer disks; they have just finished adjusting their production process so that it is operating in control. Suppose random samples of five disks could be taken during the first hour of operation, during the second hour of operation, and so on, until 20 samples have been selected. Table 20.2 provides the diameter of each disk sampled as well as the mean $\bar{x}_j$ and range R_j for each of the samples.

The estimate of the process mean μ is given by the overall sample mean.

Overall sample mean

JENSEN

$$\bar{\bar{x}} = \frac{\bar{x}_1 + \bar{x}_2 \cdots + \bar{x}_k}{k}$$

(20.4)

where

$\bar{x}_j$ = mean of the jth sample $j = 1, 2, \ldots, k$
k = number of samples

Table 20.2 Data for the Jensen Computer Supplies problem

Sample number	Observations					Sample mean $\bar{x}_j$	Sample range R_j
1	12.0056	12.0086	12.0144	12.0009	12.0030	12.0065	0.0135
2	11.9882	12.0085	11.9884	12.0250	12.0031	12.0026	0.0368
3	11.9897	11.9898	11.9995	12.0130	11.9969	11.9978	0.0233
4	12.0153	12.0120	11.9989	11.9900	11.9837	12.0000	0.0316
5	12.0059	12.0113	12.0011	11.9773	11.9801	11.9951	0.0340
6	11.9977	11.9961	12.0050	12.0014	12.0060	12.0012	0.0099
7	11.9910	11.9913	11.9976	11.9831	12.0044	11.9935	0.0213
8	11.9991	11.9853	11.9830	12.0083	12.0094	11.9970	0.0264
9	12.0099	12.0162	12.0228	11.9958	12.0004	12.0090	0.0270
10	11.9880	12.0015	12.0094	12.0102	12.0146	12.0047	0.0266
11	11.9881	11.9887	12.0141	12.0175	11.9863	11.9989	0.0312
12	12.0043	11.9867	11.9946	12.0018	11.9784	11.9932	0.0259
13	12.0043	11.9769	11.9944	12.0014	11.9904	11.9935	0.0274
14	12.0004	12.0030	12.0082	12.0045	12.0234	12.0079	0.0230
15	11.9846	11.9938	12.0065	12.0089	12.0011	11.9990	0.0243
16	12.0145	11.9832	12.0188	11.9935	11.9989	12.0018	0.0356
17	12.0004	12.0042	11.9954	12.0020	11.9889	11.9982	0.0153
18	11.9959	11.9823	11.9964	12.0082	11.9871	11.9940	0.0259
19	11.9878	11.9864	11.9960	12.0070	11.9984	11.9951	0.0206
20	11.9969	12.0144	12.0053	11.9985	11.9885	12.0007	0.0259

For the JCS data in Table 20.2, the overall sample mean is $\bar{\bar{x}} = 11.9995$. This value will be the centre line for the chart. The range of each sample, denoted R_j, is simply the difference between the largest and smallest values in each sample. The average range for k samples is computed as follows.

Average range

$$\bar{R} = \frac{R_1 + R_2 + \cdots + R_k}{k} \tag{20.5}$$

where

$$R_j = \text{range of the } j\text{th sample, } j = 1, 2, \ldots, k$$
$$k = \text{number of samples}$$

For the JCS data in Table 20.2, the average range is $\bar{R} = 0.0253$.

In the preceding section we showed that the upper and lower control limits for the chart are

$$\bar{x} \pm \frac{3\sigma}{\sqrt{n}} \tag{20.6}$$

Hence, to construct the control limits for the chart, we need to estimate μ and σ, the mean and standard deviation of the process. An estimate of μ is given by $\bar{\bar{x}}$. An estimate of σ can be developed by using the range data.

It can be shown that an estimator of the process standard deviation σ is the average range divided by d_2, a constant that depends on the sample size n. That is,

$$\text{Estimator of } \sigma = \frac{\bar{R}}{d_2} \tag{20.7}$$

Values for d_2 are shown in Table 20.3. For instance, when $n = 5$, $d_2 = 2.326$, and the estimate of σ is the average range divided by 2.326. If we substitute $\bar{R}/d_2$ for σ in expression (20.6), we can write the control limits for the $\bar{x}$ chart as

Control limits for an $\bar{x}$ chart: process mean and standard deviation unknown

$$\bar{\bar{x}} \pm \frac{3\bar{R}/d_2}{\sqrt{n}} = \bar{\bar{x}} \pm \frac{3\bar{R}}{d_2\sqrt{n}} = \bar{\bar{x}} \pm A_2\bar{R} \tag{20.8}$$

Note that $A_2 = 3/(d_2 \sqrt{n})$ is a constant that depends only on the sample size; Values for A_2 are provided in Table 20.3. For $n = 5$, $A_2 = 0.577$: thus, the control limits for the $\bar{x}$ chart are

$$11.9995 \pm (0.577)(0.0253) = 11.9995 \pm 0.0146$$

Hence, UCL $= 12.014$ and LCL $= 11.985$.

Figure 20.5 shows the $\bar{x}$ chart for the Jensen Computer Supplies problem. We used the data in Table 20.2 and MINITAB's control chart routine to construct the chart. The centre line is shown at the overall sample mean $\bar{\bar{x}} = 11.999$. The upper control limit

Figure 20.5 $\bar{x}$ chart for the Jensen Computer Supplies problem

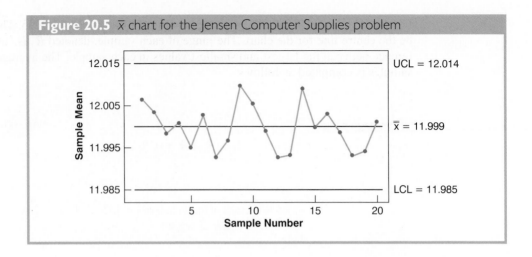

(UCL) is 12.014 and the lower control (LCL) is 11.985. The $\bar{x}$ chart shows the 20 sample means plotted over time. Because all 20 sample means are within the control limits, the indication is that the mean of the Jensen manufacturing process is in control. This chart can now be used to monitor the process mean on an ongoing basis.

R chart

Let us now consider a range chart (R chart) that can be used to control the variability of a process. To develop the R chart, we need to think of the range of a sample as a random variable with its own mean and standard deviation. The average range $\bar{R}$ provides an estimate of the mean of this random variable. Moreover, it can be shown that an estimate of the standard deviation of the range is

$$\hat{\sigma}_R = \frac{d_3}{d_2}\bar{R} \tag{20.9}$$

where d_2 and d_3 are constants that depend on the sample size: values of d_2 and d_3 are also provided in Table 20.3. Thus, the UCL for the R chart is given by

$$\bar{R} + 3\hat{\sigma}_R = \bar{R}\left(1 + 3\frac{d_3}{d_2}\right) \tag{20.10}$$

and the LCL is

$$\bar{R} - 3\hat{\sigma}_R = \bar{R}\left(1 - 3\frac{d_3}{d_2}\right) \tag{20.11}$$

If we let

$$D_4 = 1 + 3\frac{d_3}{d_2} \tag{20.12}$$

$$D_3 = 1 - 3\frac{d_3}{d_2} \tag{20.13}$$

Table 20.3 Factors for $\bar{x}$ and R control charts

Observations in sample, n	d_2	A_2	d_3	D_3	D_4
2	1.128	1.880	0.853	0	3.267
3	1.693	1.023	0.888	0	2.574
4	2.059	0.729	0.880	0	2.282
5	2.326	0.577	0.864	0	2.114
6	2.534	0.483	0.848	0	2.004
7	2.704	0.419	0.833	0.076	1.924
8	2.847	0.373	0.820	0.136	1.864
9	2.970	0.337	0.808	0.184	1.816
10	3.078	0.308	0.797	0.223	1.777
11	3.173	0.285	0.787	0.256	1.744
12	3.258	0.266	0.778	0.283	1.717
13	3.336	0.249	0.770	0.307	1.693
14	3.407	0.235	0.763	0.328	1.672
15	3.472	0.223	0.756	0.347	1.653
16	3.532	0.212	0.750	0.363	1.637
17	3.588	0.203	0.744	0.378	1.622
18	3.640	0.194	0.739	0.391	1.608
19	3.689	0.187	0.734	0.403	1.597
20	3.735	0.180	0.729	0.415	1.585
21	3.778	0.173	0.724	0.425	1.575
22	3.819	0.167	0.720	0.434	1.566
23	3.858	0.162	0.716	0.443	1.557
24	3.895	0.157	0.712	0.451	1.548
25	3.931	0.153	0.708	0.459	1.541

Source: Adapted from Table 27 of ASTM STP 15D, ASTM *Manual on Presentation of Data and Control Chart Analysis*. © 1976 American Society for Testing and Materials, Philadelphia, PA. Reprinted with permission.

we can write the control limits for the R chart as

Control limits for an R chart

$$UCL = \bar{R}\,D_4 \tag{20.14}$$
$$LCL = \bar{R}\,D_3 \tag{20.15}$$

Values for D_3 and D_4 are also provided in Table 20.3. Note that for $n = 5$, $D_3 = 0$, and $D_4 = 2.115$. Thus, with $\bar{R} = 0.0253$, the control limits are

$$UCL = 0.0253(2.115) = 0.0535$$
$$LCL = 0.0253(0) = 0$$

Figure 20.6 shows the R chart for the Jensen Computer Supplies problem. We used the data in Table 20.2 and MINITAB's control chart routine to construct the chart. The centre line is shown at the overall mean of the 20 sample ranges, $\bar{R} = 0.02527$. The UCL is 0.05344 and the LCL is 0.0. The R chart shows the 20 sample ranges plotted over time.

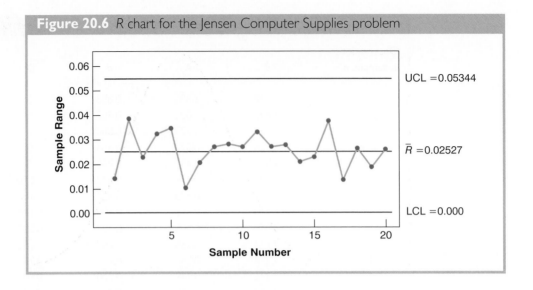

Figure 20.6 *R chart for the Jensen Computer Supplies problem*

Because all 20 sample ranges are within the control limits, we confirm that the process was in control during the sampling period.

p chart

Consider the case in which the output quality is measured by either non-defective or defective items. The decision to continue or to adjust the production process will be based on p, the proportion of defective items found in a sample. The control chart used for proportion-defective data is called a p chart.

To illustrate the construction of a p chart, consider the use of automated mail-sorting machines in a post office. These automated machines scan the post codes on letters and divert each letter to its proper carrier route. Even when a machine is operating properly, some letters are diverted to incorrect routes. Assume that when a machine is operating correctly, or in a state of control, 3 per cent of the letters are incorrectly diverted. Thus π, the proportion of letters incorrectly diverted when the process is in control, is 0.03.

The sampling distribution of P, as presented in Chapter 7, can be used to determine the variation that can be expected in P values for a process that is in control. Recall that the expected value or mean of P is π, the proportion defective when the process is in control. With samples of size n, the formula for the standard deviation of P, called the standard error of the proportion, is

---

Standard error of the proportion

$$\sigma_p = \sqrt{\frac{\pi(1 - \pi)}{n}}$$ (20.16)

We also learned in Chapter 7 that the sampling distribution of P can be approximated by a normal distribution whenever the sample size is large. The sample size can be considered large whenever the following two conditions are satisfied.

$$n\pi \geq 5$$
$$n(1 - \pi) \geq 5$$

Figure 20.7 Sampling distribution of P

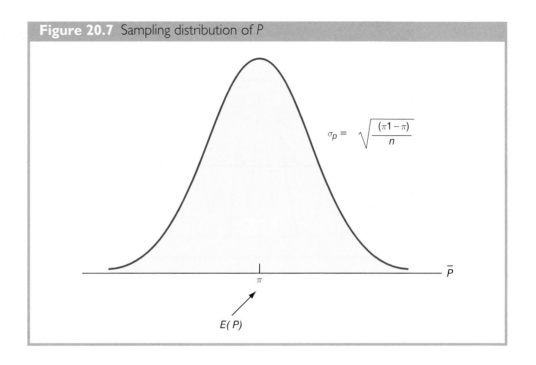

$$\sigma_p = \sqrt{\frac{(\pi 1 - \pi)}{n}}$$

In summary, whenever the sample size is large, the sampling distribution of P can be approximated by a normal distribution with mean π and standard deviation σ_p. This distribution is shown in Figure 20.7.

To establish control limits for a p chart, we follow the same procedure we used to establish control limits for an $\bar{x}$ chart. That is, the limits for the control chart are set at three standard deviations, or standard errors, above and below the proportion defective when the process is in control. Thus, we have the following control limits.

Control limits for a p chart

$$\text{UCL} = \pi + 3\sigma_p \tag{20.17}$$
$$\text{LCL} = \pi - 3\sigma_p \tag{20.18}$$

With $\pi = 0.03$ and samples of size $n = 200$, equation (20.16) shows that the standard error is

$$\sigma_p = \sqrt{\frac{0.03(1 - 0.03)}{200}} = 0.0121$$

Hence, the control limits are UCL $0.03 + 3(0.0121) = 0.0663$, and LCL $= 0.03 - 3(0.0121) = -0.0063$. Whenever equation (20.18) provides a negative value for LCL, LCL is set equal to zero in the control chart.

Figure 20.8 is the control chart for the mail-sorting process. The points plotted show the sample proportion defective found in samples of letters taken from the process. All points are within the control limits, providing no evidence to conclude that the sorting process is out of control. In fact, the p chart indicates that the process is in control and should continue to operate.

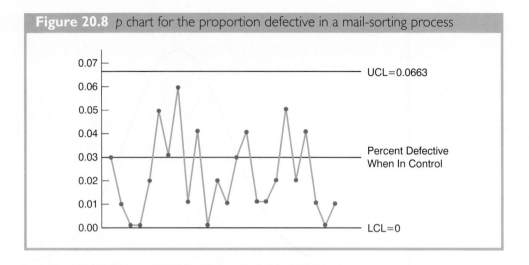

Figure 20.8 p chart for the proportion defective in a mail-sorting process

If the proportion of defective items for a process that is in control is not known, that value is first estimated by using sample data. Suppose, for example, that k different samples, each of size n, are selected from a process that is in control. The fraction or proportion of defective items in each sample is then determined. Treating all the data collected as one large sample, we can compute the proportion of defective items for all the data; that value can then be used to estimate π, the proportion of defective items observed when the process is in control. Note that this estimate of π also enables us to estimate the standard error of the proportion; upper and lower control limits can then be established.

np chart

An *np* chart is a control chart developed for the number of defective items in a sample. In this case, n is the sample size and π is the probability of observing a defective item when the process is in control. Whenever the sample size is large, that is when $n\pi \geq 5$ and $n(1 - \pi) \geq 5$, the distribution of the number of defective items observed in a sample of size n can be approximated by a normal distribution with mean $n\pi$ and standard deviation

$$\sqrt{n\pi(1 - \pi)}$$

Thus, for the mail-sorting example, with $n = 200$ and $\pi = 0.03$, the number of defective items observed in a sample of 200 letters can be approximated by a normal distribution with a mean of $200(0.03) = 6$ and a standard deviation of

$$\sqrt{200 \times 0.03 \times 0.97} = 2.4125$$

The control limits for an *np* chart are set at three standard deviations above and below the expected number of defective items observed when the process is in control. Thus, we have the following control limits.

Control limits for an *np* chart

$$UCL = n\pi + 3\sqrt{n\pi(1 - \pi)} \tag{20.19}$$
$$LCL = n\pi - 3\sqrt{n\pi(1 - \pi)} \tag{20.20}$$

For the mail-sorting process example, with $\pi = 0.03$ and $n = 200$, the control limits are UCL = $6 + 3(2.4125) = 13.2375$, and LCL = $6 - 3(2.4125) = -.2375$. When LCL is negative, LCL is set equal to zero in the control chart. Hence, if the number of letters diverted to incorrect routes is greater than 13, the process is concluded to be out of control.

The information provided by an *np* chart is equivalent to the information provided by the *p* chart; the only difference is that the *np* chart is a plot of the number of defective items observed whereas the *p* chart is a plot of the proportion of defective items observed. Thus, if we were to conclude that a particular process is out of control on the basis of a *p* chart, the process would also be judged to be out of control on the basis of an *np* chart.

Interpretation of control charts

The location and pattern of points in a control chart enable us to determine, with a small probability of error, whether a process is in statistical control. A primary indication that a process may be out of control is a data point outside the control limits, such as point 5 in Figure 20.4. Finding such a point is statistical evidence that the process is out of control; in such cases, corrective action should be taken as soon as possible.

In addition to points outside the control limits, certain patterns of the points within the control limits can be warning signals of quality control problems. For example, assume that all the data points are within the control limits but that a large number of points are on one side of the centre line. This pattern may indicate that an equipment problem, a change in materials, or some other assignable cause of a shift in quality has occurred. Careful investigation of the production process should be undertaken to determine whether quality has changed.

Another pattern to watch for in control charts is a gradual shift, or trend, over time. For example, as tools wear out, the dimensions of machined parts will gradually deviate from their designed levels. Gradual changes in temperature or humidity, general equipment deterioration, dirt build-up or operator fatigue may also result in a trend pattern in control charts. Six or seven points in a row that indicate either an increasing or decreasing trend should be cause for concern, even if the data points are all within the control limits.

When such a pattern occurs, the process should be reviewed for possible changes or shifts in quality. Corrective action to bring the process back into control may be necessary.

Exercises

Methods

1   A process that is in control has a mean of $\pi = 2.5$ and a standard deviation of $\sigma = 0.8$.

   a.   Construct an $\bar{x}$ chart if samples of size 4 are to be used.
   b.   Repeat part (a) for samples of size 8 and 16.
   c.   What happens to the limits of the control chart as the sample size is increased? Discuss why this is reasonable.

2   Twenty-five samples, each of size 5, were selected from a process that was in control. The sum of all the data collected was 307.3 kg.

a. What is an estimate of the process mean (in terms of kg per unit) when the process is in control?

b. Develop the control chart for this process if samples of size 5 will be used. Assume that the process standard deviation is 0.5 when the process is in control, and that the mean of the process is the estimate developed in part (a).

3 Twenty-five samples of 100 items each were inspected when a process was considered to be operating satisfactorily. In the 25 samples, a total of 135 items were found to be defective.

a. What is an estimate of the proportion defective when the process is in control?

b. What is the standard error of the proportion if samples of size 100 will be used for statistical process control?

c. Compute the upper and lower control limits for the control chart.

4 A process sampled 20 times with a sample of size 8 resulted in $\bar{\bar{x}} = 28.5$ and $\bar{R} = 1.6$. Compute the upper and lower control limits for the $\bar{x}$ and R charts for this process.

Applications

5 Temperature is used to measure the output of a production process. When the process is in control, the mean of the process is $\mu = 128.5$ and the standard deviation is $\sigma = 0.4$.

a. Construct an $\bar{x}$ chart if samples of size 6 are to be used.

b. Is the process in control for a sample providing the following data?

     128.8   128.2   129.1   128.7   128.4   129.2

c. Is the process in control for a sample providing the following data?

     129.3   128.7   128.6   129.2   129.5   129.0

6 A quality control process monitors the weight per carton of laundry detergent. Control limits are set at UCL = 0.570 kg and LCL = 0.564 kg. Samples of size 5 are used for the sampling and inspection process. What are the process mean and process standard deviation for the manufacturing operation?

7 The Guttman Tyre and Rubber Company periodically tests its tyres for tread wear under simulated road conditions. To study and control the manufacturing process, 20 samples, each containing three radial tyres, were chosen from different shifts over several days of operation, with the following results. Assuming that these data were collected when the manufacturing process was believed to be operating in control, develop the R and $\bar{x}$ charts.

Sample	Tread wear (mm)		
1	8	11	7
2	7	5	9
3	6	8	9
4	4	6	5
5	10	7	9
6	10	11	9
7	5	4	7
8	8	7	7
9	10	9	8
10	7	4	8
11	7	8	10

TYRES

Sample	Tread wear (mm)		
12	6	5	6
13	4	6	8
14	11	9	4
15	5	6	7
16	8	11	8
17	7	9	8
18	10	7	8
19	5	7	7
20	6	9	7

8 A company is concerned with monitoring the pH value of a liquid. Measurements are taken at intervals, three times per day so that over a 24 hour data period we have data as follows:

PH

Sample	Measurement			Sample	Measurement		
	1	2	3		1	2	3
1	6.0	5.8	6.1	13	6.1	6.9	7.4
2	5.2	6.4	6.9	14	6.2	5.2	6.8
3	5.5	5.8	5.2	15	4.9	6.6	6.6
4	5.0	5.7	6.5	16	7.0	6.4	6.1
5	6.7	6.5	5.5	17	5.4	6.5	6.7
6	5.8	5.2	5.0	18	6.6	7.0	6.8
7	5.6	5.1	5.2	19	4.7	6.2	7.1
8	6.0	5.8	6.0	20	6.7	5.4	6.7
9	5.5	4.9	5.7	21	6.8	6.5	5.2
10	4.3	6.4	6.3	22	5.9	6.4	6.0
11	6.2	6.9	5.0	23	6.7	6.3	4.6
12	6.7	7.1	6.2	24	7.4	6.8	6.3

Compute R and $\bar{x}$ charts for the data and hence determine if the underlying production process is in control.

9 An automotive industry supplier produces pistons for several models of cars. Twenty samples, each consisting of 200 pistons, were selected when the process was known to be operating correctly. The numbers of defective pistons found in the samples follow.

8	10	6	4	5	7	8	12	8	15
14	10	10	7	5	8	6	10	4	8

a. What is an estimate of the proportion defective for the piston manufacturing process when it is in control?

b. Construct a p chart for the manufacturing process, assuming each sample has 200 pistons.

c. With the results of part (b), what conclusion should be made if a sample of 200 has 20 defective pistons?

d. Compute the upper and lower control limits for an np chart.

e. Answer part (c) using the results of part (d).

20.2 Acceptance sampling

In acceptance sampling, the items of interest can be incoming shipments of raw materials or purchased parts as well as finished goods from final assembly. Suppose we want to decide whether to accept or reject a group of items on the basis of specified quality characteristics. In quality control terminology, the group of items is a **lot**, and **acceptance sampling** is a statistical method that enables us to base the accept-reject decision on the inspection of a sample of items from the lot.

The general steps of acceptance sampling are shown in Figure 20.9. After a lot is received, a sample of items is selected for inspection. The results of the inspection are compared with specified quality characteristics. If the quality characteristics are satisfied, the lot is accepted and sent to production or shipped to customers. If the lot is rejected, managers must decide on its disposal. In some cases, the decision may be to keep the lot and remove the unacceptable or nonconforming items. In other cases, the lot may be returned to the supplier at the supplier's expense; the extra work and cost placed on the supplier can motivate the supplier to provide high-quality lots. Finally, if the rejected lot consists of finished goods, the goods must be scrapped or reworked to meet acceptable quality standards.

The statistical procedure of acceptance sampling is based on the hypothesis testing methodology presented in Chapter 9. The null and alternative hypotheses are stated as follows.

$$H_0: \text{Good-quality lot}$$
$$H_1: \text{Poor-quality lot}$$

Figure 20.9 Acceptance sampling procedure

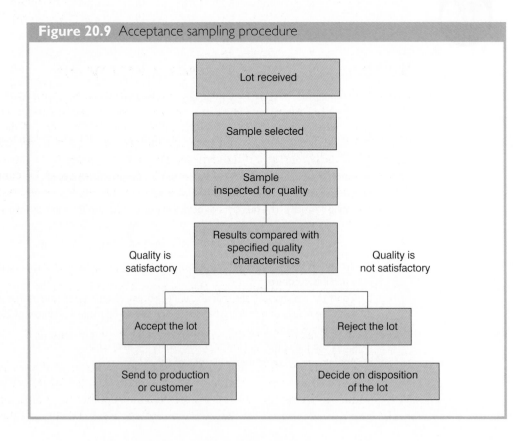

Table 20.4 The outcomes of acceptance sampling

	State of the lot	
Decision	H_0 True Good-quality lot	H_0 False Poor-quality lot
Accept the lot	Correct decision	Type II error (accepting a poor-quality lot)
Reject the lot	Type I error (rejecting a good-quality lot)	Correct decision

Table 20.4 shows the results of the hypothesis testing procedure. Note that correct decisions correspond to accepting a good-quality lot and rejecting a poor-quality lot. However, as with other hypothesis testing procedures, we need to be aware of the possibilities of making a Type I error (rejecting a good-quality lot) or a Type II error (accepting a poor-quality lot).

The probability of a Type I error provides a measure of the risk for the producer of the lot and is known as the **producer's risk**. For example, a producer's risk of 0.05 indicates a 5 per cent chance that a good-quality lot will be erroneously rejected. The probability of a Type II error, on the other hand, provides a measure of the risk for the consumer of the lot and is known as the **consumer's risk**. For example, a consumer's risk of 0.10 means a 10 per cent chance that a poor-quality lot will be erroneously accepted and thus used in production or shipped to the customer. Specific values for the producer's risk and the consumer's risk can be controlled by the person designing the acceptance sampling procedure. To illustrate how to assign risk values, let us consider the problem faced by KALI.

KALI: an example of acceptance sampling

KALI, manufactures home appliances that are marketed under a variety of trade names. However, KALI does not manufacture every component used in its products. Several components are purchased directly from suppliers. For example, one of the components that KALI purchases for use in home air conditioners is an overload protector, a device that turns off the compressor if it overheats. The compressor can be seriously damaged if the overload protector does not function properly, and therefore KALI is concerned about the quality of the overload protectors. One way to ensure quality would be to test every component received through an approach known as 100 per cent inspection. However, to determine proper functioning of an overload protector, the device must be subjected to time-consuming and expensive tests, and KALI cannot justify testing every overload protector it receives.

Instead, KALI uses an acceptance sampling plan to monitor the quality of the overload protectors. The acceptance sampling plan requires that KALI's quality control inspectors select and test a sample of overload protectors from each shipment. If very few defective units are found in the sample, the lot is probably of good quality and should be accepted. However, if a large number of defective units are found in the sample, the lot is probably of poor quality and should be rejected.

An acceptance sampling plan consists of a sample size n and an acceptance criterion c. The **acceptance criterion** is the maximum number of defective items that can be found in the sample and still indicate an acceptable lot. For example, for the KALI problem let us assume that a sample of 15 items will be selected from each incoming shipment or lot. Furthermore, assume that the manager of quality control states that the lot can be accepted only if no defective items are found. In this case, the acceptance sampling plan established by the quality control manager is $n = 15$ and $c = 0$.

This acceptance sampling plan is easy for the quality control inspector to implement. The inspector simply selects a sample of 15 items, performs the tests, and reaches a conclusion based on the following decision rule.

- *Accept the lot* if zero defective items are found.
- *Reject the lot* if one or more defective items are found.

Before implementing this acceptance sampling plan, the quality control manager wants to evaluate the risks or errors possible under the plan. The plan will be implemented only if both the producer's risk (Type I error) and the consumer's risk (Type II error) are controlled at reasonable levels.

Computing the probability of accepting a lot

The key to analysing both the producer's risk and the consumer's risk is a 'what-if' type of analysis. That is, we will assume that a lot has some known percentage of defective items and compute the probability of accepting the lot for a given sampling plan. By varying the assumed percentage of defective items, we can examine the effect of the sampling plan on both types of risks.

Let us begin by assuming that a large shipment of overload protectors has been received and that 5 per cent of the overload protectors in the shipment are defective. For a shipment or lot with 5 per cent of the items defective, what is the probability that the $n = 15$, $c = 0$ sampling plan will lead us to accept the lot? Because each overload protector tested will be either defective or non-defective and because the lot size is large, the number of defective items in a sample of 15 has a *binomial distribution*. The binomial probability function, which was presented in Chapter 5, follows.

Binomial probability function for acceptance sampling

$$p(x) = \frac{n!}{x!(n - x)!}\, \pi^x(1 - \pi)^{(n-x)} \qquad (20.21)$$

where

$$n = \text{the sample size}$$
$$\pi = \text{the proportion of defective items in the lot}$$
$$x = \text{the number of defective items in the sample}$$
$$p(x) = \text{the probability of } x \text{ defective items in the sample}$$

For the KALI acceptance sampling plan, $n = 15$: thus, for a lot with 5 per cent defective ($\pi = 0.05$), we have

$$p(x) = \frac{15!}{x!(15 - x)!}\, (0.05)^x(1 - 0.05)^{(15 - x)} \qquad (20.22)$$

Using equation (20.22), $p(0)$ will provide the probability that zero overload protectors will be defective and the lot will be accepted. In using equation (20.22), recall that $0! = 1$. Thus, the probability computation for $p(0)$ is

$$p(0) = \frac{15!}{0!(15 - 0)!}(0.05)^0 (1 - 0.05)^{(15 - 0)}$$

$$\frac{15!}{0!(15)!}(0.05)^0 (0.95)^{15} = (0.95)^{15} = 0.4633$$

We now know that the $n = 15$, $c = 0$ sampling plan has a 0.4633 probability of accepting a lot with 5 per cent defective items. Hence, there must be a corresponding $1 - 0.4633 = 0.5367$ probability of rejecting a lot with 5 per cent defective items.

Tables of binomial probabilities (see Table 5, Appendix B) can help reduce the computational effort in determining the probabilities of accepting lots. From this table, we can determine that if the lot contains 10 per cent defective items, there is a 0.2059 probability that the $n = 15$, $c = 0$ sampling plan will indicate an acceptable lot. The probability that the $n = 15$, $c = 0$ sampling plan will lead to the acceptance of lots with 1 per cent, 2 per cent, 3 per cent, . . . defective items is summarized in Table 20.5.

Using the probabilities in Table 20.5, a graph of the probability of accepting the lot versus the percent defective in the lot can be drawn as shown in Figure 20.10. The resulting graph, or curve, is called the **operating characteristic (OC) curve** for the $n = 15$, $c = 0$ acceptance sampling plan. This links with what was covered in Chapter 9.

Perhaps we should consider other sampling plans, ones with different sample sizes n or different acceptance criteria c. First consider the case in which the sample size remains $n = 15$ but the acceptance criterion increases from $c = 0$ to $c = 1$. That is, we will now accept the lot if zero or one defective component is found in the sample. For a lot with 5 per cent defective items ($\pi = 0.05$), Table 5, in Appendix B shows that with $n = 5$ and $\pi = 0.05$, $p(0) = 0.4633$ and $p(1) = 0.3658$. Thus, there is a $0.4633 + 0.3658 = 0.8291$ probability that the $n = 15$, $c = 1$ plan will lead to the acceptance of a lot with 5 per cent defective items.

Continuing these calculations we obtain Figure 20.11, which shows the operating characteristic curves for four alternative acceptance sampling plans for the KALI

Table 20.5 Probability of accepting the lot for the Kali problem with $n = 15$ and $c = 0$

Percent defective in the lot	Probability of accepting the lot
1	0.8601
2	0.7386
3	0.6333
4	0.5421
5	0.4633
10	0.2059
15	0.0874
20	0.0352
25	0.0134

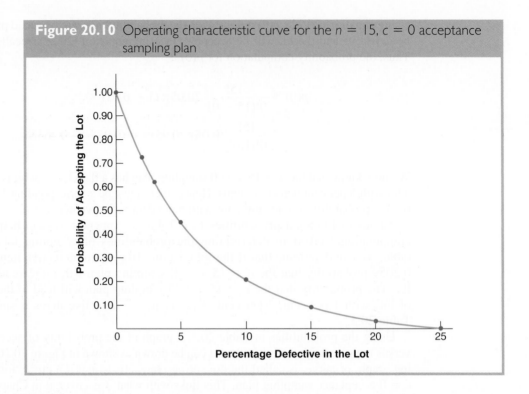

Figure 20.10 Operating characteristic curve for the $n = 15$, $c = 0$ acceptance sampling plan

problem. Samples of size 15 and 20 are considered. Note that regardless of the proportion defective in the lot, the $n = 15$, $c = 1$ sampling plan provides the highest probabilities of accepting the lot. The $n = 20$, $c = 0$ sampling plan provides the lowest probabilities of accepting the lot: however, that plan also provides the highest probabilities of rejecting the lot.

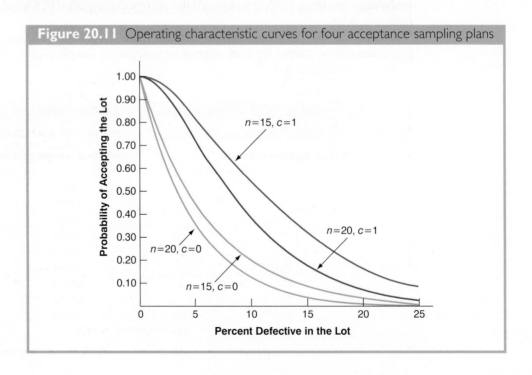

Figure 20.11 Operating characteristic curves for four acceptance sampling plans

Selecting an acceptance sampling plan

Now that we know how to use the binomial distribution to compute the probability of accepting a lot with a given proportion defective, we are ready to select the values of n and c that determine the desired acceptance sampling plan for the application being studied. To develop this plan, managers must specify two values for the fraction defective in the lot. One value, denoted p_0, will be used to control for the producer's risk, and the other value, denoted p_1, will be used to control for the consumer's risk.

We will use the following notation.

α = the producer's risk; the probability of rejecting a lot with p_0 defective items

β = the consumer's risk; the probability of accepting a lot with p_1 defective items

Suppose that for the KALI problem, the managers specify that $p_0 = 0.03$ and $p_1 = 0.15$. From the OC curve for $n = 15$, $c = 0$ in Figure 20.12, we see that $p_0 = 0.03$ provides a producer's risk of approximately $1 - 0.63 = 0.37$, and $p_1 = 0.15$ provides a consumer's risk of approximately 0.09. Thus, if the managers are willing to tolerate both a 0.37 probability of rejecting a lot with 3 per cent defective items (producer's risk) and a 0.09 probability of accepting a lot with 15 per cent defective items (consumer's risk), the $n = 5$, $c = 0$ acceptance sampling plan would be acceptable.

Suppose, however, that the managers request a producer's risk of $\alpha = 0.10$ and a consumer's risk of $\beta = 0.20$. We see that now the $n = 15$, $c = 0$ sampling plan has a better than-desired consumer's risk but an unacceptably large producer's risk. The fact that $\alpha = 0.37$ indicates that 37 per cent of the lots will be erroneously rejected when only 3 per cent of the items in them are defective. The producer's risk is too high, and a different acceptance sampling plan should be considered.

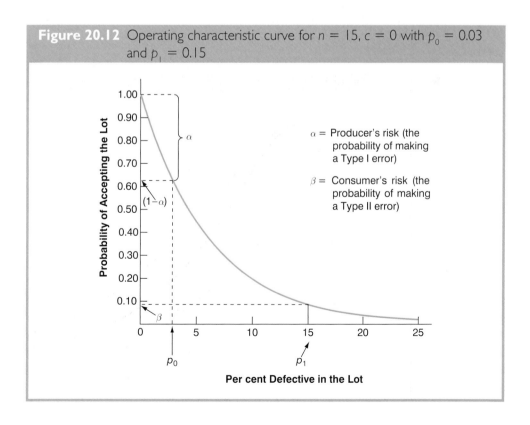

Figure 20.12 Operating characteristic curve for $n = 15$, $c = 0$ with $p_0 = 0.03$ and $p_1 = 0.15$

Using $p_0 = 0.03$, $\alpha = 0.10$, $p_1 = 0.15$, and $\beta = 0.20$, Figure 20.11 shows that the acceptance sampling plan with $n = 20$ and $c = 1$ comes closest to meeting both the producer's and the consumer's risk requirements. Exercise 13 at the end of this section will ask you to compute the producer's risk and the consumer's risk for the $n = 20$, $c = 1$ sampling plan.

As we have seen, several computations and several operating characteristic curves may need to be considered to determine the sampling plan with the desired producer's and consumer's risk. Fortunately, tables of sampling plans are published. For example, the ISO 2859-0 standard provides information helpful in designing acceptance sampling plans. More advanced texts on quality control, such as those listed in the bibliography, describe the use of such tables. Such texts also discuss the role of sampling costs in determining the optimal sampling plan.

Multiple sampling plans

The acceptance sampling procedure we presented for the KALI problem is a *single-sample* plan. It is so-called because only one sample or sampling stage is used. After the number of defective components in the sample is determined, a decision must be made to accept

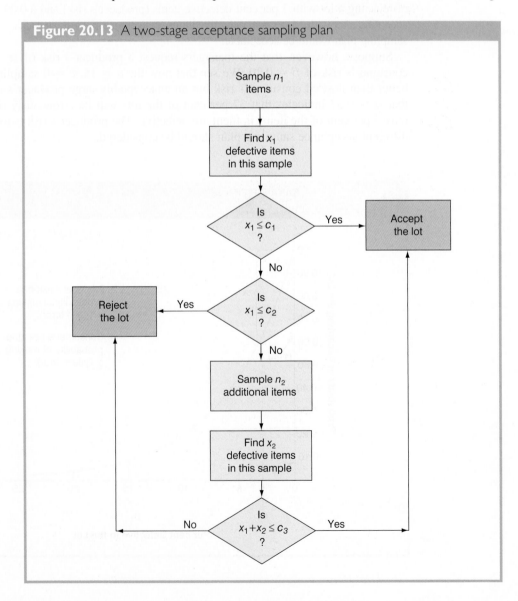

Figure 20.13 A two-stage acceptance sampling plan

or reject the lot. An alternative to the single-sample plan is a **multiple sampling plan**, in which two or more stages of sampling are used. At each stage a decision is made among three possibilities: stop sampling and accept the lot, stop sampling and reject the lot, or continue sampling. Although more complex, multiple sampling plans often result in a smaller total sample size than single-sample plans with the same α and β probabilities.

The logic of a two-stage, or double-sample, plan is shown in Figure 20.13. Initially a sample of n_1 items is selected. If the number of defective components x_1 is less than or equal to c_1, accept the lot. If x_1 is greater than or equal to c_2, reject the lot. If x_1 is between c_1 and c_2 $(c_1 < x_1 < c_2)$, select a second sample of n_2 items. Determine the combined, or total, number of defective components from the first sample (x_1) and the second sample (x_2). If $x_1 + x_2 \leq c_3$, accept the lot; otherwise reject the lot. The development of the double-sample plan is more difficult because the sample sizes n_1 and n_2 and the acceptance numbers c_1, c_2, and c_3 must meet both the producer's and consumer's risks desired.

Exercises

Methods

10 For an acceptance sampling plan with $n = 25$ and $c = 0$, find the probability of accepting a lot that has a detect rate of 2 per cent. What is the probability of accepting the lot it the detect rate is 6 per cent?

11 Consider an acceptance sampling plan with $n = 20$ and $c = 0$. Compute the producer's risk for each of the following cases.

a. The lot has a detect rate of 2 per cent.
b. The lot has a detect rate of 6 per cent.

12 A production company is considering two possible plans for the acceptance sampling of some raw material. Both are attribute inspection plans, the first specifying sample size 10 and acceptance if the number of substandard items is no greater than 1, and the second specifying sample size 25 and acceptance if the number of substandard items is no greater than 3. Plot on the same graph the operating characteristic for each plan. If the production process can work well on raw material 5 per cent substandard, but cannot work if the proportion gets close to 15 per cent, which plan should the company choose?

Applications

13 Refer to the KALI problem presented in this section. The quality control manager requested a producer's risk of 0.10 when p_0 was 0.03 and a consumer's risk of 0.20 when p_1 was 0.15.

Consider the acceptance sampling plan based on a sample size of 20 and an acceptance number of 1. Answer the following questions.
a. What is the producer's risk for the $n = 20$, $c = 1$ sampling plan?
b. What is the consumer's risk for the $n = 20$, $c = 1$ sampling plan?
c. Does the $n = 20$, $c = 1$ sampling plan satisfy the risks requested by the quality control manager? Discuss.

14 A domestic manufacturer of watches purchases quartz crystals from a Swiss firm. The crystals are shipped in lots of 1000. The acceptance sampling procedure uses 20 randomly selected crystals.

a. Construct operating characteristic curves for acceptance numbers of 0, 1 and 2.

b. It p_0 is 0.01 and $p_1 = 0.08$, what are the producer's and consumer's risks for each sampling plan in part (a)?

15 A company wishes to design and implement a single attribute sampling plan so that a good lot with a defective rate of 2 per cent will be accepted 95 per cent of the time while a bad lot with a defective rate of 10 per cent will be accepted 15 per cent of the time. How would you advise the company?

For additional online summary questions and answers go to the companion website at www.cengage.co.uk/aswsbe2

Summary

In this chapter we discussed how statistical methods can be used to assist in the control of quality. We first presented the $\bar{x}$, R, p and np control charts as graphical aids in monitoring process quality. Control limits are established for each chart; samples are selected periodically, and the data points plotted on the control chart. Data points outside the control limits indicate that the process is out of control and that corrective action should be taken. Patterns of data points within the control limits can also indicate potential quality control problems and suggest that corrective action may be warranted.

We also considered the technique known as acceptance sampling. With this procedure, a sample is selected and inspected. The number of detective items in the sample provides the basis for accepting or rejecting the lot. The sample size and the acceptance criterion can be adjusted to control both the producer's risk (Type I error) and the consumer's risk (Type II error). In addition to this single scheme, a double sample scheme was also considered.

Key terms

Acceptance criterion
Acceptance sampling
Assignable causes
Common causes
Consumer's risk
Control chart
Lot
Multiple sampling plan

np chart
Operating characteristic curve
p chart
Producer's risk
Quality control
R chart
$\bar{x}$ chart

Key formulae

Standard error of the mean

$$\sigma_{\bar{x}} = \frac{\sigma}{\sqrt{n}} \tag{20.1}$$

Control limits for an $\bar{x}$ chart: process mean and standard deviation known

$$\text{UCL} = \mu + 3\sigma_{\bar{x}} \tag{20.2}$$

$$\text{LCL} = \mu - 3\sigma_{\bar{x}} \tag{20.3}$$

Overall sample mean

$$\bar{\bar{x}} = \frac{\bar{x}_1 + \bar{x}_2 \ldots + \bar{x}_k}{k} \tag{20.4}$$

Average range

$$\bar{R} = \frac{R_1 + R_2 + \cdots + R_k}{k}$$ (20.5)

Control limits for an $\bar{X}$ chart: process mean and standard deviation unknown

$$\bar{\bar{x}} \pm \frac{3\bar{R}/d_2}{\sqrt{n}} = \bar{\bar{x}} \pm \frac{3\bar{R}}{d_2\sqrt{n}} = \bar{\bar{x}} \pm A_2\bar{R}$$ (20.8)

Control limits for an R chart

$$UCL = \bar{R}\,D_4$$ (20.14)

$$LCL = \bar{R}\,D_3$$ (20.15)

Standard error of the proportion

$$\sigma_p = \sqrt{\frac{\pi(1 - \pi)}{n}}$$ (20.16)

Control limits for a p chart

$$UCL = \pi + 3\sigma_p$$ (20.17)

$$LCL = \pi - 3\sigma_p$$ (20.18)

Control limits for an np chart

$$UCL = n\pi + 3\sqrt{n\pi(1 - \pi)}$$ (20.19)

$$LCL = n\pi - 3\sqrt{n\pi(1 - \pi)}$$ (20.20)

Binomial probability function for acceptance sampling

$$p(x) = \frac{n!}{x!(n - x)!}\,\pi^x(1 - \pi)^{(n-x)}$$ (20.21)

Case problem ISN Company

Reflecting its total quality management commitment, the ISN Company operates a strict inspection procedure for purchased (incoming) parts. A check on a shipment of 2000 bushings Part No 3128-1 supplied by one vendor yielded the following results:

Sample #	Dimension (cm) for each of five bushings tested				
1	0.429	0.401	0.411	0.424	0.409
2	0.419	0.414	0.417	0.411	0.419
3	0.419	0.417	0.419	0.419	0.419
4	0.417	0.417	0.404	0.419	0.414
5	0.419	0.414	0.417	0.417	0.411
6	0.419	0.417	0.422	0.417	0.401
7	0.404	0.411	0.409	0.419	0.419
8	0.411	0.417	0.414	0.419	0.417
9	0.424	0.417	0.417	0.417	0.417
10	0.422	0.419	0.424	0.419	0.427

The company has asked you to carry out a statistical evaluation of these results. Note that the specified dimension of the part is 0.417 ± 0.010 cm. The new quality control supervisor would specifically like for you to check how the percentages of non-conforming parts (breaching the upper or lower specification limits) compare with theoretical results from the normal distribution. What would you advise the supervisor and why?

Managerial report

1 Perform an analysis of the test data for the ISN Company.

2 Prepare a report for the quality control supervisor, summarizing your findings and recommendations.

Dimension testing. © Robert Kyllo.

Software Section
for Chapter 20

Control charts using MINITAB

In this section we describe the steps required to generate MINITAB control charts using the Jensen sample data shown in Table 20.2. The sample number appears in column C1. The first observation is in column C2, the second observation is in column C3 and so on. In particular, the steps involved in producing both the $\bar{x}$ chart and R chart simultaneously in MINITAB are as follows:.

Step 1   **Stat > Control Charts > Variables Charts for**   [Main menu bar]
          Subgroups > Xbar-R

Step 2   Select **Observations are in one row of columns**   [**Xbar-R** panel]
          Enter C2-C6 in the box below
          Select **Xbar-R Options**
          Select the **Tests** tab   [options panel]
          Choose **One point > 3.0 standard deviations from centre line***.
          Click **OK**

The $\bar{x}$ chart and the R chart will be shown together on the MINITAB output. MINITAB provides access to a variety of control charts. For example, the $\bar{x}$ and the R chart can be selected separately. Additional options include the p chart, the np chart, and others.

Control charts using PASW

To show how PASW can be used to develop generate control charts, we again use the Jensen data provided in Table 20.2. First, the data must be entered into a PASW work-sheet. In 'Data View' mode, the sample number is entered in rows 1 to 20 of the leftmost column.

This is automatically labelled by the system V00001. Similarly the observations are entered in the five adjacent columns to the right and these columns are labelled by the system V00002 to V00006 respectively. TheV00001 variable name can be changed for

---

*MINITAB provides several additional tests for detecting special causes of variation and out-of-control conditions. The user may select several of these tests simultaneously.

example to 'sample' in 'Variable View' mode. The following steps can be used to create $\overline{X}$ and R control charts for the data.

Step 1 **Analyze > Quality Control > Control**   [Main menu bar]
          Charts > X-Bar, R, s

Step 2 Click on **Cases are subgroups**   [**X-Bar, R, s** panel]
          Click on **Define**
          Enter V2-V6 in the **Samples** box
          Click on **Xbar and range** (the default setting) – if necessary – under **Charts**
          Click **OK**

The $\overline{x}$ chart and the R chart will be shown together on the PASW output. As with MINITAB a number of alternative formats are available with PASW as well as p, np and other chart options.

Chapter 21

Decision Analysis

Learning objectives

After reading this chapter and doing the exercises, you should be able to:

1 Describe a problem situation in terms of decisions to be made, chance events and consequences.

2 Understand how the decision alternatives and chance outcomes are combined to generate the consequence.

3 Analyze a simple decision analysis problem from both a payoff table and decision tree point of view.

4 Determine the potential value of additional information.

5 Use new information and revised probability values in the decision analysis approach to problem solving.

6 Understand what a decision strategy is.

7 Evaluate the contribution and efficiency of additional decision-making information.

8 Use a Bayesian approach to computing revised probabilities.

9 Know the definition of the following terms:

decision alternatives
consequence
chance event
states of nature
payoff table
decision tree
expected value approach
expected value of perfect information (EVPI)
decision strategy
expected value of sample information (EVSI)
Bayesian revision
prior probabilities
posterior probabilities

Decision analysis can be used to develop an optimal decision strategy when a decision-maker is faced with various decision alternatives and an uncertain or risk-filled pattern of future events. We begin the study of decision analysis by considering decision problems that involve reasonably few decision alternatives and reasonably few future events. Payoff tables are introduced to provide a structure for decision problems. We then show how decision trees can be used to represent problems involving sequences of decisions enabling the optimal sequence of decisions (optimal decision strategy) to be identified. In the last section, we show how Bayes' theorem, presented in Chapter 4, can be used to compute branch probabilities for decision trees. The EXCEL add-in TreePlan can be used to set up the decision trees and solve the decision problems presented in this chapter. The TreePlan software and a manual for using TreePlan are on the website, http://www.cengage.co.uk/aswsbe/. An example showing how to use TreePlan is provided in Section 21.5.

21.1 Problem formulation

The first step in the decision analysis process is problem formulation. We begin with a verbal statement of the problem. We then identify the decision alternatives, the uncertain future events, referred to as **chance events**, and the **consequences** associated with each decision alternative and each chance event outcome. Let us begin by considering a construction project of the PDC development group in Palma.

PDC has purchased land for the site of a new luxury apartment complex. The group plans to price the individual apartment units between €300 000 and €1 400 000.

Statistics in Practice

Military hardware procurement in Greece

In order to maintain its defences at an adequate level in the increasingly unstable security environment of the 1990s, the Greek army identified the procurement of a fleet of HAVs (heavily armoured vehicles) as a top military priority.

A senior officer and two groups of lower-ranking officers were involved in the HAV selection process which was based on a comprehensive list of attributes that had been initially agreed. These covered such issues as firepower, mobility, communications and sustainability. Personnel issues were also considered. On the basis of relevant performance, cost and maintenance criteria, three different HAVs (HAV1, HAV2, HAV3) were short-listed for consideration. A decision tree

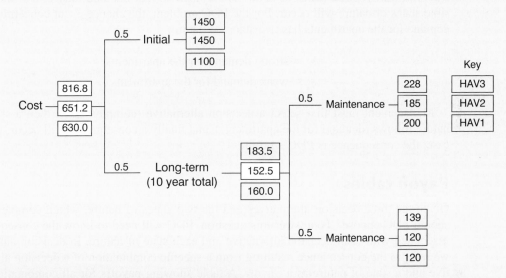

Greek military police standing outside their armoured vehicle. © Marc Garanger/CORBIS.

methodology – illustrated above – was adopted for assessing the merits of the different vehicle choices.

By formally incorporating the senior officer's specific preferences on the various decision factors into the analysis, a clear ranking order emerged – with HAV2 being judged overwhelmingly the favourite option. Coincidentally this was the model the senior officer favoured most before the analysis and was actually the one that was finally procured.

Source: Basileiou D and Owen W (1996) 'Top Tank' OR Insight 9 (1) 8–12

PDC commissioned preliminary architectural drawings for three different-sized projects: one with 30 apartments, one with 60 apartments, and one with 90 apartments. The financial success of the project depends upon the size of the apartment complex and the chance event concerning the demand for the apartments. Specifically, PDC needs to select the size of the new luxury apartment project that will lead to the largest profit given the uncertainty concerning the demand for the apartments.

In this case, the three decision alternatives are:

$$d_1 = \text{a small complex with 30 apartments}$$
$$d_2 = \text{a medium complex with 60 apartments}$$
$$d_3 = \text{a large complex with 90 apartments}$$

A factor in selecting the best decision alternative is the uncertainty associated with the chance event concerning the demand for the apartments. When asked about the possible demand for the apartments, PDC's CEO acknowledged a wide range of possibilities, but decided that it would be adequate to consider two possible chance event outcomes: a strong demand and a weak demand.

In decision analysis, the possible outcomes for a chance event are referred to as the **states of nature**. The states of nature are defined so that one and only one of the possible states of nature will occur. For the PDC problem, the chance event concerning the demand for the apartments has two states of nature:

$$s_1 = \text{strong demand for the apartments}$$
$$s_2 = \text{weak demand for the apartments}$$

Management must first select a decision alternative (complex size), then a state of nature follows (demand for the apartments), and finally a consequence will occur. In this case, the consequence is PDC's profit.

Payoff tables

Given the three decision alternatives and the two states of nature, which complex size should PDC choose? To answer this question, PDC will need to know the consequence associated with each decision alternative and each state of nature. In decision analysis, we refer to the consequence resulting from a specific combination of a decision alternative and a state of nature as a **payoff**. A table showing payoffs for all combinations of decision alternatives and states of nature is a **payoff table**.

Because PDC wants to select the complex size that provides the largest profit, profit is used as the consequence. The payoff table with profits expressed in millions of euros is shown in Table 21.1. Note, for example, that if a medium complex is built and demand turns out to be strong, a profit of €14 million will be realized. We will use the notation V_{ij} to denote the payoff associated with decision alternative i and state of nature j. Using Table 21.1, $V_{31} = 20$ indicates a payoff of €20 million occurs if the decision is to build a large complex (d_3) and the strong demand state of nature (s_1) occurs. Similarly, $V_{32} = -9$ indicates a loss of €9 million if the decision is to build a large complex (d_3) and the weak demand state of nature (s_2) occurs.

Table 21.1 Payoff table for the PDC apartment project (payoffs in € millions)

	State of nature	
Decision alternative	Strong demand s_1	Weak demand s_2
Small complex, d_1	8	7
Medium complex, d_2	14	5
Large complex, d_3	20	−9

Decision trees

A **decision tree** shows graphically the sequential nature of the decision-making process. Figure 21.1 presents a decision tree for the PDC problem, demonstrating the natural or logical progression that will occur over time. First, PDC must make a decision regarding the size of the apartment complex (d_1, d_2 or d_3). Then, after the decision is implemented, either state of nature s_1 or s_2 will occur. The number at each end point of the tree indicates the payoff associated with a particular sequence. For example the topmost payoff of 8 indicates that an €8 million profit is anticipated if PDC constructs a small apartment complex (d_1) and demand turns out to be strong (s_1). The next payoff of 7 indicates an anticipated profit of €7 million if PDC constructs a small apartment complex (d_1) and demand turns out to be weak (s_2). Thus, the decision tree shows graphically the sequences of decision alternatives and states of nature that provide the six possible payoffs.

The decision tree in Figure 21.1 has four **nodes**, numbered 1–4, that represent the decisions and chance events. Squares are used to depict **decision nodes** and circles are used to depict **chance nodes**. Thus, node 1 is a decision node, and nodes 2, 3 and 4 are chance nodes. The **branches** leaving the decision node correspond to the decision alternatives. The branches leaving each chance node correspond to the states of nature. The payoffs are shown at the end of the states-of-nature branches. (Note that payoffs can be expressed in terms of profit, cost, time, distance, or any other measure appropriate for the decision problem being analyzed.) We now turn to the question: How can the decision-maker use the information in the payoff table or the decision tree to select the best decision alternative?

Figure 21.1 Decision tree for the PDC apartment project (payoffs in € millions)

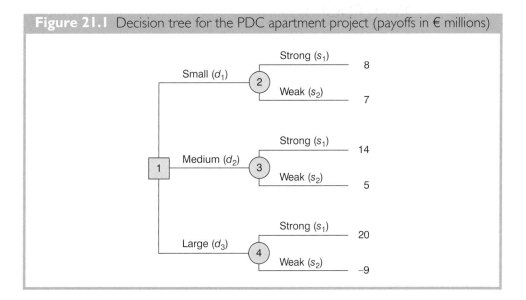

21.2 Decision-making with probabilities

Once we define the decision alternatives and the states of nature for the chance events, we can focus on determining probabilities for the states of nature. The classical method, the relative frequency method, or the subjective method of assigning probabilities discussed in Chapter 4 may be used to identify these probabilities. After determining the appropriate probabilities, we show how to use the **expected value approach** to identify the best, or recommended, decision alternative for the problem.

Expected value approach

Let

$$N = \text{the number of states of nature}$$
$$P(s_j) = \text{the probability of state of nature } s_j$$

Because one and only one of the N states of nature can occur, the probabilities must satisfy two conditions:

$$P(s_j) > 0 \qquad \text{for all states of nature} \tag{21.1}$$

$$\sum_{j=1}^{N} P(s_j) = P(s_1) + P(s_2) + \ldots + P(s_N) = 1 \tag{21.2}$$

The **expected value (EV)** of decision alternative d_i is as follows.

Expected value

$$EV(d_i) = \sum_{j=1}^{N} P(s_j) V_{ij} \tag{21.3}$$

where

$V_{ij} =$ the value of the payoff for decision alternative d_i and state of nature s_j.

In other words, the expected value of a decision alternative is the sum of weighted payoffs for the decision alternative. The weight for a payoff is the probability of the associated state of nature and therefore the probability that the payoff will occur. We now return to the PDC problem to see how the expected value approach can be applied.

It is believed PDC is optimistic about the potential for the luxury high-rise apartment complex. Suppose that this optimism leads to an initial subjective probability assessment of 0.8 that demand will be strong (s_1) and a corresponding probability of 0.2 that demand will be weak (s_2). Therefore, $P(s_1) = 0.8$ and $P(s_2) = 0.2$. Using the payoff values in Table 21.1 and equation (21.3), we compute the expected value for each of the three decision alternatives as follows:

$$EV(d_1) = 0.8(8) + 0.2(7) = 7.8$$
$$EV(d_2) = 0.8(14) + 0.2(5) = 12.2$$
$$EV(d_3) = 0.8(20) + 0.2(-9) = 14.2$$

Thus, we find that the large apartment complex, with an expected value of €14.2 million, is the recommended decision.

The calculations required to identify the decision alternative with the best expected value can be conveniently carried out on a decision tree. Figure 21.2 shows the decision tree for the PDC problem with state-of-nature branch probabilities. Working backward through the decision tree, we first compute the expected value at each chance node; that is, at each chance node, we weight each possible payoff by its probability of occurrence. By doing so, we obtain the expected values for nodes 2, 3 and 4, as shown in Figure 21.3.

Figure 21.2 PDC decision tree with state-of-nature branch probabilities

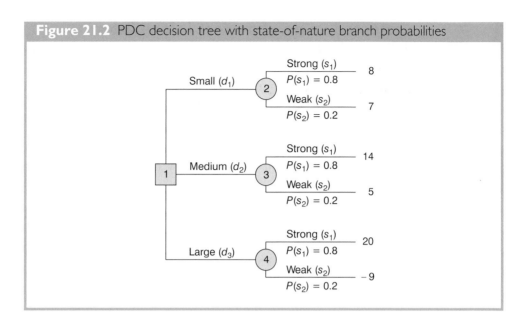

Because the decision-maker controls the branch leaving decision node 1 and because we are trying to maximize the expected profit, the best decision alternative at node 1 is d_3. Thus, the decision tree analysis leads to a recommendation of d_3 with an expected value of €14.2 million. Note that this recommendation corresponds with the expected value approach in conjunction with the payoff table.

Other decision problems may be substantially more complex than the PDC problem, but if a reasonable number of decision alternatives and states of nature are present, you can use the decision tree approach outlined here. First, draw a decision tree consisting of decision nodes, chance nodes and branches that describe the sequential nature of the problem.

If you use the expected value approach, the next step is to determine the probabilities for each of the states of nature and compute the expected value at each chance node. Then select the decision branch leading to the chance node with the best expected value. The decision alternative associated with this branch is the recommended decision.

Figure 21.3 Applying the expected value approach using decision trees

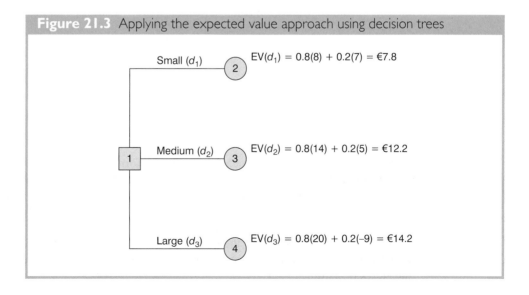

Expected value of perfect information

Suppose that PDC has the opportunity to conduct a market research study that would help evaluate buyer interest in the apartment project and provide information that management could use to improve the probability assessments for the states of nature. To determine the potential value of this information, we begin by supposing that the study could provide *perfect information* regarding the states of nature; that is, we assume for the moment that PDC could determine with certainty, prior to making a decision, which state of nature is going to occur. To make use of this perfect information, we will develop a decision strategy that PDC should follow once it knows which state of nature will occur. A decision strategy is simply a decision rule that specifies the decision alternative to be selected after new information becomes available.

To help determine the decision strategy for PDC, we refer back to PDC's payoff table in Table 21.1. Note that, if PDC knew for sure that state of nature s_1 would occur, the best decision alternative would be d_3, with a payoff of €20 million. Similarly, if PDC knew for sure that state of nature s_2 would occur, the best decision alternative would be d_1, with a payoff of €7 million. Thus, we can state PDC's optimal decision strategy if the perfect information becomes available as follows:

If s_1, select d_3 and receive a payoff of €20 million.

If s_2, select d_1 and receive a payoff of €7 million.

What is the expected value for this decision strategy? To compute the expected value with perfect information, we return to the original probabilities for the states of nature:

$$P(s_1) = 0.8 \quad \text{and} \quad P(s_2) = 0.2.$$

Thus, there is a 0.8 probability that the perfect information will indicate state of nature s_1 and the resulting decision alternative d_3 will provide a €20 million profit.

Similarly, with a 0.2 probability for state of nature s_2, the optimal decision alternative d_1 will provide a €7 million profit. Thus, using equation (21.3), the expected value of the decision strategy based on perfect information is

$$0.8(20) + 02(7) = 17.4$$

We refer to the expected value of €17.4 million as the *expected value with perfect information*.

Earlier in this section we showed that the recommended decision using the expected value approach is decision alternative d_3, with an expected value of €14.2 million.

Therefore, the expected value of the perfect information (EVPI) is €17.4 − €14.2 = €3.2 million. In other words, €3.2 million represents the additional expected value that can be obtained if perfect information were available about the states of nature. Generally speaking, a market research study will not provide 'perfect' information; however, if the market research study is a good one, the information gathered might be worth a sizeable portion of the €3.2 million. Given the EVPI of €3.2 million, PDC might seriously consider a market survey as a way to obtain more information about the states of nature.

In general, the **expected value of perfect information** is computed as follows:

Expected value of perfect information

$$\text{EVPI} = |\text{expected value with perfect information} - \text{best EV}| \qquad \textbf{(21.4)}$$

Note the role of the absolute value in equation (21.4). For minimization problems, information helps reduce or lower cost, thus the expected value with perfect information is less than or equal to the corresponding best expected value otherwise. In this case, EVPI is the magnitude or absolute value of the difference as shown in equation (21.4).

Exercises

Methods

1   The following payoff table shows profit for a decision analysis problem with two decision alternatives and three states of nature.

	States of nature		
Decision alternative	s_1	s_2	s_3
d_1	250	100	25
d_2	100	100	75

a. Construct a decision tree for this problem.
b. Suppose that the decision-maker obtains the probabilities $P(s_1) = 0.65$, $P(s_2) = 0.15$ and $P(s_3) = 0.20$. Use the expected value approach to determine the optimal decision.

2   A decision-maker faced with four decision alternatives and four states of nature develops the following profit payoff table.

	States of nature			
Decision alternative	s_1	s_2	s_3	s_4
d_1	14	9	10	5
d_2	1	10	8	7
d_3	9	10	10	11
d_4	8	10	11	13

The decision-maker obtains information that enables the following probabilities assessments:

$$P(s_1) = 0.5, P(s_2) = 0.2, P(s_3) = 0.2, \text{ and } P(s_4) = 0.1.$$

a. Use the expected value approach to determine the optimal solution.
b. Now assume that the entries in the payoff table are costs. Use the expected value approach to determine the optimal decision.

Applications

3 Holland Corporation is considering three options for managing its data processing operation: continue with its own staff, hire an outside vendor to do the managing (referred to as *outsourcing*) or use a combination of its own staff and an outside vendor. The cost of the operation depends on future demand. The annual cost of each option (in thousands of euros) depends on demand as follows.

	Demand		
Staffing options	High	Medium	Low
Own staff	650	650	600
Outside vendor	900	600	300
Combination	800	650	500

a. If the demand probabilities are 0.2, 0.5 and 0.3, which decision alternative will minimize the expected cost of the data processing operation? What is the expected annual cost associated with your recommendation?

b. What is the expected value of perfect information?

4 Magyar Air Express decided to offer direct service from Budapest to Prague. Management must decide between a full price service using the company's new fleet of jet aircraft and a discount service using smaller capacity commuter planes. It is clear that the best choice depends on the market reaction to the service Magyar Air offers. Management developed estimates of the contribution to profit for each type of service based upon two possible levels of demand for service to Prague: strong and weak. The following table shows the estimated quarterly profits (in thousands of euros).

	Demand for service	
Service	Strong	Weak
Full price	960	-490
Discount	670	320

a. What is the decision to be made, what is the chance event, and what is the consequence for this problem? How many decision alternatives are there? How many outcomes are there for the chance event?

b. Suppose that management of Magyar Air Express believes that the probability of strong demand is 0.7 and the probability of weak demand is 0.3. Use the expected value approach to determine an optimal decision.

c. Suppose that the probability of strong demand is 0.8 and the probability of weak demand is 0.2. What is the optimal decision using the expected value approach?

5 A South African company is considering whether it should tender for two contracts (MS1 and MS2) on offer from a government department for the supply of certain components. The company has three options: (i) tender for MS1 only (ii) tender for MS2 only (ii) or tender for both MS1 and MS2.

If tenders are to be submitted the company will incur additional costs. These costs will have to be entirely recouped from the contract price. The risk, of course, is that if a tender is unsuccessful the company will have made a loss.

The cost of tendering for contract MS1 only is 500 000 RAND. The component supply cost if the tender is successful would be 180 000 RAND.

The cost of tendering for contract MS2 only is 140 000 RAND. The component supply cost if the tender is successful would be 120 000 RAND.

The cost of tendering for both contract MS1 and contract MS2 is 550 000 RAND. The component supply cost if the tender is successful would be 240 000 RAND.

For each contract, possible tender prices have been determined. In addition, subjective assessments have been made of the probability of getting the contract with a particular tender price as shown below. Note here that the company can only submit one tender and cannot, for example, submit two tenders (at different prices) for the same contract.

Option	Possible tender prices (RAND)	Probability of getting contract
MS1 only	1 300 000	0.20
	1 150 000	0.85
MS2 only	700 000	0.15
	650 000	0.80
	600 000	0.95
MS1 and MS2	1 900 000	0.05
	1 400 000	0.65

In the event that the company tenders for both MS1 and MS2 it will either win both contracts (at the price shown above) or no contract at all.

a. What do you suggest the company should do and why?
b. What is the downside and the upside of the suggested course of action?
c. A consultant has approached the company with an offer that in return for 200 000 RAND in cash she will ensure that if you tender 600 000 RAND for contract MS2 only your tender is guaranteed to be successful. Should you accept her offer or not and why?

6   Seneca Vineyards recently purchased land for the purpose of establishing a new vineyard. Management is considering two varieties of white grapes for the new vineyard: Chardonnay and Pinot. The Chardonnay grapes would be used to produce a dry Chardonnay wine, and the Pinot grapes would be used to produce a semi-dry Pinot wine. It takes approximately four years from the time of planting before new grapes can be harvested. This length of time creates a great deal of uncertainty concerning future demand and makes the decision concerning the type of grapes to plant difficult. Three possibilities are being considered: Chardonnay grapes only; Pinot grapes only; and both Chardonnay and Pinot grapes. Seneca management decided that for planning purposes it would be adequate to consider only two demand possibilities for each type of wine: strong or weak. With two possibilities for each type of wine it was necessary to assess four probabilities. With the help of some forecasts in industry publications management made the following probability assessments.

Chardonnay demand	Pinot demand	
	Weak	Strong
Weak	0.05	0.50
Strong	0.25	0.20

Revenue projections show an annual contribution to profit of €20 000 if Seneca Vineyards only plants Chardonnay grapes and demand is weak for Chardonnay wine, and €70 000 if they only plant Chardonnay grapes and demand is strong for Chardonnay wine. If they only plant Pinot grapes, the annual profit projection is €25 000 if demand is weak for Pinot grapes and €45 000 if demand is strong for Pinot grapes. If Seneca Vineyards plants both types of grapes, the annual profit projections are shown in the following table.

| | Pinot demand | |
Chardonnay demand	Weak	Strong
Weak	€22 000	€40 000
Strong	€26 000	€60 000

a. What is the decision to be made, what is the chance event and what is the consequence? Identify the alternatives for the decisions and the possible outcomes for the chance events.
b. Develop a decision tree.
c. Use the expected value approach to recommend which alternative Seneca Vineyards should follow in order to maximize expected annual profit.
d. Suppose management is concerned about the probability assessments when demand for Chardonnay wine is strong. Some believe it is likely for Pinot demand to also be strong in this case. Suppose the probability of strong demand for Chardonnay and weak demand for Pinot is 0.05 and that the probability of strong demand for Chardonnay and strong demand for Pinot is 0.40. How does this change the recommended decision? Assume that the probabilities when Chardonnay demand is weak are still 0.05 and 0.50.
e. Other members of the management team expect the Chardonnay market to become saturated at some point in the future causing a fall in prices. Suppose that the annual profit projections fall to €50 000 when demand for Chardonnay is strong and Chardonnay grapes only are planted. Using the original probability assessments, determine how this change would affect the optimal decision.

7 An Australian government committee is considering the economic benefits of a programme of preventative flu vaccinations. If vaccinations are not introduced then the estimated cost to the government if flu strikes in the next year is A$7 m with probability 0.1, A$10 m with probability 0.3 and A$15 m with probability 0.6. It is estimated that such a programme will cost A$7 m and that the probability of flu striking in the next year is 0.75.

One alternative open to the committee is to institute an 'early-warning' monitoring scheme (costing A$3 m) which will enable it to detect an outbreak of flu early and hence institute a rush vaccination program (costing A$10 m because of the need to vaccinate quickly before the outbreak spreads).

a. What recommendations should the committee make to the government if their objective is to maximize expected monetary value (EMV)?
b. The committee has also been informed that there are alternatives to using EMV. What are these alternatives and would they be appropriate in this case?

21.3 Decision analysis with sample information

In applying the expected value approach, we showed how probability information about the states of nature affects the expected value calculations and thus the decision recommendation.

Frequently, decision-makers have preliminary or **prior probability** assessments for the states of nature that are the best probability values available at that time. However, to make the best possible decision, the decision-maker may want to seek additional information about the states of nature. This new information can be used to revise or update the prior probabilities so that the final decision is based on more accurate probabilities for the states of nature. Most often, additional information is obtained through experiments designed to provide **sample information** about the states of nature. Raw material sampling, product testing, and market research studies are examples of experiments (or studies) that may enable management to revise or update the state-of-nature probabilities. These revised probabilities are called **posterior probabilities**.

For the PDC problem suppose that management is considering a six-month market research study designed to learn more about potential market acceptance of the PDC apartment project. Management anticipates that the market research study will provide one of the following two results:

1 Favourable report: A significant number of the individuals contacted express interest in purchasing a PDC apartment.

2 Unfavourable report: Very few of the individuals contacted express interest in purchasing a PDC apartment.

Decision tree

The decision tree for the PDC problem with sample information shows the logical sequence for the decisions and the chance events in Figure 21.4. First, PDC's management must decide whether the market research should be conducted. If it is conducted, PDC's management must be prepared to make a decision about the size of the apartment project if the market research report is favourable and, possibly, a different decision about the size of the apartment project if the market research report is unfavourable.

At each (square) decision node, the branch of the tree that is taken is based on the decision made. At each (circle) chance node, the branch of the tree that is taken is based on probability or chance. For example, decision node 1 shows that PDC must first make the decision whether to conduct the market research study. If the market research study is undertaken, chance node 2 indicates that both the favourable report branch and the unfavourable report branch are not under PDC's control and will be determined by chance. Node 3 is a decision node, indicating that PDC must make the decision to construct the small, medium or large complex if the market research report is favourable. Node 4 is a decision node showing that PDC must make the decision to construct the small, medium or large complex if the market research report is unfavourable. Node 5 is a decision node indicating that PDC must make the decision to construct the small, medium or large complex if the market research is not undertaken. Nodes 6 to 14 are chance nodes indicating that the strong demand or weak demand state-of-nature branches will be determined by chance.

Figure 21.4 The PDC decision tree including the market research study

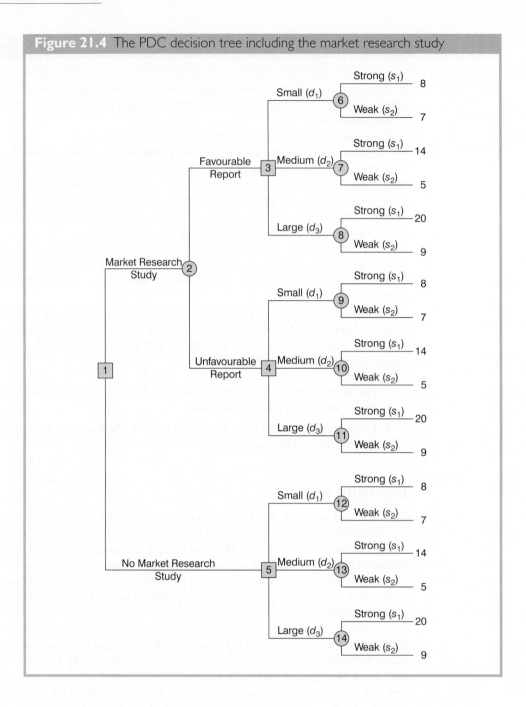

Analysis of the decision tree and the choice of an optimal strategy requires that we know the branch probabilities corresponding to all chance nodes. PDC developed the following branch probabilities.

If the market research study is undertaken

$$P(\text{Favourable report}) = P(F) = 0.77$$

$$P(\text{Unfavourable report}) = P(U) = 0.23$$

If the market research report is favourable

$$F(\text{Strong demand given a favourable report}) = P(s_1|F) = 0.94$$
$$F(\text{Weak demand given a favourable report}) = P(s_2|F) = 0.06$$

If the market research report is unfavourable

$$P(\text{Strong demand given an unfavourable report}) = P(s_1|U) = 0.35$$
$$P(\text{Weak demand given an unfavourable report}) = P(s_2|U) = 0.65$$

If the market research report is not undertaken the prior probabilities are applicable.

$$P(\text{Strong demand}) = P(s_1) = 0.80$$
$$P(\text{Weak demand}) = P(s_2) = 0.20$$

The branch probabilities are shown on the decision tree in Figure 21.5.

Decision strategy

A **decision strategy** is a sequence of decisions and chance outcomes where the decisions chosen depend on the yet to be determined outcomes of chance events. The approach used to determine the optimal decision strategy is based on a backward pass through the decision tree using the following steps:

1 At chance nodes, compute the expected value by multiplying the payoff at the end of each branch by the corresponding branch probability.

2 At decision nodes, select the decision branch that leads to the best expected value. This expected value becomes the expected value at the decision node.

Starting the backward pass calculations by computing the expected values at chance nodes 6 to 14 provides the following results.

$$EV(\text{Node } 6) = 0.94(8) + 0.06(7) = 7.94$$
$$EV(\text{Node } 7) = 0.94(14) + 0.06(5) = 13.46$$
$$EV(\text{Node } 8) = 0.94(20) + 0.06\,(-9) = 18.26$$
$$EV(\text{Node } 9) = 0.35(8) + 0.65(7) = 7.35$$
$$EV(\text{Node } 10) = 0.35(14) + 0.65(5) = 8.15$$
$$EV(\text{Node } 11) = 0.35(20) + 0.65(-9) = 1.15$$
$$EV(\text{Node } 12) = 0.80(8) + 0.20(7) = 7.80$$
$$EV(\text{Node } 13) = 0.80(14) + 0.20(5) = 12.20$$
$$EV(\text{Node } 14) = 0.80(20) + 0.20(-9) = 14.20$$

Figure 21.6 shows the reduced decision tree after computing expected values at these chance nodes.

Next move to decision nodes 3, 4 and 5. For each of these nodes, we select the decision alternative branch that leads to the best expected value. For example, at node 3 we have the choice of the small complex branch with EV(Node 6) = 7.94, the medium complex branch with EV(Node 7) = 13.46 and the large complex branch with EV(Node 8) 18.26. Thus, we select the large complex decision alternative branch and the expected value at node 3 becomes EV(Node 3) = 18.26.

Figure 21.5 The PDC decision tree with branch probabilities

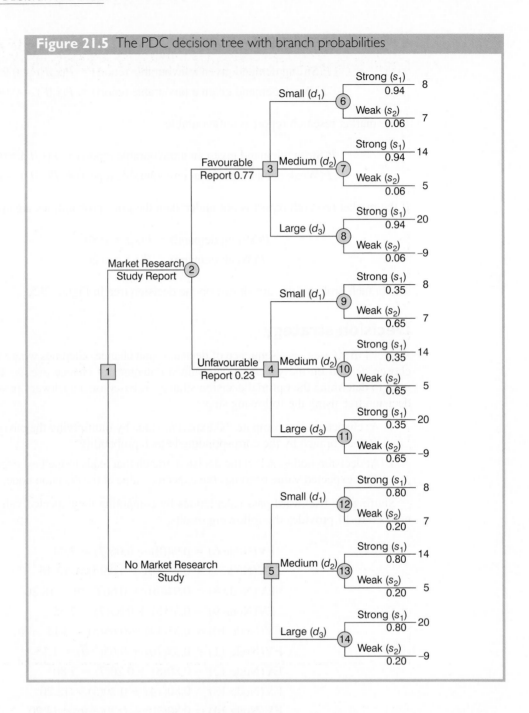

For node 4, we select the best expected value from nodes 9, 10 and 11. The best decision alternative is the medium complex branch that provides EV(Node 4) = 8.15. For node 5, we select the best expected value from nodes 12, 13 and 14. The best decision alternative is the large complex branch that provides EV(Node 5) = 14.20. Figure 21.7 shows the reduced decision tree after choosing the best decisions at nodes 3, 4 and 5.

The expected value at chance node 2 can now be computed as follows:

$$EV(\text{Node 2}) = 0.77EV(\text{Node 3}) + 0.23EV(\text{Node 4})$$
$$= 0.77(18.26) + 0.23(8.15) = 15.93$$

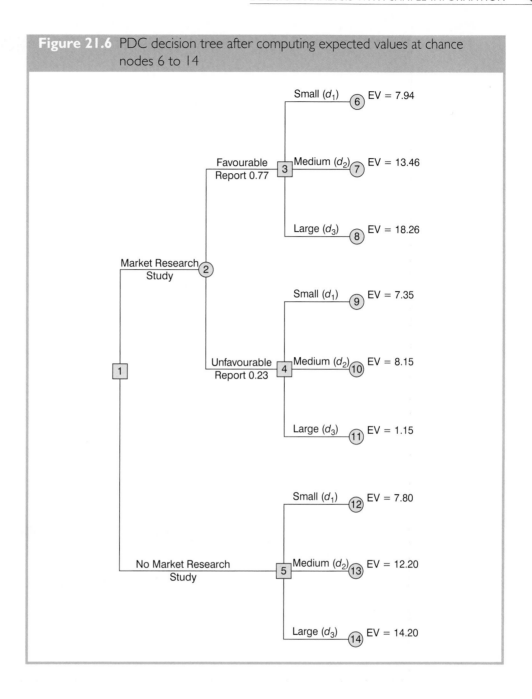

Figure 21.6 PDC decision tree after computing expected values at chance nodes 6 to 14

This calculation reduces the decision tree to one involving only the two decision branches from node 1 (see Figure 21.8).

Finally, the decision can be made at decision node 1 by selecting the best expected values from nodes 2 and 5. This action leads to the decision alternative to conduct the market research study, which provides an overall expected value of 15.93.

The optimal decision for PDC is to conduct the market research study and then carry out the following decision strategy:

If the market research is favourable, construct the large apartment complex.

If the market research is unfavourable, construct the medium apartment complex.

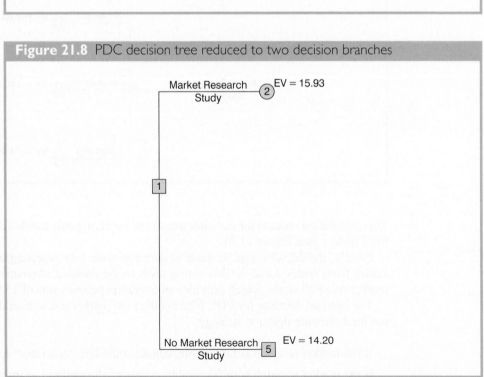

Figure 21.7 PDC decision tree after choosing best decisions at nodes 3, 4 and 5

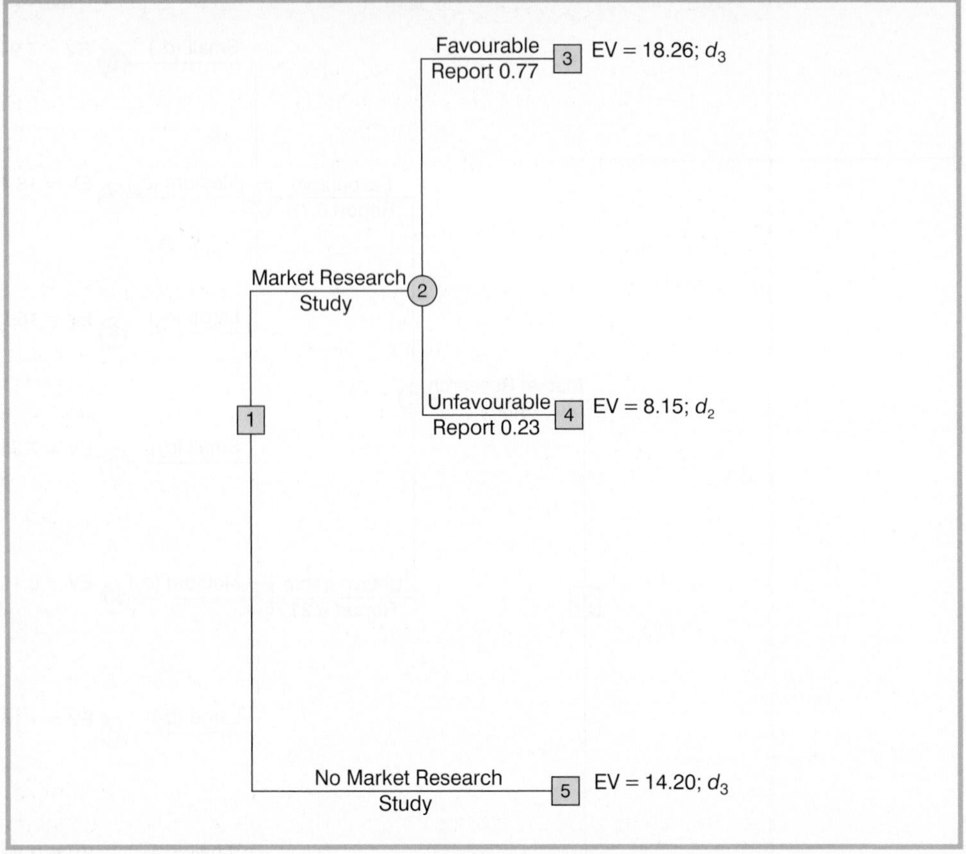

Favourable
Report 0.77   3   EV = 18.26; d_3

Market Research
Study   ②

Unfavourable
Report 0.23   4   EV = 8.15; d_2

1

No Market Research
Study   5   EV = 14.20; d_3

Figure 21.8 PDC decision tree reduced to two decision branches

Market Research
Study   ②   EV = 15.93

1

No Market Research
Study   5   EV = 14.20

The analysis of the PDC decision tree illustrates the methods that can be used to analyse more complex sequential decision problems. First, draw a decision tree consisting of decision and chance nodes and branches that describe the sequential nature of the problem.

Determine the probabilities for all chance outcomes. Then, by working backward through the tree, compute expected values at all chance nodes and select the best decision branch at all decision nodes. The sequence of optimal decision branches determines the optimal decision strategy for the problem.

Expected value of sample information

In the PDC problem, the market research study is the sample information used to determine the optimal decision strategy. The expected value associated with the market research study is €15.93. In Section 21.3 we showed that the best expected value if the market research study is *not* undertaken is €14.20. Thus, we can conclude that the difference, €15.93 − €14.20 = €1.73, is the **expected value of sample information (EVSI)**. In other words, conducting the market research study adds €1.73 million to the PDC expected value. The EVSI = €1.73 million suggests PDC should be willing to pay up to €1.73 million to conduct the market research study. In general, the expected value of sample information is as follows:

Expected value of sample information

$$\text{EVSI} = |\text{expected value with sample information} - \text{best EV}| \qquad \textbf{(21.5)}$$

Note the role of the absolute value in equation (21.5). For minimization problems the expected value with sample information is always less than or equal to the best expected value without sample information. Thus, by taking the absolute value of the difference as shown in equation (21.5), we can handle both the maximization and minimization cases with one equation.

Exercises

Methods

8 Consider a variation of the PDC decision tree shown in Figure 21.5. The company must first decide whether to undertake the market research study. If the market research study is conducted, the outcome will either be favourable (F) or unfavourable (U). Assume there are only two decision alternatives d_1 and d_2 and two states of nature s_1 and s_2. The payoff table showing profit is as follows:

| | State of nature | |
Decision alternative	s_1	s_2
d_1	100	300
d_2	400	200

a. Show the decision tree.
b. Use the following probabilities. What is the optimal decision strategy?

$$P(F) = 0.56 \quad P(s_1|F) = 0.57 \quad P(s_1|U) = 0.18 \quad P(s_1) = 0.40$$
$$P(U) = 0.44 \quad P(s_2|F) = 0.43 \quad P(s_2|U) = 0.82 \quad P(s_2) = 0.60$$

Applications

9   A property investor has the opportunity to purchase land currently zoned for residential development. If the county board approves a request to re-zone the property for commercial development within the next year, the investor will be able to lease the land to a large discount firm that wants to open a new store on the property. However, if the zoning change is not approved, the investor will have to sell the property at a loss. Profits (in thousands of euros) are shown in the following payoff table.

	State of nature	
	Re-zoning approved	Re-zoning not approved
Decision alternative	s_1	s_2
Purchase, d_1	600	−200
Do not purchase, d_2	0	0

a. If the probability that the re-zoning will be approved is 0.5, what decision is recommended?
   What is the expected profit?
b. The investor can purchase an option to buy the land. Under the option, the investor maintains the rights to purchase the land any time during the next three months while learning more about possible resistance to the re-zoning proposal from area residents. Probabilities are as follows.

$$\text{Let } H = \text{high resistance to re-zoning}$$
$$L = \text{low resistance to re-zoning}$$
$$P(H) = 0.55 \quad P(s_1|H) = 0.18 \quad P(s_2|H) = 0.82$$
$$P(L) = 0.45 \quad P(s_1|L) = 0.89 \quad P(s_2|L) = 0.11$$

What is the optimal decision strategy if the investor uses the option period to learn more about the resistance from area residents before making the purchase decision?
c. If the option will cost the investor an additional €10 000, should the investor purchase the option? Why or why not? What is the maximum that the investor should be willing to pay for the option?

10 Dante Development Corporation is considering bidding on a contract for a new office building complex. Figure 21.9 shows the decision tree prepared by one of Dante's analysts. At node 1, the company must decide whether to bid on the contract. The cost of preparing the bid is €200 000. The upper branch from node 2 shows that the company has a 0.8 probability of winning the contract if it submits a bid. If the company wins the bid, it will have to pay €2 000 000 to become a partner in the project. Node 3 shows that the company will then consider doing a market research study to forecast demand for the office units prior to beginning construction. The cost of this study is €150 000. Node 4 is a chance node showing the possible outcomes of the market research study.

Figure 21.9 Decision tree for the Dante Development Corporation

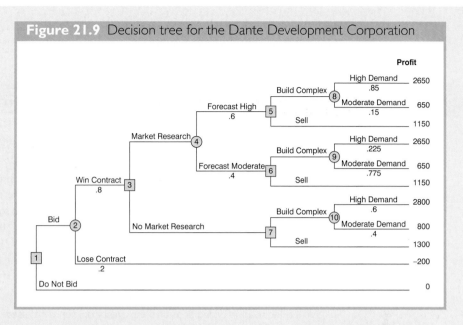

Nodes 5, 6 and 7 are similar in that they are the decision nodes for Dante to either build the office complex or sell the rights in the project to another developer. The decision to build the complex will result in an income of €5 000 000 if demand is high and €3 000 000 if demand is moderate. If Dante chooses to sell its rights in the project to another developer, income from the sale is estimated to be €3 500 000. The probabilities shown at nodes 4, 8 and 9 are based on the projected outcomes of the market research study.

a. Verify Dante's profit projections shown at the ending branches of the decision tree by calculating the payoffs of €2 650 000 and €650 000 for the first two outcomes.

b. What is the optimal decision strategy for Dante, and what is the expected profit for this project?

c. What would the cost of the market research study have to be before Dante would change its decision about conducting the study?

11 A Turkish company is trying to decide whether to bid for a certain contract or not. They estimate that merely preparing the bid will cost YTL10 000. If their company bid then they estimate that there is a 50 per cent chance that their bid will be put on the 'short-list', otherwise their bid will be rejected. Once 'short-listed' the company will have to supply further detailed information (entailing costs estimated at YTL5000). After this stage their bid will either be accepted or rejected.

The company estimate that the labour and material costs associated with the contract are YTL127 000. They are considering three possible bid prices, namely YTL155 000, YTL170 000 and YTL190 000. They estimate that the probability of these bids being accepted (once they have been short-listed) is 0.90, 0.75 and 0.35 respectively.

What should the company do and what is the expected monetary value of your suggested course of action?

12 Hale's TV Productions is considering producing a pilot for a comedy series in the hope of selling it to a major television network. The network may decide to reject the series, but it may also decide to purchase the rights to the series for either one or two years. At this point in time, Hale may either produce the pilot and wait for the network's decision or transfer the

rights for the pilot and series to a competitor for €100 000. Hale's decision alternatives and profits (in thousands of euros) are as follows:

| | State of nature | | |
Decision alternative	Reject, s_1	1 Year, s_2	2 Years, s_3
Produce pilot, d_1	-100	50	150
Sell to competitor, d_2	100	100	100

The probabilities for the states of nature are $P(s_1) = 0.2$, $P(s_2) = 0.3$, and $P(s_3) = 0.5$. For a consulting fee of €5000, an agency will review the plans for the comedy series and indicate the overall chances of a favourable network reaction to the series. Assume that the agency review will result in a favourable (F) or an unfavourable (U) review and that the following probabilities are relevant.

$$P(F) = 0.69 \quad P(s_1|F) = 0.09 \quad P(s_2|F) = 0.26 \quad P(s_3|F) = 0.65$$
$$P(U) = 0.31 \quad P(s_1|U) = 0.45 \quad P(s_2|U) = 0.39 \quad P(s_3|U) = 0.16$$

a. Construct a decision tree for this problem.
b. What is the recommended decision if the agency opinion is not used? What is the expected value?
c. What is the expected value of perfect information?
d. What is Hale's optimal decision strategy assuming the agency's information is used?
e. What is the expected value of the agency's information?
f. Is the agency's information worth the €5000 fee? What is the maximum that Hale should be willing to pay for the information?
g. What is the recommended decision?

13 Larson's Department Store faces a buying decision for a seasonal product for which demand can be high, medium or low. The purchaser for Larson's can order 1, 2 or 3 lots of the product before the season begins but cannot reorder later. Profit projections (in thousands of euros) are shown.

| | State of nature | | |
	High demand s_1	Medium demand s_2	Low demand s_3
Order 1 lot, d_1	60	60	50
Order 2 lots, d_2	80	80	30
Order 3 lots, d_3	100	70	10

a. If the prior probabilities for the three states of nature are 0.3, 0.3 and 0.4, respectively, what is the recommended order quantity?
b. At each pre-season sales meeting, the head of sales provides a personal opinion regarding potential demand for this product. Because of the CEO's enthusiasm and optimistic nature, the predictions of market conditions have always been either 'excellent' (E) or 'very good' (V). Probabilities are as follows. What is the optimal decision strategy?

$$P(E) = 0.7 \quad P(s_1|E) = 0.34 \quad P(s_2|E) = 0.32 \quad P(s_3|E) = 0.34$$
$$P(V) = 0.3 \quad P(s_1|V) = 0.20 \quad P(s_2|V) = 0.26 \quad P(s_3|V) = 0.54$$

c. Compute EVPI and EVSI. Discuss whether the firm should consider a consulting expert who could provide independent forecasts of market conditions for the product.

21.4 Computing branch probabilities using Bayes' theorem

In Section 21.3 the branch probabilities for the PDC decision tree chance nodes were specified in the problem description. No computations were required to determine these probabilities. In this section we show how **Bayes' theorem,** a topic covered in Chapter 4, can be used to compute branch probabilities for decision trees.

The PDC decision tree is shown again in Figure 21.10. As before, it is assumed

$$F = \text{Favourable market research report}$$
$$U = \text{Unfavourable market research report}$$
$$s_1 = \text{Strong demand (state of nature 1)}$$
$$s_2 = \text{Weak demand (state of nature 2)}$$

At chance node 2, we need to know the branch probabilities $P(F)$ and $P(U)$. At chance nodes 6, 7 and 8, we need to know the branch probabilities $P(s_1|F)$, the probability of state of nature 1 given a favourable market research report, and $P(s_2|F)$, the probability of state of nature 2 given a favourable market research report. $P(s_1|F)$ and $P(s_2|F)$ are referred to as *posterior probabilities* because they are conditional probabilities based on the outcome of the sample information. At chance nodes 9, 10 and 11, we need to know the branch probabilities $P(s_1|U)$ and $P(s_2|U)$; note that these are also posterior probabilities, denoting the probabilities of the two states of nature *given* that the market research report is unfavourable.

Finally at chance nodes 12, 13 and 14, we need to know the prior probabilities for the states of nature, $P(s_1)$ and $P(s_2)$, if the market research study is not undertaken. In addition, we must know the **conditional probability** of the market research outcomes (the sample information) *given* each state of nature. For example, we need to know the conditional probability of a favourable market research report given that strong demand exists for the PDC project; note that this conditional probability of F given state of nature s_1 is written $P(F|s_1)$. To carry out the probability calculations, we will need conditional probabilities for all sample outcomes given all states of nature, that is, $P(F|s_1)$, $P(F|s_2)$, $P(U|s_1)$ and $P(U|s_2)$. In the PDC problem, we assume that the following assessments are available for these conditional probabilities.

State of nature	Market research	
	Favourable, F	Unfavourable, U
Strong demand, s_1	$P(F|s_1) = 0.90$	$P(U|s_1) = 0.10$
Weak demand, s_2	$P(F|s_2) = 0.25$	$P(U|s_2) = 0.75$

Note that these should provide a reasonable degree of confidence in the market research study. If the true state of nature is s_1, the probability of a favourable market research report is 0.90 and the probability of an unfavourable market research report is 0.10. If the true state of nature is s_2, the probability of a favourable market research report is 0.25 and the probability of an unfavourable market research report is 0.75. The reason for a 0.25 probability of a potentially misleading favourable market research report for state of nature s_2 is that when some potential buyers first hear about the new apartment project, their enthusiasm may lead them to overstate their real interest in it. A potential buyer's initial favourable response can change quickly to a 'no thank you' when later faced with the reality of signing a purchase contract and making a down-payment.

Figure 21.10 The PDC decision tree

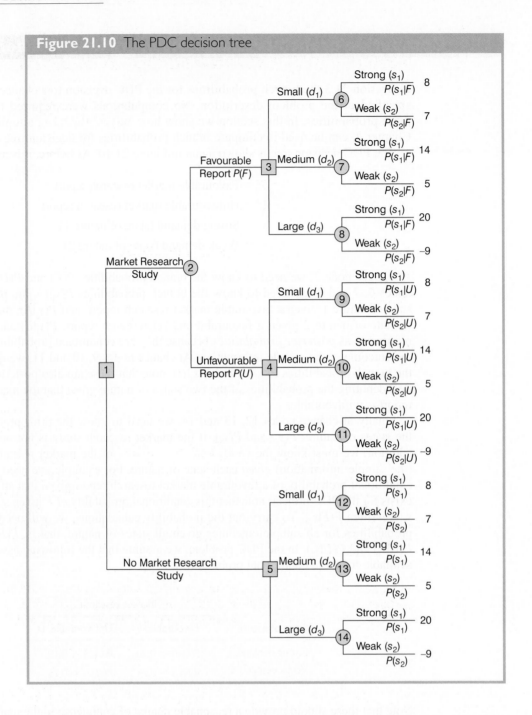

We present a tabular approach as a convenient method for carrying out the probability computations. The computations for the PDC problem based on a favourable market research report (F) are summarized in Table 21.2. The steps used to develop this table are as follows.

Step 1 In column 1 enter the states of nature. In column 2 enter the *prior probabilities* for the states of nature. In column 3 enter the *conditional probabilities* of a favourable market research report (F) given each state of nature.

Table 21.2 Branch probabilities for the PDC apartment project based on a favourable market research report

States of nature s_j	Prior probabilities $P(s_j)$	Conditional probabilities $P(F\|s_j)$	Joint probabilities $P(F \cap s_j)$	Posterior probabilities $P(s_j\|F)$
s_1	0.8	0.90	0.72	0.94
s_2	0.2	0.25	0.05	0.06
	1.0		$P(F) = 0.77$	1.00

Step 2 In column 4 compute the **joint probabilities** by multiplying the prior probability values in column 2 by the corresponding conditional probability values in column 3.

Step 3 Sum the joint probabilities in column 4 to obtain the probability of a favourable market research report, $P(F)$.

Step 4 Divide each joint probability in column 4 by $P(F) = 0.77$ to obtain the revised or posterior probabilities, $P(s_1|F)$ and $P(s_2|F)$.

Table 21.2 shows that the probability of obtaining a favourable market research report is $P(F) = 0.77$. In addition, $P(s_1|F) = 0.94$ and $P(s_2|F) = 0.06$. In particular, note that a favourable market research report will prompt a revised or posterior probability of 0.94 that the market demand of the apartment will be strong, s_1.

The tabular probability computation procedure must be repeated for each possible sample information outcome. Thus, Table 21.3 shows the computations of the branch probabilities of the PDC problem based on an unfavourable market research report. Note that the probability of obtaining an unfavourable market research report is $P(U) = 0.23$. If an unfavourable report is obtained, the posterior probability of a strong market demand, s_1, is 0.35 and of a weak market demand, s_2, is 0.65. The branch probabilities from Tables 21.2 and 21.3 were shown on the PDC decision tree in Figure 21.5.

The discussion in this section shows an underlying relationship between the probabilities on the various branches in a decision tree. To assume different prior probabilities, $P(s_1)$ and $P(s_2)$, without determining how these changes would alter $P(F)$ and $P(U)$, as well as the posterior probabilities $P(s_1|F)$, $P(s_2|F)$, $P(s_1|U)$, and $P(s_2|U)$, would be inappropriate.

Table 21.3 Branch probabilities for the PDC apartment project based on an unfavourable market research report

States of nature s_j	Prior probabilities $P(s_j)$	Conditional probabilities $P(U\|s_j)$	Joint probabilities $P(U \cap s_j)$	Posterior Probabilities $P(s_j\|U)$
s_1	0.8	0.10	0.08	0.35
s_2	0.2	0.75	0.15	0.65
	1.0		$P(U) = 0.23$	1.00

Exercises

Methods

14 Suppose that you are given a decision situation with three possible states of nature: s_1, s_2, and s_3. The prior probabilities are $P(s_1) = 0.2$, $P(s_2) = 0.5$ and $P(s_3) = 0.3$. With sample information I, $P(I|s_1) = 0.1$, $P(I|s_2) = 0.05$ and $P(I|s_3) = 0.2$. Compute the revised or posterior probabilities: $P(s_1|I)$, $P(s_2|I)$, and $P(s_3|I)$.

15 In the following profit payoff table for a decision problem with two states of nature and three decision alternatives, the prior probabilities for s_1 and s_2 are $P(s_1) = 0.8$ and $P(s_2) = 0.2$.

	State of nature	
Decision alternative	s_1	s_2
d_1	15	10
d_2	10	12
d_3	8	20

a. What is the optimal decision?
b. Find the EVPI.
c. Suppose that sample information I is obtained, with $P(I|s_1) = 0.20$ and $P(I|s_2) = 0.75$.

Find the posterior probabilities $P(s_1|I)$ and $P(s_2|I)$. Recommend a decision alternative based on these probabilities.

Applications

16 To save on expenses, Rene and Jacques agreed to form a carpool for travelling to and from work. Rene preferred to use the somewhat longer but more consistent Boulevard Peripherique. Although Jacques preferred the quicker expressway, he agreed with Rene that they should take Boulevard Peripherique if the expressway had a traffic jam. The following payoff table provides the one-way time estimate in minutes for travelling to and from work.

	State of nature	
Decision alternative	Expressway open s_1	Expressway jammed s_2
Boulevard Peripherique, d_1	30	30
Expressway, d_2	25	45

Based on their experience with traffic problems, Rene and Jacques agreed on a 0.15 probability that the expressway would be jammed.

In addition, they agreed that weather seemed to affect the traffic conditions on the expressway.

Let

$$C = \text{clear}$$
$$O = \text{overcast}$$
$$R = \text{rain}$$

The following conditional probabilities apply.

$$P(C|s_1) = 0.8 \quad P(O|s_1) = 0.2 \quad P(R|s_1) = 0.0$$
$$P(C|s_2) = 0.1 \quad P(O|s_2) = 0.3 \quad P(R|s_2) = 0.6$$

a. Use Bayes' theorem for probability revision to compute the probability of each weather condition and the conditional probability of the expressway open s_1 or jammed s_2 given each weather condition.

b. Show the decision tree for this problem.

c. What is the optimal decision strategy, and what is the expected travel time?

17 The Granaldi Manufacturing Company must decide whether to manufacture a component part at its Milan plant or purchase the component part from a supplier. The resulting profit is dependent upon the demand for the product. The following payoff table shows the projected profit (in thousands of euros).

| Decision alternative | State of nature | | |
	Low demand s_1	Medium demand s_2	High demand s_3
Manufacture, d_1	−20	40	100
Purchase, d_2	10	45	70

The state-of-nature probabilities are $P(s_1) = 0.35$, $P(s_2) = 0.35$, and $P(s_3) = 0.30$.

a. Use a decision tree to recommend a decision.

b. Use EVPI to determine whether Granaldi should attempt to obtain a better estimate of demand.

c. A test market study of the potential demand for the product is expected to report either a favourable (F) or unfavourable (U) situation. The relevant conditional probabilities are as follows:

$$P(F|s_1) = 0.10 \quad P(U|s_1) = 0.90$$
$$P(F|s_2) = 0.40 \quad P(U|s_2) = 0.60$$
$$P(F|s_3) = 0.60 \quad P(U|s_3) = 0.40$$

What is the probability that the market research report will be favourable?

d. What is Granaldi's optimal decision strategy?

e. What is the expected value of the market research information?

For additional online summary questions and answers go to the companion website at www.cengage.co.uk/aswsbe2

Summary

Decision analysis can be used to determine the best decision alternative or an optimal decision strategy when a decision-maker is faced with an uncertain and riskfilled pattern of future events.

We showed how payoff tables and decision trees could be used to structure decision problems and described the relationships between the decisions, chance events and their consequences. We showed how having perfect information about chance events (states of nature) could be valued. Given corresponding probability assessments for the states of nature, the expected value approach could be used to identify the best decision alternative or decision strategy.

Where sample information about the chance events could be made available, it was shown how relevant probabilities could be straightforwardly revised following a Bayesian approach to arrive at the optimal decision strategy.

Key terms

Bayes' theorem
Branch
Chance event
Chance nodes
Conditional probabilities
Consequence
Decision nodes
Decision strategy
Decision tree
Expected value (EV)
Expected value approach
Expected value of perfect information (EVPI)

Expected value of sample information (EVSI)
Joint probabilities
Node
Payoff
Payoff table
Posterior (revised) probabilities
Prior probabilities
Sample information
States of nature

Key formulae

Expected value

$$EV(d_i) = \sum_{j=1}^{N} P(s_j) V_{ij}$$ (21.3)

Expected value of perfect information

$$EVPI = |\text{expected value with perfect information} - \text{best EV}|$$ (21.4)

Expected value of sample information

$$EVSI = |\text{expected value with sample information} - \text{best EV}|$$ (21.5)

Case problem 1 Stock-ordering at Mintzas

The Mintzas Trading Company sells typewriters, adding machines, desk calculators, duplicating equipment, paper, ink, etc. In addition to these items for businesses, it also carries a line of transcription equipment. Sales of such kit form an appreciable portion of Mintzas' total sales. The kit is ordered from the same distributor who handles the office machines. Because the manager felt it was too expensive to order each item separately, he had established a policy of placing an order to the distributor once a month. It took approximately one month for the order to be filled (i.e. kits ordered the first of this month would not be available to sell until the first of the next month).

Because the transcription equipment has a high average unit cost, the manager of Mintzas is interested in keeping the inventory to a minimum. He estimates that every piece of transcription equipment that he has remaining in stock at the end of the month costs him €10. This figure is obtained by estimating the cost of capital tied up in stock, cost of handling and the alternative uses of space required.

On the other hand, each piece of equipment which he sells contributes approximately €50 profit and fixed costs. The manager estimates that of the customers who come into the store to buy transcription equipment when he has none in stock, half of them will buy the machines elsewhere rather than wait until Mintzas receives a shipment.

Court reporter typing up the court transcript on her laptop computer. © Lisa F. Young.

WEEKLY SALES RECORD	
Number of transcription machines demanded during week	Weeks
4	1
5	4
6	5
7	6
8	9
9	12
10	12
11	10
12	9
13	11
14	9
15	7
16	4
17	1
18	2
19	2

It is almost impossible to predict what the weekly or monthly sales of transcription kit will be. Fluctuations from week to week are apparently random and there also appears to be no increasing or decreasing sales trend over the two year period during which sales were kept. A summary of the sales record is given in Table 1. From this he wondered if he could develop a policy regarding the number of kits to order each month.

Management Report

1. Estimate Mintzas' existing stock management costs in terms of order costs and holding costs.

2. Using mean sales as a guide, consider relevant order quantity options and cost these out – as in 1-applying the methods of Chapter 21. Discuss your results.

3. What advice would you give to the company?

Case problem 2 Production Strategies

Fellini Foods is an Italian company, currently manufacturing a canned long spaghetti in Italy, importing it to the UK and selling it to grocery and delicatessen outlets. HMR Products Limited has been offered the opportunity to purchase the brand name / franchise and in particular to take over production arrangements for the product. As part of HMR's strategic assessment of this opportunity, two different production options are currently under review:

A. Continue to import from Italy

Cost £344 per tonne delivered, plus import duty of 2.5 per cent.
Maximum capacity 6500 tonnes per annum.

B. Can at Greatham using pasta supplied by Fellini Foods Limited

1 Fellini Foods would charge £63 000 per annum fixed costs.

2 Variable cost per tonne of finished product is:

Cost ex Fellini Foods	£139.65
Greatham costs	£133.35
Total variable cost	£273.00

3 Greatham fixed costs are £315.00

4 Figures above assume one shift production at Greatham with a maximum capacity of 7000 tonnes per annum.

Tinned spaghetti in tomato sauce. © Monkey Business Images.

5 Additional capacity can be obtained by:

(a) Overtime working – this gives a capability of an additional 1000 tonnes per annum, has no effect on fixed costs but increases the basic Greatham variable cost by £14 per tonne.

(b) Evening shift – this gives a capability of an additional 2000 tonnes per annum, increases the basic Greatham variable cost by £9 per tonne and involves additional fixed costs of £47 250 per annum.

Note: (a) and (b) can be done separately or together.

(c) Introducing two and/or three shift working. The second shift has a capacity of 7000 tonnes per annum. The third shift has a capacity of 5000 tonnes per annum, once a second shift is introduced, no overtime or evening shift working is feasible.

The second and third shift will each incur an additional fixed cost of £160 000 per annum.

Variable costs are as follows:

First 7000 tonnes @ €300 per tonne
Next 7000 tonnes @ €320 per tonne
Remaining tonnage @ €350 per tonne

HMR is considering three very different demand outcomes for the product, allowing for expected market volatility over the next three years: Low (6000 tonnes), Medium (12 000 tonnes) and High (18 000 tonnes)

Prior probabilities for these outcomes have been assessed as follows:

P(Low Demand) = 0.25
P(Medium Demand) = 0.5
P(High Demand) = 0.25

Management Report

Perform an analysis of the problem facing HMR and prepare a report that summarizes your findings and recommendations. Your report should include in particular:

1 A decision tree.

2 A risk profile of your recommended strategy.

21.5 Solving the PDC problem using TreePlan

TreePlan is an EXCEL add-in that can be used to develop decision trees for decision analysis problems. The software package is provided on this textbook's companion website (http://www.cengage.co.uk/aswsbe/). A manual containing additional information on starting and using TreePlan is also included. In the following example, we show how to use TreePlan to build a decision tree and solve the PDC problem presented in Section 21.1. The decision tree for the PDC problem is shown in Figure 21.11.

Getting started: an initial decision tree

We begin by assuming that TreePlan has been installed and an EXCEL workbook is open. To build a TreePlan version of the PDC decision tree the steps are as follows:

Step 1 Select cell A1   [Main menu bar]
      Add-Ins > Decision Tree
      Choose **New Tree**
      Click **OK**

A decision tree with one decision node and two branches appears as follows:

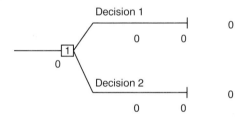

Figure 21.11 PDC decision tree

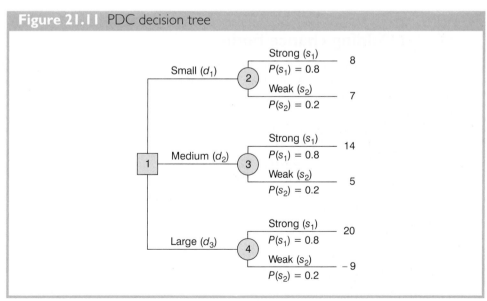

---

*TreePlan was developed by Professor Michael R. Middleton at the University of San Francisco and modified for use by Professor James E. Smith at Duke University. The TreePlan website is located at http://www.treeplan.com.

Adding a branch

The PDC problem includes three decision alternatives (small, medium and large apartment complexes), so we must add another decision branch to the tree.

Step 1   Select cell B5   [Main menu bar]
        Add-Ins > Decision Tree
        Choose **Add branch**
        Click **OK**

A revised tree with three decision branches now appears in the EXCEL worksheet.

Naming the decision alternatives

The decision alternatives can be named by selecting the cells containing the labels Decision 1, Decision 2 and Decision 3, and then entering the corresponding PDC names Small, Medium and Large. After naming the alternatives, the PDC tree with three decision branches appears as follows:

TreePlan Tryout for Evaluation

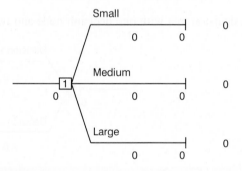

Adding chance nodes

The chance event for the PDC problem is the demand for the apartments, which may be either strong or weak. Thus, a chance node with two branches must be added at the end of each decision alternative branch.

Step 1   Select cell F3   [Main menu bar]
        Add-Ins > Decision Tree
        Choose **Change to event node**
        Select **Two** in the **Branches** section
        Click **OK**

The tree now appears as follows:

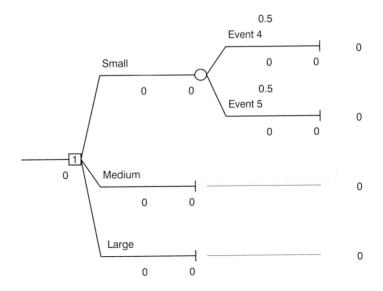

We next select the cells containing Event 4 and Event 5 and rename them Strong and Weak to provide the proper names for the PDC states of nature. After doing so we can copy the subtree for the chance node in cell F5 to the other two decision branches to complete the structure of the PDC decision tree.

Step 1 Select cell F5   [Main menu bar]
      Add-Ins > Decision Tree > Copy subtree
      Click **OK**

Step 2 Select cell F13   [Main menu bar]
      Add-Ins > Decision Tree > Paste subtree
      Click **OK**

This copy/paste procedure places a chance node at the end of the Medium decision branch.

Repeating the same copy/paste procedure for the Large decision branch completes the structure of the PDC decision tree as shown in Figure 21.12.

Inserting probabilities and payoffs

TreePlan provides the capability of inserting probabilities and payoffs into the decision tree. In Figure 21.12, we see that TreePlan automatically assigned an equal probability 0.5 to each of the states of nature. For PDC, the probability of strong demand is 0.8 and the probability of weak demand is 0.2. We can select cells H1, H6, H11, H16, H21 and H26 and insert the appropriate probabilities. The payoffs for the chance outcomes are inserted in cells H4, H9, H14, H19, H24 and H29. After inserting the PDC probabilities and payoffs, the PDC decision tree appears as shown in Figure 21.13.

Note that the payoffs also appear in the right-hand margin of the decision tree. The payoffs in the right margin are computed by a formula that adds the payoffs on all of the branches leading to the associated terminal node. For the PDC problem, no payoffs are associated with the decision alternative branches so we leave the default values of zero in cells D6, D16 and D26. The PDC decision tree is now complete.

Figure 21.12 The PDC decision tree developed by TreePlan

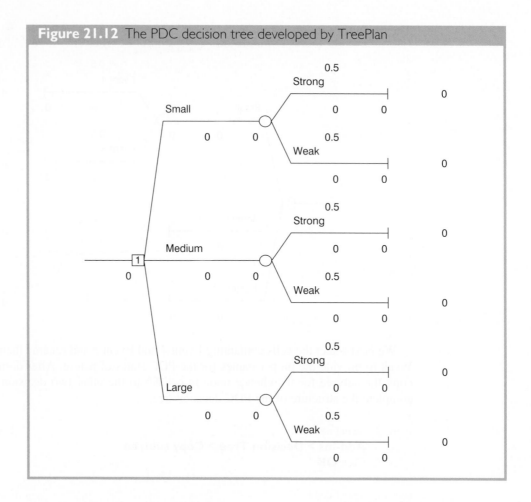

Interpreting the result

When probabilities and payoffs are inserted, TreePlan automatically makes the backward pass computations necessary to compute expected values and determine the optimal solution.

Optimal decisions are identified by the number in the corresponding decision node. In the PDC decision tree in Figure 21.13, cell B15 contains the decision node. Note that a 3 appears in this node, which tells us that decision alternative branch 3 provides the optimal decision. Thus, decision analysis recommends PDC construct the large apartment complex.

The expected value of this decision appears at the beginning of the tree in cell A16. Thus, we see the optimal expected value is €14.2 million. The expected values of the other decision alternatives are displayed at the end of the corresponding decision branch. Thus, referring to cells E6 and E16, we see that the expected value of the Small complex is €7.8 million and the expected value of the medium complex is €12.2 million.

Other options

TreePlan defaults to a maximization objective. If you would like a minimization objective, follow the command sequence:

Step 1 **Tools > Decision Tree > Options**   [Main menu bar]

Step 3 Choose **Minimize (costs)**   [Options panel]
            Click **OK**

Figure 21.13 The PDC decision tree with branch probabilities and payoffs

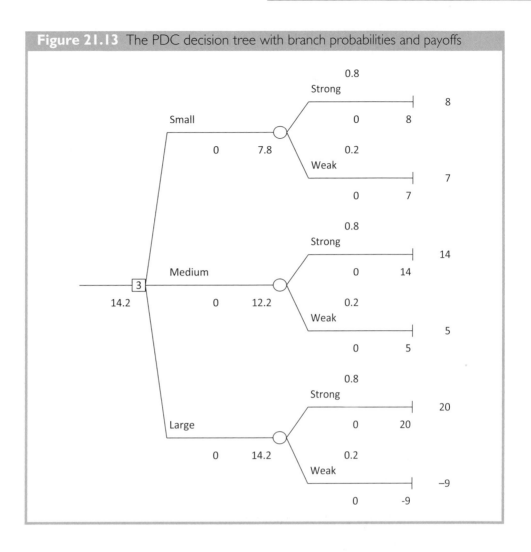

In using a TreePlan decision tree, we can modify probabilities and payoffs and quickly observe the impact of the changes on the optimal solution. Using this 'what if' type of sensitivity analysis, we can identify changes in probabilities and payoffs that would change the optimal decision. Also, because TreePlan is an EXCEL add-in, most of EXCEL's capabilities are available. For instance, we could use boldface to highlight the name of the optimal decision alternative on the final decision tree solution. A variety of other options provided by TreePlan are contained in the TreePlan manual. Computer software packages such as TreePlan make it easier to do a thorough analysis of a decision problem.

Appendix A References and Bibliography

General

Barlow J. F. (2005) *Excel Models for Business & Operations Management.* 2nd ed. Wiley.

Bowerman, B. L., and O'Connell, R. T. (1996) *Applied Statistics: Improving Business Processes.* Irwin.

Coakes, S. J., and Steed, L. G. (1996) *SPSS Analysis Without Anguish.* Wiley.

Field A (2005) *Discovering Statistics Using SPSS for Windows,* 2nd ed. Sage.

Freedman, D., Pisani, R., and Purves, R. (1997) *Statistics,* 3rd ed. W. W. Norton.

Green, S. B., Saltkind, N. J., and Akey, T. M. (2003) *Using SPSS for Windows,* 3rd ed. Prentice Hall.

Hare, C. T., and Bradow, R. L. (1977) *Light duty diesel emissions correction factors for ambient conditions.* SAE Paper 770717.

Hogg, R. V., and Craig, A. T. (1994) *Introduction to Mathematical Statistics,* 5th ed. Prentice Hall.

Hogg, R. V., and Tanis, E. A. (2001) *Probability and Statistical Inference,* 6th ed. Prentice Hall.

Joiner, B., Cryer, J., and Ryan, B. F. (2004) *Minitab Handbook,* 5th ed. Brooks/Cole.

Kinnear, P. R., and Gray, C. D. (2008) *SPSS 16 Made Simple.* Psychology Press.

MacPherson, G. (2001) *Applying and Interpreting Statistics: A Comprehensive Guide,* 2nd ed. Springer.

Miller, I., and Miller, M. (1998) *John E. Freund's Mathematical Statistics.* Prentice Hall.

Moore, D. S., and McCabe, G. P. (2003) *Introduction to the Practice of Statistics,* 4th ed. Freeman.

OECD (1982) Forecasting car ownership and use: a report by the Road Research Group. OECD.

Rossman, A. J. (1994) Televisions, physicians, and life expectancy. *Journal of Statistics Education,* vol 2 (2).

Tanur, J. M. (2002) *Statistics: A Guide to the Unknown,* 4th ed. Brooks/Cole.

Tukey, J.W. (1977) *Exploratory Data Analysis.* Addison-Wesley.

Wisniewski, M. (2002) *Quantitative Methods for Decision-Maklers.* Prentice Hall.

Experimental Design

Cochran, W. G., and Cox, G. M. (1992) *Experimental Designs,* 2nd ed. Wiley.

Hicks, C. R., and Turner, K. V. (1999) *Fundamental Concepts in the Design of Experiments,* 5th ed. Oxford University Press.

Montgomery, D. C. (2005) *Design and Analysis of Experiments,* 6th ed. Wiley.

Winer, B. J., Michels, K. M., and Brown, D. R. (1991) *Statistical Principles in Experimental Design,* 3rd ed. McGraw-Hill.

Wu, C. F. Jeff, and Hamada, M. (2000) *Experiments: Planning, Analysis, and Parameter Optimization.* Wiley.

Forecasting

Bowerman, B. L., and O'Connell, R. T. (2000) *Forecasting and Time Series: An Applied Approach,* 3rd ed. Brooks/Cole.

Box, G. E. P., Reinsel, G. C., and Jenkins, G. (1994) *Time Series Analysis: Forecasting and Control,* 3rd ed. Prentice Hall.

Makridakis, S., Wheelwright, S. C., and Hyndman, R. J. (1977) *Forecasting: Methods and Applications,* 3rd ed. Wiley.

Index Numbers

Allen, R. G. D. (1982) *Index Numbers in Theory and Practice,* Palgrave.

Consumer Price Indices Technical Manual (2005) UK Office for National Statistics.

Richardson, I. *Producer Price Indices: Principles and Procedures.* UK Office for National Statistics.

Non-parametric Methods

Conover, W. J. (1999) *Practical Nonparametric Statistics,* 3rd ed. Wiley.

Gibbons, J. D., and Chakraborti, S. (2003) *Nonparametric Statistical Inference,* 4th ed. Marcel Dekker.

Siegel, S., and Castellan, N. J. (1988) *Nonparametric Statistics for the Behavioral Sciences,* 2nd ed. McGraw-Hill.

Sprent, P., and Smeeton, N. C. (2000) *Applied Nonparametric Statistical Methods,* 3rd ed. CRC.

Probability

Hogg, R. V., and Tanis, E. A. (2005) *Probability and Statistical Inference,* 7th ed. Prentice Hall.

Ross, S. M. (2002) *Introduction to Probability Models,* 8th ed. Academic Press.

Wackerly, D. D., Mendenhall, W., and Scheaffer, R. L. (2002) *Mathematical Statistics with Applications,* 6th ed. Duxbury Press.

Quality Control

Deming, W. E. (1982) *Quality, Productivity, and Competitive Position.* MIT.

Evans, J. R., and Lindsay, W. M. (2004) *The Management and Control of Quality,* 6th ed. South-Western.

Gryna, F. M., and Juran, I. M. (1993) *Quality Planning and Analysis: From Product Development Through Use,* 3rd ed. McGraw-Hill.

Ishikawa, K. (1991) *Introduction to Quality Control.* Kluwer Academic.

Montgomery, D. C. (2004) *Introduction to Statistical Quality Control,* 5th ed. Wiley.

Regression Analysis

Belsley, D. A. (1991) *Conditioning Diagnostics: Collinearity and Weak Data in Regression.* Wiley.

Chatterjee, S., and Price, B. (1999) *Regression Analysis by Example,* 3rd ed. Wiley.

Draper, N. R., and Smith, H. (1998) *Applied Regression Analysis,* 3rd ed. Wiley.

Graybill, F. A., and Iyer, H. (1994) *Regression Analysis: Concepts and Applications.* Duxbury Press.

Hosmer, D. W., and Lemeshow, S. (2000) *Applied Logistic Regression,* 2nd ed. Wiley.

Kleinbaum, D. G., Kupper, L. L., and Muller, K. E. (1997) *Applied Regression Analysis and Other Multivariate Methods,* 3rd ed. Duxbury Press.

Kutner, M. H., Nachtschiem, C. J., Wasserman, W., and Neter, J. (1996) *Applied Linear Statistical Models,* 4th ed. Irwin.

Mendenhall, M., and Sincich, T. (2002) *A Second Course in Statistics: Regression Analysis,* 6th ed. Prentice Hall.

Myers, R. H. (1990) *Classical and Modern Regression with Applications,* 2nd ed. PWS.

Decision Analysis

Chernoff, H., and Moses, L. E. (1987) *Elementary Decision Theory.* Dover.

Clemen, R. T., and Reilly, T. (2001) *Making Hard Decisions with Decision Tools.* Duxbury Press.

Goodwin, P., and Wright, G. (2004) *Decision Analysis for Management Judgment.* 3rd ed. Wiley.

Pratt, J. W., Raiffa, H., and Schlaifer, R. (1995) *Introduction to Statistical Decision Theory.* MIT Press.

Raiffa, H. (1997) *Decision Analysis: Introductory Readings on Choices Under Uncertainty.* McGraw-Hill.

Sampling

Cochran, W. G. (1977) *Sampling Techniques*, 3rd ed. Wiley.

Deming, W. E. (1984) *Some Theory of Sampling.* Dover.

Hansen, M. H., Hurwitz, W. N., Madow, W. G., and Hanson, M. N. (1993) *Sample Survey Methods and Theory.* Wiley.

Kish, L. (1995) *Survey Sampling.* Wiley.

Levy, P. S., and Lemeshow, S. (1999) *Sampling of Populations: Methods and Applications*, 3rd ed. Wiley.

Scheaffer, R. L., Mendenhall, W., and Ott, L. (2006) *Elementary Survey Sampling*, 6th ed. Brooks/Cole.

Appendix B Tables

Table I Cumulative Probabilities for the Standard Normal Distribution

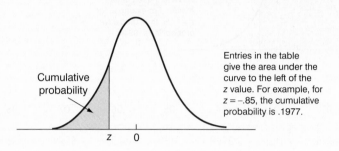

Entries in the table give the area under the curve to the left of the z value. For example, for $z = -.85$, the cumulative probability is .1977.

z	.00	.01	.02	.03	.04	.05	.06	.07	.08	.09
−3.0	.0013	.0013	.0013	.0012	.0012	.0011	.0011	.0011	.0010	.0010
−2.9	.0019	.0018	.0018	.0017	.0016	.0016	.0015	.0015	.0014	.0014
−2.8	.0026	.0025	.0024	.0023	.0023	.0022	.0021	.0021	.0020	.0019
−2.7	.0035	.0034	.0033	.0032	.0031	.0030	.0029	.0028	.0027	.0026
−2.6	.0047	.0045	.0044	.0043	.0041	.0040	.0039	.0038	.0037	.0036
−2.5	.0062	.0060	.0059	.0057	.0055	.0054	.0052	.0051	.0049	.0048
−2.4	.0082	.0080	.0078	.0075	.0073	.0071	.0069	.0068	.0066	.0064
−2.3	.0107	.0104	.0102	.0099	.0096	.0094	.0091	.0089	.0087	.0084
−2.2	.0139	.0136	.0132	.0129	.0125	.0122	.0119	.0116	.0113	.0110
−2.1	.0179	.0174	.0170	.0166	.0162	.0158	.0154	.0150	.0146	.0143
−2.0	.0228	.0222	.0217	.0212	.0207	.0202	.0197	.0192	.0188	.0183
−1.9	.0287	.0281	.0274	.0268	.0262	.0256	.0250	.0244	.0239	.0233
−1.8	.0359	.0351	.0344	.0336	.0329	.0322	.0314	.0307	.0301	.0294
−1.7	.0446	.0436	.0427	.0418	.0409	.0401	.0392	.0384	.0375	.0367
−1.6	.0548	.0537	.0526	.0516	.0505	.0495	.0485	.0475	.0465	.0455
−1.5	.0668	.0655	.0643	.0630	.0618	.0606	.0594	.0582	.0571	.0559
−1.4	.0808	.0793	.0778	.0764	.0749	.0735	.0721	.0708	.0694	.0681
−1.3	.0968	.0951	.0934	.0918	.0901	.0885	.0869	.0853	.0838	.0823
−1.2	.1151	.1131	.1112	.1093	.1075	.1056	.1038	.1020	.1003	.0985
−1.1	.1357	.1335	.1314	.1292	.1271	.1251	.1230	.1210	.1190	.1170
−1.0	.1587	.1562	.1539	.1515	.1492	.1469	.1446	.1423	.1401	.1379
−.9	.1841	.1814	.1788	.1762	.1736	.1711	.1685	.1660	.1635	.1611
−.8	.2119	.2090	.2061	.2033	.2005	.1977	.1949	.1922	.1894	.1867
−.7	.2420	.2389	.2358	.2327	.2296	.2266	.2236	.2206	.2177	.2148
−.6	.2743	.2709	.2676	.2643	.2611	.2578	.2546	.2514	.2483	.2451
−.5	.3085	.3050	.3015	.2981	.2946	.2912	.2877	.2843	.2810	.2776
−.4	.3446	.3409	.3372	.3336	.3300	.3264	.3228	.3192	.3156	.3121
−.3	.3821	.3783	.3745	.3707	.3669	.3632	.3594	.3557	.3520	.3483
−.2	.4207	.4168	.4129	.4090	.4052	.4013	.3974	.3936	.3897	.3859
−.1	.4602	.4562	.4522	.4483	.4443	.4404	.4364	.4325	.4286	.4247
−.0	.5000	.4960	.4920	.4880	.4840	.4801	.4761	.4721	.4681	.4641

Table I (Continued)

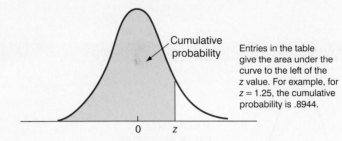

Cumulative probability

Entries in the table give the area under the curve to the left of the z value. For example, for z = 1.25, the cumulative probability is .8944.

z	.00	.01	.02	.03	.04	.05	.06	.07	.08	.09
0.0	.5000	.5040	.5080	.5120	.5160	.5199	.5239	.5279	.5319	.5359
0.1	.5398	.5438	.5478	.5517	.5557	.5596	.5636	.5675	.5714	.5753
0.2	.5793	.5832	.5871	.5910	.5948	.5987	.6026	.6064	.6103	.6141
0.3	.6179	.6217	.6255	.6293	.6331	.6368	.6406	.6443	.6480	.6517
0.4	.6554	.6591	.6628	.6664	.6700	.6736	.6772	.6808	.6844	.6879
0.5	.6915	.6950	.6985	.7019	.7054	.7088	.7123	.7157	.7190	.7224
0.6	.7257	.7291	.7324	.7357	.7389	.7422	.7454	.7486	.7517	.7549
0.7	.7580	.7611	.7642	.7673	.7704	.7734	.7764	.7794	.7823	.7852
0.8	.7881	.7910	.7939	.7967	.7995	.8023	.8051	.8078	.8106	.8133
0.9	.8159	.8186	.8212	.8238	.8264	.8289	.8315	.8340	.8365	.8389
1.0	.8413	.8438	.8461	.8485	.8508	.8531	.8554	.8577	.8599	.8621
1.1	.8643	.8665	.8686	.8708	.8729	.8749	.8770	.8790	.8810	.8830
1.2	.8849	.8869	.8888	.8907	.8925	.8944	.8962	.8980	.8997	.9015
1.3	.9032	.9049	.9066	.9082	.9099	.9115	.9131	.9147	.9162	.9177
1.4	.9192	.9207	.9222	.9236	.9251	.9265	.9279	.9292	.9306	.9319
1.5	.9332	.9345	.9357	.9370	.9382	.9394	.9406	.9418	.9429	.9441
1.6	.9452	.9463	.9474	.9484	.9495	.9505	.9515	.9525	.9535	.9545
1.7	.9554	.9564	.9573	.9582	.9591	.9599	.9608	.9616	.9625	.9633
1.8	.9641	.9649	.9656	.9664	.9671	.9678	.9686	.9693	.9699	.9706
1.9	.9713	.9719	.9726	.9732	.9738	.9744	.9750	.9756	.9761	.9767
2.0	.9772	.9778	.9783	.9788	.9793	.9798	.9803	.9808	.9812	.9817
2.1	.9821	.9826	.9830	.9834	.9838	.9842	.9846	.9850	.9854	.9857
2.2	.9861	.9864	.9868	.9871	.9875	.9878	.9881	.9884	.9887	.9890
2.3	.9893	.9896	.9898	.9901	.9904	.9906	.9909	.9911	.9913	.9913
2.4	.9918	.9920	.9922	.9925	.9927	.9929	.9931	.9932	.9934	.9936
2.5	.9938	.9940	.9941	.9943	.9945	.9946	.9948	.9949	.9951	.9952
2.6	.9953	.9955	.9956	.9957	.9959	.9960	.9961	.9962	.9963	.9964
2.7	.9965	.9966	.9967	.9968	.9969	.9970	.9971	.9972	.9973	.9974
2.8	.9974	.9975	.9976	.9977	.9977	.9978	.9979	.9979	.9980	.9981
2.9	.9981	.9982	.9982	.9983	.9984	.9984	.9985	.9985	.9986	.9986
3.0	.9986	.9987	.9987	.9988	.9988	.9989	.9989	.9989	.9990	.9990

Table 2 *t* distribution

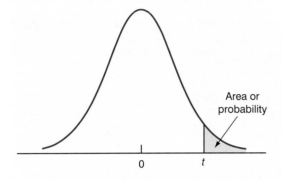

Area or probability

Entries in the table give *t* values for an area or probability in the upper tail of the *t* distribution. For example, with 10 degrees of freedom and a .05 area in the upper tail, $t_{.05} = 1.812$.

	Area in upper tail					
Degrees of freedom	.20	.10	.05	.025	.01	.005
1	1.376	3.078	6.314	12.706	31.821	63.656
2	1.061	1.886	2.920	4.303	6.965	9.925
3	.978	1.638	2.353	3.182	4.541	5.841
4	.941	1.533	2.132	2.776	3.747	4.604
5	.920	1.476	2.015	2.571	3.365	4.032
6	.906	1.440	1.943	2.447	3.143	3.707
7	.896	1.415	1.895	2.365	2.998	3.499
8	.889	1.397	1.860	2.306	2.896	3.355
9	.883	1.383	1.833	2.262	2.821	3.250
10	.879	1.372	1.812	2.228	2.764	3.169
11	.876	1.363	1.796	2.201	2.718	3.106
12	.873	1.356	1.782	2.179	2.681	3.055
13	.870	1.350	1.771	2.160	2.650	3.012
14	.868	1.345	1.761	2.145	2.624	2.977
15	.866	1.341	1.753	2.131	2.602	2.947
16	.865	1.337	1.746	2.120	2.583	2.921
17	.863	1.333	1.740	2.110	2.567	2.898
18	.862	1.330	1.734	2.101	2.552	2.878
19	.861	1.328	1.729	2.093	2.539	2.861
20	.860	1.325	1.725	2.086	2.528	2.845
21	.859	1.323	1.721	2.080	2.518	2.831
22	.858	1.321	1.717	2.074	2.508	2.819
23	.858	1.319	1.714	2.069	2.500	2.807
24	.857	1.318	1.711	2.064	2.492	2.797
25	.856	1.316	1.708	2.060	2.485	2.787
26	.856	1.315	1.706	2.056	2.479	2.779
27	.855	1.314	1.703	2.052	2.473	2.771
28	.855	1.313	1.701	2.048	2.467	2.763

(*continued*)

Table 2 (Continued)

Degrees of freedom	Area in upper tail					
	.20	.10	.05	.025	.01	.005
29	.854	1.311	1.699	2.045	2.462	2.756
30	.854	1.310	1.697	2.042	2.457	2.750
31	.853	1.309	1.696	2.040	2.453	2.744
32	.853	1.309	1.694	2.037	2.449	2.738
33	.853	1.308	1.692	2.035	2.445	2.733
34	.852	1.307	1.691	2.032	2.441	2.728
35	.852	1.306	1.690	2.030	2.438	2.724
36	.852	1.306	1.688	2.028	2.434	2.719
37	.851	1.305	1.687	2.026	2.431	2.715
38	.851	1.304	1.686	2.024	2.429	2.712
39	.851	1.304	1.685	2.023	2.426	2.708
40	.851	1.303	1.684	2.021	2.423	2.704
41	.850	1.303	1.683	2.020	2.421	2.701
42	.850	1.302	1.682	2.018	2.418	2.698
43	.850	1.302	1.681	2.017	2.416	2.695
44	.850	1.301	1.680	2.015	2.414	2.692
45	.850	1.301	1.679	2.014	2.412	2.690
46	.850	1.300	1.679	2.013	2.410	2.687
47	.849	1.300	1.678	2.012	2.408	2.685
48	.849	1.299	1.677	2.011	2.407	2.682
49	.849	1.299	1.677	2.010	2.405	2.680
50	.849	1.299	1.676	2.009	2.403	2.678
51	.849	1.298	1.675	2.008	2.402	2.676
52	.849	1.298	1.675	2.007	2.400	2.674
53	.848	1.298	1.674	2.006	2.399	2.672
54	.848	1.297	1.674	2.005	2.397	2.670
55	.848	1.297	1.673	2.004	2.396	2.668
56	.848	1.297	1.673	2.003	2.395	2.667
57	.848	1.297	1.672	2.002	2.394	2.665
58	.848	1.296	1.672	2.002	2.392	2.663
59	.848	1.296	1.671	2.001	2.391	2.662
60	.848	1.296	1.671	2.000	2.390	2.660
61	.848	1.296	1.670	2.000	2.389	2.659
62	.847	1.295	1.670	1.999	2.388	2.657
63	.847	1.295	1.669	1.998	2.387	2.656
64	.847	1.295	1.669	1.998	2.386	2.655
65	.847	1.295	1.669	1.997	2.385	2.654
66	.847	1.295	1.668	1.997	2.384	2.652
67	.847	1.294	1.668	1.996	2.383	2.651
68	.847	1.294	1.668	1.995	2.382	2.650
69	.847	1.294	1.667	1.995	2.382	2.649

Table 2 (Continued)

Degrees of freedom	Area in upper tail					
	.20	.10	.05	.025	.01	.005
70	.847	1.294	1.667	1.994	2.381	2.648
71	.847	1.294	1.667	1.994	2.380	2.647
72	.847	1.293	1.666	1.993	2.379	2.646
73	.847	1.293	1.666	1.993	2.379	2.645
74	.847	1.293	1.666	1.993	2.378	2.644
75	.846	1.293	1.665	1.992	2.377	2.643
76	.846	1.293	1.665	1.992	2.376	2.642
77	.846	1.293	1.665	1.991	2.376	2.641
78	.846	1.292	1.665	1.991	2.375	2.640
79	.846	1.292	1.664	1.990	2.374	2.639
80	.846	1.292	1.664	1.990	2.374	2.639
81	.846	1.292	1.664	1.990	2.373	2.638
82	.846	1.292	1.664	1.989	2.373	2.637
83	.846	1.292	1.663	1.989	2.372	2.636
84	.846	1.292	1.663	1.989	2.372	2.636
85	.846	1.292	1.663	1.988	2.371	2.635
86	.846	1.291	1.663	1.988	2.370	2.634
87	.846	1.291	1.663	1.988	2.370	2.634
88	.846	1.291	1.662	1.987	2.369	2.633
89	.846	1.291	1.662	1.987	2.369	2.632
90	.846	1.291	1.662	1.987	2.368	2.632
91	.846	1.291	1.662	1.986	2.368	2.631
92	.846	1.291	1.662	1.986	2.368	2.630
93	.846	1.291	1.661	1.986	2.367	2.630
94	.845	1.291	1.661	1.986	2.367	2.629
95	.845	1.291	1.661	1.985	2.366	2.629
96	.845	1.290	1.661	1.985	2.366	2.628
97	.845	1.290	1.661	1.985	2.365	2.627
98	.845	1.290	1.661	1.984	2.365	2.627
99	.845	1.290	1.660	1.984	2.364	2.626
100	.845	1.290	1.660	1.984	2.364	2.626
∞	.842	1.282	1.645	1.960	2.326	2.576

Table 3 Chi-squared distribution

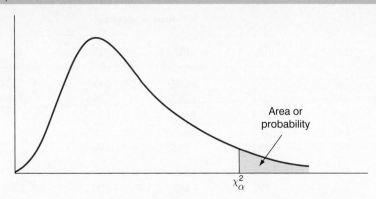

Entries in the table give χ^2_α values, where α is the area or probability in the upper tail of the chi-squared distribution. For example, with 10 degrees of freedom and a .01 area in the upper tail, $\chi^2_{.01} = 23.209$.

Degrees of freedom	Area in upper tail									
	.995	.99	.975	.95	.90	.10	.05	.025	.01	.005
1	.000	.000	.001	.004	.016	2.706	3.841	5.024	6.635	7.879
2	.010	.020	.051	.103	.211	4.605	5.991	7.378	9.210	10.597
3	.072	.115	.216	.352	.584	6.251	7.815	9.348	11.345	12.838
4	.207	.297	.484	.711	1.064	7.779	9.488	11.143	13.277	14.860
5	.412	.554	.831	1.145	1.610	9.236	11.070	12.832	15.086	16.750
6	.676	.872	1.237	1.635	2.204	10.645	12.592	14.449	16.812	18.548
7	.989	1.239	1.690	2.167	2.833	12.017	14.067	16.013	18.475	20.278
8	1.344	1.647	2.180	2.733	3.490	13.362	15.507	17.535	20.090	21.955
9	1.735	2.088	2.700	3.325	4.168	14.684	16.919	19.023	21.666	23.589
10	2.156	2.558	3.247	3.940	4.865	15.987	18.307	20.483	23.209	25.188
11	2.603	3.053	3.816	4.575	5.578	17.275	19.675	21.920	24.725	26.757
12	3.074	3.571	4.404	5.226	6.304	18.549	21.026	23.337	26.217	28.300
13	3.565	4.107	5.009	5.892	7.041	19.812	22.362	24.736	27.688	29.819
14	4.075	4.660	5.629	6.571	7.790	21.064	23.685	26.119	29.141	31.319
15	4.601	5.229	6.262	7.261	8.547	22.307	24.996	27.488	30.578	32.801
16	5.142	5.812	6.908	7.962	9.312	23.542	26.296	28.845	32.000	34.267
17	5.697	6.408	7.564	8.672	10.085	24.769	27.587	30.191	33.409	35.718
18	6.265	7.015	8.231	9.390	10.865	25.989	28.869	31.526	34.805	37.156
19	6.844	7.633	8.907	10.117	11.651	27.204	30.144	32.852	36.191	38.582
20	7.434	8.260	9.591	10.851	12.443	28.412	31.410	34.170	37.566	39.997
21	8.034	8.897	10.283	11.591	13.240	29.615	32.671	35.479	38.932	41.401
22	8.643	9.542	10.982	12.338	14.041	30.813	33.924	36.781	40.289	42.796
23	9.260	10.196	11.689	13.091	14.848	32.007	35.172	38.076	41.638	44.181
24	9.886	10.856	12.401	13.848	15.659	33.196	36.415	39.364	42.980	45.558
25	10.520	11.524	13.120	14.611	16.473	34.382	37.652	40.646	44.314	46.928
26	11.160	12.198	13.844	15.379	17.292	35.563	38.885	41.923	45.642	48.290
27	11.808	12.878	14.573	16.151	18.114	36.741	40.113	43.195	46.963	49.645

Table 3 (*Continued*)

Degrees of freedom	Area in upper tail									
	.995	.99	.975	.95	.90	.10	.05	.025	.01	.005
28	12.461	13.565	15.308	16.928	18.939	37.916	41.337	44.461	48.278	50.994
29	13.121	14.256	16.047	17.708	19.768	39.087	42.557	45.722	49.588	52.335
30	13.787	14.953	16.791	18.493	20.599	40.256	43.773	46.979	50.892	53.672
35	17.192	18.509	20.569	22.465	24.797	46.059	49.802	53.203	57.342	60.275
40	20.707	22.164	24.433	26.509	29.051	51.805	55.758	59.342	63.691	66.766
45	24.311	25.901	28.366	30.612	33.350	57.505	61.656	65.410	69.957	73.166
50	27.991	29.707	32.357	34.764	37.689	63.167	67.505	71.420	76.154	79.490
55	31.735	33.571	36.398	38.958	42.060	68.796	73.311	77.380	82.292	85.749
60	35.534	37.485	40.482	43.188	46.459	74.397	79.082	83.298	88.379	91.952
65	39.383	41.444	44.603	47.450	50.883	79.973	84.821	89.177	94.422	98.105
70	43.275	45.442	48.758	51.739	55.329	85.527	90.531	95.023	100.425	104.215
75	47.206	49.475	52.942	56.054	59.795	91.061	96.217	100.839	106.393	110.285
80	51.172	53.540	57.153	60.391	64.278	96.578	101.879	106.629	112.329	116.321
85	55.170	57.634	61.389	64.749	68.777	102.079	107.522	112.393	118.236	122.324
90	59.196	61.754	65.647	69.126	73.291	107.565	113.145	118.136	124.116	128.299
95	63.250	65.898	69.925	73.520	77.818	113.038	118.752	123.858	129.973	134.247
100	67.328	70.065	74.222	77.929	82.358	118.498	124.342	129.561	135.807	140.170

Table 4 F distribution

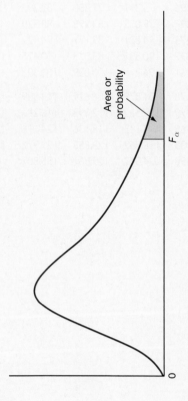

Area or probability

F_α

Entries in the table give F_α values, where α is the area or probability in the upper tail of the F distribution. For example, with 4 numerator degrees of freedom, 8 denominator degrees of freedom, and a .05 area in the upper tail, $F_{.05} = 3.84$.

Denominator degrees of freedom	Area in upper tail																			
		1	2	3	4	5	6	7	8	9	10	15	20	25	30	40	60	100	1000
1	.10	39.86	49.50	53.59	55.83	57.24	58.20	58.91	59.44	59.86	60.19	61.22	61.74	62.05	62.26	62.53	62.79	63.01	63.30
	.05	161.45	199.50	215.71	224.58	230.16	233.99	236.77	238.88	240.54	241.88	245.95	248.02	249.26	250.10	251.14	252.20	253.04	254.19
	.025	647.79	799.48	864.15	899.60	921.83	937.11	948.20	956.64	963.28	968.63	984.87	993.08	998.09	1001.40	1005.60	1009.79	1013.16	1017.76
	.01	4052.18	4999.34	5403.53	5624.26	5763.96	5858.95	5928.33	5980.95	6022.40	6055.93	6156.97	6208.66	6239.86	6260.35	6286.43	6312.97	6333.92	6362.80
2	.10	8.53	9.00	9.16	9.24	9.29	9.33	9.35	9.37	9.38	9.39	9.42	9.44	9.45	9.46	9.47	9.47	9.48	9.49
	.05	18.51	19.00	19.16	19.25	19.30	19.33	19.35	19.37	19.38	19.40	19.43	19.45	19.46	19.46	19.47	19.48	19.49	19.49
	.025	38.51	39.00	39.17	39.25	39.30	39.33	39.36	39.37	39.39	39.40	39.43	39.45	39.46	39.46	39.47	39.48	39.49	39.50
	.01	98.50	99.00	99.16	99.25	99.30	99.33	99.36	99.38	99.39	99.40	99.43	99.45	99.46	99.47	99.48	99.48	99.49	99.50
3	.10	5.54	5.46	5.39	5.34	5.31	5.28	5.27	5.25	5.24	5.23	5.20	5.18	5.17	5.17	5.16	5.15	5.14	5.13
	.05	10.13	9.55	9.28	9.12	9.01	8.94	8.89	8.85	8.81	8.79	8.70	8.66	8.63	8.62	8.59	8.57	8.55	8.53
	.025	17.44	16.04	15.44	15.10	14.88	14.73	14.62	14.54	14.47	14.42	14.25	14.17	14.12	14.08	14.04	13.99	13.96	13.91
	.01	34.12	30.82	29.46	28.71	28.24	27.91	27.67	27.49	27.34	27.23	26.87	26.69	26.58	26.50	26.41	26.32	26.24	26.14
4	.10	4.54	4.32	4.19	4.11	4.05	4.01	3.98	3.95	3.94	3.92	3.87	3.84	3.83	3.82	3.80	3.79	3.78	3.76
	.05	7.71	6.94	6.59	6.39	6.26	6.16	6.09	6.04	6.00	5.96	5.86	5.80	5.77	5.75	5.72	5.69	5.66	5.63
	.025	12.22	10.65	9.98	9.60	9.36	9.20	9.07	8.98	8.90	8.84	8.66	8.56	8.50	8.46	8.41	8.36	8.32	8.26
	.01	21.20	18.00	16.69	15.98	15.52	15.21	14.98	14.80	14.66	14.55	14.20	14.02	13.91	13.84	13.75	13.65	13.58	13.47

Numerator degrees of freedom

Table 4 (Continued)

Denominator degrees of freedom	Area in upper tail	Numerator degrees of freedom																	
		1	2	3	4	5	6	7	8	9	10	15	20	25	30	40	60	100	1000
5	.10	4.06	3.78	3.62	3.52	3.45	3.40	3.37	3.34	3.32	3.30	3.24	3.21	3.19	3.17	3.16	3.14	3.13	3.11
	.05	6.61	5.79	5.41	5.19	5.05	4.95	4.88	4.82	4.77	4.74	4.62	4.56	4.52	4.50	4.46	4.43	4.41	4.37
	.025	10.01	8.43	7.76	7.39	7.15	6.98	6.85	6.76	6.68	6.62	6.43	6.33	6.27	6.23	6.18	6.12	6.08	6.02
	.01	16.26	13.27	12.06	11.39	10.97	10.67	10.46	10.29	10.16	10.05	9.72	9.55	9.45	9.38	9.29	9.20	9.13	9.03
6	.10	3.78	3.46	3.29	3.18	3.11	3.05	3.01	2.98	2.96	2.94	2.87	2.84	2.81	2.80	2.78	2.76	2.75	2.72
	.05	5.99	5.14	4.76	4.53	4.39	4.28	4.21	4.15	4.10	4.06	3.94	3.87	3.83	3.81	3.77	3.74	3.71	3.67
	.025	8.81	7.26	6.60	6.23	5.99	5.82	5.70	5.60	5.52	5.46	5.27	5.17	5.11	5.07	5.01	4.96	4.92	4.86
	.01	13.75	10.92	9.78	9.15	8.75	8.47	8.26	8.10	7.98	7.87	7.56	7.40	7.30	7.23	7.14	7.06	6.99	6.89
7	.10	3.59	3.26	3.07	2.96	2.88	2.83	2.78	2.75	2.72	2.70	2.63	2.59	2.57	2.56	2.54	2.51	2.50	2.47
	.05	5.59	4.74	4.35	4.12	3.97	3.87	3.79	3.73	3.68	3.64	3.51	3.44	3.40	3.38	3.34	3.30	3.27	3.23
	.025	8.07	6.54	5.89	5.52	5.29	5.12	4.99	4.90	4.82	4.76	4.57	4.47	4.40	4.36	4.31	4.25	4.21	4.15
	.01	12.25	9.55	8.45	7.85	7.46	7.19	6.99	6.84	6.72	6.62	6.31	6.16	6.06	5.99	5.91	5.82	5.75	5.66
8	.10	3.46	3.11	2.92	2.81	2.73	2.67	2.62	2.59	2.56	2.54	2.46	2.42	2.40	2.38	2.36	2.34	2.32	2.30
	.05	5.32	4.46	4.07	3.84	3.69	3.58	3.50	3.44	3.39	3.35	3.22	3.15	3.11	3.08	3.04	3.01	2.97	2.93
	.025	7.57	6.06	5.42	5.05	4.82	4.65	4.53	4.43	4.36	4.30	4.10	4.00	3.94	3.89	3.84	3.78	3.74	3.68
	.01	11.26	8.65	7.59	7.01	6.63	6.37	6.18	6.03	5.91	5.81	5.52	5.36	5.26	5.20	5.12	5.03	4.96	4.87
9	.10	3.36	3.01	2.81	2.69	2.61	2.55	2.51	2.47	2.44	2.42	2.34	2.30	2.27	2.25	2.23	2.21	2.19	2.16
	.05	5.12	4.26	3.86	3.63	3.48	3.37	3.29	3.23	3.18	3.14	3.01	2.94	2.89	2.86	2.83	2.79	2.76	2.71
	.025	7.21	5.71	5.08	4.72	4.48	4.32	4.20	4.10	4.03	3.96	3.77	3.67	3.60	3.56	3.51	3.45	3.40	3.34
	.01	10.56	8.02	6.99	6.42	6.06	5.80	5.61	5.47	5.35	5.26	4.96	4.81	4.71	4.65	4.57	4.48	4.41	4.32
10	.10	3.29	2.92	2.73	2.61	2.52	2.46	2.41	2.38	2.35	2.32	2.24	2.20	2.17	2.16	2.13	2.11	2.09	2.06
	.05	4.96	4.10	3.71	3.48	3.33	3.22	3.14	3.07	3.02	2.98	2.85	2.77	2.73	2.70	2.66	2.62	2.59	2.54
	.025	6.94	5.46	4.83	4.47	4.24	4.07	3.95	3.85	3.78	3.72	3.52	3.42	3.35	3.31	3.26	3.20	3.15	3.09
	.01	10.04	7.56	6.55	5.99	5.64	5.39	5.20	5.06	4.94	4.85	4.56	4.41	4.31	4.25	4.17	4.08	4.01	3.92
11	.10	3.23	2.86	2.66	2.54	2.45	2.39	2.34	2.30	2.27	2.25	2.17	2.12	2.10	2.08	2.05	2.03	2.01	1.98
	.05	4.84	3.98	3.59	3.36	3.20	3.09	3.01	2.95	2.90	2.85	2.72	2.65	2.60	2.57	2.53	2.49	2.46	2.41
	.025	6.72	5.26	4.63	4.28	4.04	3.88	3.76	3.66	3.59	3.53	3.33	3.23	3.16	3.12	3.06	3.00	2.96	2.89
	.01	9.65	7.21	6.22	5.67	5.32	5.07	4.89	4.74	4.63	4.54	4.25	4.10	4.01	3.94	3.86	3.78	3.71	3.61
12	.10	3.18	2.81	2.61	2.48	2.39	2.33	2.28	2.24	2.21	2.19	2.10	2.06	2.03	2.01	1.99	1.96	1.94	1.91
	.05	4.75	3.89	3.49	3.26	3.11	3.00	2.91	2.85	2.80	2.75	2.62	2.54	2.50	2.47	2.43	2.38	2.35	2.30
	.025	6.55	5.10	4.47	4.12	3.89	3.73	3.61	3.51	3.44	3.37	3.18	3.07	3.01	2.96	2.91	2.85	2.80	2.73
	.01	9.33	6.93	5.95	5.41	5.06	4.82	4.64	4.50	4.39	4.30	4.01	3.86	3.76	3.70	3.62	3.54	3.47	3.37

(continued)

Table 4 (Continued)

Denominator degrees of freedom	Area in upper tail	Numerator degrees of freedom																	
		1	2	3	4	5	6	7	8	9	10	15	20	25	30	40	60	100	1000
13	.10	3.14	2.76	2.56	2.43	2.35	2.28	2.23	2.20	2.16	2.14	2.05	2.01	1.98	1.96	1.93	1.90	1.88	1.85
	.05	4.67	3.81	3.41	3.18	3.03	2.92	2.83	2.77	2.71	2.67	2.53	2.46	2.41	2.38	2.34	2.30	2.26	2.21
	.025	6.41	4.97	4.35	4.00	3.77	3.60	3.48	3.39	3.31	3.25	3.05	2.95	2.88	2.84	2.78	2.72	2.67	2.60
	.01	9.07	6.70	5.74	5.21	4.86	4.62	4.44	4.30	4.19	4.10	3.82	3.66	3.57	3.51	3.43	3.34	3.27	3.18
14	.10	3.10	2.73	2.52	2.39	2.31	2.24	2.19	2.15	2.12	2.10	2.01	1.96	1.93	1.91	1.89	1.86	1.83	1.80
	.05	4.60	3.74	3.34	3.11	2.96	2.85	2.76	2.70	2.65	2.60	2.46	2.39	2.34	2.31	2.27	2.22	2.19	2.14
	.025	6.30	4.86	4.24	3.89	3.66	3.50	3.38	3.29	3.21	3.15	2.95	2.84	2.78	2.73	2.67	2.61	2.56	2.50
	.01	8.86	6.51	5.56	5.04	4.69	4.46	4.28	4.14	4.03	3.94	3.66	3.51	3.41	3.35	3.27	3.18	3.11	3.02
15	.10	3.07	2.70	2.49	2.36	2.27	2.21	2.16	2.12	2.09	2.06	1.97	1.92	1.89	1.87	1.85	1.82	1.79	1.76
	.05	4.54	3.68	3.29	3.06	2.90	2.79	2.71	2.64	2.59	2.54	2.40	2.33	2.28	2.25	2.20	2.16	2.12	2.07
	.025	6.20	4.77	4.15	3.80	3.58	3.41	3.29	3.20	3.12	3.06	2.86	2.76	2.69	2.64	2.59	2.52	2.47	2.40
	.01	8.68	6.36	5.42	4.89	4.56	4.32	4.14	4.00	3.89	3.80	3.52	3.37	3.28	3.21	3.13	3.05	2.98	2.88
16	.10	3.05	2.67	2.46	2.33	2.24	2.18	2.13	2.09	2.06	2.03	1.94	1.89	1.86	1.84	1.81	1.78	1.76	1.72
	.05	4.49	3.63	3.24	3.01	2.85	2.74	2.66	2.59	2.54	2.49	2.35	2.28	2.23	2.19	2.15	2.11	2.07	2.02
	.025	6.12	4.69	4.08	3.73	3.50	3.34	3.22	3.12	3.05	2.99	2.79	2.68	2.61	2.57	2.51	2.45	2.40	2.32
	.01	8.53	6.23	5.29	4.77	4.44	4.20	4.03	3.89	3.78	3.69	3.41	3.26	3.16	3.10	3.02	2.93	2.86	2.76
17	.10	3.03	2.64	2.44	2.31	2.22	2.15	2.10	2.06	2.03	2.00	1.91	1.86	1.83	1.81	1.78	1.75	1.73	1.69
	.05	4.45	3.59	3.20	2.96	2.81	2.70	2.61	2.55	2.49	2.45	2.31	2.23	2.18	2.15	2.10	2.06	2.02	1.97
	.025	6.04	4.62	4.01	3.66	3.44	3.28	3.16	3.06	2.98	2.92	2.72	2.62	2.55	2.50	2.44	2.38	2.33	2.26
	.01	8.40	6.11	5.19	4.67	4.34	4.10	3.93	3.79	3.68	3.59	3.31	3.16	3.07	3.00	2.92	2.83	2.76	2.66
18	.10	3.01	2.62	2.42	2.29	2.20	2.13	2.08	2.04	2.00	1.98	1.89	1.84	1.80	1.78	1.75	1.72	1.70	1.66
	.05	4.41	3.55	3.16	2.93	2.77	2.66	2.58	2.51	2.46	2.41	2.27	2.19	2.14	2.11	2.06	2.02	1.98	1.92
	.025	5.98	4.56	3.95	3.61	3.38	3.22	3.10	3.01	2.93	2.87	2.67	2.56	2.49	2.44	2.38	2.32	2.27	2.20
	.01	8.29	6.01	5.09	4.58	4.25	4.01	3.84	3.71	3.60	3.51	3.23	3.08	2.98	2.92	2.84	2.75	2.68	2.58
19	.10	2.99	2.61	2.40	2.27	2.18	2.11	2.06	2.02	1.98	1.96	1.86	1.81	1.78	1.76	1.73	1.70	1.67	1.64
	.05	4.38	3.52	3.13	2.90	2.74	2.63	2.54	2.48	2.42	2.38	2.23	2.16	2.11	2.07	2.03	1.98	1.94	1.88
	.025	5.92	4.51	3.90	3.56	3.33	3.17	3.05	2.96	2.88	2.82	2.62	2.51	2.44	2.39	2.33	2.27	2.22	2.14
	.01	8.18	5.93	5.01	4.50	4.17	3.94	3.77	3.63	3.52	3.43	3.15	3.00	2.91	2.84	2.76	2.67	2.60	2.50
20	.10	2.97	2.59	2.38	2.25	2.16	2.09	2.04	2.00	1.96	1.94	1.84	1.79	1.76	1.74	1.71	1.68	1.65	1.61
	.05	4.35	3.49	3.10	2.87	2.71	2.60	2.51	2.45	2.39	2.35	2.20	2.12	2.07	2.04	1.99	1.95	1.91	1.85
	.025	5.87	4.46	3.86	3.51	3.29	3.13	3.01	2.91	2.84	2.77	2.57	2.46	2.40	2.35	2.29	2.22	2.17	2.09
	.01	8.10	5.85	4.94	4.43	4.10	3.87	3.70	3.56	3.46	3.37	3.09	2.94	2.84	2.78	2.69	2.61	2.54	2.43

Table 4 (Continued)

Denominator degrees of freedom	Area in upper tail	Numerator degrees of freedom																	
		1	2	3	4	5	6	7	8	9	10	15	20	25	30	40	60	100	1000
21	.10	2.96	2.57	2.36	2.23	2.14	2.08	2.02	1.98	1.95	1.92	1.83	1.78	1.74	1.72	1.69	1.66	1.63	1.59
	.05	4.32	3.47	3.07	2.84	2.68	2.57	2.49	2.42	2.37	2.32	2.18	2.10	2.05	2.01	1.96	1.92	1.88	1.82
	.025	5.83	4.42	3.82	3.48	3.25	3.09	2.97	2.87	2.80	2.73	2.53	2.42	2.36	2.31	2.25	2.18	2.13	2.05
	.01	8.02	5.78	4.87	4.37	4.04	3.81	3.64	3.51	3.40	3.31	3.03	2.88	2.79	2.72	2.64	2.55	2.48	2.37
22	.10	2.95	2.56	2.35	2.22	2.13	2.06	2.01	1.97	1.93	1.90	1.81	1.76	1.73	1.70	1.67	1.64	1.61	1.57
	.05	4.30	3.44	3.05	2.82	2.66	2.55	2.46	2.40	2.34	2.30	2.15	2.07	2.02	1.98	1.94	1.89	1.85	1.79
	.025	5.79	4.38	3.78	3.44	3.22	3.05	2.93	2.84	2.76	2.70	2.50	2.39	2.32	2.27	2.21	2.14	2.09	2.01
	.01	7.95	5.72	4.82	4.31	3.99	3.76	3.59	3.45	3.35	3.26	2.98	2.83	2.73	2.67	2.58	2.50	2.42	2.32
23	.10	2.94	2.55	2.34	2.21	2.11	2.05	1.99	1.95	1.92	1.89	1.80	1.74	1.71	1.69	1.66	1.62	1.59	1.55
	.05	4.28	3.42	3.03	2.80	2.64	2.53	2.44	2.37	2.32	2.27	2.13	2.05	2.00	1.96	1.91	1.86	1.82	1.76
	.025	5.75	4.35	3.75	3.41	3.18	3.02	2.90	2.81	2.73	2.67	2.47	2.36	2.29	2.24	2.18	2.11	2.06	1.98
	.01	7.88	5.66	4.76	4.26	3.94	3.71	3.54	3.41	3.30	3.21	2.93	2.78	2.69	2.62	2.54	2.45	2.37	2.27
24	.10	2.93	2.54	2.33	2.19	2.10	2.04	1.98	1.94	1.91	1.88	1.78	1.73	1.70	1.67	1.64	1.61	1.58	1.54
	.05	4.26	3.40	3.01	2.78	2.62	2.51	2.42	2.36	2.30	2.25	2.11	2.03	1.97	1.94	1.89	1.84	1.80	1.74
	.025	5.72	4.32	3.72	3.38	3.15	2.99	2.87	2.78	2.70	2.64	2.44	2.33	2.26	2.21	2.15	2.08	2.02	1.94
	.01	7.82	5.61	4.72	4.22	3.90	3.67	3.50	3.36	3.26	3.17	2.89	2.74	2.64	2.58	2.49	2.40	2.33	2.22
25	.10	2.92	2.53	2.32	2.18	2.09	2.02	1.97	1.93	1.89	1.87	1.77	1.72	1.68	1.66	1.63	1.59	1.56	1.52
	.05	4.24	3.39	2.99	2.76	2.60	2.49	2.40	2.34	2.28	2.24	2.09	2.01	1.96	1.92	1.87	1.82	1.78	1.72
	.025	5.69	4.29	3.69	3.35	3.13	2.97	2.85	2.75	2.68	2.61	2.41	2.30	2.23	2.18	2.12	2.05	2.00	1.91
	.01	7.77	5.57	4.68	4.18	3.85	3.63	3.46	3.32	3.22	3.13	2.85	2.70	2.60	2.54	2.45	2.36	2.29	2.18
26	.10	2.91	2.52	2.31	2.17	2.08	2.01	1.96	1.92	1.88	1.86	1.76	1.71	1.67	1.65	1.61	1.58	1.55	1.51
	.05	4.23	3.37	2.98	2.74	2.59	2.47	2.39	2.32	2.27	2.22	2.07	1.99	1.94	1.90	1.85	1.80	1.76	1.70
	.025	5.66	4.27	3.67	3.33	3.10	2.94	2.82	2.73	2.65	2.59	2.39	2.28	2.21	2.16	2.09	2.03	1.97	1.89
	.01	7.72	5.53	4.64	4.14	3.82	3.59	3.42	3.29	3.18	3.09	2.81	2.66	2.57	2.50	2.42	2.33	2.25	2.14
27	.10	2.90	2.51	2.30	2.17	2.07	2.00	1.95	1.91	1.87	1.85	1.75	1.70	1.66	1.64	1.60	1.57	1.54	1.50
	.05	4.21	3.35	2.96	2.73	2.57	2.46	2.37	2.31	2.25	2.20	2.06	1.97	1.92	1.88	1.84	1.79	1.74	1.68
	.025	5.63	4.24	3.65	3.31	3.08	2.92	2.80	2.71	2.63	2.57	2.36	2.25	2.18	2.13	2.07	2.00	1.94	1.86
	.01	7.68	5.49	4.60	4.11	3.78	3.56	3.39	3.26	3.15	3.06	2.78	2.63	2.54	2.47	2.38	2.29	2.22	2.11
28	.10	2.89	2.50	2.29	2.16	2.06	2.00	1.94	1.90	1.87	1.84	1.74	1.69	1.65	1.63	1.59	1.56	1.53	1.48
	.05	4.20	3.34	2.95	2.71	2.56	2.45	2.36	2.29	2.24	2.19	2.04	1.96	1.91	1.87	1.82	1.77	1.73	1.66
	.025	5.61	4.22	3.63	3.29	3.06	2.90	2.78	2.69	2.61	2.55	2.34	2.23	2.16	2.11	2.05	1.98	1.92	1.84
	.01	7.64	5.45	4.57	4.07	3.75	3.53	3.36	3.23	3.12	3.03	2.75	2.60	2.51	2.44	2.35	2.26	2.19	2.08

(continued)

Table 4 (Continued)

Denominator degrees of freedom	Area in upper tail	Numerator degrees of freedom																	
		1	2	3	4	5	6	7	8	9	10	15	20	25	30	40	60	100	1000
29	.10	2.89	2.50	2.28	2.15	2.06	1.99	1.93	1.89	1.86	1.83	1.73	1.68	1.64	1.62	1.58	1.55	1.52	1.47
	.05	4.18	3.33	2.93	2.70	2.55	2.43	2.35	2.28	2.22	2.18	2.03	1.94	1.89	1.85	1.81	1.75	1.71	1.65
	.025	5.59	4.20	3.61	3.27	3.04	2.88	2.76	2.67	2.59	2.53	2.32	2.21	2.14	2.09	2.03	1.96	1.90	1.82
	.01	7.60	5.42	4.54	4.04	3.73	3.50	3.33	3.20	3.09	3.00	2.73	2.57	2.48	2.41	2.33	2.23	2.16	2.05
30	.10	2.88	2.49	2.28	2.14	2.05	1.98	1.93	1.88	1.85	1.82	1.72	1.67	1.63	1.61	1.57	1.54	1.51	1.46
	.05	4.17	3.32	2.92	2.69	2.53	2.42	2.33	2.27	2.21	2.16	2.01	1.93	1.88	1.84	1.79	1.74	1.70	1.63
	.025	5.57	4.18	3.59	3.25	3.03	2.87	2.75	2.65	2.57	2.51	2.31	2.20	2.12	2.07	2.01	1.94	1.88	1.80
	.01	7.56	5.39	4.51	4.02	3.70	3.47	3.30	3.17	3.07	2.98	2.70	2.55	2.45	2.39	2.30	2.21	2.13	2.02
40	.10	2.84	2.44	2.23	2.09	2.00	1.93	1.87	1.83	1.79	1.76	1.66	1.61	1.57	1.54	1.51	1.47	1.43	1.38
	.05	4.08	3.23	2.84	2.61	2.45	2.34	2.25	2.18	2.12	2.08	1.92	1.84	1.78	1.74	1.69	1.64	1.59	1.52
	.025	5.42	4.05	3.46	3.13	2.90	2.74	2.62	2.53	2.45	2.39	2.18	2.07	1.99	1.94	1.88	1.80	1.74	1.65
	.01	7.31	5.18	4.31	3.83	3.51	3.29	3.12	2.99	2.89	2.80	2.52	2.37	2.27	2.20	2.11	2.02	1.94	1.82
60	.10	2.79	2.39	2.18	2.04	1.95	1.87	1.82	1.77	1.74	1.71	1.60	1.54	1.50	1.48	1.44	1.40	1.36	1.30
	.05	4.00	3.15	2.76	2.53	2.37	2.25	2.17	2.10	2.04	1.99	1.84	1.75	1.69	1.65	1.59	1.53	1.48	1.40
	.025	5.29	3.93	3.34	3.01	2.79	2.63	2.51	2.41	2.33	2.27	2.06	1.94	1.87	1.82	1.74	1.67	1.60	1.49
	.01	7.08	4.98	4.13	3.65	3.34	3.12	2.95	2.82	2.72	2.63	2.35	2.20	2.10	2.03	1.94	1.84	1.75	1.62
100	.10	2.76	2.36	2.14	2.00	1.91	1.83	1.78	1.73	1.69	1.66	1.56	1.49	1.45	1.42	1.38	1.34	1.29	1.22
	.05	3.94	3.09	2.70	2.46	2.31	2.19	2.10	2.03	1.97	1.93	1.77	1.68	1.62	1.57	1.52	1.45	1.39	1.30
	.025	5.18	3.83	3.25	2.92	2.70	2.54	2.42	2.32	2.24	2.18	1.97	1.85	1.77	1.71	1.64	1.56	1.48	1.36
	.01	6.90	4.82	3.98	3.51	3.21	2.99	2.82	2.69	2.59	2.50	2.22	2.07	1.97	1.89	1.80	1.69	1.60	1.45
1000	.10	2.71	2.31	2.09	1.95	1.85	1.78	1.72	1.68	1.64	1.61	1.49	1.43	1.38	1.35	1.30	1.25	1.20	1.08
	.05	3.85	3.00	2.61	2.38	2.22	2.11	2.02	1.95	1.89	1.84	1.68	1.58	1.52	1.47	1.41	1.33	1.26	1.11
	.025	5.04	3.70	3.13	2.80	2.58	2.42	2.30	2.20	2.13	2.06	1.85	1.72	1.64	1.58	1.50	1.41	1.32	1.13
	.01	6.66	4.63	3.80	3.34	3.04	2.82	2.66	2.53	2.43	2.34	2.06	1.90	1.79	1.72	1.61	1.50	1.38	1.16

Table 5 Binomial probabilities

Entries in the table give the probability of **x** successes in **n** trials of a binomial experiment, where π is the probability of a success on one trial. For example, with six trials and $\pi = 0.05$, the probability of two successes is 0.0305.

| | | \multicolumn{9}{c}{π} |
n	x	.01	.02	.03	.04	.05	.06	.07	.08	.09
2	0	.9801	.9604	.9409	.9216	.9025	.8836	.8649	.8464	.8281
	1	.0198	.0392	.0582	.0768	.0950	.1128	.1302	.1472	.1638
	2	.0001	.0004	.0009	.0016	.0025	.0036	.0049	.0064	.0081
3	0	.9703	.9412	.9127	.8847	.8574	.8306	.8044	.7787	.7536
	1	.0294	.0576	.0847	.1106	.1354	.1590	.1816	.2031	.2236
	2	.0003	.0012	.0026	.0046	.0071	.0102	.0137	.0177	.0221
	3	.0000	.0000	.0000	.0001	.0001	.0002	.0003	.0005	.0007
4	0	.9606	.9224	.8853	.8493	.8145	.7807	.7481	.7164	.6857
	1	.0388	.0753	.1095	.1416	.1715	.1993	.2252	.2492	.2713
	2	.0006	.0023	.0051	.0088	.0135	.0191	.0254	.0325	.0402
	3	.0000	.0000	.0001	.0002	.0005	.0008	.0013	.0019	.0027
	4	.0000	.0000	.0000	.0000	.0000	.0000	.0000	.0000	.0001
5	0	.9510	.9039	.8587	.8154	.7738	.7339	.6957	.6591	.6240
	1	.0480	.0922	.1328	.1699	.2036	.2342	.2618	.2866	.3086
	2	.0010	.0038	.0082	.0142	.0214	.0299	.0394	.0498	.0610
	3	.0000	.0001	.0003	.0006	.0011	.0019	.0030	.0043	.0060
	4	.0000	.0000	.0000	.0000	.0000	.0001	.0001	.0002	.0003
	5	.0000	.0000	.0000	.0000	.0000	.0000	.0000	.0000	.0000
6	0	.9415	.8858	.8330	.7828	.7351	.6899	.6470	.6064	.5679
	1	.0571	.1085	.1546	.1957	.2321	.2642	.2922	.3164	.3370
	2	.0014	.0055	.0120	.0204	.0305	.0422	.0550	.0688	.0833
	3	.0000	.0002	.0005	.0011	.0021	.0036	.0055	.0080	.0110
	4	.0000	.0000	.0000	.0000	.0001	.0002	.0003	.0005	.0008
	5	.0000	.0000	.0000	.0000	.0000	.0000	.0000	.0000	.0000
	6	.0000	.0000	.0000	.0000	.0000	.0000	.0000	.0000	.0000
7	0	.9321	.8681	.8080	.7514	.6983	.6485	.6017	.5578	.5168
	1	.0659	.1240	.1749	.2192	.2573	.2897	.3170	.3396	.3578
	2	.0020	.0076	.0162	.0274	.0406	.0555	.0716	.0886	.1061
	3	.0000	.0003	.0008	.0019	.0036	.0059	.0090	.0128	.0175
	4	.0000	.0000	.0000	.0001	.0002	.0004	.0007	.0011	.0017
	5	.0000	.0000	.0000	.0000	.0000	.0000	.0000	.0001	.0001
	6	.0000	.0000	.0000	.0000	.0000	.0000	.0000	.0000	.0000
	7	.0000	.0000	.0000	.0000	.0000	.0000	.0000	.0000	.0000
8	0	.9227	.8508	.7837	.7214	.6634	.6096	.5596	.5132	.4703
	1	.0746	.1389	.1939	.2405	.2793	.3113	.3370	.3570	.3721
	2	.0026	.0099	.0210	.0351	.0515	.0695	.0888	.1087	.1288
	3	.0001	.0004	.0013	.0029	.0054	.0089	.0134	.0189	.0255
	4	.0000	.0000	.0001	.0002	.0004	.0007	.0013	.0021	.0031
	5	.0000	.0000	.0000	.0000	.0000	.0000	.0001	.0001	.0002
	6	.0000	.0000	.0000	.0000	.0000	.0000	.0000	.0000	.0000

(continued)

						π				
n	x	.01	.02	.03	.04	.05	.06	.07	.08	.09
	7	.0000	.0000	.0000	.0000	.0000	.0000	.0000	.0000	.0000
	8	.0000	.0000	.0000	.0000	.0000	.0000	.0000	.0000	.0000
9	0	.9135	.8337	.7602	.6925	.6302	.5730	.5204	.4722	.4279
	1	.0830	.1531	.2116	.2597	.2985	.3292	.3525	.3695	.3809
	2	.0034	.0125	.0262	.0433	.0629	.0840	.1061	.1285	.1507
	3	.0001	.0006	.0019	.0042	.0077	.0125	.0186	.0261	.0348
	4	.0000	.0000	.0001	.0003	.0006	.0012	.0021	.0034	.0052
	5	.0000	.0000	.0000	.0000	.0000	.0001	.0002	.0003	.0005
	6	.0000	.0000	.0000	.0000	.0000	.0000	.0000	.0000	.0000
	7	.0000	.0000	.0000	.0000	.0000	.0000	.0000	.0000	.0000
	8	.0000	.0000	.0000	.0000	.0000	.0000	.0000	.0000	.0000
	9	.0000	.0000	.0000	.0000	.0000	.0000	.0000	.0000	.0000
10	0	.9044	.8171	.7374	.6648	.5987	.5386	.4840	.4344	.3894
	1	.0914	.1667	.2281	.2770	.3151	.3438	.3643	.3777	.3851
	2	.0042	.0153	.0317	.0519	.0746	.0988	.1234	.1478	.1714
	3	.0001	.0008	.0026	.0058	.0105	.0168	.0248	.0343	.0452
	4	.0000	.0000	.0001	.0004	.0010	.0019	.0033	.0052	.0078
	5	.0000	.0000	.0000	.0000	.0001	.0001	.0003	.0005	.0009
	6	.0000	.0000	.0000	.0000	.0000	.0000	.0000	.0000	.0001
	7	.0000	.0000	.0000	.0000	.0000	.0000	.0000	.0000	.0000
	8	.0000	.0000	.0000	.0000	.0000	.0000	.0000	.0000	.0000
	9	.0000	.0000	.0000	.0000	.0000	.0000	.0000	.0000	.0000
	10	.0000	.0000	.0000	.0000	.0000	.0000	.0000	.0000	.0000
12	0	.8864	.7847	.6938	.6127	.5404	.4759	.4186	.3677	.3225
	1	.1074	.1922	.2575	.3064	.3413	.3645	.3781	.3837	.3827
	2	.0060	.0216	.0438	.0702	.0988	.1280	.1565	.1835	.2082
	3	.0002	.0015	.0045	.0098	.0173	.0272	.0393	.0532	.0686
	4	.0000	.0001	.0003	.0009	.0021	.0039	.0067	.0104	.0153
	5	.0000	.0000	.0000	.0001	.0002	.0004	.0008	.0014	.0024
	6	.0000	.0000	.0000	.0000	.0000	.0000	.0001	.0001	.0003
	7	.0000	.0000	.0000	.0000	.0000	.0000	.0000	.0000	.0000
	8	.0000	.0000	.0000	.0000	.0000	.0000	.0000	.0000	.0000
	9	.0000	.0000	.0000	.0000	.0000	.0000	.0000	.0000	.0000
	10	.0000	.0000	.0000	.0000	.0000	.0000	.0000	.0000	.0000
	11	.0000	.0000	.0000	.0000	.0000	.0000	.0000	.0000	.0000
	12	.0000	.0000	.0000	.0000	.0000	.0000	.0000	.0000	.0000
15	0	.8601	.7386	.6333	.5421	.4633	.3953	.3367	.2863	.2430
	1	.1303	.2261	.2938	.3388	.3658	.3785	.3801	.3734	.3605
	2	.0092	.0323	.0636	.0988	.1348	.1691	.2003	.2273	.2496
	3	.0004	.0029	.0085	.0178	.0307	.0468	.0653	.0857	.1070
	4	.0000	.0002	.0008	.0022	.0049	.0090	.0148	.0223	.0317
	5	.0000	.0000	.0001	.0002	.0006	.0013	.0024	.0043	.0069
	6	.0000	.0000	.0000	.0000	.0000	.0001	.0003	.0006	.0011
	7	.0000	.0000	.0000	.0000	.0000	.0000	.0000	.0001	.0001
	8	.0000	.0000	.0000	.0000	.0000	.0000	.0000	.0000	.0000
	9	.0000	.0000	.0000	.0000	.0000	.0000	.0000	.0000	.0000
	10	.0000	.0000	.0000	.0000	.0000	.0000	.0000	.0000	.0000
	11	.0000	.0000	.0000	.0000	.0000	.0000	.0000	.0000	.0000

Table 5 *(Continued)*

		π								
n	x	.01	.02	.03	.04	.05	.06	.07	.08	.09
	12	.0000	.0000	.0000	.0000	.0000	.0000	.0000	.0000	.0000
	13	.0000	.0000	.0000	.0000	.0000	.0000	.0000	.0000	.0000
	14	.0000	.0000	.0000	.0000	.0000	.0000	.0000	.0000	.0000
	15	.0000	.0000	.0000	.0000	.0000	.0000	.0000	.0000	.0000
18	0	.8345	.6951	.5780	.4796	.3972	.3283	.2708	.2229	.1831
	1	.1517	.2554	.3217	.3597	.3763	.3772	.3669	.3489	.3260
	2	.0130	.0443	.0846	.1274	.1683	.2047	.2348	.2579	.2741
	3	.0007	.0048	.0140	.0283	.0473	.0697	.0942	.1196	.1446
	4	.0000	.0004	.0016	.0044	.0093	.0167	.0266	.0390	.0536
	5	.0000	.0000	.0001	.0005	.0014	.0030	.0056	.0095	.0148
	6	.0000	.0000	.0000	.0000	.0002	.0004	.0009	.0018	.0032
	7	.0000	.0000	.0000	.0000	.0000	.0000	.0001	.0003	.0005
	8	.0000	.0000	.0000	.0000	.0000	.0000	.0000	.0000	.0001
	9	.0000	.0000	.0000	.0000	.0000	.0000	.0000	.0000	.0000
	10	.0000	.0000	.0000	.0000	.0000	.0000	.0000	.0000	.0000
	11	.0000	.0000	.0000	.0000	.0000	.0000	.0000	.0000	.0000
	12	.0000	.0000	.0000	.0000	.0000	.0000	.0000	.0000	.0000
	13	.0000	.0000	.0000	.0000	.0000	.0000	.0000	.0000	.0000
	14	.0000	.0000	.0000	.0000	.0000	.0000	.0000	.0000	.0000
	15	.0000	.0000	.0000	.0000	.0000	.0000	.0000	.0000	.0000
	16	.0000	.0000	.0000	.0000	.0000	.0000	.0000	.0000	.0000
	17	.0000	.0000	.0000	.0000	.0000	.0000	.0000	.0000	.0000
	18	.0000	.0000	.0000	.0000	.0000	.0000	.0000	.0000	.0000
20	0	.8179	.6676	.5438	.4420	.3585	.2901	.2342	.1887	.1516
	1	.1652	.2725	.3364	.3683	.3774	.3703	.3526	.3282	.3000
	2	.0159	.0528	.0988	.1458	.1887	.2246	.2521	.2711	.2818
	3	.0010	.0065	.0183	.0364	.0596	.0860	.1139	.1414	.1672
	4	.0000	.0006	.0024	.0065	.0133	.0233	.0364	.0523	.0703
	5	.0000	.0000	.0002	.0009	.0022	.0048	.0088	.0145	.0222
	6	.0000	.0000	.0000	.0001	.0003	.0008	.0017	.0032	.0055
	7	.0000	.0000	.0000	.0000	.0000	.0001	.0002	.0005	.0011
	8	.0000	.0000	.0000	.0000	.0000	.0000	.0000	.0001	.0002
	9	.0000	.0000	.0000	.0000	.0000	.0000	.0000	.0000	.0000
	10	.0000	.0000	.0000	.0000	.0000	.0000	.0000	.0000	.0000
	11	.0000	.0000	.0000	.0000	.0000	.0000	.0000	.0000	.0000
	12	.0000	.0000	.0000	.0000	.0000	.0000	.0000	.0000	.0000
	13	.0000	.0000	.0000	.0000	.0000	.0000	.0000	.0000	.0000
	14	.0000	.0000	.0000	.0000	.0000	.0000	.0000	.0000	.0000
	15	.0000	.0000	.0000	.0000	.0000	.0000	.0000	.0000	.0000
	16	.0000	.0000	.0000	.0000	.0000	.0000	.0000	.0000	.0000
	17	.0000	.0000	.0000	.0000	.0000	.0000	.0000	.0000	.0000
	18	.0000	.0000	.0000	.0000	.0000	.0000	.0000	.0000	.0000
	19	.0000	.0000	.0000	.0000	.0000	.0000	.0000	.0000	.0000
	20	.0000	.0000	.0000	.0000	.0000	.0000	.0000	.0000	.0000

(continued)

Table 5 (Continued)

n	x	π .10	.15	.20	.25	.30	.35	.40	.45	.50
2	0	.8100	.7225	.6400	.5625	.4900	.4225	.3600	.3025	.2500
	1	.1800	.2550	.3200	.3750	.4200	.4550	.4800	.4950	.5000
	2	.0100	.0225	.0400	.0625	.0900	.1225	.1600	.2025	.2500
3	0	.7290	.6141	.5120	.4219	.3430	.2746	.2160	.1664	.1250
	1	.2430	.3251	.3840	.4219	.4410	.4436	.4320	.4084	.3750
	2	.0270	.0574	.0960	.1406	.1890	.2389	.2880	.3341	.3750
	3	.0010	.0034	.0080	.0156	.0270	.0429	.0640	.0911	.1250
4	0	.6561	.5220	.4096	.3164	.2401	.1785	.1296	.0915	.0625
	1	.2916	.3685	.4096	.4219	.4116	.3845	.3456	.2995	.2500
	2	.0486	.0975	.1536	.2109	.2646	.3105	.3456	.3675	.3750
	3	.0036	.0115	.0256	.0469	.0756	.1115	.1536	.2005	.2500
	4	.0001	.0005	.0016	.0039	.0081	.0150	.0256	.0410	.0625
5	0	.5905	.4437	.3277	.2373	.1681	.1160	.0778	.0503	.0312
	1	.3280	.3915	.4096	.3955	.3602	.3124	.2592	.2059	.1562
	2	.0729	.1382	.2048	.2637	.3087	.3364	.3456	.3369	.3125
	3	.0081	.0244	.0512	.0879	.1323	.1811	.2304	.2757	.3125
	4	.0004	.0022	.0064	.0146	.0284	.0488	.0768	.1128	.1562
	5	.0000	.0001	.0003	.0010	.0024	.0053	.0102	.0185	.0312
6	0	.5314	.3771	.2621	.1780	.1176	.0754	.0467	.0277	.0156
	1	.3543	.3993	.3932	.3560	.3025	.2437	.1866	.1359	.0938
	2	.0984	.1762	.2458	.2966	.3241	.3280	.3110	.2780	.2344
	3	.0146	.0415	.0819	.1318	.1852	.2355	.2765	.3032	.3125
	4	.0012	.0055	.0154	.0330	.0595	.0951	.1382	.1861	.2344
	5	.0001	.0004	.0015	.0044	.0102	.0205	.0369	.0609	.0938
	6	.0000	.0000	.0001	.0002	.0007	.0018	.0041	.0083	.0156
7	0	.4783	.3206	.2097	.1335	.0824	.0490	.0280	.0152	.0078
	1	.3720	.3960	.3670	.3115	.2471	.1848	.1306	.0872	.0547
	2	.1240	.2097	.2753	.3115	.3177	.2985	.2613	.2140	.1641
	3	.0230	.0617	.1147	.1730	.2269	.2679	.2903	.2918	.2734
	4	.0026	.0109	.0287	.0577	.0972	.1442	.1935	.2388	.2734
	5	.0002	.0012	.0043	.0115	.0250	.0466	.0774	.1172	.1641
	6	.0000	.0001	.0004	.0013	.0036	.0084	.0172	.0320	.0547
	7	.0000	.0000	.0000	.0001	.0002	.0006	.0016	.0037	.0078
8	0	.4305	.2725	.1678	.1001	.0576	.0319	.0168	.0084	.0039
	1	.3826	.3847	.3355	.2670	.1977	.1373	.0896	.0548	.0312
	2	.1488	.2376	.2936	.3115	.2965	.2587	.2090	.1569	.1094
	3	.0331	.0839	.1468	.2076	.2541	.2786	.2787	.2568	.2188
	4	.0046	.0185	.0459	.0865	.1361	.1875	.2322	.2627	.2734
	5	.0004	.0026	.0092	.0231	.0467	.0808	.1239	.1719	.2188
	6	.0000	.0002	.0011	.0038	.0100	.0217	.0413	.0703	.1094
	7	.0000	.0000	.0001	.0004	.0012	.0033	.0079	.0164	.0313
	8	.0000	.0000	.0000	.0000	.0001	.0002	.0007	.0017	.0039

Table 5 (Continued)

		π								
n	x	.10	.15	.20	.25	.30	.35	.40	.45	.50
9	0	.3874	.2316	.1342	.0751	.0404	.0207	.0101	.0046	.0020
	1	.3874	.3679	.3020	.2253	.1556	.1004	.0605	.0339	.0176
	2	.1722	.2597	.3020	.3003	.2668	.2162	.1612	.1110	.0703
	3	.0446	.1069	.1762	.2336	.2668	.2716	.2508	.2119	.1641
	4	.0074	.0283	.0661	.1168	.1715	.2194	.2508	.2600	.2461
	5	.0008	.0050	.0165	.0389	.0735	.1181	.1672	.2128	.2461
	6	.0001	.0006	.0028	.0087	.0210	.0424	.0743	.1160	.1641
	7	.0000	.0000	.0003	.0012	.0039	.0098	.0212	.0407	.0703
	8	.0000	.0000	.0000	.0001	.0004	.0013	.0035	.0083	.0176
	9	.0000	.0000	.0000	.0000	.0000	.0001	.0003	.0008	.0020
10	0	.3487	.1969	.1074	.0563	.0282	.0135	.0060	.0025	.0010
	1	.3874	.3474	.2684	.1877	.1211	.0725	.0403	.0207	.0098
	2	.1937	.2759	.3020	.2816	.2335	.1757	.1209	.0763	.0439
	3	.0574	.1298	.2013	.2503	.2668	.2522	.2150	.1665	.1172
	4	.0112	.0401	.0881	.1460	.2001	.2377	.2508	.2384	.2051
	5	.0015	.0085	.0264	.0584	.1029	.1536	.2007	.2340	.2461
	6	.0001	.0012	.0055	.0162	.0368	.0689	.1115	.1596	.2051
	7	.0000	.0001	.0008	.0031	.0090	.0212	.0425	.0746	.1172
	8	.0000	.0000	.0001	.0004	.0014	.0043	.0106	.0229	.0439
	9	.0000	.0000	.0000	.0000	.0001	.0005	.0016	.0042	.0098
	10	.0000	.0000	.0000	.0000	.0000	.0000	.0001	.0003	.0010
12	0	.2824	.1422	.0687	.0317	.0138	.0057	.0022	.0008	.0002
	1	.3766	.3012	.2062	.1267	.0712	.0368	.0174	.0075	.0029
	2	.2301	.2924	.2835	.2323	.1678	.1088	.0639	.0339	.0161
	3	.0853	.1720	.2362	.2581	.2397	.1954	.1419	.0923	.0537
	4	.0213	.0683	.1329	.1936	.2311	.2367	.2128	.1700	.1208
	5	.0038	.0193	.0532	.1032	.1585	.2039	.2270	.2225	.1934
	6	.0005	.0040	.0155	.0401	.0792	.1281	.1766	.2124	.2256
	7	.0000	.0006	.0033	.0115	.0291	.0591	.1009	.1489	.1934
	8	.0000	.0001	.0005	.0024	.0078	.0199	.0420	.0762	.1208
	9	.0000	.0000	.0001	.0004	.0015	.0048	.0125	.0277	.0537
	10	.0000	.0000	.0000	.0000	.0002	.0008	.0025	.0068	.0161
	11	.0000	.0000	.0000	.0000	.0000	.0001	.0003	.0010	.0029
	12	.0000	.0000	.0000	.0000	.0000	.0000	.0000	.0001	.0002
15	0	.2059	.0874	.0352	.0134	.0047	.0016	.0005	.0001	.0000
	1	.3432	.2312	.1319	.0668	.0305	.0126	.0047	.0016	.0005
	2	.2669	.2856	.2309	.1559	.0916	.0476	.0219	.0090	.0032
	3	.1285	.2184	.2501	.2252	.1700	.1110	.0634	.0318	.0139
	4	.0428	.1156	.1876	.2252	.2186	.1792	.1268	.0780	.0417
	5	.0105	.0449	.1032	.1651	.2061	.2123	.1859	.1404	.0916
	6	.0019	.0132	.0430	.0917	.1472	.1906	.2066	.1914	.1527
	7	.0003	.0030	.0138	.0393	.0811	.1319	.1771	.2013	.1964
	8	.0000	.0005	.0035	.0131	.0348	.0710	.1181	.1647	.1964

(continued)

Table 5 (Continued)

n	x					π				
		.10	.15	.20	.25	.30	.35	.40	.45	.50
	10	.0000	.0000	.0001	.0007	.0030	.0096	.0245	.0515	.0916
	11	.0000	.0000	.0000	.0001	.0006	.0024	.0074	.0191	.0417
	12	.0000	.0000	.0000	.0000	.0001	.0004	.0016	.0052	.0139
	13	.0000	.0000	.0000	.0000	.0000	.0001	.0003	.0010	.0032
	14	.0000	.0000	.0000	.0000	.0000	.0000	.0000	.0001	.0005
	15	.0000	.0000	.0000	.0000	.0000	.0000	.0000	.0000	.0000
18	0	.1501	.0536	.0180	.0056	.0016	.0004	.0001	.0000	.0000
	1	.3002	.1704	.0811	.0338	.0126	.0042	.0012	.0003	.0001
	2	.2835	.2556	.1723	.0958	.0458	.0190	.0069	.0022	.0006
	3	.1680	.2406	.2297	.1704	.1046	.0547	.0246	.0095	.0031
	4	.0700	.1592	.2153	.2130	.1681	.1104	.0614	.0291	.0117
	5	.0218	.0787	.1507	.1988	.2017	.1664	.1146	.0666	.0327
	6	.0052	.0301	.0816	.1436	.1873	.1941	.1655	.1181	.0708
	7	.0010	.0091	.0350	.0820	.1376	.1792	.1892	.1657	.1214
	8	.0002	.0022	.0120	.0376	.0811	.1327	.1734	.1864	.1669
	9	.0000	.0004	.0033	.0139	.0386	.0794	.1284	.1694	.1855
	10	.0000	.0001	.0008	.0042	.0149	.0385	.0771	.1248	.1669
	11	.0000	.0000	.0001	.0010	.0046	.0151	.0374	.0742	.1214
	12	.0000	.0000	.0000	.0002	.0012	.0047	.0145	.0354	.0708
	13	.0000	.0000	.0000	.0000	.0002	.0012	.0045	.0134	.0327
	14	.0000	.0000	.0000	.0000	.0000	.0002	.0011	.0039	.0117
	15	.0000	.0000	.0000	.0000	.0000	.0000	.0002	.0009	.0031
	16	.0000	.0000	.0000	.0000	.0000	.0000	.0000	.0001	.0006
	17	.0000	.0000	.0000	.0000	.0000	.0000	.0000	.0000	.0001
	18	.0000	.0000	.0000	.0000	.0000	.0000	.0000	.0000	.0000
20	0	.1216	.0388	.0115	.0032	.0008	.0002	.0000	.0000	.0000
	1	.2702	.1368	.0576	.0211	.0068	.0020	.0005	.0001	.0000
	2	.2852	.2293	.1369	.0669	.0278	.0100	.0031	.0008	.0002
	3	.1901	.2428	.2054	.1339	.0716	.0323	.0123	.0040	.0011
	4	.0898	.1821	.2182	.1897	.1304	.0738	.0350	.0139	.0046
	5	.0319	.1028	.1746	.2023	.1789	.1272	.0746	.0365	.0148
	6	.0089	.0454	.1091	.1686	.1916	.1712	.1244	.0746	.0370
	7	.0020	.0160	.0545	.1124	.1643	.1844	.1659	.1221	.0739
	8	.0004	.0046	.0222	.0609	.1144	.1614	.1797	.1623	.1201
	9	.0001	.0011	.0074	.0271	.0654	.1158	.1597	.1771	.1602
	10	.0000	.0002	.0020	.0099	.0308	.0686	.1171	.1593	.1762
	11	.0000	.0000	.0005	.0030	.0120	.0336	.0710	.1185	.1602
	12	.0000	.0000	.0001	.0008	.0039	.0136	.0355	.0727	.1201
	13	.0000	.0000	.0000	.0002	.0010	.0045	.0146	.0366	.0739
	14	.0000	.0000	.0000	.0000	.0002	.0012	.0049	.0150	.0370
	15	.0000	.0000	.0000	.0000	.0000	.0003	.0013	.0049	.0148
	16	.0000	.0000	.0000	.0000	.0000	.0000	.0003	.0013	.0046
	17	.0000	.0000	.0000	.0000	.0000	.0000	.0000	.0002	.0011
	18	.0000	.0000	.0000	.0000	.0000	.0000	.0000	.0000	.0002
	19	.0000	.0000	.0000	.0000	.0000	.0000	.0000	.0000	.0000
	20	.0000	.0000	.0000	.0000	.0000	.0000	.0000	.0000	.0000

Table 6 Poisson probabilities

Entries in the table give the probability of x occurrences for a Poisson process with a mean μ. For example, when $\mu = 2.5$, the probability of four occurrences is .1336.

					μ					
x	0.1	0.2	0.3	0.4	0.5	0.6	0.7	0.8	0.9	1.0
0	.9048	.8187	.7408	.6703	.6065	.5488	.4966	.4493	.4066	.3679
1	.0905	.1637	.2222	.2681	.3033	.3293	.3476	.3595	.3659	.3679
2	.0045	.0164	.0333	.0536	.0758	.0988	.1217	.1438	.1647	.1839
3	.0002	.0011	.0033	.0072	.0126	.0198	.0284	.0383	.0494	.0613
4	.0000	.0001	.0002	.0007	.0016	.0030	.0050	.0077	.0111	.0153
5	.0000	.0000	.0000	.0001	.0002	.0004	.0007	.0012	.0020	.0031
6	.0000	.0000	.0000	.0000	.0000	.0000	.0001	0002	.0003	.0005
7	.0000	.0000	.0000	.0000	.0000	.0000	.0000	.0000	.0000	.0001

					μ					
x	1.1	1.2	1.3	1.4	1.5	1.6	1.7	1.8	1.9	2.0
0	.3329	.3012	.2725	.2466	.2231	.2019	.1827	.1653	.1496	.1353
1	.3662	.3614	.3543	.3452	.3347	.3230	.3106	.2975	.2842	.2707
2	.2014	.2169	.2303	.2417	.2510	.2584	.2640	.2678	.2700	.2707
3	.0738	.0867	.0998	.1128	.1255	.1378	.1496	.1607	.1710	.1804
4	.0203	.0260	.0324	.0395	.0471	.0551	.0636	.0723	.0812	.0902
5	.0045	.0062	.0084	.0111	.0141	.0176	.0216	.0260	.0309	.0361
6	.0008	.0012	.0018	.0026	.0035	.0047	.0061	.0078	.0098	.0120
7	.0001	.0002	.0003	.0005	.0008	.0011	.0015	.0020	.0027	.0034
8	.0000	.0000	.0001	.0001	.0001	.0002	.0003	.0005	.0006	.0009
9	.0000	.0000	.0000	.0000	.0000	.0000	.0001	.0001	.0001	.0002

					μ					
x	2.1	2.2	2.3	2.4	2.5	2.6	2.7	2.8	2.9	3.0
0	.1225	.1108	.1003	.0907	.0821	.0743	.0672	.0608	.0550	.0498
1	.2572	.2438	.2306	.2177	.2052	.1931	.1815	.1703	.1596	.1494
2	.2700	.2681	.2652	.2613	.2565	.2510	.2450	.2384	.2314	.2240
3	.1890	.1966	.2033	.2090	.2138	.2176	.2205	.2225	.2237	.2240
4	.0992	.1082	.1169	.1254	.1336	.1414	.1488	.1557	.1622	.1680
5	.0417	.0476	.0538	.0602	.0668	.0735	.0804	.0872	.0940	.1008
6	.0146	.0174	.0206	.0241	.0278	.0319	.0362	.0407	.0455	.0504
7	.0044	.0055	.0068	.0083	.0099	.0118	.0139	.0163	.0188	.0216
8	.0011	.0015	.0019	.0025	.0031	.0038	.0047	.0057	.0068	.0081
9	.0003	.0004	.0005	.0007	.0009	.0011	.0014	.0018	.0022	.0027
10	.0001	.0001	.0001	.0002	.0002	.0003	.0004	.0005	.0006	.0008
11	.0000	.0000	.0000	.0000	.0000	.0001	.0001	.0001	.0002	.0002
12	.0000	.0000	.0000	.0000	.0000	.0000	.0000	.0000	.0000	.0001

(continued)

Table 6 (Continued)

					μ					
x	3.1	3.2	3.3	3.4	3.5	3.6	3.7	3.8	3.9	4.0
0	.0450	.0408	.0369	.0344	.0302	.0273	.0247	.0224	.0202	.0183
1	.1397	.1304	.1217	.1135	.1057	.0984	.0915	.0850	.0789	.0733
2	.2165	.2087	.2008	.1929	.1850	.1771	.1692	.1615	.1539	.1465
3	.2237	.2226	.2209	.2186	.2158	.2125	.2087	.2046	.2001	.1954
4	.1734	.1781	.1823	.1858	.1888	.1912	.1931	.1944	.1951	.1954
5	.1075	.1140	.1203	.1264	.1322	.1377	.1429	.1477	.1522	.1563
6	.0555	.0608	.0662	.0716	.0771	.0826	.0881	.0936	.0989	.1042
7	.0246	.0278	.0312	.0348	.0385	.0425	.0466	.0508	.0551	.0595
8	.0095	.0111	.0129	.0148	.0169	.0191	.0215	.0241	.0269	.0298
9	.0033	.0040	.0047	.0056	.0066	.0076	.0089	.0102	.0116	.0132
10	.0010	.0013	.0016	.0019	.0023	.0028	.0033	.0039	.0045	.0053
11	.0003	.0004	.0005	.0006	.0007	.0009	.0011	.0013	.0016	.0019
12	.0001	.0001	.0001	.0002	.0002	.0003	.0003	.0004	.0005	.0006
13	.0000	.0000	.0000	.0000	.0001	.0001	.0001	.0001	.0002	.0002
14	.0000	.0000	.0000	.0000	.0000	.0000	.0000	.0000	.0000	.0001

					μ					
x	4.1	4.2	4.3	4.4	4.5	4.6	4.7	4.8	4.9	5.0
0	.0166	.0150	.0136	.0123	.0111	.0101	.091	.0082	.0074	.0067
1	.0679	.0630	.0583	.0540	.0500	.0462	.0427	.0395	.0365	.0337
2	.1393	.1323	.1254	.1188	.1125	.1063	.1005	.0948	.0894	.0842
3	.1904	.1852	.1798	.1743	.1687	.1631	.1574	.1517	.1460	.1404
4	.1951	.1944	.1933	.1917	.1898	.1875	.1849	.1820	.1789	.1755
5	.1600	.1633	.1662	.1687	.1708	.1725	.1738	.1747	.1753	.1755
6	.1093	.1143	.1191	.1237	.1281	.1323	.1362	.1398	.1432	.1462
7	.0640	.0686	.0732	.0778	.0824	.0869	.0914	.0959	.1002	.1044
8	.0328	.0360	.0393	.0428	.0463	.0500	.0537	.0575	.0614	.0653
9	.0150	.0168	.0188	.0209	.0232	.0255	.0280	.0307	.0334	.0363
10	.0061	.0071	.0081	.0092	.0104	.0118	.0132	.0147	.0164	.0181
11	.0023	.0027	.0032	.0037	.0043	.0049	.0056	.0064	.0073	.0082
12	.0008	.0009	.0011	.0014	.0016	.0019	.0022	.0026	.0030	.0034
13	.0002	.0003	.0004	.0005	.0006	.0007	.0008	.0009	.0011	.0013
14	.0001	.0001	.0001	.0001	.0002	.0002	.0003	.0003	.0004	.0005
15	.0000	.0000	.0000	.0000	.0001	.0001	.0001	.0001	.0001	.0002

					μ					
x	5.1	5.2	5.3	5.4	5.5	5.6	5.7	5.8	5.9	6.0
0	.0061	.0055	.0050	.0045	.0041	.0037	.0033	.0030	.0027	.0025
1	.0311	.0287	.0265	.0244	.0225	.0207	.0191	.0176	.0162	.0149

Table 6 (Continued)

	μ									
x	5.1	5.2	5.3	5.4	5.5	5.6	5.7	5.8	5.9	6.0
2	.0793	.0746	.0701	.0659	.0618	.0580	.0544	.0509	.0477	.0446
3	.1348	.1293	.1239	.1185	.1133	.1082	.1033	.0985	.0938	.0892
4	.1719	.1681	.1641	.1600	.1558	.1515	.1472	.1428	.1383	.1339
5	.1753	.1748	.1740	.1728	.1714	.1697	.1678	.1656	.1632	.1606
6	.1490	.1515	.1537	.1555	.1571	.1587	.1594	.1601	.1605	.1606
7	.1086	.1125	.1163	.1200	.1234	.1267	.1298	.1326	.1353	.1377
8	.0692	.0731	.0771	.0810	.0849	.0887	.0925	.0962	.0998	.1033
9	.0392	.0423	.0454	.0486	.0519	.0552	.0586	.0620	.0654	.0688
10	.0200	.0220	.0241	.0262	.0285	.0309	.0334	.0359	.0386	.0413
11	.0093	.0104	.0116	.0129	.0143	.0157	.0173	.0190	.0207	.0225
12	.0039	.0045	.0051	.0058	.0065	.0073	.0082	.0092	.0102	.0113
13	.0015	.0018	.0021	.0024	.0028	.0032	.0036	.0041	.0046	.0052
14	.0006	.0007	.0008	.0009	.0011	.0013	.0015	.0017	.0019	.0022
15	.0002	.0002	.0003	.0003	.0004	.0005	.0006	.0007	.0008	.0009
16	.0001	.0001	.0001	.0001	.0001	.0002	.0002	.0002	.0003	.0003
17	.0000	.0000	.0000	.0000	.0000	.0001	.0001	.0001	.0001	.0001

	μ									
x	6.1	6.2	6.3	6.4	6.5	6.6	6.7	6.8	6.9	7.0
0	.0022	.0020	.0018	.0017	.0015	.0014	.0012	.0011	.0010	.0009
1	.0137	.0126	.0116	.0106	.0098	.0090	.0082	.0076	.0070	.0064
2	.0417	.0390	.0364	.0340	.0318	.0296	.0276	.0258	.0240	.0223
3	.0848	.0806	.0765	.0726	.0688	.0652	.0617	.0584	.0552	.0521
4	.1294	.1249	.1205	.1162	.1118	.1076	.1034	.0992	.0952	.0912
5	.1579	.1549	.1519	.1487	.1454	.1420	.1385	.1349	.1314	.1277
6	.1605	.1601	.1595	.1586	.1575	.1562	.1546	.1529	.1511	.1490
7	.1399	.1418	.1435	.1450	.1462	.1472	.1480	.1486	.1489	.1490
8	.1066	.1099	.1130	.1160	.1188	.1215	.1240	.1263	.1284	.1304
9	.0723	.0757	.0791	.0825	.0858	.0891	.0923	.0954	.0985	.1014
10	.0441	.0469	.0498	.0528	.0558	.0588	.0618	.0649	.0679	.0710
11	.0245	.0265	.0285	.0307	.0330	.0353	.0377	.0401	.0426	.0452
12	.0124	.0137	.0150	.0164	.0179	.0194	.0210	.0227	.0245	.0264
13	.0058	.0065	.0073	.0081	.0089	.0098	.0108	.0119	.0130	.0142
14	.0025	.0029	.0033	.0037	.0041	.0046	.0052	.0058	.0064	.0071
15	.0010	.0012	.0014	.0016	.0018	.0020	.0023	.0026	.0029	.0033
16	.0004	.0005	.0005	.0006	.0007	.0008	.0010	.0011	.0013	.0014
17	.0001	.0002	.0002	.0002	.0003	.0003	.0004	.0004	.0005	.0006
18	.0000	.0001	.0001	.0001	.0001	.0001	.0001	.0002	.0002	.0002
19	.0000	.0000	.0000	.0000	.0000	.0000	.0000	.0001	.0001	.0001

(continued)

Table 6 (Continued)

					μ					
x	7.1	7.2	7.3	7.4	7.5	7.6	7.7	7.8	7.9	8.0
0	.0008	.0007	.0007	.0006	.0006	.0005	.0005	.0004	.0004	.0003
1	.0059	.0054	.0049	.0045	.0041	.0038	.0035	.0032	.0029	.0027
2	.0208	.0194	.0180	.0167	.0156	.0145	.0134	.0125	.0116	.0107
3	.0492	.0464	.0438	.0413	.0389	.0366	.0345	.0324	.0305	.0286
4	.0874	.0836	.0799	.0764	.0729	.0696	.0663	.0632	.0602	.0573
5	.1241	.1204	.1167	.1130	.1094	.1057	.1021	.0986	.0951	.0916
6	.1468	.1445	.1420	.1394	.1367	.1339	.1311	.1282	.1252	.1221
7	.1489	.1486	.1481	.1474	.1465	.1454	.1442	.1428	.1413	.1396
8	.1321	.1337	.1351	.1363	.1373	.1382	.1388	.1392	.1395	.1396
9	.1042	.1070	.1096	.1121	.1144	.1167	.1187	.1207	.1224	.1241
10	.0740	.0770	.0800	.0829	.0858	.0887	.0914	.0941	.0967	.0993
11	.0478	.0504	.0531	.0558	.0585	.0613	.0640	.0667	.0695	.0722
12	.0283	.0303	.0323	.0344	.0366	.0388	.0411	.0434	.0457	.0481
13	.0154	.0168	.0181	.0196	.0211	.0227	.0243	.0260	.0278	.0296
14	.0078	.0086	.0095	.0104	.0113	.0123	.0134	.0145	.0157	.0169
15	.0037	.0041	.0046	.0051	.0057	.0062	.0069	.0075	.0083	.0090
16	.0016	.0019	.0021	.0024	.0026	.0030	.0033	.0037	.0041	.0045
17	.0007	.0008	.0009	.0010	.0012	.0013	.0015	.0017	.0019	.0021
18	.0003	.0003	.0004	.0004	.0005	.0006	.0006	.0007	.0008	.0009
19	.0001	.0001	.0001	.0002	.0002	.0002	.0003	.0003	.0003	.0004
20	.0000	.0000	.0001	.0001	.0001	.0001	.0001	.0001	.0001	.0002
21	.0000	.0000	.0000	.0000	.0000	.0000	.0000	.0000	.0001	.0001

					μ					
x	8.1	8.2	8.3	8.4	8.5	8.6	8.7	8.8	8.9	9.0
0	.0003	.0003	.0002	.0002	.0002	.0002	.0002	.0002	.0001	.0001
1	.0025	.0023	.0021	.0019	.0017	.0016	.0014	.0013	.0012	.0011
2	.0100	.0092	.0086	.0079	.0074	.0068	.0063	.0058	.0054	.0050
3	.0269	.0252	.0237	.0222	.0208	.0195	.0183	.0171	.0160	.0150
4	.0544	.0517	.0491	.0466	.0443	.0420	.0398	.0377	.0357	.0337
5	.0882	.0849	.0816	.0784	.0752	.0722	.0692	.0663	.0635	.0607
6	.1191	.1160	.1128	.1097	.1066	.1034	.1003	.0972	.0941	.0911
7	.1378	.1358	.1338	.1317	.1294	.1271	.1247	.1222	.1197	.1171
8	.1395	.1392	.1388	.1382	.1375	.1366	.1356	.1344	.1332	.1318
9	.1256	.1269	.1280	.1290	.1299	.1306	.1311	.1315	.1317	.1318
10	.1017	.1040	.1063	.1084	.1104	.1123	.1140	.1157	.1172	.1186
11	.0749	.0776	.0802	.0828	.0853	.0878	.0902	.0925	.0948	.0970
12	.0505	.0530	.0555	.0579	.0604	.0629	.0654	.0679	.0703	.0728

Table 6 (Continued)

					μ					
x	8.1	8.2	8.3	8.4	8.5	8.6	8.7	8.8	8.9	9.0
13	.0315	.0334	.0354	.0374	.0395	.0416	.0438	.0459	.0481	.0504
14	.0182	.0196	.0210	.0225	.0240	.0256	.0272	.0289	.0306	.0324
15	.0098	.0107	.0116	.0126	.0136	.0147	.0158	.0169	.0182	.1094
16	.0050	.0055	.0060	.0066	.0072	.0079	.0086	.0093	.0101	.0109
17	.0024	.0026	.0029	.0033	.0036	.0040	.0044	.0048	.0053	.0058
18	.0011	.0012	.0014	.0015	.0017	.0019	.0021	.0024	.0026	.0029
19	.0005	.0005	.0006	.0007	.0008	.0009	.0010	.0011	.0012	.0014
20	.0002	.0002	.0002	.0003	.0003	.0004	.0004	.0005	.0005	.0006
21	.0001	.0001	.0001	.0001	.0001	.0002	.0002	.0002	.0002	.0003
22	.0000	.0000	.0000	.0000	.0001	.0001	.0001	.0001	.0001	.0001

					μ					
x	9.1	9.2	9.3	9.4	9.5	9.6	9.7	9.8	9.9	10
0	.0001	.0001	.0001	.0001	.0001	.0001	.0001	.0001	.0001	.0000
1	.0010	.0009	.0009	.0008	.0007	.0007	.0006	.0005	.0005	.0005
2	.0046	.0043	.0040	.0037	.0034	.0031	.0029	.0027	.0025	.0023
3	.0140	.0131	.0123	.0115	.0107	.0100	.0093	.0087	.0081	.0076
4	.0319	.0302	.0285	.0269	.0254	.0240	.0226	.0213	.0201	.0189
5	.0581	.0555	.0530	.0506	.0483	.0460	.0439	.0418	.0398	.0378
6	.0881	.0851	.0822	.0793	.0764	.0736	.0709	.0682	.0656	.0631
7	.1145	.1118	.1091	.1064	.1037	.1010	.0982	.0955	.0928	.0901
8	.1302	.1286	.1269	.1251	.1232	.1212	.1191	.1170	.1148	.1126
9	.1317	.1315	.1311	.1306	.1300	.1293	.1284	.1274	.1263	.1251
10	.1198	.1210	.1219	.1228	.1235	.1241	.1245	.1249	.1250	.1251
11	.0991	.1012	.1031	.1049	.1067	.1083	.1098	.1112	.1125	.1137
12	.0752	.0776	.0799	.0822	.0844	.0866	.0888	.0908	.0928	.0948
13	.0526	.0549	.0572	.0594	.0617	.0640	.0662	.0685	.0707	.0729
14	.0342	.0361	.0380	.0399	.0419	.0439	.0459	.0479	.0500	.0521
15	.0208	.0221	.0235	.0250	.0265	.0281	.0297	.0313	.0330	.0347
16	.0118	.0127	.0137	.0147	.0157	.0168	.0180	.0192	.0204	.0217
17	.0063	.0069	.0075	.0081	.0088	.0095	.0103	.0111	.0119	.0128
18	.0032	.0035	.0039	.0042	.0046	.0051	.0055	.0060	.0065	.0071
19	.0015	.0017	.0019	.0021	.0023	.0026	.0028	.0031	.0034	.0037
20	.0007	.0008	.0009	.0010	.0011	.0012	.0014	.0015	.0017	.0019
21	.0003	.0003	.0004	.0004	.0005	.0006	.0006	.0007	.0008	.0009
22	.0001	.0001	.0002	.0002	.0002	.0002	.0003	.0003	.0004	.0004
23	.0000	.0001	.0001	.0001	.0001	.0001	.0001	.0001	.0002	.0002
24	.0000	.0000	.0000	.0000	.0000	.0000	.0000	.0001	.0001	.0001

(continued)

Table 6 (Continued)

| | | | | | μ | | | | | |
x	11	12	13	14	15	16	17	18	19	20
0	.0000	.0000	.0000	.0000	.0000	.0000	.0000	.0000	.0000	.0000
1	.0002	.0001	.0000	.0000	.0000	.0000	.0000	.0000	.0000	.0000
2	.0010	.0004	.0002	.0001	.0000	.0000	.0000	.0000	.0000	.0000
3	.0037	.0018	.0008	.0004	.0002	.0001	.0000	.0000	.0000	.0000
4	.0102	.0053	.0027	.0013	.0006	.0003	.0001	.0001	.0000	.0000
5	.0224	.0127	.0070	.0037	.0019	.0010	.0005	.0002	.0001	.0001
6	.0411	.0255	.0152	.0087	.0048	.0026	.0014	.0007	.0004	.0002
7	.0646	.0437	.0281	.0174	.0104	.0060	.0034	.0018	.0010	.0005
8	.0888	.0655	.0457	.0304	.0194	.0120	.0072	.0042	.0024	.0013
9	.1085	.0874	.0661	.0473	.0324	.0213	.0135	.0083	.0050	.0029
10	.1194	.1048	.0859	.0663	.0486	.0341	.0230	.0150	.0095	.0058
11	.1194	.1144	.1015	.0844	.0663	.0496	.0355	.0245	.0164	.0106
12	.1094	.1144	.1099	.0984	.0829	.0661	.0504	.0368	.0259	.0176
13	.0926	.1056	.1099	.1060	.0956	.0814	.0658	.0509	.0378	.0271
14	.0728	.0905	.1021	.1060	.1024	.0930	.0800	.0655	.0514	.0387
15	.0534	.0724	.0885	.0989	.1024	.0992	.0906	.0786	.0650	.0516
16	.0367	.0543	.0719	.0866	.0960	.0992	.0963	.0884	0.772	.0646
17	.0237	.0383	.0550	.0713	.0847	.0934	.0963	.0936	.0863	.0760
18	.0145	.0256	.0397	.0554	.0706	.0830	.0909	.0936	.0911	.0844
19	.0084	.0161	.0272	.0409	.0557	.0699	.0814	.0887	.0911	.0888
20	.0046	.0097	.0177	.0286	.0418	.0559	.0692	.0798	.0866	.0888
21	.0024	.0055	.0109	.0191	.0299	.0426	.0560	.0684	.0783	.0846
22	.0012	.0030	.0065	.0121	.0204	.0310	.0433	.0560	.0676	.0769
23	.0006	.0016	.0037	.0074	.0133	.0216	.0320	.0438	.0559	.0669
24	.0003	.0008	.0020	.0043	.0083	.0144	.0226	.0328	.0442	.0557
25	.0001	.0004	.0010	.0024	.0050	.0092	.0154	.0237	.0336	.0446
26	.0000	.0002	.0005	.0013	.0029	.0057	.0101	.0164	.0246	.0343
27	.0000	.0001	.0002	.0007	.0016	.0034	.0063	.0109	.0173	.0254
28	.0000	.0000	.0001	.0003	.0009	.0019	.0038	.0070	.0117	.0181
29	.0000	.0000	.0001	.0002	.0004	.0011	.0023	.0044	.0077	.0125
30	.0000	.0000	.0000	.0001	.0002	.0006	.0013	.0026	.0049	.0083
31	.0000	.0000	.0000	.0000	.0001	.0003	.0007	.0015	.0030	.0054
32	.0000	.0000	.0000	.0000	.0001	.0001	.0004	.0009	.0018	.0034
33	.0000	.0000	.0000	.0000	.0000	.0001	.0002	.0005	.0010	.0020
34	.0000	.0000	.0000	.0000	.0000	.0000	.0001	.0002	.0006	.0012
35	.0000	.0000	.0000	.0000	.0000	.0000	.0000	.0001	.0003	.0007
36	.0000	.0000	.0000	.0000	.0000	.0000	.0000	.0001	.0002	.0004
37	.0000	.0000	.0000	.0000	.0000	.0000	.0000	.0000	.0001	.0002
38	.0000	.0000	.0000	.0000	.0000	.0000	.0000	.0000	.0000	.0001
39	.0000	.0000	.0000	.0000	.0000	.0000	.0000	.0000	.0000	.0001

Table 7 Critical values for the Durbin-Watson test for autocorrelation

Entries in the table give the critical values for a one-tailed Durbin-Watson test for autocorrelation. For a two-tailed test, the level of significance is doubled.

Significance points of d_L and d_U: $\alpha = .05$
Number of independent variables

k	1		2		3		4		5	
n	d_L	d_U	d_L	d_U	d_L	d_U	d_L	d_U	d_L	d_U
15	1.08	1.36	0.95	1.54	0.82	1.75	0.69	1.97	0.56	2.21
16	1.10	1.37	0.98	1.54	0.86	1.73	0.74	1.93	0.62	2.15
17	1.13	1.38	1.02	1.54	0.90	1.71	0.78	1.90	0.67	2.10
18	1.16	1.39	1.05	1.53	0.93	1.69	0.82	1.87	0.71	2.06
19	1.18	1.40	1.08	1.53	0.97	1.68	0.86	1.85	0.75	2.02
20	1.20	1.41	1.10	1.54	1.00	1.68	0.90	1.83	0.79	1.99
21	1.22	1.42	1.13	1.54	1.03	1.67	0.93	1.81	0.83	1.96
22	1.24	1.43	1.15	1.54	1.05	1.66	1.96	1.80	0.86	1.94
23	1.26	1.44	1.17	1.54	1.08	1.66	0.99	1.79	0.90	1.92
24	1.27	1.45	1.19	1.55	1.10	1.66	1.01	1.78	0.93	1.90
25	1.29	1.45	1.21	1.55	1.12	1.66	1.04	1.77	0.95	1.89
26	1.30	1.46	1.22	1.55	1.14	1.65	1.06	1.76	0.98	1.88
27	1.32	1.47	1.24	1.56	1.16	1.65	1.08	1.76	1.01	1.86
28	1.33	1.48	1.26	1.56	1.18	1.65	1.10	1.75	1.03	1.85
29	1.34	1.48	1.27	1.56	1.20	1.65	1.12	1.74	1.05	1.84
30	1.35	1.49	1.28	1.57	1.21	1.65	1.14	1.74	1.07	1.83
31	1.36	1.50	1.30	1.57	1.23	1.65	1.16	1.74	1.09	1.83
32	1.37	1.50	1.31	1.57	1.24	1.65	1.18	1.73	1.11	1.82
33	1.38	1.51	1.32	1.58	1.26	1.65	1.19	1.73	1.13	1.81
34	1.39	1.51	1.33	1.58	1.27	1.65	1.21	1.73	1.15	1.81
35	1.40	1.52	1.34	1.58	1.28	1.65	1.22	1.73	1.16	1.80
36	1.41	1.52	1.35	1.59	1.29	1.65	1.24	1.73	1.18	1.80
37	1.42	1.53	1.36	1.59	1.31	1.66	1.25	1.72	1.19	1.80
38	1.43	1.54	1.37	1.59	1.32	1.66	1.26	1.72	1.21	1.79
39	1.43	1.54	1.38	1.60	1.33	1.66	1.27	1.72	1.22	1.79
40	1.44	1.54	1.39	1.60	1.34	1.66	1.29	1.72	1.23	1.79
45	1.48	1.57	1.43	1.62	1.38	1.67	1.34	1.72	1.29	1.78
50	1.50	1.59	1.46	1.63	1.42	1.67	1.38	1.72	1.34	1.77
55	1.53	1.60	1.49	1.64	1.45	1.68	1.41	1.72	1.38	1.77
60	1.55	1.62	1.51	1.65	1.48	1.69	1.44	1.73	1.41	1.77
65	1.57	1.63	1.54	1.66	1.50	1.70	1.47	1.73	1.44	1.77
70	1.58	1.64	1.55	1.67	1.52	1.70	1.49	1.74	1.46	1.77
75	1.60	1.65	1.57	1.68	1.54	1.71	1.51	1.74	1.49	1.77
80	1.61	1.66	1.59	1.69	1.56	1.72	1.53	1.74	1.51	1.77
85	1.62	1.67	1.60	1.70	1.57	1.72	1.55	1.75	1.52	1.77
90	1.63	1.68	1.61	1.70	1.59	1.73	1.57	1.75	1.54	1.78
95	1.64	1.69	1.62	1.71	1.60	1.73	1.58	1.75	1.56	1.78
100	1.65	1.69	1.63	1.72	1.61	1.74	1.59	1.76	1.57	1.78

(continued)

Table 7 (Continued)

Significance points of d_L and d_U: $\alpha = .025$

Number of independent variables

n	k 1 d_L	d_U	2 d_L	d_U	3 d_L	d_U	4 d_L	d_U	5 d_L	d_U
15	0.95	1.23	0.83	1.40	0.71	1.61	0.59	1.84	0.48	2.09
16	0.98	1.24	0.86	1.40	0.75	1.59	0.64	1.80	0.53	2.03
17	1.01	1.25	0.90	1.40	0.79	1.58	0.68	1.77	0.57	1.98
18	1.03	1.26	0.93	1.40	0.82	1.56	0.72	1.74	0.62	1.93
19	1.06	1.28	0.96	1.41	0.86	1.55	0.76	1.72	0.66	1.90
20	1.08	1.28	0.99	1.41	0.89	1.55	0.79	1.70	0.70	1.87
21	1.10	1.30	1.01	1.41	0.92	1.54	0.83	1.69	0.73	1.84
22	1.12	1.31	1.04	1.42	0.95	1.54	0.86	1.68	0.77	1.82
23	1.14	1.32	1.06	1.42	0.97	1.54	0.89	1.67	0.80	1.80
24	1.16	1.33	1.08	1.43	1.00	1.54	0.91	1.66	0.83	1.79
25	1.18	1.34	1.10	1.43	1.02	1.54	0.94	1.65	0.86	1.77
26	1.19	1.35	1.12	1.44	1.04	1.54	0.96	1.65	0.88	1.76
27	1.21	1.36	1.13	1.44	1.06	1.54	0.99	1.64	0.91	1.75
28	1.22	1.37	1.15	1.45	1.08	1.54	1.01	1.64	0.93	1.74
29	1.24	1.38	1.17	1.45	1.10	1.54	1.03	1.63	0.96	1.73
30	1.25	1.38	1.18	1.46	1.12	1.54	1.05	1.63	0.98	1.73
31	1.26	1.39	1.20	1.47	1.13	1.55	1.07	1.63	1.00	1.72
32	1.27	1.40	1.21	1.47	1.15	1.55	1.08	1.63	1.02	1.71
33	1.28	1.41	1.22	1.48	1.16	1.55	1.10	1.63	1.04	1.71
34	1.29	1.41	1.24	1.48	1.17	1.55	1.12	1.63	1.06	1.70
35	1.30	1.42	1.25	1.48	1.19	1.55	1.13	1.63	1.07	1.70
36	1.31	1.43	1.26	1.49	1.20	1.56	1.15	1.63	1.09	1.70
37	1.32	1.43	1.27	1.49	1.21	1.56	1.16	1.62	1.10	1.70
38	1.33	1.44	1.28	1.50	1.23	1.56	1.17	1.62	1.12	1.70
39	1.34	1.44	1.29	1.50	1.24	1.56	1.19	1.63	1.13	1.69
40	1.35	1.45	1.30	1.51	1.25	1.57	1.20	1.63	1.15	1.69
45	1.39	1.48	1.34	1.53	1.30	1.58	1.25	1.63	1.21	1.69
50	1.42	1.50	1.38	1.54	1.34	1.59	1.30	1.64	1.26	1.69
55	1.45	1.52	1.41	1.56	1.37	1.60	1.33	1.64	1.30	1.69
60	1.47	1.54	1.44	1.57	1.40	1.61	1.37	1.65	1.33	1.69
65	1.49	1.55	1.46	1.59	1.43	1.62	1.40	1.66	1.36	1.69
70	1.51	1.57	1.48	1.60	1.45	1.63	1.42	1.66	1.39	1.70
75	1.53	1.58	1.50	1.61	1.47	1.64	1.45	1.67	1.42	1.70
80	1.54	1.59	1.52	1.62	1.49	1.65	1.47	1.67	1.44	1.70
85	1.56	1.60	1.53	1.63	1.51	1.65	1.49	1.68	1.46	1.71
90	1.57	1.61	1.55	1.64	1.53	1.66	1.50	1.69	1.48	1.71
95	1.58	1.62	1.56	1.65	1.54	1.67	1.52	1.69	1.50	1.71
100	1.59	1.63	1.57	1.65	1.55	1.67	1.53	1.70	1.51	1.72

Table 7 (Continued)

Significance points of d_L and d_U: $\alpha = .01$
Number of independent variables

k	1		2		3		4		5	
n	d_L	d_U	d_L	d_U	d_L	d_U	d_L	d_U	d_L	d_U
15	0.81	1.07	0.70	1.25	0.59	1.46	0.49	1.70	0.39	1.96
16	0.84	1.09	0.74	1.25	0.63	1.44	0.53	1.66	0.44	1.90
17	0.87	1.10	0.77	1.25	0.67	1.43	0.57	1.63	0.48	1.85
18	0.90	1.12	0.80	1.26	0.71	1.42	0.61	1.60	0.52	1.80
19	0.93	1.13	0.83	1.26	0.74	1.41	0.65	1.58	0.56	1.77
20	0.95	1.15	0.86	1.27	0.77	1.41	0.68	1.57	0.60	1.74
21	0.97	1.16	0.89	1.27	0.80	1.41	0.72	1.55	0.63	1.71
22	1.00	1.17	0.91	1.28	0.83	1.40	0.75	1.54	0.66	1.69
23	1.02	1.19	0.94	1.29	0.86	1.40	0.77	1.53	0.70	1.67
24	1.04	1.20	0.96	1.30	0.88	1.41	0.80	1.53	0.72	1.66
25	1.05	1.21	0.98	1.30	0.90	1.41	0.83	1.52	0.75	1.65
26	1.07	1.22	1.00	1.31	0.93	1.41	0.85	1.52	0.78	1.64
27	1.09	1.23	1.02	1.32	0.95	1.41	0.88	1.51	0.81	1.63
28	1.10	1.24	1.04	1.32	0.97	1.41	0.90	1.51	0.83	1.62
29	1.12	1.25	1.05	1.33	0.99	1.42	0.92	1.51	0.85	1.61
30	1.13	1.26	1.07	1.34	1.01	1.42	0.94	1.51	0.88	1.61
31	1.15	1.27	1.08	1.34	1.02	1.42	0.96	1.51	0.90	1.60
32	1.16	1.28	1.10	1.35	1.04	1.43	0.98	1.51	0.92	1.60
33	1.17	1.29	1.11	1.36	1.05	1.43	1.00	1.51	0.94	1.59
34	1.18	1.30	1.13	1.36	1.07	1.43	1.01	1.51	0.95	1.59
35	1.19	1.31	1.14	1.37	1.08	1.44	1.03	1.51	0.97	1.59
36	1.21	1.32	1.15	1.38	1.10	1.44	1.04	1.51	0.99	1.59
37	1.22	1.32	1.16	1.38	1.11	1.45	1.06	1.51	1.00	1.59
38	1.23	1.33	1.18	1.39	1.12	1.45	1.07	1.52	1.02	1.58
39	1.24	1.34	1.19	1.39	1.14	1.45	1.09	1.52	1.03	1.58
40	1.25	1.34	1.20	1.40	1.15	1.46	1.10	1.52	1.05	1.58
45	1.29	1.38	1.24	1.42	1.20	1.48	1.16	1.53	1.11	1.58
50	1.32	1.40	1.28	1.45	1.24	1.49	1.20	1.54	1.16	1.59
55	1.36	1.43	1.32	1.47	1.28	1.51	1.25	1.55	1.21	1.59
60	1.38	1.45	1.35	1.48	1.32	1.52	1.28	1.56	1.25	1.60
65	1.41	1.47	1.38	1.50	1.35	1.53	1.31	1.57	1.28	1.61
70	1.43	1.49	1.40	1.52	1.37	1.55	1.34	1.58	1.31	1.61
75	1.45	1.50	1.42	1.53	1.39	1.56	1.37	1.59	1.34	1.62
80	1.47	1.52	1.44	1.54	1.42	1.57	1.39	1.60	1.36	1.62
85	1.48	1.53	1.46	1.55	1.43	1.58	1.41	1.60	1.39	1.63
90	1.50	1.54	1.47	1.56	1.45	1.59	1.43	1.61	1.41	1.64
95	1.51	1.55	1.49	1.57	1.47	1.60	1.45	1.62	1.42	1.64
100	1.52	1.56	1.50	1.58	1.48	1.60	1.46	1.63	1.44	1.65

This table is reprinted by permission of Oxford University Press on behalf of The Biometrika Trustees from J. Durbin and G. S. Watson, "Testing for serial correlation in least square regression II," *Biometrika* 38 (1951), 159–178.

Table 8 T_L Values for the Mann-Whitney-Wilcoxon test

Reject the hypothesis of identical populations if the sum of the ranks for the n_1 items is *less* than the value T_L shown in the following table or if the sum of the ranks for the n_1 items is *greater* than the value T_U where

$$T_U = n_1(n_1 + n_2 + 1) - T_L$$

$\alpha = .10$				n_2					
n_1	2	3	4	5	6	7	8	9	10
2	3	3	3	4	4	4	5	5	5
3	6	7	7	8	9	9	10	11	11
4	10	11	12	13	14	15	16	17	18
5	16	17	18	20	21	22	24	25	27
6	22	24	25	27	29	30	32	34	36
7	29	31	33	35	37	40	42	44	46
8	38	40	42	45	47	50	52	55	57
9	47	50	52	55	58	61	64	67	70
10	57	60	63	67	70	73	76	80	83

$\alpha = 0.05$				n_2					
n_1	2	3	4	5	6	7	8	9	10
2	3	3	3	3	3	3	4	4	4
3	6	6	6	7	8	8	9	9	10
4	10	10	11	12	13	14	15	15	16
5	15	16	17	18	19	21	22	23	24
6	21	23	24	25	27	28	30	32	33
7	28	30	32	34	35	37	39	41	43
8	37	39	41	43	45	47	50	52	54
9	46	48	50	53	56	58	61	63	66
10	56	59	61	64	67	70	73	76	79

Table 9 Critical values for the studentized range

$\alpha = .05$

Degrees of Freedom	Number of populations																		
	2	3	4	5	6	7	8	9	10	11	12	13	14	15	16	17	18	19	20
1	18.0	27.0	32.8	37.1	40.4	43.1	45.4	47.4	49.1	50.6	52.0	53.2	54.3	55.4	56.3	57.2	58.0	58.8	59.6
2	6.08	8.33	9.80	10.9	11.7	12.4	13.0	13.5	14.0	14.4	14.7	15.1	15.4	15.7	15.9	16.1	16.4	16.6	16.8
3	4.50	5.91	6.82	7.50	8.04	8.48	8.85	9.18	9.46	9.72	9.95	10.2	10.3	10.5	10.7	10.8	11.0	11.1	11.2
4	3.93	5.04	5.76	6.29	6.71	7.05	7.35	7.60	7.83	8.03	8.21	8.37	8.52	8.66	8.79	8.91	9.03	9.13	9.23
5	3.64	4.60	5.22	5.67	6.03	6.33	6.58	6.80	6.99	7.17	7.32	7.47	7.60	7.72	7.83	7.93	8.03	8.12	8.21
6	3.46	4.34	4.90	5.30	5.63	5.90	6.12	6.32	6.49	6.65	6.79	6.92	7.03	7.14	7.24	7.34	7.43	7.51	7.59
7	3.34	4.16	4.68	5.06	5.36	5.61	5.82	6.00	6.16	6.30	6.43	6.55	6.66	6.76	6.85	6.94	7.02	7.10	7.17
8	3.26	4.04	4.53	4.89	5.17	5.40	5.60	5.77	5.92	6.05	6.18	6.29	6.39	6.48	6.57	6.65	6.73	6.80	6.87
9	3.20	3.95	4.41	4.76	5.02	5.24	5.43	5.59	5.74	5.87	5.98	6.09	6.19	6.28	6.36	6.44	6.51	6.58	6.64
10	3.15	3.88	4.33	4.65	4.91	5.12	5.30	5.46	5.60	5.72	5.83	5.93	6.03	6.11	6.19	6.27	6.34	6.40	6.47
11	3.11	3.82	4.26	4.57	4.82	5.03	5.20	5.35	5.49	5.61	5.71	5.81	5.90	5.98	6.06	6.13	6.20	6.27	6.33
12	3.08	3.77	4.20	4.51	4.75	4.95	5.12	5.27	5.39	5.51	5.61	5.71	5.80	5.88	5.95	6.02	6.09	6.15	6.21
13	3.06	3.73	4.15	4.45	4.69	4.88	5.05	5.19	5.32	5.43	5.53	5.63	5.71	5.79	5.86	5.93	5.99	6.05	6.11
14	3.03	3.70	4.11	4.41	4.64	4.83	4.99	5.13	5.25	5.36	5.46	5.55	5.64	5.71	5.79	5.85	5.91	5.97	6.03
15	3.01	3.67	4.08	4.37	4.59	4.78	4.94	5.08	5.20	5.31	5.40	5.49	5.57	5.65	5.72	5.78	5.85	5.90	5.96
16	3.00	3.65	4.05	4.33	4.56	4.74	4.90	5.03	5.15	5.26	5.35	5.44	5.52	5.59	5.66	5.73	5.79	5.84	5.90
17	2.98	3.63	4.02	4.30	4.52	4.70	4.86	4.99	5.11	5.21	5.31	5.39	5.47	5.54	5.61	5.67	5.73	5.79	5.84
18	2.97	3.61	4.00	4.28	4.49	4.67	4.82	4.96	5.07	5.17	5.27	5.35	5.43	5.50	5.57	5.63	5.69	5.74	5.79
19	2.96	3.59	3.98	4.25	4.47	4.65	4.79	4.92	5.04	5.14	5.23	5.31	5.39	5.46	5.53	5.59	5.65	5.70	5.75
20	2.95	3.58	3.96	4.23	4.45	4.62	4.77	4.90	5.01	5.11	5.20	5.28	5.36	5.43	5.49	5.55	5.61	5.66	5.71
24	2.92	3.53	3.90	4.17	4.37	4.54	4.68	4.81	4.92	5.01	5.10	5.18	5.25	5.32	5.38	5.44	5.49	5.55	5.59
30	2.89	3.49	3.85	4.10	4.30	4.46	4.60	4.72	4.82	4.92	5.00	5.08	5.15	5.21	5.27	5.33	5.38	5.43	5.47
40	2.86	3.44	3.79	4.04	4.23	4.39	4.52	4.63	4.73	4.82	4.90	4.98	5.04	5.11	5.16	5.22	5.27	5.31	5.36
60	2.83	3.40	3.74	3.98	4.16	4.31	4.44	4.55	4.65	4.73	4.81	4.88	4.94	5.00	5.06	5.11	5.15	5.20	5.24
120	2.80	3.36	3.68	3.92	4.10	4.24	4.36	4.47	4.56	4.64	4.71	4.78	4.84	4.90	4.95	5.00	5.04	5.09	5.13
*	2.77	3.31	3.63	3.86	4.03	4.17	4.29	4.39	4.47	4.55	4.62	4.68	4.74	4.80	4.85	4.89	4.93	4.97	5.01

(continued)

Table 9 (Continued)

$\alpha = .01$

Degrees of Freedom	Number of populations																		
	2	3	4	5	6	7	8	9	10	11	12	13	14	15	16	17	18	19	20
1	90.0	135.0	164.0	186.0	202.0	216.0	227.0	237.0	246.0	253.0	260.0	266.0	272.0	277.0	282.0	286.0	290.0	294.0	298.0
2	14.0	19.0	22.3	24.7	26.6	28.2	29.5	30.7	31.7	32.6	33.4	34.1	34.8	35.4	36.0	36.5	37.0	37.5	37.9
3	8.26	10.6	12.2	13.3	14.2	15.0	15.6	16.2	16.7	17.1	17.5	17.9	18.2	18.5	18.8	19.1	19.3	19.5	19.8
4	6.51	8.12	9.17	9.96	10.6	11.1	11.5	11.9	12.3	12.6	12.8	13.1	13.3	13.5	13.7	13.9	14.1	14.2	14.4
5	5.70	6.97	7.80	8.42	8.91	9.32	9.67	9.97	10.2	10.5	10.7	10.9	11.1	11.2	11.4	11.6	11.7	11.8	11.9
6	5.24	6.33	7.03	7.56	7.97	8.32	8.61	8.87	9.10	9.30	9.49	9.65	9.81	9.95	10.1	10.2	10.3	10.4	10.5
7	4.95	5.92	6.54	7.01	7.37	7.68	7.94	8.17	8.37	8.55	8.71	8.86	9.00	9.12	9.24	9.35	9.46	9.55	9.65
8	4.74	5.63	6.20	6.63	6.96	7.24	7.47	7.68	7.87	8.03	8.18	8.31	8.44	8.55	8.66	8.76	8.85	8.94	9.03
9	4.60	5.43	5.96	6.35	6.66	6.91	7.13	7.32	7.49	7.65	7.78	7.91	8.03	8.13	8.23	8.32	8.41	8.49	8.57
10	4.48	5.27	5.77	6.14	6.43	6.67	6.87	7.05	7.21	7.36	7.48	7.60	7.71	7.81	7.91	7.99	8.07	8.15	8.22
11	4.39	5.14	5.62	5.97	6.25	6.48	6.67	6.84	6.99	7.13	7.25	7.36	7.46	7.56	7.65	7.73	7.81	7.88	7.95
12	4.32	5.04	5.50	5.84	6.10	6.32	6.51	6.67	6.81	6.94	7.06	7.17	7.26	7.36	7.44	7.52	7.59	7.66	7.73
13	4.26	4.96	5.40	5.73	5.98	6.19	6.37	6.53	6.67	6.79	6.90	7.01	7.10	7.19	7.27	7.34	7.42	7.48	7.55
14	4.21	4.89	5.32	5.63	5.88	6.08	6.26	6.41	6.54	6.66	6.77	6.87	6.96	7.05	7.12	7.20	7.27	7.33	7.39
15	4.17	4.83	5.25	5.56	5.80	5.99	6.16	6.31	6.44	6.55	6.66	6.76	6.84	6.93	7.00	7.07	7.14	7.20	7.26
16	4.13	4.78	5.19	5.49	5.72	5.92	6.08	6.22	6.35	6.46	6.56	6.66	6.74	6.82	6.90	6.97	7.03	7.09	7.15
17	4.10	4.74	5.14	5.43	5.66	5.85	6.01	6.15	6.27	6.38	6.48	6.57	6.66	6.73	6.80	6.87	6.94	7.00	7.05
18	4.07	4.70	5.09	5.38	5.60	5.79	5.94	6.08	6.20	6.31	6.41	6.50	6.58	6.65	6.72	6.79	6.85	6.91	6.96
19	4.05	4.67	5.05	5.33	5.55	5.73	5.89	6.02	6.14	6.25	6.34	6.43	6.51	6.58	6.65	6.72	6.78	6.84	6.89
20	4.02	4.64	5.02	5.29	5.51	5.69	5.84	5.97	6.09	6.19	6.29	6.37	6.45	6.52	6.59	6.65	6.71	6.76	6.82
24	3.96	4.54	4.91	5.17	5.37	5.54	5.69	5.81	5.92	6.02	6.11	6.19	6.26	6.33	6.39	6.45	6.51	6.56	6.61
30	3.89	4.45	4.80	5.05	5.24	5.40	5.54	5.65	5.76	5.85	5.93	6.01	6.08	6.14	6.20	6.26	6.31	6.36	6.41
40	3.82	4.37	4.70	4.93	5.11	5.27	5.39	5.50	5.60	5.69	5.77	5.84	5.90	5.96	6.02	6.07	6.12	6.17	6.21
60	3.76	4.28	4.60	4.82	4.99	5.13	5.25	5.36	5.45	5.53	5.60	5.67	5.73	5.79	5.84	5.89	5.93	5.98	6.02
120	3.70	4.20	4.50	4.71	4.87	5.01	5.12	5.21	5.30	5.38	5.44	5.51	5.56	5.61	5.66	5.71	5.75	5.79	5.83
∞	3.64	4.12	4.40	4.60	4.76	4.88	4.99	5.08	5.16	5.23	5.29	5.35	5.40	5.45	5.49	5.54	5.57	5.61	5.65

Appendix C Summation Notation

Summations

Definition

$$\sum_{i=1}^{n} x_i = x_1 + x_2 + \cdots + x_n \tag{C.1}$$

Example for $x_1 = 5$, $x_2 = 8$, $x_3 = 14$:

$$\sum_{i=1}^{3} x_i = x_1 + x_2 + x_3$$
$$= 5 + 8 + 14 = 27$$

Result 1
For a constant c:

$$\sum_{i=1}^{n} c = (c + c + \cdots + c) = nc \tag{C.2}$$
$$n \text{ times}$$

Example for $c = 5$, $n = 10$:

$$\sum_{i=1}^{10} 5 = 10(5) = 50$$

Example for $c = \bar{x}$

$$\sum_{i=1}^{n} \bar{x} = n\bar{x}$$

Result 2

$$\sum_{i=1}^{n} cx_i = cx_1 + cx_2 + \cdots + cx_n$$
$$= c(x_1 + x_2 + \cdots + x_n) = c\sum_{i=1}^{n} x_i \tag{C.3}$$

Example for $x_1 = 5$, $x_2 = 8$, $x_3 = 14$, $c = 2$:

$$\sum_{i=1}^{3} 2x_i = 2\sum_{i=1}^{3} x_i = 2(27) = 54$$

Result 3

$$\sum_{i=1}^{n}(ax_i + by_i) = a\sum_{i=1}^{n}x_i + b\sum_{i=1}^{n}y_i \qquad \text{(C.4)}$$

Example for $x_1 = 5$, $x_2 = 8$, $x_3 = 14$, $a = 2$, $y_1 = 7$, $y_2 = 3$, $y_3 = 8$, $b = 4$:

$$\sum_{i=1}^{3}(2x_i + 4y_i) = 2\sum_{i=1}^{3}x_i + 4\sum_{i=1}^{3}y_i$$
$$= 2(27) + 4(18)$$
$$= 54 + 72 = 126$$

Double Summations

Consider the following data involving the variable x_{ij}, where i is the subscript denoting the row position and j is the subscript denoting the column position:

		Column		
		1	2	3
Row	1	$x_{11} = 10$	$x_{12} = 8$	$x_{13} = 6$
	2	$x_{21} = 7$	$x_{22} = 4$	$x_{23} = 12$

Definition

$$\sum_{i=1}^{n}\sum_{j=1}^{m}x_{ij} = (x_{11} + x_{12} + \cdots + x_{1m}) + (x_{21} + x_{22} + \cdots + x_{2m})$$
$$+ (x_{31} + x_{32} + \cdots + x_{3m}) + \cdots + (x_{n1} + x_{n2} + \cdots + x_{nm}) \quad \text{(C.5)}$$

Example:

$$\sum_{i=1}^{2}\sum_{j=1}^{3}x_{ij} = x_{11} + x_{12} + x_{13} + x_{21} + x_{22} + x_{23}$$
$$= 10 + 8 + 6 + 7 + 4 + 12 = 47$$

Definition

$$\sum_{i=1}^{n}x_{ij} = x_{1j} + x_{2j} + \cdots + x_{nj} \qquad \text{(C.6)}$$

Example:

$$\sum_{i=1}^{2}x_{i2} = x_{12} + x_{22}$$
$$= 8 + 4 = 12$$

Shorthand Notation

Sometimes when a summation is for all values of the subscript, we use the following shorthand notations:

$$\sum_{i=1}^{n} x_i = \sum_i x_i \tag{C.7}$$

$$\sum_{i=1}^{n} \sum_{j=1}^{m} x_{ij} = \sum \sum x_{ij} \tag{C.8}$$

$$\sum_{i=1}^{n} x_{ij} = \sum_i x_{ij} \tag{C.9}$$

Appendix D Answers to Even-numbered Exercises

Chapter 1

Solutions

2 a. 10
   b. 4
   c. Country is a qualitative variable; hot list ranking, number of rooms and room rate are quantitative variables.
   d. Country is nominal; room rate and hot list ranking are ordinal; number of rooms and room rate are ratio.

4 a. 8
   b. All brands of audio systems manufactured.
   c. Average output power = 1130/8 = 141.25 watts

6 Questions a, c, and d provide quantitative data. Questions b and e provide qualitative data.

8 a. Quantitative; ratio.
   b. Qualitative; nominal.
   c. Qualitative; ordinal.
   d. Qualitative; nominal.
   e. Quantitative; ratio.

10 Cross-sectional data.

12 a. (48/120)100 per cent = 40 per cent in the sample died from some form of heart disease. This can be used as an estimate of the percentage of all males 60 or older who die of heart disease.
   b. The data on cause of death is qualitative.

Chapter 2

Solutions

2 a. $1 - (0.22 + 0.18 + 0.40) = 0.20$
   b. $0.20(200) = 40$
   c/d.

Class	Frequency	Percentage frequency
A	0.22(200) = 44	0.22(100) = 22
B	0.18(200) = 36	0.18(100) = 18
C	0.40(200) = 80	0.40(100) = 40
D	0.20(200) = 40	0.20(100) = 20
Total	200	100

4 a. Qualitative
   b.

Show	Frequency	Relative frequency	Percentage frequency
The X Factor	19	0.38	38
Coronation Street	12	0.24	24
A Touch of Frost	7	0.14	14
Strictly Come Dancing	12	0.24	24
	50	1.00	100

   c.

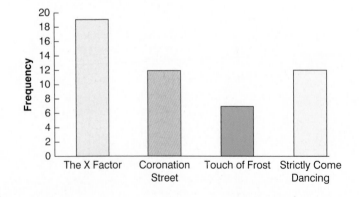

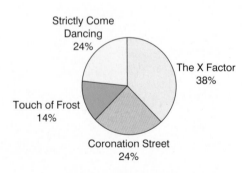

870

d. Most popular was *The X Factor*. Tied second were *Coronation Street* and *Strictly Come Dancing*.

6   a/b.

Starting time	Frequency	Percentage frequency
7:00	3	$100 \times (3/20) = 15$
7:30	4	$100 \times (4/20) = 20$
8:00	4	$100 \times (4/20) = 20$
8:30	7	$100 \times (7/20) = 35$
9:00	2	$100 \times (2/20) = 10$
	20	100

c.

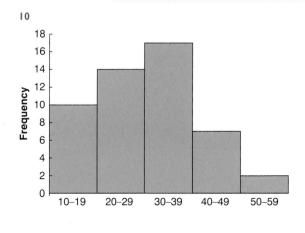

d.

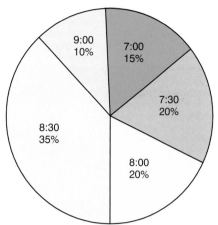

e. The most preferred starting time is 8:30 a.m. Starting times of 7:30 and 8:00 a.m. are next.

8   a/b.

Class	Frequency	Relative frequency	Percentage frequency
12–14	2	$100 \times (2/40) = 0.050$	$100 \times 0.050 = 5.0$
15–17	8	$100 \times (8/40) = 0.200$	$100 \times 0.200 = 20.0$
18–20	11	$100 \times (11/40) = 0.275$	$100 \times 0.275 = 27.5$
21–23	10	$100 \times (10/40) = 0.250$	$100 \times 0.250 = 25.5$
24–26	9	$100 \times (9/40) = 0.225$	$100 \times 0.225 = 22.5$
Total	40	1.000	100.0

10

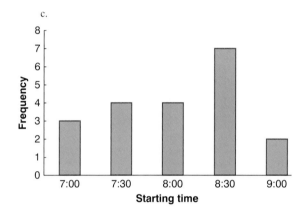

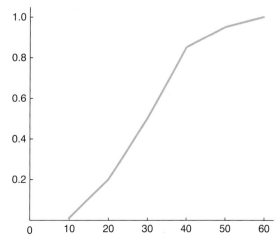

12 Unordered stem-and-leaf:

```
5 | 8  7
6 | 4  8  5
7 | 0  2  5  6  5  8  2
8 | 3  0  2  5
```

Ordered stem and leaf

```
5 | 7  8
6 | 4  5  8
7 | 0  2  2  5  5  6  8
8 | 0  2  3  5
```

14 a/b.

Waiting time	Frequency	Relative frequency
0–4	4	0.20
5–9	8	0.40
10–14	5	0.25
15–19	2	0.10
20–24	1	0.05
Totals	20	1.00

c/d.

Waiting time	Cumulative frequency	Cumulative relative frequency
Less than or equal to 4	4	0.20
Less than or equal to 9	12	0.60
Less than or equal to 14	17	0.85
Less than or equal to 19	19	0.95
Less than or equal to 24	20	1.00

e. 12/20 = 0.60

16 a.

Closing Price (€)	Frequency	Relative frequency
0–9.99	9	0.225
10–19.99	10	0.250
20–29.99	5	0.125
30–39.99	11	0.275
40–49.99	2	0.050
50–59.99	2	0.050
60–69.99	0	0.000
70–79.99	1	0.025
Totals	40	1.000

b.

Closing Price (€)	Cumulative frequency	Cumulative Relative frequency
Less than or equal to 9.99	9	0.225
Less than or equal to 19.99	19	0.475
Less than or equal to 29.99	24	0.600
Less than or equal to 39.99	35	0.875
Less than or equal to 49.99	37	0.925
Less than or equal to 59.99	39	0.975
Less than or equal to 69.99	39	0.975
Less than or equal to 79.99	40	1.000

c.

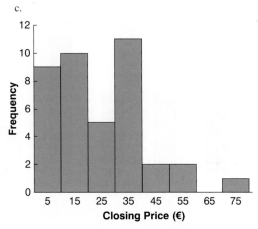

d. Over 87 per cent of shares trade for less than €40 a share and 60 per cent trade for less than €30 per share.

18 a/b.

Computer usage (hours)	Frequency	Relative frequency
0.0–2.9	5	(5/50) = 0.10
3.0–5.9	28	(28/50) = 0.56
6.0–8.9	8	(8/50) = 0.16
9.0–11.9	6	(6/50) = 0.12
12.0–14.9	3	(3/50) = 0.06
Total	50	1.00

c.

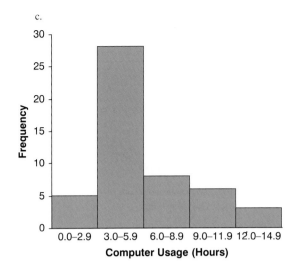

d.

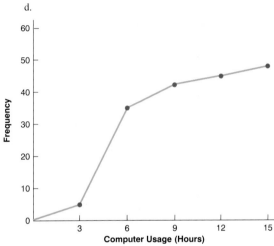

e. The majority of the computer users are in the 3 to 6 hour range. Usage is somewhat skewed toward the right with three users in the 12 to 15 hour range.

20 a.

	Y		
	1	2	Total
A	5	0	5
X **B**	11	2	13
C	2	10	12
Total	18	12	30

b.

	Y		
	1	2	Total
A	100.0	0.0	100.0
X **B**	84.6	15.4	100.0
C	16.7	83.3	100.0

c.

	Y	
	1	2
A	27.8	0.0
X **B**	61.1	16.7
C	11.1	83.3
Total	100.0	100.0

d. $X = A$ values are always paired with $Y = 1$

$X = B$ values are most often paired with $Y = 1$

$X = C$ values are most often paired with $Y = 2$

22 a. The cross-tabulation of condition of the greens by gender is below.

	Green Condition		
Gender	Too fast	Fine	Total
Male	35	65	100
Female	40	60	100
Total	75	125	200

The female golfers have the highest percentage saying the greens are too fast: 40 per cent.

b. 10 per cent of the women think the greens are too fast. 20 per cent of the men think the greens are too fast. So, for the low handicappers, the men have a higher percentage who think the greens are too fast.

c. 43 per cent of the woman think the greens are too fast. 50 per cent of the men think the greens are too fast. So, for the high handicappers, the men have a higher percentage who think the greens are too fast.

d. This is an example of Simpson's Paradox. At each handicap level a smaller percentage of the women think the greens are too fast. But, when the cross-tabulations are aggregated, the result is reversed and we find a higher percentage of women who think the greens are too fast. The hidden variable explaining the reversal is handicap level. Fewer people with low handicaps think the greens are too fast, and there are more men with low handicaps than women.

24 a.

	House type			
		Semi-		
	Detached	detached	Terraced	Total
Bedrooms 2	1	2	3	6
3	7	1	3	11
4	22	1	3	26
5	4	0	1	5
6	2	0	0	2
Total	36	4	10	50

b. See 'Total' column above.

c. See 'Total' row above.

d. The frequency distributions in (b) and (c) are formed from the row and column totals respectively of the cross-tabulation.

Chapter 3

Solutions

2   Mean = 16, median = 16.5

4   Mean

$$= \frac{53 + 55 + 70 + 58 + 64 + 57 + 53 + 69 + 57 + 68 + 53}{11}$$

$$= 59.7$$

Ordered data set: 53 53 53 55 57 57 58 64 68 69 70

Median is the 6th ordered value = 57

Mode is the most frequent value = 53 (occurs 3 times)

6   a.   Mean $= \dfrac{88.3 + 4.3 + \cdots + 63.6}{30} = 46.0$

     b.   Ordered data set is: 0.0 0.0 0.0 0.0 4.3 4.4 4.6 7.0 7.6 9.2 17.5 28.8 29.1 34.9 45.0 52.9 53.3 56.6 63.6 64.5 65.1 67.9 70.4 81.7 85.4 88.3 94.2 98.9 99.2 145.6

        Median is midway between the 15th and 16th ordered values $= \dfrac{45.0 + 52.9}{2} = 48.95$

     c.   Index for lower quartile is $\dfrac{25}{100}30 = 7.5$, rounded up to 8. Lower quartile = 7.0

        Index for upper quartile is $\dfrac{75}{100}30 = 22.5$, rounded up to 23. Upper quartile = 70.4

     d.   Index for 40th percentile is $\dfrac{40}{100}30 = 12$, so 40th percentile is midway between the 12th and 13th ordered values $= \dfrac{28.8 + 29.1}{2} = 28.95$. This implies that at least 40 per cent of values are less than or equal 28.95, and at least 60 per cent are greater than or equal to 28.95.

8   a.   Mean = 34.75. The modal age is 29; it appears three times.

     b.   Median = 34.5. Data suggest at-home workers are slightly younger.

     c. $Q_1 = 25.5$. $Q_3 = 43.5$

     d.   32nd percentile = 27. Approximately 32 per cent are aged under 27.

10 $\bar{x} = \dfrac{\Sigma x_i}{n} = \dfrac{10 + 20 + 12 + 17 + 16}{5} = \dfrac{75}{5} = 15$

      $s^2 = \dfrac{\Sigma(x_i - \bar{x})^2}{n - 1} = \dfrac{(-5)^2 + (5)^2 + (-3)^2 + (2)^2 + (1)^2}{4}$

      $= \dfrac{64}{4} = 16$

      $s = \sqrt{16} = 4$

12   a.   Range = Maximum − minimum = 45 − 34 = 11

      b.   Mean $= \dfrac{\Sigma x_i}{n} = \dfrac{41 + 34 + 42 + 45 + 35 + 37}{6}$

         $= \dfrac{234}{6} = 39$

       Variance $= \dfrac{\Sigma(x_i - \bar{x})^2}{n - 1}$

         $= \dfrac{(2)^2 + (-5)^2 + (3)^2 + (6)^2 + (4)^2 + (-2)^2}{5}$

         $= \dfrac{94}{5} = 18.8$

      c.   Standard deviation $= \sqrt{18.8} = 4.34$

      d.   Coefficient of variation $= 100 \times \dfrac{4.34}{39} = 11.1\%$

14   Dawson Supply: range = 2, s = 0.67

      J.C. Clark: range = 8, s = 2.58

16 $s^2 = 0.0021$. Production should not be shut down, because the variance is less than 0.005.

18   520, $z = \dfrac{520 - 500}{100} = +0.20$

      650, $z = \dfrac{650 - 500}{100} = +1.50$

      500, $z = \dfrac{500 - 500}{100} = 0.00$

      450, $z = \dfrac{450 - 500}{100} = -0.50$

      280, $z = \dfrac{280 - 500}{100} = -2.20$

20   a. Approximately 95 per cent. b. Almost all.

      c. Approximately 68 per cent.

22   a.   IQ scores of 85 and 115 are respectively 1 standard below and above the mean. Using the empirical rule, approximately 68 per cent of scores are within 1 standard deviation from the mean.

      b.   Similarly, approximately 95 per cent of scores are within 2 standard deviations from the mean.

      c.   Approximately (100% − 95%)/2 = 2.5% of scores are over 130.

      d.   An IQ score of 45 is 3 standard deviations above the mean. Yes, the empirical rule suggests that almost all IQ scores are less than 145.

24   a.   Mean = 3.99, median = 4.185

      b. $Q_1 = 4.00$, $Q_3 = 4.50$

      c.   s = 0.8114

      d.   The distribution is markedly skewed to the left.

      e.   Allison One, z = 0.16; Omni Audio, z = −2.06

      f.   The lowest rating is for the Bose 501 Series, z = −2.28. This is not an outlier, so there are no outliers.

26

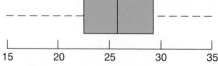

28   Lower limit = 30, upper limit = 62, 65 is an outlier.

30 a. Ordered data set is: 5.0 5.1 6.2 8.6 9.0 9.2 9.6 11.2 11.4 12.2 12.3 12.8 14.5 14.7 15.8 16.6 17.3 17.3 19.2 19.6 22.9 30.3 31.1 41.6 52.7

Five number summary: 5 (min) 9.6 (7th) 14.5 (13th) 19.2 (19th) 52.7 (max)

b. IQR $= Q_3 - Q_1 = 19.2 - 9.6 = 9.6$

Lower Limit:
$$Q_1 - 1.5 \,(\text{IQR}) = 9.6 - 1.5(9.6) = -4.8$$

Upper Limit:
$$Q_3 + 1.5(\text{IQR}) = 19.2 + 1.5(9.6) = 33.6$$

c. The data value 41.6 is an outlier (larger than the upper limit) and so is the data value 52.7. The financial analyst should first verify that these values are correct. Perhaps a typing error has caused 25.7 to be typed as 52.7 (or 14.6 to be typed as 41.6). If the outliers are correct, the analyst might consider these companies with an unusually large return on equity as good investment candidates.

d.

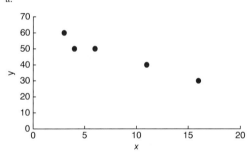

32 a.

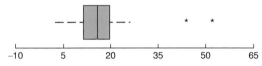

b. The scatter diagram suggests a negative relationship between the two variables.

c/d. $\Sigma x_i = 40 \quad \bar{x} = \dfrac{40}{5} = 8 \quad \Sigma y_i = 230 \quad \bar{y} = \dfrac{230}{5} = 46$

$\Sigma(x_i - \bar{x})(y_i - \bar{y}) = -240 \quad \Sigma(x_i - \bar{x})^2 = 118$
$\Sigma(y_i - \bar{y})^2 = 520$

$$s_{XY} = \frac{\Sigma(x_i - \bar{x})(y_i - \bar{y})}{n - 1} = \frac{-240}{5 - 1} = -60$$

$$s_X = \sqrt{\frac{\Sigma(x_i - \bar{x})^2}{n - 1}} = \sqrt{\frac{118}{5 - 1}} = 5.4314$$

$$s_Y = \sqrt{\frac{\Sigma(y_i - \bar{y})^2}{n - 1}} = \sqrt{\frac{520}{5} - 1} = 11.4018$$

$$r_{XY} = \frac{s_{XY}}{s_X s_Y} = \frac{-60}{(5.4314)(11.4018)} = -0.969$$

There is a strong negative linear relationship. The sample covariance and sample correlation coefficient are both negative. The sample correlation coefficient is quite close to -1.

34 a.

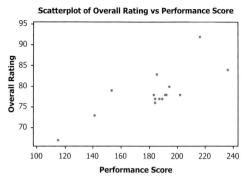

b. The scatter diagram suggests a positive relationship between the two variables.

c. $s_{XY} = 123.12$. The covariance is positive, confirming the impression given by the scatter diagram.

d. $r_{XY} = 0.78$. The correlation coefficient indicates quite a strong positive linear relationship between the two variables.

e. High performance scores tend to be paired with high overall ratings, and low performance scores tend to be paired with high overall ratings.

36 a. $\bar{x} = \dfrac{\Sigma w_i x_i}{\Sigma w_i} = \dfrac{6(3.2) + 3(2) + 2(2.5) + 8(5)}{6 + 3 + 2 + 8}$

$= \dfrac{70.2}{19} = 3.69$

b. $\dfrac{3.2 + 2 + 2.5 + 5}{4} = \dfrac{12.7}{4} = 3.175$

38 a. 17.62 per cent

b. 1.25 per cent

Chapter 4

Solutions

2 $^6C_3 = \dfrac{6!}{3!} = \dfrac{6 \times 5 \times 4}{3 \times 2 \times 1} = 20$

ABC	ACE	BCD	BEF
ABD	ACF	BCE	CDE
ABE	ADE	BCF	CDF
ABF	ADF	BDE	CEF
ACD	AEF	BDF	DEF

4 a.

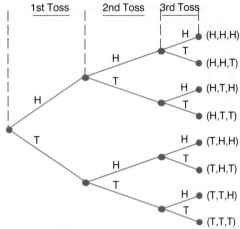

1st Toss | 2nd Toss | 3rd Toss

H — H — H (H,H,H)
H — H — T (H,H,T)
H — T — H (H,T,H)
H — T — T (H,T,T)
T — H — H (T,H,H)
T — H — T (T,H,T)
T — T — H (T,T,H)
T — T — T (T,T,T)

b. Let: H be head and T be tail

(H,H,H) (T,H,H)
(H,H,T) (T,H,T)
(H,T,H) (T,T,H)
(H,T,T) (T,T,T)

c. The outcomes are equally likely, so the probability of each outcome is 1/8.

6 $P(E_1) = 20/50 = 0.40$, $P(E_2) = 13/50 = 0.26$, $P(E_3) = 17/50 = 0.34$

The relative frequency method was used.

8 a. There are four outcomes possible for this 2-step experiment; planning commission positive – council approves; planning commission positive – council disapproves; planning commission negative – council approves; planning commission negative – council disapproves.

b. Let p = positive, n = negative, a = approves, and d = disapproves

Planning Commission Council

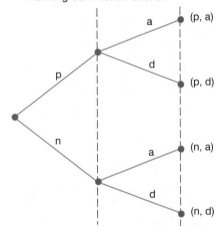

p — a (p, a)
p — d (p, d)
n — a (n, a)
n — d (n, d)

10 a. Choose a person at random. Have the person taste the four blends and state which is preferred 100 times.

b. Assign a probability of 1/4 to each blend. We use the classical method of equally likely outcomes here.

c.

Blend	Probability
1	0.20
2	0.30
3	0.35
4	0.15
Total	1.00

The relative frequency method was used.

12 a. $P(E_2) = 1/4$

b. $P(\text{any 2 outcomes}) = 1/4 + 1/4 = 1/2$

c. $P(\text{any 3 outcomes}) = 1/4 + 1/4 + 1/4 = 3/4$

14 a. (6) (6) = 36 sample points

b.

	Die 2					
	1	2	3	4	5	6
1	2	3	4	5	6	7
2	3	4	5	6	7	8
3	4	5	6	7	8	9
4	5	6	7	8	9	10
5	6	7	8	9	10	11
6	7	8	9	10	11	12

Die 1 ← Total for Both

c. 6/36 = 1/6

d. 10/36 = 5/18

e. No. $P(\text{odd}) = 18/36 = P(\text{even}) = 18/36$ or 1/2 for both.

f. Classical. A probability of 1/36 is assigned to each experimental outcome.

16 a. $P(0) = 0.10$

b. $P(\text{4 or 5}) = 0.25$

c. $P(0, 1, \text{or } 2) = 0.55$

18 a. $P(A) = 0.40$, $P(B) = 0.40$, $P(C) = 0.60$

b. $P(A \cup B) = P(E_1, E_2, E_3, E_4) = 0.80$. Yes

$P(A \cup B) = P(A) + P(B)$.

c. $\bar{A} = \{E_3, E_4, E_5\}$ $\bar{C} = \{E_1, E_4\}$ $P(\bar{A}) = 0.60$

$P(\bar{C}) = 0.40$

d. $A \cup \bar{B} = \{E_1, E_2, E_5\}$ $P(A \cup \bar{B}) = 0.60$

e. $P(B \cup C) = P(E_2, E_3, E_4, E_5) = 0.80$

20 Let: **B** = rented a car for business reasons

P = rented a car for personal reasons

a. $P(B \cup P) = P(B) + P(P) - P(B \cap P)$

$= 0.54 + 0.458 - 0.30 = 0.698$

b. $P(\text{Neither}) = 1 - 0.698 = 0.302$

22 a. $P(A \cap B) = 0$

b. $P(A \mid B) = \dfrac{P(A \cap B)}{P(B)} = 0/0.4 = 0$

c. No. $P(A \mid B) \neq P(A)$; ∴ the events, although mutually exclusive, are not independent.

d. Mutually exclusive events are dependent.

24 a.

	Reason for applying			
	Quality	Cost/ convenience	Other	Total
Full Time	0.218	0.204	0.039	0.461
Part Time	0.208	0.307	0.024	0.539
	0.426	0.511	0.063	1.00

   b. It is most likely a student will cite cost or convenience as the first reason – probability 0.511.

   c. $P(\text{Quality} \mid \text{full time}) = 0.218/0.461 = 0.473$

   d. $P(\text{Quality} \mid \text{part time}) = 0.208/0.539 = 0.386$

   e. For independence, we must have $P(A)P(B) = P(A \cap B)$.
      From the table, $P(A \cap B) = 0.218$, $P(A) = 0.461$, $P(B) = 0.426$

      $P(A)P(B) = (0.461)(0.426) = 0.196$

      Since $P(A)P(B) \neq P(A \cap B)$, the events are not independent.

26 a. $P(A \cap B) = P(A)P(B) = (0.55)(0.35) = 0.19$

   b. $P(A \cup B) = P(A) + P(B) - P(A \cap B) = 0.55 + 0.35 - 0.19 = 0.71$

   c. $P(\text{shutdown}) = 1 - P(A \cup B) = 1 - 0.71 = 0.29$

28 a. $P(B \cap A_1) = P(A_1)P(B \mid A_1) = (0.20)(0.50) = 0.10$
      $P(B \cap A_2) = P(A_2)P(B \mid A_2) = (0.50)(0.40) = 0.20$
      $P(B \cap A_3) = P(A_3)P(B \mid A_3) = (0.30)(0.30) = 0.09$

   b. $P(A_2 \mid B) = \dfrac{0.20}{0.10 \ 0.20 \ 0.09} = 0.51$

   c.

Events	$P(A_i)$	$P(B \mid A_i)$	$P(A_i \cap B)$	$P(A_i \mid B)$
A_1	0.20	0.50	0.10	0.26
A_2	0.50	0.40	0.20	0.51
A_3	0.30	0.30	0.09	0.23
	1.00		0.39	1.00

30 M = missed payment
   D_1 = customer defaults
   D_2 = customer does not default
   $P(D_1) = 0.05 \ P(D_2) = 0.95 \ P(M \mid D_2) = 0.2$
   $P(M \mid D_1) = 1$

   a.

$$P(D_1 \mid M) = \frac{P(D_1)P(M \mid D_1)}{P(D_1)P(M \mid D_1) \ P(D_2)P(M \mid D_2)}$$

$$\frac{(0.05)(1)}{(0.05)(1) \ (0.95)(0.2)} = \frac{0.05}{0.24} = 0.21$$

   b. Yes, the probability of default is greater than 0.20.

32 C = Company belongs to the Consumer industry
   B = Company belongs to the Banking industry
   PE9 = P/E value greater than 9
   PE1519 = P/E value falls in the 15–19 range

   a. $P(\text{PE9} \cup C) = 46/100 = 0.46$

   b. $P(B \mid \text{PE1519}) = 12/30 = 0.4$

Chapter 5

Solutions

2 a. Let x = time (in minutes) to assemble the product.

   b. It may assume any positive value: $x > 0$.

   c. Continuous

4 $x = 0, 1, 2, \ldots, 12$.

6 a. values: $0, 1, 2, \ldots, 20$
           discrete

   b. values: $0, 1, 2, \ldots$
           discrete

   c. values: $0, 1, 2, \ldots, 50$
           discrete

   d. values: $0 \leq x \leq 8$
           continuous

   e. values: $x > 0$
           continuous

8 a.

x	$p(x)$
1	3/20 = 0.15
2	5/20 = 0.25
3	8/20 = 0.40
4	4/20 = 0.20
Total	1.00

   b.

   c. $p(x) \geq 0$ for $x = 1, 2, 3, 4$.
      $\Sigma p(x) = 1$

10 a.

Duration of call	
x	p(x)
1	0.25
2	0.25
3	0.25
4	0.25
	1.00

b.

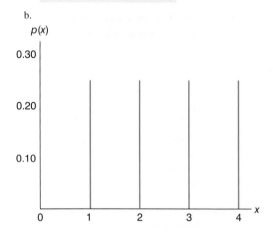

p(x)

c. $p(x) \geq 0$ and $p(1) + p(2) + p(3) + p(4) =$
$$0.25 + 0.25 + 0.25 + 0.25 = 1.00$$

d. $p(3) = 0.25$

e. P(overtime) $= p(3) + p(4) = 0.25 + 0.25 = 0.50$

12 a. Yes, since $p(x) \geq 0$ for $x = 1, 2, 3$ and
$\Sigma p(x) = p(1) + p(2) + p(3) = 1/6 + 2/6 +$
$3/6 = 1$

b. $p(2) = 2/6 = 0.333$

c. $p(2) + p(3) = 2/6 + 3/6 = 0.833$

14 a.

x	p(x)	xp(x)
3	0.25	0.75
6	0.50	3.00
9	0.25	2.25
	1.00	6.00

$E(x) = \mu = 6.00$

b.

x	x − μ	(x − μ)²	p(x)	(x − μ)² p(x)
3	−3	9	0.25	2.25
6	0	0	0.50	0.00
9	3	9	0.25	2.25
				4.50

$Var(x) = \sigma^2 = 4.50$

c. $\sigma = \sqrt{4.50} = 2.12$

16 a/b.

x	p(x)	xp(x)	x − μ	(x − μ)²	(x − μ)²p(x)
0	0.10	0.00	−2.45	6.0025	0.600250
1	0.15	0.15	−1.45	2.1025	0.315375
2	0.30	0.60	−0.45	0.2025	0.060750
3	0.20	0.60	0.55	0.3025	0.060500
4	0.15	0.60	1.55	2.4025	0.360375
5	0.10	0.50	2.55	6.5025	0.650250
		2.45			2.047500

$E(x) = \mu = 2.45$
$\sigma^2 = 2.0475$
$\sigma = 1.4309$

18 a.

x	p(x)	xp(x)
0	0.90	0.00
400	0.04	16.00
1000	0.03	30.00
2000	0.01	20.00
4000	0.01	40.00
6000	0.01	60.00
	1.00	166.00

$E(x) = 166$. If the company charged a premium of
€166.00 they would break even.

b.

Gain to policy holder	p (Gain)	Gain x p (Gain)
−260.00	0.90	−234.00
140.00	0.04	5.60
740.00	0.03	22.20
1740.00	0.01	17.40
3740.00	0.01	37.40
5740.00	0.01	57.40
		−94.00

$E(\text{gain}) = -94.00$. The policy holder is more concerned
that the big accident will break him than with the
expected annual loss of €94.00.

20 a. $E(x) = \Sigma xp(x) = 300(0.20) + 400(0.30)$
$+ 500(0.35) + 600(0.15) = 445$
The monthly order quantity should be 445 units.

b. Cost: 445 @ €50 = €22 250
Revenue: 300 @ €70 = €21 000
€1 250 Loss

22 a. $p(0) = 0.3487$
b. $p(2) = 0.1937$

c. $P(x \leq 2) = 0.9298$

d. $P(x \geq 1) = 0.6513$

e. $E(X) = n\pi = 10(0.1) = 1$

f. $Var(X) = 0.9$

$\sigma = \sqrt{0.9} = 0.9487$

24 a. Probability of a defective part being produced must be 0.03 for each part selected; parts must be selected independently.

b. Let: D = defective

G = not defective

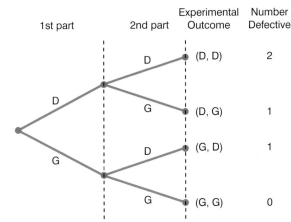

1st part	2nd part	Experimental Outcome	Number Defective
D	D	(D, D)	2
	G	(D, G)	1
G	D	(G, D)	1
	G	(G, G)	0

c. Two outcomes result in exactly one defect.

d. P (no defects) = 0.9409

P (1 defect) = 0.0582

P (2 defects) = 0.0009

26 a. $p(0) + p(1) + p(2) = 0.0115 + 0.0576$
$+ 0.1369 = 0.2060$

b. $p(4) = 0.2182$

c. $1 - [p(0) + p(1) + p(2) + p(3)]$
$= 1 - 0.2060 - 0.2054 = 0.5886$

d. $\mu = n\pi = 20(0.20) = 4$

28 a.

$$p(x) = \frac{3^x e^{-3}}{x!}$$

b.

$$p(2) = \frac{3^2 e^{-3}}{2!} = \frac{9(0.0498)}{2} = 0.2241$$

c.

$$p(1) = \frac{3^1 e^{-3}}{1!} = 3(0.0498) = 0.1494$$

d. $P(x \geq 2) = 1 - p(0) - p(1)$

$= 1 - 0.0498 - 0.1494 = 0.8008$

30 a. $\mu = 48(5/60) = 4$

$p(3) = 0.1952$

b. $\mu = 48(15/60) = 12$

$p(10) = 0.1048$

c. $\mu = 48(5/60) = 4.$

$p(0) = 0.0183$

d. $\mu = 48(3/60) = 2.4$

$p(0) = 0.0907$

32 a. $p(0) = e^{-10} = 0.000045$

b. $p(0) + p(1) + p(2) + p(3) = 0.010245$

c. 2.5 arrivals/15 sec. period Use $\mu = 2.5$

$p(0) = 0.0821$

d. 0.9179

34

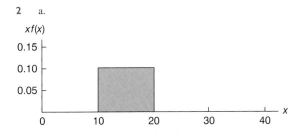

$$p(3) = \frac{\binom{4}{3}\binom{15 - 4}{10 - 3}}{\binom{15}{10}} = \frac{4 \times 330}{3003} = 0.4396$$

36 a. $p(2) = 0.3768$

b. $p(2) = 0.2826$

c. $p(5) = 0.0377$

d. $p(0) = 0.0087$

Chapter 6

Solutions

2 a.

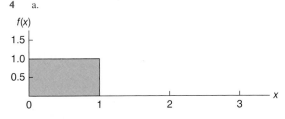

b. $P(X < 15) = 0.10(5) = 0.50$

c. $P(12 \leq X \leq 18) = 0.10(6) = 0.60$

d. $E(X) = \dfrac{10 + 20}{2} = 15$

e. $Var(X) = \dfrac{(20 - 10)^2}{12} = 8.33$

4 a.

f(x) graph

b. $P(0.25 < X < 0.75) = 1(0.50) = 0.50$

c. $P(X \leq 0.30) = 1(0.30) = 0.30$

d. $P(X > 0.60) = 1(0.40) = 0.40$

6 a. $P(300 \leq X \leq 301) = 0.5(1) = 0.5$

b. $P(X \geq 300.4) = 0.5(0.6) = 0.3$

c. $P(X < 299.6) + P(x > 300.4) = 0.5(0.6) + 0.5(0.6)$

Therefore, the probability is $0.30 + 0.30 = 0.60$.

8

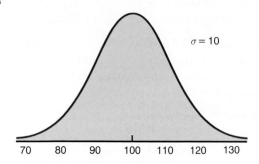

$\sigma = 10$

70 80 90 100 110 120 130

10

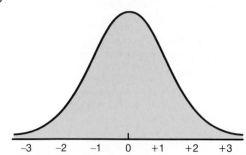

-3 -2 -1 0 +1 +2 +3

a. 0.3413

b. 0.4332

c. 0.4772

d. 0.4938

12 a. $P(0 \leq Z \leq 0.83) = P(Z \leq 0.83) - P(0 \leq Z) =$
0.7967 - 0.5000 = 0.2967

b. $P(-1.57 \leq Z \leq 0) = P(Z \leq 0) - P(-1.57 \leq Z)$
0.5000 - 0.0582 = 0.4418

c. 1 - 0.6700 = 0.3300

d. 1 - 0.4090 = 0.5910

e. 0.8849

f. 0.2389

14 a. $z = 1.96$.

b. $z = 0.61$.

c. $z = 1.12$.

d. $z = 0.44$.

16 a. Look in the table for an area of 1.0000 - 0.0100 =
0.9900. The area value in the table closest to 0.9900
provides the value $z = 2.33$.

b. Look in the table for an area of 1.0000 - 0.0250 =
0.9750. This corresponds to $z = 1.96$.

c. Look in the table for an area of 1.0000 - 0.0500 =
0.9500. Since 0.9500 is exactly halfway between 0.9495
($z = 1.64$) and 0.9505 ($z = 1.65$), we select $z = 1.645$.
However, $z = 1.64$ or $z = 1.65$ are also acceptable
answers.

d. Look in the table for an area of 1.0000 - 0.1000 =
0.9000. The area value in the table closest to 0.9000
provides the value $z = 1.28$.

18 We require the value $\mu + \sigma z$
where $\mu = 20000$, $\sigma = 1500$ $z = 0.84$, approximately.
Hence the minimum holding (£) is $20000 + 1500 \times 0.84 =$
21 260.

20 a. $\mu = n\pi = 100(0.20) = 20$
$\sigma^2 = n(1 - \pi) = 100(0.20)(0.80) = 16$
$\sigma = \sqrt{16} = 4$

b. Yes since $n\pi = 20$ and $n(1 - \pi) = 80$

c. $P(23.5 \leq X \leq 24.5)$
$$z = \frac{24.5 - 20}{4} = +1.13 \quad \text{Area} = 0.8708$$
$$z = \frac{23.5 - 20}{4} = +0.88 \quad \text{Area} = 0.8106$$
$P(23.5 \leq X \leq 24.5) = 0.8708 - 0.8106 = 0.0602$

d. $P(17.5 \leq X \leq 22.5)$
$$z = \frac{17.5 - 20}{4} = -0.63 \quad \text{Area} = 0.2643$$
$$z = \frac{22.5 - 20}{4} = +0.63 \quad \text{Area} = 0.7357$$
$P(17.5 \leq X \leq 22.5) = 0.7357 - 0.2643 = 0.4714$

e. $P(X \leq 15.5)$
$$z = \frac{15.5 - 20}{4} = -1.13 \quad \text{Area} = 0.1292$$
$P(X \leq 15.5) = 0.1292$

22 a. $\mu = n\pi = 120(0.75) = 90$
$\sigma = \sqrt{n\pi(1 - \pi)} = \sqrt{120 \times 0.75 \times 0.25} = 4.74$
The probability at least half the rooms are occupied is
the normal probability: $P(X \geq 59.5)$. At $x = 59.5$
$$z = \frac{59.5 - 90}{4.74} = 6.43$$
Therefore, probability is approximately 1

b. Find the normal probability: $P(X \geq 99.5)$ At $x = 99.5$
$$z = \frac{99.5 - 90}{4.74} = 2.00$$
$P(X \geq 99.5) = P(z \geq 2.00) = 1.0000 - 0.9772$
$= 0.0228$

c. Find the normal probability: $P(X \leq 80.5)$ At $x = 80.5$
$$z = \frac{80.5 - 90}{4.74} = -2.00$$
$P(X \leq 80.5) = P(Z \leq -2.00) = 1.0000 - 0.9772$
$= 0.0228$

24 a. $P(X \leq x_0) = 1 - e^{-x_0/3}$

b. $P(X \leq 2) = 1 - e^{-2/3} = 1 - 0.5134 = 0.4866$

c. $P(X \geq 3) = 1 - P(X \leq 3) = 1 - (1 - e^{-3/3})$
$= e^{-1} = 0.3679$

d. $P(X \leq 5) = 1 - e^{-5/3} = 1 - 0.1889 = 0.8111$

e. $P(2 \leq X \leq 5) = P(X \leq 5) - P(X \leq 2)$
$= 0.8111 - 0.4866 = 0.3245$

26 a.

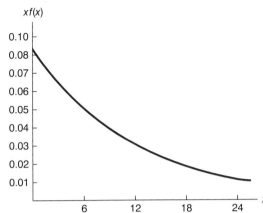

6 Finite, infinite, infinite, infinite, finite

8 a. $p = 75/150 = 0.50$ b. $p = 55/150 = 0.3667$

10 a. 0.19   b. 0.32   c. 0.79

12 a. $\bar{x} = \dfrac{4376 + 5578 + 2717 + 4929 + 4495 + 4798 + 6446 + 4119 + 4237 + 3814}{10} = \dfrac{45\ 500}{10} = 4550$

b. $s = \sqrt{\dfrac{\Sigma (x_i - \bar{x})^2}{n - 1}}$

$= \sqrt{\dfrac{(-174)^2 + (1028)^2 + (-1833)^2 + (370)^2 + (-55)^2 + (248)^2 + (1896)^2 + (-431)^2 + (-313)^2 + (-736)^2}{9}}$

$= \sqrt{\dfrac{9\ 068\ 620}{9}} = \sqrt{1\ 007\ 624} = 1004$

14 a. The sampling distribution is normal with:

$E(\bar{X}) = \mu = 200$

$\sigma_{\bar{x}} = \sigma/\sqrt{n} = 50/\sqrt{100} = 5$

For ± 5, $(\bar{x} - \mu) = 5$

$z = \dfrac{\bar{x} - \mu}{\sigma_{\bar{x}}} = \dfrac{5}{5} = 1$, cumulative probability for $z = 1$ is 0.8413, Required probability $= 2(0.8413 - 0.5000) = 0.6826$

b. For ± 10, $(\bar{x} - \mu) = 10$

$z = \dfrac{\bar{x} - \mu}{\sigma_{\bar{x}}} = \dfrac{10}{5} = 2$, cumulative probability for $z = 2$ is 0.9772, Required probability $= 2(0.9772 - 0.5000) = 0.9544$

16 a. 1.41
b. 1.41
c. 1.41
d. 1.34. Only case (d) requires the use of the fpc factor.

18 a. $\sigma_{\bar{x}} = \sigma / \sqrt{n} = 4000/\sqrt{60} = 516.40$

$z = \dfrac{52,300 - 51,800}{516.40} = +0.97$, cumulative probability for $z = 1.37$ is 0.8340, Required probability $= 2(0.8340 - 0.5000) = 0.6680$

b. $P(x \le 12) = 1 - e^{-12/12} = 1 - 0.3679 = 0.6321$
c. $P(x \le 6) = 1 - e^{-6/12} = 1 - 0.6065 = 0.3935$
d. $P(x \ge 30) = 1 - P(x < 30)$
$= 1 - (1 - e^{-30/12})$
$= 0.0821$

Chapter 7

Solutions

2 Elements 22, 147, 229, 289 are selected. (The numbers 98 601, 83 448, 27 553 are ignored, because 601, 448 and 553 are out of the 1 to 350 range. The number 84147 is ignored because 147 has already been selected.)

4 Elements 283, 610, 39, 254, 568, 353, 602, 421, 638, 164 are selected. (The numbers 816, 763, 980, 964 are ignored, because they are out of the 1 to 645 range.)

b. $\sigma_{\bar{x}} = \sigma / \sqrt{n} = 4000/\sqrt{120} = 365.15$

$z = \dfrac{52,300 - 51,800}{365.15} = +1.37$, cumulative probability for $z = 1.37$ is 0.9147, Required probability $= 2(0.9147 - 0.5000) = 0.8294$

20 a. Normal with $E(\bar{X}) = 95$ and $\sigma_{\bar{x}} = 2.56$
b. 0.7580
c. 0.8502
d. Part (c), larger sample size

22 a. No, $n/N = 0.01$
b. With, 1.29; without, 1.30. Including the fpc provides only a slightly different value for the standard error
c. 0.8764

24 $\sigma_P = \sqrt{\dfrac{\pi(1 - \pi)}{n}}$

For $n = 100$, $\sigma_P = \sqrt{\dfrac{(0.55)(0.45)}{100}} = 0.0497$

For $n = 200$, $\sigma_P = \sqrt{\dfrac{(0.55)(0.45)}{200}} = 0.0352$

For $n = 500$, $\sigma_P = \sqrt{\dfrac{(0.55)(0.45)}{500}} = 0.0222$

For $n = 1000$, $\sigma_P = \sqrt{\dfrac{(0.55)(0.45)}{1000}} = 0.0157$

σ_P decreases as n increases

26 a. Normal with mean = 0.30, standard deviation = 0.0458

b. 0.9708

c. 0.7242

28 a. 0.8882

b. 0.0233

30 $\sigma_p = \sqrt{\dfrac{\pi(1-\pi)}{n}} = \sqrt{\dfrac{(0.40)(0.60)}{400}} = 0.0245$

$P(p \geq 0.375) = ?$

$z = \dfrac{0.375 - 0.40}{0.0245} = -1.02$, cumulative probability for

$z = -1.02$ is 0.1539

$P(p \geq 0.375) = 1 - 0.1539 = 0.8461$

Chapter 8

Solutions

2 The confidence interval is $\bar{x} \pm z_{\alpha/2} \dfrac{\sigma}{\sqrt{n}}$. Here, $\bar{x} = 32$, $\sigma = 6$, $n = 50$

a. $z_{\alpha/2} = 1.645$

The confidence interval is $32 \pm 1.645 (6/\sqrt{50}) = 32 \pm 1.4$ or 30.6 to 33.4

b. $z_{\alpha/2} = 1.96$

$32 \pm 1.96 (6/\sqrt{50}) = 32 \pm 1.66$ or 30.34 to 33.66

c. $z_{\alpha/2} = 2.576$

The confidence interval is $32 \pm 2.576 (6/\sqrt{50}) = 32 \pm 2.19$ or 29.81 to 34.19

4 $n = 54$

6 The margin of error is $z_{\alpha/2} \dfrac{\sigma}{\sqrt{n}}$. Here, $\sigma = 4000$, $n = 60$

The margin of error is $1.96(4000/\sqrt{60}) = 1012$

8 a. 0.025   b. 0.90   c. 0.05

d. 0.01   e. 0.95   f. 0.90

10 a. $\bar{x} = \dfrac{\Sigma x_i}{n} = \dfrac{80}{8} = 10$

b.

x_i	$(x_i - \bar{x})$	$(x_i - \bar{x})^2$
10	0	0
8	−2	4
12	2	4
15	5	25
13	3	9
11	1	1
6	−4	16
5	−5	25
80		84

$s = \sqrt{\dfrac{\Sigma(x_i - \bar{x})^2}{n-1}} = \sqrt{\dfrac{84}{7}} = 3.464$

c. $t_{0.025}(s/\sqrt{n}) = 2.365(3.464/\sqrt{8}) = 2.9$

d. $\bar{x} \pm t_{0.025}(s/\sqrt{n})$

10 ± 2.9 or 7.1 to 12.9

12 90 per cent confidence, 18.42 to 20.58

95 per cent confidence, 18.21 to 20.79

14 Using Minitab, PASW or Excel, $\bar{x} = 6.34$ and $s = 2.163$

The confidence interval is $\bar{x} \pm t_{0.025}(s/\sqrt{n})$, with $df = 49$ and $t_{0.025} = 2.010$

The confidence interval is $6.34 \pm 2.010(2.163/\sqrt{50}) = 6.34 \pm 0.61$ or 5.73 to 6.95

16 a. $\bar{x} = 3.8$ minutes

b. 0.84

c. 2.96 to 4.64

d. Modest positive skewness. Considering a larger sample next time would be a good strategy.

18 The required sample size is given by $n = \dfrac{z_{0.025}^2 \, \sigma^2}{E^2}$

$= \dfrac{(1.96)^2 (40)^2}{10^2} = 61.47$

Use $n = 62$.

20 a. $n = 80$

b. $n = 32$

22 About 265

24 The required sample size is given by $n = \dfrac{z_{0.025}^2 \, \sigma^2}{E^2}$

For 95 per cent confidence, $z_{0.025} = 1.96$, $n = \dfrac{(1.96)^2 (8)^2}{2^2} = 61.47$. Use $n = 62$

For 99 per cent confidence, $z_{0.025} = 2.576$, $n = \dfrac{(2.576)^2 (8)^2}{2^2} = 106.17$. Use $n = 107$

26 a. The confidence interval is $p \pm z_{\alpha/2} \sqrt{\dfrac{p(1-p)}{n}} = 0.70$

$\pm 1.645 \sqrt{\dfrac{0.70(0.30)}{800}}$

$= 0.70 \pm 0.0267$ or 0.6733 to 0.7267

b. $0.70 \pm 1.96 \sqrt{\dfrac{0.70(0.30)}{800}} = 0.70 \pm 0.0318$ or 0.6682 to 0.7318

28 $n = 1068$

30 a. $p = 0.310$

b. 0.036

c. 0.274 to 0.346

32 a. 0.320 to 0.372

b. 0.312 to 0.380

c. The margin of error increases.

34 The required sample size is given by $n = \dfrac{z_{\alpha/2}^2 \, p(1-p)}{E^2}$

a. $n = (2.33)^2 (0.70)(0.30)/(0.03)^2 = 1266.74$. Use $n = 1267$.

b. $n = (2.33)^2 (0.50)(0.50)/(0.03)^2 = 1508.03$. Use $n = 1509$.

Chapter 9

Solutions

2 a. $H_0: \mu \le 14$

$H_1: \mu > 14$ Research hypothesis

b. There is no statistical evidence that the new bonus plan increases sales volume.

c. The research hypothesis that $\mu > 14$ is supported. We can conclude that the new bonus plan increases the mean sales volume.

4 a. $H_0: \mu \ge 320$

$H_1: \mu < 320$ Research hypothesis that mean cost is less than €320.

b. Unable to conclude that new method reduces costs.

c. Conclude $\mu < 320$. Consider implementing new method on basis of lower mean cost per hour.

6 a. $H_0: \mu \le 5000$

$H_1: \mu > 5000$ Research hypothesis that the plan increases average sales.

b. Claiming $\mu > 5000$ when plan does not increase sales. Mistake could be implementing plan when it does not help.

c. Concluding $\mu \le 5000$ when plan really would increase sales. Could lead to not implementing a plan that would increase sales.

8 a. $z = \dfrac{\bar{x} - \mu_0}{\sigma/\sqrt{n}} = \dfrac{19.4 - 20}{2/\sqrt{50}} = -2.12$

b. The cumulative probability for $z = -2.12$ is 0.0170.

p-value = 0.0170

c. p-value < 0.05, reject H_0

d. Reject H_0 if $z \le -1.645$

$-2.12 < -1.645$, so reject H_0

10 Reject H_0 if $z \ge 1.645$

a. $z = 2.42$, reject H_0

b. $z = 0.97$, do not reject H_0

c. $z = 1.74$, reject H_0

12 a. $H_0: \mu \ge 181\ 900$

$H_1: \mu < 181\ 900$

b. $z = -2.93$

c. 0.0017

d. Reject H_0. Conclude that the mean for the North-east is less than the national mean.

14 a. $H_0: \mu = 8$

$H_1: \mu \ne 8$

b. $z = \dfrac{\bar{x} - \mu_0}{\sigma/\sqrt{n}} = \dfrac{8.4 - 8}{3.2/\sqrt{120}} = 1.37$

Cumulative probability for $z = 1.37$ is 0.9147

p-value = 2(1.0000 − 0.9147) = 0.1706

c. p-value > α = 0.05. Do not reject H_0. Cannot conclude that the population mean waiting time differs from 8 minutes.

d. $\bar{x} \pm z_{0.025}\ (\sigma/\sqrt{n})$

$8.4 \pm 1.96\ (3.2/\sqrt{120})$

8.4 ± 0.57 (7.83 to 8.97)

Yes; $\mu = 8$ is in the interval. Do not reject H_0.

16 a. $H_0: \mu = 500$

$H_1: \mu \ne 500$

b. p-value = 0.0286.

p-value < 0.05, reject H_0. Over-filling occurring, recommend readjustment of the mechanism.

c. p-value = 0.2714.

p-value $\le$ 0.05, do not reject H_0. No action required.

d. Reject H_0 if $\bar{x} \ge 508.95$ or if $\bar{x} \le 491.05$. Same conclusions as in b and c.

18 a. $t = \dfrac{\bar{x} - \mu_0}{s/\sqrt{n}} = \dfrac{17 - 18}{4.5/\sqrt{48}} = -1.54$

b. Degrees of freedom = $n - 1 = 47$. Area in lower tail is between 0.05 and 0.10. p-value (two-tailed) is between 0.10 and 0.20. (Actual p-value = 0.1304.)

c. p-value > 0.05, do not reject H_0.

d. With $df = 47$, $t_{0.025} = 2.012$

Reject H_0 if $t \le -2.012$ or $t \ge 2.012$

$t = -1.54$; do not reject H_0

20 a. $H_0: \mu \ge 300$

$H_1: \mu < 300$

b. $t = \dfrac{\bar{x} - \mu_0}{s/\sqrt{n}} = \dfrac{299.5 - 300}{1.9/\sqrt{30}} = -1.44$

Degrees of freedom = $n - 1 = 29$

Using t table, p-value is between 0.05 and 0.10. (Actual p-value = 0.080.)

c. p-value > 0.10, do not reject H_0

22 a. $H_0: \mu \le 4\ 671\ 264$

$H_1: \mu > 4\ 671\ 264$

b. £1 132 069

c. p-value is between 0.005 and 0.01. Reject H_0 at $\alpha = 0.05$, conclude that the mean valuation of mid-fielders is higher than that of strikers.

24 a. $H_0: \mu = 2$

$H_1: \mu \ne 2$

b. $\bar{x} = 2.2$

c. $s = 0.516$

d. p-value is between 0.20 and 0.40

e. p-value > 0.05; do not reject H_0. No reason to change from the two hours for cost-estimating purposes.

26 a. p-value = 0.0026, reject H_0

b. p-value = 0.1151, do not reject H_0

c. p-value = 0.0228, reject H_0

d. p-value = 0.7881, do not reject H_0

28 a. $H_0: \pi \leq 0.10$, $H_1: \pi \leq 0.10$, where π is the population proportion of customers who will use the coupons.

b. $p = 0.13$

c. p-value $= 0.20$, do not reject H_0. Eagle should not go national on this evidence.

30 a. $H_0: \pi = 0.64$

$H_1: \pi \neq 0.64$

b. $p = \dfrac{52}{100} = 0.52$

$z = \dfrac{p - \pi_0}{\sqrt{\dfrac{\pi_0(1 - \pi_0)}{n}}} = \dfrac{0.52 - 0.64}{\sqrt{\dfrac{0.64(1 - 0.64)}{100}}} = -2.50$

Cumulative probability for $z = -2.50$ is 0.0062

p-value $= 2(0.0062) = 0.0124$

c. p-value < 0.05; reject H_0. Proportion differs from the reported 0.64.

d. Yes. Since $p = 0.52$, it indicates that fewer than 64 per cent of the shoppers believe the supermarket brand is as good as the name brand.

32 a. $H_0: \pi = 0.63$

$H_1: \pi \neq 0.63$

b. p-value < 0.002

c. Reject H_0. Conclude that support changed.

34 Reject H_0 if $z \leq -1.96$ or if $z \geq 1.96$

$\sigma_{\bar{x}} = \dfrac{\sigma}{\sqrt{n}} = \dfrac{10}{\sqrt{200}} = 0.71$

The critical values are:

$c_1 = 20 - 1.96\,(10/\sqrt{200}) = 18.61$

$c_2 = 20 + 1.96\,(10/\sqrt{200}) = 21.39$

a. $\mu = 18$

$z = \dfrac{18.61 - 18}{10/\sqrt{200}} = 0.86$

$\beta = P(Z > 0.86) = 1 - 0.8051 = 0.1949$

b. $\mu = 22.5$

$z = \dfrac{21.39 - 22.5}{10/\sqrt{200}} = -1.57$

$\beta = P(Z < -1.57) = 0.0582$

c. $\mu = 21$

$z = \dfrac{21.39 - 21}{10/\sqrt{200}} = 0.55$

$\beta = P(Z < 0.55) = 0.7088$

36 At $\mu = 17$, $\beta = 0.1151$

At $\mu = 18$, $\beta = 0.0015$

Increasing the sample size reduces the probability of making a Type II error.

38 a. A Type II error would be 'accepting' that the mean level of tax-deferred investments is no greater than €100, when in fact it is greater than €100.

b. 0.739   c. 0.421   d. 0.476   e. 0.106

40 The required sample size is given by:

$n = \dfrac{(z_\alpha + z_\beta)^2\,\sigma^2}{(\mu_0 - \mu_1)^2} = \dfrac{(1.96 + 1.645)^2\,(10)^2}{(20 - 22)^2} = 325$

42 $n = 32$

Chapter 10

Solutions

2 a. $z = \dfrac{(\bar{x}_1 - \bar{x}_2) - D_0}{\sqrt{\dfrac{\sigma_1^2}{n_1} + \dfrac{\sigma_2^2}{n_2}}} = \dfrac{(25.2 - 22.8) - 0}{\sqrt{\dfrac{5.2^2}{40} + \dfrac{6^2}{50}}} = 2.03$

b. p-value $= 1 - 0.9788 = 0.0212$

c. p-value < 0.05, reject H_0.

4 a. 4.6 years

b. 1.3 years

c. 3.3 to 5.9 years

6 a. \$67.03

b. \$17.08

c. \$49.95 to \$84.11

8 a. $t = \dfrac{(\bar{x}_1 - \bar{x}_2) - 0}{\sqrt{\dfrac{s_1^2}{n_1} + \dfrac{s_2^2}{n_2}}} = \dfrac{(13.6 - 10.1) - 0}{\sqrt{\dfrac{5.2^2}{35} + \dfrac{8.5^2}{40}}} = 2.18$

b. $df = \dfrac{\left(\dfrac{s_1^2}{n_1} + \dfrac{s_2^2}{n_2}\right)^2}{\dfrac{1}{n_1 - 1}\left(\dfrac{s_1^2}{n_1}\right)^2 + \dfrac{1}{n_2 - 1}\left(\dfrac{s_2^2}{n_2}\right)^2}$

$= \dfrac{\left(\dfrac{5.2^2}{35} + \dfrac{8.5^2}{40}\right)^2}{\dfrac{1}{34}\left(\dfrac{5.2^2}{35}\right)^2 + \dfrac{1}{39}\left(\dfrac{8.5^2}{40}\right)^2} = 65.7$

Use $df = 65$.

c. Using t table, area in tail is between 0.01 and 0.025. Therefore two-tailed p-value is between 0.02 and 0.05. (Actual p-value $= 0.0329$.)

d. p-value < 0.05, reject H_0.

10 -0.52 to 1.28

12 a. $H_0: \mu_1 - \mu_2 \leq 0$

$H_1: \mu_1 - \mu_2 > 0$

b. 38

c. $t = 1.80$, $df = 25$

Using t table, p-value is between 0.025 and 0.05. Exact p-value $= 0.0420$.

d. Reject H_0; conclude higher mean score if college grad.

14 $H_0: \mu_1 - \mu_2 = 0$

$H_1: \mu_1 - \mu_2 \neq 0$

$$z = \frac{(\bar{x}_1 - \bar{x}_2) - D_0}{\sqrt{\dfrac{\sigma_1^2}{n_1} + \dfrac{\sigma_2^2}{n_2}}} = \frac{(4.1 - 3.4) - 0}{\sqrt{\dfrac{(2.2)^2}{120} + \dfrac{(1.5)^2}{100}}} = 2.79$$

p-value $= 2(1 - 0.9974) = 0.0052$. p-value < 0.05, reject H_0. A difference exists with system B having the lower mean checkout time.

16 a. $H_0: \mu_1 - \mu_2 \geq 120$
   $H_1: \mu_1 - \mu_2 < 120$

 b. p-value is between 0.01 and 0.025, reject H_0. The improvement is less than the stated average of 120 points.

 c. 32 to 118

 d. This is a wide interval. A larger sample should be used to reduce the margin of error.

18 a. $11 - 8 = 3, 7 - 8 = -1, 9 - 6 = 3, 12 - 7 = 5,$
   $13 - 10 = 3, 15 - 15 = 0, 15 - 14 = 1$

 b. $\bar{d} = \Sigma d_i / n = (3 - 1 + 3 + 5 + 3 + 0 + 1)/7 = 14/7 = 2$

 c. $s_d = \sqrt{\dfrac{\Sigma(d_i - \bar{d})^2}{n-1}} = \sqrt{\dfrac{26}{7-1}} = 2.08$

 d. $\bar{d} = 2 \sqrt{\dfrac{\Sigma(d_i - \bar{d})^2}{n-1}}$

 e. With 6 degrees of freedom $t_{0.025} = 2.447$. The confidence interval is $2 \pm 2.447 (2.082/\sqrt{7}) = 2 \pm 1.93$, or 0.07 to 3.93.

20 p-value is between 0.10 and 0.20. Do not reject H_0; we cannot conclude that seeing the commercial improves the mean potential to purchase.

22 p-value is between 0.01 and 0.025. Reject H_0. Conclude that the population of readers spends more time, on average, watching television than reading.

24 a. $p = \dfrac{n_1 p_1 + n_2 p_2}{n_1 + n_2} = \dfrac{200(0.22) + 300(0.16)}{200 + 300} = 0.1840$

   $$z = \frac{p_1 - p_2}{\sqrt{p(1-p)\left(\dfrac{1}{n_1} + \dfrac{1}{n_2}\right)}}$$

   $$= \frac{0.22 - 0.16}{\sqrt{0.1840(1 - 0.1840)\left(\dfrac{1}{200} + \dfrac{1}{300}\right)}} = 1.70$$

   p-value $= 1 - 0.9554 = 0.0446$

 b. p-value < 0.05; reject H_0.

26 0.062 to 0.178, higher proportion of Jordanians holding the view.

28 $H_0: \pi_1 - \pi_2 = 0$
   $H_1: \pi_1 - \pi_2 \neq 0$

   $p_1 = 0.74, \quad n_1 = 1103$

   $p_2 = 0.66, \quad n_2 = 1065$

   $p = \dfrac{n_1 p_1 + n_2 p_2}{n_1 + n_2} = \dfrac{1519}{2168} = 0.700$

$$z = \frac{p_1 - p_2}{\sqrt{p(1-p)\left(\dfrac{1}{n_1} + \dfrac{1}{n_2}\right)}}$$

$$= \frac{0.74 - 0.66}{\sqrt{0.70(1 - 0.70)\left(\dfrac{1}{1103} + \dfrac{1}{1065}\right)}} = 4.06$$

p-value < 0.002. p-value < 0.05, reject H_0. There is a difference between the proportions of students agreeing with the statement (higher proportion in 2001).

Chapter 11

Solutions

2 a. $15.76 \leq \sigma^2 \leq 46.95$

 b. $14.46 \leq \sigma^2 \leq 53.33$

 c. $3.8 \leq \sigma \leq 7.3$

4 a. $n = 18, s^2 = 0.36$

   $\chi_{0.05}^2 = 27.587$ and $\chi_{0.95}^2 = 8.672$ (17 degrees of freedom)

   $$\frac{17(0.36)}{27.587} \leq \sigma^2 \leq \frac{17(0.36)}{8.672}$$

   $0.22 \leq \sigma^2 \leq 0.71$

 b. $0.47 \leq \sigma \leq 0.84$

6 a. $s^2 = 900$

 b. $567 \leq \sigma^2 \leq 1690$

 c. $23.8 \leq \sigma \leq 41.1$

8 a. $\bar{x} = \dfrac{\Sigma x_i}{n} = 260.16$

 b. $s^2 = \dfrac{\Sigma(x_i - \bar{x})^2}{n-1} = 4996.8$

   $s = \sqrt{4996.8} = 70.69$

 c. $\chi_{0.025}^2 = 32.852$ and $\chi_{0.975}^2 = 8.907$ (19 degrees of freedom)

   $$\frac{(20 - 1)(4996.8)}{32.852} \leq \sigma^2 \leq \frac{(20 - 1)(4996.8)}{8.907}$$

   $2890 \leq \sigma^2 \leq 10659$

   $53.76 \leq \sigma \leq 103.24$

10 a. $\chi^2 = 24.39$, Degrees of freedom $= n - 1 = 14$

   p-value is between 0.025 and 0.05. p-value ≤ 0.10, reject H_0. Conclude that variance exceeds maximum variance requirement.

 b. $\chi_{0.05}^2 = 23.685$ and $\chi_{0.95}^2 = 6.571$ (14 degrees of freedom)

   $0.00257 \leq \sigma^2 \leq 0.00928$

12 a. Try $n = 15$

   $\chi_{0.025}^2 = 26.119$ and $\chi_{0.975}^2 = 5.629$ (14 degrees of freedom)

   $5.86 \leq \sigma \leq 12.62$, therefore a sample size of 15 was used.

 b. $n = 25$; expect the width of the interval to be smaller.

   $\chi_{0.05}^2 = 39.364$ and $\chi_{0.975}^2 = 12.401$ (24 degrees of freedom), $6.25 \leq \sigma \leq 11.13$

14 a. $F = \dfrac{s_1^2}{s_2^2} = \dfrac{5.8}{2.4} = 2.4$, degrees of freedom 15 and 20

Using F table, p-value is between 0.025 and 0.05. (Actual p-value $= 0.0334$.)

p-value < 0.05, reject H_0. Conclude $\sigma_1^2 > \sigma_2^2$

b. $F_{0.05} = 2.20$, reject H_0 if $F \geq 2.20$

$2.4 > 2.20$, reject H_0. Conclude $\sigma_1^2 > \sigma_2^2$

16 a. $H_0 : \sigma_1^2 \leq \sigma_2^2$ (population 1 is four-year-old automobiles), $H_1 : \sigma_1^2 > \sigma_2^2$

b. $F = 2.89$, degrees of freedom 25 and 24, p-value is less than 0.01.

Reject H_0. Conclude that four-year-old cars have a larger variance in annual repair costs compared to two-year-old cars. This is expected due to the fact that older cars are more likely to have more expensive repairs that lead to greater variance in the annual repair costs.

18 $F = 2.15$, degrees of freedom 25 and 25, one-tailed p-value is between 0.05 and 0.025.

p-value ≤ 0.05, reject H_0. Conclude that the small cap fund is riskier than the large cap fund.

20 a. $H_0 : \sigma_1^2 \leq \sigma_2^2$ (Population 1 is wet roads), $H_1 : \sigma_1^2 > \sigma_2^2$

$F = \dfrac{s_1^2}{s_2^2} = \dfrac{32^2}{16^2} = 4.00$, degrees of freedom 15 and 15

Using F table, p-value is less than 0.01. (Actual p-value $= 0.0054$.)

p-value < 0.05, reject H_0. Conclude that there is greater variability in stopping distances on wet roads.

b. Drive carefully on wet roads because of the uncertainty in stopping distances.

22 $H_0 : \sigma_1^2 = \sigma_2^2, H_1 : \sigma_1^2 \neq \sigma_2^2$

$F = 8.28$, degrees of freedom 24 and 21, two-tailed p-value is less than 0.02.

p-value < 0.05, reject H_0. The process variances are significantly different. Machine 1 offers the greater opportunity for process quality improvements.

Chapter 12

Solutions

2 Expected frequencies: $e_1 = 300(0.25) = 75$, $e_2 = 300(0.25) = 75$, $e_3 = 300(0.25) = 75$, $e_4 = 300(0.25) = 75$. Actual frequencies: $f_1 = 85, f_2 = 95, f_3 = 50, f_4 = 70$.

$$\chi^2 = \frac{(85-75)^2}{75} + \frac{(95-75)^2}{75} + \frac{(50-75)^2}{75} + \frac{(70-75)^2}{75}$$

$$= \frac{100}{75} + \frac{400}{75} + \frac{625}{75} + \frac{25}{75}$$

$$= \frac{1150}{75}$$

$$= 15.33$$

$k - 1 = 3$ degrees of freedom. Chi-squared table shows p-value less than 0.005. (Actual p-value $= 0.0016$.) p-value < 0.05, reject H_0, conclude that the population proportions are not the same.

4 $\chi^2 = 16.3$, $df = 3$, p-value < 0.005, reject H_0. Conclude that the ratings differ. A comparison of observed and expected frequencies shows telephone service has more excellent and good ratings.

6 $\chi^2 = 21.7$, $df = 6$, p-value < 0.005, reject H_0. The park manager should not plan on the same number attending each day. Plan on a larger staff for Sundays.

8 $H_0 = $ The column variable is independent of the row variable

$H_1 = $ The column variable is not independent of the row variable

Expected Frequencies:

	A	B	C
P	28.5	39.9	45.6
Q	21.5	30.1	34.4

$$\chi^2 = \frac{(20-28.5)^2}{28.5} + \frac{(44-39.9)^2}{39.9} + \frac{(50-45.6)^2}{45.6}$$

$$+ \frac{(30-21.5)^2}{21.5} + \frac{(26-30.1)^2}{30.1} + \frac{(30-34.4)^2}{34.4}$$

$$= 7.86$$

Degrees of freedom $= (2-1)(3-1) = 2$. Using χ^2 table, $\chi^2 = 7.86$ provides a p-value between 0.01 and 0.025. (Actual p-value $= 0.0196$.) p-value < 0.05, reject H_0. Conclude that the column variable is not independent of the row variable.

10 $\chi^2 = 100.4$, $df = 2$, p-value is between 0.025 and 0.05, reject H_0. Conclude that the type of ticket purchased is not independent of the type of flight.

12 a. Observed Frequency (f_{ij})

	Pharm	Consumer	Computer	Telecom	Total
Correct	207	136	151	178	672
Incorrect	3	4	9	12	28
Total	210	140	160	190	700

Expected Frequency (e_{ij})

	Pharm	Consumer	Computer	Telecom	Total
Correct	201.6	134.4	153.6	182.4	672
Incorrect	8.4	5.6	6.4	7.6	28
Total	210	140	160	190	700

Chi-squared $(f_{ij} - e_{ij})^2/e_{ij}$

	Pharm	Consumer	Computer	Telecom	Total
Correct	0.14	0.02	0.04	0.11	0.31
Incorrect	3.47	0.46	1.06	2.55	7.53
					$\chi^2 = 7.85$

Degrees of freedom $= (2 - 1)(4 - 1) = 3$. Using χ^2 table, $\chi^2 = 7.85$ shows p-value is between 0.025 and 0.05. (Actual p-value $= 0.0493$.) p-value < 0.05, reject H_0. Conclude that fulfilment of orders is not independent of industry.

b. The pharmaceutical industry is doing the best with 207 of 210 (98.6 per cent) correctly filled orders.

14 $\chi^2 = 8.10$, $df = 23$, p-value is between 0.01 and 0.025, reject H_0. Conclude that shift and quality are not independent.

16 $\chi^2 = 9.76$, $df = 4$, p-value is between 0.025 and 0.05, reject H_0. We can conclude that industry type and P/E ratio are related. Banking tends to have lower P/E ratios.

18 First estimate μ from the sample data. Sample size $= 120$.

$$\bar{x} = \frac{0(39) + 1(30) + 2(30) + 3(18) + 4(3)}{120} = \frac{156}{120} = 1.3$$

Therefore, we use Poisson probabilities with $\mu = 1.3$ to compute expected frequencies.

$$\chi^2 = \frac{(6.30)^2}{32.70} + \frac{(-12.51)^2}{42.51} + \frac{(2.37)^2}{27.63} + \frac{(6.02)^2}{11.98} + \frac{(-2.17)^2}{5.16} = 9.04$$

x	Observed frequency	Poisson probability	Expected frequency	Difference $(f_i - e_i)$
0	39	0.2725	32.70	6.30
1	30	0.3543	42.51	−12.51
2	30	0.2303	27.63	2.37
3	18	0.0998	11.98	6.02
4	3	0.0431	5.16	−2.17

Degrees of freedom $= 5 - 1 - 1 = 3$. Using χ^2 table, $\chi^2 = 9.04$ shows p-value is between 0.025 and 0.05. (Actual p-value $= 0.0287$.) p-value < 0.05, reject H_0. Conclude that the data do not follow a Poisson probability distribution.

20 $\bar{x} = 24.5$, $s = 3$, $n = 30$. Use 6 classes.

Interval	Observed frequency	Expected frequency
less than 21.56	5	5
21.56–23.20	4	5
23.21–24.49	3	5
24.50–25.78	7	5
25.79–27.40	7	5
27.41 upwards	4	5

$\chi^2 = 2.8$. Degrees of freedom $= 6 - 2 - 1 = 3$. Using χ^2 table, $\chi^2 = 2.8$ shows p-value is greater than 0.10. (Actual p-value $= 0.4235$.) p-value > 0.10, do not reject H_0. The assumption of a normal distribution cannot be rejected.

22 $\chi^2 = 4.30$, $df = 2$. p-value greater than 0.10. Do not reject H_0.

Chapter 13

Solutions

2   a. $\bar{\bar{x}} = (153 + 169 + 158)/3 = 160$

$$SSTR = \sum_{j=1}^{k} n_j(\bar{x}_j - \bar{\bar{x}})^2 = 4(153 - 160)^2$$
$$+ 4(169 - 160)^2$$
$$+ 4(158 - 160)^2 = 536$$

MSTR = SSTR/$(k - 1)$ = 536/2 = 268

b. $SSE = \sum_{j=1}^{k}(n_j - 1)s_j^2 = 3(96.67) + 3(97.33)$
$$+ 3(82.00) = 828.00$$

MSE = SSE/$(n_\tau - k)$ = 828.00/(12 − 3) = 92.00

c. F = MSTR/MSE = 268/92 = 2.91

Using F table (2 degrees of freedom numerator and 9 denominator), p-value is greater than 0.10

Actual p-value = 0.1060

Because p-value > α = 0.05, we cannot reject the null hypothesis.

d.

Source of variation	Degrees of freedom	Sum of squares	Mean square	F
Treatments	2	536	268	2.91
Error	9	828	92	
Total	11	1364		

4   a.

Source of variation	Degrees of freedom	Sum of squares	Mean square	F
Treatments	1200	3	400	80
Error	300	60	5	
Total	1500	63		

b.   Using F table (3 degrees of freedom numerator and 60 denominator), p-value is less than 0.01

Because p-value ≤ α = 0.05, we reject the null hypothesis that the means of the four populations are equal.

6

	Manufacturer 1	Manufacturer 2	Manufacturer 3
Sample Mean	23	28	21
Sample Variance	6.67	4.67	3.33

$\bar{\bar{x}} = (23 + 28 + 21)/3 = 24$

$$SSTR = \sum_{j=1}^{k} n_j(\bar{x}_j - \bar{\bar{x}})^2 = 4(23 - 24)^2$$
$$+ 4(28 - 24)^2 + 4(21 - 24)^2 = 104$$

MSTR = SSTR/$(k - 1)$ = 104/2 = 52

$$SSE = \sum_{j=1}^{k}(n_j - 1)s_j^2 = 5(0.8) + 5(0.03)$$
$$+ 5(0.4) = 7.50$$

MSE = SSE/$(n_\tau - k)$ = 44.01/(12 − 3) = 4.89
F = MSTR/MSE = 52/4.89 = 10.63

Using F table (2 degrees of freedom numerator and 9 denominator), p-value is less than 0.01

Actual p-value = 0.0043

Because p-value < α = 0.05, we reject the null hypothesis that the mean time needed to mix a batch of material is the same for each manufacturer.

8

	Marketing managers	Marketing research	Advertising
Sample Mean	5	4.5	6
Sample Variance	0.8	0.3	0.4

$\bar{\bar{x}} = (5 + 4.5 + 6)/3 = 5.17$

SSTR = 7.00

MSTR = SSTR/$(k - 1)$ = 7.00/2 = 3.5

SSE = 7.50

MSE = SSE/$(n_\tau - k)$ = 7.50/(18 − 3) = 0.5
F = MSTR/MSE = 3.5/0.50 = 7.00

Using F table (2 degrees of freedom numerator and 15 denominator), p-value is less than 0.01

Actual p-value = 0.0071

Because p-value ≤ α = 0.05, we reject the null hypothesis that the mean perception score is the same for the three groups of specialists.

10   Because p-value ≤ α = 0.05, we reject the null hypothesis that the mean service ratings are equal.

```
One-way ANOVA: small, medium, large

Source   DF      SS      MS      F      P
Factor    3    226.1   113.0   3.70   0.042
Error    21    640.8    30.5
Total    23    866.9
```

12 a. $\bar{\bar{x}} = 62$

SSTR = 1448

MSTR = SSTR/$(k-1)$ = 1448/2 = 724

SSE = 828

MSE = SSE/$(n_r - k)$ = 828/(12 − 3) = 92

F = MSTR/MSE = 724/92 = 7.87

Using F table (2 degrees of freedom numerator and 9 denominator), p-value is between 0.01 and 0.025

Actual p-value = 0.0106

Because p-value $\le \alpha = 0.05$, we reject the null hypothesis that the means of the three populations are equal.

b.

$$LSD = t_{\alpha/2}\sqrt{MSE(\frac{1}{n_i} + \frac{1}{n_j})} = t_{0.025}\sqrt{92(\frac{1}{4} + \frac{1}{4})}$$

$$= 2.262\sqrt{46} = 15.34$$

$|\bar{x}_1 - \bar{x}_2| = |51 - 77| = 26 >$ LSD; significant difference

$|\bar{x}_1 - \bar{x}_3| = |51 - 58| = 7 <$ LSD; no significant difference

$|\bar{x}_2 - \bar{x}_3| = |77 - 58| = 19 >$ LSD; significant difference

14 $\bar{x}_1 - \bar{x}_2 \pm$ LSD

23 − 28 ± 3.54

− 5 ± 3.54 = −8.54 to − 1.46

16 a.

	Machine 1	Machine 2	Machine 3	Machine 4
Sample Mean	7.1	9.1	9.9	11.4
Sample Variance	1.21	0.93	0.70	1.02

$\bar{\bar{x}} = (7.1 + 9.1 + 9.9 + 11.4)/4 = 9.38$

$SSTR = \sum_{j=1}^{k} n_j (\bar{x}_j - \bar{\bar{x}}^2) = 6(7.1 - 9.38)^2$

$+ 6(9.1 - 9.38)^2 + 6(9.9 - 9.38)^2$

$+ 6(11.4 - 9.38)^2 = 57.77$

MSTR = SSTR/$(k-1)$ = 57.77/3 = 19.26

$SSE = \sum_{j=1}^{k}(n_j - 1) s_j^2 = 5(1.21) + 5(0.93)$

$+ 5(0.70) + 5(0.07) + 5(1.02) = 19.30$

MSE = SSE/$(n_r - k)$ = 19.30/(24 − 4) = 0.97

F = MSTR/MSE = 19.26/0.97 = 19.86

Using F table (3 degrees of freedom numerator and 20 denominator), p-value is less than 0.01

Actual p-value = 0.0000 (to 4 decimal places)

Because p-value $\le \alpha = 0.05$, we reject the null hypothesis that the mean time between breakdowns is the same for the four machines.

b. Note: $t_{\alpha/2}$ is based upon 20 degrees of freedom

$$LSD = t_{\alpha/2}\sqrt{MSE(\frac{1}{n_i} + \frac{1}{n_j})} = t_{0.025}\sqrt{0.97(\frac{1}{6} + \frac{1}{6})}$$

$$= 2.086\sqrt{0.3233} = 1.19$$

$|\bar{x}_2 - \bar{x}_4| = |9.1 - 11.4| = 2.3 >$ LSD:

significant difference

18 $n_1 = 8, n_2 = 8, n_3 = 8$

$t_{\alpha/2}$ is based upon 21 degrees of freedom

$$LSD = t_{0.025}\sqrt{30.5\,(\frac{1}{8} + \frac{1}{8})} = 2.080\sqrt{7.6250} = 5.74$$

Comparing Small and Medium

92.20 − 89.65 = 2.55 < LSD; no significant difference

Comparing Small and Large

92.20 − 84.80 = 7.40 > LSD; significant difference

Comparing Medium and Large

89.65 − 84.80 = 4.85 LSD; no significant difference

20 a.

Source of variation	Degrees of freedom	Sum of squares	Mean square	F
Treatments	2	1488	744	5.50
Error	15	2030	135.3	
Total	17	3518		

b.

$$LSD = t_{\alpha/2}\sqrt{MSE(\frac{1}{n_i} + \frac{1}{n_j})} = 2.131\sqrt{135.3(\frac{1}{6} + \frac{1}{6})}$$

$$= 14.31$$

$|156 - 142| = 14 < 14.31$; no significant difference

$|156 - 134| = 22 > 14.31$; significant difference

$|142 - 134| = 8 < 14.31$; no significant difference

22 a. $H_0: \mu_1 = \mu_2 = \mu_3 = \mu_4 = \mu_5$

H_1: Not all the population means are equal

b. Using F table (4 degrees of freedom numerator and 30 denominator), p-value is less than 0.01

Actual p-value = 0.0000

Because p-value $\le \alpha = 0.05$, we reject H_0

24

Source of variation	Degrees of freedom	Sum of squares	Mean square	F
Treatments	2	1200	600	43.99
Error	44	600	13.64	
Total	46	1800		

Using F table (2 degrees of freedom numerator and 44 denominator), p-value is less than 0.01

Actual p-value = 0.0000 (to 4 decimal places)

Because p-value $\leq \alpha = 0.05$, we reject the hypothesis that the treatment means are equal.

26 a.

Source of variation	Degrees of freedom	Sum of squares	Mean square	F
Treatments	2	4 560	2280	9.87
Error	27	6 240	231.11	
Total	29	10 800		

b. Using F table (2 degrees of freedom numerator and 27 denominator), p-value is less than 0.01

Actual p-value = 0.0006

Because p-value $\leq \alpha = 0.05$, we reject the null hypothesis that the means of the three assembly methods are equal.

28

	50°	60°	70°
Sample Mean	33	29	28
Sample Variance	32	17.5	9.5

$\bar{\bar{x}} = (33 + 29 + 28)/3 = 30$

$\text{SSTR} = 70$

$\text{MSTR} = \text{SSTR}/(k - 1) = 70/2 = 35$

$\text{SSE} = 236$

$\text{MSE} = \text{SSE}/(n_T - k) = 236/(15 - 3) = 19.67$

$F = \text{MSTR/MSE} = 35/19.67 = 1.78$

Using F table (2 degrees of freedom numerator and 12 denominator), p-value is greater than 0.10

Actual p-value = 0.2104

Because p-value $> \alpha = 0.05$, we cannot reject the null hypothesis that the mean yields for the three temperatures are equal.

30

	Paint 1	Paint 2	Paint 3	Paint 4
Sample Mean	13.3	139	136	144
Sample Variance	47.5	0.50	21	54.5

$\bar{\bar{x}} = (133 + 139 + 136 + 144)/3 = 138$

$\text{SSTR} = \sum_{j=1}^{k} n_j(x_j - \bar{\bar{x}})^2 = 5(133 - 138)^2$

$+ 5(139 - 138)^2 + 5(136 - 138)^2$

$+ 5(144 - 138)^2 = 338$

$\text{MSTR} = \text{SSTR}/(k - 1) = 330/3 = 110$

$\text{SSE} = \sum_{j=1}^{k}(n_j - 1)s_j^2 = 4(47.5) + 4(50)$

$+ 4(21) + 4(54.5) = 692$

$\text{MSE} = \text{SSE}/(n_T - k) = 692/(20 - 4) = 43.25$

$F = \text{MSTR/MSE} = 110/43.25 = 2.54$

Using F table (3 degrees of freedom numerator and 16 denominator), p-value is between 0.05 and 0.10

Actual p-value = 0.0931

Because p-value $> \alpha = 0.05$, we cannot reject the null hypothesis that the mean drying times for the four paints are equal.

32 Note: degrees of freedom for $t_{\alpha/2}$ are 18

$\text{LSD} = t_{\alpha/2}\sqrt{\text{MSE}(\frac{1}{n_i} + \frac{1}{n_j})} = t_{0.025}\sqrt{5.09\ (\frac{1}{7} + \frac{1}{7})}$

$= 2.101\sqrt{1.4543} = 2.53$

$|\bar{x}_1 - \bar{x}_2| = |17.0 - 20.4| = 3.4 > 2.53;$
significant difference

$|\bar{x}_1 - \bar{x}_3| = |17.0 - 25.0| = 8 > 2.53;$
significant difference

$|\bar{x}_2 - \bar{x}_3| = |20.4 - 25| = 4.6 > 2.53;$
significant difference

34 Treatment Means:

$\bar{x}_{.1} = 13.6 \quad \bar{x}_{.2} = 11.0 \quad \bar{x}_{.3} = 10.6$

Block Means:

$\bar{x}_{1.} = 9 \quad \bar{x}_{2.} = 7.67 \quad \bar{x}_{3.} = 15.67 \quad \bar{x}_{4.} = 18.67 \quad \bar{x}_{5.} = 7.67$

Overall Mean:

$\bar{\bar{x}} = 176/15 = 11.73$

Step 1

$\text{SST} = \sum\sum_j (x_{ij} - \bar{\bar{x}})^2 = (10 - 11.73)^2 + (9 - 11.73)^2$

$+ \cdots + (8 - 11.73)^2 = 354.93$

Step 2

$\text{SSTR} = b\sum_{j=1} (\bar{x}_j - \bar{\bar{x}})^2 = 5\ [(13.6 - 11.73)^2$

$+(11.0 - 11.73)^2 + (10.6 - 11.73)^2] = 26.53$

Step 3

$\text{SSBL} = k\sum_i (\bar{x}_i - \bar{\bar{x}})^2 = 3\ [(9 - 11.73)^2$

$+ (7.67 - 11.73)^2 + (15.67 - 11.73)^2$

$+ (18.67 - 11.73)^2 + (7.67 - 11.73)^2]$

$= 312.32$

Step 4

$\text{SSE} = \text{SST} - \text{SSTR} - \text{SSBL} = 354.93 - 26.53$

$- 312.32 = 16.08$

Source of variation	Degrees of freedom	Sum of squares	Mean square	F
Treatments	2	26.53	13.27	6.60
Blocks	4	312.32	78.08	
Error	8	16.08	2.01	
Total	14	354.93		

Using F table (2 degrees of freedom numerator and 8 denominator), p-value is between 0.01 and 0.025

Actual p-value $= 0.0203$

Because p-value $\leq \alpha = 0.05$, we reject the null hypothesis that the means of the three treatments are equal.

36

Source of variation	Degrees of freedom	Sum of squares	Mean square	F
Treatments	3	900	300	12.60
Blocks	7	400	57.14	
Error	21	500	23.81	
Total	31	1800		

Using F table (3 degrees of freedom numerator and 21 denominator), p-value is less than 0.01

Actual p-value $= 0.0001$

Because p-value $\leq \alpha = 0.05$, we reject the null hypothesis that the means of the treatments are equal.

38

Source of variation	Degrees of freedom	Sum of squares	Mean square	F
Cloth(block)	2	0.389	0.195	9.00
Silicone	4	16.103	4.026	
Error	8	3.577	0.447	
Total	14	20.069		

Using F table (4 degrees of freedom numerator and 8 denominator), p-value is less than 0.01

Actual p-value $= 0.005$

Because p-value $\leq \alpha = 0.05$, we reject the null hypothesis that the mean indices corresponding to the five silicone solution strengths are equal. Note that the highest mean index is for the 15 per cent solution.

40

		Factor B			Factor A
		Level 1	Level 2	Level 3	Means
Factor A	Level 1	$\bar{x}_{11} = 150$	$\bar{x}_{12} = 78$	$\bar{x}_{13} = 84$	$\bar{x}_{1.} = 104$
	Level 2	$\bar{x}_{21} = 110$	$\bar{x}_{22} = 116$	$\bar{x}_{23} = 128$	$\bar{x}_{2.} = 118$
Factor B	Means	$\bar{x}_{.1} = 130$	$\bar{x}_{.2} = 97$	$\bar{x}_{.3} = 106$	$\bar{\bar{x}} = 111$

Step 1

$$\text{SST} \sum_i \sum_j \sum_k (x_{ijk} - \bar{\bar{x}})^2$$
$$(135 - 111)^2 + (165 - 111)^2$$
$$+ .. + (136 - 111)^2 = 9028$$

Step 2

$$\text{SSA} = br\sum_i (\bar{x}_{i.} - \bar{\bar{x}})^2 = 3(2)[(104 - 111)^2$$
$$+ (118 - 111)^2] = 588$$

Step 3

$$\text{SSB} = ar\sum_j (\bar{x}_{.j} - \bar{\bar{x}})^2 = 2(2)[(130 - 111)^2$$
$$+ (97 - 111)^2 + (106 - 111)]$$
$$= 2328$$

Step 4

$$\text{SSAB} = r\sum_i \sum_j (\bar{x}_{ij} - \bar{x}_{i.} - \bar{x}_{.j} - \bar{\bar{x}})^2$$
$$= 2[(150 - 104 - 130 - 111)^2$$
$$+ (78 - 104 - 97 + 111)^2] +$$
$$+ (128 - 118 - 106 + 111)^2 = 4392$$

Step 5

$$\text{SSE} = \text{SST} - \text{SSA} - \text{SSB} - \text{SSAB} =$$
$$9028 - 588 - 2328 - 4392 = 1720$$

Source of variation	Degrees of freedom	Sum of squares	Mean square	F
Factor A	1	588	588	2.05
Factor B	2	2328	1164	4.06
Interaction	2	4392	2196	7.66
Error	6	1720	286.67	
Total	11	9028		

Factor A: $F = 2.05$

Using F table (1 degree of freedom numerator and 6 denominator), p-value is greater than 0.10

Actual p-value $= 0.2022$

Because p-value $> \alpha = 0.05$, Factor A is not significant

Factor B: $F = 4.06$

Using F table (2 degrees of freedom numerator and 6 denominator), p-value is between 0.05 and 0.10

Actual p-value $= 0.0767$

Because p-value $> \alpha = 0.05$, Factor B is not significant

Interaction: $F = 7.66$

Using F table (2 degrees of freedom numerator and 6 denominator), p-value is between 0.01 and 0.025

Actual p-value $= 0.0223$

Because p-value $\leq \alpha = 0.05$, Interaction is significant

42

		Factor B		Factor A
		Small	Large	Means
Factor A	A	$\bar{x}_{11} = 10$	$\bar{x}_{12} = 10$	$\bar{x}_1 = 10$
	B	$\bar{x}_{21} = 18$	$\bar{x}_{22} = 28$	$\bar{x}_2 = 23$
	C	$\bar{x}_{31} = 14$	$\bar{x}_{32} = 16$	$\bar{x}_3 = 15$
Factor B	Means	$\bar{x}_{.1} = 14$	$\bar{x}_{.2} = 18$	$\bar{\bar{x}} = 16$

$SST = 544$
$SSA = 344$
$SSB = 48$
$SSAB = 56$
$SSE = 96$

Source of variation	Degrees of freedom	Sum of squares	Mean square	F
Factor A	2	344	172	172/16 = 10.75
Factor B	1	48	48	48/16 = 3.00
Interaction	2	56	28	28/16 = 1.75
Error	6	96	16	
Total	11	544		

Using F table for Factor A (2 degrees of freedom numerator and 6 denominator), p-value is between 0.01 and 0.025

Actual p-value = 0.0104

Because p-value $\leq \alpha = 0.05$, Factor A is significant; there is a difference due to the type of advertisement design

Using F table for Factor B (1 degree of freedom numerator and 6 denominator), p-value is greater than 0.01

Actual p-value = 0.1340

Because p-value $> \alpha = 0.05$, Factor B is not significant; there is not a significant difference due to size of advertisement

Using F table for Interaction (2 degrees of freedom numerator and 6 denominator), p-value is greater than 0.10

Actual p-value = 0.2519

Because p-value $> \alpha = 0.05$, Interaction is not significant.

44

$\bar{x}_{1.} = (1.13 + 1.56 + 2.00)/3 = 1.563$

$\bar{x}_{2.} = (0.48 + 1.68 + 2.86)/3 = 1.673$

$\bar{x}_{.1} = (1.13 + 0.48)/2 = 0.805$

$\bar{x}_{.2} = (1.56 + 1.68)/2 = 1.620$

$\bar{x}_{.3} = (2.00 + 2.86)/2 = 2.43$

$\bar{\bar{x}} = 1.618$

$SST = 327.50$ (given in problem statement)
$SSA = 0.4538$
$SSB = 66.0159$
$SSAB = 14.2525$
$SSE = 246.7778$

Source of variation	Degrees of freedom	Sum of squares	Mean square	F
Factor A	1	0.4538	0.4538	0.2648
Factor B	2	66.1059	33.0080	19.2608
Interaction	2	14.2525	7.1263	4.1583
Error	144	246.7778	1.7137	
Total	149	327.5000		

Factor A: Actual p-value = 0.6076. Because p-value $> \alpha = 0.05$, Factor A is not significant. Factor B: Actual p-value = 0.0000. Because p-value $\leq \alpha = 0.05$, Factor B is significant. Interaction: Actual p-value = 0.0176. Because p-value $\leq \alpha = 0.05$, Interaction is significant.

Chapter 14

Solutions

2 a.

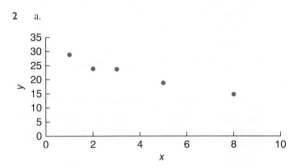

b. There appears to be a linear relationship between x and y.

c. Many different straight lines can be drawn to provide a linear approximation of the relationship between x and y; in part d we will determine the equation of a straight line that 'best' represents the relationship according to the least squares criterion.

d. Summations needed to compute the slope and y-intercept are:

$\sum x_i = 19 \quad \sum y_i = 116 \quad \sum (x_i - \bar{x})(y_i - \bar{y}) = -57.8$

$\sum (x_i - \bar{x})^2 = 30.8$

$b_1 = \dfrac{\sum (x_i - \bar{x})(y_i - \bar{y})}{\sum (x_i - \bar{x})^2} = \dfrac{57.8}{30.8} = -1.8766$

$b_0 = \bar{y} - b_1 \bar{x} = 23.2 - (-1.8766)(3.8) = 30.3311$

$\hat{y} = 30.33 - 1.88x$

e. $\hat{y} = 30.33 - 1.88(6) = 19.05$

4 a.

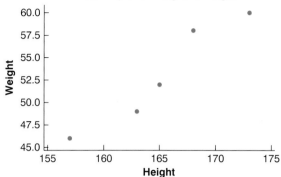

Scatterplot of Weight vs Height

b. There appears to be a linear relationship between x and y.

c. Many different straight lines can be drawn to provide a linear approximation of the relationship between x and y; in

d. Summations needed to compute the slope and y-intercept are:

$$\Sigma x_i = 826 \quad \Sigma y_i = 265 \quad \Sigma(x_i - \bar{x})(y_i - \bar{y})$$
$$= 135 \quad \Sigma(x_i - \bar{x})^2 = 140.8$$
$$b_1 = \frac{\Sigma(x_i - \bar{x})(y_i - \bar{y})}{\Sigma(x_i - \bar{x})^2} = \frac{135}{140.8} = 0.9588$$
$$b_0 = \bar{y} - b_1\bar{x} = 53 - (0.9588)(165.2) = -105.39$$
$$\hat{y} = -105.39 + 0.9588x$$

e. $\hat{y} = -105.39 + 0.9588(160) = 48.02$ kg

6 a.

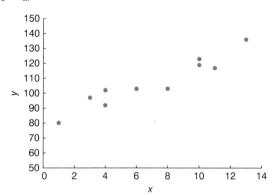

b. Summations needed to compute the slope and y-intercept are:

$$\Sigma x_i = 70 \quad \Sigma y_i = 1080 \quad \Sigma(x_i - \bar{x})(y_i - \bar{y}) = 568$$
$$\Sigma(x_i - \bar{x})^2 = 142$$

$$b_1 = \frac{\Sigma(x_i - \bar{x})(y_i - \bar{y})}{\Sigma(x_i - \bar{x})^2} = \frac{568}{142} = 4$$

$$b_0 = \bar{y} - b_1\bar{x} = 108 - (4)(7) = 80$$
$$\hat{y} = 80 + 4x$$

c. $\hat{y} = 80 + 4(9) = 116$

8 a. The estimated regression equation and the mean for the dependent variable are:

$$\hat{y} = 30.33 - 1.88x \quad \bar{y} = 23.2$$

The sum of squares due to error and the total sum of squares are

$$SSE = \Sigma(y_i - \hat{y}_i)^2 = 6.33$$
$$SST = \Sigma(y_i - \bar{y})^2 = 114.80$$

Thus, SSR = SST − SSE = 114.80 − 6.33 = 108.47

b. r^2 = SSR/SST = 108.47/114.80 = 0.945

The least squares line provided an excellent fit; 94.5 per cent of the variability in y has been explained by the estimated regression equation.

c. $r = \sqrt{0.945} = 0.9721$

Note: the sign for r is negative because the slope of the estimated regression equation is negative. ($b_1 = -1.88$)

10 a. The estimated regression equation and the mean for the dependent variable are:

$$\hat{y} = -75.586 + 0.115x \quad \bar{y} = 784.215$$

The sum of squares due to error and the total sum of squares are

$$SSE = 10083.87$$
$$SST = 1481257$$

Thus, SSR = 1471173.13

b. r^2 = SSR/SST = 1471173.13/1481257 = 0.99

We see that 99 per cent of the variability in y has been explained by the least squares line.

c. $r = \sqrt{0.99} = +0.99$

12 a. Let x = speed (ppm) and y = price (€)

The summations needed in this problem are:

$$\Sigma x_i = 188 \qquad \Sigma y_i = 953.97 \qquad \Sigma(x_i - x)(y_i - y)$$
$$= 324.864 \qquad \Sigma(x_i - \bar{x})^2 = 83.6$$
$$b_1 = \frac{\Sigma(x_i - \bar{x})(y_i - \bar{y})}{\Sigma(x_i - \bar{x})^2} = \frac{324.864}{83.6} = 3.886$$
$$b_0 = \bar{y} - b_1\bar{x} = 95.397 - (3.886)(18.8) = 22.341$$
$$\hat{y} = 22.341 + 3.886x$$

b. The sum of squares due to error and the total sum of squares are:

$$SSE = 3746.309$$
$$SST = 5008.708$$

Thus, SSR = 1262.399

$$r^2 = SSR/SST = 1262.399/5008.708 = 0.252$$

Approximately 25 per cent of the variability in price is explained by the speed.

c. $r = \sqrt{0.252} = +0.50$

It reflects a weak linear relationship.

14 a. s^2 = MSE = SSE/$(n - 2)$ = 6.33/3 = 2.11

b. $s = \sqrt{MSE} = \sqrt{2.11} = 1.453$

c. $\sum(x_i - \bar{x})^2 = 30.8$

$$s_{b_1} = \frac{s}{\sqrt{\sum(x - \bar{x})^2}} = \frac{1.453}{\sqrt{30.8}} = 0.262$$

d. $t = \dfrac{b_1}{s_{b_1}} = \dfrac{-1.88}{0.262} = -7.18$

Using t table (3 degrees of freedom), area in tail is less than 0.005; p-value is less than 0.01

Actual p-value = 0.0056

Because p-value $\leq \alpha$, we reject H_0: $\beta_1 = 0$

e. MSR = SSR/1 = 108.47

   F = MSR/MSE = 108.47/2.11 = 51.41

Using F table (1 degree of freedom numerator and 3 denominator), p-value is less than 0.01

Actual p-value = 0.0056

Because p-value $\leq \alpha$, we reject H_0: $\beta_1 = 0$

Source of variation	Degrees of freedom	Sum of squares	Mean square	F
Regression	1	108.47	108.47	51.41
Error	3	6.33	2.11	
Total	4	114.80		

16 SSE = 233 333.33 SST = 5 648 333.33 SSR = 5 415 000

MSE = SSE/$(n - 2)$ = 233 333.33/$(6 - 2)$ = 58 333.33

MSR = SSR/1 = 5 415 000

F = MSR/MSE = 5 415 000/58 333.25 = 92.83

Source of variation	Degrees of freedom	Sum of squares	Mean square	F
Regression	1	5 415 000.00	5 415 000	92.83
Error	4	233 333.33	58 333.33	
Total	5	5 648 333.33		

Using F table (1 degree of freedom numerator and 4 denominator), p-value is less than 0.01

Actual p-value = 0.0006

Because p-value $\leq \alpha = 0.05$ we reject H_0: $\beta_1 = 0$.
Production volume and total cost are related.

18 a. $s = 2.033$

   $\bar{x} = 3$ $\sum(x_i - \bar{x})^2 = 10$

   $\hat{y} = 0.2 + 2.6x = 0.2 + 2.6(4) = 10.6$

$$s_{\hat{y}_p} = s\sqrt{\frac{1}{n} + \frac{(x_p - \bar{x})^2}{\sum(x_i - x)^2}} = 2.033\sqrt{\frac{1}{5} + \frac{(4 - 3)^2}{10}} = 1.11$$

   $\hat{y}_p \pm t_{\alpha/2}s_{\hat{y}_p}$

   $10.6 \pm 3.182 (1.11) = 10.6 \pm 3.53$ or 7.07 to 14.13

b.

$$\hat{y}_p \pm t_{\alpha/2}s\sqrt{1 + \frac{1}{n} + \frac{(x_p - \bar{x})^2}{\sum(x_i - \bar{x})^2}} = 10.6 \pm 3.182(2.033)\sqrt{1 + \frac{1}{5} + \frac{(4 - 3)^2}{10}}$$

   $10.6 \pm 3.182 (2.32) = 10.6 \pm 7.38$ or 3.22 to 17.98

20 $s = 1.33$

   $\bar{x} = 5.2$ $\sum(x_i - x)^2 = 22.8$

$$s_{\hat{y}_p} = s\sqrt{\frac{1}{n} + \frac{(x_p - \bar{x})^2}{\sum(x_i - \bar{x})^2}} = 1.33\sqrt{\frac{1}{5} + \frac{(3 - 5.2)^2}{22.8}}$$

   $= 0.85$

   $\hat{y} = 0.75 + 0.51x = 0.75 + 0.51(3) = 2.28$

   $\hat{y}_p \pm t_{\alpha/2}\, s_{\hat{y}_p}$

   $2.28 \pm 3.182 (0.85) = 2.28 \pm 2.70$

   or -0.40 to 4.98

$$s\sqrt{1 + \frac{1}{n} + \frac{(x_p - \bar{x})^2}{\sum(x_i - x)^2}} \qquad \sqrt{1 + \frac{1}{5} + \frac{(3 - 5.2)^2}{22.8}} = 1.58$$

   $\hat{y}_p \pm t_{\alpha/2}(1.58) =$

   $2.28 \pm 3.182 (1.58) = 2.28 \pm 5.03$

   or -2.27 to 7.31

22 a. 9

   b. $\hat{y} = 20.0 + 7.21x$

   c. 1.3626

   d. SSE = SST $-$ SSR = 51 984.1 $-$ 41 587.3 = 10 396.8

   MSE = 10 396.8/7 = 1485.3

   F = MSR/MSE = 41 587.3/1485.3 = 28.00

   Using F table (1 degree of freedom numerator and 7 denominator), p-value is less than 0.01

   Actual p-value = 0.0011

   Because p-value $\leq \alpha$, we reject H_0: $\beta_1 = 0$.

   e. $\hat{y} = 20.0 + 7.21(50) = 380.5$ or €380 500

24 a. $\hat{y} = 80.0 + 50.0x$

   b. 30

   c. F = MSR/MSE = 6828.6/82.1 = 83.17

   Using F table (1 degree of freedom numerator and 28 denominator), p-value is less than 0.01

   Actual p-value = 0.0001

   Because p-value $< \alpha = 0.05$, we reject H_0: $\beta_1 = 0$.
   Branch office sales are related to the salespersons.

   d. $\hat{y} = 80 + 50 (12) = 680$ or €680 000

26 a. $\hat{y} = 2.32 + 0.64x$

   b.

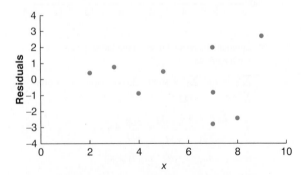

The assumption that the variance is the same for all values of x is questionable. The variance appears to increase for larger values of x.

28 a. $\hat{y} = 80 + 4x$

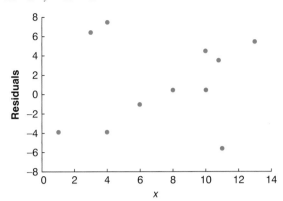

b. The assumptions concerning the error term appear reasonable.

30 a. The MINITAB output is shown below:

```
The regression equation is
Y = 13.0 + 0.425 X

Predictor   Coef    SE Coef    T       p
Constant    13.002   2.396     5.43    0.002
X           0.4248   0.2116    2.01    0.091

S = 3.181   R-sq = 40.2%   R-sq(adj) = 30.2%

Analysis of Variance

SOURCE           DF     SS      MS      F     p
Regression        1    40.78   40.78   4.03  0.091
Residual Error    6    60.72   10.12
Total             7   101.50

Unusual Observations

                             Stdev.              St.
Obs.   X      Y      Fit     Fit    Residual  Resid

 7    12.0   24.00   18.10   1.20    5.90     2.00R
 8    22.0   19.00   22.35   2.78   -3.35    -2.16RX

R denotes an observation with a large stan-
dardized residual.
X denotes an observation whose X value gives
it a large influence.
```

The standardized residuals are: $-1.00, -0.41, 0.01, -0.48, 0.25, 0.65, -2.00, -2.16$

The last two observations in the data set appear to be outliers since the standardized residuals for these observations are 2.00 and -2.16, respectively.

b. Using MINITAB, we obtained the following leverage values:

0.28, 0.24, 0.16, 0.14, 0.13, 0.14, 0.14, 0.76

MINITAB identifies an observation as having high leverage if $h_i > 6/n$: for these data, $6/n = 6/8 = 0.75$. Since the leverage for the observation $x = 22, y = 19$ is 0.76, MINITAB would identify observation 8 as a high leverage point. Thus, we conclude that observation 8 is an influential observation.

c.

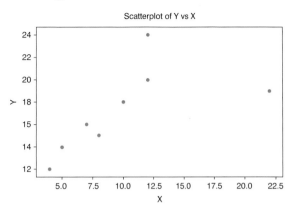

The scatter diagram indicates that the observation $x = 22, y = 19$ is an influential observation.

Chapter 15

Solutions

2 a. The estimated regression equation is
$$\hat{y} = 45.06 + 1.94x_1$$
An estimate of Y when $x_1 = 45$ is
$$\hat{y} = 45.06 + 1.94(45) = 132.36$$

b. The estimated regression equation is
$$\hat{y} = 85.22 + 4.32x_2$$
An estimate of y when $x_2 = 15$ is
$$\hat{y} = 85.22 + 4.32(15) = 150.02$$

c. The estimated regression equation is
$$\hat{y} = -18.37 + 2.01x_1 + 4.74x_2$$
An estimate of y when $x_1 = 45$ and $x_2 = 15$ is
$$\hat{y} = -18.37 + 2.01(45) + 4.74(15) = 143.18$$

4 a. $\hat{y} = 25 + 10(15) + 8(10) = 255$: sales estimate: €255 000

b. Sales can be expected to increase by €10 for every dollar increase in inventory investment when advertising expenditure is held constant. Sales can be expected to increase by €8 for every dollar increase in advertising expenditure when inventory investment is held constant.

6 a. The MINITAB output is shown below:

```
The regression equation is
Return = 247 - 32.8 Safety    34.6 ExpRatio
Predictor    Coef    SE Coef     T       P
Constant    247.4    110.4      2.24    0.039
Safety      -32.84    13.95    -2.35    0.031
ExpRatio     34.59    14.13     2.45    0.026

S = 16.98   R-Sq = 58.2%   R-Sq(adj) = 53.3%

Analysis of Variance

Source           DF     SS      MS      F      P
Regression        2   6823.2  3411.6  11.84  0.001
Residual Error   17   4899.7   288.2
Total            19  11723.0
```

b. $\hat{y} = 247 - 32.8(7.5) + 34.6(2) = 70.2$

8 a. $R^2 = \dfrac{SSR}{SST} = \dfrac{14\ 052.2}{15\ 182.9} = 0.926$

b. adj $R^2 = 1 - (1 - R^2)\dfrac{n-1}{n-p-1}$

$= 1 - (1 - 0.926)\dfrac{10-1}{10-2-1} = 0.905$

c. Yes: after adjusting for the number of independent variables in the model, we see that 90.5 per cent of the variability in y has been accounted for.

10 a. $R^2 = \dfrac{SSR}{SST} = \dfrac{12\ 000}{16\ 000} = 0.75$

b. adj $R^2 = 1 - (1 - R^2)\dfrac{n-1}{n-p-1}$

$= 1 - (0.25)\dfrac{9}{7} = 0.68$

c. The adjusted coefficient of determination shows that 68 per cent of the variability has been explained by the two independent variables: thus, we conclude that the model does not explain a large proportion of variability.

12 a. $MSR = SSR/p = 6216.375/2 = 3108.188$

$MSE = \dfrac{SSE}{n-p-1} = \dfrac{507.75}{10-2-1} = 72.536$

b. $F = MSR/MSE = 3108.188/72.536 = 42.85$

Using F table (2 degrees of freedom numerator and 7 denominator), p-value is less than 0.01

Because p-value $\leq \alpha = 0.05$, the overall model is significant.

c. $t = 0.5906/0.0813 = 7.26$

Using t table (7 degrees of freedom), area in tail is less than 0.005: p-value is less than 0.01.

Because p-value $\leq \alpha$, β_1 is significant.

d. $t = 0.4980/0.0567 = 8.78$

Using t table (7 degrees of freedom), area in tail is less than 0.005: p-value is less than 0.01.

Because p-value $\leq \alpha$, β_2 is significant.

14 a. In the two independent variable case the coefficient of x_1 represents the expected change in y corresponding to a one unit increase in x_1 when x_2 is held constant. In the single independent variable case the coefficient of x_1 represents the expected change in y corresponding to a one unit increase in x_1.

b. Yes. If x_1 and x_2 are correlated one would expect a change in x_1 to be accompanied by a change in x_2.

16 a. $F = 28.38$

Using F table (2 degrees of freedom numerator And 7 denominator), p-value is less than 0.01.

Actual p-value $= 0.002$

Because p-value $\leq \alpha$, there is a significant relationship.

b. $t = 7.53$

Using t table (7 degrees of freedom), area in tail is less than 0.005: p-value is less than 0.01.

Actual p-value $= 0.001$

Because p-value $\leq \alpha$, β_1 is significant and x_1 should not be dropped from the model.

c. $t = 4.06$

Actual p-value $= 0.010$

Because p-value $\leq \alpha$, β_2 is significant and x_2 should not be dropped from the model.

18 a. Using Minitab, the 95 per cent confidence interval is 132.16 to 154.16.

b. Using Minitab, the 95 per cent prediction interval is 111.13 to 175.18.

20 a. $E(Y) = \beta_0 + \beta_1 x_1 + \beta_2 x_2$ where

$x_2 = 0$ if level 1 and 1 if level 2

b. $E(Y) = \beta_0 + \beta_1 x_1 + \beta_2(0) = \beta_0 + \beta_1 x_1$

c. $E(Y) = \beta_0 + \beta_1 x_1 + \beta_2(1) = \beta_0 + \beta_1 x_1 + \beta_2$

d. $\beta_2 = E(Y \mid \text{level 2}) - E(y \mid \text{level 1})$

β_1 is the change in $E(Y)$ for a 1 unit change in x_1 holding x_2 constant.

22 a. €15 300

b. Estimate of sales $= 10.1 - 4.2(2) + 6.8(8)$

$+ 15.3(0) = 56.1$ or €56 100

c. Estimate of sales $= 10.1 - 4.2(1) + 6.8(3)$

$+ 15.3(1) = 41.6$ or €41 600

24 a. Relevant correlation details are as follows:

```
              Tar     Nicotine
Nicotine     0.977
             0.000

CO           0.957     0.926
             0.000     0.000

Cell Contents: Pearson correlation
               P-Value
```

Clearly the independent variables Tar and Nicotine are highly correlated. Not surprisingly when both predictors are fitted in the model a VIF of 21.6 is observed for each variable indicating the presence of serious multicollinearity.

b. Because of its slightly higher correlation with CO only the Tar variable is therefore considered as a predictor in the relevant regression model. Selected Minitab output is shown below:

```
The regression equation is
CO = 2.74 + 0.80 Tar

Predictor    Coef    SE Coef       T       P
Constant   2.7433    0.6752     4.06   0.000
Tar        0.8010    0.08360   15.92   0.000

S = 1.40   R-Sq = 91.7%   R-Sq(adj) = 91.3%
```

c. The p-value corresponding to $t = 15.92$ is $0.000 < \alpha = 0.05$: thus, Tar is a statistically significant predictor.

26 a. The Minitab output is shown below:

```
The regression equation is
Y - 0.20 + 2.60 X

Predictor     Coef    SE Coef       T        P
Constant     0.200     2.132     0.09    0.931
X           2.6000     0.6429     4.04    0.027

S = 2.033   R-Sq = 84.5%   R-Sq(adj) = 79.3%

Analysis of Variance

Source           DF      SS      MS      F       P
Regression        1  67.600  67.600  16.35  0.027
Residual Error    3  12.400   4.133
Total             4  80.000
```

b. Using Minitab we obtained the following values:

x_i	y_i	$\hat{y}_j$	Standardized residual
1	3	2.8	0.16
2	7	5.4	0.94
3	5	8.0	−1.65
4	11	10.6	0.24
5	14	13.2	0.62

The point (3, 5) does not appear to follow the trend of remaining data: however, the value of the standardized residual for this point, −1.65, is not large enough for us to conclude that (3, 5) is an outlier.

c. Using Minitab, we obtained the following values:

x_i	y_i	Studentized deleted residual
1	3	0.13
2	7	0.91
3	5	−4.42
4	11	0.19
5	14	0.54

$$t_{0.025} = 4.303$$

($n - p - 2 = 5 - 1 - 2 = 2$ degrees of freedom)

Since the studentized deleted residual for (3, 5) is −4.42 < −4.303, we conclude that the third observation is an outlier.

28 a. The Minitab output appears in the solution to part (b) of exercise 5; the estimated regression equation is:

Revenue = 83.2 + 2.29 TVAdv + 1.30 NewsAdv

b. Using Minitab we obtained the following values:

$\hat{y}_i$	Standardized residual
96.63	−1.62
90.41	−1.08
94.34	1.22
92.21	−0.37

$\hat{y}_i$	Standardized residual
94.39	1.10
94.24	−0.40
94.42	−1.12
93.35	1.08

With the relatively few observations, it is difficult to determine if any of the assumptions regarding the error term have been violated. For instance, an argument could be made that there does not appear to be any pattern in the plot; alternatively an argument could be made that there is a curvilinear pattern in the plot.

c. The values of the standardized residuals are greater than −2 and less than +2; thus, using test, there are no outliers. As a further check for outliers, we used Minitab to compute the following studentized deleted residuals:

Observation	Studentized deleted residual
1	−2.11
2	−1.10
3	1.31
4	−.33
5	1.13
6	−.36
7	−1.16
8	1.10

d. Using Minitab we obtained the following values:

Observation	h_i	D_i
1	0.63	1.52
2	0.65	0.70
3	0.30	0.22
4	0.23	0.01
5	0.26	0.14
6	0.14	0.01
7	0.66	0.81
8	0.13	0.06

The critical average value is

$$\frac{3(p + 1)}{n} = \frac{3(2 + 1)}{8} = 1.125$$

Since none of the values exceed 1.125, we conclude that there are no influential observations. However, using Cook's distance measure, we see that $D_1 > 1$ (rule of thumb critical value); thus, we conclude the first observation is influential. Final conclusion: observation 1 is an influential observation.

30 a. $E(Y) = \dfrac{e^{\beta_0 + \beta_1 X}}{1 + e^{\beta_0 + \beta_1 X}}$

b. It is an estimate of the probability that a customer that does not have a Simmons credit card will make a purchase.

```
Logistic Regression Table
                                                         Odds              95% CI
Predictor      Coef      SE Coef      z        P       Ratio    Lower      Upper
Constant     -0.9445     0.3150    -3.00    0.003
Card          1.0245     0.4235     2.42    0.016     2.79     1.21       6.39

Log-Likelihood = 64.265
Test that all slopes are zero: G = 6.072, DF = 1, P-Value = 0.014
```

Thus, the estimated logit is $\hat{g}(x) = -0.9445 + 1.0245x$.

d. For customers that do not have a Simmons credit card $(x = 0)$

$\hat{g}(0) = -0.9445 + 1.245(0) = 0.9445$

and

$\hat{y} = \dfrac{e^{\hat{g}(0)}}{1 + e^{\hat{g}(0)}} = \dfrac{e^{0.9445}}{1 + e^{0.9445}} = \dfrac{0.3889}{1 + 0.3889} = 0.28$

$t_{0.025} = 2.776$

$(n - p - 2 = 8 - 2 - 2 = 4 \text{ degrees of freedom})$

Since none of the studentized deleted residuals is less than -2.776 or greater than 2.776, we conclude that there are no outliers in the data.

For customers that have a Simmons credit card $(x = 1)$

```
Logistic Regression Table
                                                         Odds              95% CI
Predictor      Coef      SE Coef      z        P       Ratio    Lower      UPPer
Constant     -2.6335     0.7985    -3.30    0.001
Balance       0.22018    0.09002    2.45    0.014     1.25     1.04       1.49

Log-Likelihood = 25.813
Test that all slop
pes are zero: G = 9.460, DF = 1, P-Value = 0.002
```

Thus, the estimated logistic regression equation is

$E(y) = \dfrac{e^{2.6355 + 0.22018x}}{1 + e^{2.6355 + 0.22018x}}$

c. Significant result: the p-value corresponding to the G test statistic is 0.0002.

d. For an average monthly balance of €1000, $x = 10$

$E(y) = \dfrac{e^{2.6355 + 0.22018x}}{1 + e^{2.6355 + 0.22018x}} = \dfrac{e^{2.6355 + 0.22018(10)}}{1 + e^{2.6355 + 0.22018(10)}}$

$= \dfrac{e^{0.4317}}{1 + e^{0.4317}} = \dfrac{0.6494}{1.6494} = 0.39$

Thus, an estimate of the probability that customers with an average monthly balance of €1000 will sign up for direct payroll deposit is 0.39.

c. A portion of the Minitab binary logistic regression output follows:

$\hat{g}(1) = -0.9445 + 1.245(1) = 0.0800$

and

$\hat{y} = \dfrac{e^{\hat{g}(1)}}{1 + e^{\hat{g}(1)}} = \dfrac{e^{0.08}}{1 + e^{0.08}} = \dfrac{1.0833}{1 + 1.0833}$

$= 0.52$

e. Using the Minitab output shown in part (c), the estimated odds ratio is 2.79. We can conclude that the estimated odds of making a purchase for customers who have a Simmons credit card are 2.79 times greater than the estimated odds of making a purchase for customers that do not have a Simmons credit card.

32 a. $E(Y) = \dfrac{e^{\beta_0 + \beta_1 X}}{1 + e^{\beta_0 + \beta_1 X}}$

b. A portion of the Minitab binary logistic regression output follows:

e. Repeating the calculations in part (d) using various values for x, a value of $x = 12$ or an average monthly balance of approximately €1200 is required to achieve this level of probability.

f. Using the Minitab output shown in part (b), the estimated odds ratio is 1.25. Because values of x are measured in hundreds of euros, the estimated odds of signing up for payroll direct deposit for customers that have an average monthly balance of €600 is 1.25 times greater than the estimated odds of signing up for payroll direct deposit for customers that have an average monthly balance of €500. Moreover, this interpretation is true for any €100 increment in the average monthly balance.

Chapter 16

Solutions

2   a.   The MINITAB output is shown below:

```
The regression equation is
Y = 9.32 + 0.424 X

Predictor    Coef     SE Coef      T       P
Constant    9.315     4.196      2.22    0.113
X           0.4242    0.1944     2.18    0.117

S = 3.531   R-sq = 61.4%   R-sq(adj) = 48.5%

Analysis of Variance

SOURCE            DF    SS      MS      F      P
Regression         1   59.39   59.39   4.76   0.117
Residual Error     3   37.41   12.47
Total              4   96.80
```

The high p-value (0.117) indicates a weak relationship; note that 61.4 per cent of the variability in y has been explained by x.

   b.   The MINITAB output is shown below:

```
The regression equation is
Y = -8.10 + 2.41 X - 0.0480 XSQ

Predictor    Coef     SE Coef      T       P

Constant   -8.101    4.104      -1.97    0.187
X           2.4127   0.4409      5.47    0.032
XSQ        -0.04797  0.01050    -4.57    0.045

S = 1.279   R-sq = 96.6%   R-sq(adj) = 93.2%

Analysis of Variance

SOURCE            DF    SS       MS      F      P
Regression         2   93.529   46.765  28.60  0.034
Residual Error     2    3.271    1.635
Total              4   96.800
```

At the 0.05 level of significance, the relationship is significant; the fit is excellent.

   c. $\hat{y} = -8.101 + 2.4127(20) - 0.04797(20)^2$
$= 20.965$

4   a.   The MINITAB output is shown below:

```
The regression equation is
Y = 943 + 8.71 X

Predictor    Coef     SE Coef      T       P
Constant   943.05    59.38      15.88    0.000
X            8.714    1.544       5.64    0.005

S = 32.29   R-sq = 88.8%   R-sq(adj) = 86.1%

Analysis of Variance

SOURCE            DF    SS       MS      F      P
Regression         1   33 223   33 223  31.86  0.005
Residual Error     4    4172     1043
Total              5   37 395
```

   b.   The p-value of $0.005 < \alpha = 0.01$; reject H_0

6   a.   The scatter diagram for LifeExp against People per Dr suggests the existence of a possible nonlinear relationship between the two variables:

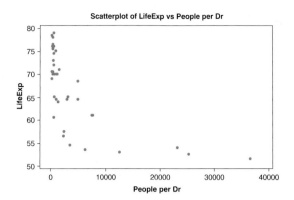

Scatterplot of LifeExp vs People per Dr

   b.   However when the People per Dr variable is replaced by its logarithm in the scatter diagram, a linear model now seems plausible:

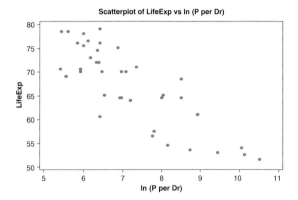

Scatterplot of LifeExp vs ln (P per Dr)

   c.   The situation is exactly analogous for the scatter diagram of LifeExp with People per TV variables. Correspondingly we have the two simple regression models:

$$\widehat{\text{LifeExp}} = 77.887 - 4.26 \ln(\text{P per TV}) \quad R^2 = 0.731$$
$$\widehat{\text{LifeExp}} = 102.873 - 4.974 \ln(\text{P per Dr}) \quad R^2 = 0.693$$

Neither of these relationships is causal but the first with the ln(P per TV) predictor has a slightly better R^2 value which might favour it in this instance.

8   a.   The scatter diagram is shown below:

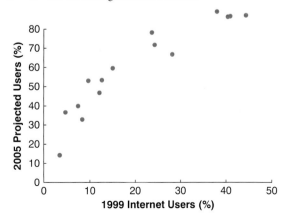

It appears that a simple linear regression model is not appropriate because there is curvature in the plot.

b. The MINITAB output is shown below:

```
The regression equation is
2005% = 17.1 + 3.15 1999% - 0.0445 1999%Sq

Predictor    Coef     SE Coef     T       p

Constant    17.099    4.639     3.69    0.003
1999%        3.1462   0.4971    6.33    0.000
1999%Sq     -0.04454  0.01018  -4.37    0.001

S = 5.667   R-sq = 89.7%   R-sq(adj) = 88.1%

Analysis of Variance

SOURCE            DF     SS      MS      F       p

Regression         2  3646.3  1823.2  56.78  0.000
Residual Error    13   417.4    32.1
Total             15  4063.8
```

c. The MINITAB output is shown below:

```
The regression equation is
Log2000% = 1.17 + 0.449 Log1999%

Predictor       Coef     SE Coef     T       p

Constant     1.17420    0.07468   15.72   0.000
Log1999%     0.44895    0.05978    7.51   0.000

S = 0.08011   R-sq = 80.1%   R-sq(adj) = 78.7%

Analysis of Variance

SOURCE            DF      SS       MS      F       p

Regression         1  0.36199  0.36199  56.40  0.000
Residual Error    14  0.08985  0.00642
Total             15  0.45184
```

d. The estimated regression in part (b) is preferred because it explains a higher percentage of the variability in the dependent variable.

10 a. $SSE = SST - SSR = 1805 - 1760 = 45$

$MSR = 1760/4 = 440$ $MSE = 45/25 = 1.8$

$F = 440/1.8 = 244.44$

$F_{0.05} = 2.76$ (4 degrees of freedom numerator and 25 denominator)

Since $244.44 > 2.76$, variables x_1 and x_4 contribute significantly to the model

b. $SSE(x_1, x_2, x_3, x_4) = 45$

c. $SSE(x_2, x_3) = 1805 - 1705 = 100$

d. $F = \dfrac{(100 - 45)/2}{1.8} = 15.28$

$F_{0.05} = 3.39$ (2 numerator and 25 denominator DF) Since $15.28 > 3.39$ we conclude that x_1 and x_3 contribute significantly to the model.

12 For the first model featuring the five predictors $x_1, x_2, x_3,$ x_4 and x_5, the significant F ratio from the ANOVA table (p-value $= 0.005 < \alpha = 0.05$) suggests that the overall model is a significant fit to the data. Yet none of the individual t tests associated with each of the regression slopes beforehand are significant except that for x_4 (p-value $= 0.005 < \alpha = 0.05$). From the VIF's which are all close to 1, multicollinearity would not appear to be a problem

for the data. The R Square of 66.3 per cent indicates that the multiple regression model explains 66.3 per cent of the variation in the response variable and this might be regarded as quite favourable. On the down side the model suffers from a single outlier according to MINITAB but for a sample of size 20 this does not seem unreasonable. The Durbin-Watson statistic is 1.72 but for a two-sided Durbin-Watson test the relevant d_L and d_U values (based on $n = 20$ and $k = 5$ predictors) are 0.70 and 1.87. As $d_L < 1.72 < d_U$ we deduce the test is inconclusive.

The second model is a simple regression with just x_4 as the predictor. The model is significant according to both the overall F test and the specific t tests associated with the regression slope for x_4. As would be expected the R square value has dropped – in this case to 51.7 per cent. Again there is an outlier (albeit for observation 12 now instead of observation 1 previously but with a corresponding standardized residual of -2.07 this does not look too serious.)

To check if the earlier five predictor model is a significant improvement on this one predictor model, a partial F test can be undertaken. The relevant calculation using equation (16,11) is as follows (note that $p = 5, q = 1$):

$$F = \frac{\dfrac{SSE(x_1, x_2, \ldots, x_q) - SSE(x_1, x_2, \ldots, x_q, x_{q+1}, \ldots, x_p)}{p - q}}{\dfrac{SSE(x_1, x_2, \ldots, x_q, x_{q+1}, \ldots, x_p)}{n - p - 1}}$$

$$= \frac{\dfrac{23.002 - 16.032}{5 - 1}}{\dfrac{16.032}{14}}$$

$$= 1.52 < 3.11 = F_{0.05}(4,14)$$

Hence we are not able to reject $H_0: \beta_1 = \beta_2 = \beta_3 = \beta_5 = 0$ at a 5 per cent significance level and deduce that the five predictor model is not a significant improvement on the corresponding 1 predictor equivalent.

Note that because of the 'ln' transformation on y the relationship between y and x_4 has an essentially exponential character despite the fact that we have effectively used a linear modelling formulation for the analysis.

14 Let Health $= 1$ if health-drugs

Health $= 0$ if energy-international or other

```
The regression equation is
P/E = 10.8 + 0.430 Sales% + 10.6 Health

Predictor     Coef     SE Coef     T       p

Constant    10.817    3.143     3.44    0.004
Sales%       0.4297   0.1813    2.37    0.034
Health      10.600    2.750     3.85    0.002

S = 5.012   R-sq = 55.4%   R-sq(adj) = 48.5%

Analysis of Variance

SOURCE            DF      SS       MS      F       p

Regression         2   405.69   202.85   8.07   0.005
Residual Error    13   326.59    25.12
Total             15   732.28

Source      DF     Seq SS

Sales%       1      32.36

Health       1     373.33
```

16 a. From the best subsets regression summary, the five predictor model with an R Square of 86.2 per cent is almost as good on all measures as the full six predictor model represented by the bottom line of the table. The same five predictor model is described in detail after the correlation matrix and can be seen from the ANOVA F statistic to be significant overall. Corresponding t statistics are also significant (though technically the p-value (of 0.054) associated with the regression slope for the *pop* variable is just slightly above the test size of 5 per cent).

   b. Clearly multicollinearity is a problem here. This is confirmed by significant correlations between predictor variables e.g. *pr* and *con*. Also the sign of the coefficient of the *con* predictor in the detailed regression output is opposite to that of the corresponding correlation between *con* and *ao*.

   c. Yes. In these circumstances the two predictor model from Stepwise now looks technically more appealing.

18 a. Because of the significant correlations between predictors in the summary table at the beginning of the output the possibility of multicollinearity looms large for any subsequent regression modelling. From the Best Subsets table the three predictor model with an R Square of 86.2 per cent compares well with the four predictor baseline model. This is essentially the same model picked out by the Stepwise (backward elimination) analysis afterwards. According to this, all predictors accept Support look as if they could be usefully retained in for regression modelling analysis. Following on, the detailed output for a three predictor regression model shows that Retired, Unemployment and Total Staff are all significant predictors of the response variable, Crimes. Because of relatively low VIF values the model does not seem to suffer from multicollinearity problems mentioned earlier. With an R Square of 84.6 per cent it compares with the three predictor model described earlier fairly well. The Durbin-Watson statistic of 2.22 is not problematic since if $n = 26$ and $k = 3$ then $d_L = 1.04$, $d_U = 1.54$. As $1.54 = d_U < 2.22 < 4-d_U = 2.46$ we deduce there is no evidence of first order serial correlation of residuals present.

   b. For the relevant two predictor model the root mean square error $s = 89.041$. Correspondingly the error sums of squares would be $23 \times 89.041^2 = 182\,350.9$. By comparing this model with the last three predictor model

in a. A partial F test statistic from equation (16.11) can be calculated as follows:

$$F = \frac{\dfrac{SSE(x_1, x_2) - SSE(x_1, x_2, x_3)}{p - q}}{\dfrac{SSE(x_1, x_2, x_3)}{n - p - 1}}$$

$$= \frac{\dfrac{182\,350.9 - 150\,468}{3 - 2}}{\dfrac{150\,468}{22}}$$

$$= 26.66 > 4.30 = F_{.95}\,(1,22)$$

Hence we reject H_0: $\beta_3 = 0$ at the 5 per cent significance level and deduce that the three predictor model is a significant improvement on the corresponding two predictor one.

   c. Following on from b. the three predictor model based on Retired, Unemployment and Total Staff would be preferred.

Chapter 17

Solutions

2 a. From the price relative we see the percentage increase was $(132 - 100) = 32$ per cent.

   b. Divide the current cost by the price relative and multiply by 100.

$$2000 \text{ cost} = \frac{€10.75}{132}\,(100) = €8.14$$

4 a. A, 110: B, 106: C,113

   b. 110

   c. 109. This implies a 9 per cent increase over the two-year period.

6 $I = \dfrac{0.19(500) + 1.80(50) + 4.20(100) + 13.20(40)}{0.15(500) + 1.60(50) + 4.50(100) + 12.00(40)}\,(100)$

   $= 104$

Paasche index

8 164

10 a. Price Relatives A $= (3.95 / 2.50)\,100 = 158$

   B $= (9.90 / 8.75)\,100 = 113$

   C $= (0.95 /0.99)\,100 = 96$

   b.

Item	Price relatives	Base-period price	Quantity	Weight $P_{i0}Q_i$	Weighted price relatives
A	158	2.50	25	62.5	9 875
B	113	8.75	15	131.3	14 837
C	96	0.99	60	59.4	5 702
				253.2	30 414

$I = \dfrac{30414}{253.2} = 120$

Cost of raw materials is up 20 per cent for the chemical.

12 Price relatives: Beer, 108.1: Wine, 106.8: Soft drink, 102.1. Index 105.4.

14 a. Deflated wage rate for 1999 = $3.60 \times \dfrac{100}{165.2} = 2.179$
        (£2.18)

    Deflated wage rate for 2008 = $5.73 \times \dfrac{100}{217.7} = 2.632$
    (£2.63)

    b. per cent change in actual wage rates =

        $100 \times \dfrac{(5.73 - 3.60)}{3.60} = 59.2\%$

    c. per cent change in deflated wage rates =
        $100 \times (2.632 - 2.179)/2.179 = 20.8\%$

16 2000, 3807; 2001, 3629; 2002, 3732; 2003, 3876; 2004, 3835; 2005, 3664; 2006, 3672; 2007, 3862; 2008, 3868 (1987 values). Volume falls noticeably 2000–2001 and 2004–2005, falls slightly 2003–2004. Rises in other years. Volume in 2008 only slightly above 2001 level.

18 $I = \dfrac{300(18.00) + 400(4.90) + 850(15.00)}{350(18.00) + 220(4.90) + 730(15.00)} (100)$

    $= \dfrac{20\,110}{18\,328} (100) = 110$

20 Quantity relatives are 85, 80, 80 respectively. Weighted aggregate quantity index is 83.

Chapter 18

Solutions

2 a.

b. MSE(4-Week) = 77.18/8 = 9.65
    MSE(5-Week) = 51.84/7 = 7.41

c. For the limited data provided, the five-week moving average provides the smallest MSE.

4

Week	Sales	Forecast	Error	Squared error
1	17			
2	21	17	4.00	16.00
3	19	18.2	0.80	0.64
4	23	18.44	4.56	20.79
5	18	19.808	−1.81	3.27
6	16	19.2656	−3.27	10.66
7	20	18.28592	1.71	2.94
8	18	18.80014	−0.80	0.64
9	22	18.5601	3.44	11.83
10	20	19.59207	0.41	0.17
11	15	19.71445	−4.71	22.23
12	22	18.30011	3.70	13.69
			MSE	9.35

6 a. $F_{13} = 0.2Y_{12} + 0.16Y_{11} + 0.64(0.2Y_{10} + 0.8F_{10})$

       $= 0.2Y_{12} + 0.16Y_{11} + 0.128Y_{10} + 0.512F_{10}$

    $F_{13} = 0.2Y_{12} + 0.16Y_{11} + 0.128Y_{10} + 0.512$

    $(0.2Y_9 + 0.8F_9) = 0.2Y_{12} + 0.16Y_{11} + 0.128Y_{10}$
    $+ 0.1024Y_9 + 0.4096F_9$

Week	Time-series value	4-week moving average forecast	(Error)^2	5-week moving average forecast	(Error)^2
1	17				
2	21				
3	19				
4	23				
5	18	20.00	4.00		
6	16	20.25	18.06	19.60	12.96
7	20	19.00	1.00	19.40	0.36
8	18	19.25	1.56	19.20	1.44
9	22	18.00	16.00	19.00	9.00
10	20	19.00	1.00	18.80	1.44
11	15	20.00	25.00	19.20	17.64
12	22	18.75	10.56	19.00	9.00
			77.18		51.84

$$F_{13} = 0.2Y_{12} + 0.16Y_{11} + 0.128Y_{10} + 0.1024Y_9$$
$$+ 0.4096(0.2Y_8 + 0.8F_8) = 0.2Y_{12} + 0.16Y_{11}$$
$$+ 0.128Y_{10} + 0.1024Y_9 + 0.08192Y_8$$
$$+ 0.32768F_8$$

b. The more recent data receives the greater weight or importance in determining the forecast. The moving averages method weights the last n data values equally in determining the forecast.

8 a.

Month	Time-series value	3-month moving average forecast	(Error)²	$\alpha = 0.2$ forecast	(Error)²
1	240				
2	350			240.00	12 100.00
3	230			262.00	1 024.00
4	260	273.33	177.69	255.60	19.36
5	280	280.00	0.00	256.48	553.19
6	320	256.67	4 010.69	261.18	3 459.79
7	220	286.67	4 444.89	272.95	2 803.70
8	310	273.33	1 344.69	262.36	2 269.57
9	240	283.33	1 877.49	271.89	1 016.97
10	310	256.67	2 844.09	265.51	1 979.36
11	240	286.67	2 178.09	274.41	1 184.05
12	230	263.33	1 110.89	267.53	1 408.50
			17 988.52		27 818.49

MSE(3-Month) = 17 988.52/9 = 1998.72

MSE($\alpha = 0.2$) = 27 818.49/11 = 2528.95

Based on the above MSE values, the three- month moving averages appears better. However, exponential smoothing was penalized by including month two which was difficult for any method to forecast. Using only the errors for months 4 to 12, the MSE for exponential smoothing is revised to

MSE($\alpha = 0.2$) = 14 694.49/9 = 1632.72

Thus, exponential smoothing was better considering months 4 to 12.

b. Using exponential smoothing,

$$F_{13} = \alpha Y_{12} + (1 - \alpha)F_{12} = 0.20(230)$$
$$+ 0.80(267.53) = 260$$

10 The following values are needed to compute the slope and intercept:

$$\Sigma t = 15 \quad \Sigma t^2 = 15 \quad \Sigma Y_t = 55 \quad \Sigma tY_t = 186$$

$$b_1 = \frac{\Sigma tY_t - (\Sigma t \Sigma Y_t)}{\Sigma t^2 - (\Sigma t)^2/n} = \frac{186 - (15)(55)15}{5/55 - (15)^2/5} = 2.1$$

$$b_0 = \bar{Y} - b_1\bar{t} = 11 - 2.1(3) = 4.7$$
$$T_t = 4.7 + 2.1t$$

Forecast: $T_6 = 4.7 + 2.1(6) = 17.3$

12 A linear trend model is not appropriate. A nonlinear model would provide a better approximation.

14

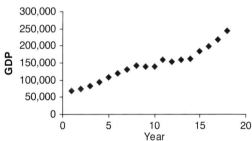

The graph shows a roughly linear trend.

b. The following values are needed to compute the slope and intercept:

$$\Sigma t = 171 \quad \Sigma t^2 = 2109 \quad \Sigma Y_t = 2572117$$
$$\Sigma_{tY_t} = 28734175$$

Computation of slope:

$$b_1 = \frac{\Sigma tY_t - (\Sigma t \Sigma Y_t)}{\Sigma t^2 - (\Sigma t)^2/n} = \frac{28734175 - (171)(2572117)18}{18/2109 - (171)^2/18}$$
$$= 8873.2$$

Computation of intercept:

$$b_0 = \bar{Y} - b_1\bar{t} = 142895.4 - 8873.2)(9.5) = 58600$$

Equation for linear trend: $T_t = 58\,600 + 8873.2t$

c. 2008: $T_t = 58\,600 + 8873.2(19) = 227\,190.8$

2009: $T_t = 58\,600 + 8873.2(20) = 236\,064$

2010: $T_t = 58\,600 + 8873.2(21) = 244\,937.2$

16 a.

Year	Quarter	Y_t	Four-quarter moving average	Centred moving average
1	1	4		
	2	2		
			3.50	
	3	3		3.750
			4.00	
	4	5		4.125
			4.25	
2	1	6		4.500
			4.75	
	2	3		5.000
			5.25	
	3	5		5.375
			5.50	
	4	7		5.875
			6.25	
3	1	7		6.375
			6.50	
	2	6		6.625
			6.75	
	3	6		
	4	8		

b.

Year	Quarter	Y_t	Centred moving average	Seasonal-irregular component
1	1	4		
	2	2		
	3	3	3.750	0.8000
	4	5	4.125	1.2121
2	1	6	4.500	1.3333
	2	3	5.000	0.6000
	3	5	5.375	0.9302
	4	7	5.875	1.1915
3	1	7	6.375	1.0980
	2	6	6.625	0.9057
	3	6		
	4	8		

Quarter	Seasonal-irregular component values	Seasonal index	Adjusted seasonal index
1	1.3333, 1.0980	1.2157	1.2050
2	0.60000, 0.9057	0.7529	0.7463
3	0.80000, 0.9032	0.8651	0.8675
4	1.2121, 1.1915	1.2018	1.1912
		4.0355	

Note: Adjustment for seasonal index = 4.000/4.0355 = 0.9912

18 a. Yes, there is a seasonal effect over the 24-hour period.

Time period	Seasonal index
12–4 a.m.	1.696
4–8 a.m.	1.458
8–12	0.711
12–4 p.m.	0.326
4–8 p.m.	0.448
8–12	1.362

b.

Time period	Forecast
12–4 p.m.	166 761.13
4–8 p.m.	146 052.99

Chapter 19

Solutions

2 There are $n = 27$ cases in which a value different from 150 is obtained. Use the normal approximation with $\mu = n\pi = 0.5(27) = 13.5$ and $\sigma = \sqrt{0.25n} = \sqrt{0.25(27)} = 2.6$. Use $x = 22$ as the number of plus signs and obtain the following test statistic:

$$z = \frac{x - \mu}{\sigma} = \frac{22 - 13.5}{2.6} = 3.27$$

Last entry in normal distribution table is $z > 3.09$, cumulative probability = 0.9990. For $z = 3.27$, p-value is less than $(1 - 0.9990) = 0.0010$. p-value < 0.01, reject H_0. Conclusion: The median is greater than 150.

4 a. $H_0: \pi = 0.50$, $H_1: \pi \neq 0.50$, where π = probability the shares held will be worth more after the split

 b. $n = 18$, $x = 14$. p-value = 0.0155, reject H_0. The results support the conclusion that stock splits are beneficial for shareholders.

6 H_0: Median $= 152\,000$, H_1: Median $\neq 152\,000$

$\mu = 0.5n = 0.5(225) = 112.5$, $\sigma = \sqrt{0.25n} = \sqrt{0.25(225)} = 7.5$

For $x = 122$, $z = (202 - 112.5)/7.5 = 1.27$

Area in tail $= 1 - 0.8980 = 0.1020$. p-value $= 2(0.1020) = 0.2040$. p-value > 0.05, do not reject H_0. We are unable to conclude that the median annual income needed differs from that reported in the survey.

8   a. $n = 43$, $x = 32$. p-value < 0.001, reject H_0. Conclusion: Median selling price in Greater Manchester is lower than national median.

b. $n = 43$, $x = 32$. p-value $= 0.014$, reject H_0. Conclusion: Median selling price in Oxfordshire is higher than national median.

10

Without relaxant	With relaxant	Difference	Rank of absolute difference	Signed rank
15	10	5	9	9
12	10	2	3	3
22	12	10	10	10
8	11	−3	6.5	−6.5
10	9	1	1	1
7	5	2	3	3
8	10	−2	3	−3
10	7	3	6.5	6.5
14	11	3	6.5	6.5
9	6	3	6.5	6.5
				Total 36

$$\mu_T = 0,\ \sigma_T = \sqrt{\frac{n(n+1)(2n+1)}{6}} = \sqrt{\frac{10(11)(21)}{6}} = 19.62$$

$$z = \frac{T - \mu_T}{\sigma_T} = \frac{36}{19.62} = 1.83$$

Using a one-tailed test, p-value $= 1 - 0.9664 = 0.0336$. p-value < 0.05, reject H_0. Conclusion: There is a significant difference, in favour of the relaxant.

12 $T = 32$, $\sigma_T = 19.6$, $z = 1.63$. p-value $= 0.0516$, do not reject H_0. Insufficient evidence to conclude that the campaign had an effect on sales.

14 H_0: There is no difference between the distributions of P/E ratios

H_1: There is a difference between the distributions of P/E ratios

Company	Japan P/E ratio	Rank	Company	US P/E ratio	Rank
Sumitomo Corp.	153	20	Gannet	19	6
Kinden	21	8	Motorola	24	11.5
Heiwa	18	5	Schlumberger	24	11.5
NCR Japan	125	19	Oracle Systems	43	16
Suzuki Motor	31	13	Gap	22	10
Fuji Bank	213	21	Winn-dixie	14	2
Sumitomo Chemical	64	17	Ingersoll-Rand	21	8
Seibu Railway	666	22	Am. Elec. Power	14	2
Shiseido	33	14	Hercules	21	8
Toho Gas	68	18	Times Mirror	38	15
		Total 157	WellPoint Health	15	4
			No. States Power	14	2
					Total 96

$$\mu_T = \frac{1}{2} n_1(n_1 + n_2 + 1) = \frac{1}{2} 10(10 + 12 + 1) = 115$$

$$\sigma_T = \sqrt{\frac{1}{12} n_1 n_2 (n_1 + n_2 + 1)}$$

$$= \sqrt{\frac{1}{12}(10)(12)(10 + 12 + 1)} = 15.17$$

$$T = 157, z = \frac{157 - 115}{15.17} = 2.77, p\text{-value} =$$

$2(1 - 0.9972) = 0.0056$. p-value < 0.01, reject H_0. We conclude that there is a significant difference in P/E ratios for the two countries.

16   $n_1 = 10, n_2 = 8. T = 82, T_L = 73, T_U = 117$. Do not reject H_0. Cannot conclude that the cost of kitchen re-modelling differs from the cost of master bedroom re-modelling.

18

A	B	C
11.5	5.0	17.0
2.5	11.5	20.0
8.0	2.5	15.0
10.0	4.0	8.0
8.0	6.0	16.0
18.0	1.0	19.0
	13.0	14.0
58.0	43.0	109.0

$$W = \frac{12}{(20)(21)} \left[\frac{(58)^2}{6} + \frac{(43)^2}{7} + \frac{(109)^2}{7} \right] - 3(21) = 9.06$$

Degrees of freedom $= 2$. Using χ^2 table, $\chi^2 = 9.06$ shows p-value is between 0.01 and 0.025 (actual p-value $= 0.0108$). p-value > 0.01, do not reject H_0. We cannot conclude that there is a significant difference in test preparation programmes.

20   $W = 4.13, df = 2, p\text{-value} > 0.10$, do not reject H_0. Insufficient evidence to conclude that the ranks differ between the three cruise lines.

22

a. $\sum d_i^2 = 52$

$$r_s = 1 - \frac{6\sum d_i^2}{n(n^2 - 1)} = 1 - \frac{6(52)}{10(99)} = 0.68$$

b. $\sigma_{r_s} = \sqrt{\frac{1}{n - 1}} = \sqrt{\frac{1}{9}} = 0.33, z = \frac{r_s - 0}{\sigma_{r_s}} = \frac{0.68}{0.33} = 2.05$

p-value $= 2(1 - 0.9798) = 0.0404$. p-value < 0.05, reject H_0. Conclude that significant rank correlation exists.

24   a. $\sum d_i^2 = 54, r_s = 0.67$

b. $z = 2.02, p\text{-value} = 0.0434$

c.   Reject H_0. Conclude there is a significant positive rank correlation.

26   $\sum d_i^2 = 136, r_s = 0.76, z = 2.83, p\text{-value} = 0.0046$, reject H_0. Conclude that there is a significant rank correlation between the two exams.

Chapter 20

Solutions

2   a. $\mu = \frac{677.5}{25(5)} = 5.42$

b.   UCL $= \mu + 3(\sigma/\sqrt{n}) = 5.42 + 3(0.5/\sqrt{5}) = 6.09$
LCL $= \mu - 3(\sigma/\sqrt{n}) = 5.42 - 3(0.5/\sqrt{5}) = 4.75$

4   R Chart:
UCL $= \bar{R}D_4 = 1.6(1.864) = 2.98$
LCL $= \bar{R}D_3 = 1.6(0.136) = 0.22$

$\bar{x}$ Chart:
UCL $= \bar{x} + A_2\bar{R} = 28.5 + 0.373(1.6) = 29.10$
LCL $= \bar{x} - A_2\bar{R} = 28.5 - 0.373(1.6) = 27.90$

6   Process Mean $= \frac{0.570 + 0.564}{2} = 0.567$

UC $\sigma = 0.0022$

Process Standard Deviation $= \frac{0.570 + 0.564}{6}$
$\sqrt{5} = (0.001)\sqrt{5} = 0.002$

When $\pi = 0.06$, the probability of accepting the lot is

$$p(x) = \frac{25!}{0!(25 - 0)!} 0.06^0 (1 - 0.06)^{(25)} = 0.2129$$

8   Using MINITAB, the following $\bar{x}$ and R charts are obtained:

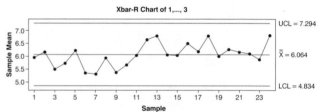

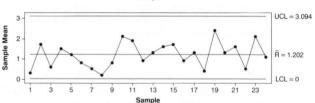

These show - in each case - the process to be in control.

10   $p(x) = \frac{n!}{x!(n - x)!} \pi^x (1 - \pi)^{n-x}$

When $\pi = 0.02$, the probability of accepting the lot is:

$$p(0) = \frac{25!}{0!(25 - 0)!} 0.02^0 (1 - 0.02)^{25-0} = 0.6035$$

When $\pi = 0.02$, the probability of accepting the lot is:

$$p(0) = \frac{25!}{0!(25 - 0)!} 0.06^0 (1 - 0.06)^{25-0} = 0.2129$$

12 At $p_0 = 0.05$, the $n = 10$ and $c = 1$ plan provides
P (Accept lot) = $p(0) + p(1) = 0.5987 + 0.3151 = 0.9139$
Producer's risk: $\alpha = 1 - 0.9139 = 0.0861$
At $p_0 = 0.05$, the $n = 25$ and $c = 3$ plan provides
P (Accept lot) = $p(0) + p(1) + p(2 + p(3) = 0.2774 +$
$0.3650 + 0.2305 + 0.0930 = 0.9659$
Producer's risk: $\alpha = 1 - 0.9659 = 0.0341$
At $p_0 = 0.15$, the $n = 10$ and $c = 1$ plan provides
P (Accept lot) = $p(0) + p(1) = 0.1969 + 0.3474 = 0.5443$
Producer's risk: $\alpha = 1 - 0.5443 = 0.4557$
At $p_0 = 0.15$, the $n = 25$ and $c = 3$ plan provides
P (Accept lot) = $p(0) + p(1) + p(2 + p(3) = 0.0172$
$0.0759 + 0.1607 + 0.2174 = 0.4711$
Producer's risk: $\alpha = 1 - 0.9659 = 0.5289$
So plan (n=25, c=3) favoured when p_0 close to 5 per cent
but plan (n =10,c=1) favoured when p_0 close to 15 per cent.

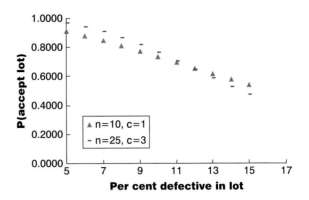

14 a. P (Accept) shown for π values below:

c	$\pi = 0.01$	$\pi = 0.05$	$\pi = 0.08$	$\pi = 0.10$	$\pi = 0.15$
0	0.8179	0.3585	0.1887	0.1216	0.0388
1	0.9831	0.7359	0.5169	0.3918	0.1756
2	0.9990	0.9246	0.7880	0.6770	0.4049

The operating characteristic curves would show the P (Accept) versus p for each value of c.

b. P (Accept)

c	At $p_0 = 0.01$	Producer's Risk	At $p_1 = .08$	Consumer's Risk
0	0.8179	0.1821	0.1887	0.1887
1	0.9831	0.0169	0.5169	0.5169
2	0.9990	0.0010	0.7880	0.7880

Chapter 21

Solutions

2   a. EV(d_1) = 0.5(14) + 0.2(9) + 0.2(10)
     + 0.1(5) = 11.3
     EV(d_2) = 0.5(11) + 0.2(10) + 0.2(8)
     + 0.1(7) = 9.8
     EV(d_3) = 0.5(9) + 0.2(10) + 0.2(10)
     + 0.1(11) = 9.6
     EV(d_4) = 0.5(8) + 0.2(10) + 0.2(11)
     + 0.1(13) = 9.5
     Recommended decision. d_1

    b. The best decision in this case is the one with the
        smallest expected value: thus, d_4, with an expected cost
        of 9.5, is the recommended decision.

4   a. The decision to be made is to choose the type of service
        to provide. The chance event is the level of demand for
        the Magyar Air service. The consequence is the amount
        of quarterly profit. There are two decision alternatives
        (full price and discount service). There are two
        outcomes for the chance event (strong demand and weak
        demand).

    b. EV(Full) = 0.7(960) + 0.3(−490) = 525
        EV (Discount) = 0.7(670) + 0.3(320) = 565
        Optimal Decision. Discount service

    c. EV(Full) = 0.8(960) + 0.2(−490) = 670
        EV (Discount) = 0.8(670) + 0.2(320) = 600
        Optimal Decision. Full price service

6   a. The decision is to choose what type of grapes to
        plant, the chance event is demand for the wine and the
        consequence is the expected annual profit contribution.
        There are three decision alternatives (Chardonnay, Pinot
        and both). There are four chance outcomes. (W,W):
        (W,S): (S,W): and (S,S). For instance, (W,S) denotes
        the outcomes corresponding to weak demand for
        Chardonnay and strong demand for Pinot.

b. In constructing a decision tree, it is only necessary to show two branches when only a single grape is planted. But, the branch probabilities in these cases are the sum of two probabilities. For example, the probability that demand for Chardonnay is strong is given by.

$$P \text{ (Strong demand for Chardonnay)} = P(S,W) + P(S,S)$$
$$= 0.25 + 0.20$$
$$= 0.45$$

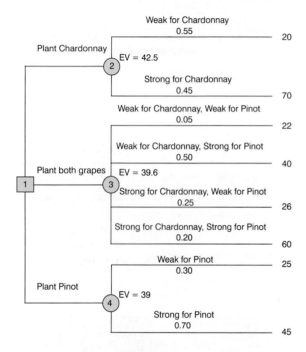

c. EV (Plant Chardonnay) = 0.55(20) + 0.45(70) = 42.5
   EV (Plant both grapes) = 0.05(22) + 0.50(40) + 0.25(26) + 0.20(60) = 39.6
   EV (Plant Pinot) = 0.30(25) + 0.70(45) = 39.0
   Optimal decision: Plant Chardonnay grapes only.

d. This changes the expected value in the case where both grapes are planted and when Pinot only is planted.
   EV (Plant both grapes) = 0.05(22) + 0.50(40) + 0.05(26) + 0.40(60) = 46.4
   EV (Plant Pinot) = 0.10(25) + 0.90(45) = 43.0
   We see that the optimal decision is now to plant both grapes. The optimal decision is sensitive to this change in probabilities.

e. Only the expected value for node 2 in the decision tree needs to be recomputed.
   EV (Plant Chardonnay) = 0.55(20) + 0.45(50) = 33.5
   This change in the payoffs makes planting Chardonnay only less attractive. It is now best to plant both types of grapes. The optimal decision is sensitive to a change in the payoff of this magnitude.

8 a.

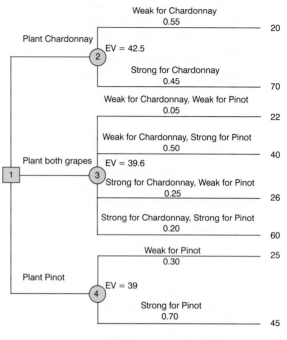

b. EV (node 6) = 0.57(100) + 0.43(300) = 186
   EV (node 7) = 0.57(400) + 0.43(200) = 314
   EV (node 8) = 0.18(100) + 0.82(300) = 264
   EV (node 9) = 0.18(400) + 0.82(200) = 236
   EV (node 10) = 0.40(100) + 0.60(300) = 220
   EV (node 11) = 0.40(400) + 0.60(200) = 280
   EV (node 3) = Max(186,314) = 314 d_2
   EV (node 4) = Max(264,236) = 264 d_1
   EV (node 5) = Max(220,280) = 280 d_2
   EV (node 2) − 0.56(314) + 0.44(264) − 292
   EV (node 1) = Max(292,280) = 292
   Market Research
   If Favourable, decision d_2
   If Unfavourable, decision d_1

10 a. Outcome 1 (€ in 000s)

Bid	−€200
Contract	−2000
Market Research	−150
High Demand	+5000
	€2650

Outcome 2 (€ in 000s)

Bid	−€200
Contract	−2000
Market Research	−150
Moderate Demand	+3000
	€650

b. EV (node 8) = 0.85(2650) + 0.15(650) = 2350

    EV (node 5) = Max(2350, 1150) = 2350
                Decision: Build

    EV (node 9) = 0.225(2650) + 0.775(650) = 1100

    EV (node 6) = Max(1100, 1150) = 1150
                Decision: Sell

    EV (node 10) = 0.6(2800) + 0.4(800) = 2000

    EV (node 7) = Max(2000, 1300) = 2000
                Decision: Build

    EV (node 4) = 0.6 EV(node 5) + 0.4 EV(node 6)
                = 0.6(2350) + 0.4(1150) = 1870

    EV (node 3) = Max (EV(node 4), EV (node 7))
                = Max (1870, 2000) = 2000
                Decision: No Market Research

    EV (node 2) = 0.8 EV(node 3) + 0.2 (− 200)
                = 0.8(2000) + 0.2(−200) = 1560

    EV (node 1) = Max (EV(node 2), 0) = Max
                (1560, 0) = 1560
                Decision: Bid on Contract

  Decision Strategy:

    Bid on the Contract

    Do not do the Market Research

    Build the Complex

    Expected Value is €1 560 000

c. Compare Expected Values at nodes 4 and 7.

    EV(node 4) = 1870 Includes €150 cost for research

    EV (node 7) = 2000

    Difference is 2000 − 1870 = €130

  Market research cost would have to be lowered €130 000 to €20 000 or less to make undertaking the research desirable.

12 a.

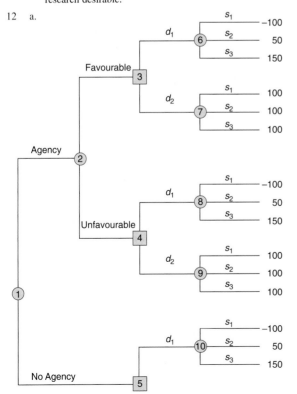

b. Using Node 5,

    EV (node 10) = 0.20(−100) + 0.30(50) + 0.50(150)
               = 70

    EV (node 11) = 100
    Decision: d_1 Sell
    Expected Value €100

c. Expected value with perfect information = 0.20(100) + 0.30(100) + 0.50(150) = €125

    EVPI = €125 − €100 = €25

d. EV (node 6) = 0.09(−100) + 0.26(50) + 0.65(150)
             = 101.5

    EV (node 7) = 100

    EV (node 8) = 0.45(−100) + 0.39(50) + 0.16(150)
             = −1.5

    EV (node 9) = 100

    EV (node 3) = Max(101.5,100) = 101.5 Produce

    EV (node 4) = Max(−1.5,100) = 100 Sell

    EV (node 2) = 0.69(101.5) + 0.31(100) = 101.04
    If Favourable, Produce
    If Unfavourable, Sell EV = €101.04

e. EVSI = €101.04 − 100 = €1.04 or €1,040.

f. No, maximum Hale should pay is €1040.

g. No agency; sell the pilot.

14

State of nature	$P(s_j)$	$P(I \mid s_j)$	$P(I \cap s_j)$	$P(s_j \mid I)$
s_1	0.2	0.10	0.020	0.1905
s_2	0.5	0.05	0.025	0.2381
s_3	0.3	0.20	0.060	0.5714
	1.0	$P(I) = 0.105$		1.0000

16

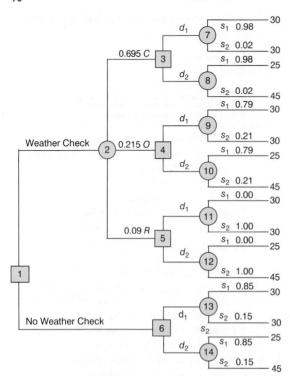

a, b. The revised probabilities are shown on the branches of the decision tree.

EV (node 7) = 30

EV (node 8) = 0.98(25) + 0.02(45) = 25.4

EV (node 9) = 30

EV (node 10) = 0.79(25) + 0.21(45) = 29.2

EV (node 11) = 30

EV (node 12) = 0.00(25) + 1.00(45) = 45.0

EV (node 13) = 30

EV (node 14) = .85(25) + 0.15(45) = 28.0

EV (node 3) = Min(30,25.4) = 25.4 Expressway EV (node 4)

   = Min(30,29.2) = 29.2 Expressway EV (node 5)

   = Min(30,45) = 30.0 Boulevard Peripherique

EV (node 6) = Min(30,28) = 28.0 Expressway

EV (node 2) = 0.695(25.4) + 0.215(29.2)

+ 0.09(30.0) = 26.6

EV (node 1) = Min(26.6,28) = 26.6
Weather check

c. Check the weather, take the expressway unless there is rain. If rain, take Boulevard Peripherique. Expected time. 26.6 minutes.

Glossary

Acceptance criterion The maximum number of defective items that can be found in the sample and still indicate an acceptable lot.

Acceptance sampling A statistical method in which the number of defective items found in a sample is used to determine whether a lot should be accepted or rejected.

Addition law A probability law used to compute the probability of the union of two events. It is $P(A \cup B) = P(A) + P(B) - P(A \cap B)$. For mutually exclusive events, $P(A \cap B) = 0$; in this case the addition law reduces to $P(A \cup B) - P(A) + P(B)$.

Adjusted multiple coefficient of determination A measure of the goodness of fit of the estimated multiple regression equation that adjusts for the number of independent variables in the model and thus avoids overestimating the impact of adding more independent variables.

Aggregate price index A composite price index based on the prices of a group of items.

Alternative hypothesis The hypothesis concluded to be true if the null hypothesis is rejected.

ANOVA table A table used to summarize the analysis of variance computations and results. It contains columns showing the source of variation, the sum of squares, the degrees of freedom, the mean square, and the F value(s).

Assignable causes Variations in process outputs that are due to factors such as machine tools wearing out, incorrect machine settings, poor-quality raw materials, operator error, and so on. Corrective action should be taken when assignable causes of output variation are detected.

Autocorrelation Correlation in the errors that arises when the error terms at successive points in time are related.

Autoregressive model A time series model whereby a regression relationship based on past time series values is used to predict the future time series values.

Bar graph, Bar chart A graphical device for depicting qualitative data that have been summarized in a frequency, relative frequency, or percentage frequency distribution.

Basic requirements for assigning probabilities Two requirements that restrict the manner in which probability assignments can be made: (1) for each experimental outcome E_i we must have $0 \leq P(E_i) \leq 1$; (2) considering all experimental outcomes, we must have $P(E_1) + P(E_2) + \cdots + P(E_n) = 1.0$.

Bayes' theorem A theorem that enables the use of sample information to revise prior probabilities.

Binomial experiment An experiment having the four properties stated at the beginning of Section 5.4.

Binomial probability distribution A probability distribution showing the probability of x successes in n trials of binomial experiments.

Binomial probability function The function used to compute binomial probabilities.

Blocking The process of using the same or similar experimental units for all treatments. The purpose of blocking is to remove a source of variation from the error term and hence provide a more powerful test for a difference in population or treatment means.

Bound on the sampling error A number added to and subtracted from a point estimate to create an approximate 95 per cent confidence interval. It is given by two times the standard error of the point estimator.

Box plot A graphical summary of data based on a five-number summary.

Branch Lines showing the alternatives from decision nodes and the outcomes from chance nodes.

Causal forecasting methods Forecasting methods that relate a time series to other variables that are believed to explain or cause its behaviour.

Census A survey to collect data on the entire population.

Central limit theorem A theorem that enables one to use the normal probability distribution to approximate the sampling distribution of $\bar{X}$ when the sample size is large.

Chance event An uncertain future event affecting the consequence, or payoff, associated with a decision.

Chance nodes Nodes indicating points where an uncertain event will occur.

Chebyshev's theorem A theorem that can be used to make statements about the proportion of data values that must be within a specified number of standard deviations of the mean.

Class midpoint The value halfway between the lower and upper class limits in a frequency distribution.

Classical method A method of assigning probabilities that is appropriate when all the experimental outcomes are equally likely.

Cluster sampling A probabilistic method of sampling in which the population is first divided into clusters and then one or more clusters are selected for sampling. In single-stage cluster sampling, every element in each selected cluster is sampled; in two-stage cluster sampling, a sample of the elements in each selected cluster is collected.

Coefficient of determination A measure of the goodness of fit of the estimated regression equation. It can be interpreted as the proportion of the variability in the dependent variable y that is explained by the estimated regression equation.

Coefficient of variation A measure of relative variability computed by dividing the standard deviation by the mean and multiplying by 100.

Common causes Normal or natural variations in process outputs that are due purely to chance. No corrective action is necessary when output variations are due to common causes.

Comparisonwise Type I error rate The probability of a Type I error associated with a single pairwise comparison.

Complement of A The event consisting of all sample points that are not in A.

Completely randomized design An experimental design in which the treatments are randomly assigned to the experimental units.

Conditional probability The probability of an event given that another event already occurred. The conditional probability of A given B is $P(A \mid B) = P(A \cap B)/P(B)$.

Confidence coefficient The confidence level expressed as a decimal value. For example, 0.95 is the confidence coefficient for a 95 per cent confidence level.

Confidence interval The interval estimate of the mean value of Y for a given value of X.

Confidence level The confidence associated with an interval estimate. For example, if an interval estimation procedure provides intervals such that 95 per cent of the intervals formed using the procedure will include the population parameter, the interval estimate is said to be constructed at the 95 per cent confidence level.

Consequence The result obtained when a decision alternative is chosen and a chance event occurs. A measure of the consequence is often called a payoff.

Consumer Price Index A price index that uses the price changes in a market basket of consumer goods and services to measure the changes in consumer prices over time.

Consumer's risk The risk of accepting a poor-quality lot; a Type II error.

Contingency table A table used to summarize observed and expected frequencies for a test of independence.

Continuity correction factor A value of 0.5 that is added to or subtracted from a value of X when the continuous normal distribution is used to approximate the discrete binomial distribution.

Continuous random variable A random variable that may assume any numerical value in an interval or collection of intervals.

Control chart A graphical tool used to help determine whether a process is in control or out of control.

Convenience sampling A non-probabilistic method of sampling whereby elements are selected on the basis of convenience.

Cook's distance measure A measure of the influence of an observation based on both the leverage of observation i and the residual for observation i.

Correlation coefficient A measure of association between two variables that takes on values between -1 and $+1$. Values near $+1$ indicate a strong positive relationship, values near -1 indicate a strong negative relationship. Values near zero indicate the lack of a relationship. Pearson's product-moment correlation coefficient measures linear association between two variables.

Covariance A measure of linear association between two variables. Positive values indicate a positive relationship; negative values indicate a negative relationship.

Critical value A value that is compared with the test statistic to determine whether H_0 should be rejected.

Cross-sectional data Data collected at the same or approximately the same point in time.

Cross-tabulation A tabular summary of data for two variables. The classes for one variable are represented by the rows; the classes for the other variable are represented by the columns.

Cumulative frequency distribution A tabular summary of quantitative data showing the number of items with values less than or equal to the upper class limit of each class.

Cumulative percentage frequency distribution A tabular summary of quantitative data showing the percentage of items with values less than or equal to the upper class limit of each class.

Cumulative relative frequency distribution A tabular summary of quantitative data showing the fraction or proportion of items with values less than or equal to the upper class limit of each class.

Cyclical component The component of the time series that results in periodic above-trend and below-trend behaviour of the time series lasting more than one year.

Data The facts and figures collected, analyzed, and summarized for presentation and interpretation.

Decision nodes Nodes indicating points where a decision is made.

Data set All the data collected in a particular study.

Decision strategy A strategy involving a sequence of decisions and chance outcomes to provide the optimal solution to a decision problem.

Decision tree A graphical representation of the decision problem that shows the sequential nature of the decision-making process.

Degrees of freedom A parameter of the t distribution. When the t distribution is used in the computation of an interval estimate of a population mean, the appropriate t distribution has $n - 1$ degrees of freedom, where n is the size of the simple random sample. (Also a parameter of the χ^2 distribution.)

Delphi method A qualitative forecasting method that obtains forecasts through group consensus.

Dependent variable The variable that is being predicted or explained. It is denoted by Y.

Descriptive statistics Tabular, graphical and numerical summaries of data.

Deseasonalized time series A time series from which the effect of season has been removed by dividing each original time series observation by the corresponding seasonal index.

Discrete random variable A random variable that may assume either a finite number of values or an infinite sequence of values.

Discrete uniform probability distribution A probability distribution for which each possible value of the random variable has the same probability.

Dot plot A graphical device that summarizes data by the number of dots above each data value on the horizontal axis.

Dummy variable A variable used to model the effect of qualitative independent variables. A dummy variable may take only the value zero or one.

Durbin-Watson test A test to determine whether first-order correlation is present.

Element The entity on which data are collected.

Empirical rule A rule that can be used to compute the percentage of data values that must be within one, two and three standard deviations of the mean for data that exhibit a bell-shaped distribution.

Estimated logistic regression equation The estimate of the logistic regression equation based on sample data: that is $\hat{y}$ = estimate of

$$P(Y = 1 \mid x_1, x_2, \ldots x_p) = \frac{e^{\beta + \beta_1 x_1 + \cdots + \beta_p x_p}}{1 + e^{\beta_0 + \beta_1 x_1 + \cdots + \beta_p x_p}}$$

Estimated logit An estimate of the logit based on sample data: that is,

$$\hat{g}(x_1, x_2, \ldots x_p) = b_0 + b_1 x_1 + b_2 x_2 + \cdots + b_p x_p$$

Estimated multiple regression equation The estimate of the multiple regression equation based on sample data and the least squares method: it is

$$\hat{y} = b_0 + b_1 x_1 + b_2 x_2 + \cdots + b_p x_p$$

Estimated regression equation The estimate of the regression equation developed from sample data by using the least squares method. For simple linear regression, the estimated regression equation is

$$\hat{y} = b_0 + b_1 x$$

Event A collection of sample points.

Expected value A measure of the central location of a random variable. For a chance node, it is the weighted average of the payoffs. The weights are the state-of-nature probabilities.

Expected value approach An approach to choosing a decision alternative that is based on the expected value

of each decision alternative. The recommended decision alternative is the one that provides the best expected value.

Expected value of perfect information (EVPI) The expected value of information that would tell the decision-maker exactly which state of nature is going to occur (i.e. perfect information).

Expected value of sample information (EVSI) The difference between the expected value of an optimal strategy based on sample information and the 'best' expected value without any sample information.

Experiment A process that generates well-defined outcomes.

Experimental units The objects of interest in the experiment.

Experimentwise Type I error rate The probability of making a Type I error on at least one of several pairwise comparisons.

Exploratory data analysis Methods that use simple arithmetic and easy-to-draw graphs to summarize data quickly.

Exponential probability distribution A continuous probability distribution that is useful in computing probabilities for the time it takes to complete a task.

Exponential smoothing A forecasting technique that uses a weighted average of past time series values as the forecast.

Factor Another word for the independent variable of interest.

Factorial experiment An experimental design that allows statistical conclusions about two or more factors.

Finite population correction factor The term $\sqrt{(N - n)/(N - 1)}$ that is used in the formulae for $\sigma_{\bar{x}}$ and σ_p when a finite population, rather than an infinite population, is being sampled. The generally accepted rule of thumb is to ignore the finite population correction factor whenever $n/N \leq 0.05$.

Five-number summary An exploratory data analysis technique that uses five numbers to summarize the data: smallest value, first quartile, median, third quartile, and largest value.

Forecast A prediction of future values of a time series.

Frame A list of the sampling units for a study. The sample is drawn by selecting units from the frame.

Frequency distribution A tabular summary of data showing the number (frequency) of items in each of several non-overlapping classes.

General linear model A model of the form $Y = \beta_0 + \beta_1 z_1 + \beta_2 z_2 + \cdots + \beta_p z_p + \varepsilon$, where each of the independent variables $z_j (j = 1, 2, \ldots, p)$ is a function of $x_1, x_2, \ldots, x_k$, the variables for which data have been collected.

Goodness of fit test A statistical test conducted to determine whether to reject a hypothesized probability distribution for a population.

Grouped data Data available in class intervals as summarized by a frequency distribution. Individual values of the original data are not available.

High leverage points Observations with extreme values for the independent variables.

Histogram A graphical presentation of a frequency distribution, relative frequency distribution, or percentage frequency distribution of quantitative data constructed by placing the class intervals on the horizontal axis and the frequencies, relative frequencies, or percentage frequencies on the vertical axis.

Hypergeometric probability distribution A probability distribution showing the probability of x successes in n trials from a population with r successes and $N - r$ failures.

Hypergeometric probability function The function used to compute hypergeometric probabilities.

Independent events Two events A and B where $P(A \mid B) = P(A)$ or $P(B \mid A) = P(B)$; that is, the events have no influence on each other.

Independent samples Where, e.g. two groups of workers are selected and each group uses a different method to collect production time data.

Independent variable The variable that is doing the predicting or explaining. It is denoted by X.

Influential observation An observation that has a strong influence or effect on the regression results.

Interaction The effect of two independent variables acting together.

Interquartile range (IQR) A measure of variability, defined to be the difference between the third and first quartiles.

Intersection of A and B The event containing the sample points belonging to both A and B. The intersection is denoted $A \cap B$.

Interval estimate An estimate of a population parameter that provides an interval believed to contain the value of the parameter.

Interval scale The scale of measurement for a variable if the data demonstrate the properties of ordinal data and the interval between values is expressed in terms of a fixed unit of measure. Interval data are always numeric.

Irregular component The component of the time series that reflects the random variation of the time series values beyond what can be explained by the trend, cyclical and seasonal components.

ith residual The difference between the observed value of the dependent variable and the value predicted using the estimated regression equation; for the ith observation the ith residual is $y_i - \hat{y}_i$.

Joint probability The probability of two events both occurring; that is, the probability of the intersection of two events.

Judgment sampling A non-probabilistic method of sampling whereby element selection is based on the judgment of the person doing the study.

Kruskal-Wallis test A non-parametric test for identifying differences among three or more populations on the basis of independent samples.

Laspeyres price index A weighted aggregate price index in which the weight for each item is its base-period quantity.

Least squares method The method used to develop the estimated regression equation. It minimizes the sum of squared residuals (the deviations between the observed values of the dependent variable, y, and the estimated values of the dependent variable, $\hat{y}_i$).

Level of significance The probability of making a Type I error when the null hypothesis is true as an equality.

Leverage A measure of how far the values of the independent variables are from their mean values.

Logistic regression equation The mathematical equation relating $E(Y)$, the probability that $Y = 1$, to the values of the independent variables: that is,

$$E(Y) = P(Y = 1 \mid x_1, x_2, \ldots x_p) = \frac{e^{\beta_0 + \beta_1 x_1 + \cdots + \beta_p x_p}}{1 + e^{\beta_0 + \beta_1 x_1 + \cdots + \beta_p x_p}}$$

Logit The natural logarithm of the odds in favour of $Y = 1$: that is, $g(x_1, x_2, \ldots x_p) = \beta_0 + \beta_1 x_1 + \beta_2 x_2 + \cdots + \beta_p x_p$

Lot A group of items such as incoming shipments of raw materials or purchased parts as well as finished goods from final assembly.

Mann-Whitney-Wilcoxon (MWW) test A non-parametric statistical test for identifying differences between two populations based on the analysis of two independent samples.

Margin of error The value added to and subtracted from a point estimate in order to construct an interval estimate of a population parameter.

Marginal probability The values in the margins of a joint probability table that provide the probabilities of each event separately.

Matched Samples Where, e.g. only a sample of workers is selected and each worker uses first one and then the other method, with each worker providing a pair of data values.

Mean A measure of central location computed by summing the data values and dividing by the number of observations.

Mean squared error (MSE) A measure of the accuracy of a forecasting method. This measure is the average of the sum of the squared differences between the forecast values and the actual time series values.

Median A measure of central location provided by the value in the middle when the data are arranged in ascending order.

Mode A measure of location, defined as the value that occurs with greatest frequency.

Moving averages A method of forecasting or smoothing a time series that uses the average of the most recent n data values in the time series as the forecast for the next period.

Multicollinearity The term used to describe the correlation among the independent variables.

Multinomial population A population in which each element is assigned to one and only one of several categories. The multinomial distribution extends the binomial distribution from two to three or more outcomes.

Multiple coefficient of determination A measure of the goodness of fit of the estimated multiple regression equation. It can be interpreted as the proportion of the variability in the dependent variable that is explained by the estimated regression equation.

Multiple comparison procedures Statistical procedures that can be used to conduct statistical comparisons between pairs of population means.

Multiple regression analysis Regression analysis involving two or more independent variables.

Multiple regression equation The mathematical equation relating the expected value or mean value of the dependent variable to the values of the independent variables; that is

$$E(Y) = \beta_0 + \beta_1 x_1 + \beta_2 x_2 + \cdots + \beta_p x_p.$$

Multiple regression model The mathematical equation that describes how the dependent variable Y is related to the independent variables $x_1, x_2, \ldots x_p$ and an error term ε.

Multiple sampling plan A form of acceptance sampling in which more than one sample or stage is used. On the basis of the number of defective items found in a sample, a decision will be made to accept the lot, reject the lot, or continue sampling.

Multiplication law A probability law used to compute the probability of the intersection of two events. It is $P(A \cap B) = P(B)P(A \mid B)$ or $P(A \cap B) = P(A)P(B \mid A)$. For independent events it reduces to $P(A \cap B) = P(A) P(B)$.

Multiplicative time series model A model whereby the separate components of the time series are multiplied together to identify the actual time series value. When the four components of trend, cyclical, seasonal and irregular are assumed present, we obtain $Y_t = T_t \times C_t \times S_t \times I_t$. When the cyclical component is not modelled, we obtain $Y_t = T_t \times S_t \times I_t$.

Mutually exclusive events Events that have no sample points in common: that is, $A \cap B$ is empty and $P(A \cap B) = 0$.

Node An intersection or junction point of an influence diagram or a decision tree.

Nominal scale The scale of measurement for a variable when the data use labels or names to identify an attribute of an element. Nominal data may be non-numeric or numeric.

Non-parametric methods Statistical methods that require relatively few assumptions about the population probability distributions and about the level of measurement. These methods can be applied when nominal or ordinal data are available.

Non-probabilistic sampling Any method of sampling for which the probability of selecting a sample of any given configuration cannot be computed.

Non-sampling error All types of errors other than sampling error, such as measurement error, interviewer error, and processing error.

Normal probability distribution A continuous probability distribution. Its probability density function is bell shaped and determined by its mean μ and standard deviation σ.

Normal probability plot A graph of the standardized residuals plotted against values of the normal scores. This plot helps determine whether the assumption that the error term has a normal probability distribution appears to be valid.

np **chart** A control chart used to monitor the quality of the output of a process in terms of the number of defective items.

Null hypothesis The hypothesis tentatively assumed true in the hypothesis testing procedure.

Observation The set of measurements obtained for a particular element.

Odds in favour of an event occurring The probability the event will occur divided by the probability the event will not occur.

Odds ratio The odds that $Y = 1$ given that one of the independent variables increased by one unit (odds_1) divided by the odds that $Y = 1$ given no change in the values for the independent variables (odds_0): that is, Odds ratio = $\text{odds}_1/\text{odds}_0$.

Ogive A graph of a cumulative distribution.

One-tailed test A hypothesis test in which rejection of the null hypothesis occurs for values of the test statistic in one tail of its sampling distribution.

Operating characteristic curve A graph showing the probability of accepting the lot as a function of the percentage defective in the lot. This curve can be used to help determine whether a particular acceptance sampling plan meets both the producer's and the consumer's risk requirements.

Ordinal scale The scale of measurement for a variable if the data exhibit the properties of nominal data and the order or rank of the data is meaningful. Ordinal data may be non-numeric or numeric.

Outlier A data point or observation that does not fit the trend shown by the remaining data, often unusually small or unusually large.

p chart A control chart used when the quality of the output of a process is measured in terms of the proportion defective.

p-value A probability, computed using the test statistic, that measures the support (or lack of support) provided by the sample for the null hypothesis. For a lower tail test, the p-value is the probability of obtaining a value for the test statistic at least as small as that provided by the sample. For an upper tail test, the p-value is the probability of obtaining a value for the test statistic at least as large as that provided by the sample. For a two-tailed test, the p-value is the probability of obtaining a value for the test statistic at least as unlikely as that provided by the sample.

Paasche price index A weighted aggregate price index in which the weight for each item is its current-period quantity.

Parameter A numerical characteristic of a population, such as a population mean μ, a population standard deviation σ, a population proportion π and so on.

Partitioning The process of allocating the total sum of squares and degrees of freedom to the various components.

Payoff A measure of the consequence of a decision such as profit, cost or time. Each combination of a decision alternative and a state of nature has an associated payoff (consequence).

Payoff table A tabular representation of the payoffs for a decision problem.

Percentage frequency distribution A tabular summary of data showing the percentage of items in each of several non-overlapping classes.

Percentile A value such that at least p per cent of the observations are less than or equal to this value and at least $(100 - p)$ per cent of the observations are greater than or equal to this value. The 50th percentile is the median.

Pie chart A graphical device for presenting data summaries based on subdivision of a circle into sectors that correspond to the relative frequency for each class.

Point estimate The value of a point estimator used in a particular instance as an estimate of a population parameter.

Point estimator The sample statistic, such as $\bar{X}$, S, or P, that provides the point estimate of the population parameter.

Poisson probability distribution A probability distribution showing the probability of x occurrences of an event over a specified interval of time or space.

Poisson probability function The function used to compute Poisson probabilities.

Pooled estimator of π A weighted average of p_1 and p_2.

Population The set of all elements of interest in a particular study.

Population parameter A numerical value used as a summary measure for a population (e.g. the population

mean μ, the population variance σ^2, and the population standard deviation σ).

Posterior probabilities Revised probabilities of events based on additional information.

Posterior (revised) probabilities The probabilities of the states of nature after revising the prior probabilities based on sample information.

Power The probability of correctly rejecting H_0 when it is false.

Power curve A graph of the probability of rejecting H_0 for all possible values of the population parameter not satisfying the null hypothesis. The power curve provides the probability of correctly rejecting the null hypothesis.

Prediction interval The interval estimate of an individual value of Y for a given value of X.

Price relative A price index for a given item that is computed by dividing a current unit price by a base-period unit price and multiplying the result by 100.

Prior probabilities The probabilities of the states of nature prior to obtaining sample information.

Probabilistic sampling Any method of sampling for which the probability of each possible sample can be computed.

Probability A numerical measure of the likelihood that an event will occur.

Probability density function A function used to compute probabilities for a continuous random variable. The area under the graph of a probability density function over an interval represents probability.

Probability distribution A description of how the probabilities are distributed over the values of the random variable.

Probability function A function, denoted by $p(x)$, that provides the probability that X assumes a particular value for a discrete random variable.

Producer Price Index A price index designed to measure changes in prices of goods sold in primary markets (i.e. first purchase of a commodity in non-retail markets).

Producer's risk The risk of rejecting a good-quality lot; a Type I error.

Qualitative data Labels or names used to identify an attribute of each element. Qualitative data use either the nominal or ordinal scale of measurement and may be non-numeric or numeric.

Qualitative independent variable An independent variable with qualitative data.

Qualitative variable A variable with qualitative data.

Quality control A series of inspections and measurements that determine whether quality standards are being met.

Quantitative data Numeric values that indicate how much or how many of something.

Quantitative variable A variable with quantitative data.

Quantity index An index designed to measure changes in quantities over time.

Quartiles The 25th, 50th and 75th percentiles, referred to as the first quartile, the second quartile (median), and third quartile, respectively. The quartiles can be used to divide a data set into four parts, with each part containing approximately 25 per cent of the data.

R chart A control chart used when the quality of the output of a process is measured in terms of the range of a variable.

Random variable A numerical description of the outcome of an experiment.

Randomized block design An experimental design employing blocking.

Range A measure of variability, defined to be the largest value minus the smallest value.

Ratio scale The scale of measurement for a variable if the data demonstrate all the properties of interval data and the ratio of two values is meaningful. Ratio data are always numeric.

Regression equation The equation that describes how the mean or expected value of the dependent variable is related to the independent variable; in simple linear regression, $E(Y) = \beta_0 + \beta_1 x$

Regression model The equation describing how Y is related to X and an error term; in simple linear regression, the regression model is $y = \beta_0 + \beta_1 x + \varepsilon$

Relative frequency distribution A tabular summary of data showing the fraction or proportion of data items in each of several non-overlapping classes.

Relative frequency method A method of assigning probabilities that is appropriate when data are available to estimate the proportion of the time the experimental outcome will occur if the experiment is repeated a large number of times.

Replications The number of times each experimental condition is repeated in an experiment.

Residual analysis The analysis of the residuals used to determine whether the assumptions made about the regression model appear to be valid. Residual analysis is also used to identify outliers and influential observations.

Residual plot Graphical representation of the residuals that can be used to determine whether the assumptions made about the regression model appear to be valid.

Sample A subset of the population.

Sample information New information obtained through research or experimentation that enables an updating or revision of the state-of-nature probabilities.

Sample point An element of the sample space. A sample point represents an experimental outcome.

Sample space The set of all experimental outcomes.

Sample statistic A numerical value used as a summary measure for a sample (e.g. the sample mean $\overline{X}$, the sample variance S^2, and the sample standard deviation S).

Sample survey A survey to collect data on a sample.

Sampled population The population from which the sample is taken.

Sampling distribution A probability distribution consisting of all possible values of a sample statistic.

Sampling error The error that occurs because a sample, and not the entire population, is used to estimate a population parameter.

Sampling unit The units selected for sampling. A sampling unit may include several elements.

Sampling with replacement Once an element has been included in the sample, it is returned to the population. A previously selected element can be selected again and therefore may appear in the sample more than once.

Sampling without replacement Once an element has been included in the sample, it is removed from the population and cannot be selected a second time.

Scatter diagram A graphical presentation of the relationship between two quantitative variables. One variable is shown on the horizontal axis and the other variable is shown on the vertical axis.

Scenario writing A qualitative forecasting method that consists of developing a conceptual scenario of the future based on a well-defined set of assumptions.

Seasonal component The component of the time series that shows a periodic pattern over one year or less.

Serial correlation Same as autocorrelation.

σ (sigma) known The condition existing when historical data or other information provide a good estimate or value for the population standard deviation prior to taking a sample. The interval estimation procedure uses this known value of σ in computing the margin of error.

σ (sigma) unknown The condition existing when no good basis exists for estimating the population standard deviation prior to taking the sample. The interval estimation procedure uses the sample standard deviation S in computing the margin of error.

Sign test A non-parametric statistical test for identifying differences between two populations based on the analysis of nominal data.

Simple linear regression Regression analysis involving one independent variable and one dependent variable in which the relationship between the variables is approximated by a straight line.

Simple random sampling Finite population: a sample selected such that each possible sample of size n has the same probability of being selected. Infinite population: a sample selected such that each element comes from the same population and the elements are selected independently.

Simpson's paradox Conclusions drawn from two or more separate cross-tabulations that can be reversed when the data are aggregated into a single cross-tabulation.

Single-factor experiment An experiment involving only one factor with k populations or treatments.

Skewness A measure of the shape of a data distribution. Data skewed to the left result in negative skewness; a symmetric data distribution results in zero skewness; and data skewed to the right result in positive skewness.

Smoothing constant A parameter of the exponential smoothing model that provides the weight given to the most recent time series value in the calculation of the forecast value.

Spearman rank-correlation coefficient A correlation measure based on rank-ordered data for two variables.

Standard deviation A measure of variability computed by taking the positive square root of the variance.

Standard error The standard deviation of a point estimator.

Standard error of the estimate The square root of the mean square error, denoted by s. It is the estimate of σ, the standard deviation of the error term ε.

Standard normal probability distribution A normal distribution with a mean of zero and a standard deviation of one.

Standardized residual The value obtained by dividing a residual by its standard deviation.

States of nature The possible outcomes for chance events that affect the payoff associated with a decision alternative.

Statistical inference The process of using data obtained from a sample to make estimates or test hypotheses about the characteristics of a population.

Statistics The art and science of collecting, analyzing, presenting and interpreting data.

Stem-and-leaf display An exploratory data analysis technique that simultaneously rank orders quantitative data and provides insight about the shape of the distribution.

Stratified random sampling A probabilistic method of selecting a sample in which the population is first divided into strata and a simple random sample is then taken from each stratum.

Studentized deleted residuals Standardized residuals that are based on a revised standard error of the estimate obtained by deleting observation i from the data set and then performing the regression analysis and computations.

Subjective method A method of assigning probabilities on the basis of judgment.

Systematic sampling A method of choosing a sample by randomly selecting the first element and then selecting every kth element thereafter.

t distribution A family of probability distributions that can be used to develop an interval estimate of a population mean whenever the population standard deviation σ is unknown and is estimated by the sample standard deviation s.

Target population The population about which inferences are made.

Test statistic A statistic whose value helps determine whether a null hypothesis can be rejected.

Time series A set of observations on a variable measured at successive points in time or over successive periods of time.

Time series data Data collected over several time periods.

Treatments Different levels of a factor.

Tree diagram A graphical representation that helps in visualizing a multiple-step experiment.

Trend The long-run shift or movement in the time series observable over several periods of time.

Trend line A line that provides an approximation of the relationship between two variables.

Two-tailed test A hypothesis test in which rejection of the null hypothesis occurs for values of the test statistic in either tail of its sampling distribution.

Type I error The error of rejecting H_0 when it is true.

Type II error The error of accepting H_0 when it is false.

Unbiasedness A property of a point estimator that is present when the expected value of the point estimator is equal to the population parameter it estimates.

Uniform probability distribution A continuous probability distribution for which the probability that the random variable will assume a value in any interval is the same for each interval of equal length.

Union of A and B The event containing all sample points belonging to A or B or both. The union is denoted $A \cup B$.

Variable A characteristic of interest for the elements.

Variable selection procedures Methods for selecting a subset of the independent variables for a regression model.

Variance A measure of variability based on the squared deviations of the data values about the mean.

Variance inflation factor A measure of how correlated an independent variable is with all other independent predictors in a multiple regression model.

Venn diagram A graphical representation for showing symbolically the sample space and operations involving events in which the sample space is represented by a rectangle and events are represented as circles within the sample space.

Weighted aggregate price index A composite price index in which the prices of the items in the composite are weighted by their relative importance.

Weighted mean The mean obtained by assigning each observation a weight that reflects its importance.

Weighted moving averages A method of forecasting or smoothing a time series by computing a weighted average of past data values. The sum of the weights must equal one.

Wilcoxon signed-rank test A non-parametric statistical test for identifying differences between two populations based on the analysis of two matched or paired samples.

$\bar{x}$ **chart** A control chart used when the quality of the output of a process is measured in terms of the mean value of a variable such as a length, weight, temperature and so on.

z-**score** A value computed by dividing the deviation about the mean $(x_i - \bar{x})$ by the standard deviation s. A z-score is referred to as a standardized value and denotes the number of standard deviations x_i is from the mean.

Index